Y0-CAZ-395

CHEMICAL ANALYSIS OF ANTIBIOTIC RESIDUES IN FOOD

WILEY SERIES IN MASS SPECTROMETRY

Series Editors

Dominic M. Desiderio
Departments of Neurology and Biochemistry
University of Tennessee Health Science Center

Nico M. M. Nibbering
Vrije Universiteit Amsterdam, The Netherlands

John R. de Laeter • *Applications of Inorganic Mass Spectrometry*
Michael Kinter and Nicholas E. Sherman • *Protein Sequencing and Identification Using Tandem Mass Spectrometry*
Chhabil Dass • *Principles and Practice of Biological Mass Spectrometry*
Mike S. Lee • *LC/MS Applications in Drug Development*
Jerzy Silberring and Rolf Eckman • *Mass Spectrometry and Hyphenated Techniques in Neuropeptide Research*
J. Wayne Rabalais • *Principles and Applications of Ion Scattering Spectrometry: Surface Chemical and Structural Analysis*
Mahmoud Hamdan and Pier Giorgio Righetti • *Proteomics Today: Protein Assessment and Biomarkers Using Mass Spectrometry, 2D Electrophoresis, and Microarray Technology*
Igor A. Kaltashov and Stephen J. Eyles • *Mass Spectrometry in Biophysics: Confirmation and Dynamics of Biomolecules*
Isabella Dalle-Donne, Andrea Scaloni, and D. Allan Butterfield • *Redox Proteomics: From Protein Modifications to Cellular Dysfunction and Diseases*
Silas G. Villas-Boas, Ute Roessner, Michael A.E. Hansen, Jorn Smedsgaard, and Jens Nielsen • *Metabolome Analysis: An Introduction*
Mahmoud H. Hamdan • *Cancer Biomarkers: Analytical Techniques for Discovery*
Chabbil Dass • *Fundamentals of Contemporary Mass Spectrometry*
Kevin M. Downard (Editor) • *Mass Spectrometry of Protein Interactions*
Nobuhiro Takahashi and Toshiaki Isobe • *Proteomic Biology Using LC-MS: Large Scale Analysis of Cellular Dynamics and Function*
Agnieszka Kraj and Jerzy Silberring (Editors) • *Proteomics: Introduction to Methods and Applications*
Ganesh Kumar Agrawal and Randeep Rakwal (Editors) • *Plant Proteomics: Technologies, Strategies, and Applications*
Rolf Ekman, Jerzy Silberring, Ann M. Westman-Brinkmalm, and Agnieszka Kraj (Editors) • *Mass Spectrometry: Instrumentation, Interpretation, and Applications*
Christoph A. Schalley and Andreas Springer • *Mass Spectrometry and Gas-Phase Chemistry of Non-Covalent Complexes*
Riccardo Flamini and Pietro Traldi • *Mass Spectrometry in Grape and Wine Chemistry*
Mario Thevis • *Mass Spectrometry in Sports Drug Testing: Characterization of Prohibited Substances and Doping Control Analytical Assays*
Sara Castiglioni, Ettore Zuccato, and Roberto Fanelli • *Illicit Drugs in the Environment: Occurrence, Analysis, and Fate Using Mass Spectrometry*
Ángel Garciá and Yotis A. Senis (Editors) • *Platelet Proteomics: Principles, Analysis, and Applications*
Luigi Mondello • *Comprehensive Chromatography in Combination with Mass Spectrometry*
Jian Wang, James MacNeil, and Jack F. Kay • *Chemical Analysis of Antibiotic Residues in Food*

CHEMICAL ANALYSIS OF ANTIBIOTIC RESIDUES IN FOOD

Edited by

JIAN WANG
JAMES D. MACNEIL
JACK F. KAY

A JOHN WILEY & SONS, INC., PUBLICATION

Published by John Wiley & Sons, Inc., Hoboken, New Jersey
Published simultaneously in Canada

Library of Congress Cataloging-in-Publication Data:
Chemical analysis of antibiotic residues in food / edited by Jian Wang, James D. MacNeil, Jack F. Kay.
p. ; cm.
Includes bibliographical references and index.
ISBN 978-0-470-49042-6 (cloth)
1. Veterinary drug residues–Analysis. 2. Antibiotic residues–Analysis. 3. Food of animal origin–Safety measures. I. Wang, Jian, 1969– II. MacNeil, James D. III. Kay, Jack F.
[DNLM: 1. Anti-Bacterial Agents–analysis. 2. Chemistry Techniques, Analytical–methods. 3. Drug Residues. 4. Food Safety. QV 350]
RA1270.V47C44 2011
615.9′54–dc22

2010054065

Printed in the United States of America

ePDF ISBN: 978-1-118-06718-5
oBook ISBN: 978-1-118-06720-8
ePub ISBN: 978-1-118-06719-2

10 9 8 7 6 5 4 3 2 1

CONTENTS

PREFACE

Food safety is of great importance to consumers. To ensure the safety of the food supply and to facilitate international trade, government agencies and international bodies establish standards, guidelines, and regulations that food producers and trade partners need to meet, respect, and follow. A primary goal of national and international regulatory frameworks for the use of veterinary drugs, including antimicrobials, in food-producing animals is to ensure that authorized products are used in a manner that will not lead to non-compliance residues. However, analytical methods are required to rapidly and accurately detect, quantify, and confirm antibiotic residues in food to verify that regulatory standards have been met and to remove foods that do not comply with these standards from the marketplace.

The current developments in analytical methods for antibiotic residues include the use of portable rapid tests for on-site use or rapid screening methods, and mass spectrometric (MS)-based techniques for laboratory use. This book, *Chemical Analysis of Antibiotic Residues in Food*, combines disciplines that include regulatory standards setting, pharmacokinetics, advanced MS technologies, regulatory analysis, and laboratory quality management. It includes recent developments in antibiotic residue analysis, together with information to provide readers with a clear understanding of both the regulatory environment and the underlying science for regulations. Other topics include the choice of marker residues and target animal tissues for regulatory analysis, general guidance for method development and method validation, estimation of measurement uncertainty, and laboratory quality assurance and quality control. Furthermore, it also includes information on the developing area of environmental issues related to veterinary use of antimicrobials. For the bench analyst, it provides not only information on sources of methods of analysis but also an understanding of which methods are most suitable for addressing the regulatory requirements and the basis for those requirements.

The main themes in this book include antibiotic chemical properties (Chapter 1), pharmacokinetics, metabolism, and distribution (Chapter 2); food safety regulations (Chapter 3); sample preparation (Chapter 4); screening methods (Chapter 5); chemical analysis focused mainly on LC-MS (Chapters 6 and 7), method development and validation (Chapter 8), measurement uncertainty (Chapter 9), and quality assurance and quality control (Chapter 10).

The editors and authors of this book are internationally recognized experts and leading scientists with extensive firsthand experience in preparing food safety regulations and in the chemical analysis of antibiotic residues in food. This book represents the cutting-edge state of the science in this area. It has been deliberately written and organized with a balance between practical use and theory to provide readers or analytical laboratory staff with a reference book for the analysis of antibiotic residues in food.

Jian Wang
James D. MacNeil
Jack F. Kay

Canadian Food Inspection Agency, Calgary, Canada
St. Mary's University, Halifax, Canada
University of Strathclyde, Glasgow, Scotland

ACKNOWLEDGMENT

The editors are grateful to Dr. Dominic M. Desiderio, the editor of *Mass Spectrometry Reviews*, for the invitation to contribute a book on antibiotic residues analysis; to individual chapter authors, leading scientists in the field, for their great contributions as the result of their profound knowledge and many years of firsthand experience; and to the editors' dear family members for their unending support and encouragement during this book project.

EDITORS

Dr. Jian Wang received his PhD at the University of Alberta in Canada in 2000, and then worked as a Post Doctoral Fellow at the Agriculture and Agri-Food Canada in 2001. He has been working as a leading Research Scientist at the Calgary Laboratory with the Canadian Food Inspection Agency since 2002. His scientific focus is on the method development using liquid chromatography-tandem mass spectrometry (LC-MS/MS) and UPLC/QqTOF for analyses of chemical contaminant residues, including antibiotics, pesticides, melamine, and cyanuric acid in various foods. He also develops statistical approaches to estimating the measurement uncertainty based on method validation and quality control data using the SAS program.

Dr. James D. MacNeil received his PhD from Dalhousie University, Halifax, NS, Canada in 1972 and worked as a government scientist until his retirement in 2007. During 1982–2007 he was Head, Centre for Veterinary Drug Residues, now part of the Canadian Food Inspection Agency. Dr. MacNeil has served as a member of the Joint FAO/WHO Expert Committee on Food Additives (JECFA), cochair of the working group on methods of Analysis and Sampling, Codex Committee on Veterinary Drugs in Foods (CCRVDF), is the former scientific editor for "Drugs, Cosmetics & Forensics" of J.AOAC Int., worked on IUPAC projects, has participated in various consultations on method validation and is the author of numerous publications on veterinary drug residue analysis. He is a former General Referee for methods for veterinary drug residues for AOAC International and was appointed scientist emeritus by CFIA in 2008. Dr. MacNeil holds an appointment as an adjunct professor in the Department of Chemistry, St. Mary's University.

Dr. Jack F. Kay received his PhD from the University of Strathclyde, Glasgow, Scotland in 1980 and has been involved with veterinary drug residue analyses since 1991. He works for the UK Veterinary Medicines Directorate to provide scientific advice on residue monitoring programmes and manages the research and development (R&D) program. Dr. Kay helped draft Commission Decision 2002/657/EC and is an International Standardization Organization (ISO)-trained assessor for audits to ISO 17025. He served as cochair of the CCRVDF ad hoc Working Group on Methods of Sampling and Analysis and steered Codex Guideline CAC/GL 71–2009 to completion after Dr. MacNeil retired. Dr. Kay now cochairs work to extend this to cover multi-residue method performance criteria. He assisted JECFA in preparing an initial consideration of setting MRLs in honey, and is now developing this further for the CCRVDF. He also holds an Honorary Senior Research Fellowship at the Department of Mathematics and Statistics at the University of Strathclyde.

CONTRIBUTORS

Bjorn Berendsen, Department of Veterinary Drug Research, RIKILT—Institute of Food Safety, Unit Contaminants and Residues, Wageningen, The Netherlands

Alistair Boxall, Environment Department, University of York, Heslington, York, United Kingdom

Andrew Cannavan, Food and Environmental Protection Laboratory, FAO/IAEA Agriculture & Biotechnology Laboratories, Joint FAO/IAEA Division of Nuclear Techniques in Food and Agriculture, International Atomic Energy Agency, Vienna, Austria

Martin Danaher, Food Safety Department, Teagasc, Ashtown Food Research Centre, Ashtown, Dublin 15, Ireland

Leslie Dickson, Canadian Food Inspection Agency, Saskatoon Laboratory, Centre for Veterinary Drug Residues, Saskatoon, Saskatchewan, Canada

Rick Fedeniuk, Canadian Food Inspection Agency, Saskatoon Laboratory, Centre for Veterinary Drug Residues, Saskatoon, Saskatchewan, Canada

Lynn G. Friedlander, Residue Chemistry Team, Division of Human Food Safety, FDA/CVM/ONADE/HFV-151, Rockville, Maryland

Kevin J. Greenlees, Office of New Animal Drug Evaluation, HFV-100, USFDA Center for Veterinary Medicine, Rockville, Maryland

Jack F. Kay, Veterinary Medicines Directorate, New Haw, Surrey, United Kingdom; *also* Department of Mathematics and Statistics, University of Strathclyde, Glasgow, United Kingdom (honorary position)

Bruno Le Bizec, Food Safety, LABERCA (Laboratoire d'Etude des Résidus et Contaminants dans les Aliments), ONIRIS—Ecole Nationale Vétérinaire, Agroalimentaire et de l'Alimentation Nantes, Atlantique, Nantes, France

Peter Lees, Veterinary Basic Sciences, Royal Veterinary College, University of London, Hatfield, Hertfordshire, United Kingdom

James D. MacNeil, Scientist Emeritus, Canadian Food Inspection Agency, Dartmouth Laboratory, Dartmouth, Nova Scotia, Canada; *also* Department of Chemistry, St. Mary's University, Halifax, Nova Scotia, Canada

Ross A Potter, Veterinary Drug Residue Unit Supervisor, Canadian Food Inspection Agency (CFIA), Dartmouth Laboratory, Dartmouth, Nova Scotia, Canada

Philip Thomas Reeves, Australian Pesticides and Veterinary Medicines Authority, Regulatory Strategy and Compliance, Canberra, ACT (Australian Capital Territory), Australia

Jacques Stark, DSM Food Specialities, Delft, The Netherlands

Sara Stead, The Food and Environment Research Agency, York, North Yorkshire, United Kingdom

Alida A. M. (Linda) Stolker, Department of Veterinary Drug Research, RIKILT—Institute of Food Safety Unit Contaminants and Residues, Wageningen, The Netherlands

Jonathan A. Tarbin, The Food and Environment Research Agency, York, North Yorkshire, United Kingdom

Pierre-Louis Toutain, UMR181 Physiopathologie et Toxicologie Experimentales INRA, ENVT, Ecole Nationale Veterinaire de Toulouse, Toulouse, France

Sherri B. Turnipseed, Animal Drugs Research Center, US Food and Drug Administration, Denver, Colorado

Jian Wang, Canadian Food Inspection Agency, Calgary Laboratory, Calgary, Alberta, Canada

1

ANTIBIOTICS: GROUPS AND PROPERTIES

PHILIP THOMAS REEVES

1.1 INTRODUCTION

The introduction of the sulfonamides in the 1930s and benzylpenicillin in the 1940s completely revolutionized medicine by reducing the morbidity and mortality of many infectious diseases. Today, antimicrobial drugs are used in food-producing animals to treat and prevent diseases and to enhance growth rate and feed efficiency. Such use is fundamental to animal health and well-being and to the economics of the livestock industry, and has seen the development of antimicrobials such as ceftiofur, florfenicol, tiamulin, tilmicosin, tulathromycin, and tylosin specifically for use in food-producing animals.[1,2] However, these uses may result in residues in foods and have been linked to the emergence of antibiotic-resistant strains of disease-causing bacteria with potential human health ramifications.[3] Antimicrobial drug resistance is not addressed in detail in this text, and the interested reader is referred to an excellent overview by Martinez and Silley.[4]

Many factors influence the residue profiles of antibiotics in animal-derived edible tissues (meat and offal) and products (milk and eggs), and in fish and honey. Among these factors are the approved uses, which vary markedly between antibiotic classes and to a lesser degree within classes. For instance, in some countries, residues of quinolones in animal tissues, milk, honey, shrimp, and fish are legally permitted (maximum residue limits [MRLs] have been established). By comparison, the approved uses of the macrolides are confined to the treatment of respiratory disease and for growth promotion (in some countries) in meat-producing animals (excluding fish), and to the treatment of American foulbrood disease in honeybees. As a consequence, residues of macrolides are legally permitted only in edible tissues derived from these food-producing species, and in honey in some countries. Although a MRL for tylosin in honey has not been established, some countries apply a safe working residue level, thereby permitting the presence of trace concentrations of tylosin to allow for its use. Substantial differences in the approved uses of antimicrobial agents also occur between countries. A second factor that influences residue profiles of antimicrobial drugs is their chemical nature and physicochemical properties, which impact pharmacokinetic behavior. Pharmacokinetics (PK), which describes the timecourse of drug concentration in the body, is introduced in this chapter and discussed further in Chapter 2.

Analytical chemists take numerous parameters into account when determining antibiotic residues in food of animal origin, some of which are discussed here.

1.1.1 Identification

A substance needs to be identified by a combination of the appropriate identification parameters including the name or other identifier of the substance, information related to molecular and structural formula, and composition of the substance.

International nonproprietary names (INNs) are used to identify pharmaceutical substances or active pharmaceutical ingredients. Each INN is a unique name that is internationally consistent and is recognized globally. As of October 2009, approximately 8100 INNs had been designated, and this number is growing every year by some 120–150 new INNs.[5] An example of an INN is tylosin, a macrolide antibiotic.

Chemical Analysis of Antibiotic Residues in Food, First Edition. Edited by Jian Wang, James D. MacNeil, and Jack F. Kay.

International Union of Pure and Applied Chemistry (IUPAC) names are based on a method that involves selecting the longest continuous chain of carbon atoms, and then identifying the groups attached to that chain and systematically indicating where they are attached. Continuing with tylosin as an example, the IUPAC name is [(2*R*,3*R*,4*E*, 6*E*,9*R*,11*R*,12*S*,13*S*,14*R*)-12-{[3,6-dideoxy-4-*O*-(2,6-dideoxy-3-*C*-methyl-α-L-ribohexopyranosyl)-3-(dimethylamino)-β-D-glucopyranosyl]oxy}-2-ethyl-14-hydroxy-5, 9,13-trimethyl-8,16-dioxo-11-(2-oxoethyl)oxacyclohexadeca-4, 6-dien-3-yl]methyl 6-deoxy-2,3-di-*O*-methyl-β-D-allopyranoside.

The Chemical Abstract Service (CAS) Registry Number is the universally recognized unique identifier of chemical substances. The CAS Registry Number for tylosin is 1401-69-0.

Synonyms are used for establishing a molecule's unique identity. For the tylosin example, there are numerous synonyms, one of which is Tylan.

1.1.2 Chemical Structure

For the great majority of drugs, action on the body is dependent on chemical structure, so that a very small change can markedly alter the potency of the drug, even to the point of loss of activity.[6] In the case of antimicrobial drugs, it was the work of Ehrlich in the early 1900s that led to the introduction of molecules selectively toxic for microbes and relatively safe for the animal host. In addition, the presence of different sidechains confers different pharmacokinetic behavior on a molecule. Chemical structures also provide the context to some of the extraction, separation, and detection strategies used in the development of analytical methods. Certain antibiotics consist of several components with distinct chemical structures. Tylosin, for example, is a mixture of four derivatives produced by a strain of *Streptomyces fradiae*. The chemical structures of the antimicrobial agents described in this chapter are presented in Tables 1.2–1.15.

1.1.3 Molecular Formula

By identifying the functional groups present in a molecule, a molecular formula provides insight into numerous properties. These include the molecule's water and lipid solubility, the presence of fracture points for gas chromatography (GC) determinations, sources of potential markers such as chromophores, an indication as to the molecule's UV absorbance, whether derivatization is likely to be required when quantifying residues of the compound, and the form of ionization such as protonated ions or adduct ions when using electrospray ionization. The molecular formulas of the antimicrobial agents described in this chapter are shown in Tables 1.2–1.15.

1.1.4 Composition of the Substance

Regulatory authorities conduct risk assessments on the chemistry and manufacture of new and generic antimicrobial medicines (formulated products) prior to granting marketing approvals. Typically, a compositional standard is developed for a new chemical entity or will already exist for a generic drug. A compositional standard specifies the minimum purity of the active ingredient, the ratio of isomers to diastereoisomers (if relevant), and the maximum permitted concentration of impurities, including those of toxicological concern. The risk assessment considers the manufacturing process (the toxicological profiles of impurities resulting from the synthesis are of particular interest), purity, and composition to ensure compliance with the relevant standard. The relevant test procedures described in pharmacopoeia and similar texts apply to the active ingredient and excipients present in the formulation. The overall risk assessment conducted by regulatory authorities ensures that antimicrobial drugs originating from different manufacturing sources, and for different batches from the same manufacturing source, have profiles that are consistently acceptable in terms of efficacy and safety to target animals, public health, and environmental health.

1.1.5 pK_a

The symbol pK_a is used to represent the negative logarithm of the acid dissociation constant K_a, which is defined as $[H^+][B]/[HB]$, where B is the conjugate base of the acid HB. By convention, the acid dissociation constant (pK_a) is used for weak bases (rather than the pK_b) as well as weak organic acids. Therefore, a weak acid with a high pK_a will be poorly ionized, and a weak base with a high pK_a will be highly ionized at blood pH. The pK_a value is the principal property of an electrolyte that defines its biological and chemical behavior. Because the majority of drugs are weak acids or bases, they exist in both ionized and un-ionized forms, depending on pH. The proportion of ionized and un-ionized species at a particular pH is calculated using the Henderson–Hasselbalch equation. In biological terms, pK_a is important in determining whether a molecule will be taken up by aqueous tissue components or lipid membranes and is related to the partition coefficient log P. The pK_a of an antimicrobial drug has implications for both the fate of the drug in the body and the action of the drug on microorganisms. From a chemical perspective, ionization will increase the likelihood of a species being taken up into aqueous solution (because water is a very polar solvent). By contrast, an organic molecule that does not readily ionize will often tend to stay in a non-polar solvent. This partitioning behavior affects the efficiency of extraction and clean-up of analytes and is an important consideration when developing enrichment methods. The pK_a values for many of the antimicrobial agents described in this chapter are presented in Tables 1.2–1.15. The consequences of pK_a for the biological and chemical properties of antimicrobial agents are discussed later in this text.

1.1.6 UV Absorbance

The electrons of unsaturated bonds in many organic drug molecules undergo energy transitions when UV light is absorbed. The intensity of absorption may be quantitatively expressed as an extinction coefficient ε, which has significance in analytical application of spectrophotometric methods.

1.1.7 Solubility

From an *in vitro* perspective, solubility in water and in organic solvents determines the choice of solvent, which, in turn, influences the choice of extraction procedure and analytical method. Solubility can also indirectly impact the timeframe of an assay for compounds that are unstable in solution. From an *in vivo* perspective, the solubility of a compound influences its absorption, distribution, metabolism, and excretion. Both water solubility and lipid solubility are necessary for the absorption of orally administered antimicrobial drugs from the gastrointestinal tract. This is an important consideration when selecting a pharmaceutical salt during formulation development. Lipid solubility is necessary for passive diffusion of drugs in the distributive phase, whereas water solubility is critical for the excretion of antimicrobial drugs and/or their metabolites by the kidneys.

1.1.8 Stability

In terms of residues in food, stability is an important parameter as it relates to (1) residues in biological matrices during storage, (2) analytical reference standards, (3) analytes in specified solvents, (4) samples prepared for residue analysis in an interrupted assay run such as might occur with the breakdown of an analytical instrument, and (5) residues being degraded during chromatography as a result of an incompatible stationary phase.

Stability is also an important property of formulated drug products since all formulations decompose with time.[7] Because instabilities are often detectable only after considerable storage periods under normal conditions, stability testing utilizes high-stress conditions (conditions of temperature, humidity, and light intensity, which are known to be likely causes of breakdown). Adoption of this approach reduces the amount of time required when determining shelf life. Accelerated stability studies involving the storage of products at elevated temperatures are commonly conducted to allow unsatisfactory formulations to be eliminated early in development and for a successful product to reach market sooner. The concept of accelerated stability is based on the Arrhenius equation:

$$k = Ae^{(-E_a/RT)}$$

where k is the rate constant of the chemical reaction; A, a pre-exponential factor; E_a, activation energy; R, gas constant; and T, absolute temperature.

In practical terms, the Arrhenius equation supports the generalization that, for many common chemical reactions at room temperature, the reaction rate doubles for every 10°C increase in temperature. Regulatory authorities generally accept accelerated stability data as an interim measure while real-time stability data are being generated.

1.2 ANTIBIOTIC GROUPS AND PROPERTIES

1.2.1 Terminology

Traditionally, the term *antibiotic* refers to substances produced by microorganisms that at low concentration kill or inhibit the growth of other microorganisms but cause little or no host damage. The term *antimicrobial agent* refers to any substance of natural, synthetic, or semisynthetic origin that at low concentration kills or inhibits the growth of microorganisms but causes little or no host damage. Neither antibiotics nor antimicrobial agents have activity against viruses. Today, the terms *antibiotic* and *antimicrobial agent* are often used interchangeably.

The term *microorganism* or *microbe* refers to (for the purpose of this chapter) prokaryotes, which, by definition, are single-cell organisms that do not possess a true nucleus. Both typical bacteria and atypical bacteria (rickettsiae, chlamydiae, mycoplasmas, and actinomycetes) are included. Bacteria range in size from 0.75 to 5 μm and most commonly are found in the shape of a sphere (coccus) or a rod (bacillus). Bacteria are unique in that they possess peptidoglycan in their cell walls, which is the site of action of antibiotics such as penicillin, bacitracin, and vancomycin. Differences in the composition of bacterial cell walls allow bacteria to be broadly classified using differential staining procedures. In this respect, the Gram stain developed by Christian Gram in 1884 (and later modified) is by far the most important differential stain used in microbiology.[8] Bacteria can be divided into two broad groups—Gram-positive and Gram-negative—using the Gram staining procedure. This classification is based on the ability of cells to retain the dye methyl violet after washing with a decolorizing agent such as absolute alcohol or acetone. Gram-positive cells retain the stain, whereas Gram-negative cells do not. Examples of Gram-positive bacteria are *Bacillus, Clostridium, Corynebacterium, Enterococcus,*

Erysipelothrix, Pneumococcus, Staphylococcus, and *Streptococcus*. Examples of Gram-negative bacteria are *Bordetella, Brucella, Escherichia coli, Haemophilus, Leptospira, Neisseria, Pasteurella, Proteus, Pseudomonas, Salmonella, Serpulina hyodysenteriae, Shigella*, and *Vibrio*. Differential sensitivity of Gram-positive and Gram-negative bacteria to antimicrobial drugs is discussed later in this chapter.

1.2.2 Fundamental Concepts

From the definitions above, it is apparent that a critically important element of antimicrobial therapy is the selective toxicity of a drug for invading organisms rather than mammalian cells. The effectiveness of antimicrobial therapy depends on a triad of bacterial susceptibility, the drug's disposition in the body, and the dosage regimen. An additional factor that influences therapeutic outcomes is the competence of host defence mechanisms. This property is most relevant when clinical improvement relies on the inhibition of bacterial cell growth rather than bacterial cell death. Irrespective of the mechanism of action, the use of antimicrobial drugs in food-producing species may result in residues.

The importance of antibacterial drug pharmacokinetics (PK) and pharmacodynamics (PD) in determining clinical efficacy and safety was appreciated many years ago when the relationship between the magnitude of drug response and drug concentration in the fluids bathing the infection site(s) was recognized. PK describes the timecourse of drug absorption, distribution, metabolism, and excretion (what the *body does to the drug*) and therefore the relationship between the dose of drug administered and the concentration of non-protein-bound drug at the site of action. PD describes the relationship between the concentration of non-protein-bound drug at the site of action and the drug response (ultimately the therapeutic effect) (what the *drug does to the body*).[9]

In conceptualizing the relationships between the host animal, drug, and target pathogens, the chemotherapeutic triangle (Fig. 1.1) alludes to antimicrobial drug PK and PD. The relationship between the host animal and the drug reflects the PK properties of the drug, whereas drug action against the target pathogens reflects the PD properties of the drug. The clinical efficacy of antimicrobial therapy is depicted by the relationship between the host animal and target pathogens.

1.2.3 Pharmacokinetics of Antimicrobial Drugs

The pharmacokinetics of antimicrobial drugs is discussed in Chapter 2. The purpose of the following discussion, then, is to introduce the concept of pharmacokinetics and, in particular, to address the consequences of an antimicrobial drug's pK_a value for both action on the target pathogen and fate in the body.

The absorption, distribution, metabolism, and excretion of an antimicrobial drug are governed largely by the drug's chemical nature and physicochemical properties. Molecular size and shape, lipid solubility, and the degree of ionization are of particular importance, although the degree of ionization is not an important consideration for amphoteric compounds such as fluoroquinolones, tetracyclines, and rifampin.[10] The majority of antimicrobial agents are weak acids and bases for which the degree of ionization depends

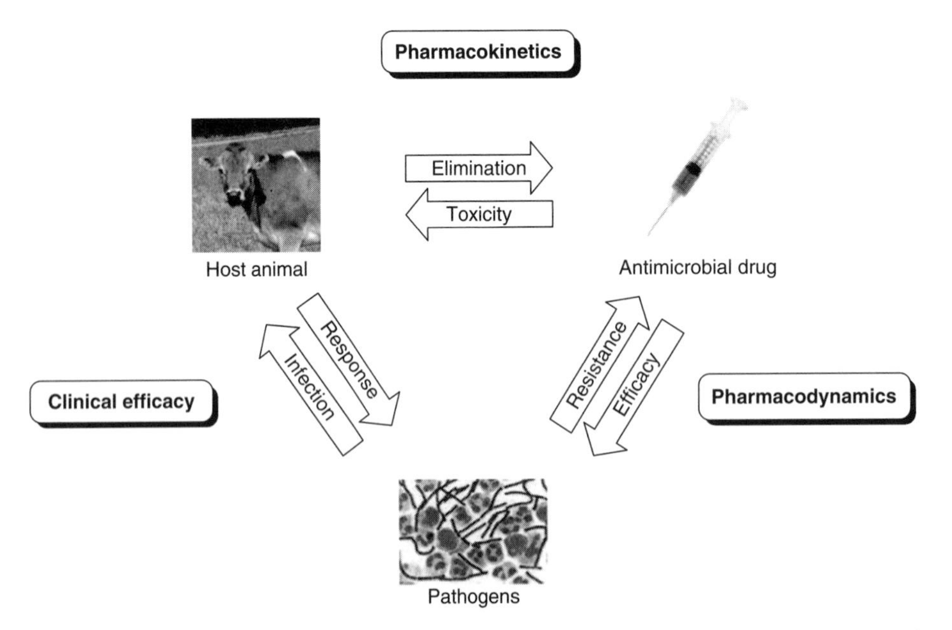

Figure 1.1 Schematic of the chemotherapeutic triangle depicting the relationships between the host animal, antimicrobial drug, and target pathogens.

on the pK_a of the drug and the pH of the biological environment. Only the un-ionized form of these drugs is lipid-soluble and able to cross cell membranes by passive diffusion. Two examples from Baggot and Brown[11] are presented here to demonstrate the implications of pK_a for the distributive phase of drug disposition. However, the same principles of passive diffusion apply to the absorption, metabolism, and excretion of drugs in the body and to the partitioning of drugs into microorganisms.

The first example relates to the sodium salt of a weak acid (with pK_a 4.4) that is infused into the mammary glands of dairy animals to treat mastitis. The pH of the normal mammary gland can be as low as 6.4, and at this pH, the Henderson–Hasselbalch equation predicts that the ratio of un-ionized to ionized drug is 1 : 100. Mastitic milk is more alkaline (with pH $\sim$ 7.4) and the ratio of un-ionized to ionized drug, as calculated by the Henderson–Hasselbalch equation, is 1 : 1000. This is identical to the ratio for plasma, which also has a pH of 7.4. This example demonstrates that, when compared to the normal mammary gland, the mastitic gland will have more drug "trapped" in the ionized form. The second example involves the injection of a lipid-soluble, organic base that diffuses from the systemic circulation (with pH 7.4) into ruminal fluid (pH 5.5–6.5) during the distributive phase of a drug. Again, the ionized form becomes trapped in the acidic fluid of the rumen; the extent of trapping will be determined by the pK_a of the organic base. In summary, weakly acidic drugs are trapped in alkaline environments and, vice versa, weakly basic drugs are trapped in acidic fluids.

A second PK issue is the concentration of antimicrobial drug at the site of infection. This value reflects the drug's distributive behavior and is critically important in terms of efficacy. Furthermore, the optimization of dosage regimens is dependent on the availability of quality information relating to drug concentration at the infection site. It raises questions regarding the choice of sampling site for measuring the concentration of antimicrobial drugs in the body and the effect, if any, that the extent of plasma protein binding has on the choice of sampling site. These matters are addressed below.

More often than not, the infection site (the biophase) is remote from the circulating blood that is commonly sampled to measure drug concentration. Several authors[12–14] have reported that plasma concentrations of free (non-protein-bound) drug are generally the best predictors of the clinical success of antimicrobial therapy. The biophase in most infections comprises extracellular fluid (plasma + interstitial fluids). Most pathogens of clinical interest are located extracellularly and as a result, plasma concentrations of free drug are generally representative of tissue concentrations; however, there are some notable exceptions:

1. Intracellular microbes such as *Lawsonia intracellularis*, the causative agent of proliferative enteropathy in pigs, are not exposed to plasma concentrations of antimicrobial drugs.
2. Anatomic barriers to the passive diffusion of antimicrobial drugs are encountered in certain tissues, including the central nervous system, the eye, and the prostate gland.
3. Pathological barriers such as abscesses impede the passive diffusion of drugs.
4. Certain antimicrobial drugs are preferentially accumulated inside cells. Macrolides, for instance, are known to accumulate within phagocytes.[15]
5. Certain antimicrobial drugs are actively transported into infection sites. The active transport of fluoroquinolones and tetracyclines by gingival fibroblasts into gingival fluid is an example.[16]

With regard to the effect of plasma protein binding on the choice of sampling site, Toutain and coworkers[14] reported that plasma drug concentrations of antimicrobial drugs that are >80% bound to plasma protein are unlikely to be representative of tissue concentrations. Those antimicrobial drugs that are highly bound to plasma protein include clindamycin, cloxacillin, doxycycline, and some sulfonamides.[17,18]

The most useful PK parameters for studying antimicrobial drugs are discussed in Chapter 2.

1.2.4 Pharmacodynamics of Antimicrobial Drugs

The PD of antimicrobial drugs against microorganisms comprises three main aspects: spectrum of activity, bactericidal and bacteriostatic activity, and the type of killing action (i.e., concentration-dependent, time-dependent, or co-dependent). Each of these is discussed below. Also described are the PD indices—minimum inhibitory concentration (MIC) and minimum bactericidal concentration (MBC)—and the mechanisms of action of antimicrobial drugs.

1.2.4.1 Spectrum of Activity

Antibacterial agents may be classified according to the class of target microorganism. Accordingly, antibacterial agents that inhibit only bacteria are described as narrow- or medium-spectrum, whereas those that also inhibit mycoplasma, rickettsia, and chlamydia (so-called atypical bacteria) are described as broad-spectrum. The spectrum of activity of common antibacterial drugs is shown in Table 1.1.

A different classification describes those antimicrobial agents that inhibit only Gram-positive or Gram-negative bacteria as narrow-spectrum, and those that are active against a range of both Gram-positive and Gram-negative bacteria as broad-spectrum. However, this distinction is not always absolute.

TABLE 1.1 Spectrum of Activity of Common Antibacterial Drugs

	Class of Microorganism				
Antibacterial Drug	Bacteria	Mycoplasma	Rickettsia	Chlamydia	Protozoa
Aminoglycosides	+	+	−	−	−
β-Lactams	+	−	−	−	−
Chloramphenicol	+	+	+	+	−
Fluoroquinolones	+	+	+	+	−
Lincosamides	+	+	−	−	+/−
Macrolides	+	+	−	+	+/−
Oxazolidinones	+	+	−	−	−
Pleuromutilins	+	+	−	+	−
Tetracyclines	+	+	+	+	−
Streptogramins	+	+	−	+	+/−
Sulfonamides	+	+	−	+	+
Trimethoprim	+	−	−	−	+

Notation: Presence or absence of activity against certain protozoa is indicated by plus or minus sign (+/−).

Source: Reference 2. Reprinted with permission of John Wiley & Sons, Inc. Copyright 2006, Blackwell Publishing.

The differential sensitivity of Gram-positive and Gram-negative bacteria to many antimicrobials is due to differences in cell wall composition. Gram-positive bacteria have a thicker outer wall composed of a number of layers of peptidoglycan, while Gram-negative bacteria have a lipophilic outer membrane that protects a thin peptidoglycan layer. Antibiotics that interfere with peptidoglycan syntheses more easily reach their site of action in Gram-positive bacteria. Gram-negative bacteria have protein channels (porins) in their outer membranes that allow the passage of small hydrophilic molecules. The outer membrane contains a lipopolysaccharide component that can be shed from the wall on cell death. It contains a highly heat-resistant molecule known as *endotoxin*, which has a number of toxic effects on the host animal, including fever and shock.

Antibiotic sensitivity also differs between aerobic and anaerobic organisms. Anaerobic organisms are further classified as facultative and obligate. Facultative anaerobic bacteria derive energy by aerobic respiration if oxygen is present but are also capable of switching to fermentation. Examples of facultative anaerobic bacteria are *Staphylococcus* (Gram-positive), *Escherichia coli* (Gram-negative), and *Listeria* (Gram-positive). In contrast, obligate anaerobes die in the presence of oxygen. Anaerobic organisms are resistant to antimicrobials that require oxygen-dependent mechanisms to enter bacterial cells. Anaerobic organisms may elaborate a variety of toxins and enzymes that can cause extensive tissue necrosis, limiting the penetration of antimicrobials into the site of infection, or inactivating them once they are present.

1.2.4.2 Bactericidal and Bacteriostatic Activity

The activity of antimicrobial drugs has also been described as being bacteriostatic or bactericidal, although this distinction depends on both the drug concentration at the site of infection and the microorganism involved. Bacteriostatic drugs (tetracyclines, phenicols, sulfonamides, lincosamides, macrolides) inhibit the growth of organisms at the MIC but require a significantly higher concentration, the MBC, to kill the organisms (MIC and MBC are discussed further below). By comparison, bactericidal drugs (penicillins, cephalosporins, aminoglycosides, fluoroquinolones) cause death of the organism at a concentration near the same drug concentration that inhibits its growth. Bactericidal drugs are required for effectively treating infections in immunocompromised patients and in immunoincompetent environments in the body.

1.2.4.3 Type of Killing Action

A further classification of antimicrobial drugs is based on their killing action, which may be time-dependent, concentration-dependent, or co-dependent. For time-dependent drugs, it is the duration of exposure (as reflected in time exceeding MIC for plasma concentration) that best correlates with bacteriological cure. For drugs characterized by concentration-dependent killing, it is the maximum plasma concentration and/or area under the plasma concentration–time curve that correlates with outcome. For drugs with a co-dependent killing effect, both the concentration achieved and the duration of exposure determine outcome (see Chapter 2 for further discussion).

Growth inhibition–time curves are used to define the type of killing action and steepness of the concentration–effect curve. Typically, reduction of the initial bacterial count (response) is plotted against antimicrobial drug concentration. The killing action (time-, concentration-, or co-dependent) of an antibacterial drug is determined largely by the slope of the curve. Antibacterial drugs that demonstrate time-dependent killing activity include the β-lactams, macrolides, tetracyclines, trimethoprim–sulfonamide combinations, chloramphenicol, and glycopeptides. A concentration-dependent killing action is demonstrated by the aminoglycosides, fluoroquinolones, and metronidazole. The antibacterial response is less sensitive to increasing drug concentration when the slope is steep and vice versa.

1.2.4.4 Minimum Inhibitory Concentration and Minimum Bactericidal Concentration

The most important indices for describing the PD of antimicrobial drugs are MIC and MBC. The MIC is the lowest concentration of antimicrobial agent that prevents visible growth after an 18- or 24-h incubation. It is a measure of the intrinsic antimicrobial activity (potency) of an antimicrobial drug. Because an MIC is an absolute value that is not based on comparison with a reference standard, it is critically important to standardize experimental factors that may influence the result, including the strain of bacteria, the size of the inocula, and the culture media used, according to internationally accepted methods (e.g., CLSI[19] or EUCAST[20]). The MIC is determined from culture broth containing antibiotics in serial two-fold dilutions that encompass the concentrations normally achieved *in vivo*. Positive and negative controls are included to demonstrate viability of the inocula and suitability of the medium for their growth, and that contamination with other organisms has not occurred during preparation, respectively.

After the MIC has been determined, it is necessary to decide whether the results suggest whether the organisms are susceptible to the tested antimicrobial *in vivo*. This decision requires an understanding of the PK of the drug (see Chapter 2 for discussion) and other factors. For example, *in vitro* assessments of activity may underestimate the *in vivo* activity because of a post-antibiotic effect and post-antibiotic leukocyte enhancement. The *post-antibiotic effect* (PAE) refers to a persistent antibacterial effect at subinhibitory concentrations, whereas the term *post-antibiotic leukocyte enhancement term* (PALE) refers to the increased susceptibility to phagocytosis and intracellular killing demonstrated by bacteria following exposure to an antimicrobial agent.[21]

The MIC test procedure described above can be extended to determine the MBC. The MBC is the minimal concentration that kills 99.9% of the microbial cells. Samples from the antibiotic-containing tubes used in the MIC determination in which microbial growth was not visible are plated on agar with no added antibiotic. The lowest concentration of antibiotic from which bacteria do not grow when plated on agar is the MBC.

1.2.4.5 Mechanisms of Action

Antimicrobial agents demonstrate five major mechanisms of action.[22] These mechanisms, with examples of each type, are as follows:

1. Inhibition of cell wall synthesis (β-lactam antibiotics, bacitracin, vancomycin)
2. Damage to cell membrane function (polymyxins)
3. Inhibition of nucleic acid synthesis or function (nitroimidazoles, nitrofurans, quinolones, fluoroquinolones)
4. Inhibition of protein synthesis (aminoglycosides, phenicols, lincosamides, macrolides, streptogramins, pleuromutilins, tetracyclines)
5. Inhibition of folic and folinic acid synthesis (sulfonamides, trimethoprim)

1.2.5 Antimicrobial Drug Combinations

The use of antimicrobial combinations is indicated in some situations. For instance, mixed infections may respond better to the use of two or more antimicrobial agents. A separate example is fixed combinations such as the potentiated sulfonamides (comprising a sulfonamide and a diaminopyrimidine such as trimethoprim) that display synergism of antimicrobial activity. Other examples include the sequential inhibition of cell wall synthesis; facilitation of one antibiotic's entry to a microbe by another; inhibition of inactivating enzymes; and the prevention of emergence of resistant populations.[2] Another potential advantage of using antimicrobial drugs in combination is that the dose, and therefore the toxicity, of drugs may be reduced when a particular drug is used in combination with another drug(s).

Disadvantages from combining antimicrobial drugs in therapy also arise, and to address this possibility, combinations should be justified from both pharmacokinetic and pharmacodynamic perspectives.[23] For example, with a fixed combination of an aminoglycoside and a β-lactam, the former displays a concentration-dependent killing action and should be administered once daily, while the latter displays time-dependent killing and should be administered more frequently in order to ensure that the plasma concentration is maintained above the MIC of the organism for the majority of the dosing interval. One way to achieve this is to combine an aminoglycoside and the procaine salt of benzylpenicillin. The former requires a high C_{max} : MIC ratio, while the procaine salt of benzylpenicillin gives prolonged absorption to maintain plasma concentrations above MIC for most of the interdose interval. Similarly, a bacteriostatic drug may prevent some classes of bactericidal drugs from being efficacious.[23]

1.2.6 Clinical Toxicities

Animals may experience adverse effects when treated with veterinary antimicrobial drugs. These effects may reflect the pharmacological or toxicological properties of the substances or may involve hypersensitivity reactions or anaphylaxis. The major adverse effects to the various classes of antibiotics used in animals are described later in this chapter.

1.2.7 Dosage Forms

Antimicrobials are available as a range of pharmaceutical formulation types for food-producing animals, and of these, oral and parenteral dosage forms are the most common. Pharmaceutical formulations are designed to ensure the stability of the active ingredient up to the expiry date (when the product is stored in accordance with label recommendations), to control the rate of release of the active ingredient, and to achieve a desirable PK profile for the active ingredient. When mixed with feed or drinking water, veterinary antimicrobials must be stable, and those incorporated in feed should (ideally) be evenly dispersed in the feed. Antimicrobial products, including generic products, should be manufactured in accordance with current good manufacturing practices (GMP) and following the specifications described in the licensing application approved by the relevant authority. Generic products should normally have been shown to be bioequivalent to the reference (usually the pioneer) product.

1.2.8 Occupational Health and Safety Issues

Occupational health and safety considerations are paramount for manufacturing staff and for veterinarians and farmers administering antimicrobials to food-producing animals. In the period 1985–2001, antimicrobial drugs accounted for 2% of all suspected adverse reactions to have occurred in humans that were reported to the UK Veterinary Medicines Directorate.[24] The major problem following human exposure to antimicrobial drugs is sensitization and subsequent hypersensitivity reactions, and these are well recognized with β-lactam antibiotics.[25] Dust inhalation and sensitization to active ingredients are major concerns in manufacturing sites and are addressed by containment and the use of protective personal equipment. Other conditions that occur in those occupationally exposed to antimicrobials include dermatitis, bronchial asthma, accidental needlesticks, and accidental self-administration of injectable formulations. The occupational health and safety issues associated with specific classes of antimicrobial drugs are discussed later in this chapter.

1.2.9 Environmental Issues

Subject to the type of animal production system being considered, antimicrobial agents used in the livestock industries may enter the environment (for a review, see Boxall[26]). In the case of manure or slurry, which is typically stored before being applied to land, anaerobic degradation of antimicrobials occurs to differing degrees during storage. For example, β-lactam antibiotics rapidly dissipate in a range of manure types whereas tetracyclines are likely to persist for months. Compared to the situation in manure or slurry, the degradation of antimicrobials in soil is more likely to involve aerobic organisms. In fish production systems, medicated food pellets are added directly to pens or cages to treat bacterial infections in fish.[27–29] This practice results in the sediment under cages becoming contaminated with antimicrobials.[30–32] More recently, the literature has described tetracycline[33] and chloramphenicol[34] produced by soil organisms being taken up by plants. This raises the possibility that food-producing species may consume naturally derived antimicrobials when grazing herbs and grasses. The effects of the various classes of antibiotics on the environment are introduced later in this chapter to provide a foundation for the discussion that follows in Chapter 3.

1.3 MAJOR GROUPS OF ANTIBIOTICS

There are hundreds of antimicrobial agents in human and veterinary use, most of which belong to a few major classes; however, only some of these drugs are approved for use in food-producing species. Many factors contribute to this situation, one of which is concern over the transfer of antimicrobial resistance from animals to humans. In 1969, the Swann report in the United Kingdom recommended against the use of antimicrobial drugs already approved as therapeutic agents in humans or animals for growth promotion in animals.[35] This recommendation was only partially implemented in Britain at the time. Since then, the use of additional drugs for growth promotion has been prohibited in several countries. In addition, the World Health Organization (WHO), Codex Alimentarius Commission (CAC), the World Organization for Animal Health [Office International des Epizooties (OIE)], and national authorities are now developing strategies for reducing losses resulting from antimicrobial resistance, of those antimicrobial agents considered to be of critical importance to human medicine. When implemented, the recommendations from these important initiatives are certain to further restrict the availability of antimicrobial drugs for prophylactic and therapeutic uses in food-producing species.

An antimicrobial class comprises compounds with a related molecular structure and generally with similar modes of action. Variations in the properties of antimicrobials within a class often arise as a result of the presence of different sidechains of the molecule, which confer different patterns of PK and PD behavior on the molecule.[36] The major classes of antimicrobial drugs are discussed below.

1.3.1 Aminoglycosides

Streptomycin, the first aminogylcoside, was isolated from a strain of *Streptomyces griseus* and became available in 1944. Over the next 20 years, other aminoglycosides were isolated from streptomycetes (neomycin and kanamycin) and *Micromonospora purpurea* (gentamicin). Semi-synthetic derivatives have subsequently been produced, including amikacin from kanamycin.

Aminoglycosides are bactericidal antibiotics with a concentration-dependent killing action, active against aerobic Gram-negative bacteria and some Gram-positive bacteria, but have little or no activity against anaerobic bacteria. Aminoglycosides are actively pumped into Gram-negative cells through an oxygen-dependent interaction between the negatively charged surface of the outer cell membrane and the aminoglycoside cations. This results in altered bacterial cell membrane permeability. The aminoglycosides then bind to the 30*S* ribosomal subunit and cause misreading of the messenger RNA, resulting in disruption of bacterial protein synthesis. This further affects cell membrane permeability, allowing more aminoglycoside uptake leading to more cell disruption and finally cell death.[37] Different aminoglycosides have slightly different effects. Streptomycin and its dihydro derivatives act at a single site on the ribosome, but other aminoglycosides act at several sites. The action of aminoglycosides is bactericidal and dose-dependent, and there is a significant post-antibiotic effect. While theoretically one would expect interaction with β-lactam antibiotics to enhance penetration of aminoglycosides into bacterial cells as a result of the interference with cell wall synthesis, human efficacy and toxicity studies now dispute that there is any therapeutic justification for this type of combination.[38] However, it would appear that some of the formulation types used in animals, such as a combination of an aminoglycoside and the procaine salt of benzylpenicillin (see discussion above), do provide enhanced antibacterial activity.

Bacterial resistance to aminoglycosides is mediated through bacterial enzymes (phosphotransferases, acetyltransferases, adenyltransferases), which inactivate aminoglycosides and prevent their binding to the ribosome. Genes encoding these enzymes are frequently located on plasmids, facilitating rapid transfer of resistance to other bacteria.

Aminoglycosides are not well absorbed from the gastrointestinal tract but are well absorbed after intramuscular or subcutaneous injection. Effective concentrations are achieved in synovial, pleural, peritoneal, and pericardial fluids. Intrauterine and intramammary administration is also effective, but significant tissue residues result. Aminoglycosides do not bind significantly to plasma proteins, and as they are large polar molecules, they are poorly lipid-soluble and do not readily enter cells or penetrate cellular barriers. This means that therapeutic concentrations are not easily achieved in cerebrospinal or ocular fluids. Their volumes of distribution are small, and the half-lives in plasma are relatively short (1–2 h).[39] Elimination is entirely via the kidney.

Aminoglycosides tend to be reserved for more serious infections because of their toxicity. The more toxic members such as neomycin are restricted to topical or oral use; the less toxic aminoglycosides such as gentamicin are used parenterally for treatment of Gram-negative sepsis. Oral preparations of neomycin and streptomycin preparations are available for treatment of bacterial enteritis in calves, ophthalmic preparations of framycetin are used in sheep and cattle, and neomycin preparations (some in combination with β-lactams) are used in the treatment of bovine mastitis. Systemic use of streptomycin, neomycin, and spectinomycin is often restricted in food-producing animals because of widespread resistance and because of extended persistence of residues in kidney tissues. Aminoglycosides are used to treat individual animals for therapeutic purposes rather than metaphylaxis or prophylaxis. An exception is the use of neomycin as a dry-cow treatment at the end of lactation in dairy cows. No aminoglycosides are used as antimicrobial growth promotants.

All aminoglycosides display ototoxicity and nephrotoxicity. Streptomycin is the most ototoxic but the least nephrotoxic; neomycin is the most nephrotoxic. Nephrotoxicity is associated with accumulation of aminoglycosides in the renal proximal tubule cells, where the drugs accumulate within the lysosomes and are released into the cytoplasm, causing damage to cellular organelles and cell death. Risk factors for aminoglycoside toxicity include prolonged therapy (>7–10 days), more than once daily treatment, acidosis and electrolyte disturbances, age (neonates, geriatrics) and pre-existing renal disease. As toxicity to aminoglycosides is related to the trough concentration of drug, once-daily high-dose treatment is used to allow drug concentration during the trough period to fall below the threshold that causes toxicity.[40] Once-daily dosing is effective because aminoglycosides display concentration-dependent killing activity and a long post-antibiotic effect. In the case of animals with impaired renal function, this may not apply as aminoglycosides are generally contraindicated or administered with extended dosing intervals.[41]

The limited information available suggests that aminoglycoside residues persist at trace levels in the environment (see also discussion in Chapter 3).

The Joint FAO/WHO Expert Committee on Food Additives (JECFA) has evaluated toxicological and residue depletion data for dihydrostreptomycin and streptomycin, gentamicin, kanamycin, neomycin, and spectinomycin (see list in Table 1.2). On the basis of the risk assessments carried out by the JECFA, ADIs were allocated for all of these substances except kanamycin.[42] In addition, on the basis of JECFA recommendations, CAC MRLs were established for dihydrostreptomycin and streptomycin in muscle, liver, kidney, and fat of cattle, sheep, pigs, and chickens, and in cow's milk and sheep's milk; for gentamicin in muscle, liver, kidney, and fat of cattle and pigs, and in cow's milk; for neomycin in muscle, liver, kidney, and fat of cattle, sheep, pigs, chickens, goats, ducks, and turkeys, and in cow's milk and chicken eggs; and for spectinomycin in muscle, liver, kidney, and fat of cattle, sheep, pigs, and chickens, and in cow's milk and chicken eggs.[43] Details of residue studies considered by JECFA in recommending MRLs for adoption by the CAC, after review by the Codex Committee on Residues of Veterinary Drugs in Foods (CCRVDF), are contained in monographs dealing with dihydrostreptomycin and streptomycin,[44–47] gentamicin,[48,49] neomycin,[50–53] and spectinomycin.[54,55]

1.3.2 β-Lactams

The discovery by Fleming in 1929 that cultures of *Penicillium notatum* produced an antibacterial substance and the subsequent purification of penicillin and its use by Florey, Chain, and others a decade later to successfully treat infections in human patients launched the chemotherapeutic revolution. In 1945, Fleming, Florey, and Chain were jointly awarded the Nobel Prize in Physiology or Medicine for this work.

There are a number of classes of β-lactam antibiotics, on the basis of their chemical structure. All are bactericidal and act by disrupting peptidoglycan synthesis in

TABLE 1.2 Aminoglycosides and Aminocyclitols

INN	IUPAC Name, Molecular Formula, and CAS Registry No.	Chemical Structure	pK_a
	Aminoglycosides		
Amikacin	(2*S*)-4-Amino-*N*-[(1*R*,2*S*,3S,4*R*,5*S*)-5-amino-2-[(2*S*,3*R*,4*S*,5*S*,6*R*)-4-amino-3,5-dihydroxy-6-(hydroxymethyl)oxan-2-yl]oxy-4-[(2*R*,3*R*,4*S*,5*S*,6*R*)-6-(aminomethyl)-3,4,5-trihydroxyoxan-2-yl]oxy-3-hydroxycyclohexyl]-2-hydroxybutanamide $C_{22}H_{43}N_5O_{13}$ 37517-28-5	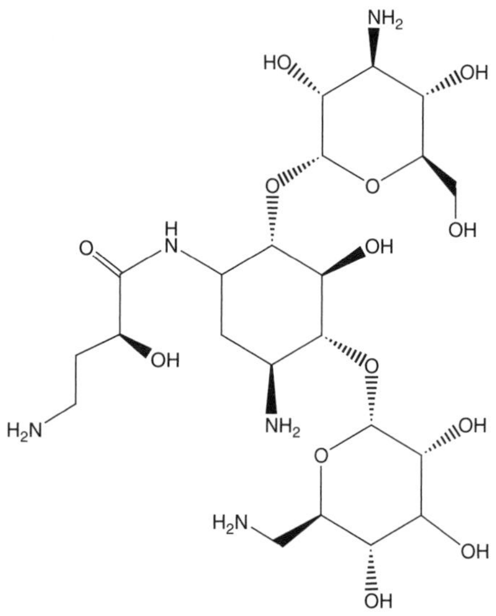	HB^+ 8.1[56]
Apramycin	(2*R*,3*R*,4*S*,5*S*,6*S*)-2-[[(2*S*,3*R*,4a*S*,6*R*,7*S*,8*R*,8a*R*)-3-Amino-2-[(1*R*,2*R*,3*S*,4*R*,6*S*)-4,6-diamino-2,3-dihydroxycyclohexyl]oxy-8-hydroxy-7-methylamino-2,3,4,4a,6,7,8,8a-octahydropyrano[2,3-*e*]pyran-6-yl]oxy]-5-amino-6-(hydroxymethyl)oxane-3,4-diol $C_{21}H_{41}N_5O_{11}$ 37321-09-08	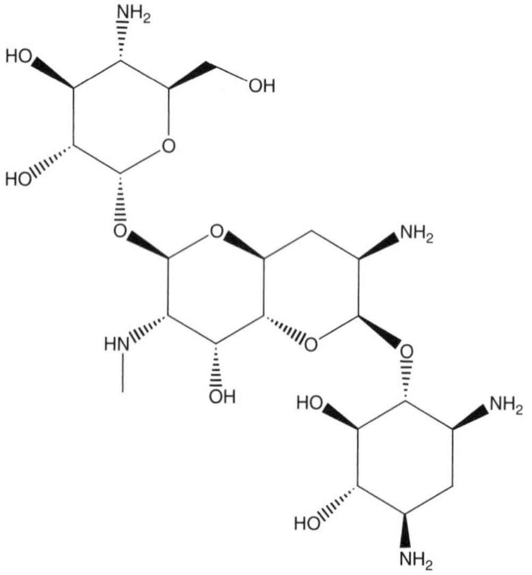	HB^+ 8.5[57]

TABLE 1.2 ***(Continued)***

INN	IUPAC Name, Molecular Formula, and CAS Registry No.	Chemical Structure	pK_a
Dihydrostreptomycin	2-[(1*S*,2*R*,3*R*,4*S*,5*R*,6*R*)-5-(Diaminomethylideneamino)-2-[(2*R*,3*R*,4*R*,5*S*)-3-[(2*S*,3*S*,4*S*,5*R*,6*S*)-4,5-dihydroxy-6-(hydroxymethyl)-3-methylaminooxan-2-yl]oxy-4-hydroxy-4-(hydroxymethyl)-5-methyloxolan-2-yl]oxy-3,4,6-trihydroxycyclohexyl]guanidine $C_{21}H_{41}N_7O_{12}$ 128-46-1	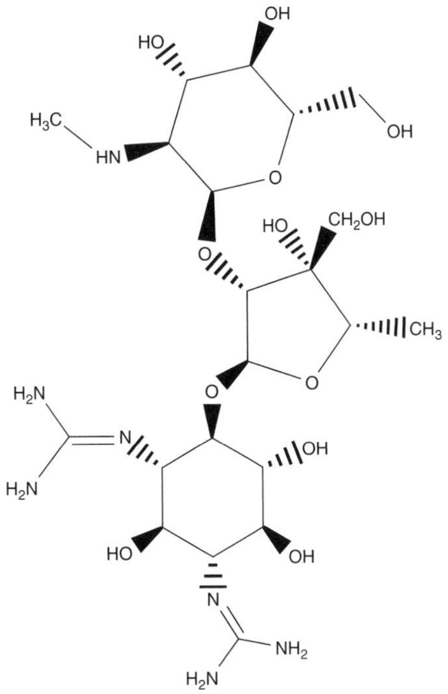	HB^+ 7.8[56]
Gentamicin	2-[4,6-Diamino-3-[3-amino-6-(1-methylaminoethyl)oxan-2-yl]oxy-2-hydroxycyclohexyl]-oxy-5-methyl-4-methylamino-oxane-3,5-diol $C_{21}H_{43}N_5O_7$ (gentamicin C_1) 1403-66-3	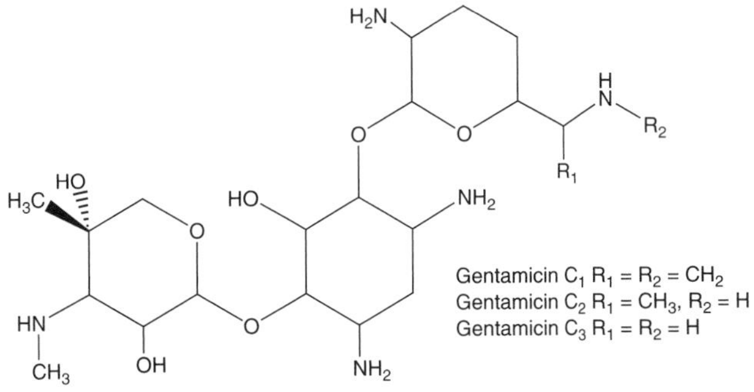 Gentamicin C_1 $R_1 = R_2 = CH_2$ Gentamicin C_2 $R_1 = CH_3$, $R_2 = H$ Gentamicin C_3 $R_1 = R_2 = H$	HB^+ 8.2[56]
Kanamycin	(2*R*,3*S*,4*S*,5*R*,6*R*)-2-(Aminomethyl)-6-[(1*R*,2*R*,3*S*,4*R*,6*S*)-4,6-diamino-3-[(2*S*,3*R*,4*S*,5*S*,6*R*)-4-amino-3,5-dihydroxy-6-(hydroxymethyl)oxan-2-yl]oxy-2-hydroxycyclohexyl]-oxyoxane-3,4,5-triol $C_{18}H_{36}N_4O_{11}$ (kanamycin A) 59-01-8	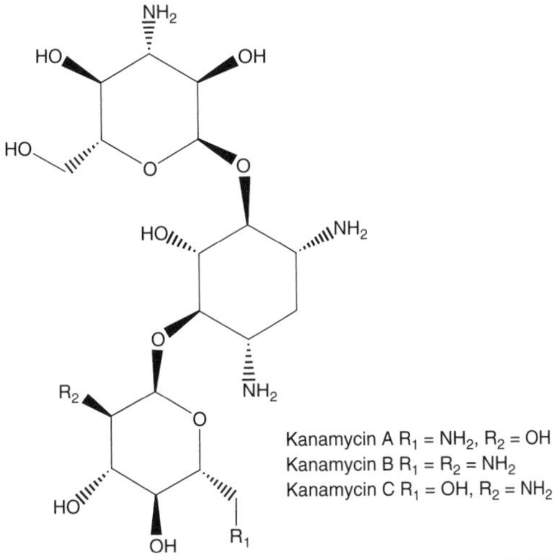 Kanamycin A $R_1 = NH_2$, $R_2 = OH$ Kanamycin B $R_1 = R_2 = NH_2$ Kanamycin C $R_1 = OH$, $R_2 = NH_2$	HB^+ 6.4[56] HB^+ 7.6[56] HB^+ 8.4[56] HB^+ 9.4[56]

(continued)

TABLE 1.2 ***(Continued)***

INN	IUPAC Name, Molecular Formula, and CAS Registry No.	Chemical Structure	pK_a
Neomycin B	(2*R*,3*S*,4*R*,5*R*,6*R*)-5-Amino-2-(aminomethyl)-6-[(1*R*,2*R*,3*S*,4*R*,6*S*)-4,6-diamino-2-[(2*S*,3*R*,4*S*,5*R*)-4-[(2*R*,3*R*,4*R*,5*S*,6*S*)-3-amino-6-(aminomethyl)-4,5-dihydroxyoxan-2-yl]oxy-3-hydroxy-5-(hydroxymethyl)-oxolan-2-yl]oxy-3-hydroxy-cyclohexyl]oxyoxane-3,4-diol $C_{23}H_{46}N_6O_{13}$ 1404-04-2	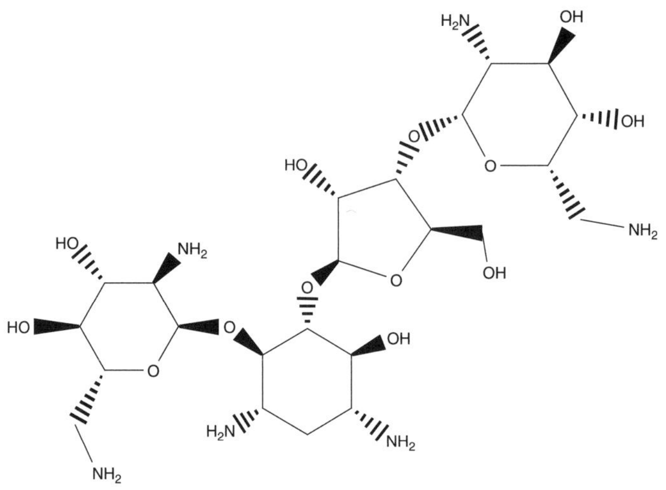	HB$^+$ 8.3[58]
Paromomycin	(2*R*,3*S*,4*R*,5*R*,6*S*)-5-Amino-6-[(1*R*,2*S*,3*S*,4*R*,6*S*)-4,6-diamino-2-[(2*S*,3*R*,4*R*,5*R*)-4-[(2*R*,3*R*,4*R*,5*R*,6*S*)-3-amino-6-(aminomethyl)-4,5-dihydroxyoxan-2-yl]oxy-3-hydroxy-5-(hydroxymethyl)oxolan-2-yl]oxy-3-hydroxy-cyclohexyl]-oxy-2-(hydroxymethyl)oxane-3,4-diol $C_{23}H_{45}N_5O_{14}$ 1263-89-4	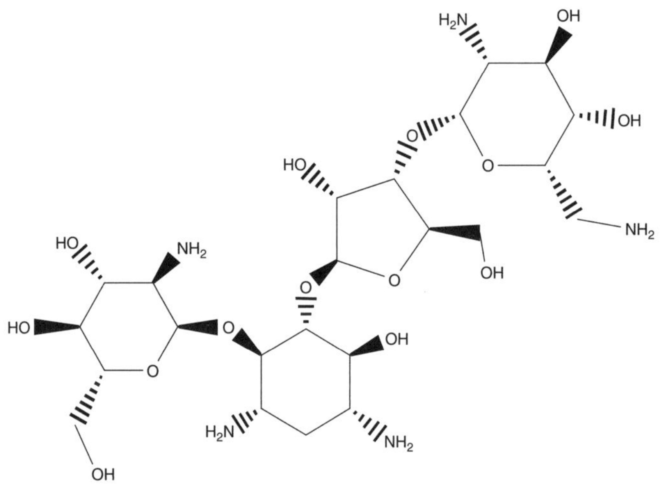	HB$^+$ 6.0[56] HB$^+$ 7.1[56] HB$^+$ 7.6[56] HB$^+$ 8.2[56] HB$^+$ 8.9[56]
Streptomycin A	2-[(1*S*,2*R*,3*R*,4*S*,5*R*,6*R*)-5-(Diaminomethylideneamino)-2-[(2*R*,3*R*,4*R*,5*S*)-3-[(2*S*,3*S*,4*S*,5*R*,6*S*)-4,5-dihydroxy-6-(hydroxymethyl)-3-methylaminooxan-2-yl]oxy-4-formyl-4-hydroxy-5-methyloxolan-2-yl]oxy-3,4,6-trihydroxycyclohexyl] guanidine $C_{21}H_{39}N_7O_{12}$ 57-92-1	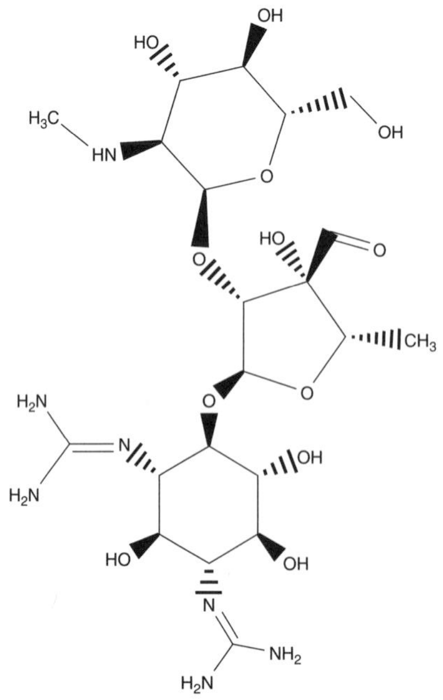	HB$^+$ 7.8[56] HB$^+$ 11.5[56] HB$^+$>12[56]

TABLE 1.2 (*Continued*)

INN	IUPAC Name, Molecular Formula, and CAS Registry No.	Chemical Structure	pK_a
Tobramycin	4-Amino-2-[4,6-diamino-3-[3-amino-6-(aminomethyl)-5-hydroxyoxan-2-yl]oxy-2-hydroxycyclohexyl]oxy-6-(hydroxymethyl)oxane-3,5-diol $C_{18}H_{37}N_5O_9$ 32986-56-4	OH, NH2, H2N, O, NH2, O, HO, NH2, O, O, HO, HO, OH, NH2	HB^+ 6.7[56] HB^+ 8.3[56] HB^+ 9.9[56]
		Aminocyclitols	
Spectinomycin	Decahydro-4α,7,9-trihydroxy-2-methyl-6,8-bis(methylamino)-4*H*-pyrano[2,3-*b*][1,4]benzodioxin-4-one $C_{14}H_{24}N_2O_7$ 1695-77-8	H3C, NH, OH, O, HO, O, H3C, NH, O, O, CH3, OH	HB^+ 7.0[56] HB^+ 8.7[56]

actively multiplying bacteria.[59] β-Lactams bind to proteins in the cell membrane [penicillin-binding proteins (PBPs)], which are enzymes that catalyze cross-linkages between the peptide chains on the *N*-acetylmuramic acid-*N*-acetylglucosamine backbone of the peptidoglycan molecule. Lack of cross-linkages results in the formation of a weak cell wall and can lead to lysis of growing cells. The differences in susceptibility of Gram-positive and Gram-negative bacteria to β-lactams are due to the larger amount of peptidoglycan in the cell wall, differences in PBPs between organisms, and the fact that it is difficult for some β-lactams to penetrate the outer lipopolysaccharide layer of the Gram-negative cell wall. Antimicrobial resistance to β-lactams is due to the action of β-lactamase enzymes that break the β-lactam ring and modification of PBPs, resulting in reduced binding affinity of the β-lactam for the peptide chain. Many Gram-negative bacteria are naturally resistant to some of the β-lactams because the β-lactam cannot penetrate the outer lipopolysaccharide membrane of the cell wall.

β-Lactams have a slower kill rate than do fluoroquinolones and aminoglycosides, and killing activity starts after a lag phase. Antimicrobial activity is usually time-dependent, not concentration-dependent. The β-lactams generally are wholly ionized in plasma and have relatively small volumes of distribution and short half-lives. They do not cross biological membranes well but are widely distributed in extracellular fluids. Elimination is generally through the kidneys.

The penicillins are characterized by their 6-aminopenicillanic acid (6-APA) core. This is a thiazolidone ring linked to a β-lactam ring and a sidechain at position C6, which allows them to be distinguished from one another. Penicillins can be separated into six groups on the basis of their activity. Benzylpenicillin (penicillin G) was the first β-lactam purified for clinical use from *Penicillium* cultures. Clinical limitations were soon recognized, with instability in the presence of gastric acids, susceptibility to β-lactamase enzymes, and ineffectiveness against many Gram-negative organisms. It also has a short terminal half-life of around 30–60 min. However, benzylpenicillin is still the best antibiotic to use against most Gram-positive organisms (except resistant staphylococci and enterococci) and some Gram-negative bacteria. Most commonly now it is administered by deep intramuscular injection as procaine penicillin, where procaine provides a depot effect as a result of slow absorption. The first modification to the 6-APA core was acylation to produce phenoxymethylpenicillin (penicillin V),[60] which is more acid-stable and active orally. This development led to the ability to produce a wide range of semi-synthetic penicillins by adding sidechains to the 6-APA core. The first group were the anti-staphylococcal penicillins such as methicillin,[61] which are resistant to staphylococcal β-lactamases. Of these, cloxacillin is commonly used to treat mastitis in dairy cows. The extended or broad-spectrum penicillins, such as ampicillin, which is active against Gram-negative bacteria, including *Escherichia coli*, was the next class of penicillins. These antibiotics are susceptible to the action of β-lactamases. However, amoxicillin and amoxicillin plus clavulanate (a β-lactamase inhibitor) are widely used in livestock and companion animals to treat Gram-negative infections, particularly those caused by enteric Enterobacteriaceae. The next development was the anti-pseudomonal penicillins such as carbenicillin. These antibiotics are not commonly used in animals. The final class is the (Gram-negative) β-lactamase resistant penicillins such as temocillin. At this time, these are not registered for use in animals.

Shortly after the development of benzypenicillin, cephalosporin C was isolated from the fungus *Cephalosporium acremonium*. Cephalosporins have a 7-aminocephalosporanic acid core that includes the β-lactam ring and were of early interest because of activity against Gram-negative bacteria. In addition, these antibiotics are less susceptible to the action of β-lactamases. Over the years the cephalosporin core molecule was also modified to provide a series of classes (generations) of semi-synthetic cephalosporins with differing activities. The first-generation cephalosporins (e.g., cephalothin) were introduced to treat β-lactamase-resistant staphylococcal infections but also demonstrated activity against Gram-negative bacteria. They are no longer used commonly in companion animals but are still used in dry-cow therapies in dairy cows. Second-generation cephalosporins (e.g., cephalexin) are active against both Gram-positive and Gram-negative organisms. Oral preparations are widely used to treat companion animals. Products are registered for use in mastitis control in dairy cows. Third-generation cephalosporins (e.g., ceftiofur) demonstrate reduced activity against Gram-positive bacteria but increased activity against Gram-negative organisms. Because of their importance in human medicine, these products should be reserved for serious infections where other therapy has failed. They are used to treat both livestock and companion animals. Fourth-generation cephalosporins (e.g., cefquinome) have increased activity against both Gram-positive and Gram-negative bacteria.[62] These are reserve drugs in human medicine but in some countries are registered for use in cattle and horses.

Other β-lactams with natural origins include carbapenems (from *Streptomyces* spp.) and monobactams. These classes of β-lactams are not registered for use in food-producing animals but are used off-label in companion animals. Carbapenems have a wide range of activity against Gram-positive and Gram-negative bacteria and are resistant to most β-lactamases. Monobactams such as aztreonam are resistant to most β-lactamases and have a narrow spectrum of activity with good activity against many Gram-negative bacteria.

β-Lactam antibiotics are largely free of toxic effects, and the margin of safety is substantial. The major adverse effect is acute anaphylaxis, which is uncommon and associated mostly with penicillins; urticaria, angioneurotic edema, and fever occur more commonly. Penicillin-induced immunity-mediated hemolytic anemia in horses has also been reported.[63] The administration of procaine penicillin has led to pyrexia, lethargy, vomiting, inappetance, and cyanosis in pigs[64] and to signs of procaine toxicity, including death in horses.[65,66]

In humans, sensitization and subsequent hypersensitivity reactions to penicillin are relatively common during treatment. By comparison, adverse reactions attributed to occupational exposure to penicillin or the ingestion of food containing residues of penicillin are now seldom reported.

The concentrations of β-lactams reportedly present in the environment are negligible. This is consistent with β-lactam antibiotics being hydrolyzed shortly after they are excreted[67] and rapidly dissipating in a range of manure types.[26]

The CAC MRLs have been established on the basis of risk assessments carried out by the JECFA for benzylpenicillin,[42,68] procaine pencillin,[69] and ceftiofur.[70] The CAC MRLs established are for benzylpenicillin in muscle, liver, kidney, and milk of all food-producing species; for procaine penicillin in muscle, liver, and kidney of pigs and chickens; and for ceftiofur (expressed as desfuroylceftiofur) in muscle, liver, kidney, and fat of cattle and pigs.[43] Details of residue studies considered by JECFA in recommending MRLs for CAC adoption are contained in monographs prepared for benzylpenicillin,[71] procaine penicillin,[72] and ceftiofur.[73,74]

From an analytical perspective, β-lactam antibiotics (Table 1.3) are stable under neutral or slightly basic conditions. These drugs degrade significantly as a result of the composition of some buffers (see Chapter 6 for further discussion).

TABLE 1.3 β-Lactams

INN	IUPAC Name, Molecular Formula, and CAS Registry No.	Chemical Structure	pK_a
	Penicillins		
Amoxicillin	(2*S*,5*R*,6*R*)-6-{[(2*R*)-2-Amino-2-(4-hydroxyphenyl)acetyl]amino}-3,3-dimethyl-7-oxo-4-thia-1-azabicyclo[3.2.0]heptane-2-carboxylic acid $C_{16}H_{19}N_3O_5S$ 26787-78-0	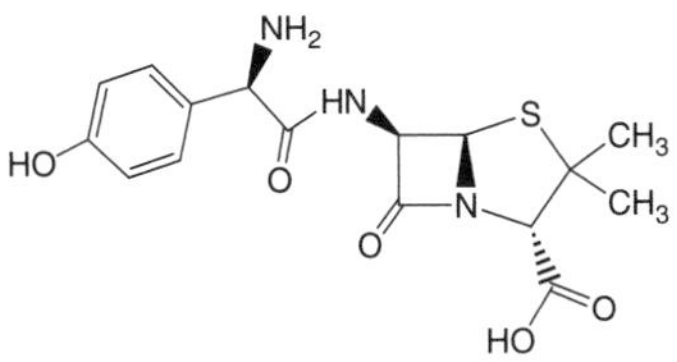	HA 2.6;[56] HB+, HA 7.3;[56] HA, HB+9.5[56]
Ampicillin	(2*S*,5*R*,6*R*)-6-{[(2*R*)-Aminophenylacetyl]amino}-3,3-dimethyl-7-oxo-4-thia-1-azabicyclo[3.2.0]heptane-2-carboxylic acid $C_{16}H_{19}N_3O_4S$ 69-53-4	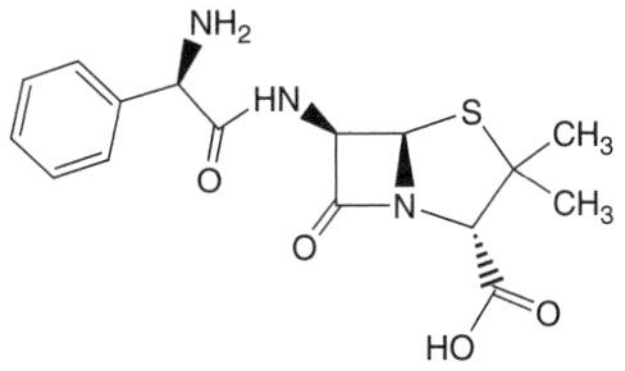	HA 2.5,[56] HB+ 7.3[56]
Benzylpenicillin (penicillin G)	(2*S*,5*R*,6*R*)-3,3-Dimethyl-7-oxo-6-[(2-phenylacetyl)amino]-4-thia-1-azabicyclo[3.2.0]heptane-2-carboxylic acid $C_{16}H_{18}N_2O_4S$ 61-33-6	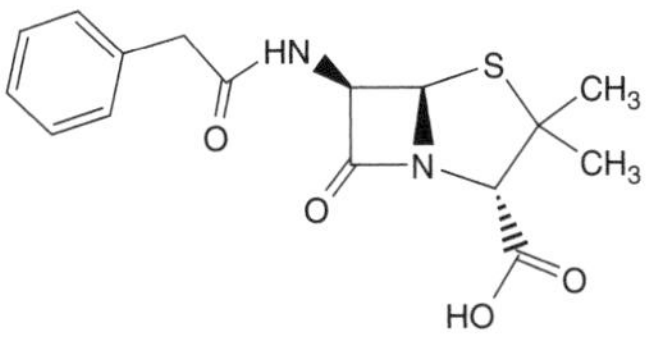	HA 2.7[56]
Carbenicillin	(2*S*,5*R*,6*R*)-6-[(3-Hydroxy-3-oxo-2-phenylpropanoyl)amino]-3,3-dimethyl-7-oxo-4-thia-1-azabicyclo[3.2.0]heptane-2-carboxylic acid $C_{17}H_{18}N_2O_6S$ 4697-36-3	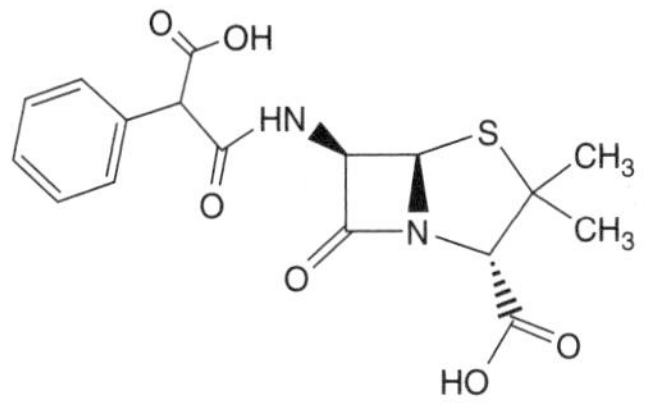	HA 2.2,[56] HA 3.3[56]
Cloxacillin	(2*S*,5*R*,6*R*)-6-[[[3-(2-Chlorophenyl)-5-methyl-4-isoxazolyl]carbonyl]amino]-3,3-dimethyl-7-oxo-4-thia-1-azabicyclo[3.2.0]heptane-2-carboxylic acid $C_{19}H_{18}ClN_3O_5S$ 61-72-3	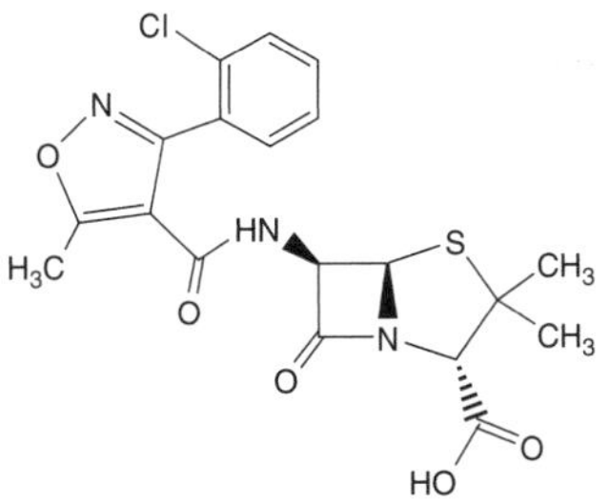	HA 2.7[56]
Dicloxacillin	(2*S*,5*R*,6*R*)-6-[[3-(2,6-Dichlorophenyl)-5-methyl-1,2-oxazole-4-carbonyl]amino]-3,3-dimethyl-7-oxo-4-thia-1-azabicyclo[3.2.0]heptane-2-carboxylic acid $C_{19}H_{17}Cl_2N_3O_5S$ 3116-76-5	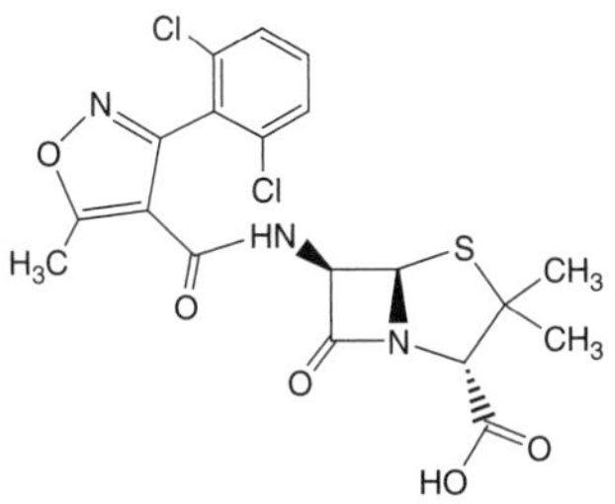	HA 2.7[56]
Mecillinam	(2*S*,5*R*,6*R*)-6-(Azepan-1-ylmethylideneamino)-3,3-dimethyl-7-oxo-4-thia-1-azabicyclo[3.2.0]heptane-2-carboxylic acid $C_{15}H_{23}N_3O_3S$ 32887-01-7	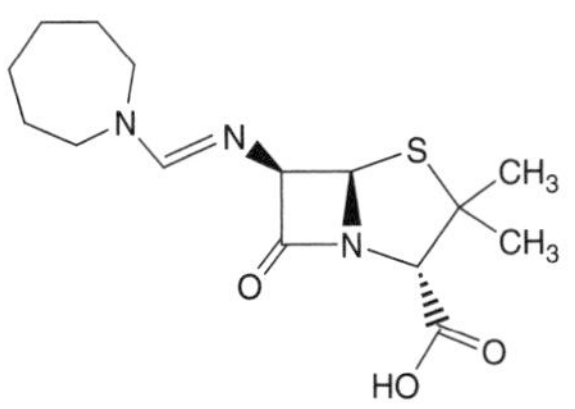	HA 2.7[56] HB+ 8.8[56]

(continued)

TABLE 1.3 (*Continued*)

INN	IUPAC Name, Molecular Formula, and CAS Registry No.	Chemical Structure	pK_a
Methicillin	(2*S*,5*R*,6*R*)-6-[(2,6-Dimethoxybenzoyl)amino]-3,3-dimethyl-7-oxo-4-thia-1-azabicyclo[3.2.0]heptane-2-carboxylic acid $C_{17}H_{20}N_2O_6S$ 61-32-5	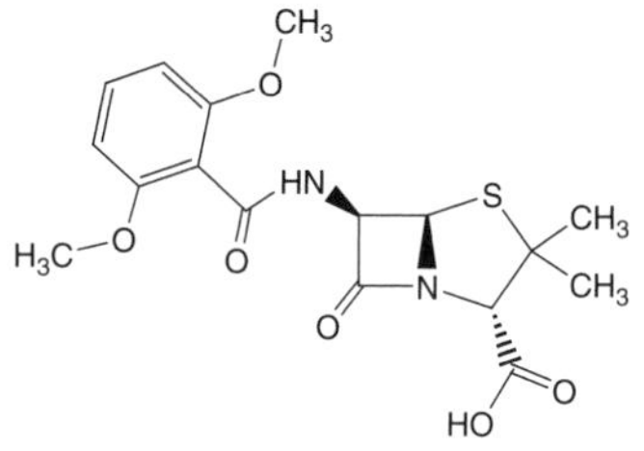	HA 2.8[56]
Nafcillin	2*S*,5*R*,6*R*)-6-[(2-Ethoxynaphthalene-1-carbonyl)amino]-3,3-dimethyl-7-oxo-4-thia-1-azabicyclo[3.2.0]heptane-2-carboxylic acid $C_{21}H_{22}N_2O_5S$ 985-16-0	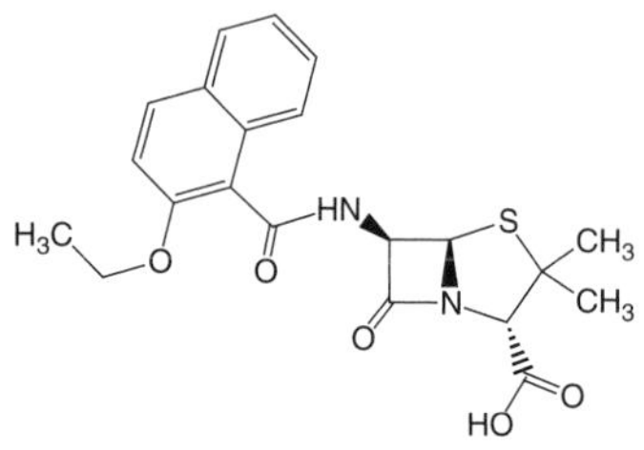	HA 2.7[56]
Oxacillin	(2*S*,5*R*,6*R*)-3,3-Dimethyl-6-[(5-methyl-3-phenyl,1,2-oxazole-4-carbonyl)amino]-7-oxo-4-thia-1-azabicyclo[3.2.0]heptane-2-carboxylic acid $C_{19}H_{19}N_3O_5S$ 66-79-5	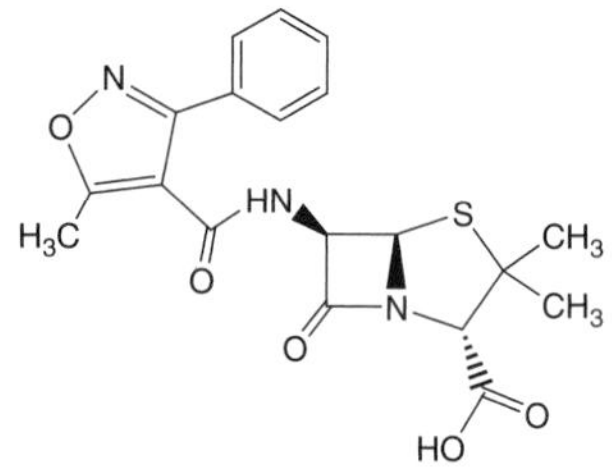	HA 2.7[56]
Penethamate	(2*S*,5*R*)-3,3-Dimethyl-7-oxo-6α-[(phenylacetyl)amino]-4-thia-1-azabicyclo[3.2.0]heptane-2β-carboxylic acid 2-(diethylamino)ethyl ester; (6α-[(phenylacetyl)amino]penicillanic acid 2-(diethylamino)ethyl)ester $C_{22}H_{31}N_3O_4S$ 3689-73-4	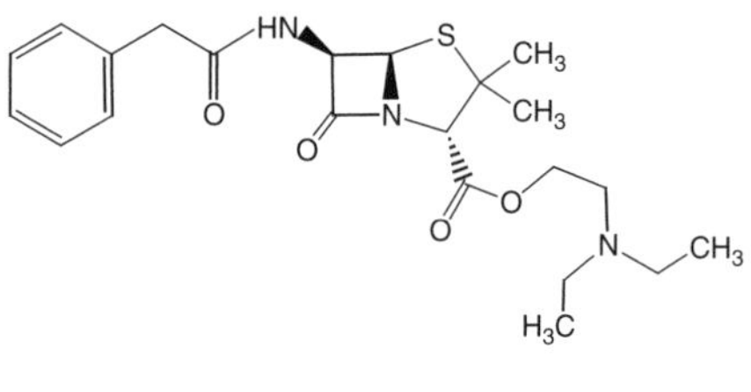	N/A[a]
Phenoxymethyl penicillin (penicillin V)	(2*S*,5*R*,6*R*)-3,3-Dimethyl-7-oxo-6-[[2-(phenoxy)acetyl]amino]-4-thia-1-azabicyclo[3.2.0]heptane-2-carboxylic acid $C_{16}H_{18}N_2O_5S$ 87-08-1	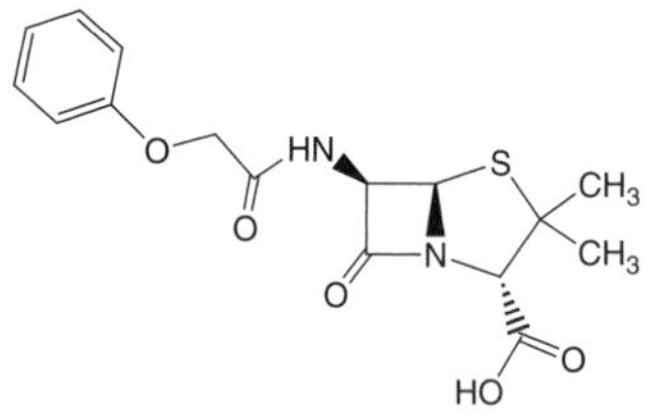	HA 2.7[56]
Temocillin	(2*S*,5*R*,6*S*)-6-[(Carboxy-3-thienylacetyl)amino]-6-methoxy-3,3-dimethyl-7-oxo-4-thia-1-azabicyclo[3.2.0] heptane-2-carboxylic acid $C_{16}H_{18}N_2O_7S_2$ 66148-78-5	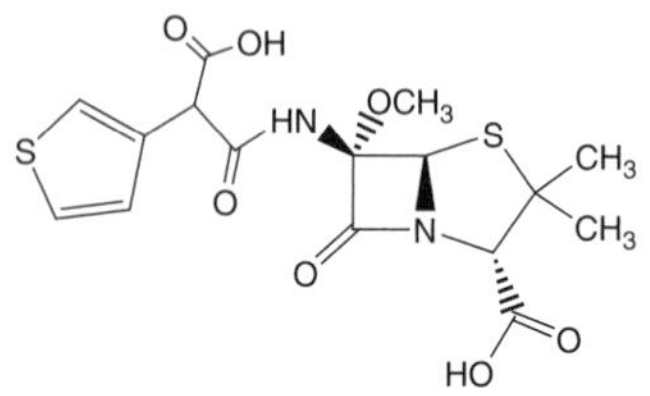	N/A[a]
Ticarcillin	(2*S*,5*R*,6*R*)-6-[[(2*R*)-3-Hydroxy-3-oxo-2-thiophen-3-ylpropanoyl]amino]-3,3-dimethyl-7-oxo-4-thia-1-azabicyclo[3.2.0]heptane-2-carboxylic acid $C_{15}H_{16}N_2O_6S_2$ 34787-01-4	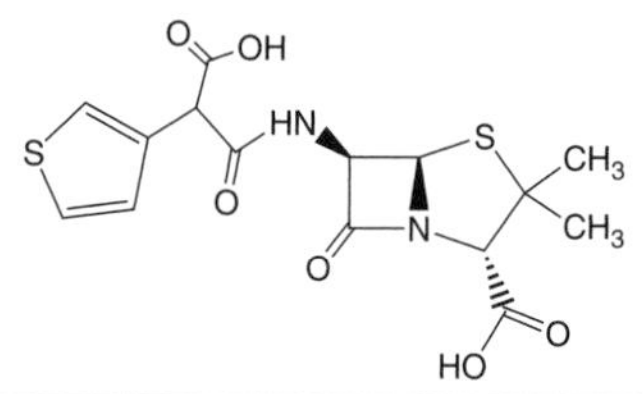	HA 2.9,[56] HB+ 3.3[56]

TABLE 1.3 (*Continued*)

INN	IUPAC Name, Molecular Formula, and CAS Registry No.	Chemical Structure	pK_a
	β-Lactamase Inhibitors		
Clavulanic acid	[2*R*-(2α,3*Z*,5α)]-3-(2-Hydroxyethylidene)-7-oxo-4-oxa-1-azabicyclo[3.2.0]heptane-2-carboxylic acid $C_8H_9NO_5$ 58001-44-8	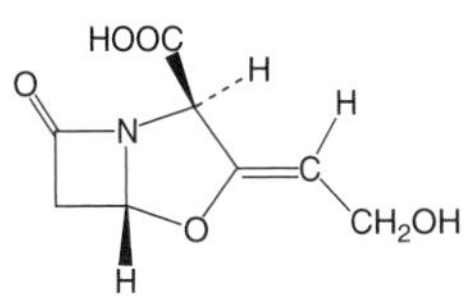	2.7[74]
	Cephalosporins		
Cefacetrile	(6*R*,7*R*)-3-(Acetyloxymethyl)-7-[(2-cyanoacetyl)amino]-8-oxo-5-thia-1-azabicyclo[4.2.0]oct-2-ene-2-carboxylic acid $C_{13}H_{13}N_3O_6S$ 10206-21-0	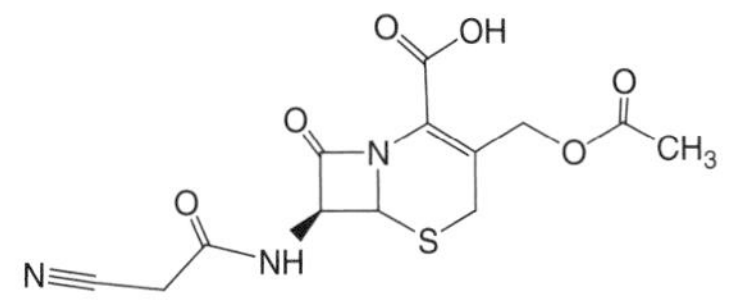	HA 2.0[56]
Cefalonium	(6*R*,7*R*)-3-[(4-Carbamoylpyridin-1-ium-1-yl)methyl]-8-oxo-7-[(2-thiophen-2-ylacetyl)amino]-5-thia-1-azabicyclo[4.2.0]oct-2-ene-2-carboxylate $C_{20}H_{18}N_4O_5S_2$ 5575-21-3	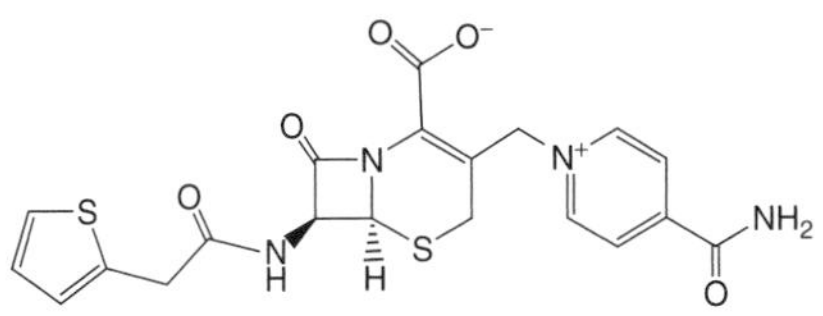	N/A[a]
Cefaprin (cephapirin)	(6*R*,7*R*)-3-(Acetyloxymethyl)-8-oxo-7-[(2-pyridin-4-ylsulfanylacetyl)amino]-5-thia-1-azabicyclo[4.2.0]oct-2-ene-2-carboxylic acid $C_{17}H_{17}N_3O_6S_2$ 21593-23-7	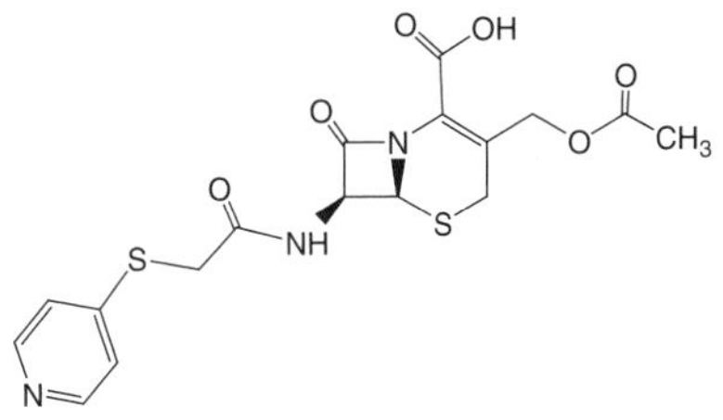	HA 1.8,[56] HB$^+$ 5.6[56]
Cefazolin	(7*R*)-3-[(5-Methyl-1,3,4-thiadiazol-2-yl)sulfanylmethyl]-8-oxo-7-[[2-(tetrazol-1-yl)acetyl]amino]-5-thia-1-azabicyclo[4.2.0]oct-2-ene-2-carboxylic acid $C_{14}H_{14}N_8O_4S_3$ 25953-19-9	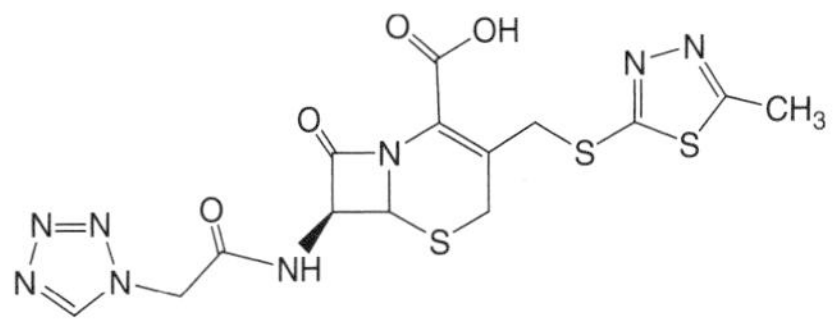	HA 2.8[56]
Cefoperazone	(6*R*,7*R*)-7-[[2-[(4-Ethyl-2,3-dioxopiperazine-1-carbonyl)amino]-2-(4-hydroxyphenyl)acetyl]amino]-3-[(1-methyltetrazol-5-yl)sulfanylmethyl]-8-oxo-5-thia-1-azabicyclo[4.2.0]oct-2-ene-2-carboxylic acid $C_{25}H_{27}N_9O_8S_2$ 62893-19-0	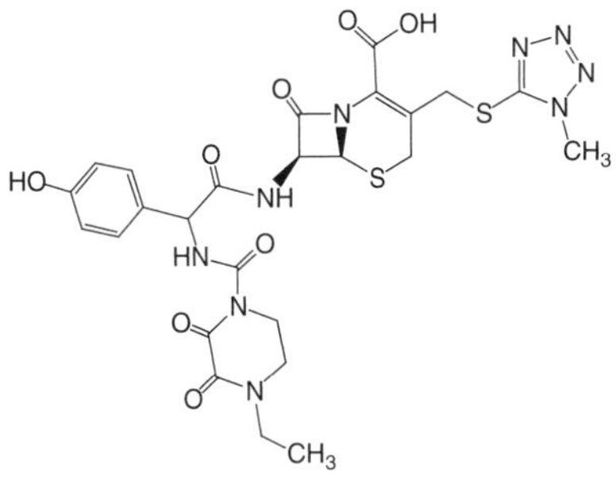	HA 2.6[56]
Cefquinome	1-[[(6*R*,7*R*)-7-[[(2*Z*)-(2-Amino-4-thiazolyl)-(methoxyimino)acetyl]amino]-2-carboxy-8-oxo-5-thia-1-azabicyclo[4.2.0]-oct-2-en-3-yl]methyl]-5,6,7,8-tetrahydroquinolinium inner salt $C_{23}H_{24}N_6O_5S_2$ 84957-30-2	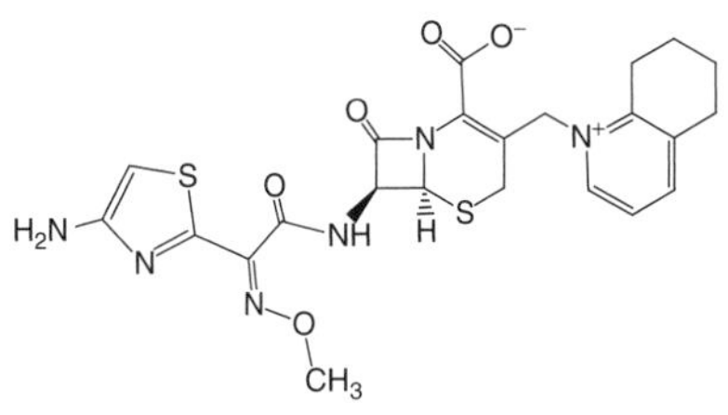	N/A[a]

(*continued*)

TABLE 1.3 (*Continued*)

INN	IUPAC Name, Molecular Formula, and CAS Registry No.	Chemical Structure	pK_a
Ceftiofur	(6*R*,7*R*)-7-[[(2*Z*)-(2-Amino-4-thiazolyl)(methoxyimino)acetyl]amino]-3-[[(2-furanylcarbonyl)thio]methyl]-8-oxo-5-thia-1-azabicyclo[4.2.0]oct-2-ene-2-carboxylic acid $C_{19}H_{17}N_5O_7S_3$ 80370-57-6	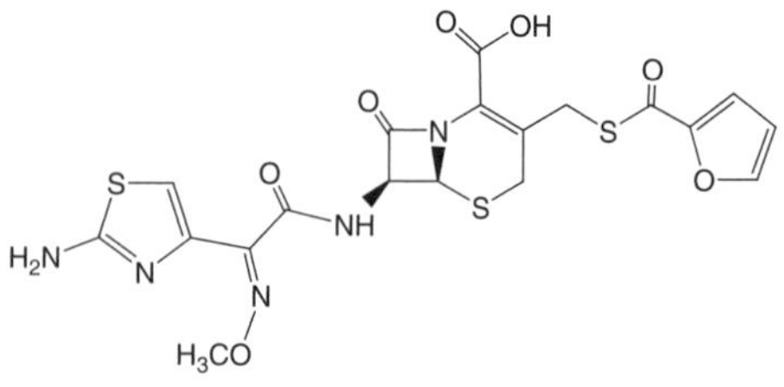	N/A[a]
Cefuroxime	(6*R*,7*R*)-3-(Carbamoyloxymethyl)-7-[[(2*E*)-2-furan-2-yl-2-methoxyiminoacetyl]amino]-8-oxo-5-thia-1-azabicyclo[4.2.0]oct-2-ene-2-carboxylic acid $C_{16}H_{16}N_4O_8S$ 55268-75-2	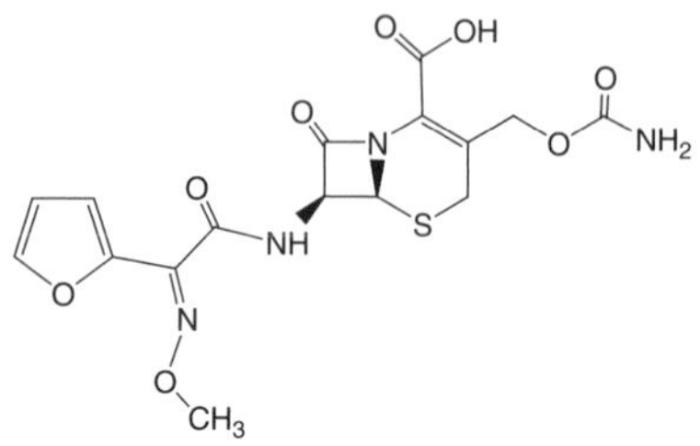	HA 2.5[56]
Cephalexin	(6*R*,7*R*)-7-[[(2*R*)-2-Amino-2-phenylacetyl]amino]-3-methyl-8-oxo-5-thia-1-azabicyclo[4.2.0]oct-2-ene-2-carboxylic acid $C_{16}H_{17}N_3O_4S$ 15686-71-2	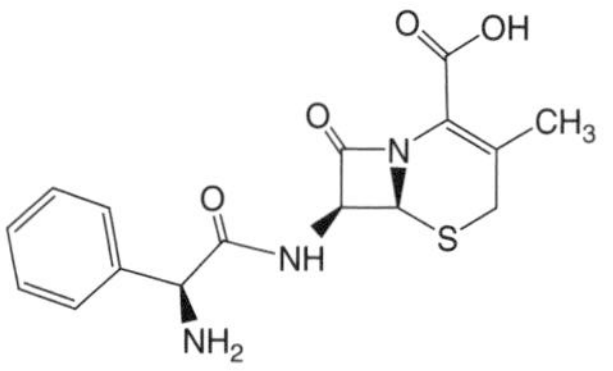	HA 2.5,[56] HB$^+$ 7.1[56]
Cephalothin	(6*R*,7*R*)-3-(Acetyloxymethyl)-8-oxo-7-[(2-thiophen-2-ylacetyl)amino]-5-thia-1-azabicyclo[4.2.0]oct-2-ene-2-carboxylic acid $C_{16}H_{16}N_2O_6S_2$ 153-61-7	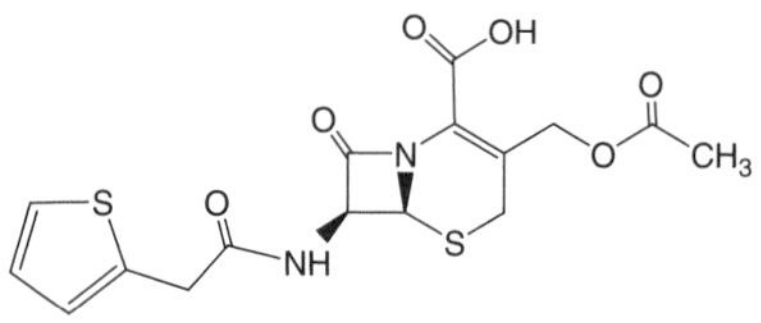	HA 2.4[56]

[a]The author was not able to find a pK_a value for the substance in the public literature (N/A = data not available).

1.3.3 Quinoxalines

The quinoxaline-1,4-di-*N*-oxides were originally investigated for potential antagonism to vitamin K activity. Quindoxin (quinoxaline-1,4-dioxide) was later used as a growth promoter in animal husbandry before being withdrawn because of its photoallergic properties. In the 1970s, three synthetic derivatives of quindoxin—carbadox, cyadox, and olaquindox—became available as antimicrobial growth promoters. These substances are active against Gram-positive and some Gram-negative bacteria as well as some chlamydiae and protozoa. Their antimicrobial activity is attributed to the inhibition of DNA synthesis by a mechanism that is not completely understood. On the basis of studies conducted in *E. coli*, Suter et al.[75] postulated that free radicals produced by the intracellular reduction of quinoxalines damage existing DNA and inhibit the synthesis of new DNA. Resistance to olaquindox has been reported in *E. coli* to be *R*-plasmid-mediated.

Carbadox is well absorbed when administered as a feed additive to pigs. Nonetheless, concentrations of carbadox in the stomach and duodenum of pigs following in-feed administration of 50 mg/kg are adequate to provide effective prophylaxis against *Brachyspira hyodysenteriae*, the causative agent in swine dysentery.[76] The major metabolites of carbadox are its aldehyde, desoxycarbadox, and quinoxaline-2-carboxylic acid. Urinary excretion accounts for two-thirds of a carbadox dose within 24 h of administration. Olaquindox is rapidly and extensively absorbed following oral administration to pigs and undergoes oxidative and/or reductive metabolism. Urinary excretion of unchanged olaquindox and a mono-*N*-oxide of olaquindox accounts for approximately 70% and 16%, respectively, of a dose within 24 h of administration.

Van der Molen et al.[77] and Nabuurs et al.[78] investigated the toxicity of quinoxalines in pigs. A dose of 50 mg/kg carbadox was demonstrated to cause increased fecal dryness, reduced appetite, dehydration, and disturbances in electrolyte homeostasis. These signs are attributable principally to hypoaldosteronism, a manifestation of carbadox-induced damage of the adrenal glands. The accidental feeding of high doses (331–363 mg/kg) of carbadox to weaner pigs resulted in inappetance, ill thrift, posterior paresis, and deaths.[79] The toxic effect of olaquindox is comparable with that of carbadox, whereas cyadox is less toxic.

Carbadox is used in feed at a dose of 10–25 mg/kg as an antimicrobial growth-promoting agent for improving weight gain and feed efficiency in pigs. The commercial product is used in starter and/or grower rations but not in finisher rations. A dose of 50–55 mg/kg carbadox is administered as a feed additive for the prevention and control of (1) swine dysentery caused by the anaerobic intestinal spirochaetal bacterium, *Brachyspira hyodysenteriae* and (2) bacterial enteritis caused by susceptible organisms. Carbadox is also used in pigs to treat nasal infections caused by *Bordetella bronchiseptica*. Olaquindox is administered as medicated feed to pigs for improving feed conversion efficiency and for the prevention of porcine proliferative enteritis caused by *Campylobacter* species. Cyadox has been used as a feed additive for pigs, calves, and poultry to promote growth.

Occupational exposure of farmworkers to the quinoxaline class of antimicrobials may result in dermal photosensitivity reactions. In general terms, photosensitivity may take the form of phototoxic reactions, whereby a drug absorbs energy from ultraviolet A light and releases it into the skin, causing cellular damage; or photoallergic reactions, whereby light causes a structural change in a drug so that it acts as a hapten, possibly binding to proteins in the skin. Olaquindox causes photoallergic reactions in humans and animals. On exposure to light, olaquindox forms a reactive oxaziridine derivative, and this imino-*N*-oxide reacts with protein to form a photoallergen. In 1999, the use of carbadox and olaquindox was banned in the European Union in response to concerns of toxicity to humans from occupational exposure.[80] More recently, the health concerns with carbadox and olaquindox identified by the JECFA were noted at the 18th Session of the CCRVDF, as was the ongoing use of these substances in some countries.[81]

In addition to the concerns relating to occupational exposure described above, the use of quinoxalines (see list in Table 1.4) in food-producing species is associated with food safety concerns. The genotoxic and carcinogenic nature of carbadox and its metabolites and the presence of relatively persistent residues in edible tissues of pigs treated with carbadox resulted in the JECFA not allocating an acceptable daily intake (ADI).[82,83] In the case of olaquindox, the JECFA[84] concluded that the substance is potentially genotoxic and that the toxicity of its metabolites is inadequately understood. For these reasons, the JECFA was unable to determine the amount of residues in food that did not cause an appreciable risk to human health, and thus MRLs were not established for these compounds by the CAC (see Chapter 3 for further discussion). Details

TABLE 1.4 Quinoxalines

INN	IUPAC Name, Molecular Formula, CAS Registry No.	Chemical Structure	pK_a
Carbadox	Methyl (2*E*)-2-[(1,4-dioxidoquinoxalin-2-yl)-methylene]hydrazine carboxylate $C_{11}H_{10}N_4O_4$ 6804-07-5	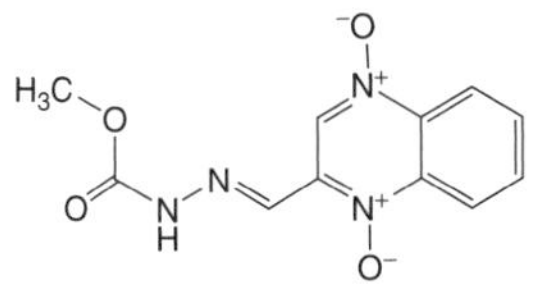 H$_3$C, O, O, N, H, N, N$^+$, $^-$O, N$^+$, O$^-$	N/A[a]
Cyadox	2-Cyano-*N*-[(*E*)-(1-hydroxy-4-oxido-quinoxalin-2-ylidene)methyl]iminoacetamide $C_{12}H_9N_5O_3$ 65884-46-0	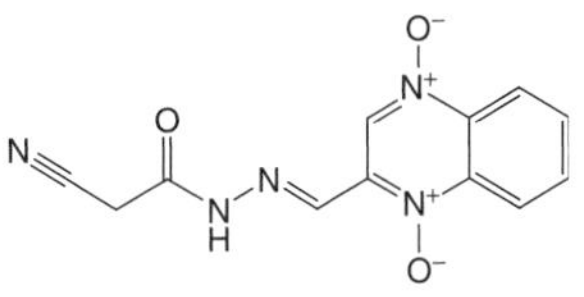 N, O, N, H, N, N$^+$, O$^-$, N$^+$, O$^-$	N/A[a]
Olaquindox	*N*-(2-Hydroxyethyl)-3-methyl-4-oxido-1-oxoquinoxalin-1-ium-2-carboxamide $C_{12}H_{13}N_3O_4$ 23696-28-8	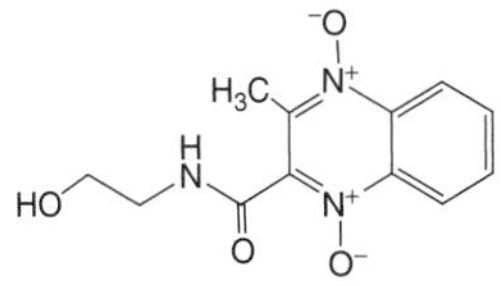 HO, H, N, O, H$_3$C, N$^+$, $^-$O, N$^+$, O$^-$	N/A[a]
Quindoxin	Quinoxaline-1,4-dioxide $C_8H_6N_2O_2$ 2423-66-7	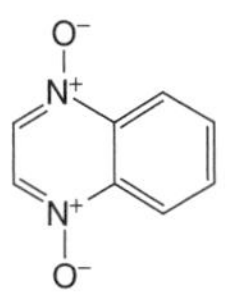 O$^-$, N$^+$, N$^+$, O$^-$	N/A[a]

[a]The author was not able to find a pK_a value for the substance in the public literature.

of residue studies on olaquindox reviewed by JECFA are available in monographs prepared for the 36th[85] and 42nd[86] meetings of the committee.

1.3.4 Lincosamides

The lincosamide class of antimicrobial drugs includes lincomycin, clindamycin, and pirlimycin; two of these drugs—lincomycin and pirlimycin—are approved for use in food-producing species. Lincosamides are derivatives of an amino acid and a sulfur-containing galactoside. Lincomycin was isolated in 1962 from the fermentation product of *Streptomyces lincolnensis* subsp. *lincolnensis*. Clindamycin is a semi-synthetic derivative of lincomycin, and pirlimycin is an analog of clindamycin.

The lincosamides inhibit protein synthesis in susceptible bacteria by binding to the 50*S* subunits of bacterial ribosomes and inhibiting peptidyltransferases; interference with the incorporation of amino acids into peptides occurs thereby. Lincosamides may be bacteriostatic or bactericidal depending on the concentration of drug at the infection site, bacterial species and bacterial strain. These drugs have activity against many Gram-positive bacteria and most obligate anaerobes but are not effective against most Gram-negative organisms. Clindamycin, which is not approved for use in food-producing animals, has a wider spectrum of activity than does lincomycin.

Resistance specific to lincosamides results from the enzymatic inactivation of these drugs. More common, however, is cross-resistance among macrolides, lincosamides, and streptogramin group B antibiotics (MLSB resistance). With this form of resistance, binding of the drug to the target is prevented on account of methylation of the adenine residues in the 23*S* ribosomal RNA of the 50*S* ribosomal subunit (the target).[87] Complete cross-resistance between lincomycin and clindamycin occurs with both forms of resistance.

Lincomycin is effective against *Staphylococcus* species, *Streptococcus* species (except *Streptococcus faecalis*), *Erysipelothrix insidiosa, Leptospira pomona*, and *Mycoplasma* species. Lincomycin hydrochloride is added to feed or drinking water to treat and control swine dysentery in pigs and to control necrotic enteritis in chickens. It is used also in medicated feed for growth promotion and to increase feed efficiency in chickens and pigs, the control of porcine proliferative enteropathies caused by *Lawsonia intracellularis* in pigs, and the treatment of pneumonia caused by *Mycoplasma* species in pigs. An injectable formulation of lincomycin is used in pigs to treat joint infections and pneumonia.

Several combination products containing lincomycin are approved for use in food-producing species. A lincomycin–spectinomycin product administered in drinking water is used for the treatment and control of respiratory disease and for improving weight gains in poultry. A product containing the same active ingredients is available for in-feed or drinking water administration to pigs for the treatment and control of enteric and respiratory disease, treatment of infectious arthritis, and increasing weight gain. Injectable combination products containing lincomycin and spectinomycin are used for the treatment of bacterial enteric and respiratory disease in pigs and calves, treatment of arthritis in pigs, and treatment of contagious foot-rot in sheep. A lincomycin–sulfadiazine combination product administered in-feed is used for the treatment of atrophic rhinitis and enzootic pneumonia in pigs. Lincomycin–neomycin combination products are used for treating acute mastitis in lactating dairy cattle.

Pirlimycin is approved as an intramammary infusion for the treatment of mastitis in lactating dairy cattle. It is active against sensitive organisms such as *Staphylococcus aureus, Streptococcus agalactiae, Streptococcus uberis, Streptococcus dysgalactiae*, and some enterococci. Pirlimycin exhibits a post-antibiotic effect *in vitro* against *Staphylococcus aureus* isolated from bovine mastitis, and exposure of pathogens to subinhibitory concentrations increases their susceptibility to phagocytosis by polymorphonuclear leukocytes. Many species of anaerobic bacteria are extremely sensitive to pirlimycin.

The use of lincosamides (see list in Table 1.5) is contraindicated in horses because of the potential risk of serious or fatal enterocolitis and diarrhea. This commonly involves overgrowth of the normal microflora by nonsusceptible bacteria such as *Clostridium* species. Oral administration of lincomycin to ruminants has also been associated with adverse side effects such as anorexia, ketosis, and diarrhea. Such use is therefore contraindicated in ruminants.

The limited information available suggests that lincomycin does not pose a risk to organisms in those environments where the drug is known to be used. A 2006 UK study that used targeted monitoring detected a maximum concentration of 21.1 μg lincomycin per liter of streamwater, which compares with the predicted no-effect concentration for lincomycin of 379.4 μg per liter.[88]

From a food safety perspective, the JECFA has allocated ADI values for lincomycin[89] and pirlimycin.[89] On the basis of JECFA recommendations, CAC MRLs for lincomycin in muscle, liver, kidney, and fat of pigs and chickens, and in cow's milk and for pirlimycin in muscle, liver, kidney, and fat of cattle and in cow's milk have also been established.[43] Details of residue studies reviewed by JECFA to develop MRL recommendations for CCRVDF may be found in monographs published for lincomycin[91–93] and pirlimycin.[94]

TABLE 1.5 Lincosamides

INN	IUPAC Name, Molecular Formula, and CAS Registry No.	Chemical Structure	pK_a
Clindamycin	(2*S*,4*R*)-*N*-[2-chloro-1-[(2*R*,3*R*,4*S*,5*R*,6*R*)-3,4,5-trihydroxy-6-methyl-sulfanyloxan-2-yl]propyl]-1-methyl-4-propylpyrrolidine-2-carboxamide $C_{18}H_{33}ClN_2O_5S$ 18323-44-9	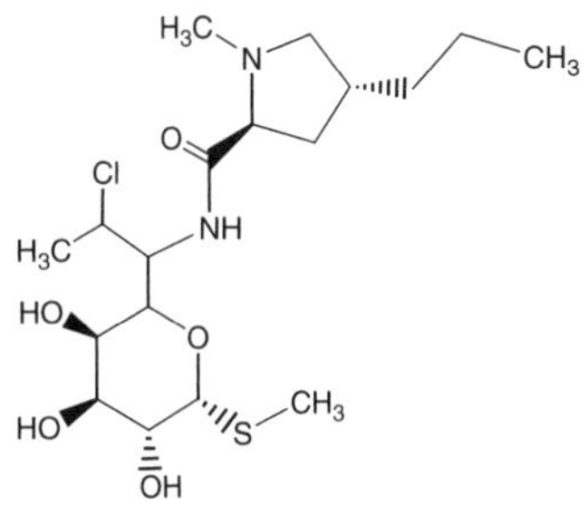	HB^+ 7.7[56]
Lincomycin	(4*R*)-*N*-[(1*R*,2*R*)-2-hydroxy-1-[(2*R*,3*R*,4*S*,5*R*,6*R*)-3,4,5-trihydroxy-6-methylsulfanyloxan-2-yl]propyl]-1-methyl-4-propylpyrrolidine-2-carboxamide $C_{18}H_{34}N_2O_6S$ 154-21-2	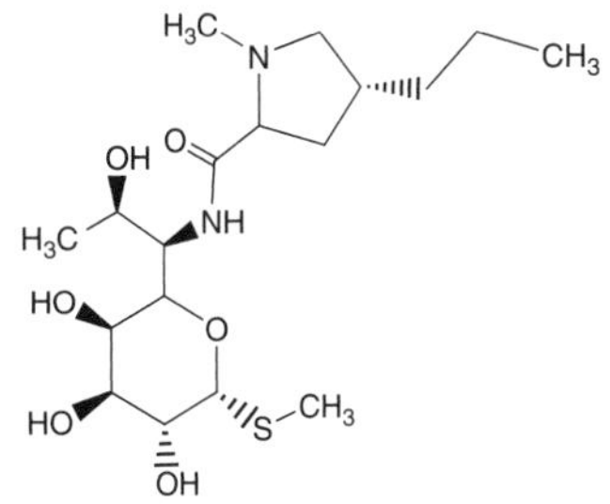	HB^+ 7.5[56]
Pirlimycin	Methyl(2*S*-*cis*)-7-chloro-6,7,8-trideoxy-6[[(4-ethyl-2-piperidinyl)-carbonyl]amino]-1-thio-L-*threo*-α-D-galactooctopyranoside $C_{17}H_{31}ClN_2O_5S$ 79548-73-5	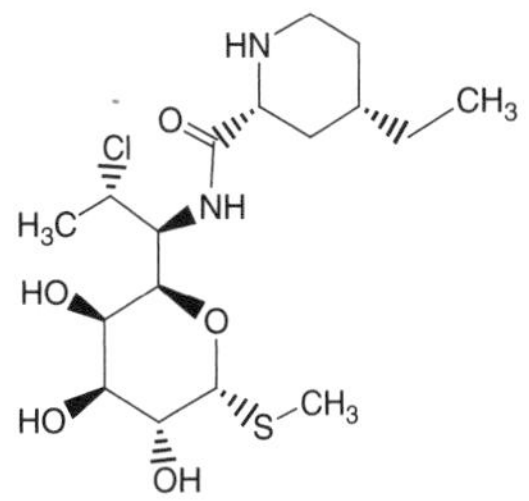	8.5[74]

1.3.5 Macrolides and Pleuromutilins

The macrolide class of antibiotics consists of natural products isolated from fungi and their semi-synthetic derivatives. The macrolide structure is characterized by a 12–16-atom lactone ring; however, none of the 12-member ring macrolides are used clinically. Erythromycin and oleandomycin are 14-member ring macrolides derived from strains of *Saccharopolyspora erythreus* (formerly *Streptomyces erythreus*) and *Streptomyces antibioticus*, respectively. Clarithromycin and azithromycin are semi-synthetic derivatives of erythromycin. Spiramycin and tylosin are 16-member ring macrolides derived from strains of *Ambofaciens streptomyces* and the actinomycete *Streptomyces fradiae*, respectively. Tilmicosin is a 16-member ring macrolide produced semi-synthetically by chemical modification of desmycosin. Tulathromycin, a semi-synthetic macrolide, is a mixture of a 13-member ring macrolide (10%) and a 15-member ring macrolide (90%) (shown in Table 1.6). Macrolide drugs are complex mixtures of closely related antibiotics that differ from one another with respect to the chemical substitutions on the various carbon atoms in the structure, and in the aminosugars and neutral sugars. Erythromycin, for example, consists primarily of erythromycin A (shown in Table 1.6), but the B, C, D, and E forms may also be present. It was not until 1981 that erythromycin A was chemically synthesized. Two pleuromutilins, tiamulin and valnemulin, are used in animals, and these compounds are semi-synthetic derivatives of the naturally occurring diterpene antibiotic, pleuromutilin.

The antimicrobial activity of the macrolides is attributed to the inhibition of protein synthesis. Macrolides bind to the 50*S* subunit of the ribosome, resulting in blockage of the transpeptidation or translocation reactions, inhibition of protein synthesis, and thus the inhibition of cell growth. These drugs are active against most aerobic and anaerobic Gram-positive bacteria, Gram-negative cocci, and also *Haemophilus, Actinobacillus, Bordetella, Pasteurella, Campylobacter*, and *Helicobacter*. However, they are not active against most Gram-negative bacilli. The macrolides display activity against atypical mycobacteria, mycobacteria, mycoplasma, chlamydia, and rickettsia species. They are predominantly bacteriostatic, however, high concentrations are slowly bactericidal against more sensitive organisms. In human medicine, erythromycin,

TABLE 1.6 Macrolides and Pleuromutilins

INN	IUPAC Name Molecular Formula, and CAS Registry No.	Chemical Structure	pK_a
		Macrolides	
Azithromycin	[2*R*-(2*R**,3*S**,4*R**,5*R**,8*R**,10*R**,11*R**,12*S**,13*S**,14*R**)]-13-[(2,6-Dideoxy-3-*C*-methyl-3-*O*-methyl-α-L-ribohexopyranosyl)oxy]-2-ethyl-3,4,10-trihydroxy-3,5,6,8,10,12,14-heptamethyl-11-[[3,4,6-trideoxy-3-(dimethylamino)-β-D-xylohexopyranosyl]oxy]1-oxa-6-azacyclopentadecan-15-one $C_{38}H_{72}N_2O_{12}$ 83905-01-5	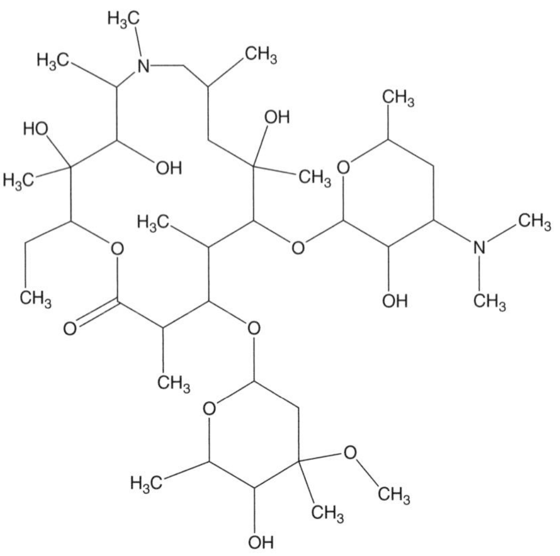	HB^+ 8.7,[56] HB^+ 9.5[56]
Carbomycin	[(2*S*,3*S*,4*R*,6*S*)-6-[(2*R*,3*S*,4*R*,5*R*,6*S*)-6-[[(3*R*,7*R*,8*S*,9*S*,10*R*,12*R*,14*E*)-7-acetyloxy-8-methoxy-3,12-dimethyl-5,13-dioxo-10-(2-oxoethyl)-4,17-dioxabicyclo[14.1.0]heptadec-14-en-9-yl]oxy]-4-(dimethylamino)-5-hydroxy-2-methyloxan-3-yl]oxy-4-hydroxy-2,4-dimethyloxan-3-yl]3-methylbutanoate $C_{42}H_{67}NO_{16}$ 4564-87-8	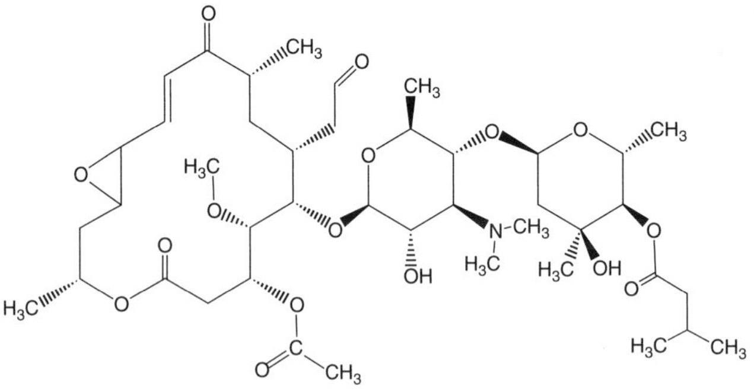	HB^+ 7.6[56]
Erythromycin A	(3*R*,4*S*,5*S*,6*R*,7*R*,9*R*,11*R*,12*R*,13*S*,14*R*)-6-[(2*S*,3*R*,4*S*,6*R*)-4-Dimethylamino-3-hydroxy-6-methyloxan-2-yl]oxy-14-ethyl-7,12,13-trihydroxy-4-[(2*R*,4*R*,5*S*,6*S*)-5-hydroxy-4-methoxy-4,6-dimethyloxan-2-yl]oxy-3,5,7,9,11,13-hexamethyl-1-oxacyclotetradecane-2,10-dione $C_{37}H_{67}NO_{13}$ 114-07-8	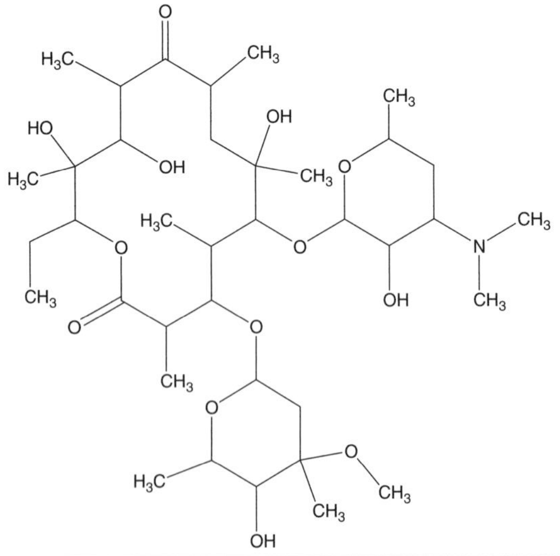	HB^+ 8.6[56]

TABLE 1.6 (*Continued*)

INN	IUPAC Name Molecular Formula, and CAS Registry No.	Chemical Structure	pK_a
Kitasamycin (Leucomycin A_1)	[(2*S*,3*S*,4*R*,6*S*)-6-[(2*R*,3*S*,4*R*,5*R*,6*S*)-6-[[(4*R*,5*S*,6*S*,7*R*,9*R*,10*R*,11*E*,13*E*,16*R*)-4-acetyloxy-10-Hydroxy-5-methoxy-9,16-dimethyl-2-oxo-7-(2-oxoethyl)-1-oxacyclohexadeca-11,13-dien-6-yl]oxy]-4-dimethylamino-5-hydroxy-2-methyloxan-3-yl]oxy-4-hydroxy-2,4-dimethyloxan-3-yl]-3-methylbutanoate $C_{40}H_{67}NO_{14}$ 1392-21-8	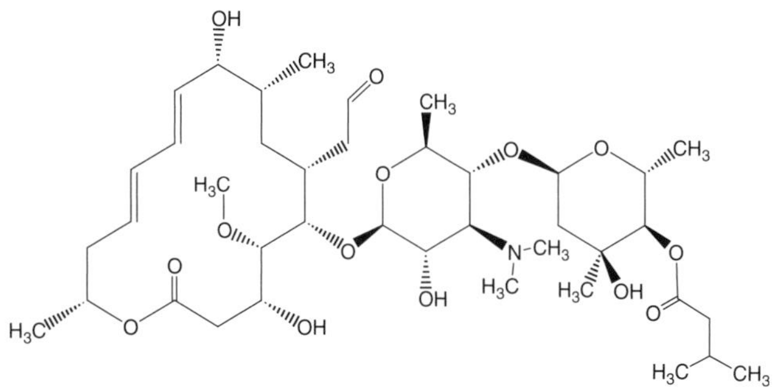	N/A[a]
Neospiramycin	2-[(1*R*,3*R*,4*R*,5*E*,7*E*,10*R*,14*R*,15*S*,16*S*)-16-[(2*S*,3*R*,4*S*,5*S*,6*R*)-4-(Dimethylamino)-3,5-dihydroxy-6-methyloxan-2-yl]oxy-4-[(2*r*,5*s*,6*r*)-5-(dimethylamino)-6-methyloxan-2-yl]oxy-14-hydroxy-15-methoxy-3,10-dimethyl-12-oxo-11-oxacyclohexadeca-5,7-dien-1-yl]acetaldehyde $C_{36}H_{62}N_2O_{11}$ 102418-06-4	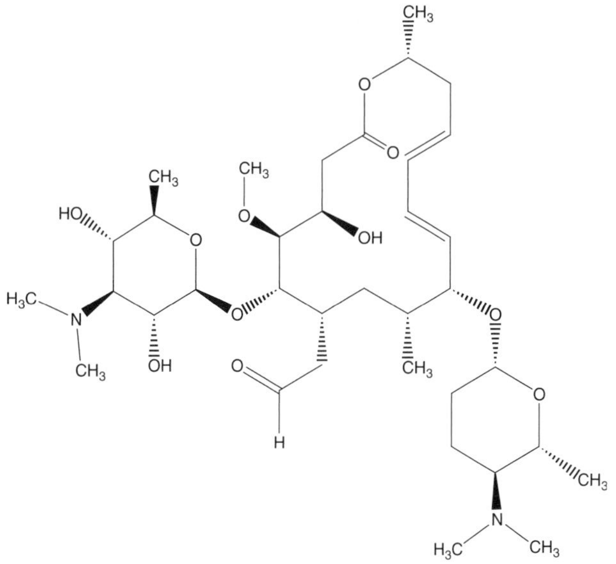	N/A[a]
Oleandomycin	(3*R*,5*R*,6*S*,7*R*,8*R*,11*R*,12*S*,13*R*,14*S*,15*S*)-14-((2*S*,3*R*,4*S*,6*R*)-4-(Dimethylamino)-3-hydroxy-6-methyltetrahydro-2*H*-pyran-2-yloxy)-6-hydroxy-12-((2*R*,4*S*,5*S*,6*S*)-5-hydroxy-4-methoxy-6-methyltetrahydro-2*H*-pyran-2-yloxy)-5,7,8,11,13,15-hexamethyl-1,9-dioxaspiro[2.13]hexadecane-4,10-dione $C_{35}H_{61}NO_{12}$ 3922-90-5	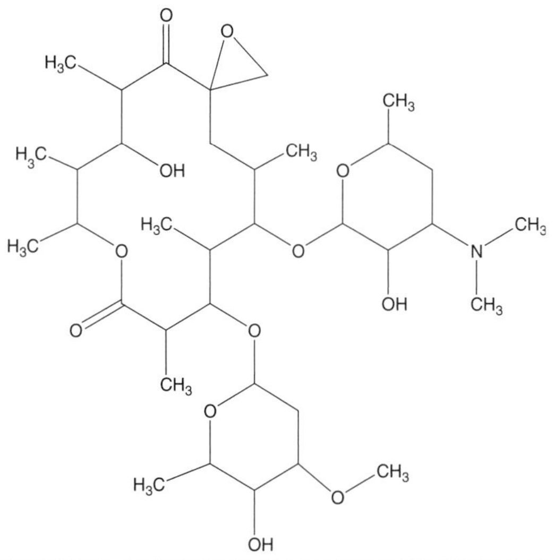	HB^+ 8.5[56]

(*continued*)

TABLE 1.6 (*Continued*)

INN	IUPAC Name Molecular Formula, and CAS Registry No.	Chemical Structure	pK_a
Roxithromycin	(3*R*,4*S*,5*S*,6*R*,7*R*,9*R*,11*S*,12*R*,13*S*,14*R*)-6-[(2*S*,3*R*,4*S*,6*R*)-4-Dimethylamino-3-hydroxy-6-methyloxan-2-yl]oxy-14-ethyl-7,12,13-trihydroxy-4-[(2*R*,4*R*,5*S*,6*S*)-5-hydroxy-4-methoxy-4,6-dimethyloxan-2-yl]oxy-10-(2-methoxyethoxy-methoxyimino)-3,5,7,9,11,13-hexamethyl-1-oxacyclotetradecan-2-one $C_{41}H_{76}N_2O_{15}$ 80214-83-1	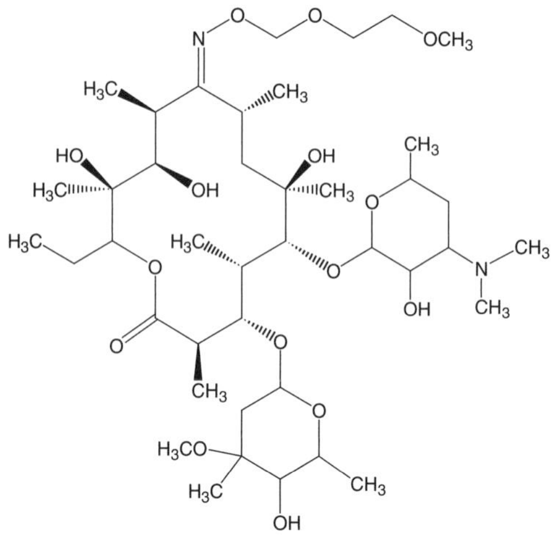	N/A[a]
Spiramycin	(4*R*,5*S*,6*R*,7*R*,9*R*,10*R*,11*E*,13*E*,16*R*)-10-{[(2*R*,5*S*,6*R*)-5-(Dimethylamino)-6-methyltetrahydro-2*H*-pyran-2-yl]oxy}-9,16-dimethyl-5-methoxy-2-oxo-7-(2-oxoethyl)oxacyclohexadeca-11,13-dien-6-yl 3,6-dideoxy-4-*O*-(2,6-dideoxy-3-*C*-methyl-α-L-*ribo*-hexopyranosyl)-3-(dimethylamino)-α-D-glucopyranoside $C_{43}H_{74}N_2O_{14}$ (spiramycin I) 8025-81-8	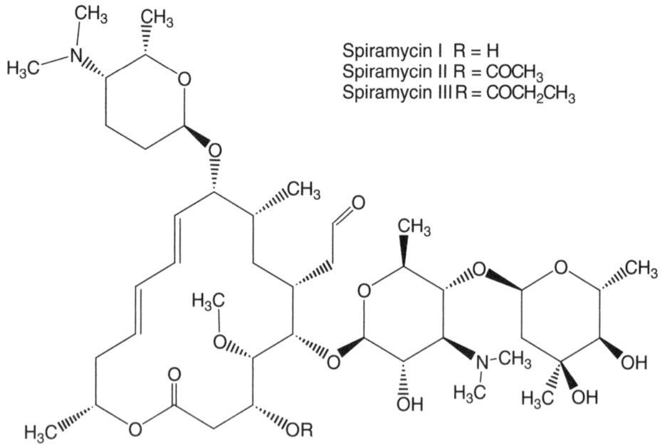 Spiramycin I R = H Spiramycin II R = $COCH_3$ Spiramycin III R = $COCH_2CH_3$	8.2[15]
Tilmicosin	(10*E*,12*E*)-(3*R*,4*S*,5*S*,6*R*,8*R*,14*R*,15*R*)-14-(6-Deoxy-2,3-di-*O*-methyl-*b*-*d*-*allo*-hexopyranosyoxymethyl)-5-(3,6-dideoxy-3-dimethylamino-*b*-*d*-glucohexapyranosyloxy)-6-[2-(*cis*-3,5-dimethylpiperidino)ethyl]-3-hydroxy-4,8,12-trimethyl-9-oxoheptadeca-10,12-dien-15-olide $C_{46}H_{80}N_2O_{13}$ 108050-54-0	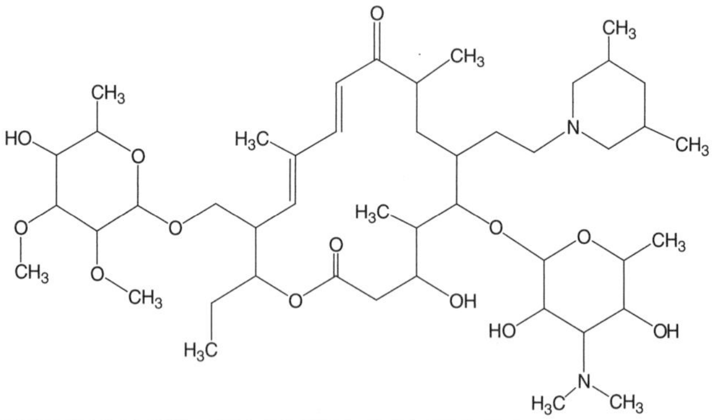	HB^+ 8.2,[56] HB^+ 9.6[56]

TABLE 1.6 ***(Continued)***

INN	IUPAC Name Molecular Formula, and CAS Registry No.	Chemical Structure	pK_a
Tulathromycin	(2*R*,3*S*,4*R*,5*R*,8*R*,10*R*,11*R*,12*S*,13*S*,14*R*)-13-[[2,6-Dideoxy-3-*C*-methyl-3-*O*-methyl-4-*C*-[(propylamino)methyl]-α-L-ribo-hexopyrano-syl]oxy]-2-ethyl-3,4,10-trihydroxy-3,5,8,10,12,14-hexamethyl-11-[[3,4,6-trideoxy-3-(dimethylamino)-β-D-xylohexopyranosyl]-oxy]-1-oxa-6-azacyclopentadecan-15-one $C_{41}H_{79}N_3O_{12}$ 217500-96-4	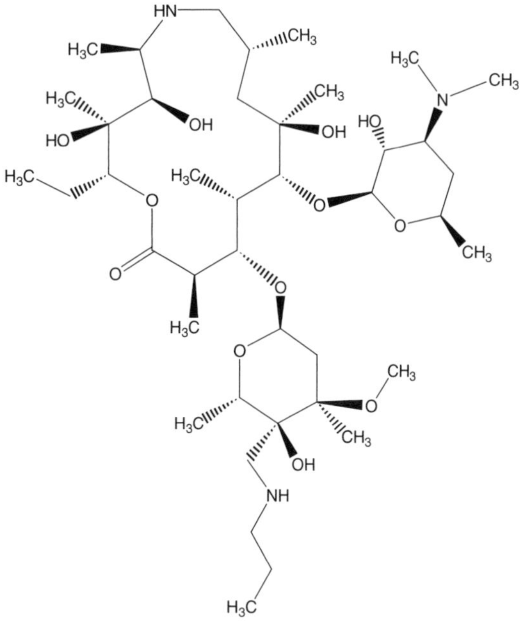	8.5[120] 9.3[120] 9.8[120] (90% isomer A)
Tylosin	[(2*R*,3*R*,4*E*,6*E*,9*R*,11*R*,12*S*,13*S*,14*R*)-12-{[3,6-Dideoxy-4-*O*-(2,6-dideoxy-3-*C*-methyl-α-L-ribo-hexopyranosyl)-3-(dimethylamino)-β-D-glucopyranosyl]oxy}-2-ethyl-14-hydroxy-5, 9,13-trimethyl-8, 16-dioxo-11-(2-oxoethyl)oxacyclohexadeca-4,6-dien-3-yl]methyl 6-deoxy-2,3-di-*O*-methyl-β-D-allopyranoside $C_{46}H_{77}NO_{17}$ 1401-69-0	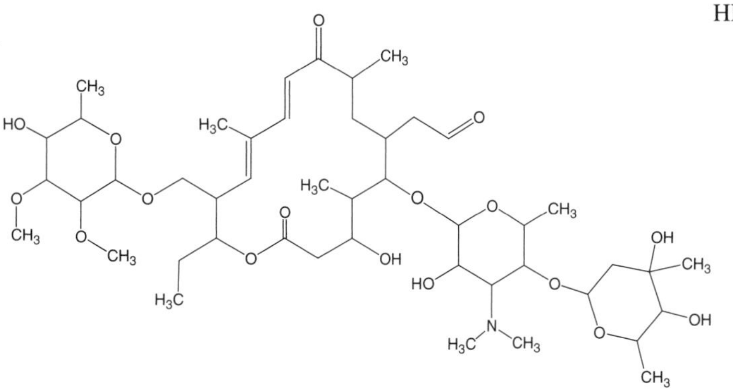	HB$^+$ 7.7[56]
Tylvalosin (acetyliso-valerylty-losin)	(4*R*,5*S*,6*S*,7*R*,9*R*,11*E*,13*E*,15*R*,16*R*)-15-{[(6-Deoxy-2,3-di-*O*-methyl-β-D-allopyranosyl)oxy]methyl}-6-({3,6-dideoxy-4-*O*-[2,6-dideoxy-3-*C*-methyl-4-*O*-(3-methylbutanoyl)-α-L-*ribo*-hexopyranosyl]-3-(dimethylamino)-β-D-glucopyranosyl}oxy)-16-ethyl-5,9,13-trimethyl-2,10-dioxo-7-(2-oxoethyl)oxacyclohexadeca-11,13-dien-4-yl acetate (2*R*,3*R*)-2,3-dihydroxybutanedioate $C_{53}H_{87}NO_{19}$ 63409-12-1	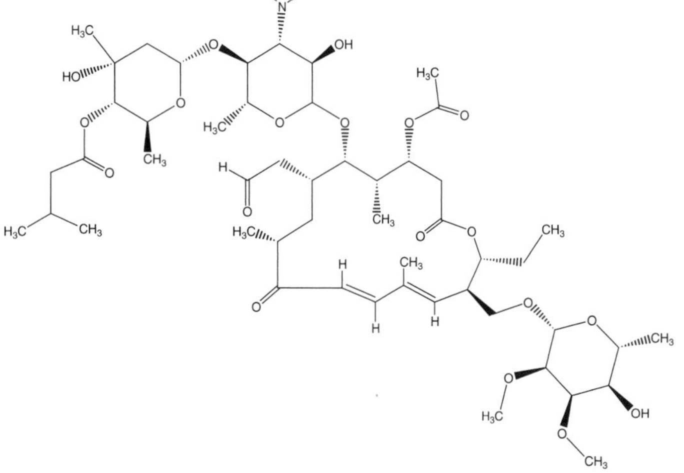	N/A[a]

(continued)

TABLE 1.6 (*Continued*)

INN	IUPAC Name Molecular Formula, and CAS Registry No.	Chemical Structure	pK_a
		Pleuromutilins	
Tiamulin	(4*R*,5*S*,6*S*,8*R*,9a*R*,10*R*)-5-Hydroxy-4,6,9,10-tetramethyl-1-oxo-6-vinyldecahydro-3a,9-propano-cyclopenta[8]annulen-8-yl {[2-(diethylamino)ethyl] sulfanyl}acetate $C_{28}H_{47}NO_4S$ 55297-95-5	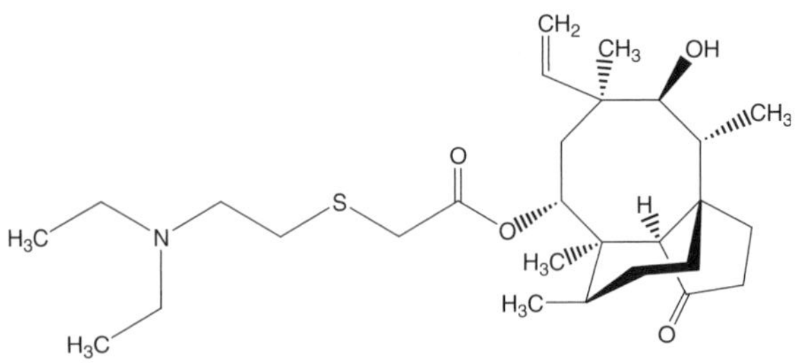	7.6[87]
Valnemulin	(3a*S*,4*R*,5*S*,6*S*,8*R*,9*R*,9a*R*,10*R*)-6-ethenyl-5-hydroxy-4,6,9,10-tetramethyl-1-oxodecahydro-3a,9-propano-3a*H*-cyclopenta[8]annulen-8-yl-[(*R*)-2-(2-amino-3-methylbutanoylamino)-1,1-dimethtylethylsulfanyl]acetate $C_{31}H_{52}N_2O_5S$ 101312-92-9	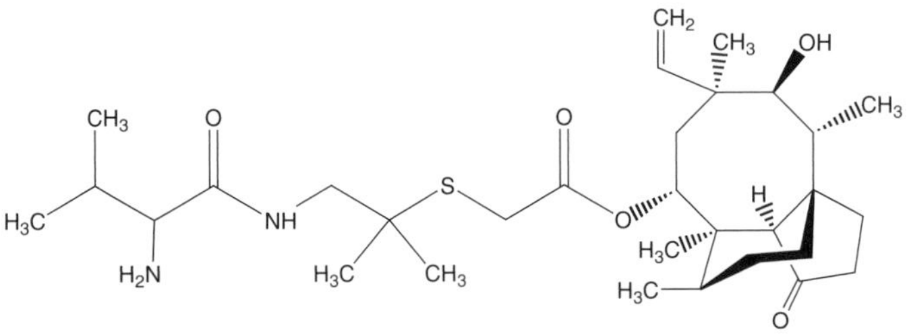	N/A[a]

[a]The author was not able to find a pK_a value for the substance in the public literature.

which is the most widely used of the macrolide class of antimicrobials, is used as an alternative to penicillin in many infections, especially in patients who are allergic to penicillin. Macrolides are significantly more active at higher pH ranges (pH 7.8–8.0).

Bacterial resistance to macrolides results from alterations in ribosomal structure with loss of macrolide binding affinity. The structural alteration very often involves methylation of ribosomal RNA and is attributed to enzymatic activity expressed by plasmids. Cross-resistance between macrolides, lincosamides, and streptogramins occurs as a result of these drugs sharing a common binding site on the ribosome.

Macrolides are used in a variety of dosage forms, including medicated feed, a water-soluble powder for the addition to drinking water, tablets, and injections for the treatment of systemic and local infections in animals. Erythromycin and/or tylosin are indicated for the prophylaxis of hepatic abscesses and the treatment of diphtheria, metritis, bacterial pneumonia, pododermatitis, and bovine respiratory disease in cattle. These drugs are also used in pigs for the prophylaxis and treatment of atrophic rhinitis, infectious arthritis, enteritis, erysipelas, respiratory syndrome, and bacterial respiratory infections, and in farrowing sows for leptospirosis. Erythromycin is indicated for the prophylaxis of enterotoxemia in lambs, while erythromycin and tylosin are used in the treatment of pneumonia and upper respiratory disease in sheep. Erythromycin is administered to chickens and turkeys for the prophylaxis of infectious coryza, chronic respiratory disease, and infectious synovitis, and to turkeys for the treatment of enteritis. Tylosin is approved in the United States for the control of American foulbrood disease in honeybees. This drug is also used in some countries to improve feed efficiency in pigs and chickens. Erythromycin is used for the treatment of *Campylobacter* enteritis and pyoderma in dogs. Although erythromycin is used in the treatment of pneumonia caused by *Rhodococcus equi* in foals, azithromycin combined with rifampicin is now more commonly used.

As mentioned above, two pleuromutilins are used in veterinary medicine. Tiamulin is available as a pre-mix and a water-soluble powder for addition to drinking water for pigs and poultry, and as an injection for pigs. It is indicated for the prophylaxis and treatment of dysentery, pneumonia, and mycoplasmal infections in pigs and poultry. In the European Union (EU), valnemulin is approved for oral administration in the treatment and prevention of swine enzootic pneumonia, swine dysentery, and proliferative ileitis in pigs.

Although the incidence of serious adverse effects to the macrolides is relatively low in animals, notable reactions do occur with some formulations and in certain animal species. For example, the irritancy of some parenteral formulations causes severe pain on intramuscular injection, thrombophlebitis at the injection site after intravenous injection, and inflammatory reactions following intramammary infusion. Macrolide-induced gastrointestinal disturbances have occurred in most species but are more serious in horses. Dosing horses with erythromycin, for instance, has resulted in fatalities from enterocolitis caused by *Clostridium difficile*.

Reports of macrolides causing adverse reactions in humans relate primarily to medicated stockfeed and parenteral formulations for injection. Farmworkers exposed to stockfeed medicated with spiramycin and tylosin have developed dermatitis and bronchial asthma.[95] In addition, accidental needlesticks with needles contaminated with tilmicosin have caused minor local reactions,[96] whereas accidental self-administration of injectable formulations of tilmicosin has resulted in serious cardiac effects and death.[97–99]

Some of the macrolides used in veterinary medicine have been detected at trace levels in the environment.[100] An investigation into the sorption behavior of a range of veterinary drugs found tylosin to be slightly mobile and slightly persistent in soil, whereas erythromycin was nonmobile and persistent.[101] Macrolides have also been shown to rapidly dissipate in a range of manure types.[102–104]

The JECFA has allocated ADIs for erythromycin,[105] spiramycin,[106] tilmicosin,[107] and tylosin,[108] with those values for erythromycin, spiramycin, and tylosin based on microbiological endpoints. The CAC also established MRLs for erythromycin in muscle, liver, kidney, and fat of chickens and turkeys, and in chicken eggs; for spiramycin in muscle, liver, kidney, and fat of cattle, pigs, and chickens; for tilmicosin in muscle, liver, kidney, fat (or fat/skin) of cattle, sheep, pigs, chickens, and turkeys; and MRLs for tylosin in muscle, liver, kidney, fat of cattle, pigs and chickens, and in chicken eggs.[43] Details of residue studies considered by JECFA are contained in monographs prepared for erythromycin,[109] spiramycin,[110–113] tilmicosin,[114,115] and tylosin.[116,117]

The properties of the macrolides from an analytical perspective were discussed in a recent review[118] and are addressed in Chapters 4–6. Some of the macrolides are pH-sensitive and degrade under acidic conditions.[119] For example, erythromycin is completely transformed to erythromycin-H_2O with the loss of one molecule of water at pH 4.[67] Erythromycin exists principally in the degraded form in aquatic environments and is measured as erythromycin-H_2O in environmental samples following pH adjustment to achieve total conversion of erythromycin to erythromycin-H_2O. Tylosin A is also unstable under acidic conditions, which accounts for its slow degradation to tylosin B in honey.[117]

1.3.6 Nitrofurans

Furans are five-membered ring heterocycles, and it is the presence of a nitro group in the 5 position of the furan ring that confers antibacterial activity on many 2-substituted furans. Although the use of nitrofurans in food-producing species is prohibited because of their carcinogenicity, nitrofurantoin, nitrofurazone, furazolidone, and nifuroxazide are used in small animals and horses.

The mechanism of antibacterial action of the furan derivatives is unknown. However, the reduced forms of nitrofurans are highly reactive and are thought to inhibit many bacterial enzyme systems, including the oxidative decarboxylation of pyruvate to acetylcoenzyme A. Nitrofurans (see list in Table 1.7) are bacteriostatic but, at high concentrations, can be bactericidal to sensitive organisms. Both chromosomal and plasmid-mediated mechanisms of resistance to nitrofurantoin occur, and these most commonly involve the inhibition of nitrofuran reductase.

Following the administration of safe doses of nitrofurantoin, effective plasma concentrations are not achieved because of its rapid elimination, and for this reason, the drug cannot be used to treat systemic infections. However, nitrofurantoin is a useful lower-urinary-tract disinfectant in small animals and occasionally in horses. The antibacterial activity observed is attributed to approximately 40% of a dose being excreted unchanged in urine, and antibacterial activity is greater in acidic urine. Nitrofurantoin has activity against several Gram-negative and some Gram-positive organisms, including many strains of *E. coli*, *Klebsiella*, *Enterobacter*, *Enterococci*, *Staphylococcus aureus* and *Staphylococcus epidermidis*, *Citrobacter*, *Salmonella*, *Shigella*, and *Corynebacterium*. It has little or no activity against most strains of *Proteus, Serratia*, or *Acinetobacter* and no activity against *Pseudomonas* species.

Nitrofurazone is used in small animals and horses as a broad-spectrum topical antibacterial agent in the prevention and treatment of bacterial skin infections and in the treatment of mixed infections in superficial wounds. It exhibits bacteriostatic activity against a variety of Gram-positive and Gram-negative microorganisms and, at high concentrations, bactericidal activity to sensitive organisms. Nitrofurazone is available as a cream, ointment, powder, soluble dressing, and topical solution. The systemic toxicity of nitrofurazone is relatively low when applied topically because absorption is not significant.

Furazolidone is occasionally used in small animals to treat enteric infections. It has activity against *Giardia, Vibrio cholera, Trichomonas*, coccidia, and many strains of *Escherichi coli, Enterobacter, Campylobacter, Salmonella*, and *Shigella*. Another nitrofuran, nifuroxazide, is used for treating acute bacterial enteritis.

There is a paucity of information describing nitrofurans in the environment. This may reflect the fact that the

TABLE 1.7 Nitrofurans

INN	IUPAC Name, Molecular Formula, and CAS Registry No.	Chemical Structure	pK_a
Furaltadone	5-(4-Morpholinomethyl)-3-(5-nitro-2-furfurylideneamino)-2-oxazolidinone $C_{13}H_{16}N_4O_6$ 139-91-3	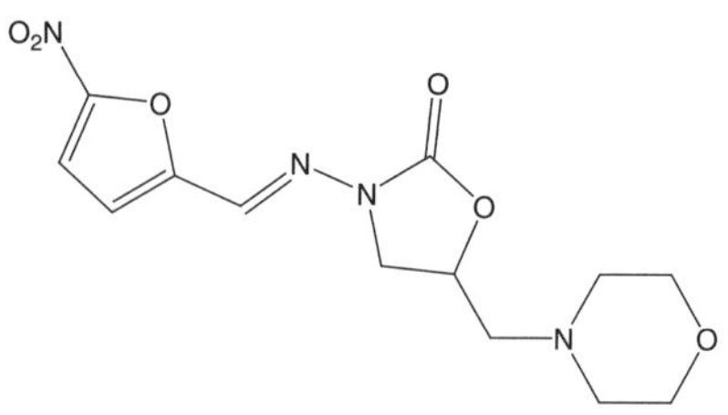	HB^+ 5.0[56]
Furazolidone	3-{[(5-Nitro-2-furyl)methylene]amino}-1,3-oxazolidin-2-one $C_8H_7N_3O_5$ 67-45-8	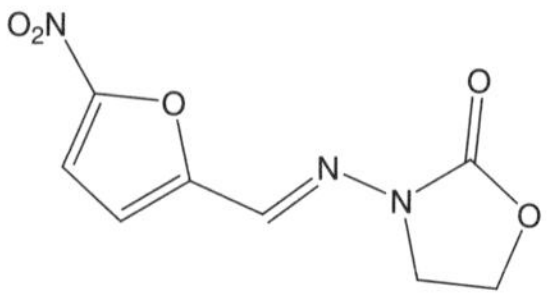	N/A[a]
Nifuroxazide	4-Hydroxybenzoic acid [(5-nitro-2-furanyl)methylene]hydrazide $C_{12}H_9N_3O_5$ 965-52-6	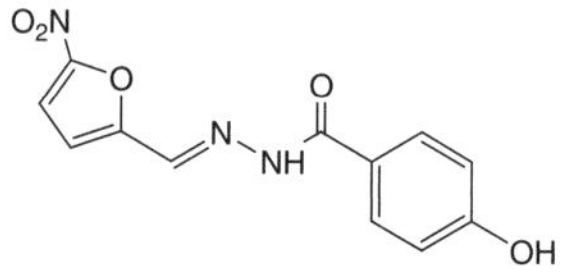	N/A[a]
Nitrofurantoin	1-[(5-Nitrofuran-2-yl)methylideneamino]imidazolidine-2,4-dione $C_8H_6N_4O_5$ 67-20-9	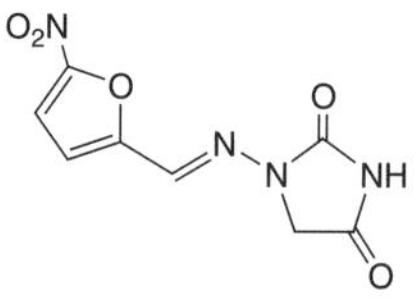	HA 7.0[56]
Nitrofurazone	[(5-Nitrofuran-2-yl)methylideneamino]urea $C_6H_6N_4O_4$ 59-87-0	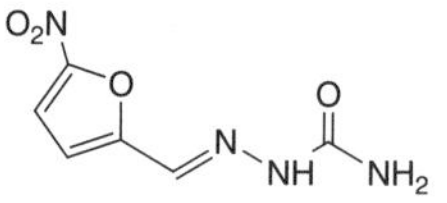	HA 9.3[56]

[a]The author was not able to find a pK_a value for the substance in the public literature.

use of these drugs in food-producing species is prohibited and is minor in small animals and horses. Consequently, the quantities of nitrofurans released into the environment will be small or negligible. Furthermore, furazolidone is unstable on exposure to light[121] and degrades very quickly in marine aquaculture sediment.[122]

Following its evaluation, the JECFA concluded that nitrofurazone was carcinogenic but not genotoxic whereas furazolidone was a genotoxic carcinogen.[121] Consequently, JECFA did not establish ADIs, and CAC MRLs have not been established for any of the nitrofurans. The carcinogenicity of the nitrofurans has led to the prohibition of their use in food-producing species in many regions, including Australia, Canada, EU, and the United States.

1.3.7 Nitroimidazoles

The chemical synthesis and biological testing of numerous nitroimidazoles occurred following the discovery in 1955 of azomycin, a 2-nitroimidazole compound, and the demonstration of its trichomonacidal properties a year later. The trichomonacidal activity of metronidazole, a 5-nitroimidazole, was reported in 1960. The chemical synthesis of other 5-nitroimidazole compounds, including dimetridazole, ipronidazole, ronidazole, and tinidazole, followed. In addition to antiprotozoal activity, these compounds display concentration-dependent activity against anaerobic bacteria. Both activities are utilized in human and veterinary medicine, although the use of nitroimidazoles in food-producing species is prohibited in Australia, Canada, the EU, and the United States.

The antimicrobial activity of the 5-nitroimidazoles involves the reduction *in vivo* of the 5-nitro group with the formation of an unstable hydroxylamine derivative that covalently binds to various cellular macromolecules. The interaction of this unstable intermediate with DNA results in a loss of helical structure and strand breakage and, in turn, the inhibition of DNA synthesis and cell death. It is via this mechanism that nitroimidazoles display antiprotozoal activity and antibacterial activity against obligate anaerobes, including penicillinase-producing strains of *Bacteroides*. They are not effective against facultative anaerobes or obligate aerobes.

The emergence of resistance to 5-nitroimidazoles is rare. When it does emerge, resistance is generally attributed to a decrease in the reduction of the 5-nitro group to form an unstable intermediate.

Metronidazole is used in dogs, cats, horses, and birds for the treatment of protozoal infections and anaerobic bacterial infections caused by susceptible organisms. The drug is effective against *Trichomonas, Entamoeba, Giardia*, and *Balantidium* species. It is used, for example, in dogs and cats with giardiasis to eliminate the shedding of giardial cysts and treat the associated diarrhea. Metronidazole is also used for the treatment of conditions such as peritonitis, empyema, and periodontal disease caused by susceptible anaerobic bacteria, and for the prevention of infection following colonic surgery. Formulations that combine metronidazole and an antimicrobial agent active against aerobic bacteria are also available. One example is a tablet for dogs and cats that combines metronidazole and erythromycin. Oral and parenteral dosage forms of metronidazole (as the sole active ingredient) are commercially available in some countries. Dimetridazole is available as a soluble powder for administration in drinking water to birds not producing meat or eggs for human consumption, and for the control of blackhead caused by *Histomonas melagridis*.

Clinical toxicity in animals treated with metronidazole at the recommended dose rate is uncommon. However, high doses lead to neurological signs including seizures, head tilt, paresis, ataxia, vertical nystagmus, tremors, and rigidity in cats, dogs, and horses. A common occurrence in animals treated with metronidazole is the voiding of reddish brown urine. This does not require medical intervention.

Residues of nitroimidazoles in the environment have not been reported. (See list of nitroimidazoles in Table 1.8).

TABLE 1.8 Nitroimidazoles

INN	IUPAC Name, Molecular Formula, and CAS Registry No.	Chemical Structure	pK_a
Dimetridazole	1,2-Dimethyl-5-nitro-1*H*-imidazole $C_5H_7N_3O_2$ 551-92-8	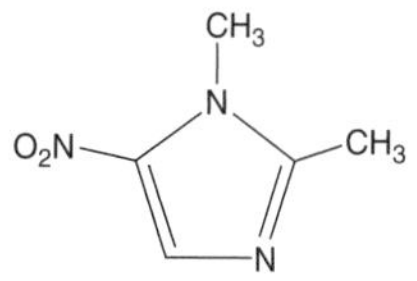 CH_3; N; O_2N; CH_3; N	N/A[a]
Ipronidazole	1-Methyl-2-(1-methylethyl)-5-nitro-1*H*-imidazole $C_7H_{11}N_3O_2$ 14885-29-1	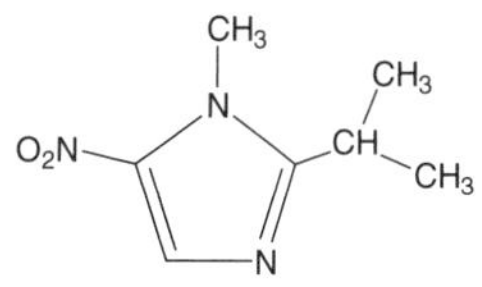 CH_3; N; CH_3; CH; O_2N; CH_3; N	HB^+ 2.7[56]
Metronidazole	2-(2-Methyl-5-nitroimidazol-1-yl)ethanol $C_6H_9N_3O_3$ 443-48-1	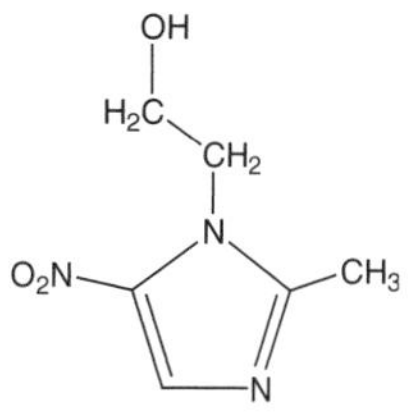 OH; H_2C; CH_2; N; O_2N; CH_3; N	HB^+ 2.6[56]
Ronidazole	1-Methyl-5-nitroimidazole-2-methanol carbamate (ester) $C_6H_8N_4O_4$ 7681-76-7	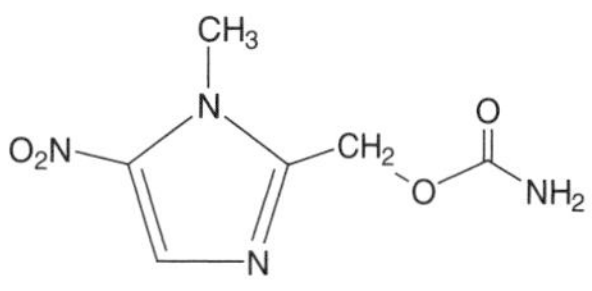 CH_3; N; O; O_2N; CH_2; O; NH_2; N	N/A[a]
Tinidazole	1-(2-Ethylsulfonylethyl)-2-methyl-5-nitroimidazole $C_8H_{13}N_3O_4S$ 19387-91-8	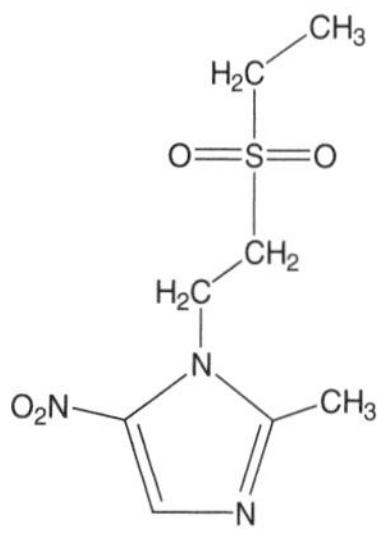 CH_3; H_2C; O=S=O; CH_2; H_2C; N; O_2N; CH_3; N	N/A[a]

[a]The author was not able to find a pK_a value for the substance in the public literature.

Although JECFA has not established ADI values for metronidazole, dimetridazole, or ipronidazole, they did allocate a temporary ADI for ronidazole in 1989[123] but it was withdrawn in 1995.[124]

1.3.8 Phenicols

In 1947, Ehrlich and coworkers reported the isolation of chloramphenicol (known at that time as *chloromycetin*) from *Streptomyces venezuelae*, a Gram-positive soil-dwelling actinomycete.[125] Today, the drug is produced for commercial use by chemical synthesis. Chloramphenicol was the first broad-spectrum antibiotic developed. It demonstrates a time-dependent bacterial effect and is bacteriostatic for most Gram-positive and many Gram-negative aerobic bacteria, although at higher concentrations, it can be bactericidal against some very sensitive organisms. Many strains of *Salmonella* species are susceptible to chloramphenicol, while most strains of *Pseudomonas aeruginosa* are resistant. The drug is also very effective against all obligate anaerobes and suppresses the growth of rickettsia and chlamydia species. Other members of the phenicol class are thiamphenicol and florfenicol. The antibacterial activity of thiamphenicol is less than that of chloramphenicol. The activity spectrum of florfenicol, which is not approved for use in humans, is similar to that of chloramphenicol but is more active.

The phenicols are transported into bacterial cells by passive or facilitated diffusion. They bind to the 50*S* subunit of the 70*S* bacterial ribosome and impair peptidyltransferase activity, thereby interfering with the incorporation of amino acids into newly formed peptides. Chloramphenicol also inhibits mitochondrial protein synthesis in mammalian bone marrow cells but does not significantly affect other intact cells.

Chloramphenicol is available as a bitter-tasting free base and as two esters—a neutral-tasting palmitate for oral administration and a water-soluble sodium succinate for injection. Other forms are available for topical and ophthalmic use. Chloramphenicol base is rapidly absorbed following oral administration to non-ruminant animals. In ruminants, however, reduction of the nitro moiety of chloramphenicol by ruminal microflora results in inactivation and very low bioavailability. Chloramphenicol sodium succinate may be injected intravenously or intramuscularly and is activated on hydrolysis to the free base. Chloramphenicol is un-ionized at physiological pH and is lipophilic; it readily crosses membranes. The drug is widely distributed to virtually all tissues and body fluids, including the central nervous system, cerebrospinal fluid, and the eye. The principal metabolic pathway for chloramphenicol is hepatic metabolism to the inactive metabolite, chloramphenicol glucuronide. Urinary excretion of unchanged chloramphenicol accounts for approximately 5–15% of a dose. Florfenicol also penetrates most body tissues but to a lesser extent than does chloramphenicol in the case of cerebrospinal fluid and the eye. In cattle, urinary excretion of unchanged florfenicol accounts for approximately 64% of a dose. Thiamphenicol does not undergo significant metabolism and is excreted unchanged in urine.

Chloramphenicol causes two distinct forms of toxicity in humans. The most serious form is an irreversible aplastic anaemia. This rare idiosyncratic response (the incidence is $\approx 1:25{,}000-60{,}000$) may have an immunological component; however, the mechanism of chloramphenicol-induced aplastic anemia remains unknown. Neither a dose–response relationship nor a threshold dose for the induction of aplastic anaemia has been established. Aplastic anemia is associated with reduced numbers of erythrocytes, leukocytes, and platelets (pancytopaenia), with resultant bleeding disorders and secondary infections. The condition tends to be irreversible and fatal. By comparison, leukemia may be a sequel of hypoplastic anemia. Because thiamphenicol and florfenicol lack the *p*-nitro moiety, they do not induce irreversible aplastic anemia in humans.

The second form of chloramphenicol toxicity in humans involves dose-dependent and reversible bone marrow suppression. With this toxicity, erythroid and myeloid precursors do not mature normally, serum iron concentration is increased, and phenylalanine concentrations are decreased. These signs of toxicity usually disappear when chloramphenicol is discontinued. Chronic dosing with thiamphenicol or florfenicol may also cause dose-dependent bone marrow suppression.

Bacteria develop resistance to chloramphenicol by four main mechanisms: (1) mutation of the 50*S* ribosomal subunit; (2) decreased membrane permeability to chloramphenicol; (3) elaboration of the inactivating enzyme, chloramphenicolacetyltransferase (CAT); and (4) increased expression of efflux pumps. Mechanism 3 is the most frequent cause of resistance to chloramphenicol. It involves CAT catalyzing the covalent binding of one or two acetyl groups derived from acetyl CoA to the hydroxyl moieties on the chloramphenicol molecule. The (di)acetylated product is unable to bind to the 50*S* subunit of the 70*S* bacterial ribosome and lacks antibacterial activity. This form of resistance may involve endogenous CAT or alternatively, CAT expressed by plasmids that are transferred during bacterial conjugation. Florfenicol is less susceptible to resistance from CAT inactivation because the hydroxyl moiety is replaced with a fluorine moiety that is less susceptible to CAT inactivation. Resistance to florfenicol in Gram-negative bacteria is attributed to increased expression of efflux pumps.[126] The findings of an Australian study indicate that cross-resistance with chloramphenicol is very important. The study found that 60% of *E. coli* isolates from pigs were resistant to florfenicol when the antimicrobial was introduced onto the Australian market in 2003.[1]

It was proposed that the past use of chloramphenicol may have selected for strains carrying *cml*A gene that had persisted for more than 20 years in the absence of selection pressure (chloramphenicol was last used in food-producing species in Australia in 1982).

Chloramphenicol is used to treat a variety of local and systemic infections in small animals and horses. Its use in food-producing species is banned in most countries because of human health implications (discussed in Chapter 3). Therapeutic uses include chronic respiratory infections, bacterial meningoencephalitis, brain abscesses, ophthalmitis and intraocular infections pododermatitis, dermal infections, and otitis externa. The drug is effective against *Salmonellosis* and *Bacteroides* sepsis. Its poor efficacy against lower-urinary-tract infections reflects the small amount of unchanged drug excreted in urine. Florfenicol is an effective therapy for bovine respiratory disease in cattle caused by *Mannheimia, Pasteurella*, and *Histophilus*. The drug is also approved in some countries for use in pigs and fish. Thiamphenicol is approved for use in Europe and Japan.

The possibility of chloramphenicol detected in food samples collected in national monitoring programs in the early 2000s being attributed to environmental exposure was the subject of a 2004 review.[127] Two aspects—natural synthesis of chloramphenicol in soil and the persistence of chloramphenicol in the environment after historical veterinary use—were considered. The review found that although the possibility of food being occasionally contaminated from environmental sources could not be completely ruled out, it was highly unlikely. More recently, Berendsen and coworkers[34] reported that non-compliant residues of chloramphenicol in animal-derived food products may, in part, be due to the natural occurrence of chloramphenicol in herbs and grasses grazed by food-producing species.

As mentioned above, in order to protect the health of consumers, few countries permit the use of chloramphenicol in food-producing animals. In addition to epidemiological studies in humans showing that treatment with chloramphenicol is associated with the induction of aplastic anemia, chloramphenicol is a genotoxin *in vivo* and may cause adverse effects in humans[127] (discussed further in Chapter 3). The use of thiamphenicol and florfenicol is permitted in food-producing species in some countries. JECFA has established an ADI for thiamphenicol[128] and recommended temporary MRLs for thiamphenicol residues that were withdrawn when additional residue data requested for evaluation were not provided.[129] Two reviews of residue studies on thiamphenicol provided for evaluation by JECFA have been published.[130,131] The CAC does not currently list MRLs for florfenicol or thiamphenicol.[43]

Properties of three phenicols are listed in Table 1.9.

1.3.9 Polyether Antibiotics (Ionophores)

The polyether ionophore class of antibiotics includes lasalocid, maduramicin, monensin, narasin, salinomycin, and semduramicin. These drugs are used exclusively in

TABLE 1.9 Phenicols

INN	IUPAC Name, Molecular Formula, and CAS Registry No.	Chemical Structure	pK_a
Chloramphenicol	2,2-Dichloro-*N*-[1,3-dihydroxy-1-(4-nitrophenyl)propan-2-yl]acetamide $C_{11}H_{12}Cl_2N_2O_5$ 56-75-7	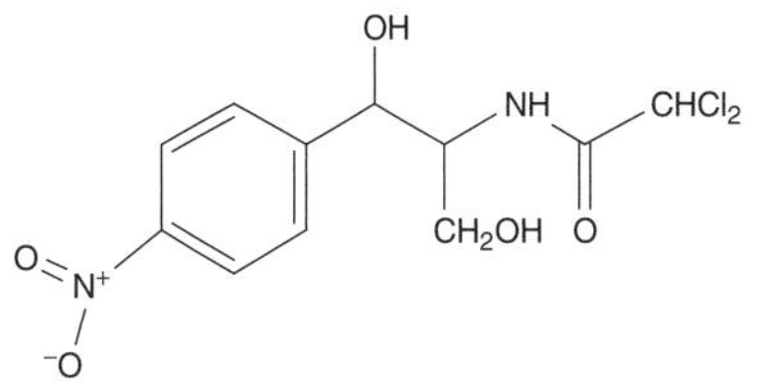 OH, NH, $CHCl_2$, CH_2OH, O, O, N^+, ^-O	N/A[a]
Florfenicol	2,2-Dichloro-*N*-[(1*S*,2*R*)-1-(fluoromethyl)-2-hydroxy-2-[4-(methylsulfonyl)phenyl]ethyl]acetamide $C_{12}H_{14}Cl_2FNO_4S$ 73231-34-2	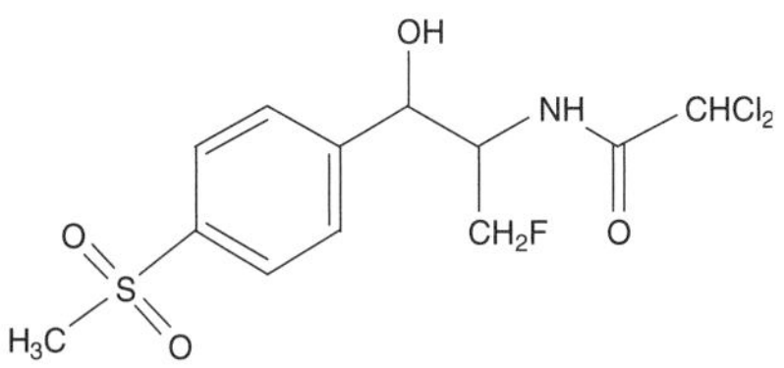 OH, NH, $CHCl_2$, CH_2F, O, O, S, H_3C, O	N/A[a]
Thiamphenicol	2,2-Dichloro-*N*-{(1*R*,2*R*)-2-hydroxy-1-(hydroxymethyl)-2-[4-(methylsulfonyl)phenyl]ethyl}acetamide $C_{12}H_{15}Cl_2NO_5S$ 15318-45-3	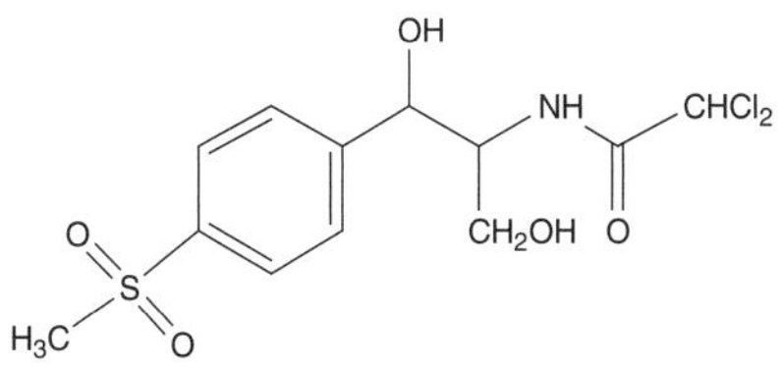 OH, NH, $CHCl_2$, CH_2OH, O, O, S, H_3C, O	N/A[a]

[a]The author was not able to find a pK_a value for the substance in the public literature.

veterinary medicine for their antibacterial and anticoccidial activities. The first ionophore to be discovered was lasalocid in 1951. This drug, which is a fermentation product of *Streptomyces lasaliensis*, is a divalent polyether ionophore. The discovery of monensin, a fermentation product of *Streptomyces cinnamonensis* and a monovalent polyether ionophore, followed in 1967. The discoveries of salinomycin, a fermentation product of *Streptomyces albus* and its methyl analoge, narasin, a fermentation product of *Streptomyces aureofaciens*, were reported in 1972 and 1975, respectively. Both salinomycin and narasin are monovalent polyether ionophores. Maduramicin, a fermentation product of *Actinomadura yumaense*, and semduramicin, a fermentation product of *Actinomadura roseorufa*, discovered in 1983 and 1988, respectively, are monovalent monoglycoside polyether ionophores.

Polyether ionophores (Table 1.10) have a distinctly different mode of action from therapeutic antibiotics.

TABLE 1.10 Polyether Antibiotics (Ionophores)

INN	IUPAC Name, Molecular Formula, and CAS Registry No.	Chemical Structure	pK_a
Lasalocid A	6-[7*R*-[5*S*-ethyl-5-(5*R*-Ethyltetrahydro-5-hydroxy-6*S*-methyl-2*H*-pyran-2*R*-yl)tetrahydro-3*S*-methyl-2*S*-furanyl]-4*S*-hydroxy-3*R*,5*S*-dimethyl-6-oxononyl]-2-hydroxy-3-methylbenzoic acid $C_{34}H_{54}O_8$ 25999-31-9	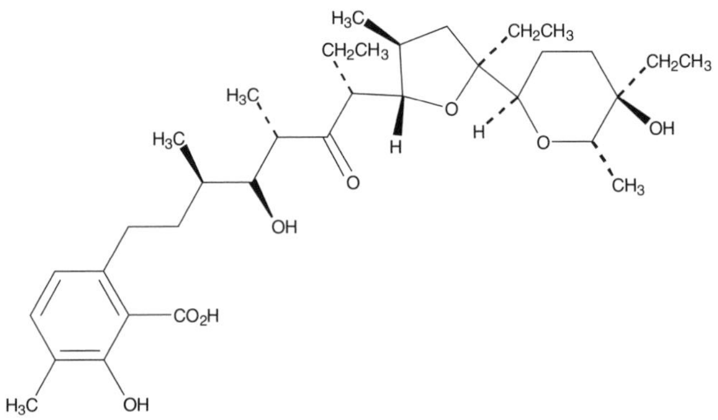	4.4[153]
Maduramicin	(2*R*,3*S*,4*S*,5*R*,6*S*)-6-[(1*R*)-1-[(2*S*,5*R*,7*S*,8*R*,9*S*)-2-[(2*S*,2′*R*,3′S,5*R*,5′*R*)-3′-[(2,6-Dideoxy-3,4-di-*O*-methyl-*b*-L-*arabino*-hexopyranosyl)oxy]-octahydro-2-methyl-5′-[(2*S*,3*S*,5*R*,6*S*)-tetrahydro-6-hydroxy-3,5,6-trimethyl-2*H*-pyran-2-yl][2,2′-bifuran]-5-yl]-9-hydroxy-2,8-dimethyl-1,6-dioxaspiro[4.5]dec-7-yl]ethyl]tetrahydro-2-hydroxy-4,5-dimethoxy-3-methyl-2*H*-pyran-2-acetic acid $C_{47}H_{80}O_{17}$ 61991-54-6	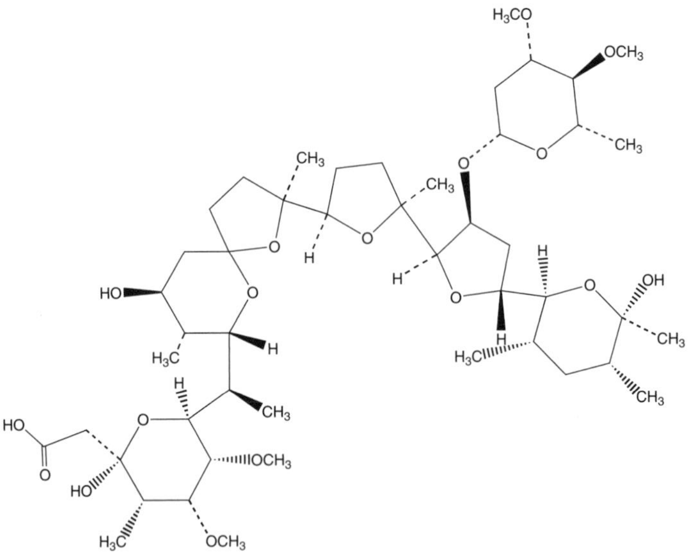	4.2[154]
Monensin	2-[5-Ethyltetrahydro-5-[tetrahydro-3-methyl-5-[tetrahydro-6-hydroxy-6-(hydroxymethyl)-3,5-dimethyl-2*H*-pyran-2-yl]-2-furyl]-2-furyl]-9-hydroxy-β-methoxy-α,γ,2,8-tetramethyl-1,6-dioxaspiro[4.5]decane-7-butyric acid $C_{36}H_{62}O_{11}$ 17090-79-8	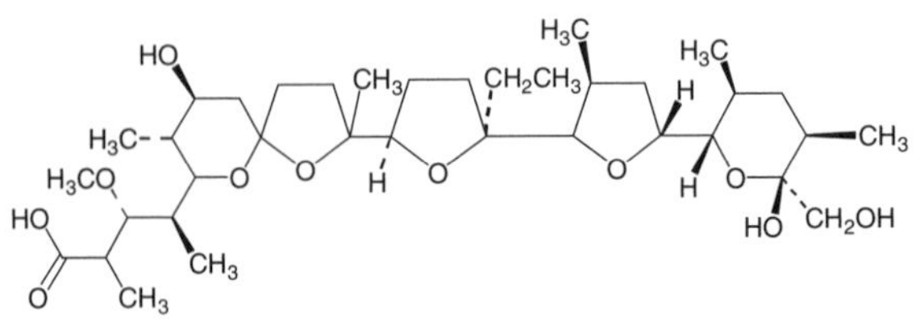	6.7[153]

TABLE 1.10 ***(Continued)***

INN	IUPAC Name Molecular Formula, and CAS Registry No.	Chemical Structure	pK_a
Narasin	(αβ,2β,3α,5α,6α)-α-Ethyl-6-[5-[5-(5α-ethyltetrahydro-5β-hydroxy-6α-methyl-2*H*-pyran-2β-yl)-3″α,4,4″,5,5″α,6″-hexahydro-3′β-hydroxy-3″β,5α,5″β-trimethylspiro]furan-2(3H),2′-[2*H*]pyan-6′(3′H),2″-[2*H*]pyran]6″α-yl]2α-hydroxy-1α,3β-dimethyl-4-oxoheptyl]-tetrahydro-3,5-dimethyl-2*H*-pyran-2-acetic acid $C_{43}H_{72}O_{11}$ 55134-13-9	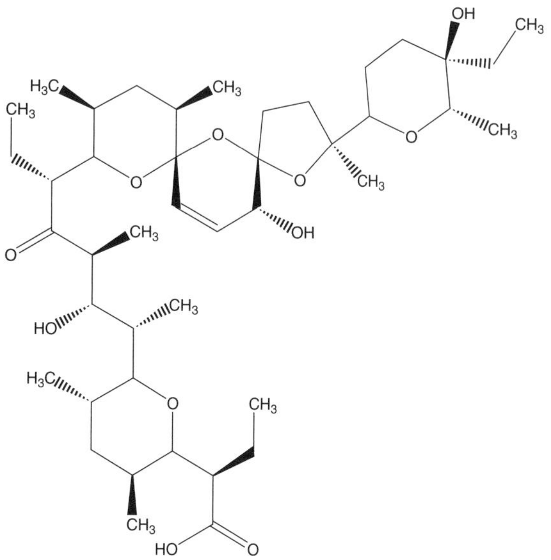	7.9[153]
Salinomycin	(2*R*)-2-((5S)-6-{5-[(10*S*,12*R*)-2-((6*S*,5*R*)-5-Ethyl-5-hydroxy-6-methylperhydro-2*H*-pyran-2-yl)-15-hydroxy-2,10,12-trimethyl-1,6,8-trioxadispiro[4.1.5.3]pentadec-13-en-9-yl](1*S*,2*S*,3*S*,5*R*)-2-hydroxy-1,3-dimethyl-4-oxoheptyl}-5-methylperhydro-2*H*-pyran-2-yl)butanoic acid $C_{42}H_{70}O_{11}$ 53003-10-4	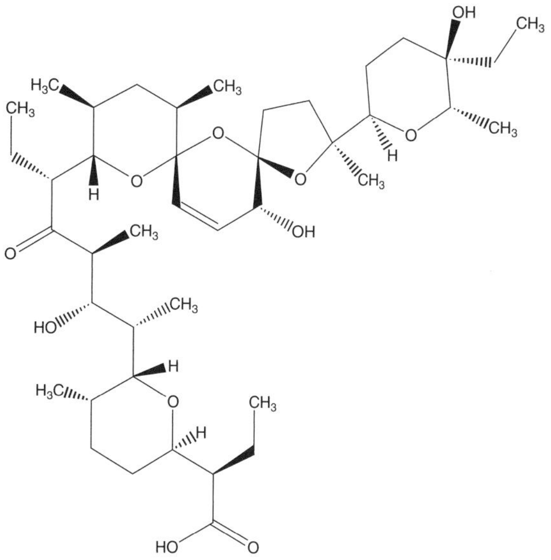	4.5[153] 6.4[153]
Semduramicin	(2*R*,3*S*,4*S*,5*R*,6*S*)-Tetrahydro-2,4-dihydroxy-6-[(1*R*)-1-[(2*S*,5*R*,7*S*,8*R*,9*S*)-9-hydroxy-2,8-dimethyl-2-[(2*R*,6*S*)-tetrahydro-5-methyl-5-[(2*R*,3*S*,5*R*)-tetrahydro-5[(2*S*,3*S*,5*R*,6*S*)-tetrahydro-6-hydroxy-3,5,6-trimethyl-2*H*-pyran-2-yl]-3-[[(2*S*,5*S*,6*R*)-tetrahydro-5-methoxy-6-methyl-2*H*-pyran-2-yl]oxy]-2-furyl{}-2-furyl]-1,6-dixoaspirol[4.5]dec-7-yl]ethyl]-5-methoxy-3-methyl-2*H*-pyran-2-acetic acid $C_{45}H_{76}O_{16}$ 113378-31-7	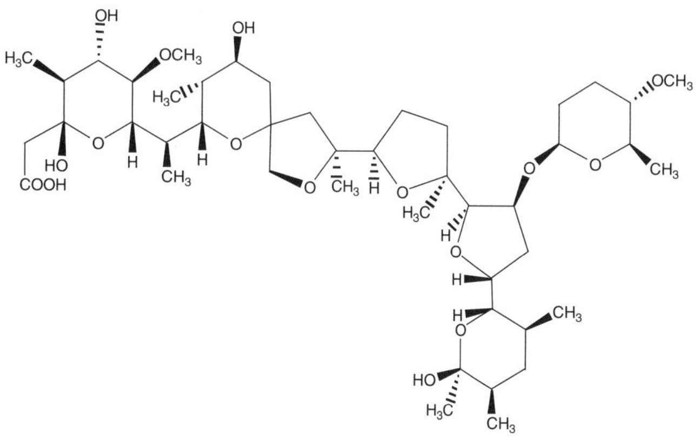	4.2[154]

Their structures involve an alkyl-rich, lipid-soluble exterior and a cagelike interior that is capable of binding and shielding monovalent metal ions (e.g., sodium, potassium) and divalent metal ions (e.g., magnesium, calcium). The ionophores are highly lipophilic and able to transport cations across cell membranes of susceptible bacteria.[132] They are most effective against Gram-positive bacteria because the peptidoglycan layer is porous, allowing them to pass through to reach the cytoplasmic membrane, where they rapidly dissolve into the membrane. The exchange of intracellular potassium for extracellular protons, and extracellular sodium for intracellular protons, disrupts ion gradients.[133] Because the potassium gradient is greater than the sodium gradient, the net effect of these exchanges is the accumulation of protons inside the bacterium.[134] The cellular response to this homeostatic disturbance is the activation of ATP-dependent processes, which in turn, exhausts cellular energy sources and leads to cell death.[133,135] Because ionophores selectively affect Gram-positive organisms, the rumen microflora shifts toward a more Gram-negative population and results in changes in the patterns of diet fermentation. The proportions of acetic acid and butyric acid in the volatile fatty acids are decreased, while the proportion of propionic acid is increased. The result is reduced energy losses per unit of feed consumed.[136] The anticoccidial activity of ionophores is thought to alter membrane integrity and internal osmolality of extracellular sporozoites and merozoites. Because coccidia have no osmoregulatory organelles, perturbances of internal osmotic conditions lead to cell death.[137]

Ionophore resistance appears to be mediated by extracellular polysaccharides (glycocalyx) that exclude ionophores from the cell membrane.[138] This is believed to involve physiological selection rather than a mutation *per se* because cattle that are not receiving ionophores can have large populations of resistant ruminal bacteria. To date, genes conferring ionophore resistance in ruminal bacteria have not been identified. Ionophore resistance is not restricted to bacteria for it is common with chicken *Eimeria* species in the United States.[137]

The use of lasalocid, monensin, and salinomycin as growth promoters was phased out in the European Union in 2006. In other regions, ionophores are used for improving production efficiency by altering the gastrointestinal microflora of animals. Ruminal fermentation is inherently inefficient, with the conversion of $\leq$12% of dietary carbon and energy to methane and heat that are unusable by the animal,[139] and $\leq$50% of dietary protein is degraded to ammonia and lost in the urine. Ionophore-induced improvements in productivity result from changes in the proportion of volatile fatty acids produced during ruminal digestion. The administration of monensin to cattle, for example, results in improvements in liveweight gain of $\leq$10%, increases in feed conversion efficiency of $\leq$7%, and decreases in food consumption of $\leq$6%. Ionophores also have a profound impact on ruminal nitrogen retention, a phenomenon referred to as a *protein-sparing effect*. Monensin is used in feedlot cattle to reduce the incidence of acute and subacute ruminal acidosis resulting from rapid fermentation of carbohydrates in the rumen and the accumulation of lactic acid. Monensin is administered by controlled-release capsules for its anti-bloat effects. The latter are mediated via a dual mechanism—the inhibition of *slime-producing bacteria* and a decrease in overall ruminal gas production.[140] Monensin is also used for decreasing the incidence of acute pneumonia caused by the eructation and inhalation of 3-methylindole, a by-product of L-tryptophan fermentation.[141] The efficacy of monensin in this condition is due to its direct inhibition of the lactobacilli producing 3-methylindole. The ionophores are also approved for use as coccidiostats in poultry, cattle, sheep, goats, and rabbits.

Ionophore toxicity has been widely reported in many species of animals, including rabbits, dogs, cats, pigeons, quail, chickens, turkeys, ostriches, goats, pigs, sheep, cattle, camels, and horses, sometimes with fatal consequences.[142] Toxicity is most often attributed to dosing errors, accidental ingestion including contaminated rations prepared by feedmills, the ingestion by ruminants of litter from ionophore-treated poultry flocks, and the concurrent administration of other agents and, in particular, tiamulin. The mechanism of ionophore toxicity generally involves cellular electrolyte imbalance, with skeletal and cardiac muscle affected most severely. Horses are particularly sensitive to ionophore toxicity; the LD_{50} for monensin in horses is 2–3 mg/kg, compared with LD_{50} values of 20 mg/kg for dogs and 200 mg/kg for chickens.[143] Food contaminated with salinomycin has resulted in polyneuropathy in cats.[144]

Kouyoumdjian and coworkers[145] reported the case of a 17-year-old male who developed myoglobinemia and renal failure and died 11 days after ingesting sodium monensin. The findings in this case were similar to those seen in animals following accidental intoxication.

Relatively few reports in the literature describe environmental concentrations, fate, and transport of monensin. Compared with tetracyclines and macrolides, monensin is not tightly adsorbed to soil and has been detected in river water and aquatic sediments in Colorado[146] and in streams in southern Ontario.[147]

The JECFA has allocated ADIs for monensin[148] and narasin.[149] The CAC has established MRLs for monensin in muscle, liver, kidney, and fat of cattle, sheep, chickens, goats, turkeys, and quails,[43] based on the residue evaluation conducted by JECFA.[150] The CAC MRLs for narasin in muscle, liver, kidney, and fat of pigs and chickens, and temporary MRLs for narasin in muscle, liver, kidney, and fat of cattle, have also been established,[43] on the basis of the JECFA evaluation.[151]

From an analytical perspective, ionophores are unstable in strongly acidic conditions. Moreover, weakly acidic extractants are not suitable for use with these substances.[152]

1.3.10 Polypeptides, Glycopeptides, and Streptogramins

The polypeptides include bacitracin A, colistin (polymyxin E), novobiocin, and polymyxin B. Bacitracin is a complex mixture of branched, cyclic decapeptides produced by *Bacillus subtilise*, which was first isolated in 1945. The polymyxins, discovered in 1947, are synthesized by various strains of *Bacillus polymyxa*. Colistin (polymyxin E) comprises a family of polymyxins and was known as colimycin when first isolated from a broth of *Bacillus polymyxa* var. *colistinus* in 1951. The polymyxins are cationic detergents. Novobiocin, first reported in 1955 as streptonivicin, is produced by the actinomycete *Streptomyces niveus*.

The glycopeptide antibiotics include avoparcin, teicoplanin, and vancomycin. Avoparcin is produced by *Amycolatopsis coloradensis*, while teicoplanin is a mixture of six closely related compounds produced by *Streptococcus teichomyetius*. Vancomycin is produced by *Streptococcus orientalis*.

The streptogramins include virginiamycin and pristinomycin. Virginiamycin is produced by a mutant strain of *Streptomyces virginiae*. It is a natural mixture of factor M and factor S, and its antibacterial activity is synergistically optimum when the $M:S$ ratio is approximately 4:1.[155–157] Pristinamycin is a combination of quinupristin, a streptogramin B, and dalfopristin, a streptogramin A, in a 30:70 ratio. Each of these compounds is a semi-synthetic derivative of naturally occurring pristinamycins produced by *Streptomyces pristinaespiralis*.

Bacitracin inhibits the synthesis of the bacterial cell wall by preventing the transport of peptidoglycan precursors through the cytoplasmic membrane. It is bactericidal to Gram-positive bacteria but exhibits little activity against Gram-negative organisms. The antibacterial activity of the polymyxins is attributed to their strong binding to phospholipids in cell membranes, which disrupts their structure and alters membrane permeability. These drugs are bactericidal and display activity against many species of Gram-negative bacteria, including *E. coli, Salmonella*, and *Pseudomonas aeruginosa* but not against *Proteus, Serratia*, or *Providencia*. The glycopeptide antibiotics inhibit cell wall synthesis by binding strongly with cell wall precursors. The antibacterial activity of the streptogramins is attributed to the inhibition of protein synthesis. This involves the M and S factors of virginiamycin binding to 50S ribosomal subunits and inhibiting the formation of peptide bonds during protein synthesis. Quinupristin and dalfopristin also inhibit protein synthesis. They bind to the 50S ribosomal subunit at different sites located in close proximity, thereby interfering with the formation of polypeptide chains.

Bacterial resistance to the polymyxins is rare; however, resistance is common in pig and chicken isolates of *Enterococcus* spp.[158] Interestingly, bacitracin administered to pigs and chickens has been shown to reduce the transfer of resistance plasmids among enteric *E. coli*. In the case of novobiocin, resistance has developed in many species of bacteria. Prior to the ban on its use in food-producing species, avoparcin was found to select for vancomycin-resistant enterococci (VRE). Bacterial resistance to vancomycin is generally uncommon, with the exception of *Enterococcus* species. Development of resistance to teicoplanin is also uncommon. In terms of pristinamycin, the mechanisms of resistance to class A streptogramins and class B streptogramins are different. With class A streptogramins, active efflux of drug from the bacterial cell as well as drug inactivation by acetyltransferases contribute to resistance. By comparison, resistance to class B streptogramins is most commonly due to methylation of the target 23S ribosomal RNA, while a less common mechanism involves enzymatic cleavage of a structural ring in the drug.

Bacitracin is used for the treatment of infections of the skin, eyes, and ears. Various topical dosage forms, including wound powders and ointments, and eye and ear ointments, are available. Bacitracin is used as a feed additive for pigs, poultry, and ruminants, except in the European Union, where use for growth promotion was banned in 1999. It improves growth rate and feed conversion efficiency of pigs, broilers, calves, sheep, and feedlot steers. Bacitracin is also used in the control of proliferative enteropathy in grower–finisher pigs, and to decrease the incidence and severity of clostridial enteritis in piglets born to sows treated during pregnancy. In the poultry industry, bacitracin is used for the prevention of necrotic enteritis in broilers and to improve the ability of broilers and layers to withstand heat stress. Novobiocin sodium is included with other agents in intramammary infusions for treating mastitis in dairy cattle.

The glycopeptides are not used in food-producing species. In humans, vancomycin is indicated for the treatment of life-threatening Gram-positive infections that are unresponsive to less toxic antibiotics. The worldwide emergence of vancomycin-resistant enterococci (VRE) is a major concern for public health and stimulated the debate concerning the use of avoparcin in agriculture and whether this contributed to VRE in humans. The agricultural use of avoparcin in many countries is now banned. A new glycopeptide antibiotic, teicoplanin, was developed against infections with resistant Gram-positive bacteria, especially bacteria resistant to vancomycin.

Virginiamycin is used to improve daily liveweight gain and feed conversion efficiency in feedlot and grazing cattle, broilers, turkeys, and pigs in several countries. However, such use was discontinued in the European Union in 1999 and in Australia in 2005. In feedlot cattle, it also reduces the incidence and severity of liver abscessation. Virginiamycin reduces the risk of fermentative lactic acidosis in cattle and sheep fed high-concentrate diets. The drug is administered to horses on high grain diets to reduce the risk of laminitis.

Polymyxins are notably nephrotoxic and neurotoxic and cause intense pain if injected. Polymyxin B is a potent histamine releaser; however, hypersensitivity reactions to all polymyxins are seen occasionally. The incidence of adverse reactions to novobiocin sodium is also relatively frequent.

The limited literature available suggests that presently none of the polypeptide antibiotics, glycopeptide antibiotics, or streptogramins pose a risk to the environment.

On the basis of risk assessments carried out in 1968,[42] JECFA has allocated ADI values to bacitracin and novobiocin. MRLs were not established by the CAC because when administered to animals, these substances should not be allowed to give rise to detectable residues in food for human consumption. More recently, JECFA allocated an ADI for colistin based on a microbiological endpoint and recommended MRLs for colistin in muscle, liver, kidney, fat, and milk of cattle and sheep; and in muscle, liver, kidney and fat of pigs, rabbits, goats, turkeys, and chickens, and in chicken eggs.[159] These MRL recommendations were adopted by the CAC.[43] Details of the residue studies for colistin considered by JECFA were published in a monograph.[160] Properties of polypeptides, glycopeptides, and streptogramens are listed in Table 1.11.

1.3.11 Phosphoglycolipids

Flavophospholipol is the only phosphoglycolipid antibiotic that is approved for use in food-producing animals (see Table 1.12 for properties of this compound). It is produced by *Streptomyces* spp., including *S bambergiensis, S. ghanaensis, S. geyirensis*, and *S. ederensis* and was discovered in the mid-1950s. The product consists of a complex of similar components in which moenomycin A predominates. Flavophospholipol has a novel mode of action in that it inhibits peptidoglycan synthesis by interfering with transglycolase activity and prevents the formation of the murein backbone of the peptidoglycan molecule.[161] It is active mainly against Gram-positive organisms and has little activity against Gram-negative bacilli as it cannot penetrate the outer lipopolysaccharide cell membrane in these organisms. Flavophospholipol is absorbed poorly from the gastrointestinal tract and if administered parenterally, is strongly bound to plasma proteins and host cell membranes. It is slowly excreted unchanged in the urine.[162] Limited information is available on acquired resistance to flavophospholipol, but it seems that many *Enterococcus* species are intrinsically resistant.

For more than 30 years, flavophospholipol has been used in many countries, including Australia and European countries, solely as a growth-promoting antimicrobial in animal feeds. However, its use was banned in the EU in 2006. The most extensive use has been in pigs and poultry, although flavophopholipol also promotes growth in ruminants. The mechanism for growth promotion of flavophopholipol is unclear. Its mode of action on the rumen microbial population appears to differ from that of the ionophore class of antibiotics in that volatile fatty acid proportions are generally unchanged.[163] An interesting characteristic of flavophospholipol is its ability to inhibit transfer of plasmids carrying antibiotic resistance genes in *E. coli, Salmonella*, and *Enterococcus* spp.[162,164] Furthermore, it has been shown to reduce the shedding of salmonella in experimentally infected animals.[161] The PK and PD profiles of flavophospholipol make it unsuitable for use as a human antibiotic.

This author is not aware of any reports that describe the presence of flavophospholipol in the environment.

The JECFA has not evaluated toxicological or residue depletion data for flavophospholipol, and CAC MRLs for the substance have not been established.

1.3.12 Quinolones

The quinolones (Table 1.13) are a family of synthetic broad-spectrum antimicrobial drugs that comprise four generations; members of the first generation have a narrow spectrum of activity compared to those in later generations. The first quinolone to be used clinically for its antimicrobial activity was nalidixic acid in 1962; this drug is a derivative of chloroquine that was discovered by Lesher and coworkers. Today, naladixic acid and other first-generation quinolones such as flumequine and oxolinic acid are used primarily in aquaculture. Successive generations of quinolones have a fluorine atom in the quinolone ring structure, typically at the C6 position. Several fluoroquinolones, including danofloxacin, difloxacin, enrofloxacin (which is deethylated to form ciprofloxacin), marbofloxacin, orbifloxacin, and sarofloxacin, are used in veterinary but not human medicine. Conversely, some fluoroquinolones that are important in human medicine are not labeled for animal use.

The activity type of the fluoroquinolone antimicrobial drugs is concentration-dependent. Because quinolones accumulate in the cytosol of macrophages and neutrophils, they are often used to treat intracellular pathogens. The preponderance of macrophages and neutrophils in infected tissues compared to healthy tissues may explain the higher concentrations of fluoroquinolones attained in infected tissues.[62] Fluoroquinolones can produce a post-antibiotic effect, suppressing bacterial growth after local drug concentrations have fallen below the MIC of the target

TABLE 1.11 Polypeptides, Glycopeptides, and Streptogramins

INN	IUPAC Name, Molecular Formula, and CAS Registry No.	Chemical Structure	pK_a
		Polypeptides	
Bacitracin A	(4*R*)-4-[[(2*S*)-2-[[2-[(1*S*)-1-Amino-2-methylbutyl]4,5-dihydro-1,3-thiazole-5-carbonyl]amino]-4-methylpentanoyl]amino]-5-[[(2*S*)-1-[[(3*S*,6*R*,9*S*,12*R*,15*S*,18*R*,21*S*)-3-(2-amino-2-oxoethyl)-18-(3-aminopropyl)-15-butan-2-yl-6-(carboxymethyl)-9-(3*H*-imidazol-4-ylmethyl)-2,5,8,11,14,17,20-heptaoxo-12-(phenylmethyl)-1,4,7,10,13,16,19-heptazacyclopentacos-21-yl]amino]-3-methyl-1-oxopentan-2-yl]amino]-5-oxopentanoic acid $C_{66}H_{103}N_{17}O_{16}S$ 1405-87-4	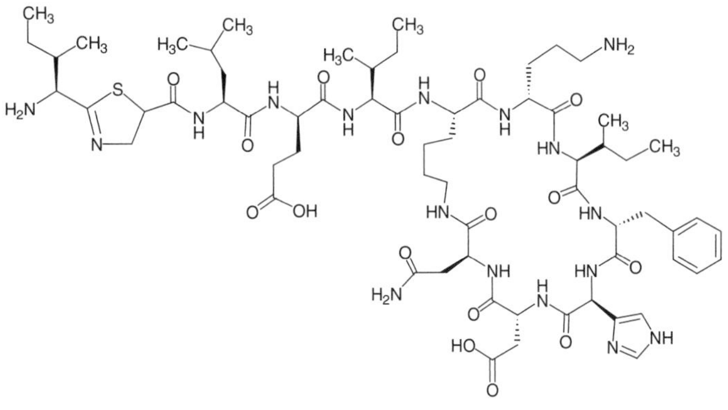	N/A[a]
Colistin (polymyxin E)	*N*-[(2*S*)-4-Amino-1-[[(2*S*,3*R*)-1-[[(2*S*)-4-amino-1-oxo-1-[[(3*S*,6*S*,9*S*,12*S*,15*R*,18*S*,21*S*)-6,9,18-tris(2-aminoethyl)-3-(1-hydroxyethyl)-12,15-bis(2-methylpropyl)-2,5,8,11,14,17,20-heptaoxo-1,4,7,10,13,16,19-heptazacyclotricos-21-yl]-amino]butan-2-yl]amino]-3-hydroxy-1-oxobutan-2-yl]amino]-1-oxobutan-2-yl]-5-methylheptanamide $C_{52}H_{98}N_{16}O_{13}$ 1066-17-7	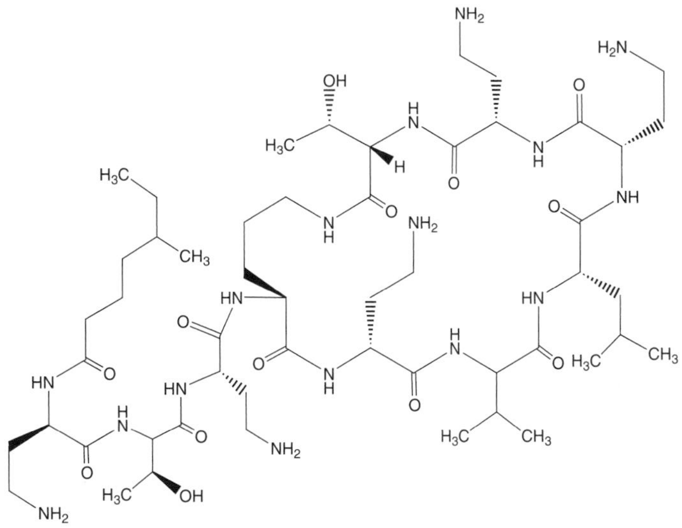	N/A[a]
Enramycin (enduracidin)	IUPAC name not available $C_{107}H_{138}Cl_2N_{26}O_{31}$ 11115-82-5	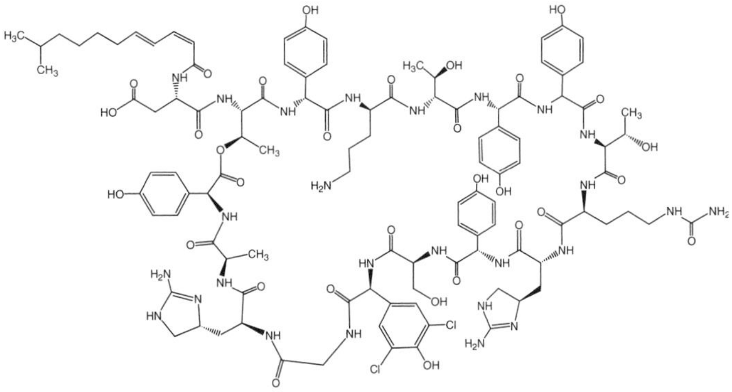	N/A[a]

(continued)

TABLE 1.11 ***(Continued)***

INN	IUPAC Name, Molecular Formula, and CAS Registry No.	Chemical Structure	pK_a
Novobiocin	*N*-[7-[[3-*O*-(Aminocarbonyl)-6-deoxy-5-*C*-methyl-4-*O*-methyl-ß-L-lyxo-hexopyranosyl]oxy]-4-hydroxy-8-methyl-2-oxo-2*H*-1-benzopyran-3-yl]-4-hydroxy-3-(3-methyl-2-butenyl)benzamide $C_{31}H_{36}N_2O_{11}$ 303-81-1	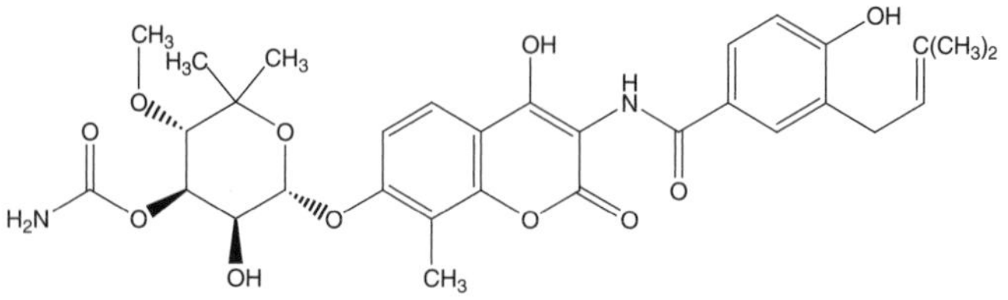	HA 4.3,[56] HA 9.1[56]
Polymyxin B	*N*-[4-Amino-1-[[1-[[4-amino-1-oxo-1-[[6,9,18-tris(2-aminoethyl)-15-benzyl-3-(1-hydroxyethyl)-12-(2-methylpropyl)-2,5,8,11,14,17,20-heptaoxo-1,4,7,10,13,16,19-heptazacyclotricos-21-yl]amino]butan-2-yl]amino]-3-hydroxy-1-oxobutan-2-yl]amino]-1-oxobutan-2-yl]-6-methyloctanamide $C_{56}H_{98}N_{16}O_{13}$ 1405-20-5	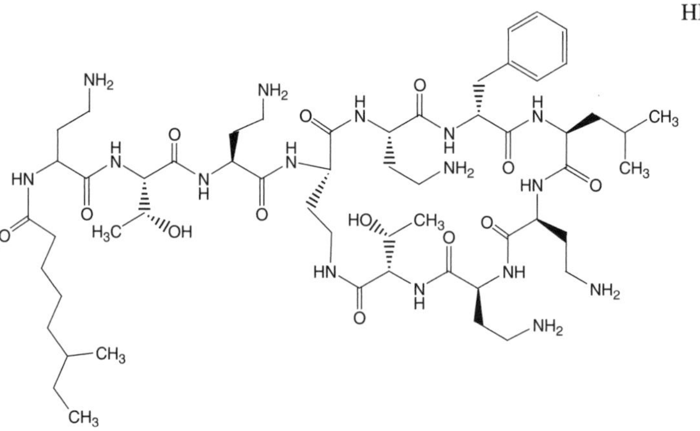	HB^+ 8.9[56]
Thiopeptin B	IUPAC name not available $C_{72}H_{104}N_{18}O_{18}S_5$ 37339-66-5	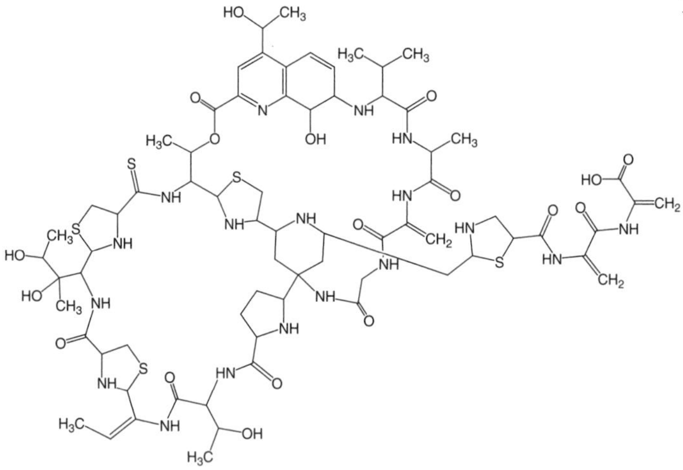	N/A[a]

TABLE 1.11 (*Continued*)

INN	IUPAC Name, Molecular Formula, and CAS Registry No.	Chemical Structure	pK_a
		Glycopeptides	
Avoparcin	IUPAC name not available $C_{89}H_{102}ClN_9O_{36}$ (α-avoparcin) $C_{89}H_{101}Cl_2N_9O_{36}$ (ß-avoparcin) 37332-99-3	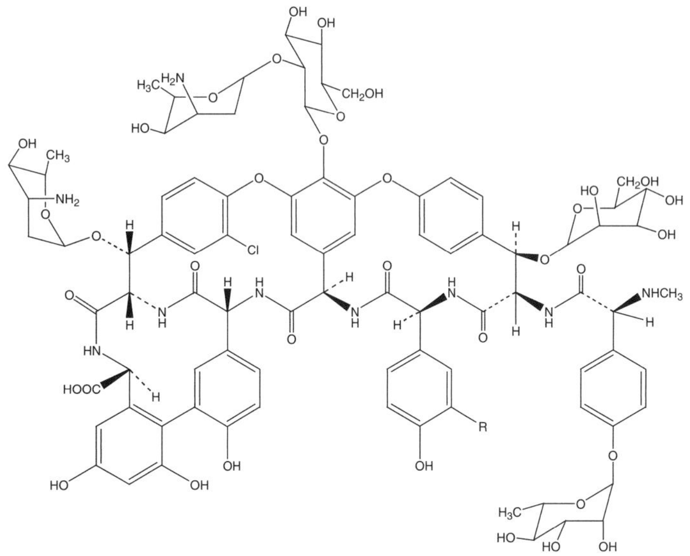 **α-avoparcin** R = H **β-avoparcin** R = Cl	N/A[a]
Teicoplanin	Ristomycin A: 34-*O*-[2-(acetylamino)-2-deoxy-β-D-glucopyranosyl]-22,31-dichloro-7-demethyl-64-*O*-demethyl-19-deoxy-56-*O*-[2-deoxy-2-[(8-methyl-1-oxononyl)amino]-β-D-glucopyranosyl]-42-O-α-D-mannopyranosyl $C_{88}H_{95}Cl_2N_9O_{33}$ (teicoplanin A_2—1) $C_{88}H_{97}Cl_2N_9O_{33}$ (teicoplanin A_2—2) $C_{88}H_{97}Cl_2N_9O_{33}$ (teicoplanin A_2—3) $C_{89}H_{99}Cl_2N_9O_{33}$ (teicoplanin A_2—4) $C_{89}H_{99}Cl_2N_9O_{33}$ (teicoplanin A_2—5) 61036-62-2	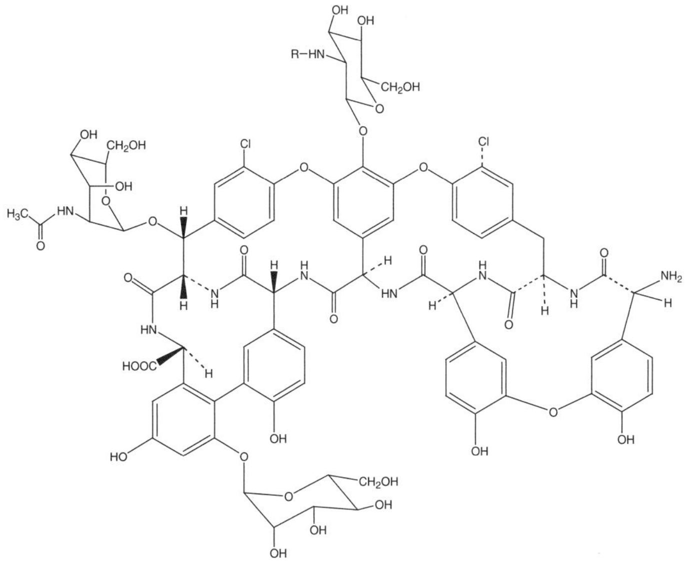 A_2-1: R = (Z)-4-decanoic acid A_2-2: R = 8-methylnonanoic acid A_2-3: R = n-decanoic acid A_2-4: R = 8-methyldecanoic acid A_2-5: R = 9-methyldecanoic acid	N/A[a]

(*continued*)

TABLE 1.11 (*Continued*)

INN	IUPAC Name, Molecular Formula, and CAS Registry No.	Chemical Structure	pK_a
Vancomycin	(3*S*,6*R*,7*R*,11*R*,23*S*,26*S*,30a*S*, 36*R*,38a*R*)-44-[2-*O*-(3-Amino-2,3,6-trideoxy-3-*C*-methyl-α-L-*lyxo*-hexopyranosyl)-β-D-glucopyranosyloxy]-3-(carbamoylmethyl)-10,19-dichloro-2,3,4,5,6,7,23,25,26, 36,37,38,38a-tetradecahydro-7,22,28,30,32-pentahydroxy-6-(*N*-methyl-D-leucyl)-2,5,24,38,39-pentaoxo-1*H*,22*H*-23,36-(epiminomethano)-8,11:18,21-dietheno-13,16:31,35-di(metheno)^{1,6,9} oxadiazacyclohexadecino[4,5-*m*]^{10,2,16} benzoxadiazacyclotetracosine-26-carboxylic acid $C_{66}H_{75}Cl_2N_9O_{24}$ 1404-90-6	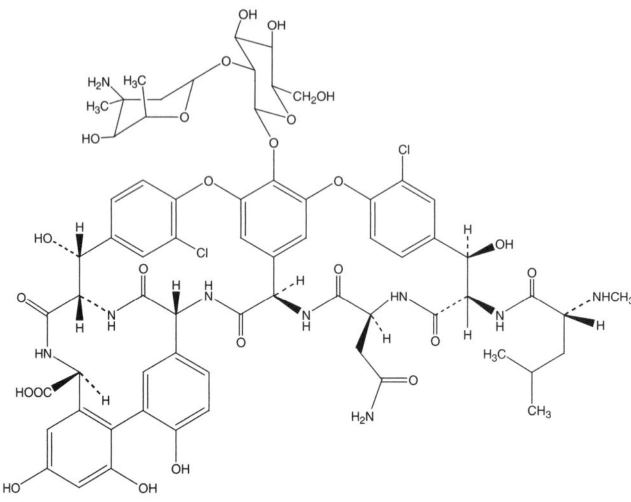	HA 2.2[56] (COOH), HB^+ 7.8[56] ($NHCH_3$) HB^+ 8.9[56] (NH_2), HA 9.6[56] (phenol), HA 10.4[56] (phenol), HA 12.0[56] (phenol)
		Streptogramins	
Quinupristin/ dalfopristin	Quinupristin: *N*-[(6*R*,9*S*,10*R*,13*S*,15a*S*,18*R*, 22*S*,24a*S*)-22-[*p*-(Dimethylamino)benzyl]-6-ethyldocosahydro-10,23-dimethyl-5,8,12,15,17,21,24-heptaoxo-13-phenyl-18-[[(3*S*)-3-quinuclidinylthio] methyl]-12*H*-pyrido[2,1-*f*]pyrrolo-[2,1-*l*]^{1,4,7,10,13,16} oxapentaazacyclononadecin-9-yl]-3-hydroxypicolinamide $C_{53}H_{67}N_9O_{10}S$ 120138-50-3 Dalfopristin: (3*R*,4*R*,5*E*,10*E*, 12*E*,14*S*,26*R*,26a*S*)-26-[[2-(Diethylamino)ethyl]sulfonyl]-8,9,14,15,24,25,26,26a-octahydro-14-hydroxy-3-isopropyl-4,12-dimethyl-3*H*-21,18-nitrilo-1*H*,22*H*-pyrrolo[2,1-*c*]^{1,8,4,19}-dioxadiazacyclotetracosine-1,7,16,22(4*H*,17*H*)-tetrone $C_{34}H_{50}N_4O_9S$ 112362-50-2 $C_{87}H_{117}N_{13}O_{19}S_2$ (combined) 126602-89-9 (combined)	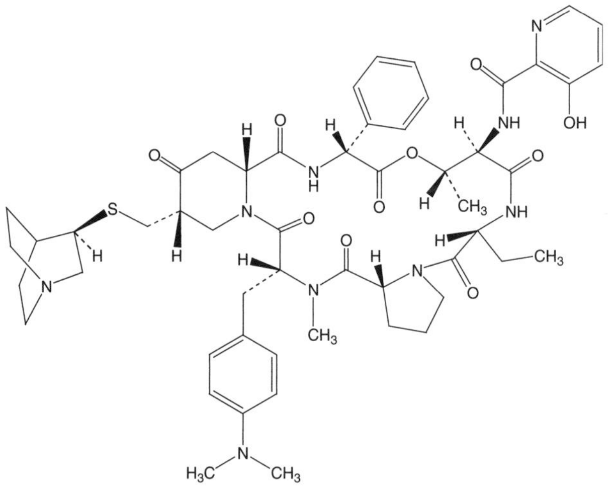 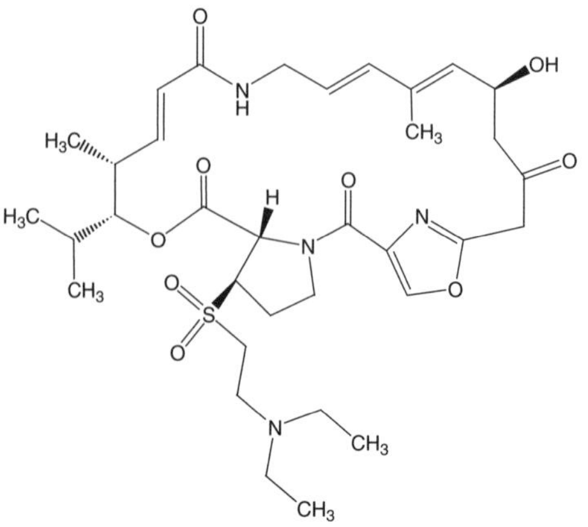	N/A[a]

TABLE 1.11 (*Continued*)

INN	IUPAC Name, Molecular Formula, and CAS Registry No.	Chemical Structure	pK_a
Virginiamycin	Virginiamycin S_1: IUPAC name not available $C_{43}H_{49}N_7O_{10}$ 23152-29-6 Virginiamycin M_1: 8,9,14,15,24,25-Hexahydro-14-hydroxy-4,12-dimethyl-3-(1-methylethyl)(3*R*,4*R*,5*E*,10*E*,12*E*,14*S*)-3*H*-21,18-nitrolo-1*H*,22*H*-pyrrolo-[2,1-*c*][1,8,4,19] dioxadiazacyclotetracosine-1,7,16,22(4*H*,17*H*)-tetrone $C_{28}H_{35}N_3O_7$ 21411-53-0	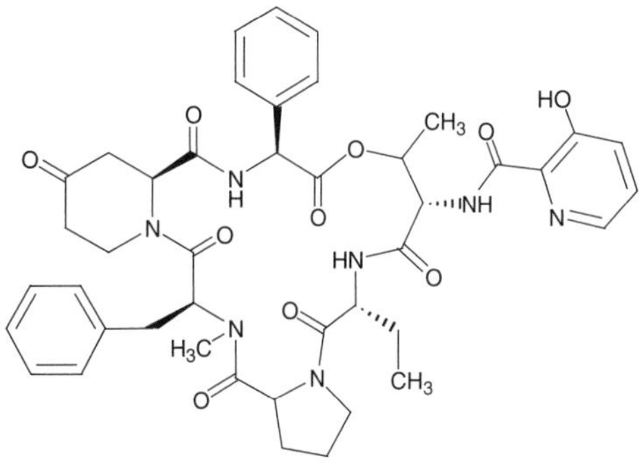 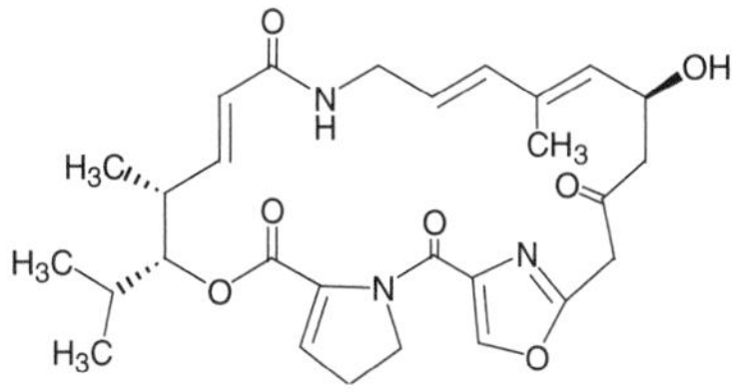	N/A[a]

[a]The author was not able to find a pK_a value for the substance in the public literature.

pathogen. The fluoroquinolones enter bacterial cells via porins and inhibit bacterial DNA gyrase in many Gram-negative bacteria, or topoisomerase IV in many Gram-positive bacteria—thereby inhibiting DNA replication and transcription. Fluoroquinolones also cause the cessation of cellular respiration and disruption of membrane integrity. Although mammalian topoisomerase II is a target for a variety of quinolone-based drugs, concentrations approximately 100-fold higher than those recommended for bacterial activity are needed for the enzyme to be inhibited.

TABLE 1.12 Phosphoglycolipids

INN	IUPAC Name, Molecular Formula, and CAS Registry No.	Chemical Structure	pK_a
Flavophospholipol (bambermycin, moenomycin A)	IUPAC name not available $C_{69}H_{107}N_4O_{35}P$ 11015-37-5	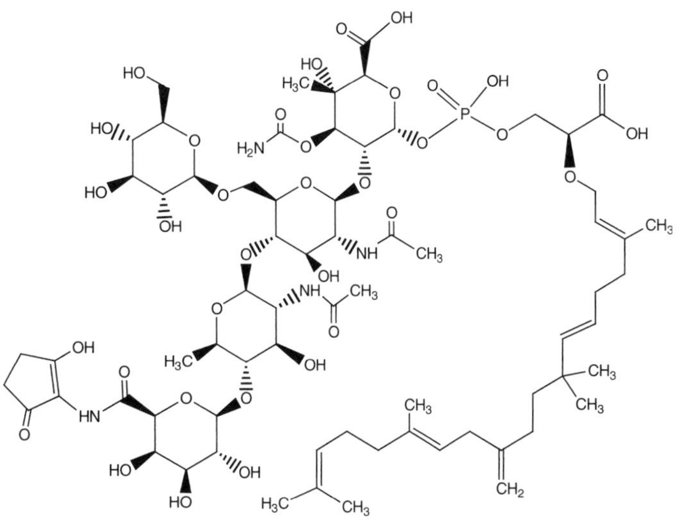	N/A[a]

[a]The author was not able to find a pK_a value for the substance in the public literature.

TABLE 1.13 Quinolones

INN	IUPAC Name, Molecular Formula, and CAS Registry No.	Chemical Structure	pK_a
Ciprofloxacin	1-Cyclopropyl-6-fluoro-1,4-dihydro-4-oxo-7-(1-piperazinyl)-3-quinoline carboxylic acid $C_{17}H_{18}FN_3O_3$ 85721-33-1	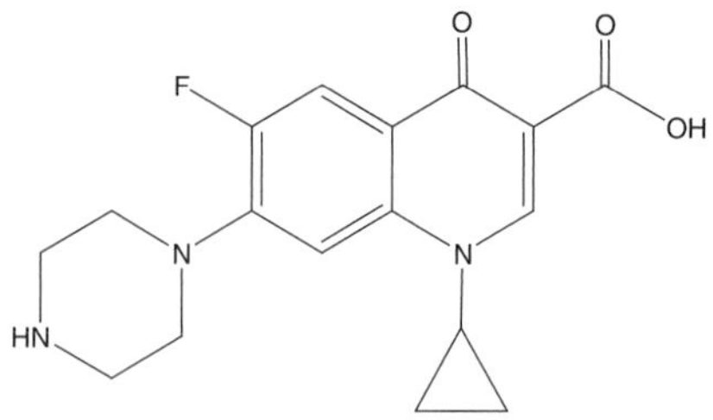	HA 6.2,[56] HB^+ 8.7[56]
Danofloxacin	(1*S*)-1-Cyclopropyl-6-fluoro-1,4-dihydro-7-(5-methyl-2,5-diazabicyclo[2.2.1]hept-2-yl)-4-oxo-3-quinoline carboxylic acid $C_{19}H_{20}FN_3O_3$ 112398-08-0	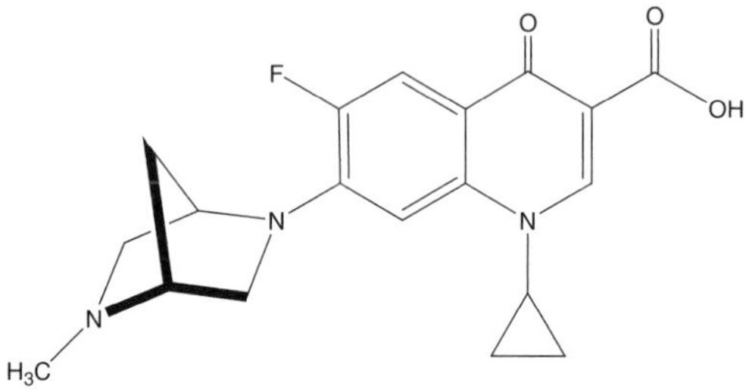	N/A[a]
Difloxacin	6-Fluoro-1-(4-fluorophenyl)-1,4-dihydro-7-(4-methyl-1-piperazinyl)-4-oxo-3-quinoline carboxylic acid $C_{21}H_{19}F_2N_3O_3$ 98106-17-3	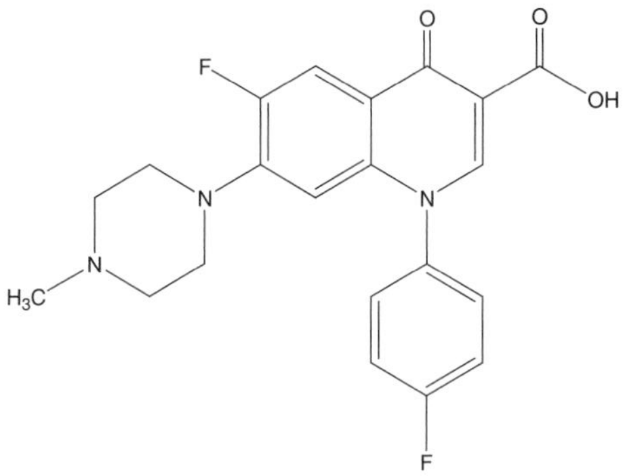	HA 6.1,[56] HB^+ 7.6[56]
Enrofloxacin	1-Cyclopropyl-7-(4-ethyl-1-piperazinyl)-6-fluoro-1,4-dihydro-4-oxo-3-quinolonecarboxylic acid $C_{19}H_{22}FN_3O_3$ 93106-60-6	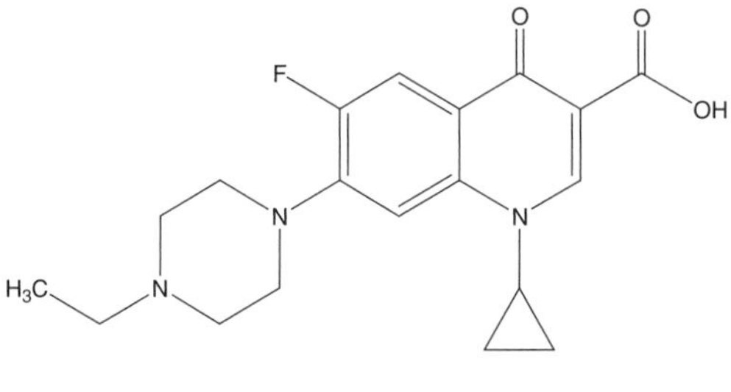	HA 6.0,[74] HB^+ 8.8[74]
Flumequine	9-Fluoro-6,7-dihydro-5-methyl-1-oxo-1*H*,5*H*-benzo[*ij*]-quinolizine-2-carboxylic acid. $C_{14}H_{12}FNO_3$ 42835-25-6	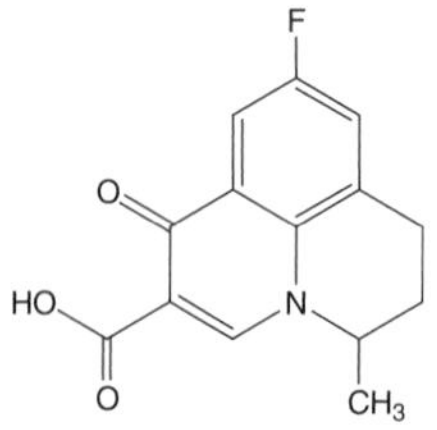	HA 6.4[56]
Marbo-floxacin	9-Fluoro-2,3-dihydro-3-methyl-10-(4-methyl-1-piperazinyl)-7-oxo-7*H*-pyridol(3,2,1-ij)(4,2,1)benzoxadiazin-6-carboxylic acid $C_{17}H_{19}N_4O_4F$ 115550-35-1	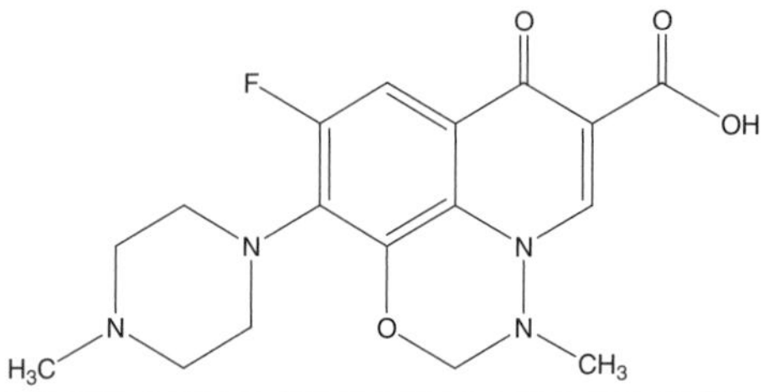	N/A[a]

TABLE 1.13 ***(Continued)***

INN	IUPAC Name, Molecular Formula, and CAS Registry No.	Chemical Structure	pK_a
Nalidixic acid	1-Ethyl-1,4-dihydro-7-methyl-4-oxo-[1,8]naphthyridine-3-carboxylic acid $C_{12}H_{12}N_2O_3$ 389-08-2	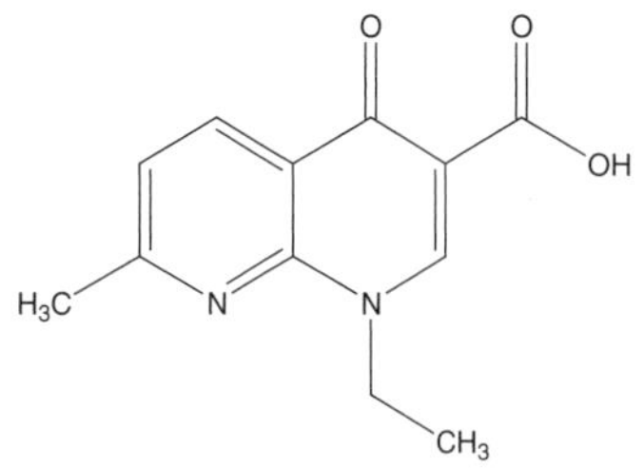	HA 6.0[56]
Norfloxacin	1-Ethyl-6-fluoro-1,4-dihydro-4-oxo-7-(1-piperazinyl)-3-quinolinecarboxylic acid $C_{16}H_{18}FN_3O_3$ 70458-96-7	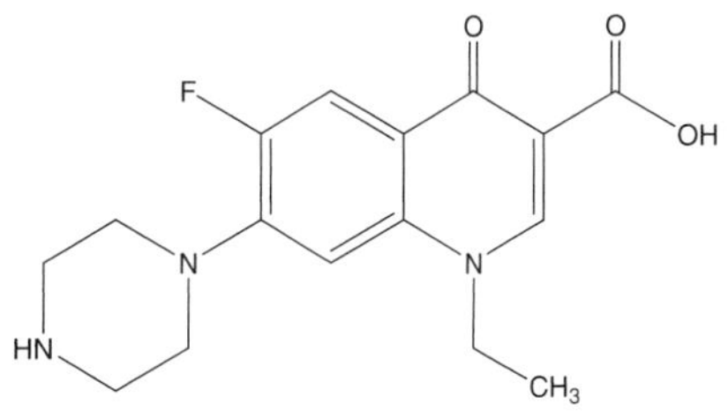	HA 6.3,[56] HB^+ 8.4
Orbifloxacin	1-Cyclopropyl-7-[(3*S*,5*R*)-3,5-dimethylpiperazin-1-yl]-5,6,8-trifluoro-4-oxo-1,4-dihydroquinoline-3-carboxylic acid $C_{19}H_{20}F_3N_3O_3$ 113617-63-3	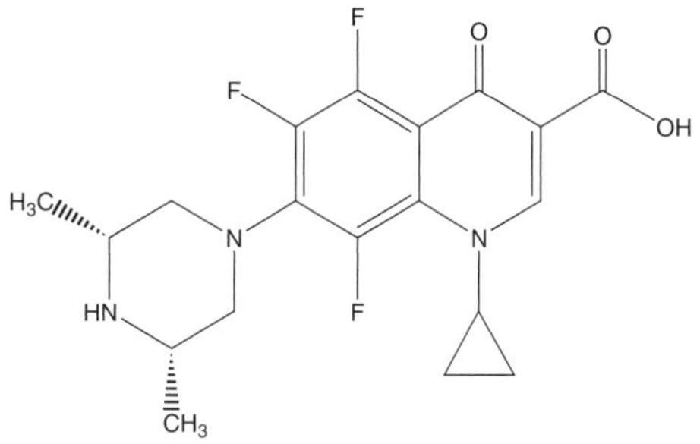	HA ~6,[56] HB^+~9[56]
Oxolinic acid	5-Ethyl-5,8-dihydro-8-oxo[1,3]dioxolo[4,5-*g*]quinoline-7-carboxylic acid $C_{13}H_{11}NO_5$ 14698-29-4	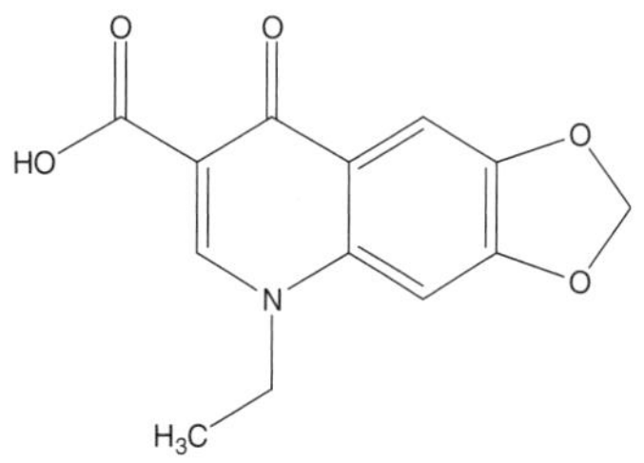	N/A[a]
Sarafloxacin	6-Fluoro-1-(4-fluorophenyl)-4-oxo-7-piperazin-1-ylquinoline-3-carboxylic acid $C_{20}H_{17}F_2N_3O_3$ 98105-99-8	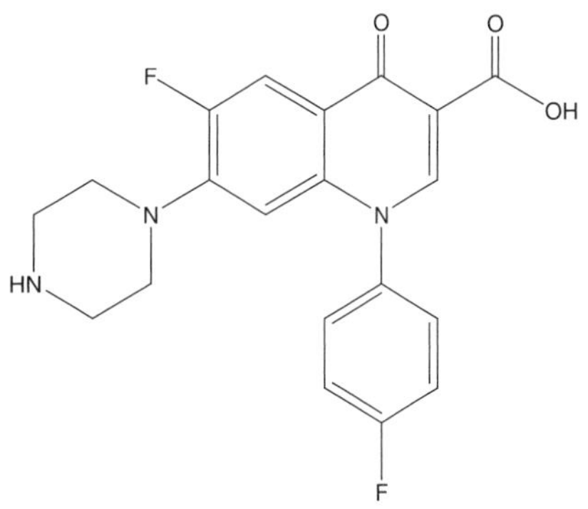	N/A[a]

[a]The author was not able to find a pK_a value for the substance in the public literature.

Resistance to quinolones can evolve rapidly; the most common mechanism involves mutation of DNA gyrase (topoisomerase II) in Gram-negative bacteria. A similar mechanism alters topoisomerase IV in Gram-positive bacteria. These mutations result in reduced binding affinity to quinolones, which decreases bactericidal activity. A second mechanism of resistance involves increased expression of efflux pumps that actively transport drug out of bacterial cells, resulting in decreased intracellular drug concentration. Plasmid-mediated resistance in Gram-negative bacteria results in the synthesis of proteins that bind to DNA gyrase, protecting it from the action of quinolones. At present, however, the clinical importance of this mechanism is unclear. The transfer of fluoroquinolone-resistant *Campylobacter* species and *Salmonella typhimurium*-type DT-104 from animals to humans is a major concern. As a consequence, some countries have established systems for monitoring and surveillance of antibiotic resistance in human and animal isolates. In many countries, approved and off-label uses of fluoroquinolones in food-producing species are either restricted or not permitted.

Fluoroquinolones are active against some Gram-negative bacteria, including *E. coli, Enterobacter* species, *Klebsiella* species, *Pasteurella* species, *Proteus* species, and *Salmonella* species. The susceptibility of *Pseudomonas aeruginosa* to fluoroquinolones is variable. These agents are also active against some Gram-positive bacteria and chlamydia, mycobacteria, and mycoplasma. In some regions, the use of fluoroquinolones is approved for the treatment of colibacillosis of chickens and turkeys, fowl cholera in turkeys, and bovine respiratory disease caused by *Mannheimia haemolytica, Pasteurella multocida, Haemophilus somnus*, and other susceptible organisms. The fluoroquinolones are administered as oral solutions to chickens and turkeys, by injection to cattle, as tablets, and by injection to dogs and cats.

Fluoroquinolone administration during rapid growth has been associated with arthropathies and cartilage erosions in weight-bearing joints in immature cats, dogs, and horses. Retinal degeneration has been associated with the administration of enrofloxacin at high doses in cats. Therefore the use of fluoroquinolones in immature animals and high doses of fluoroquinolones in cats should be avoided.

Literature relating to quinolones in the environment as a result of veterinary use is sparse. Trace amounts of flumequine, oxolinic acid, and sarafloxacin in sediment at fish farms have been detected.[27] Trace amounts of enrofloxacin in soil were detected in a UK monitoring study,[87] while studies into the sorption behavior of danofloxacin and sarafloxacin in soil showed these drugs to be non-mobile and persistent.[100]

On the basis of JECFA evaluations of toxicological and residues depletion data for oxolinic acid,[165] flumequine,[166] enrofloxacin,[167] danofloxacin,[168] and sarafloxacin,[169] CAC MRL recommendations were established for flumequine in cattle, sheep, pigs, chickens, and trout;[43] danofloxacin in cattle, pigs, and chickens;[43] and sarafloxacin in chickens and turkeys[43] but not for oxolinic acid[165] or enrofloxacin.[167] Details of residue studies reviewed by JECFA have been published for oxolinic acid,[170] flumequine,[171–174] enrofloxacin,[175] danofloxacin,[176] and sarafloxacin.[177]

1.3.13 Sulfonamides

The sulfonamides were the first effective chemotherapeutic agents to be employed systemically for the prevention and cure of bacterial infections in humans. Foerster reported the first clinical case study of prontosil in 1933; 2 years later, this compound was shown to be a pro-drug of sulfanilamide. In 1939, Domagk was awarded the Nobel Prize in Physiology or Medicine for discovering the chemotherapeutic value of prontosil. Interestingly, sulfanilamide had been prepared by Gelmo in 1906 while investigating azo dyes.

The sulfonamide class contains a large number of antibacterial drugs, including sulfadiazine, sulfamethazine (sulfadimidine), sulfathiazole, sulfamethoxazole, and many more. Potentiated sulfonamides, in which a sulfonamide and an antibacterial diaminopyrimidine such as trimethoprim are combined, demonstrate improved efficacy compared with sulfonamides alone. Relatively few sulfonamides are currently (as of 2011) approved for use in food-producing species. This is attributed to numerous factors, including toxicological concerns associated with some sulfonamides and the lack of contemporary data to support the historical uses of other sulfonamides.

The sulfonamides are structural analogues of *para*-aminobenzoic acid (PABA) and competitively inhibit dihydropteroate synthetase, the enzyme that catalyzes the synthesis of dihydrofolic acid (folic acid). Organisms susceptible to sulfonamides must synthesize their own folic acid, unlike mammalian cells, which utilize preformed folic acid. The decreased synthesis of dihydrofolic acid, in turn, causes decreased synthesis of tetrahydrofolic acid (folinic acid), which is required for the synthesis of DNA. A variety of effects may result, including suppression of protein synthesis, impairment of metabolic processes, and inhibition of growth and multiplication in susceptible organisms. Sulfonamides, which are not efficacious in the presence of purulent material, are bacteriostatic. Before such activity is exhibited, however, existing stores of folic acid, folinic acid, purines, thymidine, and amino acids are utilized by bacteria. Sulfonamides inhibit both Gram-positive and Gram-negative bacteria, some chlamydia, *Nocardia*, and *Actinomyces* species, and some protozoa including coccidia and *Toxoplasma* species. Organisms resistant to sulfonamides include *Pseudomonas, Klebsiella, Proteus, Clostridium*, and *Leptospira* species.

Although dihydrofolate reductase catalyzes the synthesis of folic acid in both bacteria and mammals, antibacterial diaminopyrimidines such as trimethoprim and ormetoprim inhibit this enzyme more efficiently in bacteria than in mammalian cells. These drugs are bacteriostatic when used alone; however, when combined with sulfonamides, the sequential blockade of dihydropteroate synthetase and dihydrofolate reductase elicits a bactericidal effect. Sulfonamide–diaminopyrimidine combinations are active against Gram-positive and Gram-negative organisms, including *Actinomyces, Bordetella, Clostridium, Corynebacterium, Fusobacterium, Haemophilus, Klebsiella, Pasteurella, Proteus, Salmonella, Shigella*, and *Campylobacter* species, as well as *E. coli*, streptococci, and staphylococci. *Pseudomonas* and *Mycobacterium* species are resistant to potentiated sulfonamides.

Resistance to sulfonamides is widespread in bacteria isolated from animals, and may involve chromosomal mutations or plasmid-mediated mechanisms. Chromosomal mutations cause impaired drug penetration, production of altered forms of dihydropteroate synthetase for which sulfonamides have a lowered affinity, or production of excessive PABA that overcomes the metabolic block imposed by the inhibition of dihydropteroate synthetase. A more common cause of bacterial resistance to sulfonamides is plasmid-mediated mechanisms, which may result in impaired drug penetration or the synthesis of sulfonamide-resistant dihydropteroate synthetase. There is cross-resistance among sulfonamides.

Resistance to bacterial diaminopyrimidines results from chromosomal mutations or plasmid-mediated mechanisms and develops very rapidly. Resistance conferred by chromosomal mutations allows bacteria to utilize exogenous sources of folinic acid or thymidine, thereby overcoming the drug-imposed blockade. Plasmid-mediated mechanisms result in the synthesis of dihydrofolate reductase characterized by a reduced affinity for antibacterial diaminopyrimidines.

Compared to most classes of antimicrobial drugs, the usage of sulfonamides and potentiated sulfonamides in veterinary medicine is high. Sulfonamides are used to treat or prevent acute systemic or local infections, including actinobacillosis, coccidiosis, mastitis, metritis, colibacillosis, pododermatitis, polyarthritis, respiratory infections, and toxoplasmosis. Sulfonamides are also used in the treatment of American foulbrood disease caused by *Paenibacillus larvae* and European foulbrood disease caused by *Melissococcus pluton* that affect honeybees. Sulfonamides in combination with pyrimethamine are used to treat protozoal diseases such as leishmaniasis and toxoplasmosis. Sulfonamides are most effective in the early stages of acute infections when organisms are rapidly multiplying.

Sulfonamides are administered to food-producing species as additives to feed and drinking water, controlled-release oral boluses, and intrauterine infusions. These drugs are applied to the brood chambers of honeybee hives mixed with confectioners' sugar or in syrup. The insoluble nature of sulfonamides is an important consideration. Highly insoluble sulfonamides such as phthalylsulfathiazole are absorbed from the gastrointestinal tract very slowly and are used to treat enteric infections. With triple sulfas for oral administration, the concentration of individual sulfonamides is limited by the drug's solubility, while efficacy reflects the additive activity of all three components. Sodium salts of sulfonamides, which are readily soluble in water, are available for intravenous administration.

The majority of adverse effects to sulfonamides are mild in nature and reversible, although idiosyncratic drug reactions may occur. Urinary tract disturbances, including sulfonamide crystalluria and hematuria, can be minimized in susceptible animals by maintaining an adequate water intake to maintain a high urine flow. Bone marrow depression and dermatologic reactions have also been associated with sulfonamide therapy in animals.

Literature describing sulfonamides and antibacterial diaminopyrimidines in the environment is sparse. The retransformation of N^4-acetylsulfamethazine to the active sulfamethazine during the storage of manure has been reported by Berger and co-workers[178] and may be an important consideration in species such as rabbits[179] and humans, in which N^4-acetylation represents a major metabolic pathway for sulfonamides. Studies into the transport of sulfonamides in runoff water[180] and the movement of sulfonamides through soil[181] indicate that these substances are not highly sorptive. This is consistent with the finding that sulfamethazine and sulfadimethoxine are non-persistent and highly mobile in soil.[101] A UK study reported that the concentrations of sulfadiazine and trimethoprim in surface water did not represent a risk to the environment.[26]

An ADI for sulfamethazine (sulfadimidine) has been allocated by JECFA.[182] The CAC has established MRLs for sulfamethazine in muscle, liver, kidney, and fat of cattle, sheep, pigs, and chickens.[43] A review of the residue studies considered by JECFA has been published.[183]

Table 1.14 lists the properties of sulfonamides and antibacterial diaminopyrimidines.

1.3.14 Tetracyclines

The tetracyclines (Table 1.15) are a large family of antibiotics, the first members of which were derived from the *Streptomyces* genus of *Actinobacteria*. Chlortetracycline was isolated from *Streptomyces aureofaciens* in 1944, and a few years later, oxytetracycline and demeclocycline were

TABLE 1.14 Sulfonamides and Antibacterial Diaminopyrimidines

INN	IUPAC Name, Molecular Formula, and CAS Registry No.	Chemical Structure	pK_a
	Sulfonamides		
Dapsone	4-[(4-Aminobenzene)sulfonyl]aniline $C_{12}H_{12}N_2O_2S$ 80-08-0		HB$^+$ 1.3,[56] HB$^+$ 2.5[56]
Phthalylsulfathiazole	2-[[[4-[(2-Thiazolylamino)sulfonyl]phenyl] amino]carbonyl]benzoic acid $C_{17}H_{13}N_3O_5S_2$ 85-73-4		N/A[a]
Sulfabenzamide	*N*-[(4-Aminophenyl)sulfonyl]benzamide $C_{13}H_{12}N_2O_3S$ 127-71-9		N/A[a]
Sulfacetamide	*N*-((4-Aminophenyl)sulfonyl)acetamide $C_8H_{10}N_2O_3S$ 144-80-9		HA 2.0,[56] HB$^+$ 5.3[47]
Sulfachloropyridazine	4-Amino-*N*-(6-chloropyridazin-3-yl)benzenesulfonamide $C_{10}H_9ClN_4O_2S$ 80-32-0		HA 6.1[56]
Sulfadiazine	4-Amino-*N*-pyrimidin-2-yl-benzenesulfonamide $C_{10}H_{10}N_4O_2S$ 68-35-9		HA 6.5[56]
Sulfadimethoxine	4-Amino-*N*-(2,6-dimethoxypyrimidin-4-yl)benzenesulfonamide $C_{12}H_{14}N_4O_4S$ 122-11-2		HB$^+$ 2.0,[56] HA 6.7[56]

TABLE 1.14 ***(Continued)***

INN	IUPAC Name, Molecular Formula, and CAS Registry No.	Chemical Structure	pK_a
Sulfadoxin	4-Amino-*N*-(5,6-dimethoxy-4-pyrimidinyl)benzenesulfonamide $C_{12}H_{14}N_4O_4S$ 2447-57-6	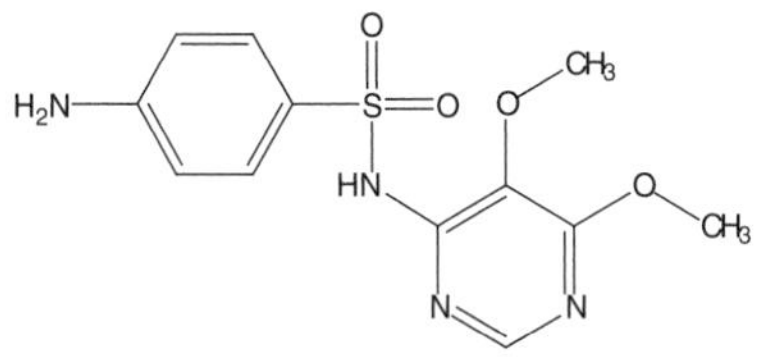	N/A[a]
Sulfaguanidine	4-Amino-*N*-[amino(imino)methyl]benzenesulfonamide $C_7H_{10}N_4O_2S$ 57-67-0	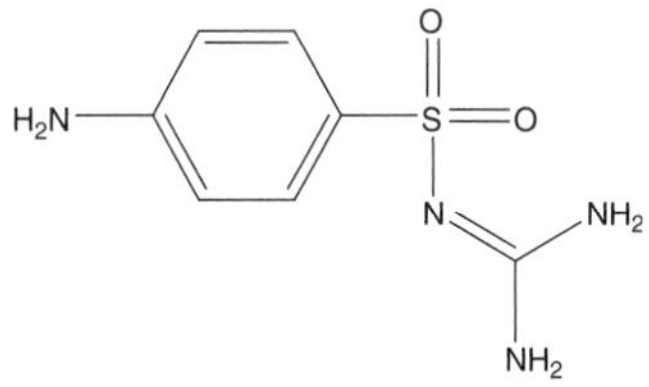	HB^+ 2.4[56]
Sulfamerazine	4-Amino-*N*-(4-methylpyrimidin-2-yl)benzenesulfonamide $C_{11}H_{12}N_4O_2S$ 127-79-7	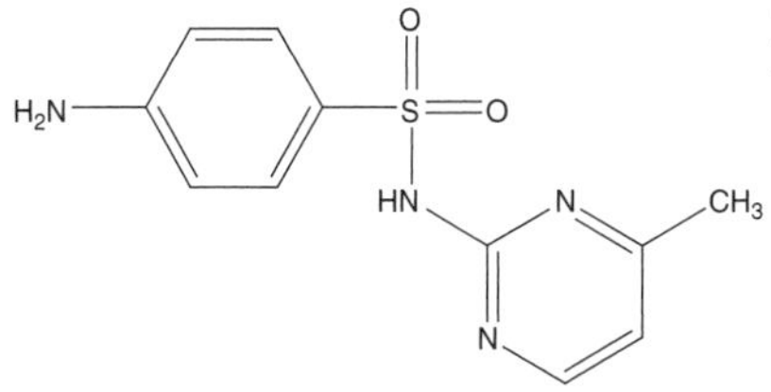	HB^+ 2.3,[56] HA 7.0[56]
Sulfamethazine (sulfadimidine)	4-Amino-*N*-(4,6-dimethylpyrimidin-2-yl)benzenesulfonamide $C_{12}H_{14}N_4O_2S$ 57-68-1	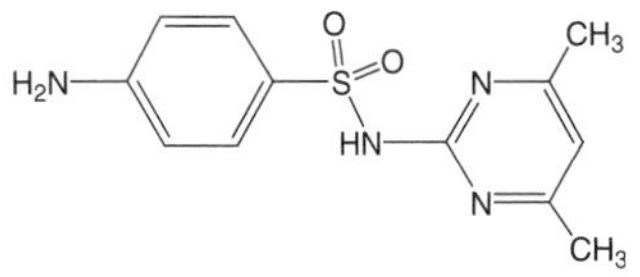	HB^+ 2.4,[56] HA 7.4[56]
Sulfamethizole	4-Amino-*N*-(5-methyl-1,3,4-thiadiazol-2-yl)benzenesulfonamide $C_9H_{10}N_4O_2S_2$ 144-82-1	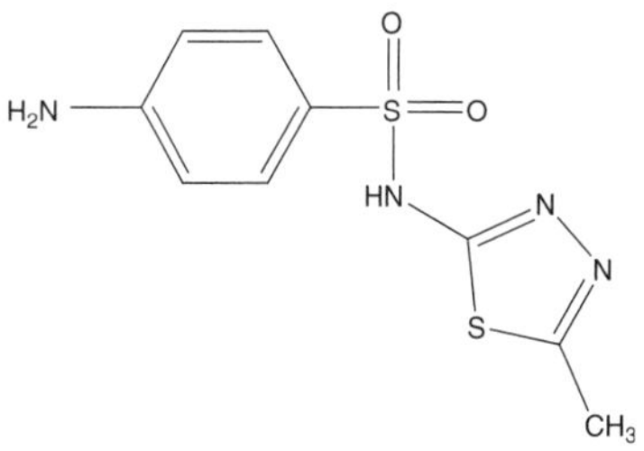	HA 5.4[56]
Sulfamethoxazole	4-Amino-*N*-(5-methyl-3-isoxazolyl)-benzenesulfonamide $C_{10}H_{11}N_3O_3S$ 723-46-6	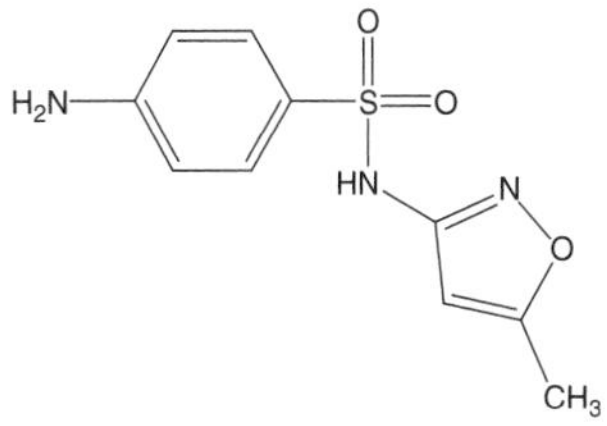	HA 5.6[56]
Sulfamethoxy-pyridazine	4-Amino-*N*-(6-methoxypyridazin-3-yl)benzenesulfonamide $C_{11}H_{12}N_4O_3S$ 80-35-3	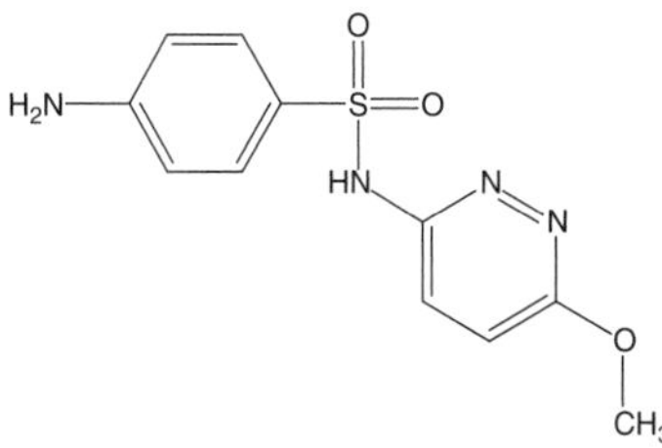	HA 7.2[56]

(continued)

TABLE 1.14 ***(Continued)***

INN	IUPAC Name, Molecular Formula, and CAS Registry No.	Chemical Structure	pK_a
Sulfamethoxydiazine (sulfameter)	4-Amino-*N*-(5-methoxy-2-pyrimidinyl)benzenesulfonamide $C_{11}H_{12}N_4O_3S$ 651-06-9	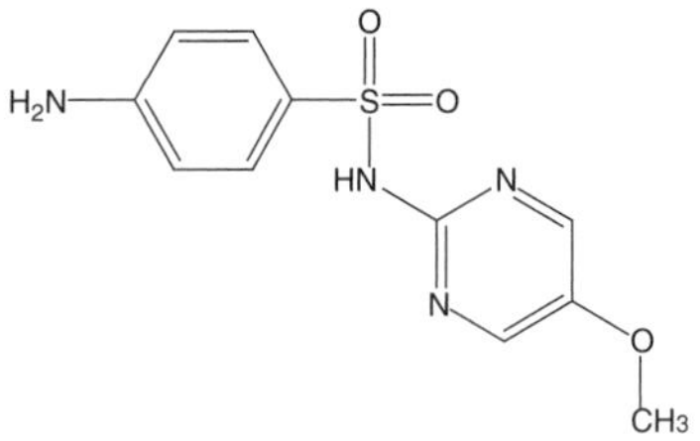	HA 6.8[56]
Sulfamonomethoxine	4-Amino-*N*-(6-methoxy-4-pyrimidinyl)benzenesulfonamide $C_{11}H_{12}N_4O_3S$ 1220-83-3	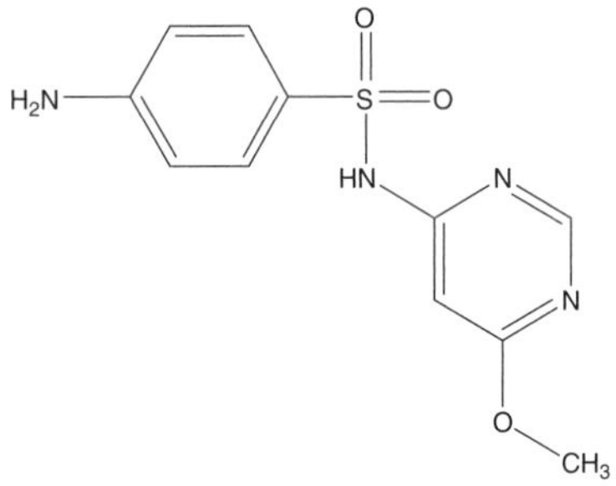	HA 5.9[56]
Sulfamoxole	4-Amino-*N*-(4,5-dimethyl1,3-oxazol-2-yl)benzenesulfonamide $C_{11}H_{13}N_3O_3S$ 729-99-7	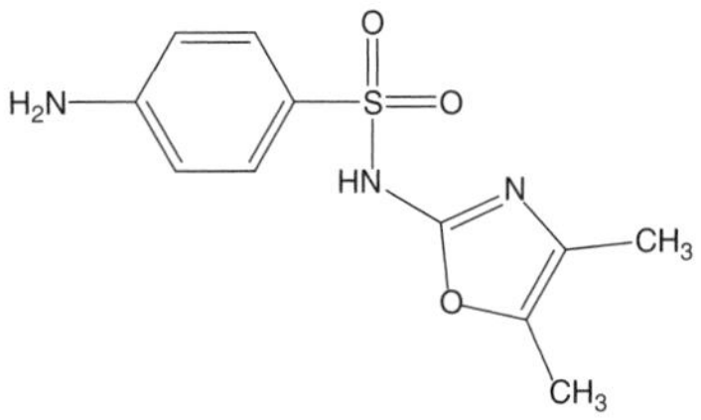	N.A.*
Sulfanilamide	4-Aminobenzenesulfonamide $C_6H_8N_2O_2S$ 63-74-1	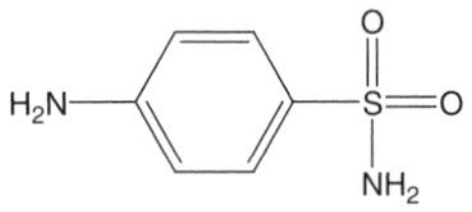	HB^+ 2.4[56]
Sulfaphenazole	4-Amino-*N*-(1-phenyl-1*H*-pyrazol-5-yl)benzenesulfonamide $C_{15}H_{14}N_4O_2S$ 526-08-9	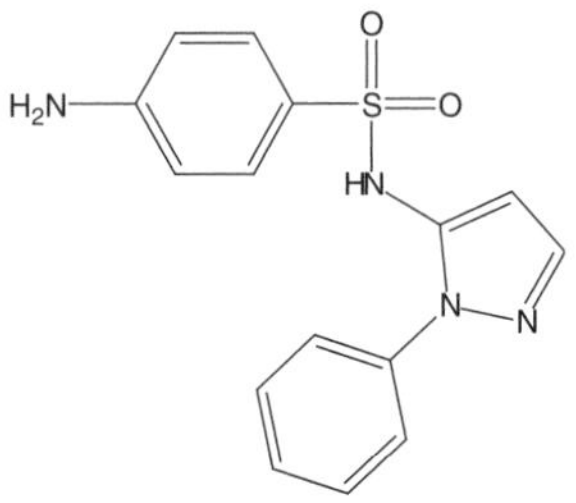	HA 5.7[56]
Sulfapyridine	4-Amino-*N*-pyridin-2-yl-benzenesulfonamide $C_{11}H_{11}N_3O_2S$ 144-83-2	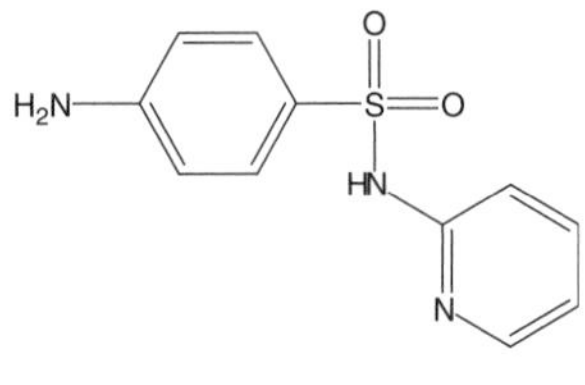	HB^+ 1.0,[56] HB^+ 2.6,[56] HA 8.4[56]
Sulfaquinoxaline	4-Amino-*N*-2-quinoxalinylbenzenesulfonamide $C_{14}H_{12}N_4O_2S$ 59-40-5	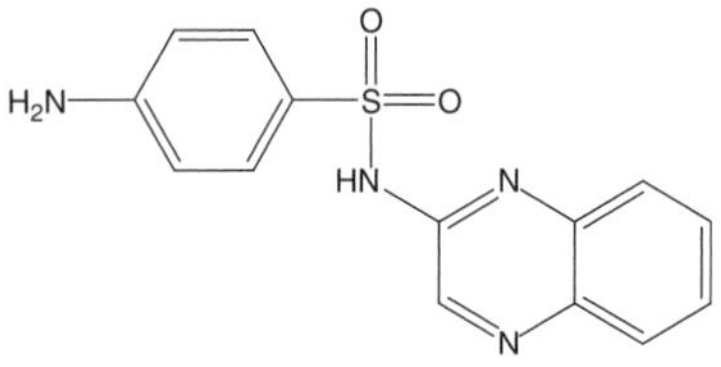	N/A[a]

TABLE 1.14 ***(Continued)***

INN	IUPAC Name, Molecular Formula, and CAS Registry No.	Chemical Structure	pK_a
Sulfathiazole	4-Amino-*N*-(1,3-thiazol-2-yl)benzenesulfonamide $C_9H_9N_3O_2S_2$ 72-14-0	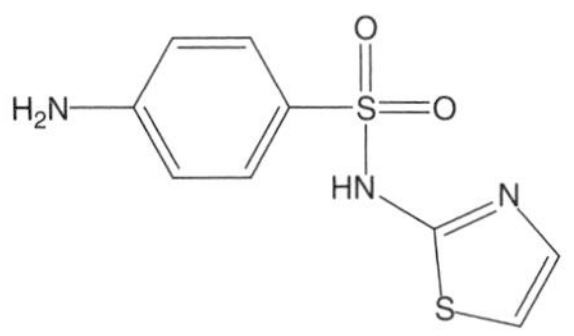	HA 7.1[56]
Sulfisomidine	4-Amino-*N*-(2,6-dimethylpyrimidin-4-yl)benzenesulfonamide $C_{12}H_{14}N_4O_2S$ 515-64-0	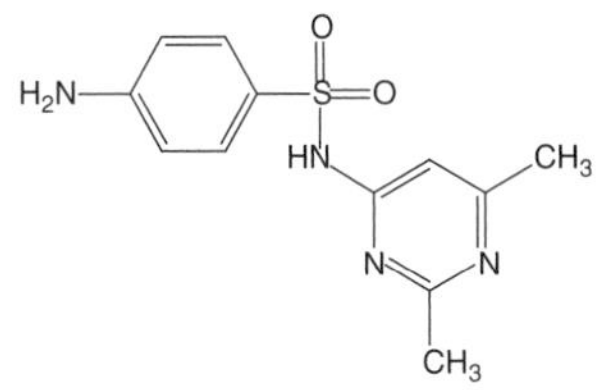	HA 7.6[56]
Sulfafurazole (sulfisoxazole)	4-Amino-*N*-(3,4-dimethyl-1,2-oxazol-5-yl)benzenesulfonamide $C_{11}H_{13}N_3O_3S$ 127-69-5	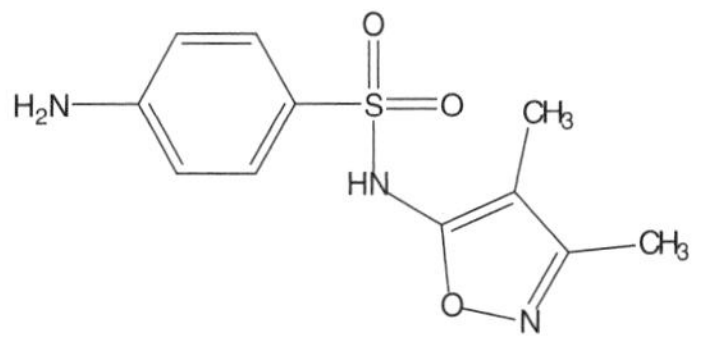	HA 5.0[56]
	Antibacterial Diaminopyrimidines		
Ormetoprim	5-(4,5-Dimethoxy-2-methylbenzyl)-2,4-diaminopyrimidine $C_{14}H_{18}N_4O_2$ 6981-18-6	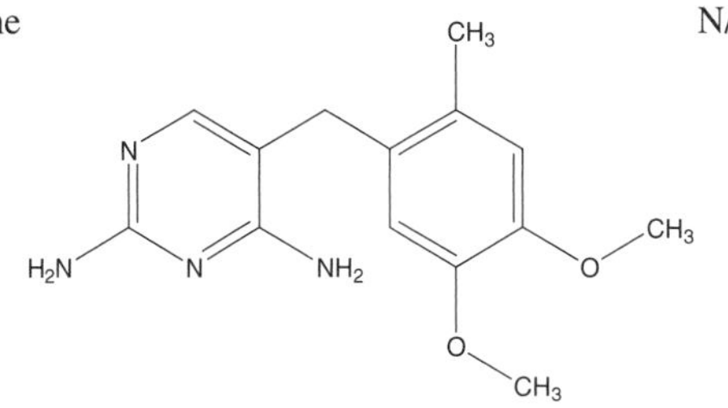	N/A[a]
Trimethoprim	5-[(3,4,5-Trimethoxyphenyl)methyl]pyrimidine-2,4-diamine $C_{14}H_{18}N_4O_3$ 738-70-5	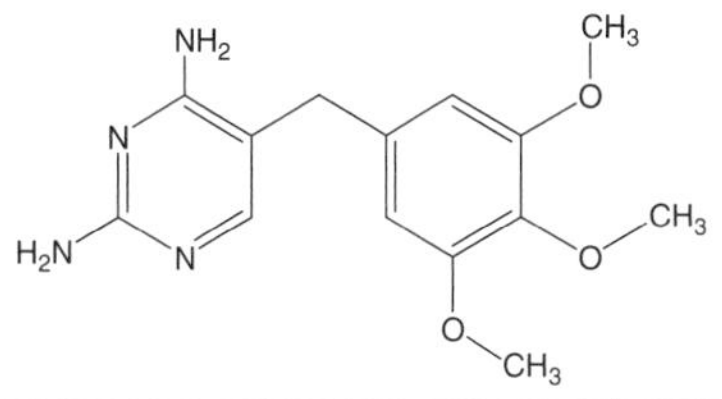	HB$^+$ 6.6[56]

[a] The author was not able to find a pK_a value for the substance in the public literature.

isolated from *Streptomyces rimosus* and *Streptomyces aureofaciens*, respectively. Tetracycline is sourced from the hydrogenolysis of chlortetracycline, while doxycycline is a semi-synthetic derivative of oxytetracycline that was developed in the early 1960s. The glycylcyclines are a new subgroup of tetracyclines that emerged with the introduction of tigecycline in 2005.

Tetracyclines are broad-spectrum antibiotics that inhibit protein synthesis—a mechanism that involves reversible binding of the drug to receptors of the 30*S* ribosomal subunit of susceptible organisms. This, in turn, blocks binding of the aminoacyl-tRNA to the acceptor site on the mRNA-ribosomal complex and prevents the addition of new amino acids to the peptide chain. Tetracyclines display bacteriostatic activity but can be bactericidal to sensitive organisms at high concentrations. They are more effective against organisms that are rapidly replicating. Tetracyclines exhibit activity against Gram-positive and Gram-negative bacteria, including some anaerobes. Susceptible organisms include *Escherichia coli, Klebsiella* species, *Pasteurella* species, *Salmonella* species, and *Streptococcus* species. Tetracyclines are also active against chlamydia, mycoplasmas, some protozoa, and several rickettsiae.

Resistance to tetracyclines is conferred on bacteria by at least three mechanisms. One mechanism involves the efflux of tetracyclines from bacterial cells and is the

TABLE 1.15 Tetracyclines

INN	IUPAC Name, Molecular Formula, and CAS Registry No.	Chemical Structure	pK_a
Chlortetracycline	(4*S*,4a*S*,5a*S*,6*S*,12a*S*,*Z*)-2-[amino(hydroxy)methylene]-7-chloro-4-(dimethylamino)-6,10,11,12a-tetrahydroxy-6-methyl-4a,5,5a,6-tetrahydrotetracene-1,3,12(2*H*,4*H*,12a*H*)-trione $C_{22}H_{23}ClN_2O_8$ 57-62-5	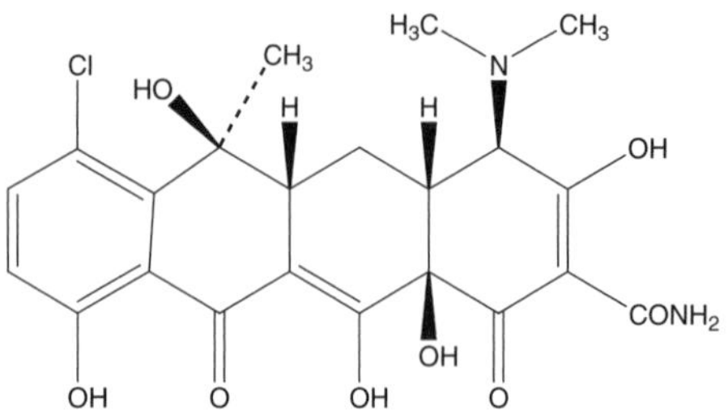	HA 3.3,[56] HA 7.4,[56] HB^+ 9.3[56]
4-epi-Chlortetracycline	4*R*,4a*S*,5a*S*,6*S*,12a*S*)-7-Chloro-4-(dimethylamino)-1,4,4a,5,5a,6,11,12a-octahydro-3,6,10,12,12a-pentahydroxy-6-methyl-1,11-dioxo-2-naphthacenecarboxamide monohydrate $C_{22}H_{23}ClN_2O_8$ 14297-93-9	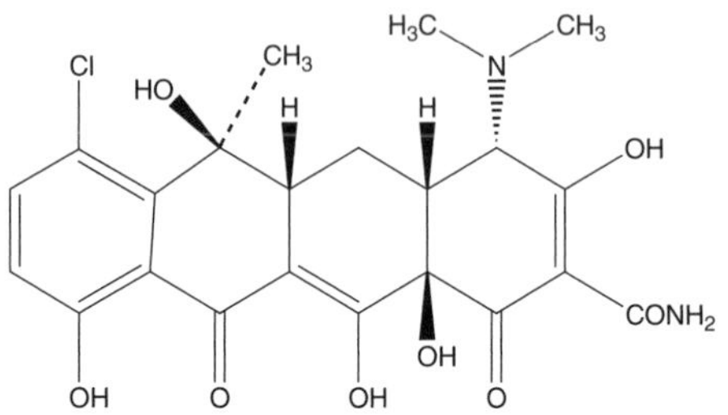	HA 3.7,[56] HA 7.7,[56] HB^+ 9.2[56]
Demeclocycline	(2*E*,4*S*,4a*S*,5a*S*,6*S*,12a*S*)-2-[Amino(hydroxy)methylidene]-7-chloro-4-(dimethylamino)-6,10,11,12a-tetrahydroxy-1,2,3,4,4a,5,5a,6,12,12a-decahydrotetracene-1,3,12-trione $C_{21}H_{21}ClN_2O_8$ 127-33-3	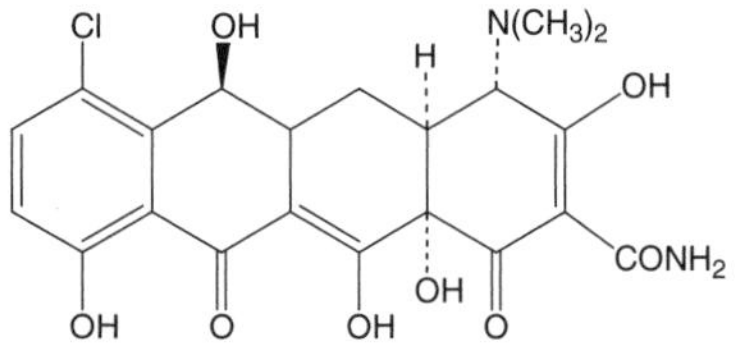	HA 3.3,[56] HA 7.2,[56] HB^+ 9.3[56]
Doxycycline	4*S*,4a*R*,5*S*,5a*R*,6*R*,12a*S*)-4-(Dimethylamino)-3,5,10,12,12a-pentahydroxy-6-methyl-1,11-dioxo-1,4,4a,5,5a,6,11,12a-octahydrotetracene-2-carboxamide $C_{22}H_{24}N_2O_8$ 564-25-0	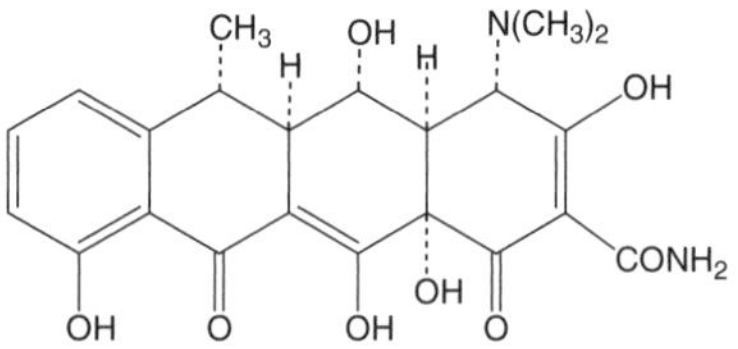	HA 3.2,[56] HA 7.6,[56] HB^+ 8.9,[56] HA 11.5[56]
Methacycline	(2*Z*,4*S*,4a*R*,5*S*,5a*R*,12a*S*)-2-(Amino-hydroxymethylidene)-4-dimethylamino-5,10,11,12a-tetrahydroxy-6-methylidene-4,4a,5,5a-tetrahydrotetracene-1,3,12-trione $C_{22}H_{22}N_2O_8$ 914-00-1	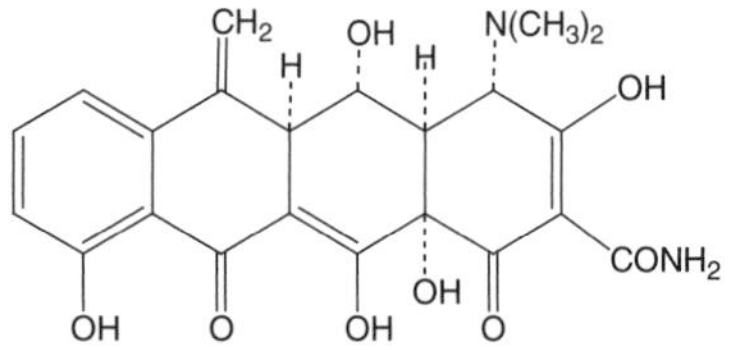	HA 3.5,[56] HA 7.6,[56] HB^+ 9.5[47]
Minocycline	2*E*,4*S*,4a*R*,5aS,12a*R*)-2-(Amino-hydroxy-methylidene)- 4,7-bis(dimethylamino)-10,11,12a-trihydroxy-4a,5,5a,6-tetrahydro-4H-tetracene- 1,3,12-trione $C_{23}H_{27}N_3O_7$ 10118-90-8	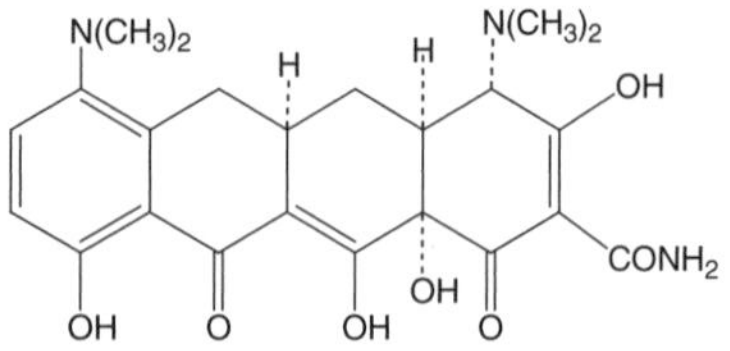	HA 2.8,[56] HA 5.0,[56] HA 7.8,[56] HB^+ 9.5[56]

TABLE 1.15 (*Continued*)

INN	IUPAC Name, Molecular Formula, and CAS Registry No.	Chemical Structure	pK_a
Oxytetracycline	[4S-(4α,4aα,5α,5aα,-6β, 12aα)]-4-(Dimethylamino)-1,4,4a,5,5a, 6,11,12a-octa-hydro-3,5,6,10,12,12a-hexahydroxy-6-methyl-1,11-dioxo-2-naphthacenecarbox-amide $C_{22}H_{24}N_2O_9$ 79-57-2	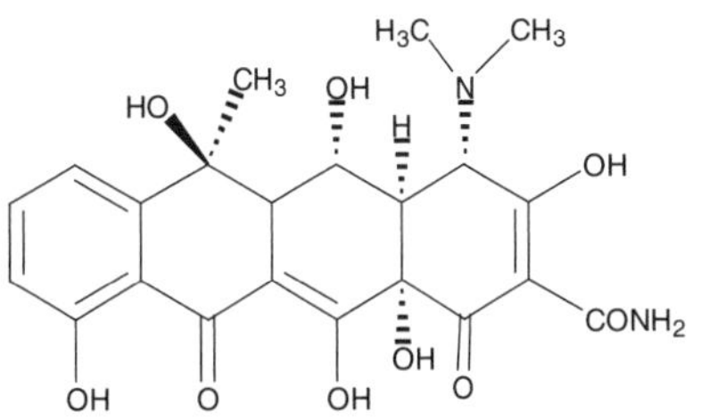	HA 3.3,[56] HA 7.3,[56] HB^+ 9.1[56]
4-epi-Oxytetracycline	[4*S*-(4α,4aα,5α,5aα,-6β,12aα)]-4-(Dimethylamino)-1,4,4a,5,5a, 6,11,12a-octahydro-3,5,6,10,12,12a-hexahydroxy-6-methyl-1,11-dioxo-2-naphthacenecarboxamide $C_{22}H_{24}N_2O_9$ 14206-58-7	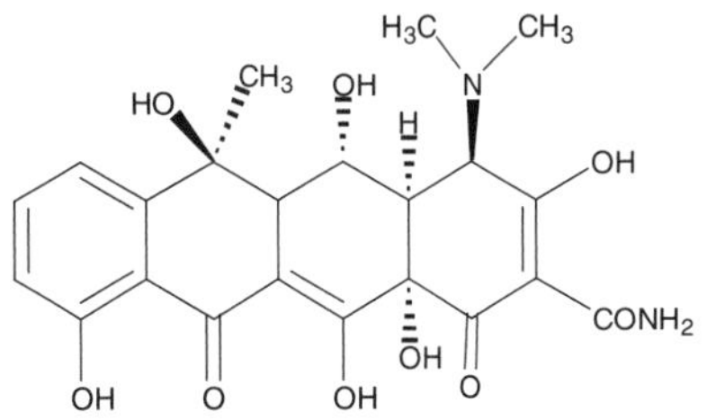	N/A[a]
Tetracycline	[4*S*-(4α,4aα,5aα,6β,-12aα)]-4-(Dimethylamino)-1,4,4a,5,5a,6,-11,12a-octahydro-3,6,10,12,12a-pentahydroxy-6-methyl-1,11-dioxo-2-naphthacenecarboxamide $C_{22}H_{24}N_2O_8$ 60-54-8	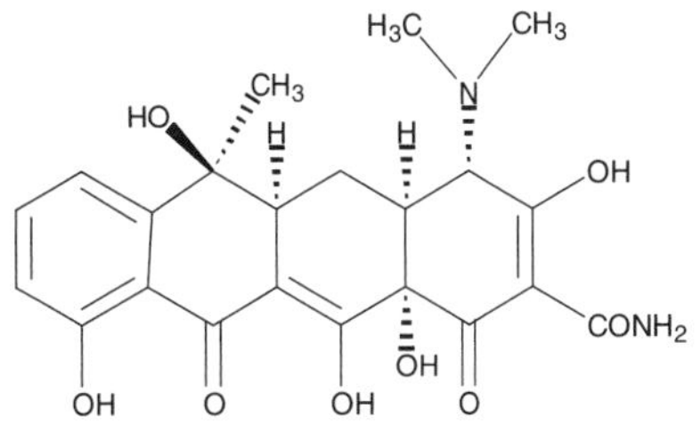	HA 3.3,[56] HA 7.7,[56] HB^+ 9.7[56]
4-epi-Tetracycline	[4*S*-(4α,4aα,5aα,6β,-12aα)]-4-(Dimethylamino)-1,4,4a,5,5a,6,-11,12a-octahydro-3,6,10,12,12a-pentahydroxy-6-methyl-1,11-dioxo-2-naphthacenecarboxamide $C_{22}H_{24}N_2O_8$ 79-85-6	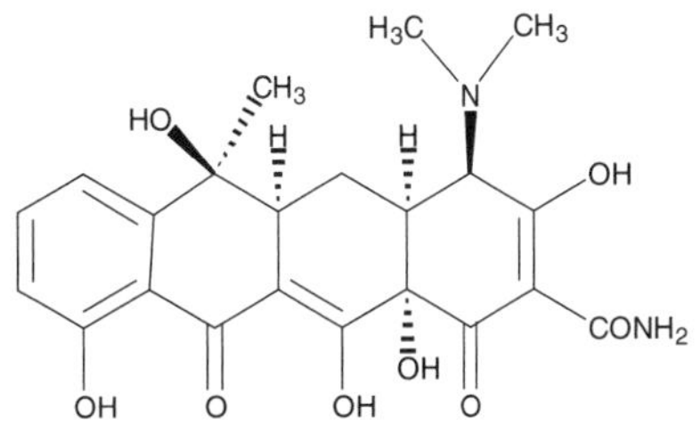	HA 4.8,[56] HA 8.0,[56] HB^+ 9.3[56]

[a]The author was not able to find a pK_a value for the substance in the public literature.

result of a resistance gene encoding for a membrane protein that actively pumps the drug out of bacterial cells. Another mechanism involves the overexpression of a gene encoding for a protein that prevents tetracyclines from binding to bacterial ribosomes. The rarest form of resistance involves the acetylation of tetracycline, which inactivates the drug.

The tetracycline derivatives are amphoteric substances that can form salts with both acids and bases. Hydrochloride is the most common salt form and is used in a variety of dosage forms, including medicated feeds, soluble powders, tablets and boluses, intrauterine infusions, intramammary infusions, and injections. Because the tetracyclines are relatively inexpensive, they tend to be used as first-line antimicrobials, especially in ruminants and pigs. Uses include those for acute uterine infections; actinobacillosis; anaplasmosis; bacterial enteritis; *Clostridium* diseases; diphtheria; infectious keratoconjunctivits; pneumonia; pododermatitis; skin and soft tissue infections in cattle; bacterial arthritis, bacterial enteritis, and vibrionic abortion in sheep; and atrophic rhinitis, bacterial enteritis, erysipelas, leptospirosis, mastitis, and pneumonia in pigs. Tetracyclines are administered to chickens for bacterial enteritis, fowl cholera, chronic respiratory disease, and infectious sinusitis; to salmon for furunculosis, bacterial haemorrhagic septicaemia, and pseudomonas disease; and to honeybees for American and European foulbrood disease, although strains of *Paenibacillus larvae* spp., the causative organism of American foulbrood disease in honeybees, are becoming increasingly resistant to oxytetracycline. Tetracyclines are also used for improved feed efficiency in cattle, chickens, pigs, sheep, and turkeys.

The rapid intravenous administration of tetracyclines causes acute toxicity in most animal species. In horses, intravenous doxycycline has caused cardiovascular dysfunction, collapse, and death. Long-acting formulations of oxytetracycline administered intramuscularly may cause local irritation at the site of injection in food-producing species. Tetracycline administration causes overgrowth of nonsusceptible organisms in several species of animals and, in horses, may result in colitis and severe diarrhea. Tooth discoloration in young animals may result when tetracyclines are administered during late pregnancy or during the period of tooth development.

A relatively small number of environmental studies in the literature relate to tetracyclines. The sorption behavior of these drugs is characterized by persistence and low mobility[101] and accounts for their superficial location in soil profiles[184] and their paucity in runoff water.[180] The enviromental fate of oxytetracycline used in aquaculture was extensively researched (cited by Boxall[26]). In these studies, oxytetracycline has been detected in wild fauna and in the sediment around fish farms. Soil samples collected from regions with intensive livestock production in Germany[184] and the UK[88] have been shown to contain tetracyclines, which possibly originate from manure.

A group ADI for tetracycline, oxytetracycline, and chlortetracycline has been allocated by JECFA.[185] The CAC has also established MRLs for tetracycline, oxytetracycline and chlortetracycline applicable to cattle, sheep, pigs, and poultry; and to fish and giant prawn for oxytetracycline only.[43] JECFA has prepared a number of reviews detailing residue studies on tetracyclines that support the development of the MRLs adopted by the CAC.[186–191]

From an analytical perspective, tetracyclines are relatively stable in acids but not in bases, and they can decompose rapidly under the influence of light and atmospheric oxygen.[152] Their decomposition is minimized by maintaining standard solutions in amber bottles and by drying samples under nitrogen in a dark room. Tetracyclines are susceptible to conformational degradation to their 4-epimers in aqueous solutions and during sample preparation. For example, Lindsey and coworkers[192] reported that the use of phosphate buffer solutions cause their degradation during the evaporation step.

1.4 RESTRICTED AND PROHIBITED USES OF ANTIMICROBIAL AGENTS IN FOOD ANIMALS

The therapeutic use of a very small number of antimicrobials in food-producing animals is prohibited because of public health concerns. The drugs involved are found on the websites of the regulatory authorities. In the United States, the provisions of the Animal Medicinal Drug Use Clarification Act (AMDUCA) of 1994 allow for the extralabel (off-label) use of drugs by veterinarians under certain conditions. However, the extralabel use of chloramphenicol, furazolidone, sulfonamide drugs in lactating dairy cattle (except for approved use of sulfadimethoxine, sulfabromomethazine, and sulfaethoxypyridazine), fluoroquinolones, and glycopeptides in food-producing animals is prohibited. The use of chloramphenicol, dapsone, furazolidone, and nitrofurans (except furazolidone) in food-producing species is banned in the EU, as is the use of antimicrobial drugs for growth promotion. In Canada, the use in food-producing animals of chloramphenicol and its salts and derivatives, and the 5-nitrofuran and nitroimidazole compounds is banned, while the sale of carbadox has been prohibited since 2006. In Australia, the use of carbadox, chloramphenicol, nitrofurans (including furazolidone and nitrofurazone), fluoroquinolones, gentamicin, [dihydro]streptomycin, and several sulfonamides is not permitted in food-producing animals. Currently, the use of carbadox is approved in pigs in the United States, while olaquindox is approved for use in pigs in Australia.

Several countries have acted to reduce the agricultural use of antimicrobial agents, and this has resulted in the discontinuation of some antimicrobial growth promotion uses. The UK banned the use of penicillin and the tetracyclines for growth promotion in the early 1970s; other European countries followed suit shortly thereafter. Sweden banned the use of all antibiotics for growth promotion in 1986. In the EU, the approval of avoparcin was withdrawn in 1998; the productivity claims for bacitracin, spiramycin, tylosin, and virginiamycin were discontinued in 1999; and the productivity claims for avilamycin, flavomycin, monensin, and salinomycin were discontinued in 2006. In the United States, the extralabel (off-label) use of drugs in treating food-producing animals for improving weight gain, feed efficiency, or other production purposes is prohibited under AMDUCA. In Australia, the registration of avoparcin was withdrawn in 2000, and the use of virginiamycin as a growth promoter was discontinued in 2005.

1.5 CONCLUSIONS

Antimicrobial drug use in food animal production is fundamental to animal health and well-being and to the economics of the livestock industry. Therefore the prudent use of antimicrobials is critically important because few new drugs are entering the market, and existing uses need to be preserved for as long as is practicable. Prudent use will minimize the development of antimicrobial resistance and maximize therapeutic effect. When introducing new products onto the market, pharmaceutical companies need to rule out the presence of cross-resistance to old products in the same class, some of which may no longer be used in animals. From a food safety perspective, responsible use of antimicrobials in food-producing species as reflected by the results of residue-monitoring programs is of paramount importance to reassure the community that the food supply is safe.

In conclusion, this chapter has discussed the major antibiotic classes used in food-producing species and in particular, the PD component depicted in the chemotherapeutic triangle (Fig. 1.1). The antimicrobial potency of antimicrobial drugs to bacterial isolates is characterized using *in vitro* MIC and/or MBC values. The killing kinetics of an antibiotic, which are the basis for determining whether the antibacterial effect of a drug is concentration-dependent, time-dependent, or co-dependent, are also established *in vitro*. While this information is fundamental to antimicrobial therapy, when considered in isolation, it is insufficient to predict effectiveness *in vivo*. Both the dosage regimen and the PK of the drug are important determinants of drug concentration at the infection site (the biophase). These important topics are discussed in Chapter 2.

ACKNOWLEDGMENTS

The author wishes to thank APVMA Science Fellow Professor Mary Barton and Professor Peter Lees for their comments on this manuscript. The assistance of Dr. Cheryl Javro with preparation of the figure and tables is gratefully acknowledged.

REFERENCES

1. Barton MD, Peng H, *Epidemiology of Antibiotic Resistant Bacteria and Genes in Piggeries*, report to Australian Pork Ltd., 2005.
2. Giguère S, Antimicrobial drug action and interaction: An introduction, in Giguère S, Prescott JF, Baggot JD, Walker RD, Dowling PM, eds., *Antimicrobial Therapy in Veterinary Medicine*, 4th ed., Blackwell, Ames, IA, 2006, pp. 3–9.
3. Joint FAO/OIE/WHO Expert Workshop, Joint FAO/OIE/WHO Expert Workshop on Non-Human Antimicrobial Usage and Antimicrobial Resistance: Scientific Assessment, Geneva, Dec. 1–5, 2003 (available at `http://www.who.int/foodsafety/publications/micro/en/report.pdf`; accessed 11/20/10).
4. Martinez M, Silley P, Antimicrobial drug resistance, in Cunningham F, Elliott J, Lees P, eds., *Handbook of Experimental Pharmacology. Comparative and Veterinary Pharmacology*, Vol. 199, Heidelberg, Springer-Verlag; 2010, pp. 227–264.
5. World Health Organisation, *The Use of Stems in the Solution of International Nonproprietary Names (INN) for Pharmaceutical Substances*, Document WHO/EMP/QSM/2009.3; 2009 (available at `http://www.who.int/medicines/services/inn/StemBook2009.pdf`; accessed 11/20/10).
6. Lees P, Cunningham FM, Elliott J, Principles of pharmacodynamics and their applications in veterinary pharmacology, *J. Vet. Pharmacol. Ther*. 2004;27:397–414.
7. Pugh J, Kinetics and product stability, in Aulton ME, ed., *Pharmaceutics The Science of Dosage Form Design*, 2nd ed., Churchill Livingstone, London, 2002, pp. 101–112.
8. Hanlon G, Fundamentals of microbiology, in Aulton ME, ed., *Pharmaceutics. The Science of Dosage Form Design*, 2nd ed., Churchill Livingstone, London, 2002, pp. 599–622.
9. Birkett DJ, *Pharmacokinetics Made Easy*, 2nd ed., McGraw-Hill Australia, Sydney, 2002.
10. Baggot JD, Principles of antimicrobial drug bioavailability and disposition, in Giguère S, Prescott JF, Baggot JD, Walker RD, Dowling PM, eds., *Antimicrobial Therapy in Veterinary Medicine*, 4th ed., Ames, IA,Blackwell, 2006, pp. 45–79.
11. Baggot JD, Brown SA, Basis for selection of the dosage form, in Hardee GE, Baggot JD, eds., *Development and Formulation of Veterinary Dosage Forms*, 2nd ed., Marcel Dekker, New York, 1998, pp. 7–143.
12. Schentag J, Swanson DF, Smith IL, Dual individualization: Antibiotic dosage calculation from the integration of in-vitro pharmacodynamics and in-vivo pharmacokinetics, *J. Antimicrob. Chemother*. 1985;15(Suppl. A):47–57.
13. Cars O, Efficacy of beta-lactam antibiotics: Integration of pharmacokinetics and pharmacodynamics, *Diagn. Microbiol. Infect. Dis*. 1997;27:29–33.
14. Toutain P-L, del Castillo JRE, Bousquet-Mélou A, The pharmacokinetic-pharmacodynamic approach to a rational dosage regimen for antibiotics, *Res. Vet. Sci*. 2002; 73(2):105–114.
15. Giguère S, Macrolides, azalides and ketolides, in Giguère S, Prescott JF, Baggot JD, Walker RD, Dowling PM, eds., *Antimicrobial Therapy in Veterinary Medicine*, 4th ed., Blackwell, Ames, IA, 2006, pp. 191–205.
16. Yang Q, Nakkula RJ, Walters JD, Accumulation of ciprofloxacin and minocycline by cultured human gingival fibroblasts, *J. Dent. Res*. 2002;81:836–840.
17. Hardman JG, Limbird LE, Molinoff PB, Ruddon RW, Gilman AG, Design and optimization of dosage regimens: Pharmacokinetic data, in Hardman JGG, Gilman A, Limbird LL, eds., *Goodman and Gilman's Pharmacological Basis of Therapeutics*, 9th ed., McGraw-Hill, New York, 1996, pp. 1712–1792.
18. McKellar QA, Sanchez Bruni SF, Jones DG, Pharmacokinetic/pharmacodynamic relationships of antimicrobial drugs used in veterinary medicine, *J. Vet. Pharmacol. Ther*. 2004;27:503–514.
19. CLSI—Clinical Laboratory Standards Institute (previously NCCLS), *Performance Standards for Antimicrobial Disk Susceptibiliy Tests: Approved Standard*, 10th ed., 2009.

20. EUCAST—European Committee on Antimicrobial Susceptibility Testing (2010) (available at http://www.eucast.org/; accessed 10/11/10).
21. Prescott JF, Walker RD, Principles of antimicrobial drug selection and use, in Prescott JF, Baggot JD, Walker RD, eds., Antimicrobial Therapy in Veterinary Medicine, 3rd ed., Iowa State Univ. Press, Ames, IA, 2000, pp. 88–104.
22. Chambers HF, General principles of antimicrobial therapy, in Brunton LL, Lazo J, Parker K, eds., Goodman and Gilman's Pharmacological Basis of Therapeutics, 11th ed., McGraw-Hill, New York, 2006, pp. 1095–1111.
23. Walker RD, Giguère S, Principles of antimicrobial drug selection and use, in Giguère S, Prescott JF, Baggot JD, Walker RD, Dowling PM, eds., Antimicrobial Therapy in Veterinary Medicine, 4th ed., Blackwell, Ames, IA, 2006, pp. 107–117.
24. Woodward KN, Veterinary pharmacovigilance. Part 4. Adverse reactions in humans to veterinary medicinal products, *J. Vet. Pharmacol. Ther*. 2005;28:185–201.
25. Woodward KN, Hypersensitivity in humans and exposure to veterinary drugs, *Vet. Hum. Toxicol*. 1991;33:168–172.
26. Boxall ABA, Veterinary medicines and the environment, in Cunningham F, Elliott J, Lees P, eds., *Handbook of Experimental Pharmacology. Comparative and Veterinary Pharmacology*, Vol. 199, Springer, Heidelberg, 2010, pp. 291–314.
27. Samuelsen OB, Lunestad BT, Husevåg B, Hølleland T, Ervik A, Residues of oxolinic acid in wild fauna following medication in fish farms, *Dis. Aquat. Organ*. 1992;12: 111–119.
28. Samuelsen OB, Torsvik V, Ervik A, Long-range changes in oxytetracycline concentration and bacterial resistance towards oxytetracycline in a fish farm sediment after medication, *Sci. Total Environ*. 1992;114:25–36.
29. Hektoen H, Berge JA, Hormazabal V, Yndestad M, Persistence of antibacterial agents in marine sediments, *Aquaculture* 1995;133:175–184.
30. Jacobsen P, Berglind L, Persistence of oxytetracycline in sediments from fish farms, *Aquaculture* 1988;70:365–370.
31. Björklund HV, Bondestam, Bylund G, Residues of oxytetracycline in wild fish and sediments from fish farms, *Aquaculture* 1990;86:359–367.
32. Björklund HV, Råbergh CMI, Bylund G, Residues of oxolinic acid and oxytetracycline in fish and sediments from fish farms, *Aquaculture* 1991;97:85–96.
33. Boxall ABA, Johnson P, Smith EJ, Sinclair CJ, Stutt E, Levy LS, Uptake of veterinary medicines from soils into plants, *J. Agric. Food Chem*. 2006;54:2288–2297.
34. Berendsen B, Stolker L, de Jong J, Nielen M, Tserendorj E, Sodnomdarjaa R, Cannavan A, Elliott C, Evidence of natural occurrence of the banned antibiotic chloramphenicol in herbs and grass, *Anal. Bioanal. Chem*. 2010;397(5):1955–1963.
35. Swann MM, Joint Committee on the Use of Antibiotics in Animal Husbandry and Veterinary Medicine, Her Majesty's Stationery Office, London, 1969.
36. World Health Organisation, *WHO Global Principles for the Containment of Antimicrobial Resistance in Animals Intended for Food*, report of a WHO consultation, Geneva, June 5–9, 2000 (available at http://whqlibdoc.who.int/hq/2000/WHO_CDS_CSR_APH_2000.4.pdf; accessed 11/20/10).
37. Begg EJ, Barclay ML, Aminoglycosides—50 years on, *Br. J. Clin. Pharmacol*. 1995;39:597–603.
38. Leibovici L, Vidal L, Paul M, Aminoglycoside drugs in clinical practice: An evidence-based approach, *J. Antimicrob. Chemother*. 2009;63(5):1081–1082.
39. Dowling PM, Aminoglycosides, in Giguère S, Prescott JF, Baggot JD, Walker RD, Dowling PM, eds., *Antimicrobial Therapy in Veterinary Medicine*, 4th ed., Blackwell, Ames, IA, 2006, pp. 207–229.
40. Marra F, Partoni N, Jewesson P, Aminoglycoside administrations as a single daily dose: An improvement to current practice or a repeat of previous errors? *Drugs* 1996;52:344–376.
41. Riviere JE, Renal impairment, in Prescott JF, Baggot JD, Walker RD, eds., Antimicrobial Therapy in Veterinary Medicine, 3rd ed., Iowa State Univ. Press, Ames, 2000, pp. 453–458.
42. World Health Organisation, *Specifications for the Identity and Purity of Food Additives and Their Toxicological Evaluation: Some Antibiotics*, 12th Report Joint FAO/WHO Expert Committee on Food Additives, WHO Technical Report Series 430, 1969 (available at http://whqlibdoc.who.int/trs/WHO_TRS_430.pdf; accessed 11/9/10).
43. Codex Veterinary Drug Residues in Food Online Database (available at http://www.codexalimentarius.net/vetdrugs/data/index.html; accessed 11/11/10).
44. Heitzman RJ, Dihydrostreptomycin and streptomycin, in *Residues of Some Veterinary Drugs in Animals and Foods*, FAO Food and Nutrition Paper 41/7, 1995, pp. 17–29 (available at ftp://ftp.fao.org/ag/agn/jecfa/vetdrug/41-7-dihydrostreptomycin_streptomycin.pdf; accessed 11/20/10).
45. Heitzman RJ, Dihydrostreptomycin and streptomycin, in *Residues of Some Veterinary Drugs in Animals and Foods*, FAO Food and Nutrition Paper 41/10, 1998, pp. 39–44 (available at ftp://ftp.fao.org/ag/agn/jecfa/vetdrug/41-10-dihydrostreptomycin_streptomycin.pdf; accessed 11/20/10).
46. Heitzman RJ, Dihydrostreptomycin and streptomycin, in *Residues of Some Veterinary Drugs in Animals and Foods*, FAO Food and Nutrition Paper 41/12, 2000, pp. 21–25 (available at ftp://ftp.fao.org/ag/agn/jecfa/vetdrug/41-12-dihydrostreptomycin_streptomycin.pdf; accessed 11/20/10).
47. Heitzman RJ, Dihydrostreptomycin and streptomycin, in *Residues of Some Veterinary Drugs in Animals and Foods*, FAO Food and Nutrition Paper 41/14, 2002, pp. 37–41 (available at ftp://ftp.fao.org/ag/agn/jecfa/vetdrug/41-14-streptomycins.pdf; accessed 11/20/10).
48. MacNeil JD, Cuerpo L, Gentamicin, in *Residues of Some Veterinary Drugs in Animals and Foods*, FAO Food

and Nutrition Paper 41/7, 1995, pp. 45–55 (available at ftp://ftp.fao.org/ag/agn/jecfa/vetdrug/41-7-gentamicin.pdf; accessed 11/20/10.

49. MacNeil JD, Gentamicin, in *Residues of Some Veterinary Drugs in Animals and Foods*, FAO Food and Nutrition Paper 41/11, 1998, pp. 61–63 (available at ftp://ftp.fao.org/ag/agn/jecfa/vetdrug/41-11-gentamicin.pdf; accessed 11/20/10).
50. Livingston RC, Neomycin, in *Residues of Some Veterinary Drugs in Animals and Foods*, FAO Food and Nutrition Paper 41/7, 1995, pp. 57–67 (available at ftp://ftp.fao.org/ag/agn/jecfa/vetdrug/41-7-nomycin.pdf; accessed 11/20/10).
51. Arnold D, Neomycin, in *Residues of Some Veterinary Drugs in Animals and Foods*, FAO Food and Nutrition Paper 41/9, 1997, pp. 73–74 (available at ftp://ftp.fao.org/ag/agn/jecfa/vetdrug/41-9-neomycin.pdf; accessed 11/20/10).
52. Livingston RC, Neomycin, in *Residues of Some Veterinary Drugs in Animals and Foods*, FAO Food and Nutrition Paper 41/12, 2000, pp. 91–95 (available at ftp://ftp.fao.org/ag/agn/jecfa/vetdrug/41-12-neomycin.pdf; accessed 11/20/10).
53. Reeves PT, Swan GE, Neomycin, in *Residues of Some Veterinary Drugs in Animals and Foods*, FAO Food and Nutrition Paper 41/15, 2003, pp. 53–63 (available at ftp://ftp.fao.org/ag/agn/jecfa/vetdrug/41-15-neomycin.pdf; accessed 11/20/10).
54. Cuerpo L, Livingston RC, Spectinomycin, in *Residues of Some Veterinary Drugs in Animals and Foods*, FAO Food and Nutrition Paper 41/7, 1995, pp. 63–77 (available at ftp://ftp.fao.org/ag/agn/jecfa/vetdrug/41-6-spectinomycin.pdf; accessed 11/20/10).
55. Ellis RL, Livingston RC, Spectinomycin, in *Residues of Some Veterinary Drugs in Animals and Foods*, FAO Food and Nutrition Paper 41/11, 1998, pp. 119–132 (available at ftp://ftp.fao.org/ag/agn/jecfa/vetdrug/41-11-spectinomycin.pdf; accessed 11/20/10).
56. Prankerd RJ, Critical compilation of pK_A values for pharmaceutical substances, in Brittain HG, ed., *Profiles of Drug Substances, Excipients, and Related Methodology*, Vol. 33, Elsevier, Amsterdam; 2007.
57. DrugBank:Apramycin (DB04626) (available at http://www.drugbank.ca/drugs/DB04626; accessed 11/23/10).
58. Riviere JE, Craigmill AL, Sundlof SF, *Handbook of Comparative Pharmacokinetics and Residues of Veterinary Antimicrobials*, CRC Press, Boca Raton, FL, 1991.
59. Kong KF, Schneper L, Mathee K, β-lactam antibiotics: From antibiosis to resistance and bacteriology, *Acta Pathol. Microbiol. Immunol. J*. 2009;118:1–36.
60. Sheehan JC, Henery-Logan KR, A general synthesis of the penicillins, *J. Biol. Chem*. 1959;81:5838–5839.
61. Fairbrother RW, Taylor G, Sodium methicillin in routine therapy, *Lancet* 1961;1:473–476.
62. Hornish RE, Kotarski SF, Cephalosporins in veterinary medicine—ceftiofur use in food animals, *Curr. Top. Med. Chem*. 2002;2:717–731.
63. Robbins RL, Wallace SS, Brunner CJ, Gardner TR, DiFranco BJ, Speirs VC, Immune-mediated haemolytic disease after penicillin therapy in a horse, *Equine Vet. J*. 1993;25:462–465.
64. Embrechts E, Procaine penicillin toxicity in pigs, *Vet. Rec*. 1982;111:314–315.
65. Nielsen IL, Jacobs KA, Huntington PJ, Chapman CB, Lloyd KC, Adverse reactions to procaine penicillin G in horses, *Austral. Vet. J*. 1988;65:181–185.
66. Chapman CB, Courage P, Nielsen IL, Sitaram BR, Huntington PJ, The role of procaine in adverse reactions to procaine penicillin in horses, *Austral. Vet. J*. 1992;69:129–133.
67. Göbel A, McArdell CS, Suter MJ-C, Giger W, Trace determinations of macrolide and sulphonamide antimicrobials, a human sulphonamide metabolite, and trimethoprim in wastewater using liquid chromatography coupled to electrospray tandem mass spectrometry, *Anal. Chem*. 2004;76:4756–4764.
68. World Health Organization, *Evaluation of Certain Veterinary Drug Residues in Food*, 36th Report Joint FAO/WHO Expert Committee on Food Additives, WHO Technical Report Series 799, 1990, pp. 37–41 (available at http://whqlibdoc.who.int/trs/WHO_TRS_799.pdf; accessed 11/20/10).
69. World Health Organization, *Evaluation of Certain Veterinary Drug Residues in Food*, 50th Report Joint FAO/WHO Expert Committee on Food Additives, WHO Technical Report Series 888, 1999;250–33 (available at http://whqlibdoc.who.int/trs/WHO_TRS_888.pdf; accessed 11/20/10).
70. World Health Organization, *Evaluation of Certain Veterinary Drug Residues in Food*, 45th Report Joint FAO/WHO Expert Committee on Food Additives., WHO Technical Report Series 864, 1996, pp. 26–32 (available at http://whqlibdoc.who.int/trs/WHO_TRS_864.pdf; accessed 11/22/10).
71. Anonymous, Benzylpenicillin, in *Residues of Some Veterinary Drugs in Animals and Foods*, FAO Food and Nutrition Paper 41/3, 1990 pp. 1–18 (available at ftp://ftp.fao.org/ag/agn/jecfa/vetdrug/41-3-benzylpenicillin.pdf; accessed 11/20/10).
72. MacNeil JD, Procaine penicillin, in *Residues of Some Veterinary Drugs in Animals and Foods*, FAO Food and Nutrition Paper 41/11, 1998, pp. 95–106 (available at ftp://ftp.fao.org/ag/agn/jecfa/vetdrug/41-11-procaine_benzylpenicillin.pdf; accessed 11/20/10).
73. MacNeil JD, Ceftiofur, in *Residues of Some Veterinary Drugs in Animals and Foods*, FAO Food and Nutrition Paper 41/10, 1997, pp. 1–8 (available at ftp://ftp.fao.org/ag/agn/jecfa/vetdrug/41-10-ceftiofur.pdf; accessed 11/20/10).
74. USP Veterinary Pharmaceutical Information Monographs—Antibiotics, *J. Vet. Pharmacol. Ther*. 2003;26 (Suppl. 2),

p. 46 (clavulanic acid), p. 89 (enrofloxacin), p. 161 (pirlimycin).

75. Suter W, Rosselet A, Knusel F, Mode of action of quindoxin and substituted quinoxaline-di-N-oxides on Escherichia coli, *Antimicrob. Agents Chemother*. 1978;13:770–783.

76. de Graaf GJ, Jager LP, Baars AJ Spierenburg TJ, Some pharmacokinetic observations of carbadox medication in pigs, *Vet. Q*. 1988;10:34–41.

77. Van der Molen EJ, Baars AJ, de Graff GJ, Jager LP, Comparative study of the effect of carbadox, olaquindox and cyadox on aldosterone, sodium and potassium plasma levels in weaned pigs, *Res. Vet. Sci*. 1989;47:11–16.

78. Nabuurs MJA, van der Molen EJ, de Graaf GJ, Jager LP, Clinical signs and performance of pigs treated with different doses of carbadox, cyadox and olaquindox, *J. Vet. Med. Ser. A* 1990;37:68–76.

79. Power SB, Donnelly WJ, McLaughlin JG, Walsh MC, Dromey MF, Accidental carbadox overdosage in pigs in an Irish weaner-producing herd, *Vet. Rec*. 1989;124:367–370.

80. Commission Regulation No. 2788/98 of December 1998 amending Council directive 70/524/EEC concerning additives in feedingstuffs as regards the withdrawal of authorization for certain growth promoters, *Off. J. Eur. Commun*. 1998;L347:32–32 (available at http://eur-lex.europa.eu/LexUriServ/LexUriServ.do?uri=OJ:L:1998:347:0031:0032:EN:PDF; accessed 11/23/10).

81. ALINORM 09/32/31, *Report 18th Session of the Codex Committee on Residues of Veterinary Drugs in Foods*, Codex Alimentarius Commission, Joint FAO/WHO Food Standards Programme, Rome, 2009 (available at http://www.codexalimentarius.net/web/archives.jsp?year=09; accessed 11/07/10).

82. World Health Organization, *Evaluation of Certain Veterinary Drug Residues in Food*, 60th Report Joint FAO/WHO Expert Committee on Food Additives, WHO Technical Report Series 918, 2003, pp. 33–41 (available at http://whqlibdoc.who.int/trs/WHO_TRS_918.pdf; accessed 11/20/10).

83. Fernández Suárez A, Arnold D, Carbadox, in *Residues of Some Veterinary Drugs in Animals and Foods*, FAO Food and Nutrition Paper 41/15, 2003, pp. 1–9 (available at ftp://ftp.fao.org/ag/agn/jecfa/vetdrug/41-15-carbadox.pdf; accessed 11/20/10).

84. World Health Organization, *Evaluation of Certain Veterinary Drug Residues in Food*, 42nd Report Joint FAO/WHO Expert Committee on Food Additives, WHO Technical Report Series 851, 1995, pp. 19–21 (available at ftp://ftp.fao.org/ag/agn/jecfa/vetdrug/41-15-carbadox.pdf; accessed 11/20/10).

85. Anonymous, Olaquindox, in *Residues of Some Veterinary Drugs in Animals and Foods*, FAO Food and Nutrition Paper 41/3, 1991, pp. 85–96 (available at ftp://ftp.fao.org/ag/agn/jecfa/vetdrug/41-3-olaquindox.pdf; accessed 11/20/10).

86. Ellis RL, Olaquindox, in *Residues of Some Veterinary Drugs in Animals and Foods*, FAO Food and Nutrition Paper 41/6, 1994, pp. 53–62 (available at ftp://ftp.fao.org/ag/agn/jecfa/vetdrug/41-6-olaquindox.pdf; accessed 11/20/10).

87. Liguère S, Lincosamides, pleuromutilins, and streptogramins, in Giguère S, Prescott JF, Baggot JD, Walker RD, Dowling PM, eds., *Antimicrobial Therapy in Veterinary Medicine*, 4th ed., Blackwell, Ames, IA, 2006, pp. 179–190.

88. Boxall ABA, Fogg L, Baird D, Telfer T, Lewis C, Gravell A, Boucard T, *Targeted Monitoring Study for Veterinary Medicines*, Environment Agency R&D Technical Report, Environment Agency, Bristol, UK, 2006.

89. World Health Organization, *Evaluation of Certain Veterinary Drug Residues in Food*, 54th Report Joint FAO/WHO Expert Committee on Food Additives, WHO Technical Report Series 900, 2001, pp. 13–29 (available at http://whqlibdoc.who.int/trs/WHO_TRS_900.pdf; accessed 11/20/10).

90. World Health Organization, *Evaluation of Certain Veterinary Drug Residues in Food*, 62nd Report Joint FAO/WHO Expert Committee on Food Additives, WHO Technical Report Series 925, 2004, pp. 26–37 (available at http://whqlibdoc.who.int/trs/WHO_TRS_925.pdf; accessed 11/20/10).

91. Röstel B, Zmudski J, MacNeil J, Lincomycin, in *Residues of Some Veterinary Drugs in Animals and Foods*, FAO Food and Nutrition Paper 41/13, 2000, pp. 59–74 (available at ftp://ftp.fao.org/ag/agn/jecfa/vetdrug/41-13-lincomycin.pdf; accessed 11/20/10).

92. Arnold D, Ellis R, Lincomycin, in *Residues of Some Veterinary Drugs in Animals and Foods*, FAO Food and Nutrition Paper 41/14, 2002, pp. 45–53 (available at ftp://ftp.fao.org/ag/agn/jecfa/vetdrug/41-14-lincomycin.pdf; accessed 11/20/10).

93. Kinabo LDB, Moulin G, Lincomycin, in *Residues of Some Veterinary Drugs in Animals and Foods*, FAO Food and Nutrition Paper 41/16, 2004, pp. 41–43 (available at ftp://ftp.fao.org/ag/agn/jecfa/vetdrug/41-16-lincomycin.pdf; accessed 11/20/10).

94. Friedlander L, Moulin G, Pirlimycin, in *Residues of Some Veterinary Drugs in Animals and Foods*, FAO Food and Nutrition Paper 41/16, 2004, pp. 55–73 (available at ftp://ftp.fao.org/ag/agn/jecfa/vetdrug/41-16-pirlimycin.pdf; accessed 11/20/10).

95. Veien NK, Hattel O, Justesen O, Nørholm A, Occupational contact dermatitis due to spiramycin and/or tylosin among farmers, *Contact Dermatitis* 1980;6:410–413.

96. McGuigan MA, Human exposures to tilmicosin (MICOTIL), *Vet. Hum. Toxicol*. 1994;36:306–308.

97. Crown LA, Smith RB, Accidental veterinary antibiotic injection into a farm worker, *Tenn. Med*. 1999;92:339–340.

98. Von Essen S, Spencer J, Hass B, List P, Seifert SA, Unintentional human exposure to tilmicosin (Micotil® 300), *J. Toxicol. Clin. Toxicol*. 2003;41:229–233.

99. Kuffner EK, Dart RC, Death following intravenous injection of Micotil® 300, *J. Toxicol. Clin. Toxicol*. 1996;34:574.

100. Kolpin DW, Furlong ET, Meyer MT, Thurman EM, Zaugg SD, Barber LB, Buxton HT, Pharmaceuticals, hormones,

and other organic wastewater contaminants in US streams 1999–2000: A national reconnaissance, *Environ Sci Technol*. 2002;36:1202–1211.

101. Pope L, Boxall ABA, Corsing C, Halling-Sorensen B, Tait A, Topp E, Exposure assessment of veterinary medicines in terrestrial systems, in Crane M, Boxall ABA, Barrett K, eds., *Veterinary Medicines in the Environment*, CRC Press, Boca Raton, FL, 2009, pp. 129–153.

102. Loke ML, Ingerslev F, Halling-Sorensen B, Tjornelund J, Stability of tylosin A in manure containing test systems determined by high performance liquid chromatography, *Chemosphere* 2000;40:759–765.

103. Teeter JS, Meyerhoff RD, Aerobic degradation of tylosin in cattle, chicken and swine excreta, *Environ. Res*. 2003;93:45–51.

104. Kolz AC, Moorman TB, Ong SK, Scoggin KD, Douglass EA, Degradation and metabolite production of tylosin in anaerobic and aerobic swine-manure lagoons, *Water Environ Res*. 2005;77:49–56.

105. World Health Organization, *Evaluation of Certain Veterinary Drug Residues in Food*, 66th Report Joint FAO/WHO Expert Committee on Food Additives, WHO Technical Report Series 939, 2006, pp. 33–44 (available at http://whqlibdoc.who.int/publications/2006/9241209399_eng.pdf; accessed 11/21/10).

106. World Health Organization, *Evaluation of Certain Veterinary Drug Residues in Food*, 43rd Report Joint FAO/WHO Expert Committee on Food Additives, WHO Technical Report Series 855, 1995, pp. 38–43 (available at http://whqlibdoc.who.int/trs/WHO_TRS_855.pdf; accessed 11/21/10).

107. World Health Organization, *Evaluation of Certain Veterinary Drug Residues in Food*, 47th Report Joint FAO/WHO Expert Committee on Food Additives, WHO Technical Report Series 876, 1998, pp. 37–44 (available at http://whqlibdoc.who.int/trs/WHO_TRS_876.pdf; accessed 11/21/10).

108. World Health Organization, *Evaluation of Certain Veterinary Drug Residues in Food*, 70th Report Joint FAO/WHO Expert Committee on Food Additives, WHO Technical Report Series 954, 2009, pp. 94–107 (available at http://whqlibdoc.who.int/trs/WHO_TRS_954_eng.pdf; accessed 11/21/10).

109. Fernández Suárez A, Ellis R, Erythromycin, in *Residue Evaluation of Certain Veterinary Drugs*, FAO JECFA Monographs 2, 2006, pp. 29–51 (available at ftp://ftp.fao.org/ag/agn/jecfa/vetdrug/2-2006-erythromycin.pdf; accessed 11/22/10).

110. Anonymous, Spiramycin, in *Residues of Some Veterinary Drugs in Animals and Foods*, FAO Food and Nutrition Paper 41/4, 1991, pp. 97–107 (available at ftp://ftp.fao.org/ag/agn/jecfa/vetdrug/41-4-spiramycin.pdf; accessed 11/22/10).

111. Ellis RL, Spiramycin, in *Residues of Some Veterinary Drugs in Animals and Foods*, FAO Food and Nutrition Paper 41/7, 1995, pp. 89–103 (available at ftp://ftp.fao.org/ag/agn/jecfa/vetdrug/41-7-spiramycin.pdf; accessed 11/22/10).

112. Marshall BL, Spiramycin, in *Residues of Some Veterinary Drugs in Animals and Foods*, FAO Food and Nutrition Paper 41/9, 1997, pp. 77–87 (available at ftp://ftp.fao.org/ag/agn/jecfa/vetdrug/41-9-spiramycin.pdf; accessed 11/22/10).

113. Livingston RC, Spiramycin, in *Residues of Some Veterinary Drugs in Animals and Foods*, FAO Food and Nutrition Paper 41/10, 1997, pp. 77–78; available at ftp://ftp.fao.org/ag/agn/jecfa/vetdrug/41-10-spiramycin.pdf; accessed 11/22/10).

114. MacNeil JD, Tilmicosin, in *Residues of Some Veterinary Drugs in Animals and Foods*, FAO Food and Nutrition Paper 41/9, 1997, pp. 105–118 (available at ftp://ftp.fao.org/ag/agn/jecfa/vetdrug/41-9-tilmicosin.pdf; accessed 11/22/10).

115. Xu S, Arnold D, Tilmicosin, in *Residue Evaluation of Certain Veterinary Drugs*, FAO JECFA Monographs 6, 2009, pp. 159–195 (available at ftp://ftp.fao.org/ag/agn/jecfa/vetdrug/6-2009-tilmicosin.pdf; accessed 11/22/10).

116. Anonymous, Tylosin, in *Residues of Some Veterinary Drugs in Animals and Foods*, FAO Food and Nutrition Paper 41/4, 1991, pp. 109–127 (available at ftp://ftp.fao.org/ag/agn/jecfa/vetdrug/41-4-tylosin.pdf; accessed 11/22/10).

117. Lewicki J, Reeves PT, Swan GE, Tylosin, in *Residue Evaluation of Certain Veterinary Drugs*, FAO JECFA Monographs 6, 2009, pp. 243–279 (available at ftp://ftp.fao.org/ag/agn/jecfa/vetdrug/6-2009-tylosin.pdf; accessed 11/22/10).

118. Wang J, Analysis of macrolide antibiotics, using liquid chromatography-mass spectrometry, in food, biological and environmental matrices, *Mass Spectr. Rev*. 2009; 28(1):50–92.

119. Horie M, Chemical analysis of macrolides, in Oka H, Nakazawa H, Harada K, MacNeil JD, eds., *Chemical Analysis for Antibiotics Used in Agriculture*, AOAC International, Arlington, VA, 1995, pp. 165–205.

120. *Draxxin Injectable Solution*, APVMA Product no. 59304, Public Release Summary, Australian Pesticides and Veterinary Medicines Authority, June 2007, p. 29 (available at http://www.apvma.gov.au/registration/assessment/docs/prs_draxxin.pdf; accessed 11/23/10).

121. World Health Organisation, *Evaluation of Certain Veterinary Drug Residues in Food*, WHO Technical Report Series 832, 1993, pp. 32–40 (available at http://whqlibdoc.who.int/trs/WHO_TRS_832.pdf; accessed 11/21/10).

122. Samuelson OB, Solheim E, Lunestad BT, Fate and microbiological effects of furazolidone in marine aquaculture sediment, *Sci. Total Environ*. 1991;108:275–283.

123. World Health Organization, *Evaluation of Certain Veterinary Drug Residues in Food*, 34th Report Joint FAO/WHO Expert Committee on Food Additives, WHO Technical Report Series 788, 1989, pp. 27–32 (available at

http://whqlibdoc.who.int/trs/WHO_TRS_788.pdf; 11/08/10).

124. World Health Organization, *Evaluation of Certain Veterinary Drug Residues in Food*, 42nd Report Joint FAO/WHO Expert Committee on Food Additives, WHO Technical Report Series 851, 1995, p. 27 (available at http://whqlibdoc.who.int/trs/WHO_TRS_851.pdf; accessed 11/08/10).

125. Ehrlich J, Bartz QR, Smith RM, Joslyn DA, Burkholder PR, Chloromycetin, a new antibiotic from a soil actinomycete, *Science* 1947;106:417.

126. Schwarz S, Kehrenberg C, Doublet B, Cloeckaert A, Molecular basis of bacterial resistance to chloramphenicol and florfenicol, *FEMS Microbiol. Rev*. 2004;28:519–542.

127. Wongtavatchai J, McLean JG, Ramos F, Arnold D, *Chloramphenicol*, WHO Food Additives Series 53, JECFA (WHO: Joint FAO/WHO Expert Committee on Food Additives), IPCS (International Programme on Chemical Safety) INCHEM. 2004; pp. 7–85 (available at http://www.inchem.org/documents/jecfa/jecmono/v53je03.htm; accessed 11/21/10).

128. World Health Organization, *Evaluation of Certain Veterinary Drug Residues in Food*, 52nd Report Joint FAO/WHO Expert Committee on Food Additives, WHO Technical Report Series 893, 2000, pp. 28–37 (available at http://whqlibdoc.who.int/trs/WHO_TRS_893.pdf; accessed 11/22/10).

129. World Health Organization, *Evaluation of Certain Veterinary Drug Residues in Food*, 58th Report Joint FAO/WHO Expert Committee on Food Additives, WHO Technical Report Series 911, 2002; pp. 35–36 (available at http://whqlibdoc.who.int/trs/WHO_TRS_911.pdf; accessed 11/22/10).

130. Francis PG, Thiamphenicol, in *Residues of Some Veterinary Drugs in Animals and Foods*, FAO Food and Nutrition Paper 41/9, 2000, pp. 89–104 (available at ftp://ftp.fao.org/ag/agn/jecfa/vetdrug/41-9-thiamphenicol.pdf; accessed 11/22/10).

131. Wells RJ, Thiamphenicol, in *Residues of Some Veterinary Drugs in Animals and Foods*, FAO Food and Nutrition Paper 41/12, 1997, pp. 119–128 (available at ftp://ftp.fao.org/ag/agn/jecfa/vetdrug/41-12-thiamphenicol.pdf; accessed 11/22/10).

132. Pressman BC, Biological applications of ionophores, *Annu. Rev. Biochem*. 1976;45:501–530.

133. Russell JB, A proposed mechanism of monensin action in inhibiting ruminal bacterial growth: Effects on ion flux and protonmotive force, *J. Anim. Sci*. 1987;64:1519–1525.

134. Chow JM, Van Kessel JAS, Russell JB, Binding of radiolabelled monensin and lasalocid to ruminal microorganisms and feed, *J. Anim. Sci*. 1994;72:1630–1635.

135. Russell JB, Strobel HJ, Effect of ionophores on ruminal fermentation, *Appl. Environ. Microbiol*. 1989;55:1–6.

136. Bergen WG, Bates DB, Ionophores: Their effect on production efficiency and mode of action, *J. Anim. Sci*. 1984;58:1465–1483.

137. Lindsay DS, Blagburn BL, Antiprotozoan drugs, in Adams HR, ed., *Veterinary Pharmacology and Therapeutics*, 8th ed., Blackwell, Ames, IA, 2001, pp. 992–1016.

138. Russell JB, Houlihan AJ, Ionophore resistance of ruminal bacteria and its potential impact on human health, *FEMS Microbiol. Rev*. 2003;27(1):65–74.

139. Blaxter K, The energy metabolism of ruminants, in Blaxter K, ed., *The Energy Metabolism of Ruminants*, Charles C. Thomas, Springfield, IL, 1962, pp. 197–200.

140. Galyean ML, Owens FN, Effects of monensin on growth, reproduction, and lactation in ruminants, in *ISI Atlas of Science: Animal and Plant Sciences*, ISI Press, Philadelphia, 1988, pp. 71–75.

141. Honeyfield DC, Carlson JR, Nocerini MR, Breeze RG, Duration and inhibition of 3-methylindole production by monensin, *J. Anim. Sci*. 1985;60:226–231.

142. Woodward KN, Veterinary pharmacovigilance. Part 3. Adverse effects of veterinary medicinal products in animals and on the environment, *J. Vet. Pharmacol. Ther*. 2005;28:171–184.

143. Dowling PM, Miscellaneous antimicrobials: Ionophores, nitrofurans, nitroimidazoles, rifamycins, oxazolidinones, and others, in Giguère S, Prescott JF, Baggot JD, Walker RD, Dowling PM, eds., Antimicrobial Therapy in Veterinary Medicine, 4th ed., Blackwell, Ames, IA, 2006, pp. 285–300.

144. Van der Linde-Sipman JS, van den Ingh T, Van Nes JJ, Verhagen H, Kersten JGTM, Benyen AC, Plekkringa R, Salinomycin-induced polyneuropathy in cats. Morphologic and epidemiologic data, *Vet. Pathol*. 1999;36:152–156.

145. Kouyoumdjian JA, Morita MPA, Sato AK, Pissolatti AF, Fatal rhabdomyolysis after acute sodium monensin (Rumensin®) toxicity, *Arq. Neuropsiquiatr*. 2001;59:596–598.

146. Kim S, Carlson K, Occurrence of ionophore antibiotics in water and sediments of a mixed-landscape watershed, *Water Res*. 2006;40:2549–2560.

147. Lissemore L, Hao C, Yang P, Sibley PK, Mabury S, Solomon KR, An exposure assessment for selected pharmaceuticals within a watershed in Southern Ontario, *Chemosphere* 2006;64:717–729.

148. World Health Organization, *Evaluation of Certain Veterinary Drug Residues in Food*, 70th Report Joint FAO/WHO Expert Committee on Food Additives, WHO Technical Report Series 954, 2009, pp. 56–71 (available at http://whqlibdoc.who.int/trs/WHO_TRS_954_eng.pdf; accessed 11/21/10).

149. World Health Organization, Evaluation of Certain Veterinary Drug Residues in Food, 70th Report Joint FAO/WHO Expert Committee on Food Additives, WHO Technical Report Series 954, 2009, pp. 71–83 (available at http://whqlibdoc.who.int/trs/WHO_TRS_954_eng.pdf; accessed 11/21/10).

150. Freidlander LG, Sanders, P, Monensin, in *Residues of Some Veterinary Drugs in Foods and Animals*, FAO JECFA Monographs 6, 2009, pp. 109–135 (available at ftp://ftp.fao.org/ag/agn/jecfa/vetdrug/6-2009-monensin.pdf; accessed 11/08/10).

151. San Martin B, Freidlander LG, Narasin, in *Residues of Some Veterinary Drugs in Foods and Animals*, FAO JECFA Monograph 6, 2009, pp. 137–158 (available at ftp://ftp.fao.org/ag/agn/jecfa/vetdrug/6-2009-narasin.pdf; accessed 11/08/10).
152. Kim S-C, Carlson K, Quantification of human and veterinary antibiotics in water and sediment using SPE/LC/MS/MS, *Anal. Bioanal. Chem*. 2007;387:1301–1315.
153. Hansen M, Krogh KA, Brandt A, Christensen JH, Halling-Sørensen B, Fate and antibacterial potency of anticoccidial drugs and their main degradation products, *Environ. Pollut*. 2009;157:474–480.
154. Hansen M, Anticoccidials in the Environment: Occurrence, Fate, Effects and Risk Assessment of Ionophores, dissertation, Univ. Copenhagen, 2009.
155. Van Dijck PJ, Vanderhaeghe H, DeSomer P, Microbiologic study of the components of Staphylomycin, *Antibiot. Chemother*. 1957;7(12):625–629.
156. Vanderhaeghe H, Parmentier G, La structure de la staphylomycie, *Bull. Soc. Chim. Biol*. 1959;69:716–718.
157. Champney WS, Tober CL, Specific inhibition of 50S ribosomal subunit formation in *Staphylococcus aureus* cells by 16-membered macrolide, lincosamide, and streptogramin B antibiotics, *Curr. Microbiol*. 2000;41:126–135.
158. Matos R, Pinto VV, Ruivo M, Lopes MFD, Study on the dissemination of the bcrABDR cluster in Enterococcus spp reveals that the BCRAB transporter is sufficient to confer high level bacitracin resistance, *Int. J. Antimicrob. Agents* 2009;34:142–147.
159. World Health Organization, *Evaluation of Certain Veterinary Drug Residues in Food*, 66th Report Joint FAO/WHO Expert Committee on Food Additives, WHO Technical Report Series 939, World Health Organization, Geneva, 2006, pp. 18–32 (available at http://whqlibdoc.who.int/publications/2006/9241209399_eng.pdf; accessed 11/9/10).
160. Freidlander LG, Arnold D, Colistin, in *Residues of Some Veterinary Drugs in Foods and Animals*, FAO JECFA Monograph 2, 2006, pp. 7–28 (available at ftp://ftp.fao.org/ag/agn/jecfa/vetdrug/2-2006-colistin.pdf; accessed 11/9/10).
161. Butaye P, Devriese LA, Haesbrouck F, Antimicrobial growth promoters used in animal feed: Effect of less well known antibiotics on Gram-positive bacteria, *Clin. Microbiol. Rev*. 2003;16:175–178.
162. Pfaller, M, Flavophospholipol use in animals: Positive implications for antimicrobial resistance based on its microbiologic properties, *Diagn. Microbiol. Infect. Dis*. 2006;52:115–121.
163. Edwards JE, McEwan NR, McKain N, Walker N, Wallace RJ, Influence of flavomycin on ruminal fermentation and microbial populations in sheep, *Microbiology* 2005;15:717–725.
164. Poole TL, McReynolds JL, Edrington TS, Byrd JA, Callaway TR, Nisbet DJ, Effect of flavophospholipol on conjugation frequency between Escherichia coli donor and recipient pairs *in vitro* and in the chicken gastrointestinal tract, *J. Antimicrob. Chemother*. 2006;58:359–366.
165. World Health Organization, *Evaluation of Certain Veterinary Drug Residues in Food*, 43rd Report Joint FAO/WHO Expert Committee on Food Additives, WHO Technical Report Series 855, 1995, pp. 36–38 (available at http://whqlibdoc.who.int/trs/WHO_TRS_855.pdf; accessed 11/9/10).
166. World Health Organization, *Evaluation of Certain Veterinary Drug Residues in Food*, 62nd Report Joint FAO/WHO Expert Committee on Food Additives, WHO Technical Report Series 925, 2004, pp. 18–20 (available at http://whqlibdoc.who.int/trs/WHO_TRS_925.pdf; accessed 11/10/10).
167. World Health Organization, *Evaluation of Certain Veterinary Drug Residues in Food*, 43rd Report Joint FAO/WHO Expert Committee on Food Additives, WHO Technical Report Series 855, 1995, pp. 17–24 (available at http://whqlibdoc.who.int/trs/WHO_TRS_855.pdf; accessed 11/9/10).
168. World Health Organization, *Evaluation of Certain Veterinary Drug Residues in Food*, 48th Report Joint FAO/WHO Expert Committee on Food Additives, WHO Technical Report Series 879, 1998, pp. 15–25 (available at http://whqlibdoc.who.int/trs/WHO_TRS_879.pdf; accessed 11/10/10).
169. World Health Organization, *Evaluation of Certain Veterinary Drug Residues in Food*, 50th Report Joint FAO/WHO Expert Committee on Food Additives, WHO Technical Report Series 888, 1999, pp. 33–43 (available at http://whqlibdoc.who.int/trs/WHO_TRS_888.pdf; accessed 11/10/10).
170. Wells R, Oxolinic acid, in *Residues of Some Veterinary Drugs in Animals and Foods*, FAO Food and Nutrition Paper 41/7, 1998, pp. 69–88 (available at ftp://ftp.fao.org/ag/agn/jecfa/vetdrug/41-7-oxolinic_acid.pdf; accessed 11/22/10).
171. Francis PG, Wells RJ, Flumequine, in *Residues of Some Veterinary Drugs in Animals and Foods*, FAO Food and Nutrition Paper 41/10, 1995, pp. 59–70 (available at ftp://ftp.fao.org/ag/agn/jecfa/vetdrug/41-10-flumequine.pdf; accessed 11/22/10).
172. Wells R, Flumequine, in *Residues of Some Veterinary Drugs in Animals and Foods*, FAO Food and Nutrition Paper 41/13, 2000, pp. 43–52 (available at ftp://ftp.fao.org/ag/agn/jecfa/vetdrug/41-13-flumequine.pdf; accessed 11/22/10).
173. Rojas JL, Soback S, Flumequine, in *Residues of Some Veterinary Drugs in Animals and Foods*, FAO Food and Nutrition Paper 41/15, 2003;43–52 (available at ftp://ftp.fao.org/ag/agn/jecfa/vetdrug/41-15-flumequine.pdf; accessed 11/22/10).
174. Rojas JL, Reeves PT, Flumequine, in *Residues of Some Veterinary Drugs in Animals and Foods*, FAO JECFA Monograph 2, 2006 pp. 1–7 (available at ftp://ftp.fao.org/ag/agn/jecfa/vetdrug/2-2006-flumequine.pdf; accessed 11/22/10).

175. Heitzman RJ, Enrofloxacin, in *Residues of Some Veterinary Drugs in Animals and Foods*, FAO Food and Nutrition Paper 41/10, 1997, pp. 31–44 (available at ftp://ftp.fao.org/ag/agn/jecfa/vetdrug/41-7-enrofloxacin.pdf; accessed 11/09/10).
176. Heitzman RJ, Danofloxacin, in *Residues of Some Veterinary Drugs in Animals and Foods*, FAO Food and Nutrition Paper 41/10, 1997, pp. 23–37 (available at ftp://ftp.fao.org/ag/agn/jecfa/vetdrug/41-10-danofloxacin.pdf; accessed 11/22/10).
177. Heitzman RJ, Sarofloxacin, in *Residues of Some Veterinary Drugs in Animals and Foods*, FAO Food and Nutrition Paper 41/10, 1998, pp. 107–117 (available at ftp://ftp.fao.org/ag/agn/jecfa/vetdrug/41-11-sarafloxacin.pdf; accessed 11/22/10).
178. Berger K, Petersen B, Buening-Pfaue H, Persistence of drugs occurring in liquid manure in the food chain, *Arch. Lebensmittelhyg*. 1986;37(4):85–108.
179. Reeves PT, Minchin RF, Ilett KF, Induction of sulfamethazine acetylation by hydrocortisone in the rabbit, *Drug Metab. Dispos*. 1988;16:104–109.
180. Kay P, Blackwell PA, Boxall ABA, Transport of veterinary antibiotics in overland flow following the application of slurry to arable land, *Chemosphere* 2005;59: 951–959.
181. Blackwell PA, Kay P, Boxall ABA, The dissipation and transport of veterinary antibiotics in a sandy loam soil, *Chemosphere* 2007;67:292–299.
182. World Health Organization, *Evaluation of Certain Veterinary Drug Residues in Food*, 42nd Report Joint FAO/WHO Expert Committee on Food Additives, WHO Technical Report Series 851, 1995, pp. 25–27 (available at http://whqlibdoc.who.int/trs/WHO_TRS_851.pdf; accessed 11/22/10).
183. Anonymous, Sulfamethazine, in *Residues of Some Veterinary Drugs in Animals and Foods*, FAO Food and Nutrition Paper 41/2, 1994, pp. 66–81 (available at ftp://ftp.fao.org/ag/agn/jecfa/vetdrug/41-2-sulfadimidine.pdf; accessed 11/22/10).
184. Hamscher G, Abu-Quare A, Sczesny S, Hoper H, Nau G, Determination of tetracyclines and tylosin in soil and water samples from agricultural areas in lower Saxony, in van Ginkel LA, Ruiter A, eds., *Proc. Euroresidue IV Conf*., May 2000, National Institute of Public Health and the Environment (RIVM), Veldhoven, The Netherlands, 2000, pp. 8–10.
185. World Health Organization, *Evaluation of Certain Veterinary Drug Residues in Food*, 45th Report Joint FAO/WHO Expert Committee on Food Additives, WHO Technical Report Series 864, 1996, pp. 38–40 (available at http://whqlibdoc.who.int/trs/WHO_TRS_864.pdf; accessed 11/11/10).
186. Anonymous, Oxytetracycline, in *Residues of Some Veterinary Drugs in Animals and Foods*, FAO Food and Nutrition Paper 41/3, 1991, pp. 97–118 (available at ftp://ftp.fao.org/ag/agn/jecfa/vetdrug/41-3-oxytetracycline.pdf; accessed 11/23/10).
187. Sinhaseni Tantiyaswasdikul P, Oxytetracycline, in *Residues of Some Veterinary Drugs in Animals and Foods*, FAO Food and Nutrition Paper 41/8, 1996, pp. 125–130 (available at ftp://ftp.fao.org/ag/agn/jecfa/vetdrug/41-8-oxytetracycline.pdf; accessed 11/23/10).
188. Wells R, Tetracycline in *Residues of Some Veterinary Drugs in Animals and Foods*, FAO Food and Nutrition Paper 41/8, 1996, pp. 131–155 (available at ftp://ftp.fao.org/ag/agn/jecfa/vetdrug/41-8-tetracycline.pdf; accessed 11/23/10).
189. Sinhaseni Tantiyaswasdikul P, Oxytetracycline, in *Residues of Some Veterinary Drugs in Animals and Foods*, FAO Food and Nutrition Paper 41/9, 1997, pp. 75–76 (available at ftp://ftp.fao.org/ag/agn/jecfa/vetdrug/41-9-oxytetracycline.pdf; accessed 11/23/10).
190. Wells R, Chlortetracycline and tetracycline, in *Residues of Some Veterinary Drugs in Animals and Foods*, FAO Food and Nutrition Paper 41/9, 1997, pp. 3–20 (available at ftp://ftp.fao.org/ag/agn/jecfa/vetdrug/41-9-chlortetracycline_tetracycline.pdf; accessed 11/23/10).
191. Roestel B, Tetracycline, oxytetracycline and chlortetracycline, in *Residues of Some Veterinary Drugs in Animals and Foods*, FAO Food and Nutrition Paper 41/11, 1998, p. 23 (available at ftp://ftp.fao.org/ag/agn/jecfa/vetdrug/41-11-chlortetracycline_oxytetracycline_tetracycline.pdf; accessed 11/23/10).
192. AliAbadi F, MacNeil JD, Oxytetracycline, in *Residues of Some Veterinary Drugs in Animals and Foods*, FAO Food and Nutrition Paper 41/14, 2002, pp. 61–67 (available at ftp://ftp.fao.org/ag/agn/jecfa/vetdrug/41-14-oxytetracycline.pdf; accessed 11/23/10).
193. Lindsey ME, Meyer M, Thurman EM, Analysis of trace levels of sulphonamide and tetracycline antimicrobials in groundwater and surface water using solid-phase extraction and liquid chromatography/mass spectrometry, *Anal. Chem*. 1998;73:4640–4646.

2

PHARMACOKINETICS, DISTRIBUTION, BIOAVAILABILITY, AND RELATIONSHIP TO ANTIBIOTIC RESIDUES

PETER LEES AND PIERRE-LOUIS TOUTAIN

2.1 INTRODUCTION

To ensure public confidence, in relation to the consumption of foodstuffs derived from animals that have received antimicrobial drugs (AMDs), regulatory authorities adopt conservative approaches and set stringent standards on data requirements. Pharmacological, toxicological, and microbiological *no observable (adverse) effect levels* [NO(A)ELs] are determined and the lowest value is used to calculate the *acceptable daily intake* (ADI), which is the amount of drug or drug metabolite that can be consumed by humans daily throughout life without appreciable risk to health. The ADI is used to calculate the *maximum residue limits* (MRLs) (termed *tolerances* in the United States) of the selected marker residue, which is usually the parent drug but can also be a drug metabolite, the total concentration of several compounds, or a chemical conversion product of the parent compound plus metabolites. In support of MRLs in edible tissues of the target animal species, determined in residue depletion studies, companies seeking a marketing authorization (MA) for a product containing one or more AMDs are required to supply target species data on the pharmacokinetics and metabolism of the active constituents of the product, when administered at recommended dose rates. The pharmacokinetic studies provide quantitative data on the absorption, distribution, metabolism, and excretion of the drugs, involving in particular plasma or blood concentration–time profiles and identification and quantification of major metabolites.

This chapter summarizes the principal pharmacokinetic properties of the major groups of AMDs, particularly in relation to residues depletion. Regulatory control strategies are reviewed and for one jurisdiction, the EU, marker residues and target tissues are indicated for each group of compounds. The chapter also briefly reviews circumstances of therapeutic use, including prophylaxis, metaphylaxis, and therapy. The requirement for conducting residue studies on generic products for which bioequivalence to a pioneer product has been demonstrated is considered. Risk assessment, characterization, management, and communication, with respect to AMD residues in food derived from food-producing species, are summarized.

2.2 PRINCIPLES OF PHARMACOKINETICS

2.2.1 Pharmacokinetic Parameters

Pharmacokinetics is the science of describing quantitatively changes in drug concentration in the body over time as a function of administered dose. Generally, it is based on subjecting serum/plasma concentration–time data to mathematical models, which provide further data on absorption, distribution, metabolism, and excretion of the drug and its metabolites. Detailed discussion of the derivation, definition, and application of pharmacokinetic terms is outside the scope of this chapter (the reader is referred to reviews by Toutain and Bousquet-Mélou[1–4]). However, it is necessary to consider the plasma and tissue pharmacokinetics of AMDs in relation to residues in food-producing species. The relevant pharmacokinetic terms are defined in Table 2.1.

Chemical Analysis of Antibiotic Residues in Food, First Edition. Edited by Jian Wang, James D. MacNeil, and Jack F. Kay.

TABLE 2.1 Definition and Characterization of Pharmacokinetic Parameters

Term	Abbreviation	Dimension (Typical Units)	Estimation/Computation	Definition/Meaning
Area under curve	AUC	ATV^{-1} (μg h ml^{-1})	From raw data with trapezoidal rule or AUC = F * dose/Cl	Integral of plasma concentration–time curve; plasma (blood) exposure; internal dose is controlled by clearance and F%
Maximum plasma concentration	C_{max}	AV^{-1} (μg ml^{-1})	Generally obtained from raw data; simple analytical solution for a monocompartmental model	Maximum plasma concentration after administration of a given dose
Time of maximum concentration	T_{max}	T(min, h)		Time of maximal plasma concentration (C_{max})
Clearance	Cl	VT^{-1} (ml/kg/min)	Cl = dose/AUC = $K_{10} * V_c$, where K_{10} is first-order rate constant of elimination from central compartment	Rate of drug elimination scaled by plasma concentration; expresses body's capacity to eliminate a drug; with F, the single determinant of plasma exposure
Volume of distribution in steady-state condition	V_{ss}	(V l/kg)	V_{ss} = dose*AUMC/(AUC)^2, where AUMC is area under first moment of plasma concentration–time curve; V_{ss} = Cl*MRT, with MRT the mean residence time	Proportionality constant between amount of drug in body in steady-state condition and corresponding steady-state plasma concentration; term used to compute a loading dose
Volume of distribution; terminal phase	V_{area}	V l/kg	V_{area} = Cl/terminal slope	Proportionality constant between amount of drug in body at a given time in terminal phase and corresponding plasma concentration; term used to compute a residual amount of drug from an observed plasma concentration located in terminal elimination phase
Terminal half-life	$T_{1/2}$	T(min, h)	Ln2/ terminal slope	Time required to halve plasma concentration during terminal phase; can be an elimination half-life when plasma decay is controlled by clearance and extent of drug distribution; can be a half-life of absorption when decay is controlled by rate of drug release from administration site (flip-flop situation)
Very late terminal half-life	$T_{1/2}$	T(h, day)	Ln2/ very late terminal slope	Time required to halve plasma concentration during a very terminal phase (called *gamma* (γ) phase in text for aminoglycosides); given a sufficiently sensitive analytical technique, it is possible to characterize a terminal half-life that has no clinical meaning but is relevant for residue depletion; this last phase may be viewed as a slow drug release from a deep but small compartment giving a situation analogous to a flip-flop from an injection site
Bioavailability	F%	Scalar (percentage)	$F = (AUC_{EV}/AUC_{IV}) * 100$; when dose administered extravascularly (EV) is equal to dose administered intravenously (IV)	Express the amount of drug that is absorbed and gains access to central compartment after dosing by a nonvascular route

Notation: A = amount; V = volume; T = time.

When a drug is administered intravenously, that is, directly into the pharmacokinetic central compartment, there is no absorption phase and the plasma or blood concentration–time profile can be used to derive, by fitting regression lines to the data using appropriate computer programs (e.g., WinNonLin) three intrinsic properties of the substance: clearance (Cl), volume of distribution(s) (V), and elimination half-life (T_{max}). These are described as PK parameters and are generally obtained in healthy animals; when a collection of these PK parameters is obtained from a set of animals, they become statistical random variables with typical values, distributions (often lognormal), and other characteristics. The goal of so-called population kinetics is precisely to evaluate these statistical parameters and to explain variability (inter- and intra-animal) with different covariables, such as age, sex, and health status. This is relevant regarding the establishment of a withholding time (WhT) that needs to take into account all these factors of variability. For example, nothing guarantees that clearance of an AMD established in healthy animals is equal or even similar to clearance that could be obtained in diseased animals.

There are many more pharmacokinetic properties, such as C_{max}, T_{max}, area under the curve (AUC), absorption half-life, terminal half-life, and bioavailability, which are not uneqivocally related to the drug substance but to a given formulation of the substance, namely, a drug product and for a given drug product, to the route of administration and circumstances of drug administration (e.g., whether administered to fed or fasted animals), and so on. From a residues perspective, this explains why a withholding time (WhT) cannot be a substance parameter but is a product characteristic depending on the route of administration, dosage regimen, and other parameters. In contrast, a MRL is a concentration that is a substance property, intrinsically independent of any kinetic characteristics of the substance, allowing authorities to fix its value as a "regulatory constant" having a universal meaning and that can be subjected to international harmonization.

The relationships between the different PK parameters and plasma drug concentrations (often reported as C_{max} and AUC) are indicated by the equations in Table 2.1. In simple terms, it will be seen that, for a given dose, if Cl_{tot} is high, AUC will be low, and this impacts on residues insofar as AUC (also termed *exposure* or *internal dose*) in plasma will relate to tissue concentrations (albeit in a possibly complex manner). Clearance is the genuine pharmacokinetic parameter expressing the body capacity to eliminate a substance and determining dose amount to administer to achieve a targeted plasma concentration (for a given bioavailability). The terminal half-life, which expresses a change in concentration in units of time, provides an easily understood parameter for describing a terminal concentration–time profile.

The terminal half-life is a hybrid parameter, and its determinants need to be acknowledged. It is either a hybrid PK parameter determined by both clearance and distribution of the substance or, when there is a flip-flop, a PK characteristic of the drug product depending on bioavailability factors (rate and extent of absorption). In this latter case, the terminal half-life does not reflect the intrinsic rate of substance elimination. Whatever the biological factors controlling the terminal half-life, $T_{1/2}$ should be considered to select an optimal interval between dose administrations. This is clear if one considers a threshold plasma concentration required for AMD efficacy; a long terminal half-life will result in a longer time for plasma concentrations to decline to the threshold concentration.

More importantly, in relation to residues of AMD, the existence or nonexistence of a so-called very late terminal phase needs to be considered. When using a sensitive analytical technique, a supplementary phase can be detected for a range of plasma concentrations that are below the microbiologically effective plasma concentration and thus without therapeutic meaning. This terminal phase decays very slowly (half-life typically higher than 24 h), and it reflects persistence of drug residues in some deep compartments. This terminal phase is actually controlled by the redistribution rate constant from tissue to plasma. Aminoglycosides provide an example of this situation with a therapeutically significant half-life of approximately 2 h, while a very late terminal phase decays with a much slower half-life (see Section 2.3.1), explaining the persistence of residues in some tissues for weeks or even months.

In addition, this very late terminal phase can lead to drug accumulation with repeated administrations, explaining that the WhT, required to fall below the MRL, can be much more prolonged after a multiple dosing regimen than after a single dose administration. It is only the remnant amount of drug during the very late terminal phase that is consistently higher after a repeated dosing regimen, while there is no therapeutically relevant accumulation in plasma concentration during the treatment itself (for further explanation, see Fig. 15 in Toutain and Bousquet-Mélou[4]).

In relation to residues of AMDs, the pharmacokinetic studies conducted in laboratory animals and target species required by regulatory authorities are designed to establish concentration–time profiles of parent drug and its biologically active and inactive metabolites in body fluids (usually blood, serum, or plasma). For laboratory animals, the purpose of comparative metabolism studies is to determine whether laboratory animals used in toxicological testing have been exposed to the metabolites that humans will be exposed to as residues in products of food animal origin.[5] For target species, metabolism studies are required to determine the nature and quantity of veterinary drugs residues.[6] This task is generally accomplished using radiolabeled drugs to cover all possible residues. Finally, target tissue concentration–time profiles are determined after administering the recommended dosage schedule of the product in the formulation intended for clinical use in order to describe the marker residue depletion to establish product WhTs.[7] Ideally, pharmacokinetic data and metabolism studies are required for each target species, for each administration route, and both minimum and maximum dose rates, in the event of variable dosage recommendations. In the latter circumstance, residue depletion studies conducted with the highest recommended dose administered for the longest recommended duration are the minimum requirement.

2.2.2 Regulatory Guidelines on Dosage Selection for Efficacy

Guidelines on dose selection of AMDs vary between jurisdictions, but all require the sponsor to demonstrate in preclinical studies the pharmacokinetic profile in the target animal species and the pharmacodynamic profile against microorganisms. The latter comprises the spectrum of activity, whether the drug is primarily bacteriostatic or bactericidal (at clinically effective dose rates), and whether the type of killing action is primarily concentration-, time- or co-dependent. AMD pharmacodynamics may be quantified using several indices, the most important of which are minimum inhibitory concentration (MIC) and minimum bactericidal concentration (MBC), while growth inhibition–time curves are used to define both the type of killing action and the steepness of the concentration–effect relationship.

The most widely used hybrid indicator of both efficacy and potency is MIC. When MIC has been determined against a sufficient number of strains (usually hundreds because of the inter-strain variability in potency) of each sensitive microbial species, the median or geometric mean MIC_{50} and MIC_{90} values are determined. It is then possible to set a provisional dose through integration of pharmacodynamic and pharmacokinetic data, using one or more of the following indices: $C_{max} : MIC_{90}$ (for some concentration-dependent drug classes, e.g., aminoglycosides), $AUC : MIC_{90}$ (for most concentration- and co-dependent drugs, e.g., fluoroquinolones, macrolides, tetracyclines), and $T > MIC_{90}$ (for most β-lactam drugs). The latter is the proportion of the inter-dose interval for which plasma/serum concentration exceeds MIC_{90} and is expressed as a percentage of the inter-dose interval.

The scientific literature is replete with proposals for numerical values of these indices, for example, $C_{max} : MIC_{90} \geq 10 : 1$ for aminoglycosides, $AUC : MIC_{90}$ ratio $\geq$ 125 h for fluoroquinolones, and $T > MIC_{90} \geq 50\%$ for β-lactams. In fact, these values provide no more than a guide to clinically effective dosage for several reasons:

1. Target numerical values, in practice, are "bug and drug"-specific.
2. The dosage required is dependent on bacterial load and level of immune competence of the host animal.
3. The dosage required depends on whether the endpoint is clinical cure, bacteriological cure, or avoidance of emergence of resistance.

For further discussion of these indices, see Lees et al.[8–10] and Toutain et al.[11–13]

The approach to final dosage schedule determination, following provisional determination using PK-PD principles as described above, varies between jurisdictions and is not considered here in detail. One example is taken. In the EU, guidance is provided by European Medicines Agency/Committee for Veterinary Medicinal Products (EMA/CVMP, previously European Medicines Evaluation Agency/Committee for Veterinary Medicinal Products, or EMEA/CVMP), which recommends the conduct, using clinically relevant disease models in the target species, of dose titration/determination studies.[14] These should be conducted separately for each administration route, each proposed dose, and each disease indication; thus the requirements can be onerous in terms of both cost and animal welfare. The dose rate, which provides a greater response than the next-lower dose but no greater response (statistically) than the next-higher dose, is selected for further study in a dose confirmation study, which is conducted again in a disease model or in clinical subjects.

The problems/disadvantages of dose-ranging studies have been discussed elsewhere.[8,10,12,13] Briefly, the dose selected may be demonstrably *effective* by appropriate statistical analyses but is unlikely to be *optimal*. Lees et al.[8] and Toutain et al.[13] have recommended as an alternative the use of PK-PD modeling approaches (not to be confused with PK-PD integration), in which computer programs, utilizing, for example, the sigmoidal E_{max} equation, are applied to evaluate the whole sweep of the concentration–effect relationship. This enables determination, using *in vitro*, *ex vivo*, and *in vivo* techniques, of drug concentrations and dosages required to achieve specific levels of inhibition of bacterial growth and bacteriostatic or bactericidal eradication of bacteria.[15,16]

2.2.3 Residue Concentrations in Relation to Administered Dose

Residue concentrations and their depletion profiles are inevitably linked to administered dose of an antimicrobial drug, albeit in a possibly complex and tissue dependent manner. This is illustrated *first* by the equation linking, for a systemically administered drug, dose to area under the plasma/blood concentration–time curve:

$$\text{Dose} = \frac{\text{Cl} \times \text{AUC}}{F} \tag{2.1}$$

where Cl = whole-body clearance, AUC = area under plasma or blood concentration–time curve, and F = bioavailability, the proportion of the administered dose absorbed, and gaining access to the central compartment. Rearrangement of Equation (2.1) leads to the following equation:

$$\text{AUC} = \frac{\text{Dose} \times F}{\text{Cl}} \quad (2.2)$$

This equation illustrates that the higher the dose and value of F and the lower the value of Cl, the greater will be the amount (exposure) of drug in plasma/blood over a measured time interval. If, in a dose-ranging study, F and Cl are held constant (i.e., if the pharmacokinetics is linear), doubling the dose increases AUC by a factor of 2. The AUC will not increase in direct proportion to administered dose, however, if Cl and/or F are not PK parameters but rather dose-dependent variables, as is the case for non-linear pharmacokinetics. For example, F may be lower at higher dosages of orally administered drugs that are highly lipid-soluble but poorly water-soluble, while Cl may be slower at high doses as a consequence of saturation of elimination pathways.

Secondly there will be a relationship between drug concentration in plasma (the driving concentration) and concentration in tissues, but the two will rarely be equal, and tissue concentrations will depend on a range of drug properties (see Section 2.2.6) and animal characteristics. It is important to note the differences in importance and consequence of total tissue concentration for pharmacological/toxicological responses on one hand and for drug and metabolite residues on the other hand. For the latter, it is the mean tissue concentration (not the separate concentrations in extracellular or intracellular fluids or intracellular distribution between several compartments) that determines intake by human consumers. In contrast, for the former, tissue concentration has limited (if any) value and may actually be misleading. This is illustrated by the macrolide, lincosamide, and pleuromutilin groups of antimicrobial drugs. Drugs of these classes, in magnitudes varying from compound to compound, achieve high overall concentrations in lung tissue, but the highest concentrations occur at intracellular sites. This circumstance is of no benefit therapeutically if the biophase where organisms are located is an extracellular site (such as epithelial lining fluid), as is the case for most bacterial species causing lung infections in farm animal species. This circumstance may be likened to an army confined to barracks and unable to contribute to the fight waging on the battlefield outside.

In relation to drug residues, a pharmacokinetic property of major significance is the very late terminal elimination half-life (see Sections 2.3.1 and 2.7.1). For many drug classes, a semi-logarithmic plot of plasma concentration versus time, after intravenous dosing, reveals a multicompartmental model, with three phases describing the compartments, of slopes λ_1, λ_2, and λ_3. These represent, respectively, rapid distribution, slow distribution, and finally, after reaching a pseudoequilibrium distribution (i.e., when the same amount of drug is exchanged from central to peripheral compartments and vice versa, from peripheral to central compartments) the decay observed during the terminal phase corresponds to the net elimination process. The third phase may be revealed only if (1) sampling is continued beyond concentrations of therapeutic relevance and (2) the analytical method is sufficiently sensitive (i.e., has a low lower limit of quantification). For most drugs the λ_2 phase is of therapeutic interest, as it determines the interval between doses required to provide clinical efficacy. On the other hand, the λ_3 phase (also named *gamma phase* in the literature) represents for some drugs the slow decline in concentration of drug beyond the therapeutically useful concentration, as drug is off-loaded from tissues. The λ_3 phase represents drug elimination from what are termed *pharmacokinetic deep compartments*. Alternatively, the λ_3 phase may represent flip-flop pharmacokinetics for a fraction of the drug that is slowly absorbed (see Sections 2.3.1 and 2.7.1). In both instances, it is the λ_3 phase value that normally determines WhTs.

There are two equations which can be used to represent terminal half-life:

$$T_{1/2} = \frac{\ln 2}{\text{Terminal} \times \text{slope}} = \frac{0.693}{\lambda_3} \quad (2.3)$$

This equation is the mathematical expression of the definition of a half-life, specifically, the time required for plasma concentrations to be divided by 2 after reaching pseudoequilibrium; as λ_3 is a hybrid parameter related to V_{area} (the volume of distribution associated with the terminal phase) and plasma clearance, Equation (2.3) can be re-written in a more mechanistically useful way as follows:

$$T_{1/2} = \frac{0.693 \times V_{\text{area}}}{\text{Cl}} \quad (2.4)$$

Equation (2.3) is conceptually useful in indicating that, when slope (λ_3) is shallow, half-life will be long and therefore WhT will also be long. Equation (2.4) is mechanistically useful in illustrating that $T_{1/2}$ depends on Cl, rate of elimination from the body, and V_{area}, on the extent of distribution within the body. Clearly, if Cl is slow, $T_{1/2}$ will be prolonged.

Volume V_{area} is a proportionality constant, indicating the relationship between a plasma concentration in the terminal phase and the corresponding total amount of drug in the body. It may be useful to perform such a computation to compare, at a given time, the total residual amount of drug in the body and the ADI. It should be stressed that V_{area} does not represent a particular physiological space, and if one wishes to discuss physiological drug repartition and WhT, the steady-state volume of distribution (V_{ss}) is the most appropriate volume of distribution to be considered because it is physiologically based and its numerical value (always lower than that of V_{area}) directly indicates equilibrium distribution mechanism, whereas V_{area} is also influenced by plasma clearance. However, interpretation of a V_{ss} to anticipate a WhT is not straightforward as a high value of V_{ss} (e.g., much greater than body water volume) may represent uptake in high concentration into intracellular fluid of most or all body cells. Alternatively, it may represent uneven distribution and a high concentration in a specific tissue or tissues. Some drugs can have prolonged WhTs because of association to a large V_{ss}, but a low V_{ss} does *not* guarantee a short WhT (see Section 2.3.1), as a drug may achieve high concentrations in one tissue, such as kidney (e.g., aminoglycosides), while its *overall distribution* is limited.

The many factors that can alter V_{ss} and/or Cl, and thereby shorten or prolong $T_{1/2}$ are discussed in detail elsewhere.[1–4,17] They include altered fluid balance, nutritional status, percentage of body fat, species, hormonal status, age of animal, and disease status. For example, renal and/or hepatic disease can reduce Cl and therefore prolong $T_{1/2}$ for the therapeutic phase, while infectious diseases may either increase or decrease Cl and/or V_{ss}. In contrast, the main factor controlling the slope of the very late terminal phase is the redistribution of the drug from a deep compartment to plasma.

2.2.4 Dosage and Residue Concentrations in Relation to Target Clinical Populations

The efficacy and safety of antimicrobial drugs depend on both pharmacodynamic (drug efficacy and potency against the disease causing organism) and pharmacokinetic (exposure of organisms in the biophase for sufficient time to provide bacteriological cure) properties. Both properties must therefore be used in selecting a rational dosage schedule for clinical use. Regulatory authorities, therefore, require pharmaceutical companies to supply pharmacokinetic data in healthy and homogeneous animals (i.e., animals selected to minimize sources of interanimal variability). The requirements are different for residue studies where regulatory authorities explicitly require that animals used to identify the nature of residues be representative of the target population.[7] For example, if there are reasons to believe that the metabolisms of non-ruminating cattle will significantly differ from those of adult cattle, two separate studies will be required to document the possible influence of age on drug metabolism and on the nature of residues.

Similarly, separate studies are recommended when the target population includes both pre-ruminant and ruminant cattle to establish product WhT. However, the health status is ignored; this is of concern especially for AMDs. According to Nouws, the disease state is the main factor affecting the WhT.[18] This author determined tissue residue concentrations and persistence of different AMDs, including β-lactams, aminoglycosides, tetracyclines, macrolides, chloramphenicol, and sulfonamides in normal and emergency-slaughtered ruminants after parenteral or intramammary administration. At that time analytical assays (microbiological assays) were rather crude and MRLs were not established. Nevertheless, it is interesting to note that, comparing with the same pharmacokinetic model normal and emergency-slaughtered cattle, Nouws concluded that to predict WhT for muscle and kidney it was necessary to multiply by a factor of 2–3 or 4–5, respectively, values obtained in normal cattle to predict values in emergency-slaughtered ruminants.[19] No recent studies using current analytical methods have been performed to update these data, but it is very likely that residue depletion of AMDs is not equivalent in healthy and diseased animals.

A solution to this difficulty would be to define the depletion profile in clinical subjects or in disease models that closely simulate clinical disease. No attempt is made to meet this ideal for a range of ethical, economic, and scientific reasons. Instead, reliance is placed on conducting residue (like pharmacokinetic) studies in healthy animals. This is justified by the series of conservative assumptions made in establishing withholding periods (see Section 2.5.4.1).

Another important topic for regulatory authorities concerns flexibility in selecting dose schedules for clinical use for a given claim in a given species. We have argued elsewhere for greater reliance on (1) PK-PD modeling approaches[8] and (2) population pharmacokinetic studies as alternatives to classical dose titration studies in disease models[20] in order to optimize dosage for clinical and bacteriological cures.[13] However, a tailored dosage regimen taking into account both PK and PD variability raises the question of the WhT that has a single value. An advance would involve proposing a range of dosage regimens and establishing corresponding lower and upper bounds for the WhT.

2.2.5 Single-Animal versus Herd Treatment and Establishment of Withholding Time (WhT)

In poultry and porcine husbandry in particular, the use of AMDs "in feed" or "in drinking water" for prophylaxis, metaphylaxis (or control in the United States) or therapy is widely practiced (see Section 2.7.3.2). Prophylaxis involves administration of AMDs to healthy animals known to be at risk (as a consequence, e.g., of close proximity of animals housed together or predictable stresses caused by transport or adverse weather conditions). Metaphylaxis involves administration of AMDs, again to animals judged to be clinically healthy, but which are in contact with animals in which clinical signs have been detected. With such group dosing procedures, the dose received by individual animals is likely to vary considerably. This is in part a simple consequence of provision to the group of medicated feed or water, but variable intake may be compounded by smaller animals losing out in competition for access to feed or water. Even worse is the disinclination of more severely diseased animals in the group to eat or drink the medicated food or water. Moreover, drug intake is discontinuous in each animal. From a residue perspective, these sources of variability in drug intake inevitably have direct consequences for variability in tissue residues, and this should be considered when WhT is established in an experimental setting where administered doses are carefully controlled.

In contrast, the therapeutic use of AMDs generally involves treating animals individually with AMDs formulated for parenteral (usually intramuscular or subcutaneous) or oral dosing, with animals dosed on a mg/kg body weight basis. Here, dosing can be more accurate even if body weight is normally assessed rather than measured under clinical conditions.

2.2.6 Influence of Antimicrobial Drug (AMD) Physicochemical Properties on Residues and WhT

With long and expensive drug development times, there is a need in the pharmaceutical industry to optimize the drug discovery process. For human drugs, "Lipinski's rule of 5" aims at predicting oral drugability of new drug candidates by computing or measuring a set of descriptors, including the substance molecular weight, octanol–water partition (expressed as log P) to assess lipophilicity/hydrophilicity, number of hydrogen-bound acceptors or donors, and so on. Considering some cutoff values (actually 5), it can be predicted if the substance is likely to have desirable pharmacokinetic properties. For veterinary drugs, the question of residues and WhT is a critical factor that should be documented very early in the development program. However, currently no systematic investigation is carried out to link residue persistence in tissues and physicochemical properties of the active ingredients, thus allowing development of a general rule comparable to the Lipinski rule. To succeed in this objective, it would be necessary to investigate the residue depletion curve after intravenous administration to establish the contribution of the substance itself versus all other factors (mainly formulation) on the WhT. Currently, it is recognized that tissue concentrations will depend on a range of properties of the drug, namely, lipid solubility, acidic/basic characteristics, which influence the passive diffuse of drug across cell membranes and, for a few drugs, active uptake by or extrusion from tissues.

A low, moderate, or high degree of lipid solubility can have profound effects on AMD pharmacokinetics and tissue residues. Table 2.2 presents a broad classification of drugs on the basis of lipid solubilities and summarizes the impact on pharmacokinetic profiles. Drugs of high lipid solubility are organic molecules, which are un-ionized or only partially ionized at physiological pH. AMDs of low lipid solubility are usually either strong acids (e.g., penicillins) or strong bases (e.g., polymyxins) and hence wholly ionized at physiological pHs. Aminoglycosides are weak bases but nevertheless highly polar and very poorly lipid-soluble, due to the presence of sugar residues in the molecules.

As drug residue science is concerned with metabolites, as well as parent molecules, it should be noted that most (especially phase II) metabolites are more polar, less lipid-soluble, and less biologically active than parent drugs. Hence, most metabolites follow the general rules on disposition (poor penetration of cell membranes) and elimination (high concentrations in urine and/or bile) applicable to poorly lipid-soluble drugs.

2.3 ADMINISTRATION, DISTRIBUTION, AND METABOLISM OF DRUG CLASSES

2.3.1 Aminoglycosides and Aminocyclitols

The principal drugs of the aminoglycoside class are streptomycin (which is not extensively used in veterinary medicine because it is less safe than dihydrostreptomycin), dihydrostreptomycin, gentamicin, amikacin, kanamycin, apramycin, tobramycin, neomycin, and paromomycin. Aminoglycosides characteristically comprise an aglycone linked to one or more sugar units (a glycosamine and/or a disaccharide). In the aminocyclitols (e.g., spectinomycin), the amino group occurs in the cyclitol ring. The pharmacokinetics of aminoglycosides are dictated by their highly polar and poorly lipid-soluble physicochemical properties; these respective solubilities in water and lipid are related to both their polycationic nature and the fact that they contain "sugar" residues, such as streptose in streptomycin and dihydrostreptomycin.

Absorption extent from the gastrointestinal tract (GIT) is very low (of the order of $\leq$1–2% of administered dose), although higher bioavailability may be achieved in neonatal animals and where there is disruption of the intestinal mucosa, caused by, for example, parvovirus infection. Within the GIT, aminoglycosides are stable and excreted unchanged in feces.

When AMDs are administered parenterally as aqueous solutions (usually intramuscularly), absorption into the

TABLE 2.2 Influence of Lipid Solubility of Antimicrobial Drugs on Pharmacokinetic Properties (ADME)[a]

Drugs with Low Lipophilicity		Drugs with Moderate to High Lipophilicity			
Strong Acids	Strong Bases or Polar Bases[b]	Weak Acids	Weak Bases	Amphoteric	Drugs with High Lipophilicity
Cephalosporins, penicillins	Aminocyclitols, aminoglycosides, polymixins	Sulfonamides	Diaminopyrimidines	Most tetracyclines Chlortetracycline Oxytetracyline	Lipophilic tetracyclines Doxycycline Minocycline Fluoroquinolones Ketolides Lincosamides Macrolides Phenicols Rifamycins Triamilides
Penetrate cell membranes poorly or not at all; limited or no significant absorption from GIT except for acid-stable aminopenicillins, which have moderate but species-variable absorption; distribution limited mainly to extracellular fluids; concentrations in intracellular fluid, CSF, milk, and ocular fluids low, but effective concentrations may be reached in synovial, peritoneal, and pleural fluids; some penicillins actively transported out of CSF into plasma; generally excreted, usually in urine, in high concentrations as the parent molecule; some drugs actively secreted into urine and/or bile; biotransformation (e.g., in the liver) usually slight or absent		Readily cross cell membranes; generally moderate to good absorption from GIT but species-dependent; effective concentrations achieved in intra- and transcellular as well as extracellular fluids except for poor penetration of sulfonamides into intracellular fluid due to acidic environment; ability to penetrate into CSF and ocular fluids depends on plasma protein binding (e.g., most sulfonamides and diaminopyrimidines penetrate well); weak acids are ion-trapped in fluids alkaline relative to plasma, such as herbivore urine; weak bases are ion-trapped in fluids acidic relative to plasma (e.g., prostatic fluid, milk, intracellular fluid, carnivore urine); commonly dependent on biotransformation for termination of activity but may also be excreted unchanged in urine and/or bile; some drugs actively secreted into bile			Cross cell membranes very readily; generally well absorbed from GIT in monogastric species; penetrate into intracellular and transcellular fluids (e.g., synovial and prostatic fluids and bronchial secretions); also penetrate well into CSF, except tetracyclines and rifampin; termination of activity dependent on a high proportion of administered dose being metabolized, for example, in the liver but also at other sites (e.g., kidney, enterocytes); some drugs actively secreted into bile

[a]Absorption, distribution, metabolism, and excretion.
[b]Polymixins are strong bases, while aminoglycosides and aminocyclitols are weak bases, but polar and poorly lipid-soluble because of the presence of sugar residues in the molecules.

systemic circulation is rapid. Maximum concentrations in plasma are achieved within 14–120 min.[21] Plasma protein binding is low (<20%), but distribution is limited largely to extracellular fluids (plasma+interstitial fluid). Aminoglycosides penetrate poorly into cells, transcellular fluids, and milk, but urine concentrations are high, as reabsorption by passive diffusion into the systemic circulation is very limited. However, they bind to brushborder vesicles and cell membrane phospholipids in cells of the proximal convoluted tubule, probably through their free amino groups. The consequences are twofold: (1) they are nephrotoxic, and this is most significant for the most ionized compounds (e.g., neomycin), which exhibit the greatest binding affinity; and (2) binding is firm, with little or no reabsorption into peritubular capillaries. Hence, concentrations in the kidney cortex in excess of MRL persist for months rather than weeks. As enzymes capable of metabolizing AMDs are located intracellularly in the liver, kidney, and enterocytes, aminoglycosides are excreted almost entirely unchanged.

Papich and Riviere report marked variability in aminoglycoside pharmacokinetics (distribution, clearance, and half-life) with altered physiologic or pathologic states, including pregnancy, obesity, dehydration, immaturity, sepsis, endotoxemia, and renal disease.[21] The latter influence is predictable from the fact that body clearance is dependent almost entirely on renal excretion. Martin-Jimenez and Riviere concluded that aminoglycoside pharmacokinetics can be predicted across species by population pharmacokinetic modeling.[22]

The volume of distribution of aminoglycosides is increased in young calves relative to adults, as a consequence of high extracellular water volume relative to body weight, because volume of distribution is proportional to plasma volume. Volumes of distribution are lower in obese animals, as aminoglycosides penetrate very poorly into adipose tissue. Overall, volume of distribution ($V_{d,\text{area}}$) is of the order of 0.15–0.45 l/kg, and the clinically relevant terminal half-life (β phase) is 1.0–2.0 h.

For some aminoglycosides, it is necessary to note, from both pharmacokinetic and residue perspectives, that they comprise a mixture of compounds. Gentamicin, for example, is a mixture of four compounds, C_1, C_{1a}, C_2, and C_{2a}; residues are usually determined as the sum of these compounds. For neomycin, the residues are determined as neomycin B and for kanamycin, as kanamycin A.

The disposition of aminoglycosides is generally best described by a three-compartment model; the α, β, and γ phases represent, respectively, distribution half-life, the clinically relevant decay phase (which dictates dose schedules for therapy), and a final slow-release elimination of drug sequestered in tissues, particularly renal cortex and liver. The γ (classified above λ_3) phase determines residue depletion profiles. The α phase is rapid (≤60 min), the β phase is also generally short (≤5 h), but the γ phase is much longer, in farm animal species ranging for gentamicin from 11.0 h in the pig, 44.9 h in cattle, to 142 h in the horse.[21] Most of the administered dose is actually eliminated during the short β (classified above λ_2) phase that correlates well with the glomerular filtration rate (GFR) as there is virtually no reabsorption and no tubular secretion in the mammalian nephron. As GFR does *not* increase in proportion to body weight, elimination half-life tends to be longer in larger animals, and clearance decreases as body weight increases.[22] The latter authors also demonstrated dependence of the slow γ phase on dose and administration route, with considerable differences also in findings from different laboratories for a given species.

The prolonged persistence of aminoglycosides in renal cortical tissue increases the possibility of non-compliant tissue residues. This problem is compounded by the inter-animal variability in depletion from renal tissue, so that an original proposed WhT for gentamicin of 18 months for adult cattle was followed by a proposal to avoid usage altogether. A WhT of 40 days has been proposed for neonatal piglets.[21] As well as in the kidney, high concentrations of aminoglycosides occur in the liver.[23,24]

Gentamicin is the most frequently studied aminoglycoside and may be taken as an example for the group. Reported bioavailability after intramuscular dosing is 93% in cattle, 87% in horses, and 60% in catfish, and is similar for different muscle sites. It is also similar, but with slower absorption, after subcutaneous dosing, while oral bioavailability is virtually 0%. Oukessou and Toutain reported lower clearance and volume of distribution and higher plasma AUC values in sheep administered a low-protein diet, compared to those receiving a high-protein diet.[25] Pharmacokinetic parameters and variables for a wide range of species and doses are given by Papich and Riviere.[21] Although 90% of administered drug is recovered from urine within 24 h in calves and adult cattle, drug residues nevertheless persist in renal cortical tissue. On the other hand, drug concentrations in skeletal muscle are low. Other aminoglycosides used in farm animal medicine are apramycin, used in feed for treatment of porcine colibacillosis, and dihydrostreptomycin, which is used in combination with penicillins, notably procaine benzylpenicillin, in parenteral formulations.

Spectinomycin is an aminocyclitol with physicochemical properties similar to those of the aminoglycosides; it is polar and poorly lipid-soluble, but does not contain aminosugars or glycosidic bonds. Unlike aminoglycosides, it is not renotoxic. Oral bioavailability is low (<10%), and volume of distribution is small. The terminal half-life in cattle is short (1.2–2.0 h) after intravenous and intramuscular dosing. It is used mainly in pigs and poultry as a powder for solution in drinking water or as a feed additive, and in cattle, poultry, and pigs by intramuscular injection.

2.3.2 β-Lactams: Penicillins and Cephalosporins

The presence of a carboxylic acid grouping confers on all β-lactams a moderate to strongly acidic character. Benzylpenicillin, for example ($pK_a = 2.7$), is for all practical purposes wholly ionized at the pH of all body fluids, with the exception of gastric juice in monogastric species. The ionised: un-ionised molecule ratio exceeds 50,000 : 1 at a blood pH of 7.4. Therefore, most β-lactams do not readily cross cell membranes, so that intra- and transcellular fluid concentrations are low relative to plasma. Absorption from the GIT varies between drugs, with bioavailability of the order of 1–2% for benzylpenicillin, which is unstable in aqueous solution, especially at extremes of pH (e.g., in gastric juice). Improved bioavailability after oral dosing occurs for phenoxymethyl penicillin, ampicillin, and amoxicillin, in ascending order. This relates to the greater stability of aminopenicillins in acid media. The aminopenicillins also contain a basic amino group and are therefore amphoteric. There is also some likely species variability in oral bioavailability with low values (5–10%) quoted for amoxicillin in the horse, compared to 64–77% in the dog and 23% in the pig.[26] Absorption is improved in species such as the horse in products containing esters such as pivampicillin, but these compounds are not used in food-producing species. Aminopenicillins are poorly absorbed in pre-ruminant calves, and bioavailability is even less in ruminating calves. Many cephalosporins also have moderate to good bioavailability after oral administration in monogastric species.

When β-lactams are formulated as aqueous solutions for parenteral (intramuscular or subcutaneous) use, they are rapidly absorbed to achieve maximum concentrations in 0.25–1 h. In many cases, therefore, the plasma concentration–time profile is very similar for intravenous and intramuscular routes.

The distribution and elimination of β-lactams is determined largely by their polar and generally non-lipophilic nature. Although individual drugs may be exceptional, the general rule is that β-lactams do not readily penetrate to intracellular sites of enzymes capable of metabolizing AMDs, so that they are excreted mainly as parent drug. Limited metabolism involves opening of the unstable, four-membered β-lactam ring to form, for example, amoxicilloic acid from amoxicillin. In the kidney the unbound fraction of parent drug is filtered at the glomerulus and excreted in urine with minimal absorption, so that by this mechanism alone the urine : plasma concentration ratio may be of the order of 100 : 1 (for free plasma drug concentration). However, many penicillins and cephalosporins are substrates for transporters in the proximal convoluted tubule, which promote the active secretion of specific organic acids from the peritubular capillaries into tubular lumen fluid. Therefore, from the combined effect of glomerular ultrafiltration and tubular secretion, the urine : plasma concentration ratios can be as high as 400 : 1.

For most penicillins in most species, binding to plasma protein ranges from low (30%) to moderate (60%). The clearance of most penicillins (non-protein-bound fraction) thus exceeds GFR and confers on most drugs rapid clearance and short terminal half-lives, of the order of 0.6–2.0 h, regardless of species. In addition, volumes of distribution (0.15–0.40 l/kg) approximate to extracellular fluid volume. Papich and Riviere provide an excellent summary of the many publications in this field.[27] Likewise, for cephalosporins, volumes of distribution generally approximate to extracellular fluid volume, and terminal half-lives are $\leq$2.0 h. An interesting exception is cefovecin, the half-life of which is 7 days in cats and 5 days in dogs. These high values are attributable mainly to a very high degree of binding to plasma protein, greatly limiting ultrafiltration at the glomerulus and presumably a lack of proximal tubular secretion, thus limiting renal excretion. Not surprisingly, this drug is not licensed for use in food-producing species. If its pharmacokinetic properties were similar to those in the dog and cat, the clearance from tissues would be unduly protracted.

It has been commonly assumed that, because of their lipophobic character and excretion in high concentrations in urine, β-lactams are metabolized slightly or not at all. This assumption is contraindicated (for amoxicillin) by the identification of metabolites, amoxicilloic acid, and amoxicillinpiperazine-2,5-lione, in tissues of pigs, after medication of drinking water with amoxicillin.[28] An important problem with frequently reported bio-analytical methods for amoxicillin is the use of a derivatization step during sample pre-treatment. Most derivatization procedures lead to the same reaction product for both amoxicillin and its amoxicilloic acid metabolite, with identical relative retention times during chromatography. This might result in an overestimation of the actual amoxicillin residue concentration.[29]

The pharmacokinetic profiles of β-lactams dictate tissue depletion profiles. Concentrations are generally high in the kidney, very low in fat, and also low in muscle. For example, in pigs, Martinez-Larrañaga et al. reported concentrations (mg/kg) of amoxicillin of 23.6 (muscle), 24.7 (fat), 49.1 (liver), and 559.7 (kidney) 2 days after oral dosing of 20 mg/kg orally for 5 days.[30] In broiler chickens administered amoxicillin in drinking water daily for 5 days, tissue concentrations (μg/kg) 1 h or less after final doses were 138 (muscle), 108 (fat), 484 (skin and fat), 2178 (liver), and 4363 (kidney).[31] For fat, the poor uptake is explained by both low bloodflow and the lipophobic character of the drugs. For muscle, the low concentration is related to poor intracellular penetration and hence restriction to the extracellular fraction of the tissue.

In food-producing species, there are compelling economic and welfare reasons for minimizing the number of AMD administrations in dosage regimens. The ideal is to achieve bacteriological and clinical cures with a single dose. For β-lactams, which are classified as time-dependent in their killing actions against most susceptible organisms, there is the additional requirement to maintain plasma drug concentrations in excess of MIC for at least half and indeed possibly for the whole of the inter-dosing interval, namely, not allowing concentrations to decrease below MIC until bacteriological cure is achieved. To attain this goal with intravenous dosing of β-lactams is, under clinical conditions, wholly impractical. One solution has been to use, instead of water-soluble sodium and potassium salts of penicillins, less soluble organic salts, such as procaine, benethamine, and benzathine benzylpenicillins.

Benzathine salts have particularly low aqueous solubility and, when injected intramuscularly, form a depot from which dissolution occurs slowly. Indeed, all benzathine benzylpenicillin salts have been banned from use in the food-producing animals in the EU, because of persistence at and erratic rate of depletion from injection sites and a consequent perceived hazard to human health. Procaine salts, on the other hand, are somewhat more water-soluble and remain in widespread use, formulated as aqueous or oily suspensions. These formulations provide flip-flop pharmacokinetics with terminal half-lives in the range of 8.9–17.0 h after intramuscular or subcutaneous dosing to calves and adult cattle. In some studies, absorption rate (reflected by terminal half-life) was slower after subcutaneous dosing compared to intramuscular dosing. There are reports, indicating differences in absorption rate for different muscle groups, that absorption from neck muscle is slower than that from gluteal muscle.[32,33] These formulations are designed for once-daily dosing regimens. Other depot formulations have been developed, for example, using aqueous suspensions of ampicillin and amoxicillin trihydrates.

Another major clinical use of β-lactams is for intramammary treatment of bovine mastitis, in both lactating and dry cows. The lactating cow (see Table 2.3 for AMD milk : plasma concentrations in lactating cows) products are administered in rapid-release formulations to achieve milk concentrations often greatly in excess of MIC_{90} values for susceptible bacteria. These products are usually rapidly cleared from the gland after two or three infusions, providing short milk withholding periods. The dry-cow formulations are administered at dry off in fixed oil formulations and sometimes containing water repellent agents, such as aluminum monostearate, to prolong presence in the gland for most or all of the dry period.

2.3.3 Quinoxalines: Carbadox and Olaquindox

Carbadox has been used as a feed additive in pigs, as a growth promoter, and therapeutically for the control of swine dysentery, enteritis, and nasal infections. Both drugs are absorbed after oral dosing, but published information on their pharmacokinetic profiles is limited. The major residue metabolite of carbadox is quinoxaline-2-carboxylic acid. After feeding carbadox to the pig (50 ppm) as a growth promoter, residues in liver and kidney exceeded 30 μg/kg for 4–5 weeks and 10 μg/kg at 62 days.[34] Carbadox is both mutagenic and carcinogenic, while olaquindox is mutagenic but probably not a carcinogen. While there is concern regarding the safety of residues, there is evidence to indicate that the residues do not possess mutagenic or carcinogenic activity. It has been suggested that any risk is likely to be to individuals handling products containing the drugs.[35] Carbadox and olaquindox were withdrawn as feed additives in the EU in 1999. In the United States, the marker residue for carbadox is quinoxaline-2-carboxylic acid and the tolerance for pig liver is 30 μg/kg. In Australia, the MRL for olaquindox in pig and poultry meat has been set at 300 μg/kg.

Anadón et al. described the residue pharmacokinetics of olaquindox in broiler chickens (Table 2.4).[36] Absorption was rapid ($T_{max} = 0.22$ h) and terminal half-life was 5.13 h. Tissue depletion rates of olaquindox illustrate well the general principles for AMDs that (1) depletion rates are tissue-dependent and (2) peak concentrations (in this case in kidney) are not necessarily at the first slaughter time.

2.3.4 Lincosamides and Pleuromutilins

Three members of the lincosamide class, lincomycin, pirlimycin, and clindamycin, and two drugs in the pleuromutilin group, tiamulin and valnemulin, are used in veterinary medicine. The binding sites are similar to those of macrolides and, like macrolides, are lipophilic weak organic bases. As predictable from their weakly basic character, they achieve high concentrations in milk and intracellular fluid, and therefore tissue concentrations generally exceed those in plasma and interstitial fluid.

Lincomycin is formulated as a pre-mix and as a soluble powder for addition to drinking water for use in poultry and pigs and is also available in a parenteral formulation for the pig. Oral administration is contraindicated in all ruminants because of the risk of bacterial overgrowth with *Clostridium* species. However, lincomycin is licensed for use parenterally in calves as a combination product

TABLE 2.3 Milk : Plasma Concentration Ratios of AMDs in Lactating Cows[a]

Drug	Drug Class	Lipid Solubility	pK_a	Milk Ultrafiltrate : Plasma Ultrafiltrate	
				Theoretical Ratio	Experimental Ratio
Bases					
Trimethoprim	2 : 4 diaminopyrimidine	High	7.3	2.32 : 1	2.90 : 1
Spiramycin	Macrolide	High	8.2	3.57 : 1	4.60 : 1
Dihydrostreptomycin	Aminoglycoside	Low	7.8	3.13 : 1	0.50 : 1
Polymixin B	Polymyxin	Very low	10	3.97 : 1	0.30 : 1
Acids					
Benzylpenicillin	Penicillin	Low	2.7	0.25 : 1	0.13 : 1–0.26 : 1
Sulfadimethoxine	Sulfonamide	Moderate/high	6	0.20 : 1	0.23 : 1
Sulfamethazine	Sulfonamide	Moderate/high	7.4	0.58 : 1	0.59 : 1

[a]Note the poor penetration into milk of strong or polar bases, strong acids, and weak acids and good penetration of weak bases, except streptomycin, which is very polar as a result of the presence of sugar residues in the molecule.

Source: Adapted from Baggot et al. (2006).[152]

TABLE 2.4 Tissue Concentrations of Olaquindox after Oral Administration in Chickens (Mean ± SEM, $n = 6$)[a]

	Drug Concentration (mg/kg)				
Tissue	Day 1	Day 3	Day 6	Day 8	Day 14
Muscle	3.33 ± 0.84	1.69 ± 0.51	0.38 ± 0.08	0.18 ± 0.04	0.03 ± 0.01
Liver	3.69 ± 0.50	2.93 ± 0.38	1.49 ± 0.33	0.88 ± 0.22	0.11 ± 0.01
Kidney	1.43 ± 0.23	2.23 ± 0.65	1.92 ± 0.28	1.34 ± 0.18	0.12 ± 0.01

[a]Administered orally directly into the crop.
Source: Data from Anadón et al (1990).[36]

with spectinomycin for treatment of lung infections. The combination has also been used in sheep, goats, and poultry. In the pig, absorption from the GIT is rapid but bioavailability is limited, in the range of 20–50%. Lincomycin is well distributed into tissues, with relatively high concentrations obtained in liver and kidney. Muscle and skin concentrations, on the other hand, are low. Elimination is primarily through hepatic metabolism and approximately 20% of the administered dose is excreted in urine as parent drug. Diffusion trapping occurs in fluids and tissues, such as milk and the prostate, which are acidic relative to plasma. Volume of distribution is in the range of 1.0–1.3 l/kg.

In the chicken, after 7 days of oral dosing, feces and urine (combined) contained approximately 80% parent drug, 10% sulfoxide metabolite, and 5% *N*-demethyllincomycin.[37] The same authors reported excretion of 11–21% of the administered dose (half as parent drug) in urine. The remainder was excreted in feces, of which 17% was parent drug and 83% as uncharacterized metabolites.

Pirlimycin is used solely by intramammary infusion for the treatment of mastitis in lactating cattle.

Tiamulin is formulated as the base for parenteral use and as the hydrogen fumarate for oral use in drinking water and pre-mix soluble formulations. Valnemulin is also formulated as a pre-mix, as the hydrochloride salt. They are used in the pig against *Mycoplasma* lung infections and swine dysentery, in poultry for both *Mycoplasma* and *Brachyspira* infections, and to a lesser degree in treatment of calf pneumonia. The absorption of tiamulin is also high when administered orally as a bolus dose, but bioavailability is said to be lower from pre-mix formulations.[38] In calves the half-life is short (25 min), and after oral dosing absorption is rapid in pre-ruminant calves. Pleuromutilins are not used in calves with functional rumens. Concentrations in milk and lung tissue exceed those in plasma severalfold.

2.3.5 Macrolides, Triamilides, and Azalides

Drugs in this class include erythromycin, tylosin, spiramycin, tylvalosin, carbomycin, oleandomycin, tilmicosin (all macrolides), azithromycin (an azalide), and tulathromycin (a triamilide). The latter drug is a regioisomeric mixture of 13-membered (10%) and 15-membered (90%) ring compounds, while erythromycin is a mixture of three related compounds, named A, B, and C. Several drugs of this class are in widespread use in food-producing species. Carbomycin, oleandomycin, and tylosin have been used as pre-mixes for addition to the feed of poultry, pigs, and cattle, either as growth promoters or for disease prophylaxis and treatment. Tylosin and tulathromycin are used in parenteral formulations as therapeutic agents for calves and pigs, and tilmicosin is used in cattle only.

The weakly basic nature (pK_a 6–9) of macrolides results in partial ionisation at physiological pHs, but the un-ionized fractions possess moderate to high lipid solubility, so that they are well absorbed orally (except for erythromycin base) and readily penetrate into intra- and transcellular fluids. However, absorption may be impaired by feed. Erythromycin base is unstable in acid gastric juice and is therefore administered either in enteric-coated formulations or as estolate or ethylsuccinate esters or the stearate salt. These esters and the salt have improved bioavailability; the esters are hydrolysed after absorption. Their weakly basic character results in diffusion trapping in acidic fluids, such as milk (Tables 2.2 and 2.3). Volumes of distribution generally exceed body water volume, sometimes by a considerable amount. For example, the reported distribution volume of tylosin in calves ranging in age from 2 to >6 weeks is 9–11 l/kg.[39] Volumes of distribution of 20 and 11 l/kg have been reported, respectively, for azithromycin and tulathromycin. A characteristic of macrolide distribution is a strong tendency to concentrate intracellularly in some tissues, notably the lung and lung macrophages. This is probably attributable to their basic nature, reflecting the Henderson–Hasselbalch diffusion trapping mechanism and being due to the acidic pH in the cell phagolysosome. Plasma protein binding is relatively low, of the order of 18–30%.

The terminal half-life of erythromycin A in calves and adult cows is relatively short (2.9–4.1 h) after intravenous administration, but much longer after intramuscular (11.9 h) or subcutaneous (18.3 to 26.9 h) dosing, as a consequence of flip-flop pharmacokinetics of commercially available formulations, that is, of a very slow process of drug absorption from its injection site. Concentrations in tissue (liver and kidney) and fluid (bile and prostatic) exceed those in plasma. Erythromycin is metabolized by demethylation in the liver by microsomal enzymes. Some 90% of the administered dose is excreted in bile, mainly as metabolites. No more than 5% is excreted in urine as the parent molecule.

After intravenous dosing, the half-life of tylosin is short in all species, 1.1, 2.1, 3.0, and 4.0 h, respectively, in calves, sheep, goats, and pigs.[40] Tylosin penetrates readily into milk and is slowly cleared from the mammary gland, so that its use in lactating cattle is not recommended. In fact, this property extends to other macrolides, due to their basic nature and lipophilic properties. For example, the half-life of tilmicosin in cows is approximately 1 h, but concentrations in milk exceed 0.8 mg/l for 8–9 days after a single subcutaneous dose of 10 mg/kg.

Tulathromycin is administered intramuscularly in pigs and subcutaneously in cattle. In both species bioavailability is of the order of 90% and volume of distribution is 12 l/kg. Tylvalosin, tilmicosin, and to an even greater extent tulathromycin among the macrolides achieve high concentrations in lung tissue. For the latter drug, lung : plasma concentrations exceeding 100 : 1 have been reported in the calf and the pig. The terminal half-life for lung tissue exceeds that of plasma; values for calves and pigs are 184 and 142 h, respectively, compared to serum half-lives of 90 and 76 h. Several macrolides, azalides, and triamilides have been shown to achieve high concentrations in leucocytes and off-loading of drug from polymorphonuclear neutrophils (PMNs) has been shown *in vitro* and proposed as a mechanism of delivery drug *in vivo* to the biophase. In fact, considering the rate of antibiotic efflux from PMN (rather slow), the total body pool of PMN (small relative to body weight), and using mass balance considerations, it is unlikely that neutrophils migrating preferentially to sites of infection provide a delivery mechanism for AMDs able to maintain, dynamically, a high local antibiotic concentration in the biophase that is extracellular water. In addition, using the microdialysis technique, it was shown that an acute inflammatory event seems to have little influence on tissue penetration. As quoted by Muller et al., "these observations are in clear contrast to reports on the increase in the target site availability of antibiotics by macrophage drug uptake and the preferential release of antibiotics at the target site, a concept which is also used as a marketing strategy by the drug industry."[41] The terminal half-life of tilmicosin is 1 h in the cow and 25 h in the pig.

2.3.6 Nitrofurans

In many countries, including the EU member states and the United States, the use of nitrofurans and furazolidone has been banned in food-producing animals, as they are genotoxic and furazolidone is carcinogenic. Therefore, from a residue perspective the interest lies in their illegal use. They are lipid-soluble weak acids, well absorbed orally, and bioavailability is enhanced when administered with feed. Some 50% of administered dose is metabolized, and the remainder is excreted in urine. In acid environments they are un-ionized, so that acidification of the urine promotes renal reabsorption and alkalinization enhances excretion.

2.3.7 Nitroimidazoles

Nitroimidazoles are antibacterial and antiprotozoal drugs, of moderate to high lipid solubility, and high bioavailability in monogastric species. The principal members of the group are metronidazole, tinidazole, ronidazole, and dimetridazole. They were used formerly in poultry and game birds to treat histomoniasis and *Spironucleus* infections and also swine dysentery in pigs. They have been classified as suspect mutagens and carcinogens,[42,43] and all the compounds in the group, except metronidazole and tinidazole have been removed from the market. All nitroimidazoles have been prohibited from use in food-producing animals in the United States and EU, where they are placed in Annex IV[44] (see Section 2.5.3).

2.3.8 Phenicols

The phenicol group of AMDs includes chloramphenicol, florfenicol, and thiamphenicol. All phenicols are relatively small organic molecules, containing neither acidic nor basic groups, and all are very lipid-soluble.

The original member of the group, chloramphenicol, has been used therapeutically for more than 60 years, and its pharmacokinetics have been studied extensively in many species, including food-producing animals. However, its toxicity profile includes a very rare but fatal form of aplastic anaemia in humans (incidence 1 : 10,000–1 : 45,000), which is not concentration-dependent. Therefore, in the United States, EU, Canada, Australia, and indeed in most jurisdictions, chloramphenicol is classified as a drug with a risk to public health and its use has been banned in food-producing animals. However, from an EU perspective, chloramphenicol continues to be used legally or illegally in some countries, and controls are still required, especially for imports of animals and their products from third-world countries (honey, crab meat, etc.). For chloramphenicol (as for nitrofurans), which has been expressly prohibited from use in food-producing animals in the EU, the concept of minimum required performance limit (MRPL) has been established in the Commission Decision 2002/657/EC.[45] MRPLs are defined as "minimum content of an analyte in a sample, which at least has to be detected and confirmed" and are the reference points for action in relation to the evaluation of consignments of food. To date, MRPLs have been established for chloramphenicol of 0.3 and 1 μg/kg for nitrofurans.[46]

After intravenous dosing, chloramphenicol clearance in ruminant species is rapid and the terminal half-life is short: 1.7 h in sheep, 1.2–4.0 h in goats (the longer time $T_{1/2}$ in goats is after a period of food deprivation), and 2.5–7.6 h in calves. The longer time period is seen in young calves (7.6 h at 1 day and 4.0 h at 14 days) than in animals aged 9 months (2.5 h).[40] The half-life is also longer in piglets (12.7–17.2 h) than in adult pigs (1.3 h). In piglets the shorter half-life (12.7 h) was obtained in colostrum-fed animals, and the longer half-life (17.2 h) occurred in colostrum-deprived piglets. In the chicken, half-life was much longer in *E. coli*–infected animals than in healthy animals: 26.2 h compared to 8.3 h.[40]

Chloramphenicol is well absorbed after oral dosing in ruminants but is rapidly inactivated by ruminal microflora, so that bioavailability is extremely low. Its distribution in the body is widespread; as predicted from its lipid-soluble character, it readily penetrates into intracellular and transcellular fluids and readily diffuses into milk. Plasma protein binding is of the order of 30–45%. The volume of distribution is of the order of 1.0–2.5 l/kg. Urinary excretion in calves is minimal. Elimination is attributable primarily to metabolism in the liver, for example, by hydrolysis (phase I) and by glucuronidation (phase II) reactions. A range of metabolites has been identified, including dehydrochloramphenicol, nitrophenylaminopropanedione, and nitrosochloramphenicol.[47] These authors reported that the latter two compounds were still detectable in kidney, liver, and muscle of chickens at 12 days post-slaughter after oral dosing with chloramphenicol at 50 mg/kg daily for 4 days. Its rapid clearance and short half-life necessitate administration with a short dosing interval.

Thiamphenicol is a semi-synthetic derivative of chloramphenicol. It can cause reversible bone marrow depression, but fatal aplastic anemia has not been reported in humans. Oral bioavailability in pre-ruminant calves is 60%. It is somewhat less lipid- and somewhat more water-soluble than chloramphenicol and therefore crosses cell membranes less readily. Hepatic metabolism is limited, and elimination is primarily as parent drug in the urine. Limited published data indicate that it has a high distribution volume in ruminants. It has been used "in feed" in pigs and chickens, but such usage is now limited.

As a successor to chloramphenicol, florfenicol is now used extensively in food-producing species, particularly calves, chickens, and young pigs. It lacks the *para*-nitro group of chloramphenicol, which seems to be an essential molecular feature for causing aplastic anaemia. Therefore, there is no public health risk relating to aplastic anaemia arising from the use of florfenicol.

Florfenicol, like chloramphenicol, is very lipid soluble and is well absorbed in calves after oral dosing (bioavailability of 79–89%) but with some reduction in bioavailability when administered with milk. Bioavailability is also high from intramuscular and subcutaneous injection sites. The terminal half-life after intravenous dosing in calves is short (2.7–3.7 h), but as a consequence of slow absorption and flip-flop pharmacokinetics it is much longer (18 h) after intramuscular dosing. The clinically recommended intramuscular dose is 20 mg/kg. When florfenicol is administered to calves subcutaneously at the higher dose rate of 40 mg/kg, terminal half-life is even longer, so that effective therapy can often be achieved with single-dose administration. In the fish species, red pacu and salmon, half-lives were 4.3 and 12.2 h, respectively. The latter value was determined at 10.8°C. In rainbow trout, the mean residence time at 10°C was 21 h.

Florfenicol is widely distributed, achieving high concentrations in muscle, kidney, urine, milk, bile, and small intestine, but with lesser penetration of the blood–brain barrier than chloramphenicol. Volume of distribution in calves is similar to body water volume (0.67–0.91 l/kg), and binding to plasma protein is low (13–19%). Approximately two-thirds of the administered dose is excreted in calf urine as parent drug. The biologically inactive metabolite, florfenicol amine, is eliminated more slowly than parent drug and is used as the marker residue with the liver as the target tissue in some jurisdictions. For example, Anadón et al. recorded highest residue concentrations in the chicken in liver, with lower and similar residue depletion profiles in kidney, muscle, and skin plus fat.[48] For florfenicol amine, residue depletion profiles were similar for kidney and liver, with much lower concentrations in muscle and skin plus fat. In the EU the marker residue is the sum of florfenicol and all metabolites expressed as florfenicol amine.

Florfenicol is available in a range of formulations: in two strengths for parenteral administration in pigs and cattle, as a solution for addition to drinking water in pigs, and as pre-mixes for incorporation into feed for pigs and fish.

2.3.9 Polyether Antibiotic Ionophores

This is a unique class of compounds with high potency against a range of critical infectious disease targets, including protozoa, bacteria, and viruses. The principal drugs in this class are lasalocid, maduramicin, monensin, narasin, semduramicin, and salinonycin. All are coccidiostats, with widespread use in poultry. As a consequence of species-based toxicity, they are not used in horses and guinea fowl, and salinomycin and narasin are not used in turkeys. For some ionophores, toxicities may be exacerbated when administered in combination with erythromycin, tiamulin, pleuromutilins, sulfonamides, and chloramphenicol, as a consequence of inhibition of ionophore metabolism.

All polyether ionophores are formulated for use in chickens in feed for the prevention of coccidiosis. Some are also licensed for use, in feed, as coccidiostats in goats (monensin), cattle (lasalocid, monensin), sheep (lasalocid), rabbits (lasalocid), turkeys (lasalocid, monensin), chukar partridges (lasalocid), and bobwhite quail (monensin, salinomycin). For individual drugs there are various restrictions on use, including narasin, for use in broiler chickens only, and monensin, which is not for use in goats producing milk for human consumption. Some ionophores are added to animal feeds as growth promoters for use in pigs and/or cattle. However, such use has been disallowed in the EU since 2006. Monensin is also licensed for improved milk production in cattle. It might be noted that there is a possibility of carryover of drugs of this class from non-target animal feed, which might give rise to residues in animal products for which no MRLs are set.[49,50]

Published data on the pharmacokinetics of ionophores are limited. Dowling reports high bioavailability in monogastric species and approximately 50% in ruminants.[51] Most ionophores are metabolized extensively in the liver, forming many metabolites that are secreted in bile and excreted in feces.

2.3.10 Polypeptides

The drug groups in this general category are polymyxins (see Section 2.3.13), glycopeptides, bacitracin, and streptogramins. The principal members of the glycopeptide group are vancomycin, teicoplanin, and avoparcin. The former two drugs are used in human therapeutics, and avoparcin has been used extensively as a growth promoter in poultry and pigs, particularly in the EU. However, it has been withdrawn from use in the EU, because of selection for vancomycin-resistant enterococci (VRE) in farm animals, which may potentially transfer resistance to human pathogens. This is of concern because vancomycin is a drug of last resort for serious human infections caused by drug-resistant Gram-positive bacteria. It might be noted that VRE cause significant problems in human hospitals in North America, where avoparcin has never been used in animals. In Australia avoparcin retains an MRL listing. Vancomycin has been used, by intravenous infusion, in horses and dogs but not in farm animal species. Its use in food-producing animals was banned in the United States in 1997.

Vancomycin is a high-molecular-weight polypeptide and teicoplanin is similar in structure and, in fact, is a complex of five related compounds. Both are poorly lipid-soluble. For both compounds, absorption after oral dosing is slight/absent, a property relating to the low lipid solubility and polypeptide structure, as they are broken down to constituent amino acids in the GIT. Teicoplanin is well absorbed after intramuscular dosing, and has a prolonged half-life in humans of 45–70 h. Vancomycin is too irritant for intramuscular administration. It is, therefore, administered intravenously in humans, where the terminal half-life is 6 h. Both drugs are poorly distributed, restricted primarily to extracellular fluids, and excreted largely unchanged by glomerular ultrafiltration.

Bacitracin has been administered orally as a growth promoter in poultry and pigs (although this use is no longer permitted in the EU) and as a therapeutic for enteritis, although it is not effective in swine dysentery. Absorption from the GIT is very low, which is fortunate as bacitracin is nephrotoxic.

Streptogramins are natural (e.g., virginiamicin) or semisynthetic (e.g., quinupristin/dalfopristin) cyclic peptides. Virginiamycin has been used as a growth promoter. It is the only member of the group used in animals and is a mixture of two compounds, virginiamycin S (a cyclic hexadepsipeptide, the minor component) and virginiamycin M (a macrolactone, the major component). The use of virginiamycin as a feed additive in pigs and poultry can result in the selection of resistance in fecal enterococci with cross-resistance to quinuspristin/dalfopristin, which has been used in human medicine to treat VRE infections. Virginiamycin is poorly absorbed after oral dosing and is excreted in bile. It was banned as a growth promoter for pigs in the EU in 2009 but is still used for this purpose in some countries. It has also been used as a therapeutic agent in swine dysentery and laminitis in horses.

2.3.11 Quinolones

First-generation quinolones were nalidixic and oxolinic acids. The latter is still used therapeutically in fish, but otherwise these drugs have been superseded by the fluoroquinolone sub-group. The principal fluoroquinolones used in food-producing species are danofloxacin, enrofloxacin, flumequine, marbofloxacin, and sarafloxacin. They contain both carboxylic acidic and basic amino groups and are therefore amphoteric; pK values for the former are 5.5–6.5 and for the latter 7.5–9.3, so that at physiological pH they exist as zwitterions (partially ionised, partially un-ionized for each group). The drugs are most lipophilic at the isoelectric point, which is close to blood pH. Lipophilicity varies between drugs but is always moderate (ciprofloxacin, marbofloxacin) or high (enrofloxacin). Two members of the group, enrofloxacin and sarofloxacin, were formerly used in poultry but have now been banned in the United States and Australia because of concerns about *Campylobacter* and *Salmonella* resistance. Many other fluoroquinolones are used extensively in human medicine.

The pharmacokinetic profiles of danofloxacin, enrofloxacin, flumequine, and marbofloxacin in food-producing species have been studied extensively.[52] Bioavailability for all drugs in all species is very high after intramuscular dosing. Some studies in calves have demonstrated flip-flop pharmacokinetics after intramuscular and subcutaneous dosing. Binding to plasma protein is relatively low to moderate and varies with species for enrofloxacin; it is low in the pig (27%) and chicken (21%) and moderate in cattle (36–60%). Volumes of distribution are of the order of 1.0–4.0 l/kg, that is, exceeding total body water volume, and elimination half-lives are in the range 2.0–8.0 h in cattle, sheep, goats, and pigs. Shorter half-lives in rabbits (1.8–2.5 h for enrofloxacin) and longer half-lives in fish (24 h in trout and 131.0 h in Atlantic salmon, both for enrofloxacin) and chickens (5.6–14.0 h for enrofloxacin) have been reported. Half-lives are also longer in reptiles (36 and 55 h for enrofloxacin in monitor lizards and alligators, respectively). For free plasma concentration of enrofloxacin, an allometric relationship seems to apply between volume of distribution and body weight, with a direct proportionality, specifically, larger distribution volumes in heavier animals.[52]

For pigs and ruminant species, dosing is generally by intramuscular injection, with once-daily dosing schedules. For example, Anadón et al.[53] reported a bioavailability of 74.5% after intramuscular dosing of enrofloxacin (2.5 mg/kg) in pigs, and tissue residue concentrations of enrofloxacin and the ciprofloxacin metabolite (mg/kg) at 5 days were 0.03, 0.08 (fat), 0.06, 0.04 (kidney), 0.06, 0.02 (liver), and 0.06, <0.003 (muscle). Products of higher strength in depot formulations have been developed for danofloxacin, enrofloxacin, and marbofloxacin and these provide, after intramuscular or subcutaneous injection, therapeutic levels for 48 h or longer; they are commonly administered as single doses. For most drugs, in most parenteral formulations, bioavailability is in the range of 75–100%.

Oral bioavailability of fluroquinolones is high in both presence and absence of feed, in most monogastric species, including the pig, but in ruminants, this administration route is not used. Nevertheless, data in adult sheep suggest good bioavailability (61%), whereas bioavailability from oral dosing was only 10% in ruminant calves. Enrofloxacin is well absorbed after oral dosing in poultry, but its use is not permitted in the EU for animals producing eggs for human consumption. Bioavailability from oral dosing of enrofloxacin in fish is reported to be 40–50%. In chickens selected for fattening, the oral bioavailability of flumequine was 57% after oral dosing.[54] These authors reported highest residue concentrations of flumequine and the metabolite 7-hydroxyflumequine in kidney, followed by liver, with lower concentrations in muscle and skin plus fat. Anadón et al. also described the differing residue depletion profiles of marbofloxacin and its *N*-desmethyl metabolite in broiler chickens.[55] In plasma at day 1 following oral administration, marbofloxacin and its *N*-desmethyl metabolite concentrations were 0.047 ± 0.003 mg/l and 0.032 ± 0.004 mg/l, respectively, but were not

TABLE 2.5 Residues of Marbofloxacin and *N*-Desmethylmarbofloxacin in Edible Tissues of Chickens Following Oral Administration of Marbofloxacin (2 mg/kg, every 24 hours, for 3 days)

Tissue	Days Post-treatment (withholding period)	Marbofloxacin (μg/g)	*N*-Desmethyl-marbofloxacin (μg/g)
Muscle	1	32 ± 3	119 ± 23
	2	18 ± 3	113 ± 23
	3	<LOD	<LOD
	5	<LOD	<LOD
Kidney	1	985 ± 72	499 ± 60
	2	420 ± 48	164 ± 32
	3	40 ± 4	69 ± 13
	5	7 ± 2	21.7 ± 4.9
Liver	1	735 ± 45	554 ± 66
	2	343 ± 38	158 ± 30
	3	28 ± 7	99 ± 15
	5	11 ± 2	51 ± 8
Skin plus fat	1	43 ± 6	266 ± 58
	2	10 ± 2	55 ± 10
	3	<LOD	<LOD
	5	<LOD	<LOD

Source: Data from Anadón et al (2002).[55]

detectable on subsequent sampling dates. Residues found in edible tissues are given in Table 2.5.

The distribution of fluoroquinolones into interstitial fluid has been shown to be predictable from free concentrations in plasma.[56] Like the macrolides group of AMDs, fluoroquinolones achieve high concentrations in leukocytes. Concentrations in lung, liver, and kidney are several times higher than those in plasma.

For enrofloxacin, there is an additional consideration in relation to pharmacokinetic and residue profiles, in that it is metabolized in the liver to a microbiologically active metabolite, ciprofloxacin, by a deethylation reaction. In cattle and calf conversion rates, from enrofloxacin to ciprofloxacin, are 25% and 41%, respectively. Residues are measured as the sum of enrofloxacin and ciprofloxacin. In poultry, pigs, and fish, much smaller amounts of ciprofloxacin are formed. Nevertheless, in chickens, ciprofloxacin residues were detectable 12 days after dosing with enrofloxacin.[57] Ciprofloxacin itself is converted to minor metabolites with no antibacterial activity. Nevertheless, metabolites are of residue concern, and tissue depletion profiles were studied in broiler chickens by Anadón et al.[58] The data in Table 2.6 illustrate the rapid conversion of ciprofloxacin to oxociprofloxacin and desethyleneciprofloxacin ($T_{\max} < 1.0$ h), the accumulation

TABLE 2.6 Residue Pharmacokinetics of Ciprofloxacin and Its Metabolites in Broiler Chickens after Oral Dosing of Ciprofloxacin[a]

Variable	Ciprofloxacin			Oxociprofloxacin			Desethyleneciprofloxacin		
	Day 1	Day 5	Day 10	Day 1	Day 5	Day 10	Day 1	Day 5	Day 10
Plasma concentration (mg/kg)	0.14 ± 0.02	N/D	N/D	0.10 ± 0.02	N/D	N/D	0.10 ± 0.02	N/D	N/D
Kidney concentration (mg/kg)	0.74 ± 0.07	0.69 ± 0.06	N/D	1.27 ± 0.13	0.63 ± 0.07	N/D	0.97 ± 0.27	0.23 ± 0.07	N/D
Liver concentration (mg/kg)	0.74 ± 0.22	0.55 ± 0.21	N/D	1.78 ± 0.72	0.75 ± 0.38	N/D	1.28 ± 0.62	0.59 ± 0.47	0.011 ± 0.008
Muscle concentration (mg/kg)	0.37 ± 0.06	0.020 ± 0.008	N/D	0.68 ± 0.20	0.32 ± 0.06	N/D	0.61 ± 0.26	0.35 ± 0.11	N/D
Skin+fat concentration (mg/kg)	0.23 ± 0.11	0.11 ± 0.06	N/D	0.51 ± 0.32	0.026 ± 0.011	N/D	0.95 ± 0.23	0.28 ± 0.08	0.010 ± 0.006
Plasma C_{max} (mg/l)		2.63 ± 0.20			1.73 ± 2.02			1.57 ± 0.14	
Plasma T_{max} (h)		0.36 ± 0.07			0.62 ± 0.08			0.75 ± 0.16	

[a]Administered at a rate of 8 mg/kg for 3 days (mean ± SD, $n = 6$).
[b]N/D not detectable.
Source: Data from Anadón et al (2001).[58]

of parent drug and both metabolites in kidney and liver, and the rates of depletion from all edible tissues.

The principal route of elimination of fluoroquinolones is via the kidney by glomerular filtration and, for some drugs, also by tubular secretion.[59] Smaller amounts are eliminated in feces.

2.3.12 Sulfonamides and Diaminopyrimidines

Sulfonamides are synthetic AMDs based on sulfanilamide as parent compound, which was introduced into medicine in 1935. Subsequently, large numbers of derivatives have been used clinically. In veterinary medicine sulfonamides are now used primarily in combination products containing the 2 : 4 diaminopyrimidines, trimethoprim and ormetoprim. The combinations are synergistic in their antibacterial actions. However, some sulfonamides (e.g., sulfadimethoxine, sulfanquinoxaline, sulfadimidine) are used alone in cattle and poultry as soluble powders or solutions for addition to drinking water or as extended-release tablets for cattle. For the latter, maintenance of therapeutic concentrations for 2–5 days has been claimed after a single dose. Overall, the potency of sulfonamides is low, so that high doses (20–100 mg/kg) are used therapeutically. This imposes a high metabolic load on the body and may saturate metabolic pathways, leading to dose dependence in clearance and terminal half-life.

As weak organic acids ($pK_a = 10.1$ for sulfanilamide, 6.1 for sulfadoxine), at physiological pHs of most body fluids they are mainly un-ionized. The un-ionized moiety is generally lipid-soluble, but this varies (lipophilicity high for sulfisoxazole, low for sulfaguanidine) between drugs. The consequence is that sulfonamides generally readily cross cell membranes and are diffusion/ion-trapped in fluids alkaline to plasma (e.g., intracellular fluid, alkaline urine). On the other hand, they penetrate poorly into fluids more acidic than plasma, such as prostatic fluid and milk (Table 2.3). Sulfonamides of high pK_a are generally the least water-soluble, and solubility in water is greater under alkaline than acidic conditions, so that the potential for precipitation to cause crystalluria and renal damage in acid urines has long been recognized, especially for those drugs of low potency and low water solubility. For sulfonamides of high pK_a, the percentage binding to plasma proteins tends to be low. Protein binding thus ranges from high (90% for sulfadimethoxine in some species) to as low as 15%. Moreover, binding to protein can vary considerably between species. Formerly used extensively were triple sulfonamide formulations, which were additive in their antimicrobial actions but obeyed the law of independent solubilities, enabling the use of lower doses of each drug in the combination.

Diaminopyrimidines are lipid-soluble weak organic bases, partially ionized at physiological pH, which, in contrast to sulfonamides, penetrate readily into cells and are poorly reabsorbed from acid urines.

As weak organic acids, sulfonamides are generally well absorbed after oral dosing in monogastric species, but rate and extent of absorption vary with species, drug (greater bioavailability of more lipid soluble drugs), and feed. For example, in horses the absorption of sulfachlorpyridazine was reduced, delayed, and exhibited two peaks when administered orally in the presence of food.[60] The double-peak phenomenon is likely due to partial binding of drug to feed by adsorption, with initial rapid absorption of an unbound fraction and subsequent absorption of the bound fraction following its release by fermentative digestion in the large intestine. Those sulfonamides with very low lipid solubility (e.g., sulfaguanidine) are excreted unchanged, with only slight absorption, in feces when administered orally. They were formerly widely used to treat GIT infections.

An influence of disease on the absorption of sulfaquinoxaline was established by Williams et al.,[61] who reported a 3.5-fold greater bioavailability in chickens infected with *Escherichia acervalina* and *E. tenella* in comparison with uninfected birds.

Both age and diet may influence sulfonamide absorption in calves. Oral absorption of sulfadiazine was very slow in calves on milk diets, and bioavailability was greater in ruminating than milk-fed pre-ruminant calves.[62] Trimethoprim, on the other hand, was well absorbed in pre-ruminant calves but not in ruminating animals, possibly as a result of inactivation by ruminal microflora.

Sulfonamides are generally well distributed into extra- and transcellular fluids, but penetration into intracellular fluid is poor to moderate, a consequence of their acidic nature and the overall acid pH within cells.

A major metabolic pathway for sulfonamides is acetylation of the amino group in the *N*-4 position of the benzene ring. It occurs primarily in the liver but also in the lung. Acetylation is of interest for several reasons: (1) it generally occurs more rapidly in herbivores than in omnivores and carnivores—acetylated derivatives are the major urinary metabolites in cattle, sheep, and pigs; (2) it is species-dependent, in that it occurs to only a slight extent in chickens and dogs; and (3) the acetylated derivatives are commonly less water-soluble (and especially so in acidic fluids) than parent compounds, potentially leading to the condition of renal crystalluria. Those sulfonamides containing a pyrimidine ring (sulfamethazine, sulfamerazine, sulfadiazine) undergo hydroxylation of a methyl group within the ring. Other metabolic pathways include glucuronidation, sulfate conjugation, aromatic hydroxylation, and deamination. All known metabolites have either much reduced or no antimicrobial activity.

Sulfonamide excretion occurs partly via the parent compounds in urine (most readily if urine pH is alkaline, as in herbivores), but predominantly through the less lipid-soluble and therefore more readily excreted metabolites described above. Some sulfonamides are also excreted via the active carrier-mediated transport system, which secretes organic acids from peritubular capillaries across proximal convoluted tubule cells and into tubular lumen fluid. Acetylated sulfonamides are usually less water-soluble than the parent compounds and are the main cause of the crystalluria that can occur, leading to tubular damage. Only small amounts are excreted in bile and milk.

A summary of the detailed information on the pharmacokinetics of sulfamethazine (also known as *sulfadimidine*) and sulfadiazine is provided by Papich and Riviere.[63] Volumes of distribution of these drugs are low to moderate in most species (0.24–0.90 l/kg), but with the buffalo ($V_d = 1.23$ l/kg) and rainbow trout ($V_d = 1.2$ l/kg at 10°C and 0.83 l/kg at 20°C) as exceptions. In cattle, the elimination half-life ranged from 3.6 to 5.9 h with some evidence of age variability.[64] In goats, sulfadimidine half-life was of the same order as that of cattle, but with a longer half-life in fasted adults (7.03 h) than in fed adults (4.75 h).[65] Similar values were reported for ewes, but with a shorter terminal half-life, after oral dosing, with a low (100 mg/kg) compared to a high (391 mg/kg) dose, of 4.3 and 14.3 h, respectively.[66] In pigs, the half-life of sulfadimidine ranged from 11.9 to 20.0 h, with little dependence on age.

For sulfadiazine, Nouws et al.[67] reported long elimination half-lives in carp of 47.1 h at 10°C and 33.0 h at 20°C. However, as for sulfadimine, the elimination half-life of sulfadiazine in calves was in the range of 3.4–7.0 h, with no apparent relationship to age.[64] Sulfadimethoxine is a long-acting sulfonamide, with an elimination half-life of 12.5 h in calves,[68] 16.2 h in pigs aged 1–2 weeks, and 9.4 h in older (11–12 weeks) animals.[69] Mengelers et al. reported similar half-lives of approximately 13 h in healthy and febrile (inoculated endobronchially with *A. pneumoniae* toxins) pigs.[70] In sheep, the elimination half-life of sulfamerizine was longer (9–14 h) in lambs aged 1 week than in older animals (4–7 h) aged 9–16 weeks.

In several species, including calves and pigs, the distribution volume of trimethoprim is of the order of 1.8–4.0 l/kg, thus significantly exceeding body water volume, and reflecting ready penetration (as a weak base) into intracellular fluid to achieve high tissue concentrations. In calves aged 1–13 weeks, elimination half-life was in the range of 0.9–4.4 h, with no apparent relationship to age.[62] After intravenous dosing in pigs, half-life was 3.3 h. In the same study longer terminal half-lives of 6.5 and 10.6 h in fasted and fed pigs, respectively, after oral dosing, indicate the likelihood of flip-flop pharmacokinetics.[71] Nouws et al. reported long elimination half-lives of trimethoprim in carp of 40.7 and 20.0 h, respectively, at 10°C and 20°C.[67]

Several groups have reported the highest tissue concentrations of sulfonamides plus metabolites in liver and kidney of various food-producing species.[72] Tissue residues of sulfonamides have been described as a cause of special concern in several jurisdictions as they (principally sulfadimidine) have been the cause of more residue violations than any other AMD group, notably in the pig.[63] For example, an early report indicated that sulfadimidine and metabolites were the most frequent cause of non-compliant residues in pig meat, associated with its use as a feed additive.[73] After in feed dosing in pigs, high concentrations of sulfadimidine and metabolites were measured in liver and kidney, with low concentrations in fat.[74] The concern has been exacerbated by reports that sulfadimidine may be carcinogenic in mouse and rat studies. Bevill reported

(1) as a major contributory factor to sulfadimidine residues its relatively long terminal half-life of 12.7 h in the pig and (2) as major causes of violations, failure to observe the WhT, improper feed mixing, and inadequate cleaning of feed-mixing equipment, leading to cross-contamination of feed.[75] Accidental exposure, such as during transport, can also lead to the presence of non-compliant residues in tissues of pigs at slaughter.[76] The high rate of sulfonamide-related violations of pig kidney of 13% in the late 1970s has since decreased considerably.

2.3.13 Polymyxins

Of several polymyxins isolated and investigated (A, B, C, D, E, and M) only compounds B and E are in veterinary use, both as sulfate salts. Polymyxin B is a mixture of B_1 and B_2. Of greater clinical use is polymyxin E, more commonly known as *colistin*, and used as colistin methanesulfonate. Their cationic structure accounts for their disruptive interaction with cell membrane phospholipids and has been described as a detergent-like action. Polymyxins, which are highly ionized molecules, are polar and very poorly lipid-soluble.

As predictable from their very low lipid solubility, clearance is relatively rapid, involving excretion by glomerular ultrafiltration and rapid excretion in urine, although ultrafiltration is somewhat limited by relatively high binding (70–90%) to plasma protein. Colistin is excreted virtually unchanged, and terminal half-life is of the order of 3–4 h. In sheep, a volume of distribution of 1.29 l/kg has been reported. Absorption from the GIT is virtually absent after oral dosing. Because of well-defined neurotoxic and renotoxic properties, polymyxin B is not administered by any route that provides measurable plasma concentrations. However, polymyxin B is used as an endotoxin-neutralizing agent in some veterinary vaccines at doses not exceeding 60 μg/dose, and this does not raise safety (including residues) issues.

For colistin sulfate, no evidence of neurotoxicity was observed in experimental animals for doses much higher than therapeutic doses, and there are many colistin products available for parenteral and intramammary administration. Colistins have, in addition to bactericidal activity against Gram-negative bacteria, direct antiendotoxaemic actions through their binding capacity for anionic lipid, a portion of the endotoxin molecule. The main clinical use of colistin sulfate in food-producing animals is oral administration for the treatment of colibacillosis in young piglets. As absorption from the GIT is very low, residues in edible tissues are not considered to be a major concern. However, it might be noted that, like aminoglycosides, systemically available polymyxins become firmly bound to renal tissue and depletion from the kidney is very slow.

2.3.14 Tetracyclines

Drugs of the tetracycline group are amphoteric, forming salts with both acids and bases. They are used as parent compounds (e.g., oxytetracycline dihydrate) or as salts (e.g., oxytetracycline hydrochloride). Their lipid solubilities range from moderate (oxytetracycline and chlortetracycline) to high (doxycycline and minocycline), so that they are able to traverse cell membranes moderately or readily. The former two drugs are natural tetracyclines, while the latter two are semi-synthetic.

After oral dosing, the bioavailability of tetracyclines varies between drugs, being lowest for oxytetracycline and chlortetracycline and highest for doxycycline, but for all drugs except doxycycline, it is relatively low (Table 2.7). This is of importance from both therapeutic and residue perspectives, because low bioavailability is associated with a high degree of inter-animal variability in both amount absorbed and the resulting plasma concentration–time profile. This can be expected to lead to high variability between animals in residue depletion.

After absorption, the tetracyclines are partially bound to plasma protein. Reported values in farm animal species are 46–51% (chlortetracycline), 28–41% (tetracycline), 21–76% (oxytetracycline), and 84–92% (doxycycline).[77] For the latter drug, high protein binding raises questions concerning effective dosage. The recommended dosage for pigs in drinking water is 10 mg/kg, which provides $AUC_{24\ h}$/MIC ratios that are claimed to be effective for several respiratory tract pathogens.[78] However, Toutain and coworkers, cited by Lees et al.,[9] taking an $AUC_{24\ h}$/MIC breakpoint of 24 h (i.e., an average plasma concentration over the dosing interval equal to the MIC) and using population pharmacokinetic data, predicted for systemic effect a dosage of 20 mg/kg, based on total plasma concentration. Allowing for 90% protein binding, the

TABLE 2.7 Oral Bioavailability of Tetracyclines (Mean Values of Studies Reported)

Drug	Species	Systemic Bioavailability (*F*%)
Chlortetracycline	Chicken	1
	Turkey	6
	Pig	6, 11, 19[a]
Oxytetracycline	Pig	3–5
	Fish	6
	Turkey	9–48
Tetracycline	Pig	5, 8, 18, 23[a]
Doxycycline	Pig	21.2
	Calf	70
	Chicken	41.3
	Turkey	25, 37, 41, 63.5[b]

[a]These *F*% values are study- and feed-dependent.
[b]These *F*% values are age-dependent.
Source: Adapted from Papich and Riviere (2009).[21]

predicted effective dose would be 200 mg/kg, which is totally unrealistic.

Tetracyclines are used extensively in food-producing species. Thus, chlortetracycline, oxytetracycline, and doxycycline are formulated for use as both in-feed and/or in-water products, in poultry, pigs, fish, and cattle for some or all of the following purposes: growth promotion, prophylaxis, metaphylaxis, and therapy. Another major use, particularly of oxytetracycline in parenteral formulations, is for therapy of a range of diseases, including calf and piglet pneumonias. Parenteral solution formulations of various strengths, ranging from 5% to 30% and containing a range of organic solvents, such as propylene glycol, 2-pyrrolidone, and *N*-methylpyrollidone, are in widespread use. When used in higher strengths ($\geq$10%), the formulations create a depot, from which slow absorption occurs, at intramuscular injection sites. After intramuscular administration, a fraction of the dose that remains in solution is rapidly absorbed to achieve maximum plasma concentrations within 1–2 h. However, as the organic solvents in the formulation are dispersed and absorbed, a larger fraction of the administered oxytetracycline precipitates. This provides a depot for subsequent slow absorption phases, giving rise to flip-flop pharmacokinetics and also an acute inflammatory reaction (see Section 2.7.1).

There have been many studies in calves and pigs confirming the retardation effect (prolonged absorption) of high-dose, high-strength solutions of oxytetracycline. Craigmill et al. reported an analysis for 41 datasets from 25 published articles on oxytetracycline in cattle.[79] Their meta-analysis for a dose of 20 mg/kg intramuscularly indicated mean values of 5.61 μg/ml ($C_{\max}$) and 21.6 h ($T_{1/2}$). The advantages of these formulations relate to convenience and economy (single injection) plus animal welfare (avoiding multiple injections) and maintenance of plasma concentrations equal to or greater than MICs of sensitive organisms for periods of 48–96 h. Nouws studied both irritation at injection sites and persistence of oxytetracycline for relatively long periods after intramuscular dosing with 10 of the then available formulations.[80] Injection site issues are considered in Section 2.7.1.

As tetracyclines have moderate to high lipophilic properties, the poor bioavailability associated with oral administration is somewhat surprising. Papich and Riviere suggest that causes may be multifactorial.[77] As zwitterions, they are mainly ionized at pHs within GIT liquor. Moreover, feed reduces bioavailability, and tetracyclines chelate with polyvalent cations. Oxytetracycline absorption has been shown, experimentally, to be reduced by feed, dairy products, Ca^{2+}, Mg^{2+}, Al^{3+}, and Fe^{2+} ions and antacids. Even though doxycycline has a similar structure, affinity for metals is different from that of oxytetracycline with greater affinity for zinc and less for calcium. Supplementation of feed for piglets with zinc can drastically reduce bioavailability of doxycycline.

Moderate variability in absorption of oxytetracycline from different intramuscular injection sites in calves was reported by Nouws and Vree.[81] Bioavailability values of 79%, 86%, and 89% were obtained for injection into the buttock, neck, and shoulder, respectively. The same group reported variable bioavailability and residue profiles with 10 formulations of oxytetracycline in pigs[80] and 5 formulations in calves, sheep, and pigs.[82]

Despite moderate to high degrees of binding to plasma protein, tetracyclines are generally well distributed to most tissues. Volumes of distribution are generally similar to body water volume (0.6–0.7 l/kg). Distribution volumes in excess of this are probably indicative of higher concentrations in intra- than extracellular fluid or binding to specific tissues, including bone. Doxycycline and minocycline traverse cell membranes better than do chlortetracycline and oxytetracycline, and doxycycline in particular concentrates intracellularly.

Systemic clearance of tetracyclines is similar to or higher than GFR. Up to 60%, depending on individual drugs, is eliminated by glomerular ultrafiltration and approximately 40% of administered dose is excreted in feces, but percentages are dependent on drug and route of administration. Bile : plasma concentration ratios may be as high as 20 : 1. For doxycycline, biliary excretion exceeds urinary excretion. Tetracyclines are also metabolized to inactive compounds, except for doxycycline, for which no metabolites have been detected in calves and pigs. In addition to possible metabolism, residue analysis, especially of chlortetracycline, can be hindered by the fact that chlortetracycline is subjected not only to epimerization but also to keto-enol tautomerism, resulting in keto-enol tautomers in the chromatogram, which influence the quantification of residues.[83] In the EU, MRLs for tetracyclines are expressed as the sum of parent compound plus the 4-epimer.

Terminal half-life varies with species, individual drug, and formulation. With the exceptions of retard, in-feed, and in-water formulations, the half-life in most species is sufficient to justify dosing once or twice daily. There are, however, exceptions; half-lives of oxytetracycline after intravenous dosing of 0.7 h (turkey) and 81.5 h (rainbow trout) have been reported.[77] As with all AMDs, there is the possibility of altered pharmacokinetics, as a consequence of disease, but the nature and direction of the change are not readily predictable. Pijpers et al.[84] reported an increase in half-life of oxytetracycline after oral dosing in pigs with pneumonia (14.1 h) compared to healthy pigs (5.9 h), with both values higher than half-life after intravenous dosing (3.7 h), described by Mevius et al.[85] In contrast, more recent studies in our laboratory indicate a lower AUC for oxytetracycline in pneumonic calves compared to healthy animals.[86] Others have reported an increased volume of distribution in diseased animals.

Bound residues of tetracyclines may occur in bones of slaughtered animals for months after treatment. Theoretically, these could reach the food chain via contaminated (mechanically deboned) meat or meat and bonemeal. The accumulation of tetracyclines in tissues is illustrated by the findings of Toutain and Raynaud for oxytetracycline in calves (Table 2.8).[87] Concentrations of oxytetracycline were relatively high in liver and kidney compared to the extrapolated zero-time concentration for serum (4.2 mg/l). The time required for residues to deplete to 0.1 mg/l in serum was 143 hr, considerably shorter than the time required for residues to deplete to 0.1 mg/kg in liver and kidney, but similar to the depletion time for muscle. The data nicely illustrate the importance of tissue elimination half-life in determining decrease to the 0.1 mg/kg concentration; despite an almost three-fold higher initial concentration in kidney compared to liver, the longer half-life for liver leads to a longer time to depletion to 0.1 mg/kg for liver. Similar data were reported for doxycycline in broiler chickens.[88] After dosing orally at 20 mg/kg for 4 days, 1- and 5-day residue concentrations (mg/kg) were as follows: kidney 1.92 and 0.17, liver 1.93 and 0.12, and muscle 1.18 and 0.06, respectively. Other tetracyclines, including tigecycline, recommended for human but not veterinary use.

2.4 SETTING GUIDELINES FOR RESIDUES BY REGULATORY AUTHORITIES

All advanced and several emerging economies have well-established, legally binding procedures for evaluating applications for marketing authorizations (MAs) for veterinary medicinal products (VMPs). The principal bodies and their legal status are indicated in Table 2.9. In the case of the EU of 27 member states, as well as the supranational

TABLE 2.8 Oxytetracycline Residues in Calves after Intramuscular Administration of a Long-Acting Formulation[a]

Tissue	B_0[b] (mg/kg)	Tissue : Serum Ratio at Zero Time[c]	$t_{1/2\beta}$[d] (h)	Delay to Depletion to 0.1 mg/kg (h)
Liver	10.7	2.4 : 1	42.4	287
Kidney	28.9	6.4 : 1	23.6	193
Muscle	3.9	0.9 : 1	26.2	138

[a]Formulation: 20 mg/kg of a 20% w/v solution.
[b]Extrapolated zero-time concentration.
[c]The initial concentration in serum B_0: 4.5 mg/l.
[d]Elimination half-life.
Source: Data from Toutain and Raynaud (1983).[87]

TABLE 2.9 Major Regulatory Authorities Granting Marketing Authorizations for Antimicrobial Drugs[a]

Country	Authority	Acronym	Legal Basis
United States of America	Food and Drug Administration Center for Veterinary Medicine	FDA/CVM[b]	Federal Food, Drug and Cosmetic Act 1996, as amended, and associated regulations
European Union of 27 member states[c]	Committee for Medicinal Products for Veterinary Use, European Medicines Agency[d]	CVMP/EMA	EC Directive 2001/82/EC and Regulation 726/2004 of European Parliament and of Council as amended by Directive 2004/28/EC (*EUDRALEX* Vol. 5)
New Zealand	New Zealand Food Safety Authority	NZFSA/ACVM	Agricultural Compounds and Veterinary Medicines Act (ACVM)
Australia	Australian Pesticides and Veterinary Medicines Authority	APVMA	Agricultural and Veterinary Code Act 1994 (Agvet Code)
Japan	Pharmaceutical Affairs and Food Sanitation Council of the Ministry of Agriculture, Forestry and Fisheries	MAFF/PAFSC	Pharmaceutical Affairs Law
Canada	Canadian Veterinary Drugs Directorate	VDD	Food and Drugs Act (R.S.C., 1985, c. F-27); last amended on 2008-06-16

[a]Legislation and registration procedures for VMPs for therapeutic and prophylactic use and for feed additives are the same in some countries (Australia and USA) but separate in others (EU and Japan).
[b]FDA establishes safety guidelines for drug use in food-producing species and the US Department of Agriculture (USDA) enforces the standards set by FDA.
[c]In the EU, marketing authorizations may be granted either by EMA (centralized procedure) or national authorities (decentralized and mutual recognition procedures), but MRLs are set by EU.
[d]Formerly the European Medicines Evaluation Agency (EMEA). Note that some cited documents refer to EMEA.

authority, there are also national authorities; MAs can be obtained through four possible channels: centralized, decentralized, mutual recognition, and a solely national channel. For products containing AMDs, all authorities require the submission of data packages that establish their quality, safety, and efficacy (QSE). Considerable progress has been made in harmonizing QSE registration requirements in the form of guidelines, at international level under the auspices of VICH (International Co-operation on Harmonization of Technical Requirements for Registration of Veterinary Medicinal Products; see Chapter 3).

2.5 DEFINITION, ASSESSMENT, CHARACTERIZATION, MANAGEMENT, AND COMMUNICATION OF RISK

2.5.1 Introduction and Summary of Regulatory Requirements

National and supranational regulatory authorities are responsible for the administration of legislation, designed to ensure that all foodstuffs obtained from animals are safe for human consumption. Safety in relation to human consumption of food containing (usually no more than trace amounts of) residues of drugs and their metabolites is based on a scientific assessment of data, which ultimately defines the risk. The risk, when defined, is used to establish a food-withholding period, which can be communicated to interested parties. Implementation of the withholding period is the responsibility of veterinarians, farmers, and others concerned with product use in clinical practice. In most countries, residue testing programs are now in place, to ensure as far as possible compliance with the statutory withholding periods. By these mechanisms, the general public is reassured that food derived from animals treated with drugs does not contain residues that might constitute a health hazard to consumers.

The health risks to human consumers of tissue residues of AMDs exceeding MRLs, or residues of AMDs for which no MRL has been determined, include direct toxicity to cells of the host, immunotoxicity (allergenicity), and the emergence of resistance in human GIT microflora and its subsequent spread. In addition, there is a requirement to achieve low concentrations of AMDs in milk to ensure noninterference with the manufacture of milk-derived products: cheese, butter, and yogurt. Concentrations of antimicrobials as low as 1 μg/kg can delay starter activity for these dairy products. Moreover, AMDs may decrease acidity and retard flavor production in butter manufacture, as well as inhibit the ripening of cheeses.

An important element of safety assessment comprises a series of studies designed to ensure that, when products are administered to food-producing species, a withholding period is established, which ensures that food from treated

TABLE 2.10 VICH Harmonized Guidelines on Various Toxicity Studies

Guideline No.	Year	Title
GL22	2001	Reproduction Toxicity Testing[a]
GL23	2001	Genotoxicity Testing[a]
GL27	2003	Pre-Approval Information for Registration of New Veterinary Medicinal Products for Food Producing Animals with Respect to Antimicrobial Resistance[b]
GL28	2002	Carcinogenicity Testing[a]
GL31	2002	Repeat-Dose (90 Day) Toxicity Testing[a]
GL32	2002	Developmental Toxicity Testing[a]
GL33	2004	General Approach to Testing[a]
GL36	2004	General Approach to Establish a Microbiological ADI[b]
GL37	2003	Repeat-Dose (Chronic) Toxicity Testing[a]

[a]Study evaluating the safety of residues of veterinary drugs in human food.
[b]Study evaluating antimicrobial potency and resistance.
Source: Guidelines accessed at http://www.vichsec.org

animals can be eaten safely by humans. Table 2.10 summarizes the range of extensive studies required to satisfy human food safety requirements. The process, which commences with basic pharmacokinetic studies, also includes metabolism studies and animal toxicity and microbiological studies (and in some instances pharmacological and immunotoxicity data are also required), designed to establish a series of NOELs (Table 2.11). VICH has issued harmonized guidelines for establishing the toxicity of VMPs (including AMDs) and also data requirements on antimicrobial properties relating to safety, as listed in Table 2.12.

The standard battery of safety studies (Tables 2.10–2.12) is designed to establish the highest dose that produces no observed effect for non-carcinogenic substances. From the experimental data from all studies, the most sensitive effect in the species most predictive of humans is designated the toxicological NOEL. Allergenicity is not a significant issue for most AMDs. The main exception is benzylpenicillin. The evaluations conducted by the Joint FAO/WHO Expert Committee on Food Additives (JECFA) at the request of the Codex Alimentarius Commission (CAC) did not establish an ADI for benzylpenicillin, but recommended that the daily intake be kept below 30 μg of parent drug per day, setting MRLs of 0.05 mg/kg for edible tissues and 0.004 mg/kg for milk.[89]

The lowest NOEL (toxicological, pharmacological, or microbiological) is used to derive an *acceptable daily intake* (ADI), expressed in milligrams per person per day throughout life, through application of the following simple formula

$$\text{ADI} = \text{NOEL} \times \text{SF}$$

TABLE 2.11 No Observable Effect Levels: Definitions, Guidelines, and Applications

NOEL (and Guidelines)	Definition/Description	Application
Toxicological (VICH, GL22,23,28,31,32,33,37)	Determined in dose–response studies as the dose, in a battery of tests, which is the most sensitive toxicological effect in the species most predictive for humans	Used to determine the toxicological ADI from the relationship $ADI_{tox} = NOEL_{tox} \times SF$
Pharmacological (VICH, GL33)	Determined in dose–response studies, in a range of tests, as the dose that is the most sensitive pharmacological effect	For some drugs (e.g., glucocorticoids) a pharmacological action may be exerted at dose rate less than the toxicological NOEL; used to determine the pharmacological ADI from the relationship $ADI_{pharm} = NOEL_{pharm} \times SF$
Microbiological (VICH, GL27,36)	GL27 outlines the risk of transfer of resistant microorganisms or resistant determinants from animal foodstuffs to humans	—
	GL36 outlines the methods and test systems for determination of the microbiological NOEL	Required for all antimicrobial drugs to establish microbiological ADI
Immunotoxicological (VICH, GL33)	For some antimicrobial drug classes (e.g., β-lactams), immunotoxicity tests are required	Used to establish potential for eliciting allergic reactions in sensitive individuals

TABLE 2.12 Classification of Studies Required by Regulatory Authorities to Satisfy Human Food Safety Requirements on AMDs Residues

Study Type	Description and Objective
Toxicology studies *in vivo* in laboratory animals and *in vitro* studies for genotoxicity	For the range of VICH approved tests and guideline numbers, see Table 2.10; these studies establish a toxicological NOEL
In vitro studies to establish spectrum of activity and potency of AMDs	Establish a microbiological NOEL
Pharmacokinetic studies in laboratory animals and target species	Establish blood/plasma concentration–time profiles and derivation of key pharmacokinetic parameters and variables
Metabolism studies in laboratory animals and target species	Identify metabolites to determine whether a metabolite or parent compound is the marker residue
Residue depletion studies in target species	Define the rate of depletion of the marker residue in edible tissues and fluids (milk, honey), using the highest recommended dose; separate studies for each administration route
Validated analytical method	Identification and quantification of marker residues in animal tissues, milk, eggs, and honey by a determinative method; if the determinative method is not sufficiently specific, a confirmatory method for structural identification is required

where SF is a safety factor. SF may also be referred to as the *uncertainty factor* (UF), which perhaps reflects more accurately the intent of terminology, that is, the management of variability, but in this text we retain the traditional terminology. The ADI typically is based on an assumed body weight of 60 kg. In the case of a toxicological NOEL, SF usually has a value of 100 or higher (see Chapter 3). It is the product of two separate 10-fold factors that allows for interspecies difference and human variability. These 10-fold factors allow for both toxicokinetic and toxicodynamic differences and were subdivided to take into account each aspect separately.[90] Values of $10^{0.6}$ (i.e., 4) for toxicokinetics and $10^{0.4}$ (i.e., 2.5) for toxicodynamics are used for species differences, and equal values of $10^{0.5}$ (i.e., 3.16) for both toxocokinetics and toxicodynamics are used for human variability.[91,92] This value of 3 is not without experimental evidence; examining different databases, it was shown that multiplying a default SF by 3 allowed a coverage of 99% for an additional uncertainty factor.[93] In 2009, MacLachlan used physiologically based pharmacokinetic (PBPK) modeling (see Section 2.5.4.1) to explore for food-producing species the possible value of a default scaling factor based on physiological differences, which can be used to improve estimates of residues from lactating dairy cattle to other food-producing species.[94]

The safety of AMD residues must also be addressed with respect to the human intestinal flora, and derivation of a microbiological ADI is required by regulatory authorities, if AMD or microbiologically active residues reach the human colon. An appropriate ADI should prevent two risks: a disruption of the colonization barrier and the possible increase of the population(s) of resistant bacteria. VICH GL 36 explains the required procedure.[6] The first step is to determine whether there is a need for a microbiological ADI. Data to assess whether a microbiological ADI is needed may be obtained experimentally or from the literature. If required, disruption of the colonization barrier and possible change of resistant bacteria should be documented. Microbiological data may be generated in humans *in vivo*, in gnotobiotic animals, or *in vitro* in species and strains of bacteria accepted as representative of the human GIT flora. Currently, the VICH guideline does not recommend any particular tests, because the reliability and validity of currently used *in vitro* and *in vivo* test systems is not fully established. Finally, the ADI is derived either from *in vitro* data, taking into account a non-observable adverse effect concentration (NOAECs) or from *in vivo* data from a NOEL divided by an uncertainty factor. Within most jurisdictions there is also a requirement to quantify the potential effect of the AMD on starter cultures used in food processing (cheese, buttermilk, sour cream, yogurt starter cultures, etc.).

The lowest ADI (usually toxicological or microbiological) is defined as the amount of drug plus metabolite residue that can be consumed daily for a lifetime, without appreciable risk to the health of the consumer. However, some residues can give rise to an acute rather than a chronic toxicological effect. β-Agonists are an example; they can induce shortlived pharmacological responses, such as tachycardia, but with no long-term consequences. Some authorities have accepted, for such drugs, an acute reference dose (ARfD) as the appropriate health standard.

On the basis of the lowest ADI (usually toxicological or microbiological), together with metabolism and residue depletion studies, the MRL of the residues is determined for each tissue, expressed in μg/kg on a freshweight basis. The range of extensive studies involved in setting ADIs is summarized in Tables 2.10 and 2.12.

Metabolism studies are required in the laboratory animal species used to determine the toxicological NOEL, as well as each food-producing animal species. ADIs are based on total residues of drug plus all metabolites, whereas MRLs comprise a single, quantifiable marker residue, most commonly the parent compound but in some instances a single metabolite or a mixture of compounds. To establish the MRL for each tissue, food consumption estimates are made on the basis of an assumed standard meal (the so-called food basket), as discussed further in Section 2.5.2.2.

The composition of the standard meal varies between regulatory authorities, so that MRLs and WhTs also differ, even when the ADIs are the same. The differences in MRLs adopted by national bodies are attributable mainly to the levels of risk each is prepared to accept, the conditions of use, and methods for establishing MRLs. These differences in national standards affect international trade in animal foods adversely, as manufacturers are required to comply with diverse standards imposed by several importing countries. The MRLs for veterinary drugs developed by the Codex Alimentarius Commission (CAC) are designed to protect the consumer, to be compatible with good veterinary practice in drug use, and to facilitate fair practices in international trade. These are objectives of the Codex Committee on Veterinary Drug Residues in Foods (CCVDRF).

Although CAC MRLs have been adopted by many countries, they are not mandatory.

MRLs are published at the following sites:

United States: `http://www.fsis.usda.gov/OPHS/red_book_2001/2001_Residue_Limits_Veterinary_Drugs_App4.pdf`

Canada: `http://www.hc-sc.gc.ca/dhp-mps/vet/mrl-lmr/mrl-lmr_versus_new-nouveau_ehtml`

EU: `http://ec.europa.eu/health/files/mrl/mrl_20101212_consol.pdf`. EU MRL Summary reports are located at `http://www.ema.europa.eu/ema/index.jsp?curl=pages/medicines/landing/vet_mrl_search.jsp&murl=menus/medicines/medicines.jsp&mid=WC0b01ac058008d7ad`.

The CCRVDF has a primary function in establishing internationally acceptable concentrations of veterinary drugs and their metabolites in food animal products. It thereby facilitates world trade in agricultural products by establishing internationally accepted standards, recommendations, codes of practice, and guidelines, based on a consensus of expert scientific opinion. Space here does not allow provision of a full list of MRLs for all regulatory authorities, but Table 2.13 summarizes marker residues, target tissues, and approved MRLs for AMDs, together with qualifying provisions, where appropriate, adopted in the EU. The MRL for each compound is contained in the European Public MRL Assessment Reports (EPMARs), published on the European Medicines Authority website. Table 2.14 lists tolerances of selected AMDs in the United States, and, for comparative purposes, the ratios of these tolerances to MRLs adopted by the EU are presented. The data illustrate both similarities and differences between the two jurisdictions.

A risk analysis framework for food safety has been developed as an approach to assessing the relationship between potential hazards and the actual human health risks.[95] The three components of risk analysis are assessment, management, and communication.

2.5.2 Risk Assessment

Risk assessment, where risk is the probability of harm to the consumer, is represented by the relationship risk = hazard × exposure. IPCS defined these terms as follows:[96]

TABLE 2.13 Marker Residues and Target Tissues for Antimicrobial Drugs for Which MRLs Have Been Defined in the EU

Pharmacological Class	Individual Drugs	Marker Residue (Par = Parent Drug)	Animal Species[a]	Target Tissues[b] and MRL (μg/kg)	Other Provisions
Sulfonamides	All sulfonamides	Par	All	M, F, L, K, Mi (all 100)	Combined total residues of all sulfonamides not >100 μg/kg
			B, C, O	Mi100	
Diaminopyrimidines	Baquiloprim	Par	B	F10, L300, K150, Mi(30)	—
			P	F40, L50, K50	—
	Trimethoprim	Par	All except E	M, F, L, K, Mi (all 50)	Not eggs[c]
			E	M, F, L, K (all 100)	—
Penicillins	Amoxicillin	Par	All	M50, F50, L50, K50, Mi 4	—
	Ampicillin	Par	All	M50, F50, L50, K50, Mi 4	—
	Benzylpenicillin (penicillin G)	Par	All	M50, F50, L50, K50, Mi 4	—
	Cloxacillin	Par	All	M300, F300, L300, K300, Mi 30	—
	Dicloxacillin	Par	All	M300, F300, L300, K300, Mi 30	—
	Nafcillin	Par	All ruminants	M300, F300, L300, K300, Mi3	Intramammary use only
	Oxacillin	Par	All	M300, F300, L300, K300, Mi30	—
	Penethamate	Benzylpenicillin	All mammalian food species	M50, F50, L50, K50, Mi4	—
	Phenoxymethyl penicillin	Par	P	M, L, K (all 25)	—
			Po	M, F, L, K (all 25)	—
Cephalosporins	Cefacetrile	Par	B	Mi125	Intramammary use only
	Cefalexin	Par	B	M200, F200, L200, K1000, Mi100	—
	Cefalonium	Par	B	Mi20	—
	Cefapirin	Sum of cephapirin + desacetylcephapirin	B	M50, F50, K100, Mi60	—
	Cefazolin	Par	B, O, C	Mi50	—
	Cefoperazone	Par	B	Mi50	—
	Cefquinome	Par	B	M50, F50, L100, K200, Mi20	—
			P	M50, F50, L100, K200	—
			E	M50, F50, L100, K200	—
	Ceftiofur	Sum of all residues retaining the β-lactam structure expressed as desfuroylceftiofur	All mammalian food-producing species	M1000, F2000, L2000, K6000, Mi100	—
Quinolones	Danofloxacin	Par	All food-producing species except B, O, C, P Po	M100, F50, L200, K200	—
			B, O, C	M200, F100, L400, K400	—
			Po	M200, F100, L400, K400	Not eggs[c]

(*continued*)

TABLE 2.13 ***(Continued)***

Pharmacological Class	Individual Drugs	Marker Residue (Par = Parent Drug)	Animal Species[a]	Target Tissues[b] and MRL (μg/kg)	Other Provisions
	Difloxacin	Par	All food-producing species except B, O, C, Po	M300, F100, L800, K600	—
			B, O, C	M400, F100, L1400, K800	—
			P	M400, F100, L800, K800	—
			Po	M300, F400, L1900, K600	Not eggs[c]
	Enrofloxacin	Sum of enrofloxacin and ciprofloxacin	All except B, O, C, P, Po, R	M100, F100, L200, K200	—
			B, O, C	M100, F100, L300, K200, Mi100	—
			P, R	M100, F100, L200, K300	—
			Po	M100, F100, L200, K300	Not eggs[c]
	Flumequine	Par	All except B, O, C, P, Po, Fi	M200, F250, L500, K1000	—
			B, P, O, C	M200, F300, L500, K1500, Mi50	—
			Po	M400, F250, L800, K1000	—
			Fi	M600	—
	Marbofloxacin	Par	B	M150, F50, L150, K150, Mi75	—
			P	M150, F50, L150, K150	—
	Oxolinic acid	Par	All except Fi	M100, F50, L150, K150	Not eggs or milk[c]
			Fi	M100	—
	Sarafloxacin	Par	Ch	F10, L100	—
			S	M30	—
Macrolides	Erythromycin	Erythromycin A	All	M200, F200, L200, K200, Mi40, Eg150	—
	Spiramycin	Sum of spiramycin + neospiramycin	B	M200, F300, L300, K300, Mi 200	—
			Ch	M200, F300, L400	—
		Spiramycin 1	P	M250, L2000, K1000	—
	Tilmicosin	Par	All except Po	M50, F50, L1000, K1000, Mi50	—
			Po	M75, F75, L1000, K250	—
	Tulathromycin	(2*R*,3*S*,4*R*,5*R*,8*R*,10*R*,11*R*,12*S*,13*S*,14*R*)-2-ethyl-3,4,10,13-tetra-hydroxyl-3,5,8,10,12,14-Hexamethyl-11-[[3,4,6-trideoxy-3-(dimethylamino)-β-D-xylo-hexopyranosyl]oxy]-1-oxa-6-azacyclopent-decan-15-one expressed as tulathromycin equivalents	B, P	F100, L3000, K3000	Not milk[c]
	Tylosin	Tylosin A	All	M100, F100, L100, K100, Mi50, Eg200	—

TABLE 2.13 ***(Continued)***

Pharmacological Class	Individual Drugs	Marker Residue (Par = Parent Drug)	Animal Species[a]	Target Tissues[b] and MRL (μg/kg)	Other Provisions
	Tylvalosin	Sum of tylvalosin + 3-*o*-acetyltylosin	P	M50, F50, L50, K50	—
			Po	F50, L50	Not eggs[c]
Phenicols	Thiamphenicol	Par	All	M50, F50, L50, K50, Mi50	Not eggs[c]
	Florfenicol	Sum of florfenicol and its metabolites measured as florfenicol amine	B	M200, L3000, K300	—
Tetracyclines	Chlortetracycline	Sum of par and its 4-epimer	All	M100, L300, K600, Mi100, Eg200	—
	Doxycyline	Par	B	M100, L300, K600	Not milk[c]
			P	M100, F300, L300, K600	—
			Po	M100, F300, L300, K600	Not eggs[c]
	Oxytetracycline	Sum of par & its 4-epimer	All	M100, L300, K600, Mi100, Eg200	—
	Tetracycline	Sum of par & its 4-epimer	All	M100, L300, K600, Mi100, Eg200	—
Naphthalene-ringed ansamycin	Rifaximin	Par	B	Mi60	—
Pleuromutilins	Tiamulin	Sum of metabolites that may be hydrolysed to 8-a-hydroxymutilin	P, R	M100, L500	—
			Ch	M100, F100, L1000	—
			T	M100, F100, L300, Eg1000	—
	Valnemulin	Par	P	M50, L500, K100	—
Lincosamides	Lincomycin	Par	All	M100, F50, L500, K1500, Mi150, Eg50	—
	Pirlimycin	Par	B	M100, F100, L1000, K400, Mi100	—
			P	M100, F50, L500, K1500	—
			Ch	M100, F50, L500, K1500, Eg50	—
Aminoglycosides	Apramycin	Par	B	M1000, F1000, L10,000, K20,000	Not milk[c]
	Dihydrostreptomycin	Par	All ruminants	M500, F500, L500, K1000, Mi200	—
			P, R	M500, F500, L500, K1000	—
	Gentamicin	Sum of gentamicin C_1 gentamicin C_{1a}, gentamicin C_2 + gentamicin C_{2a}	B	M50, F50, L200, K750, Mi100	—
			P	M50, F50, L200, K750	—
	Kanamycin	Kanamycin A	All except Fi	M100, F100, L600, K2500, Mi150	—
	Neomycin (including framomycin)	Neomycin B	All	M500, F500, L500, K5000, Mi1500, Eg500	—
	Paromomycin	Par	All	M500, L1500, K1500	Not eggs or milk[c]
	Spectinomycin	Par	All except O	M300, F500, L1000, K5000, Mi200	Not eggs[c]
			O	M300, F500, L2000, K5000, Mi200	—

(continued)

TABLE 2.13 ***(Continued)***

Pharmacological Class	Individual Drugs	Marker Residue (Par = Parent Drug)	Animal Species[a]	Target Tissues[b] and MRL (μg/kg)	Other Provisions
	Streptomycin	Par	All ruminants	M500, F500, L500, K1000, Mi200	—
			P, R	M500, F500, L500, K1000	—
Polypeptides	Bacitracin	Sum of bacitracin A, bacitracin B, + bacitracin C	B	Mi100	—
			R	M150, F150, L150, K150	—
β-Lactamase inhibitors	Clavulanic acid	Par	B	M100, F100, L200, K400, Mi200	—
			P	M100, F100, L200, K400	—
Polymyxins	Colistin	Par	All	M150, F150, L150, K200, Mi50, Eg300	—
Orthosamomycins	Avilamycin	Dichloroisoeverninic acid	P, Po, R	M50, F100, L300, K200	Not eggs[3]
Ionophores	Monensin	Monensin A	B	M2, F10, L30, K2, Mi2	—
	Lasalocid	Lasalocid A	Po	M20, F100, L100, K50, Eg150	—
Miscellaneous	Novobiocin	Par	B	Mi50	—

[a]Abbreviations used in this column: all = all food-producing species; B = bovine; C = caprine; Ch = chicken; E = equidae; Fi = fin fish; O = ovine; P = porcine; Po = poultry; R = rabbit; S = salmonidae; T = turkey.
[b]Abbreviations used in this column: M = muscle; F = fat; L = liver; K = kidney; Mi = milk; Eg = eggs (for poultry, chickens, and pigs, F = skin and fat; for fin fish and salmonidae, M = muscle and skin in natural proportions).
[c]Not for use in animals from which eggs and/or milk are produced for human consumption.
Source: Data from EU Commission Regulation 37/2010.[102]

> Hazard is the inherent property of an agent or situation having the potential to cause adverse effects when an organism, system or (sub) population is exposed.
>
> Risk is the probability of an adverse effect in an organism, system, or (sub) population caused under specified circumstances by exposure to an agent.

2.5.2.1 Hazard Assessment

For AMD residues in foodstuffs, the hazard is drug/drug metabolite residues and exposure is dietary intake.[97] As risk assessment must be applied independently to each AMD, it follows that the database identifying hazard must be extensive and should comprise, for each drug, information on structure, purity, physicochemical, pharmacological (including pharmacokinetic and metabolism data), and toxicological properties. Data may be obtained from human epidemiological and animal toxicology studies and *in vitro* assays (e.g., for mutagenicity/genotoxicity).

After hazard identification, hazard characterisation is undertaken, and this is normally based on dose–response relationships in the range of toxicological studies summarized in Section 2.5.1. It is assumed that a threshold dose for response can be identified, where the NOEL is the highest dose that causes no (adverse) detectable effect in the most sensitive animal species or strain. Other approaches have been used, however, such as determination of a benchmark dose (BMD).[98] Determination of BMD involves modeling of all dose–response data, with increased weighting applied to data at low levels of response (e.g., ED_{05}, ED_{10}, doses producing respectively 5% or 10% of maximum), with the subsequent application of safety/uncertainty factors in determining ADI. As discussed in Section 2.5.1, the ADI is likely to differ between regulatory authorities.

The NOEL procedure for determining ADI is not acceptable for drugs with effects that are characterized by non-threshold mechanisms. An example is genotoxic carcinogens, which cause genetic alterations in target cells. Mutagenicity tests are carried out to provide evidence of genotoxicity. Such compounds are assumed to be harmful at any exposure and are banned from use in food-producing animals in many countries. Other countries accept their use, provided residue concentrations are small enough to be regarded as posing negligible risk. On the other hand, non-genotoxic carcinogens act extragenetically to cause enhanced cell proliferation or sustained hyperfunction, dysfunction, or both. At least theoretically, non-genotoxic carcinogens can be regulated through the NOAEL determined ADI procedure.

A final step in hazard characterization leading to setting the ADI is for the regulatory authority to reflect on the significance and applicability of those responses revealed in high-dose-rate toxicology studies to the circumstance of

TABLE 2.14 FDA Tolerance Levels for Some AMD Residues in Edible Animal Meat Products and EU MRL: USA Tolerance

Compound	Animal Species, USA	Tolerance, USA (μg/kg)	Ratio EU MRL: USA Tolerance
Amoxicillin	Bovine	10 (M, K, L, F)[a]	5 : 1 all tissues
Ampicillin	Bovine, porcine	10 (M, K, L, F)	5 : 1 all tissues
Benzylpenicillin	Bovine	50 (M, K, L, F)	1 : 1 all tissues
	Turkey	10 (M, K, L, F)	5 : 1 all tissues
Ceftiofur	Bovine	1000 (M), 2000 (L), 8000 (K)	M1 : 1, L1 : 1, K0.75 : 1
Cephapirin	Bovine	100 (M, K, L, F)	0.5 : 1
Cloxacillin	Bovine	10 (M, K, L, F)	30 : 1
Danofloxacin	Bovine	200 (M, L)	M1 : 1, L2 : 1
Dihydrostreptomycin	Bovine, porcine	2000 (K), 500 (M, L, F)	M, L, F 1 : 1, K 0.5 : 1,
Enrofloxacin	Bovine	100 (L)	L 3 : 1
	Chicken, turkey	300(M)	M 0.33 : 1
Erythromycin	Bovine, porcine	100 (M, K, L, F)	2 : 1 all tissues
	Chicken, turkey	125 (M, K, L, F)	1.6 : 1 all tissues
Florfenicol	Bovine	300 (M), 3700 (L)	M 0.66 : 1, L0.8 : 1
	Porcine	200 (M), 2500 (L)	M 0.66 : 1, L0.8 : 1
Gentamicin	Porcine	100 (M), 300 (L), 400 (K, F)	M 0.5 : 1, L0.66 : 1, K 1.88 : 1, F 0.125 : 1
Lincomycin	Porcine	100 (M), 600 (L)	M 1 : 1, L 0.83 : 1
Neomycin	Bovine, ovine, porcine	1200 (M), 3600 (L), 7200 (K)	M 0.42 : 1, L 0.14 : 1, K0.69 : 1
Pirlimycin	Bovine	300 (M), 500 (L)	M 0.33 : 1, L 2 : 1
Spectinomycin	Bovine	250 (M), 4000 (K)	M 1.2 : 1, K 1.25 : 1
	Chicken, turkey	100 (M, K, L, F)	M 3 : 1, K50 : 1, L20 : 1, F5 : 1
Streptomycin	Bovine, porcine	500 (M, L, F), 2000 (K)	M, L, F 1 : 1, K 0.5 : 1
Sulfonamide group	Bovine, porcine, poultry	100 (M, L. K, F)	1 : 1 all tissues
Tiamulin	Porcine	600 (L)	L 0.83 : 1
Tilmicosin	Bovine, ovine	100 (M), 1200 (L)	M 0.5 : 1, L 0.83 : 1
	Porcine	100 (M), 7500 (L)	M 0.5 : 1, L0.13 : 1
Tylosin	Bovine, porcine, chicken, turkey	200 (M, L, K, F)	0.5 : 1 all tissues
Tetracycline group	Bovine, ovine, porcine, poultry	2000 (M), 6000 (L), 12000 (F, K)	M, L, K 0.05 : 1, F 0.025 : 1

[a]Tissue: F, fat; L, liver; K, kidney; M, meat

Source: Adapted from Croubels et al. (2004).[35]

residue intake, when the latter may be several orders of magnitude less than the former. Further information on the process by which an ADI is typically assigned for an AMD is provided in Chapter 3.

2.5.2.2 *Exposure Assessment*

Three factors determine exposure assessment: quantity of food consumed, residue concentration in that food, and the marker residue : total residue ratio. The "food basket" adopted by most authorities comprises:

300 g muscle (for fish, muscle and skin in natural proportions).
50 g fat (for pigs and poultry, fat and skin in natural proportions, for poultry 90 g)
100 g liver
50 g kidney (10 g poultry)
100 g eggs
20 g honey
1.5 l milk

In the EU, for example, the food basket comprises, for mammals, muscle (300 g), liver (100 g), kidney (50 g), and fat (50 g) and, if appropriate, milk, eggs, and honey (for poultry, 10 g kidney and 90 g fat). For poultry and pigs, the MRL for fat relates to fat and skin in natural proportions in the EU, while for fin fish, muscle includes muscle and skin in natural proportions. In general, a numerically greater MRL is allowed for foods likely to be consumed infrequently or in small amounts (e.g., kidney relative to muscle). In addition, residues that occur in food of plant origin or that come from the environment need to be considered when fixing the MRL.

On the basis of the food basket, the EU authority (EMA/CVMP) then requires applicants for MAs to estimate the theoretical maximum daily intake (TMDI) for persons weighing 60 kg, applying the equation:

$$\mathrm{TMDI} = \sum \left(\text{daily intake}_i \times \mathrm{MRL}_i \times \frac{\mathrm{TR}_i}{\mathrm{MR}_i} \right)$$

where daily intake$_i$ (kg) = daily consumption as defined in the model food basket; MRL$_i$ = MRL (μg/kg) for muscle, fat, liver, kidney, eggs, and honey; TR_i = total residue concentration (or pharmacological or microbiological activity where relevant); and MR_i = marker residue concentration (or pharmacological or microbiological activity where relevant) in the same tissues and commodities.

Later, JECFA proposed an alternative to TMDI, namely the estimated dietary intake (EDI),[99] which has been accepted by the Australian authority. The difference from TMDI is the replacement of MRL by median residue concentration, on the reasonable consideration that, in the chronic intake circumstance, MRL does not provide a realistic estimate of residue intake. MRL is the upper limit of a high percentile (usually 95th) of the distribution of marker residue. In contrast, the median residue concentration provides the best point estimate of the central tendency over a prolonged period.

For a 60-kg person, EDI is calculated from the equation

$$\mathrm{EDI} = \sum \left(\text{daily intake}_i \times \text{median residue concentration}_i \times \mathrm{MRL}_i \times \frac{\mathrm{TR}_i}{\mathrm{MR}_i} \right)$$

where daily intake_i (kg) = daily consumption as defined in the model food basket; median residue concentration_i = median residue concentration (μg/kg) for muscle, fat, liver, kidney, eggs, and honey; TR_i = total residue concentration; and MR_i = marker residue concentration in the same tissues and commodities.

In the United States, FDA/CVM reasonably assumes that an individual consuming 300 g of muscle will not, on a given day, also consume liver or kidney but might well consume a full allocation of milk and eggs. FDA therefore calculates a safe concentration of total residues for edible tissues (and if appropriate additionally milk and eggs) from the equation:

$$\mathrm{SC} = \frac{\mathrm{ADI} \times 60\ \mathrm{kg}}{\mathrm{FCF}}$$

where SC = safe concentration for total residues in a specified edible tissue, as defined in the model food basket; ADI = acceptable daily intake; and FCF = daily consumption of the specified edible tissue.

It is clear that all regulatory authority approaches to predicting dietary exposure are very conservative, in that all are higher than actual dietary intake. These conservative assumptions are as follows: daily consumption for a lifetime of the model food basket, the treatment of all animals at the maximum recommended dose rate and duration, slaughter of treated animals at the WhT, and the presence of residues in all edible tissues (including milk and eggs) at the MRL (TMDI calculation) or at median residue concentrations (EDI calculation). The conservative assumptions used to calculate exposure are additional to the conservative assumptions made in the toxicology studies used to determine NOELs.

2.5.3 Risk Characterization

The characterization of risk involves consideration of the information garnered when identifying and characterizing the hazard together with exposure assessment. For AMDs used in food-producing species, the outcome comprises the MRLs derived from the ADI approach together with application of the model meal.[100] In characterizing risk, two extreme examples may be cited. On one hand, residues can constitute a health hazard at any concentration, so that MRLs cannot be established and use of the AMD in food-producing animals is disallowed. On the other hand, drugs may leave residues that are not considered to pose a health risk to humans, and MRLs are therefore not required. The latter applies to many excipients used in VMPs. Most AMDs lie between the two extremes cited, but furans that are both mutagenic and carcinogenetic are banned from use in food-producing animals. Similarly, nitroimidazoles (metronidazole, ronidazole) are suspected to be mutagens and carcinogens, and their use in food-producing animals is prohibited. In the EU, under Council Regulation (EEC) 2377/90[44] and up until the entry in force of a new regulation EC 470/2009[101] laying down European Community (EC) procedure for the establishment of residue limits of pharmacologically active substances in foodstuffs of animal origin, all pharmacologically active substances were until recently included in one of the following annexes:

- Annex I—definitive MRLs have been set.
- Annex II—MRLs are not required for the protection of public health.
- Annex III—provisional MRLs have been set, but finalization has not been completed.
- Annex IV—substances are prohibited from use in VMPs for all food-producing animals.

European Community Regulations 470/2009[101] and 37/2010[102] have introduced a new system of classification, whereby all pharmacologically active substances are now listed in a single annex in alphabetical order in two tables, the first to include all compounds listed in Annexes I, II, and III and a second table listing prohibited substances from Annex IV. In the United States procedures are broadly similar [see *Code of Federal Regulations* (CFR) Title 21, Part 556, on the FDA website[103]]. In the United States a tolerance is not required if there is no reasonable expectation that residues may be present, or when the drug is metabolized or assimilated into tissues in such form that any possible residue would be indistinguishable from normal tissue constituents.

In summary, to establish MRLs for a given drug requires provision of the following data: knowledge of dosage schedule (amount, dose interval, duration) and administration route; pharmacokinetic and metabolic information in laboratory animals and each of the target food-producing species; residue depletion in each target species using radiolabeled drug; validated analytical methods for detection and quantification of residues, including marker residue; and data defining the effect of residues on food processing. It is not possible to address fully all of these aspects here, but the reader is referred to (1) Section 2.3 of this chapter on AMD pharmacokinetics and metabolism; (2) later chapters of this text and MacNeil,[104] describing analytical method requirements and validation; and (3) an excellent account in 2010 on risk characterization in relation to residues provided by Reeves.[105]

There is no consistency in procedures used to establish MRLs either between national authorities and JECFA or, indeed, between different national authorities. This is regrettable from a sponsor's perspective, as it may involve, at worst, several similar studies (with major implications for animal welfare and added expense) and variance in achieving (or not) a MA between jurisdictions. The rational procedures proposed by JECFA for MRL determination are based on three premises: (1) they can be enforced by regulatory programs that use available analytical methods, (2) they are no higher than necessary, and (3) they reflect residue concentrations expected when the product containing AMD is used in accordance with good veterinary practice. For this latter point (3), JECFA introduced the concept of the so-called estimated daily intake (EDI). They soundly assumed that EDI is a more appropriate tool in determining whether residues pose any chronic (lifetime) risks than does TDMI. In other words, the JECFA approach copes only with chronic, not acute, exposure that is a matter of concern for CVMP.[106]

The new JECFA pivotal approach to determine a MRL is to use median residue concentrations (not MRL) in animal-derived food for the calculation of the EDI from a model daily food basket. JECFA combines an iterative approach with a statistical tool, as agreed to at the 66th JECFA meeting.[107] A linear regression analysis is performed on the terminal depletion of marker residue in edible tissues after the last administration of drug. If appropriate, the data from the analysis enable calculation of the upper limits of the 95% confidence interval of the upper one-sided tolerance limit of the 95th percentile of the test animal. Then the JECFA approach defines a link between daily residue intake as expressed through the EDI and MRL as follows:

> The MRL and the median concentration are derived from the same time point of the depletion data of the marker residue. The MRL is a point on the curve describing the upper one-sided 95% confidence limit over the 95th percentile. The median is the corresponding point on the regression line for the same time point. Both figures are obtained from a statistical evaluation of the data.

Practically, the MRL is that point on the tolerance limit as defined above at or beyond which the predicted EDI using the median concentration guarantees that EDI$\leq$ADI.

In contrast, the EU approach consists of computing a TMDI using MRL, not median residue concentrations; CVMP criticized the JECFA approach for its intrinsic limitation to a chronic risk scenario and also for several technical reasons, including difficulties of using linear regression to estimate the tolerance limit of residue concentrations with its confidence interval and to the rather loose concept of good veterinary practice. (For further details, see Ref. 106. For further details on and a discussion of the USFDA approach, see Refs. 105 and 108.)

2.5.4 Risk Management

2.5.4.1 Withholding Times

In relation to the use of AMDs in food-producing species, risk is managed by setting withdrawal/withholding periods, defined as the time interval from last administration of a product to when the animal can be slaughtered to provide animal foods, milk, eggs, or honey that can be safely consumed. WhTs are set by regulatory authorities. Adherence to the WhT provides assurance that food derived from treated animals will not exceed the MRL for the drug substance. The normal procedure for assigning the WhT is to utilize the data generated in residue depletion studies conducted: (1) with non-radiolabeled drug; (2) in the product formulation proposed for marketing; and (3) administered at the highest dose rate, shortest dose interval, and longest duration specified in the product literature. Typically, a company would use a minimum of four animals of the target species at each of at least four slaughter times to define the depletion profile. The animals should be typical of target animals in clinical use, for example, young calves *or* lactating adult cattle. Extensive tables of tissue depletion pharmacokinetic data have been published by Craigmill et al.[109]

A simple approach is used in Europe when no statistical approach is possible; it consists of setting the WhT as the time when residues in all tissues from all investigated animals have been depleted to less than their respective MRLs and, to add to the observed delay, an additional, arbitrary safety span to compensate for the uncertainties of biological origin. This is the so-called pragmatic EU approach. The size of the safety span (typically 10–30%) was documented by comparing the WhT obtained by the regular EU statistical method for 62 depletion studies and the WhT that would have been fixed using the so-called pragmatic approach; it was shown that the median value of the appropriate safety span was 25% but the range was from −24% to 233%, indicating that for some studies, selecting an arbitrary safety span of 10–30% is not conservative enough.[110]

The second and preferred method is to apply appropriate statistical analysis to the dataset, based on linear regression. Both EU and USFDA authorities assume log-linear decline of residue concentrations and apply least-squares regression to derive the fitted depletion line. Then the one-sided upper tolerance limit (95% in EU and 99% in USA) with a 95% confidence level is computed. The WhT is the time when this upper one-sided 95% tolerance limit for the residue is below the MRL with 95% confidence. In other words, this definition of the WhT says that at least 95% of the population in EU (or 99% in USA) is covered in an average of 95% of cases. It should be stressed that the nominal statistical risk that is fixed by regulatory authorities should be viewed as a statistical protection of farmers who actually observe the WhT and not a supplementary safety factor to protect the consumer even if consumers indirectly benefit from this rather conservative statistical approach.

Concordet and Toutain initiated a scientific debate on the estimation of WhTs, using a regression method to estimate a 99th (USA) or a 95th (EU) percentile of the population with a 95% confidence level.[111,112] The regression approach requires a pharmacokinetic/residue study, then modeling the depletion curve as a straight line after logarithmic transformation. Five assumptions are involved in applying linear regression:

1. No experimental uncertainty on the time of slaughter (x axis)
2. Linearity of the depletion curve
3. Normality of distribution of the logarithm of residue concentrations at each slaughter time
4. Homoscedasticity, that is, assumption of constant variance
5. Independence of observation (as outlined in detail by Concordet and Toutain[112]), which may or may not be satisfied in every instance.

As an alternative to the regression method, Concordet and Toutain[111,112] proposed a simple, understandable non-parametric method that allows the control of risk, defined as a percentage less than $100/1 - \alpha$. The only assumption is that the probability of exceeding MRL decreases monotonically with time. The method requires a few assumptions, including assuming that observations are independent and slaughter times are chosen during the declining phase of residue kinetics. The latter condition is especially important with controlled, slow-release formulations. For details of the method, see Concordet and Toutain.[111] They emphasize the advantage that all animals/samples contribute to WhT estimation, including those for which concentrations are lower limit of quantification (LLOQ) of the analytical method. The limit of this approach is its low statistical power requiring many more animals than the conventional regression method.

Martinez et al.[113] compared USFDA approaches on WhTs with those of Concordet and Toutain[111,112] (see Table 4 of Martinez et al.[113]) and concluded that the USFDA regression method was, on several grounds, more appropriate for prevention of exposure to non-compliant residues. The approach of EMA/CVMP is similar to that of USFDA, except that it is based on 95/95% tolerance limits, considered preferable because of uncertainty in extrapolating to extreme percentages of the population. Fisch extended the discussion on estimating WhTs.[114] He proposed the application of Bayesian methods, using Markov chain Monte Carlo methods for circumstances in which neither regression nor non-parametric approaches apply. In an early review, Martin-Jimenez and Riviere indicated a possible role for and relevance of using population pharmacokinetic data for describing drug disposition in fluids and drug deposition as residues in edible tissues.[115] However, they pointed out that adequate strategies were not available then, and indeed, this remains the case. Nevertheless, they highlighted the prospect and value at some future time of using multicompartment models to characterize plasma–tissue relationships. They even envisaged the possibility of defining WhTs with appropriate confidence intervals for subpopulations, depending on differences in clinical or production parameters. Such models would use a Bayesian approach to harvest information from a range of protocols (efficacy, safety, residues) pooled in a single model.

The most promising research in the area of residues and WhT is that of physiologically based pharmacokinetic modeling (PBPK) as illustrated for oxytetracycline in sheep.[116] Another example is the prediction of sulfonamide concentrations in pigs for prediction of non-compliant residues.[117] The principle of PBPK modeling is to develop a model with a generally complicated structure, including explicitly critical anatomical (e.g., polygastric vs. monogastric stomach), physiological, physicochemical, and biochemical processes regarding the purpose for which the model is developed. For example, a PBPK model intended to explore the consequence of inhibition of intestinal enzymes on WhT in milk in a given species would include explicitly not only the different compartments representing the relevant tissue for food safety (muscle, adipose tissue, kidney, liver, other carcass tissues, and also milk) but also a clearance component at the intestinal level. PBPK models should be viewed as sophisticated dosimetry models that offer great flexibility in modeling exposure scenarios for which there are no or limited data in order to predict tissue concentrations between varying routes of exposure and across species. MacLachlan developed such a model in lactating cattle to explore the influence of differing physiological statuses on residues of lipophilic xenobiotics in livestock.[94] Interspecies extrapolation is a key issue for the establishment of WhT in minor species. The application of a PBPK model to the prediction of WhTs for oxytetracycline in salmon has demonstrated the applicability of PBPK modeling to the prediction of tissue residues in food animals and the establishment of WhTs.[118] More recently, it was shown that a PBPK model developed for midazolam in the chicken and then adapted to take into account species-specific physiological parameters for turkey, pheasant, and quail provided good predictions of the observed tissue residues in each species, in particular for liver and kidney.[118,119]

There are differences between regulatory authorities in procedures used to set milk withholding periods. USFDA/CVM requires use of at least 20 animals and analysis of milk samples for the marker residue in triplicate.[120] If the product is authorized for mastitis treatment, it is assumed that no more than one-third of the milk is derived from treated cows. A regression line is fitted to the log residue concentration data for each cow, and then fitted lines are used to estimate the distribution of log residue concentrations at each sampling time. Between-animal variance and measurement error variability are estimated and used to calculate a tolerance limit at each time. The WhT is set as the first time at which the upper 95% confidence limit of the 99th percentile of residue concentrations is equal to or less than the MRL.

For milk, EMA/CVMP uses a "time to safe concentration" (TTSC) approach.[121] TTSC is the first time when the upper 95% confidence limit of the 95th percentile of individual milk sampling times complies with the MRL. The method assumes a log normal distribution of individual times to safe concentration. If the data set is not suitable for analysis by the TTSC method, alternative statistical approaches may be used. Thus, the distributional assumptions of the USFDA/CVM and EMA/CVMP relate, respectively, to residue concentrations and time to safe concentration. An advantage of the latter approach is that an assumption of log linear depletion of residues is not made.

2.5.4.2 *Prediction of Withdholding Times from Plasma Pharmacokinetic Data*

Gehring et al.[122] and Riviere and Sundlof[123] have proposed that an approximation to WhT can be derived from the equation:

$$\mathrm{WhT} = 1.44 \times \ln\left(\frac{C_0}{\mathrm{MRL}}\right) \times T_{1/2}$$

where C_0 = concentration of drug in target tissue at the end of administration and $T_{1/2}$ is terminal half-life. While this equation is an approximation, it does provide a perspective on WhT relative to terminal half-life. If the marker residue is a metabolite, it is the terminal half-life of metabolite that should be used to estimate WhT. Assuming homogeneous drug distribution in the body (admittedly a tall order!), a therapeutic AMD plasma concentration of 10 mg/l and a MRL of 0.01 mg/l, then

$$\mathrm{WhT} = 1.44 \times \ln(10/0.01) \times T_{1/2} = 9.947 \times T_{1/2}$$

Thus, a WhT would be very slightly less than 10 half-lives if the ratio C_0/MRL were 1000. This "rule of 10" is based on the elimination of 99.9% of drug after 10 half-lives. If MRL is low relative to C_0, this will lead to a greater WhT value. If $T_{1/2}$ is short (e.g., for penicillin), WhT is correspondingly short. However, if tissue half-life is long (as for aminoglycosides), WhT can be very protracted. If drug dose is doubled, WhT is increased by a single half-life to 10.94 half-lives, showing that an error on the dose exerts a rather limited influence (of approximately 10%) on WhT. However, there is a circumstance in which the increase in dose increases a WhT disproportionally. This occurs when the residue depletion curve obeys a multiexponential decay and when the MRL value transects a phase having a relatively short half-life, when the dose is low but a very late terminal phase, when the dose is increased or when multiple drug administrations lead to some "stacking" as reported by KuKanich et al. for a long-acting formulation of florfenicol.[124] This can result in illegal residues when the product is administered for more than the single-label dose.

2.5.4.3 *International Trade*

An additional aspect of risk management is the requirement for compliance with regulations of animal-derived foods in international trade. Because there is no harmonized worldwide legislation on residues, barriers to international trade can and do arise through distortion of the conditions of competition in the market. Imported foods must comply with either CAC or national MRLs of the importing country. When MRLs have been established, the imported food can be subjected to a testing program. However, in some instances the MRL of the importing country may be lower than that of the exporting country. Moreover, no MRLs may have been established and then a zero-tolerance approach to residues is normally adopted. These varying circumstances can be addressed in several ways: through trade agreements between trading partners, by establishing import MRLs, and through assignment of export slaughter intervals (ESIs) to veterinary products intended for use in food-producing animals destined for overseas markets. The latter approach is unique, at present, to the Australian authority (APVMA). The ESI is defined as the interval between product administration and slaughter for export. It is determined first by a consideration of trade data for meat and edible offal and the degree of risk acceptable to major stakeholders. Then, a suite of algorithms is used to calculate the probability of lot rejection of meat consignments, when treated animals are killed at various intervals. The ESI is that time when the upper tolerance limit about the regression line for the censored data intersects with the residue concentration associated with the acceptable risk. If a MRL has not been established by the importing company, the ESI endpoint is taken as the LLOQ of the analytical method.[105]

2.5.5 Risk Communication

The aims in risk communication are first to involve or inform all participants in the food chain, including marketing companies, regulatory authorities, and consumers, of the nature of risks associated with drug residues in foods, and then to provide assurance that precautionary principles have been applied to the generation and interpretation of data and the adoption of standards that ensure a safe food supply.

In large part, the risk communication and management procedures encompass the link between prescribing veterinarian and farmer. Farm animal veterinarians, at least in the EU, have more recently assumed a significantly increased role as educators of their clients in the safe and effective use of drugs and the maintenance of adequate records. The latter should constitute an essential element in ensuring compliance with effective therapy and adherence with statutory WhTs. Quality assurance programs, instituted by veterinarians and producers, have made a significant contribution to reducing the occurrence of non-compliant residues and thereby providing assurances on food safety to the public.[125,126]

2.6 RESIDUE VIOLATIONS: THEIR SIGNIFICANCE AND PREVENTION

2.6.1 Roles of Regulatory and Non-regulatory Bodies

Residue violations may occur as a consequence of the use of drugs and pesticides or from environmental contaminants and naturally occurring toxicants in foods. Drugs (including pesticides registered for veterinary use) are the most commonly detected chemicals in animal-derived foods and, of these, a large majority of positive findings are AMDs. Dowling has outlined the roles of the US Department of Agriculture's Food Safety and Inspection Service (FSIS) and the Canadian Food Inspection Agency (CFIA) in monitoring meat, poultry, eggs, and honey for residues of chemicals, including AMDs.[51] FSIS monitors tissues through its National Residue Program (NRP). Both agencies utilize hazard analysis and critical control point (HACCP)-based systems as the basis for conducting risk analyses.

Annually, FSIS and CFIA analyze approximately 300,000 and 200,000 samples, respectively, from all market classes of food-producing animals. When a non-compliant residue is detected in a slaughter animal or food animal product, it is condemned. FSIS informs USFDA of residues violations and seeks to obtain the names of producers of products and/or to identify other parties offering animals or products for sale. Appropriate action by the federal agency may include follow-up inspections, seizure and recall of products, and, on the basis of a surveillance plan, further sampling. The action taken depends on the magnitude of the health risk, and emphasis is placed on avoidance of any repeat occurrence and/or further distribution of products.

The standard adopted by FSIS and CFIA is that the incidence of non-compliant residues should not exceed 1%, when veterinary drugs are administered according to label instructions. Any value greater than 1% is deemed to indicate that a product has not been used in accordance with label directions. As discussed by Paige et al., examples include administration to a non-approved species, administration of doses exceeding the recommended maximum, administration by a non-approved route, failure to adhere to prescribed WhTs, failure to maintain treatment records (and thereby failure to identify treated animals), and administration of drug products in error.[127]

Another cause of non-compliant residues, notably in culled dairy cows and veal calves, is the salvaging of animals for slaughter following AMD treatment. Dowling reported that the consumption of medicated feeds is a common cause of residue violations in pigs and poultry, partly as a consequence of the difficulties involved in adhering to WhTs.[51] Adherence may be both expensive and inconvenient, in that it involves replacement of medicated with non-medicated feed in the WhT. Contamination of compound feeds can also arise through inadequate processes in mills, including plant design, and inadequate pre-mix formulation. Croubels et al. list problem compounds as sulfonamides, tetracyclines, nitroimidazoles, nitrofurans, nicarbazin, and ionophore coccidiostats.[35] In some jurisdictions, products can be used in a non-approved species, under the responsibility of the prescribing veterinarian and based on estimation of a suitable WhT. Such recommendations may be based on estimations lacking sufficient accuracy. Gehring et al. have discussed the application of risk management principles to the extra-label (off-label) use of drugs.[128]

As discussed below, various residue detection programs are in use. All are designed to minimize the incidence of non-compliant residues. In the matter of prevention, the role of the Food Animal Residue Avoidance Databank (FARAD) should be noted. FARAD is a USDA-supported computerized databank, established in 1982, with the objective of minimizing residue violations, through the collection, collation, and dissemination of information relevant to residues prediction. It is a cooperative project between North Carolina State University, the University of Florida, and the USDA. FARAD collates information on approved animal drugs, extralabel drug use and environmental toxins on a searchable computer database. It incorporates data in an on-line database (VetGRAM) for more than 1000 drugs and chemicals and more than 20,000 published pharmacokinetic studies. The latter includes a range of pharmacokinetic parameters and variables (clearance, volume of distribution, half-life, maximum concentration, etc.) on drugs, pesticides, and environmental contaminants with potential for presence in livestock tissues. Mathematical models of residue depletion have been developed from the pharmacokinetic data; these predict residue depletion profiles, irrespective of administered dose. A second role of FARAD, both educational and consultative, is to provide advice on residue avoidance and mitigation for chemical contamination incidents and the extralabel use of drugs. The database also provides regulatory information on

1. Indications and directions for use of drugs in food-producing animals for therapeutic and production-enhancing purposes
2. Toxicokinetic data
3. Foreign registration and safety data
4. Tolerances of AMDs in tissues, eggs, and milk
5. Withholding times
6. Bibliographic citations

All FARAD pharmacokinetic data have been published in book form.[109] It is regularly updated. Of 912 enquiries to FARAD reported in 2003, the greatest numbers were for AMDs (338) and NSAIDs (143). Of AMD enquiries the greatest numbers related to dairy cattle, followed by pigs and beef cattle.[129]

In 1998 the concept was extended to development of a global(g) FARAD, embracing several countries and facilitating international dissemination of data on drugs used in food animals and residue avoidance. The collaboration extends to sharing data on withdrawal recommendations, interspecies extrapolations, and the extralabel use of drugs. Data may be obtained from the FARAD organizations at FARAD@ncsu.edu, FARAD@ucdavis.edu, or www.farad.org for the United States and cgfarad@umontreal.ca or www.cgforad.usask.ca for Canada. The FARAD advisory service has particular value in relation to the extralabel use of AMDs, for example, use of a dose different from that authorized or in a different food-producing species. This was legalized in the United States with passage of the Animal Medicinal Drug Use Clarification Act of 1994 (AMDUCA).[130] AMDUCA applies solely to FDA-approved animal and human drugs administered only by or under the order of a licensed veterinarian.[131] AMDUCA does not permit extralabel use of drugs in feed and also specifically prohibits the extralabel use of fluoroquinolones, glycopeptides, furazolidone, nitrofurazone, chloramphenicol, dimetridazole, ipronidazole, other nitroimidazoles (such as metronidazole), and sulfonamide drugs in lactating dairy animals (except for approved uses of sulfadimethoxine, sulfabromomethazine and sulfaethoxypyridazine). Under AMDUCA the hierarchy of use of AMDs in food-producing animals is

1. A product approved for the condition being treated that is effective as labeled
2. A product approved for a food animal that may be used in an extralabel manner
3. A product approved in non-food-producing animals or humans that may be used in an extra-label manner

If no products exist that satisfy these requirements, a compounded product may be permitted.

In the EU, the 1999 ban on the use of AMDs as growth promoters included avoparcin, ardacin, zinc bacitracin, spiramycin, tylosin, virginamycin, carbadox, and olaquindox. Also in the EU, avilamycin, flavomycin, salinomycin, and monensin were phased out in 2006.

In addition to the several classes of residue detection program and the excellent FARAD program, there are non-regulatory approaches to minimizing the incidence of non-compliant residues. Professional associations (e.g., the UK National Office of Animal Health) and academics training future veterinarians have roles to play in drawing attention to guidelines issued by the local jurisdiction. Roeber et al. stress the importance of the responsible use of drugs on farms and the implementation of quality assurance programs by producers.[126] There is a key role for veterinarians through the provision of advice to farmers on matters of good practice and injection technique to reduce wastage of meat as a result of trimming blemishes. Pharmaceutical companies will continue to make advances in product formulation and drug delivery technologies. As potential means of tackling injection site residue problems, the use of biodegradable polymers as drug carriers, injectable microspheres, and microcapsules may be mentioned. For sustained drug release at intramuscular injection sites, the use of liposomes as carriers has been reported. Finally, there is a range of residue detection programs now in use, and their specificity and sensitive increases with time.

2.6.2 Residue Detection Programs

Federal and national agencies have adopted residue detection (control) programs for both domestically produced and imported products. The control programs vary, as they are structured according to the needs of the particular country, but overall it might be noted that published results from these programs provide overwhelming evidence of a safe food supply. The methods used are of two classes, screening and confirmatory. The former detect the presence of an analyte and have the capacity for high sample throughout, checking large sample numbers for potential non-compliance, that is, positive results. The latter enable the analyte to be identified unequivocally and if necessary quantified at the concentration of interest (e.g., at 0.5 MRL). They utilize a range of methods, with many laboratories moving progressively to mass spectrophotometric methods, as described in subsequent chapters of this text. Additionally, domestic residue sampling is classified into four categories according to purpose: monitoring, enforcement, surveillance, and exploratory.[51] Another means of classifying methods is on the basis of analytical principle used, namely, physicochemical, immunochemical, and microbiological. Immunochemical assays are further subdivided into immunoassays, including enzyme-linked immunosorbent assay (ELISA) and immunoaffinity chromatography (IAC). Physicochemical methods are based on chromatographic purification following spectroscopic quantification such as ultraviolet, fluorescence, or mass spectrometric methods. Microbiological methods comprise rapid screening tests. They are inexpensive, provided an extraction procedure is not required. Domestic programs are a trade requirement, either mandatory or as an expectation of importing countries. For detailed discussion, see Croubels et al.[35] and De Brabander et al.[132]

2.6.2.1 *Monitoring Program*

These programs involve a statistically based selection of random samples, each analyzed for selected drugs, collected from healthy animals at slaughter. Animals or animal-derived foods (eggs, milk, honey) sampled are normally those that have passed ante- and postmortem inspections. The residues are assessed for compliance with the MRL. Food products are not retained prior to analysis, but may be recalled, if analyses suggest a public health concern. The data obtained are used to evaluate trends, such as drug misuse, which may lead subsequently to a targeted sampling program. The number of animals sampled is selected to provide a 95% probability of detecting one or more violations when 1% of the animal population contains a non-compliant residue concentration, that is, exceeding the MRL.[133]

Dowling summarized data obtained from United States (FSIS) and Canadian (CFIA) monitoring programs (predominantly meat, eggs, and honey) in 2003.[51] For example, FSIS identified 87 residue violations from 26,214 samples. Of these, 5608 samples analyzed for antimicrobial residues (excluding sulfonamides) yielded 36 violations. Most non-compliant residues related to neomycin, for which there were 29 violations in veal calves. In addition, there were 14 violations for sulfonamides from 5276 samples. Dowling also reported data for CFIA for 2002/03.[51] It is of interest to note for CFIA analyses, using microbial inhibition tests, the presence of inhibitors in 42 of 312 rabbit samples, while of 1055 honey samples, non-compliant residues were reported for erythromycin (2), oxytetracycline (14), sulfathiazole (2), and tetracycline (10).

In the EU, monitoring programs are conducted according to Council Directive 96/23/EC;[134] residue substances and the number of samples to be tested are defined in Annex I of Directive 96/23.[45] Included are all AMDs with MRLs (group B), including those incorporated in feedstuffs, and banned compounds listed in Annex IV of Council Regulation 2377/90 (group A).[44] The AMDs in group A are chloramphenicol, dimetridazole, metronidazole, ronidazole, and nitrofurans, including furazolidone. For banned substances in the EU, emphasis is placed on identification in a large number of matrices, including meat, urine, and even hair at a concentration as low as possible, in accordance with the principle of zero tolerance. Accordingly, the minimum required performance limit (MRPL) is set at a concentration 10–100 times lower than those generally used for MRLs. For banned substances the requirement is for a positive finding from an initial qualitative multi-residue method to be followed by confirmation of identity using a method that provides sufficient identification points, as specified in EU Commission Decision 2002/657/EC.[45]

As an aside, one might mention hair as a generally very stable matrix for most drugs and metabolites.[135] These authors detected quantitatively sulfonamide and trimethoprim in horse tail hair 3 years after oral dosing in a horse, at a distance of 45–55 cm from the follicle. For the nitrofurans a MRPL of 1 μg/kg has been set in the EU. "Bound Residues and Nitrofuran Detection" (FoodBRAND) comprises a rapid multi-residue screening test and also includes definitive multi-residue reference methods for protein-bound remnants of the nitrofurans.[132]

In addition to the tests conducted by regulatory authorities, major abattoirs often have their own quality assurance programs to ensure freedom of products from non-compliant residues, particularly in products destined for export markets, which may impose a different regulatory limit for certain residues than their national authority. Results of such tests are seldom included in totals published by national authorities, unless their inclusion is clearly stated.

2.6.2.2 *Enforcement Programs*

The object of these programs is to analyze samples from animals judged to be at high risk of having non-compliant residues. High-risk expectation may be due to historical data for a product group and includes appearance on ante- and post-mortem inspections. In North America, typically suspect animals include young calves (aged <3 weeks and weighing <68 kg), culled dairy cows, a visible injection site, animals of a production class in which a residue monitoring program has revealed a high incidence of non-compliant residues, and animals in which there is evidence of infectious disease.

Initial "in plant" tests using microbiologically based rapid screening methods of detection for AMDs, include the "Swab Test on Premises" (STOP), "Fast Antimicrobial Screen Test" (FAST), "Overnight Rapid Beef Identification Test" (ORBIT), "Calf Antibiotic and Sulfonamide Test" (CAST) and "Sulfa-on-Site" (SOS). Also used are a range of ELISAs. Carcasses from suspect animals are retained at abattoirs pending the results of tests. If the screening test gives a positive result, FSIS or CFIA carry out a confirmation test. In the absence of availability of a screening test or if a residue not detected by FAST or STOP is nevertheless suspected, tissue samples are submitted directly to FSIS or CFIA. If the confirmatory test is positive for a non-compliant residue, the carcass is classified as adulterated and condemned. Tests for use "on site" are also available for detecting AMDs in honey, such as the CAP Residue Rapid Inspection Device for chloramphenicol and the Tetrasensor Honey test, which detects four tetracyclines. Novel electrochemical and optical immunosensors, flow cytometric immunoassays, and biochip assay methods for residue analysis are currently under development.[132] Rapid tests are discussed further in Chapter 5.

Data from enforcement programs are evaluated to ascertain their effectiveness in reducing residues. In 2003, FSIS recorded 1923 residue violations from 230,351 samples; these violations included 1470 for AMDs (excluding sulfonamides), 335 for sulfonamides, and 118 for the non-steroidal anti-inflammatory drug, flunixin. A full analysis is provided by Dowling.[51] Of interest because of high incidence in the FAST test on 215,813 samples were 552, 199, and 195 violations for penicillin, sulfadimethoxine, and gentamicin, respectively, all in dairy cows and 372 neomycin positives in young veal calves. These represent high percentages of the 1820 (0.84%) total non-compliances in 1665 animals. The STOP test on a total of 14,360 samples revealed 28 positives for penicillin in dairy cows. These represent a high proportion of the 74 (0.51%) total non-compliances in 64 animals. Similar data generated by CFIA on suspect animals in 2003 in SOS, STOP, and CAST tests, with confirmatory tests, indicated the highest incidence of non-compliances for oxytetracycline (120) and benzylpenicillin (182) from a total of 346 non-compliant samples from 11,877 samples analyzed. The great majority of samples related to pork and beef. In both Canada and the United States, food producers and distributors who violate standards are placed on enhanced inspection, with the objectives of identifying causes and reducing non-compliances.

2.6.2.3 Surveillance Programs

The objective of surveillance sampling is to evaluate the occurrence and incidence of a potential residue concern in a particular animal population, when there is suspicion of non-compliant residues on the basis of herd history or clinical signs. The data obtained provide regulatory authorities with information on whether residues have been reduced by interventions. Carcasses or products may or may not be retained pending laboratory findings; this depends on the nature and weight of the evidence that initiated the surveillance and also varies according to the policies of the particular jurisprudence. Normally, the source of supply is traced and action is taken to ensure non-recurrence. The action may include seizure and disposal of produce, quarantining a farm, additional residue testing at the expense of the producer, and prevention of sale of produce until the commodity has been shown to be safe for consumption and acceptable for sale in domestic and export markets. Additional potential procedures include implementing industry codes of practice, auditing users and operators, and obtaining feedback from sellers. Dowling quotes the example of submitting 6295 market hogs for FSIS screening for sulfonamides; the SOS test revealed 10 sulfamethazine non-compliances.[51]

In the UK, the Veterinary Medicines Directorate has monitoring programs, the results of which are published periodically in the *Veterinary Record* and in the annual report of the Veterinary Residues Committee.[137]

2.6.2.4 Exploratory Programs

Exploratory residue testing is conducted on drugs that do not have MRLs. The objective is to evaluate new methods and approaches for both monitoring and sampling. Exploratory testing generates information on non-MRL drugs, but is not designed to facilitate regulatory action.

2.6.2.5 Imported Food Animal Products

In addition to residue monitoring, enforcement, surveillance, and exploratory programs, designed to evaluate residues in animal-derived foods produced in the country of origin, regulatory authorities conduct inspections on imported foods. These are actually re-inspections, the extent of which depends on previous knowledge of the particular exporting country's standards. Each importing country seeks to verify that imported foods meet the same standards as those operating under the domestic programs. As discussed by Dowling, in 2003 the United States tested various foods (processed meat, poultry, and eggs) for residues in eight compound classes of veterinary drugs and pesticides.[51] Of 2212 samples tested, two residue non-compliances were identified, both for the anthelmintic, avermectin. Interestingly, in 2003 Canadian import testing revealed 27 and 6 detections of chloramphenicol in honey imported from India and the United States, respectively.

2.6.2.6 Residue Testing in Milk

The problems and penalties associated with residues of AMDs in milk and other dairy products have given rise to several tests for monitoring non-compliant residues of AMDs in milk. These include receptor-binding assays, microbial receptor assays, microbial growth inhibition assays, enzyme assays, and chromatographic analyses. These are used to test bulk tank milk and include the following: Charm SL-Beta-lactam, DelvoTest P5 Pack Beta-lactam, IDEXX SNAP Beta-lactam, and Charm II Tablet Competitive Beta-lactam. The Delvo test is very widely used. More recently, rapid tests, providing results in 3 min, have been described, for example, Charm MRL-3 and β-STAR 1+1 (STAR = Screening Test for Antimicrobial Residues). The Parallux Milk Residue Testing System detects six β-lactams, tetracyclines, spectinomycin, neomycin, streptomycin, spiramycin, sulfonamides, and quinolones in one test within 4 min.[132] As these commercial names indicate, there is particular concern over the presence of non-compliant concentrations of AMDs of the β-lactam group (penicillins and cephalosporins). In fact, it is rare for inhibitor-positive milk to contain antibiotics other than β-lactams. Most methods therefore used *Geobacillus stearothermophilus*, which is very sensitive to β-lactams (see Chapter 5. for further discussion). Riviere and Sundlof point out that summarizing available tests is difficult because of rapid advances in methods leading to new tests.[123]

In 2003 in the United States, 4,456,141 tests for residues in milk and other dairy products yielded 3246 (0.07%) positive results. Of these, 4,354,087 tests and 3207 positives (0.07%) were for β-lactam drugs.[138] Of 66,124 tests for sulfonamides, 23 (0.03%) were positive. In the same year, of 3577 milk and cheese products tested by CFIA, none yielded positive results for AMDs or sulfonamides. The screening tests in use have good sensitivity and negative predictive values, but poor positive predictive values. Thus, a positive test on an individual cow does not necessarily indicate a bulk tank milk concentration exceeding the MRL (see Chapter 5 for further discussion).

In setting withholding periods for AMDs in milk, regulatory authorities allow the assumption that milk from a treated cow will be admixed with milk from untreated animals; that is, MRLs for milk are based on bulk tank concentrations, prompting the comment that "the solution to pollution is dilution." This is not unreasonable, as the human consumption of milk from a single animal will be a relatively rare event, at least in developed countries. An increased somatic cell count (SCC) of bulk milk is an indicator of the prevalence of mastitis in a dairy herd. Such infections are widespread and are therefore widely and routinely treated by AMDs administered by intramammary infusion (most countries) or systemically (Scandinavian countries). Cases of peracute mastitis will almost always be treated systemically in all countries. There is an expectation that a non-compliant residue is more likely to occur in milk from herds with a high SCC. The US Pasteurized Milk Ordinance requires all bulk milk tankers to be sampled and analyzed for AMD residues prior to processing.[139] Additionally, at least four samples from pasteurized milk and milk products are required to be tested from each plant at 6-month intervals, and each producer must be tested at least 4 times every 6 months. Typically, dairies include additional testing for their own purposes to ensure the freedom of milk from residues that pose a risk to the consumer, or to manufacture of products such as yogurt and cheese. Results of these tests are seldom included in the totals published by national authorities.

2.7 FURTHER CONSIDERATIONS

2.7.1 Injection Site Residues and Flip-Flop Pharmacokinetics

When drug products are administered to animals by a parenteral (other than the intravenous or oral) route, usually intramuscularly or subcutaneously, drug concentrations at the injection site are initially always high. As drug is absorbed into the circulation, the concentration falls rapidly. In the case of drugs administered as aqueous solutions, such as aminoglycosides and the sodium or potassium salt of a penicillin, the drug normally remains in solution at the injection site and complete absorption is generally very rapid. Thus, for sodium benzylpenicillin administered intramuscularly, T_{max} occurs within 10–20 min of administration. In this circumstance, depletion from the injection site is sufficiently rapid to ensure that the concentration in injection site muscle decreases to a concentration not distinguishable from non-injection-site muscle by the time of slaughter. Therefore, marker tissue will not be injection site muscle, and may or may not be non-injection-site muscle.

However, for many AMDs and indeed drugs of other classes (e.g., anthelmintics), there has long been a practice of developing slow-release (depot) formulations, administered intramuscularly, subcutaneously, or as "pour-on products," for use in farm animal species. As discussed in Section 2.2.3, these products commonly display flip-flop pharmacokinetics, in which the terminal half-life represents a slow absorption phase and is longer than the elimination half-life determined after intravenous dosing. The advantages and disadvantages of depot preparations are summarized in Table 2.15.

The potential complexity of the pharmacokinetic profile for slow-release products is illustrated by the early study of Toutain and Raynaud.[87] These workers administered a 20% w/v solution of oxytetracycline intramuscularly to young calves at a dose rate of 20 mg/kg. The data best fitted an open two-compartment model (central and peripheral) with two absorption compartments (rapid and slow release). The rapid absorption phase was attributed to immediate

TABLE 2.15 Advantages and Disadvantages of Parenteral Depot Formulations of AMDs

Advantages	Disadvantages
Provide products with long duration of action, requiring single dosing and/or or a long (48–72-h) dosing interval	Generally have much longer withholding periods than rapidly absorbed formulations, discouraging innovative developments and field use
Greater convenience and lower cost than products requiring more frequent dosing (e.g., once daily for 3–4 days)	Burden for farmers required to adhere to prolonged withholding period
Greater compliance with administration of dosage schedule	Greater risk for consumers through possible non-adherence to prolonged withholding periods
Improved consumer safety through greater compliance	Possibly pain and inflammation at injection sites, leading to a local response involving formation of fibrous granulation tissue that can enclose and create a "protected pocket" containing AMD
Improved animal welfare by minimizing stress of handling and pain on repeated injection	Problems for regulatory authorities in setting withholding periods based on slow depletion from injection sites
	Threats to international trade through persistence of residue at injection sites

availability of a small fraction (14%) of the administered dose, with an absorption half-life of 48 min. The slow absorption phase was associated with a larger fraction (37.5%) of the administered dose, with a half-life of 18.1 h. As elimination half-life after intravenous dosing was 9.04 h, the product displayed flip-flop pharmacokinetics.

Some depot products have given rise to injection site residue concerns, as discussed in official guidelines (e.g., see Ref. 140), in a review in 2007,[141] and in peer-reviewed articles.[142,143] The quantity of administered drug is generally large in sustained-release products, as it is required to provide both initially high and then well-maintained therapeutic concentrations (over $\geq$2 days). Injection site residues of slow-release formulations are likely to be less of a problem when absorption is steady, but in practice depletion is often erratic and non-exponential and therefore unpredictable. This seems to arise partly because the products, to achieve prolonged release, are formulated as suspensions in either water (e.g., benzathine and procaine benzylpenicillins) or water-repelling fixed oils (e.g., procaine benzylpenicillin) or as solutions containing organic solvents [e.g., high-strength (20–30%) oxytetracycline]. For the latter, after intramuscular or subcutaneous dosing, the rapid absorption of organic solvents leads to precipitation of the AMD, which is then slowly taken up into solution in interstitial fluid at the injection site. In addition, variable amounts of the suspension or precipitate may induce a local acute inflammatory response and/or, as a foreign body, become walled off by granulation tissue and therefore subject to very slow and erratic absorption.[80,82]

There is a lack of consistency between jurisdictions in addressing the issue of injection site residues.[141] One reason for the lack of harmonization of risk assessment is the limited data available on the probability of dietary exposure to injection sites containing residues.[144] The paucity of data extends to three areas: (1) prevalence of injection site tissue at slaughter, (2) incidence of remaining residues above MRL at injection sites, and (3) the fate of injection sites. Determination of prevalence is confounded by regional differences in animal husbandry practices, so that extrapolation cannot be made from data obtained in one region to another. Injection site tissue is trimmed out and discarded when identified, and the efficiency of this procedure (although unknown) is inevitably variable. There are only limited data available on percentages of injection sites that actually contain residues, but the proportion may be small. However, Beechinor and colleagues *have* generated good data on (1) injection site muscle residues of tilmicosin, tiamulin, and enrofloxacin in livestock in Ireland[145] and (2) prevalence and public health significance of blemishes in cuts of Irish beef and pork.[146,147]

Several regulatory approaches have been taken to address injection site residues. The one adopted in some jurisdictions (e.g., USA and Australia) is to use ARfD in place of ADI as the permissible exposure standard. This is reasonable, but the validity of this approach depends on consumption of injection site residues being a rare event. Use of ARfD will shorten the WhT, provided its value exceeds that of ADI, and if muscle is the target tissue. An additional consideration is that this approach requires a residue surveillance sampling protocol that can differentiate between injection site and non-injection-site muscle. At several meetings of the CCRVDF, through the 1990s and 2000s, this was proposed but regrettably not adopted. Thus, in 2001 a working paper discussed by the CCRVDF proposed the analysis of two muscle samples; two positive results would indicate violation of the WhT, whereas a single positive would indicate an injection site sample, in which event the ARfD could be applied and violation would occur only if the positive value exceeded ARfD. EMA/CVMP envisaged three practical problems relating to this proposal: (1) injection sites might not be easily identifiable, and moreover it was possible that only a part of an injection site would be sampled; (2) in some cases additional analytical method validation might be required; and (3) an additional analytical method might be needed if the marker residue at the injection site (normally a parent drug) differed from the marker residue in non-injection-site muscle. EMA/CVMP therefore continues to require injection site muscle to be treated as non-injection-site muscle; specifically, the former must decline below the MRL. This was the standard adopted when all benzathine benzylpenicillin–containing parenteral products were banned in the EU; the argument was that potentially serious allergenic consequences of consumption of small amounts (above the MRL of 50 μg/kg) could arise, even if consumption was a very rare event.

In the present authors' opinion, acceptable advances might be made in addressing injection site residue issues through consideration, development, and possible adoption of the proposals of Sanquer et al.[142,143] They questioned the EMA/CVMP guideline, which recommends application of the same calculation method for injection site muscle as other edible tissues. This was considered to be scientifically unsound, on the grounds that injection site residues often violate regression assumptions regarding both homoscedasticity (same variance in residue concentrations for different slaughter times) and linearity (of mean depletion curve in $\log_e$ scale). These authors applied a probabilistic approach in assessing risk of consumption of an injection site, in whole or in part, during one year, based on a 7-day AMD treatment. The analysis indicated, for EU consumers, a maximal risk of 4 days of injection site consumption (containing or not containing residues). They proposed a nonparametric approach for calculation of WhT, stating that acute risk exposure associated with injection site consumption could be more appropriately dealt with by use of ARfD or acute single-dose intake (ASDI) indices. In earlier studies, Concordet and Toutain had already proposed a nonparametric approach as an alternative to the recommended statistical approach.[111,112]

2.7.2 Bioequivalence and Residue Depletion Profiles

Many of the antimicrobial drug products licensed for use in food-producing species are generics, that is, products containing one or more drugs developed initially as pioneer products, but that have subsequently been formulated in products containing the same drug, usually (but not necessarily) in the same concentration and usually (but not necessarily) in a similar formulation. A crucial component of the data required by regulatory authorities for generic products is a study to determine whether the generic and pioneer products are bioequivalent. Bioequivalence allows applicants for MAs of AMDs, which are generic to a pioneer product, to claim essential similarity, in terms of efficacy and safety in each target species.

The assessment of bioequivalence is based on 90% confidence intervals for the ratio of the population geometric means (test/reference) for the parameters under consideration. This method is equivalent to two one-sided tests with the null hypothesis of bio-inequivalence at the 5% significance level. Two products are declared bioequivalent if upper and lower limits of the confidence interval of the mean (median) of log-transformed AUC and $C_{\max}$ each fall within the a priori bioequivalence intervals 0.80–1.25. It is then assumed that both rate (represented by $C_{\max}$) and extent (represented by AUC) of absorption are essentially similar. $C_{\max}$ is less robust than AUC, as it is a single-point estimate. Moreover, $C_{\max}$ is determined by the elimination as well as the absorption rate (Table 2.1). Because the variability (inter- and intra-animal) of $C_{\max}$ is commonly greater than that of AUC, some authorities have allowed wider confidence intervals (e.g., 0.70–1.43) for log-transformed $C_{\max}$, provided this is specified and justified in the study protocol.

Regulatory bodies in general accept that, although the pioneer and generic products are not pharmacokinetically identical, they are nevertheless deemed to be *sufficiently similar* to permit the assumption that they will be therapeutically equivalent. Therapeutic equivalence is taken to mean that the products will have the same efficacy and safety profiles in the target species. This, in turn, means that the company seeking to license the generic product will not normally have to undertake the otherwise extensive laboratory animal and target species safety studies and clinical trials to establish safety and efficacy in clinical use, assuming that the claims for the generic product are identical to those made for the pioneer product.

In relation to residues in food-producing species, it is important to recognize that demonstration of average bioequivalence does not obviate the need for separate residue depletion studies for a generic product. There are several reasons why this is so:

1. It might be noted the definitions of WhT and average bioequivalence are fundamentally different; it should be stressed that to guarantee a 90% confidence interval for the ratio of the two treatment means of AUC and $C_{\max}$, respectively, it should be entirely contained within the range 80–125%. Meeting this criterion gives no guarantee that the upper 95% confidence limit of the 95th percentile of the population is below the MRL. Indeed, bioequivalence can be determined at three levels: average, population, and individual. It is outside the scope of this chapter to discuss each of these, except to say that average bioequivalence is the most easily satisfied (least stringent) of the three. Regulatory

authorities require companies to demonstrate average bioequivalence only, and two formulations can be declared bioequivalent while their variances for AUC are different, and this may have a large impact on WhT that controls a population percentage, not a mean parameter. Only a so-called population bioequivalence could also guarantee equivalence of variance.

2. It is clear that, for a parenteral product administered intramuscularly or subcutaneously, depletion from the injection site may well be sufficiently similar to provide bioequivalence variables that fall well within the preset limits, but are nevertheless sufficiently different to yield significant, even quite large, differences in concentration at the injection site.[141]
3. It is not only at the actual injection site, however, that residue depletion is likely to differ; depletion rate is not guaranteed either for *all* edible tissues. This is because bioequivalence demonstrates essential similarity in rate and extent of absorption between a pioneer product and a generic product only for the range of therapeutically useful plasma concentrations. Bioequivalence does not guarantee the same rate of decrease of concentration in the terminal phase, which often has no therapeutic meaning. For many drugs (see Section 2.3.1 for an example of gentamicin), a rapid elimination phase is followed by a much slower decline in concentration ($T_{1/2}$ values of 1.83 and 44.9 h, respectively, for β and γ phases for gentamicin in calves). The γ phase represents the unloading and elimination of drugs from tissues, including edible tissues. It is not possible for average bioequivalence established over the first 24 h following product administration to give assurance on the same exposure in the γ phase, that is, much later (in days or even weeks). In addition, a very late terminal phase may reflect a flip-flop phenomenon undetected at the plasma level by a conventional bioequivalence trial. **For all these reasons, there is no (statistical) basis for having the same WhT for different generic products and a pioneer product.**

2.7.3 Sales and Usage Data

Sales and usage of AMDs inevitably vary considerably between and even in regions within countries. As examples, in this section we will consider recent sales data for two countries (United Kingdom and France) and sales data for a single clinical condition (bovine respiratory disease, BRD).

2.7.3.1 Sales of AMDs in the United Kingdom, 2003–2008

The United Kingdom's Veterinary Medicines Directorate (an executive agency of the Department for Environment, Food and Rural Affairs) 2009 report describes sales of AMDs, antiprotozoals, antifungals, and coccidiostats, authorized for use as veterinary medicines, annually for the period 2003–2008.[148] To illustrate trends in usage, this account summarizes data for 2003 and 2008. Prior to that, between 1998 and 2003, the total sale of veterinary therapeutic AMDs in the UK was relatively constant at approximately 434 metric tons per annum. From 435 metric tons in 2003, total sales decreased to 384 metric tons in 2008. Livestock numbers in thousands in the national herd (2003 and 2008) were as follows: pigs (5046 and 4714), cattle (10,508 and 10,107), sheep (35,812 and 33,131), and poultry (178,800 and 166,200). Therefore, the proportional decreases in total AMD sales and in animal numbers were broadly similar.

Sales figures for AMDs and coccidiostats are presented in Table 2.16. The decrease in tonnage sales of AMDs in food animals is partly accounted for by the ban on growth promoters, which took effect on January 1, 2006. For AMDs, it will be seen that by far the largest tonnage (59% of total) relates to medicated feedstuffs, followed by products formulated for oral or water medication (29%) and injectable medication (10%). For intramammary products, 56.6% was for dry-cow and 43.4% for lactating cow therapy. Of the coccidiostats, 72% comprised ionophores. It should be noted that the proportion of the 327 metric tons of AMDs, which was administered to food animals but did not enter the food chain, is unknown.

Categorized by species, by far the largest groups of AMD sales were for pigs and poultry, with smaller and

TABLE 2.16 Sales of Therapeutic AMDs and Coccidiostats in UK in 2003 and 2008 (Metric Tons of Active Ingredient)

Category	2003	2008
Therapeutic AMDs		
Total sales	435	384
Food-producing animals only	377	327
Combination of food and non-food-producing animals	28	18
Non-food-producing animals only	30	38
Growth-promoting products	36	0
Medicated feedings stuffs	307	228
Oral/water medication	87	112
Injectables	34	38
Intramammaries	5	4
Intramammaries[a]	4735	4092
Dry cow[a]	2590	2317
Lactating cow[a]	2145	1775
Coccidiostats		
Total coccidiostats	240	207
Ionophores	190	150
Non-ionophores	50	57

[a]Values in kilograms, not metric tons.

Source: Data from Veterinary Medicines Directorate (2009).[148]

TABLE 2.17 Sales of Therapeutic AMDs in UK in 2003 and 2008 (Metric Tons of Active Ingredient) by Species and Chemical Groups

Product	2003	2008
Species		
Cattle only	12	11
Pig only	70	62
Poultry only	11	31
Fish only	2	1
Pig plus poultry	261	195
Multiple edible animal species	21	28
Chemical group		
Tetracyclines	212	174
Trimethoprims/sulfonamide	89	70
Trimethoprims	15	12
Sulfonamides	74	58
β-Lactams	62	69
Cephalosporins[a]	3(3037)	6(6242)
Penicillins[b]	16	13
Other penicillins[c]	43	50
Aminoglycosides	21	18
Streptomycin	7	6
Neomycin + framycetin	5	1
Other aminoglycosides[d]	9	11
Macrolides	39	35
Fluoroquinolones[a]	1(1364)	2(1928)
Others	12	15

[a]Values in kilograms given in parentheses.
[b]Includes potassium and procaine salts of benzylpenicillin.
[c]Includes cloxacillin, amoxicillin, ampicillin, nafcillin, and penethamate hydriodide.
[d]Includes gentamicin, apramycin, kanamycin, and spectinomycin.
Source: Data from Veterinary Medicines Directorate (2009).[148]

much smaller tonnages used in cattle and fish, respectively (Table 2.17). Of the 195 metric tons of AMDs in the pig–poultry category, VMD has estimated 60% usage in pigs, 38% in poultry, and 2% were sold off-label for use in other (non-authorized) bird species (e.g., duck, turkey, game). Classified by chemical grouping, by far the largest category is the tetracycline group (45% of total sales) followed by potentiated sulfonamides and β-lactams (18% each). Most tetracyclines were sold for pigs and poultry as medicated feedstuffs under veterinary prescription. While total tonnages were small, it is of interest to note increasing trends in the sales of fluoroquinolones and cephalosporins and the decreased use, between 2003 and 2008, of neomycin and framycetin, due to the withdrawal of neomycin from the market. In 2008, the total number of AMD, antiprotozoal and coccidiostat products sold (for all species, including non-food-producing animals) was 370, made up as follows: β-lactams 131, tetracyclines 46, trimethoprim/sulfonamides 41, others 41, aminoglycosides 28, fluoroquinolones 25, macrolides 22, coccidiostats 11, and antiprotozoals 10. AMD products can be imported into the UK, when no authorized products are available; sales of imported AMDs increased markedly from 159 to 3883 kg of active ingredient, between 2003 and 2008, but remained a very small proportion of total sales.

Because of the nature of the sales data harvested, it is not possible to specify, on an interspecies basis, the precise usage/sales data for particular drug classes. The VMD report emphasizes that there is no central record of the use of AMDs in animals in the UK. However, VMD has estimated, from liveweight slaughter data for cattle, pigs, sheep, poultry, and fish (5,327,000 metric tons in 2003 and 5,516,000 metric tons in 2008) that one metric ton of AMD was used to produce 12,898 metric tons of liveweight animal in 2003 and 16,869 metric tons of liveweight animal in 2008. These data correspond to the sale of 80 g (2003) and 60 g (2008) of AMD for each metric ton of liveweight animal slaughtered.

2.7.3.2 Comparison of AMD Usage in Human and Veterinary Medicine in France, 1999–2005

Moulin et al. compared tonnages of AMDs sold in human and veterinary medicine in France for the 7-year period from 1999 to 2005.[149] Data were compiled from the registers of the French Agency for Veterinary Medicinal Products (AFSSA ANMV) for animals and the French Health Products Safety Agency (AFSSAPS) for humans. Data in tonnages of active ingredients were related to animal and human biomasses to compare usages expressed in mg/kg of body weight (Table 2.18). While approximately 60% and 40% of total tonnages were in animals and humans, respectively, in relation to unit biomass, usage was 2.4 times higher in humans than in animals.

The highest sales in humans and animals were β-lactams and tetracyclines, respectively. Tetracyclines alone, in veterinary medicine, represented approximately 50.4% of all sales, while tetracyclines, sulfonamides/trimethoprim, β-lactams, and aminoglycosides combined accounted for more than 80% of AMDs used. During the 7-year period, sales of cephalosporins and fluoroquinolones in veterinary medicine increased by 38.4% and 31.6%, respectively. Nevertheless, as percentages of total veterinary sales, their use remained relatively small: cephalosporins 0.64% and fluoroquinolones 0.33% in 2005. In animals, oral administration accounted for 88% of sales, and estimated sales for parenteral products were 10.5%. Moreover, while 92% of total tonnage was intended for food-producing animals, 64% of cephalosporins were intended for pets.

The human/animal comparison revealed that some classes were used almost exclusively either in animals (aminoglycosides, amphenicols, polymyxins, tetracyclines) or in humans (nitrofurans). Expressed as percentages of total tonnage sales within each sector (animal or human), several differences were revealed as follows (animal first, humans second): tetracyclines 50.4 and 1.7,

TABLE 2.18 AMD Consumption and Biomasses Estimated in Humans and Animals in France from 1999 to 2005

	AMD Sales (metric tons)		Population Body Mass (metric tons)		AMD Sales Relative to Biomass (mg/kg live weight)	
Year	Human	Animal	Human	Animal	Human	Animal
1999	896.20	1316.31	3,597,843	17,122,220	249.1	76.9
2002	809.44	1331.53	3,709,154	17,268,049	218.2	77.1
2005	759.67	1320.10	3,810,215	15,795,105	199.4	83.6

Source: Data from Moulin et al (2008).[149]

TABLE 2.19 Global Animal Health Sales by Product Category, Species, and Country in 2005 in $ ($billions)

Product Category	Sales	% of Total	Species	Sales	% of Total	Country	Sales	% of Total
Antiparasitics	4.875	28	Dogs + cats	5.75	33	USA	6.29	36.1
			Cattle	4.885	28	China	1.095	6.3
Biologicals	3.655	21	Pigs	2.94	17	France	1.04	6.0
Antimicrobials	2.785	16	Poultry	1.94	11	Brazil	0.909	5.2
Medicated feed additives	1.915	11	Horses	1.045	6	UK	0.825	4.7
Other pharmaceuticals	4.180	24	Other food animals + aquaculture	0.85	5	Japan	0.793	4.6
						Germany	0.74	4.3
						Others	5.718	32.8
Total	17.41	100	Total	17.41	100	Total	17.41	100

Source: Data provided by A. R. Peters.[150]

TABLE 2.20 Estimated Sales of AMDs for BRD Therapy in Four Territories

Territory	Sales of AMDs ($millions)	Percent of Global BRD Market (%)
USA	250	36.1
EU	186	26.7
China	44	6.3
UK	32.5	4.7

Source: Data provided by A. R. Peters.[150]

sulfonamides/trimethoprim 18.8 and 2.9, aminoglycosides 5.9 and 0.2, polymyxins 4.9 and 0.2, β-lactams 8.3 and 51.6, and macrolides 9.0 and 14.8.

2.7.3.3 *Global Animal Health Sales and Sales of AMDs for Bovine Respiratory Disease*

Data for AMD usage and sales in the treatment of bovine respiratory disease (BRD) have been supplied by Professor A. R. Peters.[150] BRD is a major cause of reduced productivity and economic loss globally;[151] in the United States alone the annual total cost to the cattle industry is estimated to approach $2 billion. In 2005 the total global animal health market was estimated at $17.4 billion. Table 2.19 classifies this on the basis of product category, animal species, and country. Together AMDs and medicated feed additives constitute 27% of the total; cattle, pigs, and poultry together make up 56% of the total; and the United States as a country takes 36.1% of the total. In 2006 AMD sales by country in $million were United States 1280, China 521, France 315, Spain 252, Germany 214, and UK 164.

TABLE 2.21 Analysis of US Market Share of Main Actives for BRD Therapy

Active	Product	Market Share	Withholding Period (days)
Tulathromycin	Draxxin	35	49
Enrofloxacin	Baytril	20	10–14
Ceftiofur	Excede		13
	Exenel	20	8
Tilmicosin	Micotil	18	60
Florfenicol	Nuflor	3.0	30–44
Ceftiofur	Naxcel	2.5	71
Danofloxacin	Advocin 180	1.5	8

Source: Data provided by A. R. Peters.[150]

Tables 2.20 and 2.21 present estimated data for sales of AMDs for treatment and prevention of BRD in three countries and the EU, together with an analysis of major products used in the United States. The market is dominated by the United States and EU, which together account for 63% of the global BRD market. The US market is dominated by two products containing macrolides (53% of total), three products containing ceftiofur (22.5% of total), and two products containing fluoroquinolones (21.5% of total). While florfenicol has a much smaller percentage share, it is nevertheless significant in view of the market size of $250 million; 3% of this market is $7.5 million annually.

REFERENCES

1. Toutain PL, Bousquet-Mélou A, Bioavailability and its assessment, *J. Vet. Pharmacol. Ther*. 2004;27:455–466.
2. Toutain PL, Bousquet-Mélou A, Volumes of distribution, *J. Vet. Pharmacol. Ther*. 2004;27:441–453.
3. Toutain PL, Bousquet-Mélou A, Plasma clearance, *J. Vet. Pharmacol. Ther*. 2004;27:415–425.
4. Toutain PL, Bousquet-Mélou A, Plasma terminal half-life, *J. Vet. Pharmacol. Ther*. 2004;27:427–439.
5. VICH GL 47, MRK—*Comparative Metabolism Studies. Studies to Evaluate the Metabolism and Residue Kinetics of Veterinary Drugs in Food-Producing Animals*, Comparative Metabolism Studies in Laboratory Animals, Draft 1, Nov. 2009 (available at `http://www.vichsec.org/en/guidelines3.htm`; accessed 11/30/10).
6. VICH GL 46, MRK—*Nature of Residues. Studies to Evaluate the Metabolism and Residue Kinetics of Veterinary Drugs in Food-Producing Animals: Metabolism Study to determine the Quantity and Identify the Nature of Residues*, Draft 1, Nov. 2009 (available at `http://www.vichsec.org/en/guidelines3.htm`; accessed 11/30/10).
7. VICH GL 48, MRK—*Marker Residue Depletion Studies, Studies to Evaluate the Metabolism and Residue Kinetics of Veterinary Drugs in Food-Producing Animals: Marker Residue Depletion Studies to Establish Product Withdrawal Periods*, Draft 1, Nov. 2009 (available at `http://www.vichsec.org/en/guidelines3.htm`; accessed 11/30/10).
8. Lees P, Shojaee Aliabadi F, Toutain PL, PK-PD modelling: An alternative to dose titration studies for antimicrobial drug dosage selection, *Regul. Aff. J. Pharmacol*. 2004;15:175–180.
9. Lees P, Concordet D, Shojaee Aliabadi F, Toutain PL, Drug selection and optimisation of dosage schedules to minimize antimicrobial resistance, in *Antimicrobial Resistance in Bacteria of Animal Origin*, Aarestrup FM, ed., ASM Press, Washington, DC, 2006, pp. 49–71.
10. Lees P, Svendsen O, Wiuff C, Strategies to minimise the impact of antimicrobial treatment on the selection of resistant bacteria, in Guardabassi L, Jensen LB, Kruse H, eds., *Guide to Antimicrobial Use in Animals*, Blackwell, Oxford, 2008, pp. 77–101.
11. Toutain, PL, Pharmacokinetics/pharmacodynamics integration in drug development and dosage regimen optimization for veterinary medicine, *J. Am. Acad. Pharm. Sci*. 2002;4(4): 160–188 (available at `http://www.pharmagateway.net/ArticlePage.aspx?DOI=10.1208/ps040438`; accessed 12/6/10).
12. Toutain PL, del Castillo JRE, Bousquet-Mélou A, The pharmacokinetic-pharmacodynamic approach to a rational dosage regimen for antibiotics, *Res. Vet. Sci*. 2002;73: 105–114.
13. Toutain PL, Pharmacokinetics/pharmacodynamics integration in dosage regimen optimization for veterinary medicine, in Riviere JE, Papich M, eds., *Veterinary Pharmacology and Therapeutics*, 9th ed., Wiley-Blackwell, Iowa State Univ. Press, Ames, 2009; pp. 75–98.
14. European Commission Notice to Applicants, *Veterinary Medicinal Products. Presentation and Contents of the Dossier*, 2004 (available at `http://ec.europa.eu/health/files/eudralex/vol-6/b/vol6b_04_2004_final_en.pdf`; accessed 12/06/10).
15. Aliabadi FS, Lees P, Pharmacokinetics and pharmacokinetic/pharmacodynamic integration of marbofloxacin in calf serum, exudate and transudate, *J. Vet. Pharmacol. Ther*. 2002;25:161–174.
16. Aliabadi FS, Lees P, Pharmacokinetic-pharmacodynamic integration of danofloxacin in the calf, *Res. Vet. Sci*. 2003;74: 247–259.
17. Riviere JE, Pharmacokinetics, in Riviere JE, Papich MG, eds., *Veterinary Pharmacology and Therapeutics*, 9th ed., Wiley-Blackwell, Iowa State Univ. Press, Ames, 2009; pp. 48–74.
18. Nouws JFM, *Tissue Distribution and Residues of Some Antimicrobial Drugs in Normal and Emergency Slaughtered Ruminants*, dissertation, Univ. Utrecht, The Netherlands, 1978.
19. Nouws JFM, Ziv G, Pre-slaughter withdrawal times for drugs in dairy cows, *J. Vet. Pharmacol. Ther*. 1978;1:47–56.
20. Toutain PL, Anti-inflammatory agents, in *The Merck Veterinary Manual*, 10th ed., Merck, Rayway, NJ, 2009.
21. Papich MG, Riviere JE, Tetracycline antibiotics, in Riviere JE, Papich MG, eds., *Veterinary Pharmacology and Therapeutics*, 9th ed., Wiley-Blackwell, Iowa State Univ. Press, Ames, 2009;915–944.
22. Martin-Jimenez T, Riviere JE, Mixed effects modeling of the disposition of gentamicin across domestic animal species, *J. Vet. Pharmacol. Ther*. 2001;24:321–332.
23. Cherlet M, Baere SD, Backer PD, Determination of gentamicin in swine and calf tissues by high-performance liquid chromatography combined with electrospray ionization mass spectrometry, *J. Mass Spectrom*. 2000;35:1342–1350.
24. Cherlet, M, De Baere S, De Backer P, Quantitative determination of dihydrostreptomycin in bovine tissues and milk by liquid chromatography-electrospray ionization-tandem mass spectrometry, *J. Mass Spectrom*. 2007;42:647–656.
25. Oukessou M, Toutain PL, Effect of dietary nitrogen intake on gentamicin disposition in sheep, *J. Vet. Pharmacol. Ther*. 1992;15:416–420.
26. Reyns T, De Boever S, Baert K, Croubels S, Schauvliege S, Gasthuys F, De Backer P, Disposition and oral bioavailability of amoxicillin and clavulanic acid in pigs, *J. Vet. Pharmacol. Ther*. 2007;30:550–555.
27. Papich MG, Riviere JE, β-lactam antibiotics: Penicillins, cephalosporins and related drugs, in Riviere JE, Papich MG, eds., *Veterinary Pharmacology and Therapeutics*, 9th ed., Wiley-Blackwell, Iowa State Univ. Press, IA, 2009; 865–894.

28. De Baere S, Cherlet M, Baert K, De Backer P, Quantitative analysis of amoxycillin and its major metabolites in animal tissues by liquid chromatography combined with electrospray ionization tandem mass spectrometry, *Anal. Chem*. 2002;74:1393–1401.
29. Reyns T, Cherlet M, De Baere S, De Backer P, Croubels S, Rapid method for the quantification of amoxicillin and its major metabolites in pig tissues by liquid chromatography-tandem mass spectrometry with emphasis on stability issues, *J. Chromatogr. B* 2008;861:108–116.
30. Martinez-Larranaga MR, Anadón A, Martinez MA, Diaz MJ, Frejo MT, Castellano VJ, Isea G, De La Cruz CO, Pharmacokinetics of amoxycillin and the rate of depletion of its residues in pigs, *Vet. Rec*. 2004;154:627–632.
31. De Baere S, Wassink P, Croubels S, De Boever S, Baert K, De Backer P, Quantitative liquid chromatographic-mass spectrometric analysis of amoxycillin in broiler edible tissues, *Anal. Chim. Acta* 2005;529:221–227.
32. Firth EC, Nouws JFM, Driessens F, Schmaetz P, Peperkamp K, Klein WR, Effect of the injection site on the pharmacokinetics of procaine penicillin-G in horses, *Am. J. Vet. Res*. 1986;47:2380–2384.
33. Papich MG, Korsrud GO, Boison JO, Yates WD, MacNeil JD, Janzen ED, Cohen RD, Landry DA, A study of the disposition of procaine penicillin G in feedlot steers following intramuscular and subcutaneous injection, *J. Vet. Pharmacol. Ther*. 1993;16:317–327.
34. Anadón A, Martinez-Larranaga MR, Residues of antimicrobial drugs and feed additives in animal products: Regulatory aspects, *Livestock Prod. Sci*. 1999;59:183–198.
35. Croubels S, Daeseleire E, De Baere S, De Backer P, Courtheyn D, Residues in meat and meat products, feed and drug residues, in Jenson WK, Devine C, Dikeman M, eds., *Encyclopedia of Meat Sciences*, Elsevier Science, Academic Press, Oxford, 2004;1172–1187.
36. Anadón A, Martinez-Larranaga MR, Diaz MJ, Velez C, Bringas P, Pharmacokinetic and residue studies of quinolone compounds and olaquindox in poultry, *Ann. Rech. Vet*. 1990;21(Suppl. 1):137S–144S.
37. Hornish RE, Gosline RE, Nappier JM, Comparative metabolism of lincomycin in the swine, chicken, and rat, *Drug Metab. Rev*. 1987;18:177–214.
38. Giguere S, Lincosamides, pleuromutilins and streptogramins, in Giguere S, Prescott JF, Baggot JD, Walker RD, Dowling PM, eds., *Antimicrobial Therapy in Veterinary Medicine*, 4th ed., Wiley-Blackwell, Univ. Iowa Press, Ames, 2006:179–205.
39. Burrows GE, Barto PB, Martin B, Tripp ML, Comparative pharmacokinetics of antibiotics in newborn calves: Chloramphenicol, lincomycin, and tylosin, *Am. J. Vet. Res*. 1983;44:1053–1057.
40. Papich MG, Riviere JE, Chloramphenicol and derivatives, macrolides, lincosamides and miscellaneous antimicrobials, in Riviere JE, Papich MG, eds., *Veterinary Pharmacology and Therapeutics*, 9th ed., Wiley-Blackwell, Iowa State Univ. Press, Ames; 2009:945–982.
41. Muller M, dela Pena A, Derendorf H, Issues in pharmacokinetics and pharmacodynamics of anti-infective agents: Distribution in tissue, *Antimicrob. Agents Chemother*. 2004;48: 1441–1453.
42. Dobias L, Cerna M, Rossner P, Sram R, Genotoxicity and carcinogenicity of metronidazole, *Mut. Res*. 1994;317: 177–194.
43. World Health Organization (WHO), *Evaluation of Certain Veterinary Drug Residues in Food*, 34th Report Joint FAO/WHO Expert Committee on Food Additives, WHO Technical Report Series 788, 1989, pp. 20–32 (available at `http://whqlibdoc.who.int/trs/WHO_TRS_788.pdf`; accessed 10/16/10).
44. Commission Regulation 2377/90/EC; consolidated version of the Annexes 1 to IV updated up to 22.12.2004, *Off. J. Eur. Commun*. 1990;L224:1–8.
45. Commission Decision 2002/657/EC, implementing Council Directive 96/23/EC concerning the performance of analytical methods and the interpretation of results, *Off. J. Eur. Commun*. 2002;L221:8–36.
46. Commission Decision 2003/181/EC amending Decision 2002/657/EC as regards the setting of minimum required performance limits (MRPLs) for certain residues in food of animal origin, *Off. J. Eur. Commun*. 2003;L71:17–18.
47. Anadón A, Bringas P, Martinez-Larranaga MR, Diaz MJ, Bioavailability, pharmacokinetics and residues of chloramphenicol in the chicken, *J. Vet. Pharmacol. Ther*. 1994;17: 52–58.
48. Anadón A, Martinez MA, Martinez M, Rios A, Caballero V, Ares I, Martinez-Larranaga MR, Plasma and tissue depletion of florfenicol and florfenicol-amine in chickens, *J. Agric. Food Chem*. 2008;56:11049–11056.
49. Kennedy DG, Blanchflower WJ, Hughes PJ, McCaughy WJ, The incidence and cause of lasalocid residues in eggs in Northern Ireland, *Food Addit. Contam*. 1996;13(7): 787–794.
50. Kennedy DG, Hughes PJ, Bleanchflower WJ, Ionophore residues in eggs in Northern Ireland, *Food Addit. Contam*. 1998;15(5):535–541.
51. Dowling PM, Miscellaneous antimicrobials: Ionophores, nitrofurans, nitroimidazoles, rifamycins, oxazolidinones and other, in Giguere S, Prescott JF, Baggot JD, Walker RD, Dowling PM, eds., *Antimicrobial Therapy in Veterinary Medicine*, 4th ed., Blackwell, Ames, IA, 2006;285–300.
52. Papich MG, Riviere JE, Fluoroquinolone antimicrobial drugs, in Riviere JE, Papich MG, eds., *Veterinary Pharmacology and Therapeutics*, 9th ed., Wiley-Blackwell, Iowa State Univ. Press, Ames, 2009; pp. 983–1012.
53. Anadón A, Martinez-Larranaga MR, Diaz MJ, Fernandez-Cruz ML, Martinez MA, Frejo MT, Martinez M, Iturbe J, Tafur M, Pharmacokinetic variables and tissue residues of enrofloxacin and ciprofloxacin in healthy pigs, *Am. J. Vet. Res*. 1999;60:1377–1382.
54. Anadón A, Martinez MA, Martinez M, De La Cruz C, Diaz MJ, Martinez-Larranaga MR, Oral bioavailability, tissue distribution and depletion of flumequine in the

food producing animal, chicken for fattening, *Food Chem. Toxicol*. 2008;46:662–670.

55. Anadón A, Martinez-Larranaga MR, Diaz MJ, Martinez MA, Frejo MT, Martinez M, Tafur M, Castellano VJ, Pharmacokinetic characteristics and tissue residues for marbofloxacin and its metabolite N-desmethyl-marbofloxacin in broiler chickens, *Am. J. Vet. Res*. 2002;63:927–933.
56. Bidgood TL, Papich MG, Plasma and interstitial fluid pharmacokinetics of enrofloxacin, its metabolite ciprofloxacin, and marbofloxacin after oral administration and a constant rate intravenous infusion in dogs, *J. Vet. Pharmacol. Ther*. 2005;28:329–341.
57. Anadón A, Martinez-Larranaga MR, Diaz MJ, Bringas P, Martinez MA, Fernandez-Cruz ML, Fernandez MC, Fernandez R, Pharmacokinetics and residues of enrofloxacin in chickens, *Am. J. Vet. Res*. 1995;56:501–506.
58. Anadón A, Martinez-Larranaga MR, Iturbe J, Martinez MA, Diaz MJ, Frejo MT, Martinez M, Pharmacokinetics and residues of ciprofloxacin and its metabolites in broiler chickens, *Res. Vet. Sci*. 2001;71:101–109.
59. Bregante MA, Saez P, Aramayona JJ, Fraile L, Garcia MA, Solans C, Comparative pharmacokinetics of enrofloxacin in mice, rats, rabbits, sheep, and cows, *Am. J. Vet Res*. 1999;60: 1111–1116.
60. Van Duijkeren E, Kessels BG, Sloet van Oldruitenborgh-Oosterbaan MM, Breukink HJ, Vulto AG, van Miert AS, In vitro and in vivo binding of trimethoprim and sulphachlorpyridazine to equine food and digesta and their stability in caecal contents, *J. Vet. Pharmacol. Ther*. 1996;19:281–287.
61. Williams RB, Farebrother DA, Latter VS, Coccidiosis: A radiological study of sulphaquinoxaline distribution in infected and uninfected chickens, *J. Vet. Pharmacol. Ther*. 1995;18:172–179.
62. Shoaf SE, Schwark WS, Guard CL, The effect of age and diet on sulfadiazine trimethoprim disposition following oral and subcutaneous administration to calves, *J. Vet. Pharmacol. Ther*. 1987;10:331–345.
63. Papich MG, Riviere JE, Sulfonamides and potentiated sulfonamides, in Riviere JE, Papich MG, eds., *Veterinary Pharmacology and Therapeutics*, 9th ed., Wiley-Blackwell, Iowa State Univ. Press, Ames, 2009; pp. 835–864.
64. Nouws JFM, Mevius D, Vree TB, Baakman M, Degen M, Pharmacokinetics, metabolism, and renal clearance of sulfadiazine, sulfamerazine, and sulfamethazine and of their N-4 acetyl and hydroxy metabolites in calves and cows, *Am. J. Vet. Res*. 1988;49:1059–1065.
65. Abdullah AS, Baggot JD, The effect of food deprivation on the rate of sulfamethazine elimination in goats, *Vet. Res. Commun*. 1988;12:441–446.
66. Bulgin MS, Lane VM, Archer TE, Baggot JD, Craigmill AL, Pharmacokinetics, safety and tissue residues of sustained-release sulfamethazine in sheep, *J. Vet. Pharmacol. Ther*. 1991;14:36–45.
67. Nouws JFM, Vanginneken VJT, Grondel JL, Degen M, Pharmacokinetics of sulfadiazine and trimethoprim in carp (*Cyprinus carpio* L.) acclimated at 2 different temperatures, *J. Vet. Pharmacol. Ther*. 1993;16:110–113.
68. Boxenbaum HG, Fellig J, Hanson LJ, Snyder WE, Kaplan SA, Pharmacokinetics of sulphadimethoxine in cattle, *Res. Vet. Sci*. 1977;23:24–28.
69. Righter HF, Showalter DH, Teske RH, Pharmacokinetic study of sulfadimethoxine depletion in suckling and growing-pigs, *Am. J. Vet. Res*. 1979;40:713–715.
70. Mengelers MJB, Vangogh ER, Kuiper HA, Pijpers A, Verheijden JHM, Vanmiert ASJPAM, Pharmacokinetics of sulfadimethoxine and sulfamethoxazole in combination with trimethoprim after intravenous administration to healthy and pneumonic pigs, *J. Vet. Pharmacol. Ther*. 1995;18: 243–253.
71. Nielsen P, Gyrdhansen N, Oral bioavailability of sulfadiazine and trimethoprim in fed and fasted pigs, *Res. Vet. Sci*. 1994;56:48–52.
72. Bevill RF, Koritz GD, Dittert LW, Bourne DWA, Disposition of sulfonamides in food-producing animals. 5. Disposition of sulfathiazole in tissue, urine, and plasma of sheep following intravenous administration, *J. Pharm. Sci*. 1977;66:1297–1300.
73. Sweeney RW, Bardalaye PC, Smith CM, Soma LR, Uboh CE, Pharmacokinetic model for predicting sulfamethazine disposition in pigs, *Am. J. Vet. Res*. 1993;54:750–754.
74. Mitchell AD, Paulson GD, Depletion kinetics of C-14 sulfamethazine [4-amino-N-(4,6-dimethyl-2-pyrimidinyl) benzene[U-C-14]sulfonamide] metabolism in swine, *Drug Metab. Dispos*. 1986;14:161–165.
75. Bevill RF, Sulfonamide residues in domestic animals, *J. Vet. Pharmacol. Ther*. 1989;12:241–252.
76. Elliot CT, McCaughy WJ, Crooks SRH, McEvoy JDG, Effects of short term exposure of unmedicated pigs to sulfadimidine contaminated housing, *Vet. Rec*. 1994;134(17): 450–451.
77. Papich MG, Riviere JE, Tetracyclines antibiotics, in Riviere JE, Papich MG, *Veterinary Pharmacology and Therapeutics*, 9th ed., Wiley-Blackwell, Iowa State Univ. Press, Ames, 2009; pp. 895–914.
78. Prats C, El Korchi G, Giralt M, Cristofol C, Pena J, Zorrilla I, Saborit J, Perez B, PK and PK/PD of doxycycline in drinking water after therapeutic use in pigs, *J. Vet. Pharmacol. Ther*. 2005;28:525–530.
79. Craigmill AL, Miller GR, Gehring R, Pierce AN, Riviere JE, Meta-analysis of pharmacokinetic data of veterinary drugs using the Food Animal Residue Avoidance Databank: Oxytetracycline and procaine penicillin G, *J. Vet. Pharmacol. Ther*. 2004;27:343–353.
80. Nouws JFM, Irritation, bioavailability, and residue aspects of 10 oxytetracycline formulations administered intramuscularly to pigs, *Vet. Quart*. 1984;6:80–84.
81. Nouws JFM, Vree TB, Effect of injection site on the bioavailability of an oxytetracycline formulation in ruminant calves, *Vet. Quart*. 1983;5:165–170.
82. Nouws JFM, Smulders A, Rappalini M, A comparative-study on irritation and residue aspects of 5 oxytetracycline

formulations administered intramuscularly to calves, pigs, and sheep, *Vet. Quart*. 1990;12:129–138.

83. Cherlet M, De Backer P, Croubels S, Control of the keto-enol tautomerism of chlortetracycline for its straightforward quantitation in pig tissues by liquid chromatography-electrospray ionization tandem mass spectrometry, *J. Chromatogr. A* 2006;1133:135–141.
84. Pijpers A, Schoevers EJ, van Gogh H, van Leengoed LA, Visser IJ, van Miert AS, Verheijden JH, The influence of disease on feed and water consumption and on pharmacokinetics of orally administered oxytetracycline in pigs, *J. Anim. Sci*. 1991;69:2947–2954.
85. Mevius DJ, Vellenga L, Breukink HJ, Nouws JFM, Vree TB, Driessens F, Pharmacokinetics and renal clearance of oxytetracycline in piglets following intravenous and oral-administration, *Vet. Quart*. 1986;8:274–284.
86. Potter P, Illambas J, Lees P, unpublished data.
87. Toutain PL, Raynaud JP, Pharmacokinetics of oxytetracycline in young cattle—comparison of conventional vs. long-acting formulations, *Am. J. Vet. Res*. 1983;44:1203–1209.
88. Anadón A, Martinez-Larranaga MR, Diaz MJ, Bringas P, Fernandez MC, Fernandez-Cruz ML, Iturbe J, Martinez MA, Pharmacokinetics of doxycycline in broiler chickens, *Avian Pathol*. 1994;23:79–90.
89. World Health Organization, *Evaluation of Certain Veterinary Drug Residues in Food*, 36th Report Joint FAO/WHO Expert Committee on Food Additives, WHO Technical Report Series 799, 1990; pp. 37–41 (available at `http://whqlibdoc.who.int/trs/WHO_TRS_799.pdf`; accessed 10/16/10).
90. Renwick AG, Walker R, An analysis of the risk of exceeding the acceptable or tolerable daily intake, *Regul. Toxicol. Pharmacol*. 1993;18:463–480.
91. International Programme on Chemical Safety, *Assessing Human Health Risks of Chemicals: Derivation of Guidance Values for Health-Based Exposure Limits*, Environmental Health Criteria, World Health Organisation, Geneva, 1994; p. 73.
92. International Programme on Chemical Safety, *Assessing Human Health Risks of Chemicals: Principles for the Assessment of Risk to Human Health from Exposure to Chemicals*, Environmental Health Criteria, World Health Organisation, Geneva, 1999.
93. Gaylor DW, Kodell RL, Percentiles of the product of uncertainty factors for establishing probabilistic reference doses, *Risk Anal*. 2000;20:245–250.
94. MacLachlan DJ, Influence of physiological status on residues of lipophilic xenobiotics in livestock, *Food Addit. Contam. Part A, Chem. Anal. Control Expos. Risk Assess*. 2009;26:692–712.
95. Anonymous, *Food Safety Risk Analysis. A Guide for National Food Safety Authorities*, FAO Food and Nutrition Paper 87, Food and Agriculture Organization of the United Nations, Rome, 2006.
96. *IPCS Risk Assessment Terminology*, Harmonization Project Document 1, *International Programme on Chemical Safety*, World Health Organisation, Geneva, 2004.
97. Davies L, O'Connor M, Logan S, Chronic intake, in Hamilton D, Crossley S, eds., *Pesticide Residues in Food and Drinking Water: Human Exposure and Risks*, Wiley, Chichester, UK, 2003, pp. 213–241.
98. Kroes R, Kozianowski G, Threshold of toxicological concern (TTC) in food safety assessment, *Toxicol. Lett*. 2002; 127:43–46.
99. World Health Organization, *Evaluation of Certain Veterinary Drug Residues in Food*, 66th Report Joint FAO/WHO Expert Committee on Food Additives, WHO Technical Report Series 939, 2006, pp. 15–16 (available at `http://whqlibdoc.who.int/publications/2006/9241209399_eng.pdf`; accessed 12/01/10).
100. Arnold D, Risk assessments for substances without ADI/MRL—an overview, *Joint FAO/WHO Technical Workshop on Residues of Veterinary Drugs without ADI/MRL*, Bangkok, 2004, pp. 22–32.
101. Regulation (EC) No 470/2009 of the European Parliament and of the Council of 6 May 2009 laying down Community procedures for the establishment of residue limits of pharmacologically active substances in foodstuffs of animal origin, repealing Council Regulation (EEC) No 2377/90 and amending Directive 2001/82/EC of the European Parliament and of the Council and Regulation (EC) No 726/2004 of the European Parliament and of the Council, *Off. J. Eur. Commun*. 2009;L152:11–22.
102. Commission Regulation (EU) No 37/2010 of 22 December 2009 on pharmacologically active substances and their classification regarding maximum residue limits in foodstuffs of animal origin, *Off. J. Eur. Commun*. 2010;L15:1–72.
103. Title 21—Food and Drugs, Chapter I—Food and Drug Administration, Dept. Health and Human Services; Subchapter E—Animal drugs, feeds, and related products, Part 556: Tolerances for residues of new animals drugs in food (available at `http://www.accessdata.fda.gov/scripts/cdrh/cfdocs/cfCFR/CFRSearch.cfm`; accessed 12/07/10).
104. MacNeil JD, Validation requirements for testing for residues of veterinary drugs, *Joint FAO/WHO Technical Workshop on Residues of Veterinary Drugs without ADI/MRL*, Bangkok, 2004, pp. 99–102.
105. Reeves PT, Drug residues, in Cunningham FM, Lees P, Elliott J, eds., *Handbook of Experimental Pharmacology: Comparative and Veterinary Pharmacology*, Springer-Verlag, Heidelberg, 2010, pp. 265–290.
106. Committee for Medicinal Products for Veterinary Use (CVMP), *Reflection Paper on the New Approach Developed by JECFA for Exposure and MRL Assessment of Residues of VMP*, European Medicines Agency, London, 2008.
107. World Health Organization, Evaluation of Certain Veterinary Drug Residues in Food, 66th Report Joint FAO/WHO Expert Committee on Food Additives, WHO Technical Report Series 939, 2006, pp. 16–17 (available at `http://whqlibdoc.who.int/publications/2006/9241209399_eng.pdf`; accessed 12/01/10).
108. Friedlander LG, Brynes SD, Fernandez AH, The human food safety evaluation of new animal drugs, *Vet. Clin. North Am. Food Anim. Pract*. 1999;15:1–11.

109. Craigmill AL, Riviere JE, Webb AI, *Tabulation of FARAD Comparative and Veterinary Pharmacokinetic Data*, Blackwell Press, Ames, IA, 2006.

110. Schefferlie GJ, Hekman P, The size of the safety span for pre-slaughter withdrawal periods, *J. Vet. Pharmacol. Ther.* 2009;32:249.

111. Concordet D, Toutain PL, The withdrawal time estimation of veterinary drugs: A non-parametric approach, *J. Vet. Pharmacol. Ther.* 1997;20:374–379.

112. Concordet D, Toutain PL, The withdrawal time estimation of veterinary drugs revisited, *J. Vet. Pharmacol. Ther.* 1997;20:380–386.

113. Martinez M, Friedlander L, Condon R, Meneses J, O'Rangers J, Weber N, Miller M, Response to criticisms of the US FDA parametric approach for withdrawal time estimation: Rebuttal and comparison to the nonparametric method proposed by Concordet and Toutain, *J. Vet. Pharmacol. Ther.* 2000;23:21–35.

114. Fisch RD, Withdrawal time estimation of veterinary drugs: Extending the range of statistical methods, *J. Vet. Pharmacol. Ther.* 2000;23:159–162.

115. Martin-Jimenez T, Riviere JE, Mixed effects modeling of the disposition of gentamicin across domestic animal species, *J. Vet. Pharmacol. Ther.* 2001;24:321–332.

116. Craigmill AL, A physiologically based pharmacokinetic model for oxytetracycline residues in sheep, *J. Vet. Pharmacol. Ther.* 2003;26:55–63.

117. Buur JL, Baynes RE, Craigmill AL, Riviere JE, Development of a physiologic-based pharmacokinetic model for estimating sulfamethazine concentrations in swine and application to prediction of violative residues in edible tissues, *Am. J. Vet. Res.* 2005;66:1686–1693.

118. Law F, A PBPK model for predicting the withdrawal period of oxytetracycline in cultured Chinook salmon, in Smith DJ, Gingerich WH, Beconi-Barker MG, eds., *Xenobiotics in Fish*, Kluwer Academic, New York, 1999, pp. 105–121.

119. Cortright KA, Wetzlich SE, Craigmill AL, A PBPK model for midazolam in four avian species, *J. Vet. Pharmacol. Ther.* 2009;32:552–565.

120. Guidance for Industry #3: *General Principles for Evaluating the Safety of Compounds Used in Food-Producing Animals*, US Food and Drug Administration, 2006 (available at http://www.fda.gov/downloads/AnimalVeterinary/GuidanceComplianceEnforcement/GuidanceforIndustry/UCM052180.pdf; accessed 12/07/10).

121. EMEA/CVMP/473/98—final, *Note for Guidance for the Determination of Withdrawal Periods for Milk*, European Medicines Authority, Committee for Medicinal Products for Veterinary Use, 2000 (available at http://www.ema.europa.eu/docs/en_GB/document_library/Scientific_guideline/2009/10/WC500004496.pdf; accessed 12/06/10).

122. Gehring R, Baynes RE, Craigmill AL, Riviere JE, Feasibility of using half-life multipliers to estimate extended withdrawal intervals following the extralabel use of drugs in food-producing animals, *J. Food Prot.* 2004;67:555–560.

123. Riviere JE, Sundlof SF, Chemical residues in tissues of food animals, in Riviere JE, Papich MG, eds., *Veterinary Pharmacology and Therapeutics*, 9th ed., Wiley-Blackwell, Iowa State Univ. Press, Ames, 2009, pp. 1453–1462.

124. KuKanich B, Gehring R, Webb AI, Craigmill AL, Riviere JE, Effect of formulation and route of administration on tissue residues and withdrawal times, *J. Am. Vet. Med. Assoc.* 2005;227:1574–1577.

125. Reeves PT, The safety assessment of chemical residues in animal-derived foods, *Austral. Vet. J.* 2005;83:151–153.

126. Roeber DL, Cannell RC, Belk KE, Scanga JA, Cowman GL, Smith GC, Incidence of injection-site lesions in beef top sirloin butts, *J. Anim. Sci.* 2001;79:2615–2618.

127. Paige JC, Chaudry MH, Pell FM, Federal surveillance of veterinary drugs and chemical residues (with recent data), *Vet. Clin. North Am. Food Anim. Pract.* 1999;15:45–61.

128. Gehring R, Baynes RE, Riviere JE, Application of risk assessment and management principles to the extralabel use of drugs in food-producing animals, *J. Vet. Pharmacol. Ther.* 2006;29,5–14.

129. Wang J, Gehring R, Baynes RE, Webb AI, Whitford C, Payne MA, Fitzgerald K, Craigmill AL, Riviere JE, Evaluation of the advisory services provided by the Food Animal Residue Avoidance Databank, *J. Am. Vet. Med. Assoc.* 2003;223:1596–1598.

130. Animal Medicinal Drug Use Clarification Act of 1994 (AMDUCA), US Food and Drug Administration, 1994 (available at http://www.fda.gov/AnimalVeterinary/GuidanceComplianceEnforcement/ActsRulesRegulations/ucm085377.htm; accessed 12/07/10).

131. Fajt VR, Regulatory considerations in the United States, *Vet. Clin. North Am. Food Anim. Pract.* 2003;19:695–705.

132. De Brabander HF, Noppe H, Verheyden K, Bussche JV, Wille K, Okerman L, Vanhaecke L, Reybroeck W, Ooghe S, Croubels S, Residue analysis: Future trends from a historical perspective, *J. Chromatogr. A.* 2009;1216:7964–7976.

133. CAC/GL 71–2009, *Guidelines for the Design and Implementation of National Regulatory Food Safety Quality Assurance Programme Associated with the Use of Veterinary Drugs in Food Producing Animals*, Joint FAO/WHO Food Standards Program, 2009 (available at http://www.codexalimentarius.net/web/more_info.jsp?id_sta = 11252; accessed 2/15/10).

134. Commission Directive No 96/23/EC of 29 April 1996 on measures to monitor certain substances and residues thereof in live animals and animal products and repealing Directives 85/358/EEC and 86/469/EEC and Decisions 89/187/EEC and 91/664/EEC, *Off. J. Eur. Commun.* 1996;L125:10–32.

135. Dunnett M, Lees P, Retrospective detection and deposition profiles of potentiated sulphonamides in equine hair by liquid chromatography, *Chromatographia* 2004;59:S69–S78.

136. Commission Decision of 13 March 2003 amending Decision 2002/657/EC as regards the setting of minimum required performance limits (MRPLs) for certain residues in food of animal origin, *Off. J. Eur. Commun.* 2003;L71:17–18.

137. *Annual Report on Surveillance for Veterinary Residues in Food in the UK 2009*, The Veterinary Residues Committee, UK, 2010 (available at http://www.vmd.gov.uk/vrc/Reports/vrcar2009.pdf; accessed 12/07/10).
138. *National Milk Drug Residue Data Base Fiscal Year 2003 Annual Report*, National Conf. Interstate Milk Shipments (NCIMS), US Food and Drug Administration, 2004 (available at http://www.fda.gov/Food/FoodSafety/Product-SpecificInformation/MilkSafety/MiscellaneousMilkSafetyReferences/ucm115592.htm#sample; accessed 12/06/10).
139. Pasteurized Milk Ordinance 2007, US Food and Drug Administration, page last updated 05/11/2009 (available at http://www.fda.gov/Food/FoodSafety/Product-SpecificInformation/MilkSafety/NationalConferenceonInterstateMilkShipmentsNCIMSModelDocuments/PasteurizedMilkOrdinance2007/default.htm; accessed 12/07/10).
140. EMEA/CVMP/520190/2007—consultation, *Reflection Paper on Injection Site Residues; Considerations for Risk Assessment and Residue Surveillance*, European Medicines Authority, Committee for Medicinal Products for Veterinary Use, 2008 (available at http://www.ema.europa.eu/docs/en_GB/document_library/Scientific_guideline/2009/10/WC500004430.pdf); accessed 12/06/10).
141. Reeves PT, Residues of veterinary drugs at injection sites, *J. Vet. Pharmacol. Ther*. 2007;30:1–17.
142. Sanquer A, Wackowiez G, Havrileck B, Qualitative assessment of human exposure to consumption of injection site residues, *J. Vet. Pharmacol. Ther*. 2006;29:345–353.
143. Sanquer A, Wackowiez G, Havrileck B, Critical review on the withdrawal period calculation for injection site residues, *J. Vet. Pharmacol. Ther*. 2006;29:355–364.
144. EMEA/CVMP/209865/2004, Overview of comments received on draft guideline on injection site residues (EMEA/CVMP/542/03-FINAL), European Medicines Authority, Committee for Medicinal Products for Veterinary Use, 2005 (available at http://www.ema.europa.eu/docs/en_GB/document_library/Other/2009/10/WC500004431.pdf; accessed 12/06/10).
145. Beechinor JG, *Studies on Muscle Residues of the Antibacterial Veterinary Medicines Tilmicosin, Enrofloxacin and Tiamulin in Livestock and the Risk to Consumers from Ingestion of Infection Sites*, dissertation Univ. Dublin, Ireland, 2000.
146. Beechinor JG, Buckley T, Bloomfield FJ, Prevalence of injection site blemishes in primal cuts of Irish pork, *Irish Vet. J*. 2001;54:121–122.
147. Beechinor JG, Buckley T, Bloomfield, FJ, Prevalence and public health significance of blemishes in cuts of Irish beef, *Vet. Rec*. 2001;149:43–44.
148. Anonymous, *Sales of Antimicrobial Products Authorized for Use as Veterinary Medicines, Antiprotozoals, Antifungals and Coccidiostats, in the UK in 2008*, Veterinary Medicines Directorate, Dept. Environment, Food and Rural Affairs, UK, 2009 (available at http://www.vmd.gov.uk/Publications/Antibiotic/salesanti08.pdf; accessed 12/06/10).
149. Moulin G, Cavalie P, Pellanne I, Chevance A, Laval A, Millemann Y, Colin P, Chauvin C, Gr ARah. (2008), A comparison of antimicrobial usage in human and veterinary medicine in France from 1999 to 2005, *J. Antimicrob. Chemother*. 2008;62:617–625.
150. Peters AR, personal communication.
151. Benchaoui H, Population medicine and control of epidemics, in Cunningham FM, Lees P, Elliott J, eds., *Handbook of Experimental Pharmacology: Comparative and Veterinary Pharmacology*, Springer-Verlag, Heidelberg, 2010, pp. 113–138.
152. Baggot JD, Brown SA, Basis for selection of the dosage form, in Hardee GE, Baggot JD, eds., *Development and Formulation of Veterinary Dosage Forms*, 2nd ed., Marcel Dekker, New York, 1998, p. 50.

3

ANTIBIOTIC RESIDUES IN FOOD AND DRINKING WATER, AND FOOD SAFETY REGULATIONS

KEVIN J. GREENLEES, LYNN G. FRIEDLANDER, AND ALISTAIR BOXALL

3.1 INTRODUCTION

Animals are treated routinely with antibiotics to prevent, treat, or control disease. Even under the best conditions of agricultural management, crowding and stress can lead to disease. While historically there have also been non-therapeutic uses of antibiotics, typically as production tools to improve endpoints such as feed efficiency and weight gain, there is a call to diminish these uses worldwide,[1,2] and concern for the development of resistance to antibiotics used in human medicine as a result of their use in animal agriculture has led to international efforts to evaluate that risk.[3–5] The results of the therapeutic uses are healthy animals that contribute to a healthful and plentiful food supply.

However, one consequence of the use of the antibiotics in food-producing animals is the presence of residues of the drug, however minute, in the edible tissues of the treated animal. The residues of the antibiotic could be systemically toxic to the consumer, adversely affecting organ systems, leading to morbidity and even death. Residues of the antibiotic in consumed food could have direct adverse effects on the complex microflora that inhabit the human gastrointestinal system, with potentially disastrous consequences for the consumer. Another potential consequence is exposure of the human consumer to bacteria that, having been exposed to the drug through the treated animal and having survived the exposure, are less susceptible to that antibiotic. People who develop a human disease resulting from exposure to these bacteria may find that the causative organisms are resistant to antibacterials used in human medicine and the disease refractory to standard treatments.

3.2 RESIDUES IN FOOD—WHERE IS THE SMOKING GUN?

If antibiotic residues are so dangerous, why aren't there plentiful examples of consumers becoming ill, seeking medical treatment, or dying? The answer is complex, as is the subject, and reflects a number of factors. The first is that toxicity can be the result of a single, large, acute exposure or can result from long-term exposure to much lower concentrations. Whether an acute or chronic toxicity will be observed reflects the exposure dose, the nature of the toxicity, pharmacokinetics, and the exposed population. While exceptions abound, in general long-term exposure can result in toxicity following a lower concentration in the diet than would be seen from a single acute exposure. Regulatory agencies typically anticipate that veterinary drug toxicity, including antibiotics, will generally result from long-term, low-level exposure in the diet. As will be discussed later, the bulk of the approach to establishing the safety of any given concentration of antibiotic residue for the human diet is based on chronic exposure. Adverse impacts on the human consumer resulting from years, or even decades, of exposure to residues of a veterinary antibiotic in the food would be very difficult to trace back to the source of the problem. On the other hand, there are a few examples where veterinary drug residues can cause an acute toxicity, sometimes from a single meal. In such instances, it is easier to determine the source of the problem, and two examples are discussed here. The first, clenbuterol, is not an antibiotic, but, unfortunately, its veterinary use in food-producing animals resulted in

Chemical Analysis of Antibiotic Residues in Food, First Edition. Edited by Jian Wang, James D. MacNeil, and Jack F. Kay.

clear toxicities to large numbers of human consumers. The second, chloramphenicol, is an antibiotic whose use in human medicine has been associated with severe illness and death. As a result, its use as an antibiotic in veterinary medicine for food-producing animals has been severely curtailed worldwide.

Clenbuterol is a sympathomimetic drug, meaning that it has properties that are similar to those exhibited by the chemical mediators of the sympathetic nervous system (epinephrine and norepinephrine). Clenbuterol is a β-2 adrenergic receptor agonist, similar to epinephrine in actions. Like epinephrine, the direct effects of clenbuterol are mediated by the β-2 adrenergic receptor, but the overall effects of the drug are complicated by complex feedback systems within the body. For example, while clenbuterol has little direct impact on cardiac tissue (primarily responding to β-1 adrenergic agonists), it causes a dramatic peripheral vasodilation, in response to which there can be considerable increases in heart rate as the body attempts to maintain blood pressure. As a β-2 adrenergic receptor agonist, clenbuterol has been approved for use as a bronchodilator in horses in the United States[6] and as a bronchodilator in horses and cattle, and as a tocolytic in cattle, in the EU.[7] Another effect of long-term use of the drug is to increase catabolism of fat and increase conservation of protein. This anabolic effect has made it a (unapproved) drug of choice by some in the human bodybuilding[8–10] and weight loss community[11] and by some in animal agriculture.[12] Unfortunately, as will be seen, its pharmacological properties are such that adverse effects are reported following as little as a single meal.

Concern began to be raised in the early 1990s as patients were identified in Spain complaining of a variety of symptoms, including racing heart, dizziness, nausea, headaches, and peripheral tremors. In 1992, 113 cases of clenbuterol poisoning were reported in Spain following ingestion of veal liver and possibly veal tongue.[13] Symptoms included tachycardia, muscle tremors, nervousness, myalgia, vomiting, and headache. Clenbuterol concentrations in veal liver samples ranged from 19 to 5395 μg/kg.

Clenbuterol poisonings as the result of veterinary drug residues continued to be reported. In Italy in 1997, 15 people were reported to have sought care for symptoms including tremors, vertigo, headache, hot flushes, increased heart rate, nervousness, and rapid breathing following consumption of veal that was subsequently found to contain residues of clenbuterol.[14] Incidents were reported in Hong Kong between 1998 and 2002.[15] In Portugal in 2005, 50 people were reported ill following consumption of lamb and beef containing residues of clenbuterol.[16] Symptoms included tremors, vertigo, headache, hot flushes, increased heart rate, nervousness, and rapid breathing. Interestingly, in each case the clenbuterol was consumed in well-cooked meals where the cooking process might be anticipated to interfere with the pharmacological properties of the drug. Subsequent studies showed that the ring structure of clenbuterol is remarkably resistant to thermal degradation and a key consideration in the ability of the drug to interact with the adrenergic receptor.[17] Clenbuterol continues to be approved for therapeutic uses in cattle and horses in some countries.

Chloramphenicol provides another example of public health consequences resulting from veterinary drug residues. In the United States, the antibiotic chloramphenicol was approved for use in companion animals (dogs, cats, horses) in ophthalmic, injectable, and oral dosage form. Later, the US Food and Drug Administration (US FDA) became aware of the use of chloramphenicol in cattle. Although not approved for food-producing animals, the utility of the antibiotic was recognized in treating systemic infections in cattle. At the same time, information began to become available in the human medical community of low-frequency adverse effects of chloramphenicol. Wallerstein et al.[18] reported low incidence (approximately 1 : 30,000) of blood dyscrasias in patients receiving chloramphenicol. In some patients, this led to a fatal aplastic anemia and, in some cases where there was recovery from the aplastic anemia, incidences of leukemia. Concern for the potential toxicity to consumers of food derived from cattle treated with chloramphenicol led the US FDA to withdraw the equine approvals for chloramphenicol, based on concern for the unapproved uses of this formulation in cattle. In a series of later reviews by the Joint FAO/WHO Expert Committee for Food Additives (JECFA), it was consistently determined that an acceptable daily intake for the human consumer could not be established for residues of chloramphenicol because no threshold could be predicted for the aplastic anemia.[19] Chloramphenicol is not approved for use in food-producing animals in Australia, Canada, the EU, or the United States.

Chloramphenicol has been evaluated by four separate JECFA meetings.[19] As recently as 2004, the 62nd Committee declined to establish an acceptable daily intake (ADI) for chloramphenicol.[19] The committee further considered the potential for environmental sources of chloramphenicol to result in residues of the drug in the edible tissues of food-producing animals. It concluded that it is unlikely that chloramphenicol is synthesized in detectable amounts in soil; however, it is possible that environmental contamination may lead to some of the very low concentrations of chloramphenicol that has sometimes been reported in animal tissue. Considering these and other recommendations, the Codex Committee on Veterinary Residues in Food has been considering how to provide appropriate risk mitigation information to members regarding chloramphenicol as a veterinary drug for use in food-producing animals.[20]

3.3 HOW ALLOWABLE RESIDUE CONCENTRATIONS ARE DETERMINED

Requirements to establish the safety of residues of veterinary drugs in food vary internationally. Table 3.7, later in the chapter, provides some of the national, regional, and international guidelines and online sources for these requirements.

3.3.1 Toxicology—Setting Concentrations Allowed in the Human Diet

So, how do regulatory agencies address the concern to establish the safety of antibiotic residues of veterinary drugs in food? Numerous national regulatory authorities and international bodies have established guidelines for the toxicological evaluation of residues of veterinary drugs and pesticides. The toxicology studies are evaluated to characterize the toxicity of the antibiotic in various *in vitro* and *in vivo* animal models as well as any available human data to predict the potential toxicity of residues of the veterinary drug in food. In general, the approach[21,22] is to evaluate the potential for short-term (acute, typically a single meal or a few meals) or long-term (chronic, months to years of exposure) dietary exposure to residues of the antibiotic (whether a veterinary drug or a pesticide) to have adverse effects on the human consumer. This is typically done in orally exposed mammalian animal models (e.g., rodents, dogs, swine) but can include *in vitro* models and even human exposure data. Adverse effects may range from systemic toxicity (e.g., damage to liver or kidney) to reproductive or developmental effects on offspring (e.g., increased stillbirths or abnormal limb development), immunological effects (e.g., decreased immune response), neurological effects (e.g., peripheral nerve damage), and cancer. Typically multiple doses of the antibiotic are orally administered to test animals to identify a dose that results in no observable change from background [a threshold dose, often called a *no observable effect level* (NOEL) or *no observable adverse effect level* (NOAEL)] and ideally, higher doses that characterize the dose–response relationship. Responses across all of the models are considered and the most appropriate is selected as the basis for an acceptable daily intake (ADI). The NOAEL is divided by a safety factor that for toxicological studies usually has a value of 100 or higher, depending on the toxicological properties of the drug and the amount and quality of data available. The default value of 100 has been investigated for its accuracy and limitations by many scientists.[23,24] A similar approach, often called an *acute reference dose* (ARfD), can establish an acceptable acute intake on the basis of short-term studies. In the case of antibiotics, a microbiological ADI may also be determined. The microbiological ADI is based not on classic endpoints of toxicology but on the potential interaction between residues of the veterinary drug in food and the microbial flora of the human gastrointestinal tract. The approach also typically differs from that of evaluation for a toxicological ADI in that it is based on a decision tree approach that determines the potential for microbiologically active residues to enter the human colon. If it is determined that there is still potential for the interaction of microbiologically active residues of the veterinary drug with the microbial flora, a microbiological ADI is determined. The microbiological ADI is based on a change in either the antimicrobial resistance for endogenous microflora or the microbial colonization barrier.[21,25] Regulatory agencies then determine whether the regulation of the antibiotic residues is most appropriately addressed using the chronic toxicological or microbiological ADI, or the short-term ARfD, which is typically based on whichever result is considered most protective of public health. International guidance on the use of an ARfD for residues of veterinary drugs in food may be found in OECD Guideline 124[26] and EHC 240,[21] publications of the Organization for Economic Co-operation and Development (OECD) and the World Health Organization, respectively. International guidelines on the development of a toxicological or microbiological ADI may be found through EHC 240[21] and through the VICH guidelines discussed later in this chapter (in Section 3.3.4).

The OECD developed its Guideline 124 as part of a large series of internationally agreed-on testing protocols in a wide range of areas related to the safety of chemicals, including veterinary drugs. The list of available guidelines may be found online at `http://www.oecd.org/department/0,3355,en_2649_34377_1_1_1_1_1,00.html`.[27] With 33 member countries, there is a broad international consensus available through the OECD guidelines. These guidelines are referenced by the International Cooperation on Harmonisation of Technical Requirements for Registration of Veterinary Products (VICH), a trilateral (EU–Japan–USA) program for harmonization of the technical requirements for the registration of a veterinary drug product, with observers participating from Australia, Canada, and New Zealand.

Guideline EHC 240 was developed by the Joint FAO/WHO Expert Committee on Food Additives (JECFA) and the Joint FAO/WHO Meeting on Pesticide Residues (JMPR). These independent expert bodies are made up of members and experts drawn from the international scientific community under the direction of the JECFA and JMPR Secretariats. JECFA meets on an *ad hoc* basis to evaluate toxicology data for food additives, contaminants, and naturally occurring toxicants in food and for residues of veterinary drugs in food and recommends concentrations of these chemicals in foods that should pose no risk to the consumer. JMPR carries out similar evaluations for pesticides. Maximum residue limits (MRLs) for residues in food are recommended to the FAO/WHO food-standards-setting body, the Codex Alimentarius Commission (CAC). Established in 1963, the CAC sets non-binding food standards with the goal of protecting consumer health while ensuring fair international trade. The CAC, on reaching international consensus among its attending members, sets international standards for the maximum residue of the veterinary drug (or pesticide) that may be contained in food (the MRL). While these standards are recommendations, they are also the principal food standards recognized by the 1995 World Trade Organization Agreement on Sanitary Phytosanitary Measures (often referred to as the *SPS agreement*).[28] International harmonization is discussed in Section 3.3.4.

Once the ADI (or ARfD) is established, there are some differences in how it is interpreted among national and regional authorities when determining the maximum concentration of residues in edible tissues (e.g., meat, milk, or eggs).

3.3.2 Setting Residue Concentrations for Substances Not Allowed in Food

National and regional authorities responsible for the protection of public health must consider the concentration of residues of veterinary drug residues, pesticides, and other chemicals that may be in food regardless of whether the substance is allowed for that use. In many regions, in the absence of an approval for the substance, the concentration of residues allowed in food is considered to be zero. In practical terms, this is frequently defined by the technical capability of the analytical method. Attempts to improve on "zero" include the ALARA (as low as reasonably achievable) approach, which recognizes that absolute zero is unattainable, and describes an approach that considers what is technically achievable, the resources needed to achieve that technical goal, and the benefit gained.

3.3.3 Setting Residue Concentrations Allowed in Food

Residues are evaluated to determine the extent of uptake of the veterinary drug, its distribution throughout the body, and its elimination. Normally, contemporary residue depletion studies establish tissue concentrations in a radiolabeled drug study, in which total residues and parent compound are determined at several pre-determined times between zero time and a time beyond the proposed withdrawal time. As well as total residues, which include free and bound components, the study quantifies major metabolites. These are compounds contributing 10% or more of total radioactivity or that are present at a concentration of ≥ 0.10 mg/kg. Metabolism studies enable identification of the marker residue and target tissue. The marker residue must give assurance that, when its concentration is at or below the MRL, total residues satisfy ADI requirements. Thus, because the marker residue is unlikely to be the only compound of toxicological concern, its depletion must ensure that other residues have depleted to concentrations considered to be safe prior to that of the marker residue.

Bound residues are often considered to represent residues that are not toxicologically available on oral consumption. Bound residues (and non-extractable residues) can result in several ways: incorporation of drug or metabolite(s) into macromolecules, physical encapsulation, integration of residues into tissue matrices, or incorporation into endogenous compounds. Bound residues may be of no toxicological concern (and their presence therefore discounted) if only small molecular fragments, say, one or two carbon units, are incorporated into endogenous molecules. However, bound residues must be documented when they represent a significant portion of the total residue.

Some authorities, such as the United States, establish a target tissue to monitor the safety of the entire carcass. The target tissue is chosen to monitor the safety of all edible (meat) tissues and is usually that tissue with the slowest rate of depletion. Some authorities, including the EU, consider all edible tissues as possible target tissues because control can be performed not only on the entire carcass but also on pieces of meat, isolated marketed offal, and other substances.

Both toxicological ADI and MRL/tolerance are linked through metabolism studies in the laboratory animal species used to determine the toxicological NOEL, as well as each food-producing animal species. A qualitatively similar metabolite profile between the laboratory animal species and the target species ensures the validity of the toxicology studies, by demonstration of broadly similar exposure to the same range of compounds. Comparative metabolism studies involve analysis of metabolites in blood and its fractions, excreta, kidney, fat, liver, and bile in both the target and laboratory animal test species.

The ADI provides the bright line between exposures that are safe and those that are unsafe. Because it is a primary (i.e., a non-derived) food safety standard, it is easily interpretable across a range of regulatory situations. However, with the exception of residue determinations made using bioassays, because the ADI reflects total residues of concern, it cannot provide a standard for regulatory enforcement. Maximum residue limits (MRLs) and tolerances (in the USA) provide that enforcement standard.

Both MRLs and tolerances, following adjustments for specific tissue consumptions and analytical performance capabilities, are referable back to the ADI. As a consequence, both MRLs and tolerances may be considered as derived food safety standards. However, the exact way in which this connection back to the ADI is made differs for MRLs and tolerances.

3.3.3.1 Tolerances

In the case of US tolerances, there is a direct link from the tissue-specific tolerance back to the ADI via the analytical method used in its determination.[29] Associated with this linkage are assumptions regarding daily dietary consumption or, as is it is known, partitioning of the ADI into tissue-specific safe concentrations. In the United States, it is assumed that daily consumption of animal-derived edible products may include milk, eggs, and edible tissues. However, it is assumed that not all of the edible tissues will be eaten at their maximum consumption value every day. As such, the ADI is initially partitioned between meat, milk, and eggs. Ideally, this determination reflects the entirety of the proposed product line development; that is, while the initial development emphasis may be approval of a product for use in beef cattle, future development may include use in lactating dairy cattle. Because the development timeline for the drug is first understood and is ultimately at the discretion of the drug sponsor, the requested partitioning of the ADI is left to the drug sponsor. Additionally, the market constraints differ for beef and lactating dairy cattle; specifically, it may be very reasonable to assume a withdrawal time of days or even weeks for a product intended for use in beef cattle, but the final milk discard period for the same product extended for use in lactating dairy cattle must be consistent with normal dairy production practices and must be managed correctly to ensure optimal product line development. For compounds where there is no reasonable extension to dairy cattle or laying hens, the entire ADI can be allocated to meat. Because the ADI represents the line between safe exposure and unsafe exposure, the entire ADI is used in the partitioning exercise and is available for use in the derivation of the tolerances. As tolerances are derived from the ADI through partitioning, they do not, with a few exceptions, require reevaluation as new uses of approved drugs are developed. An example of partitioning and the assignment of tissue-specific safe concentrations is given in Tables 3.1–3.3.

Once the partitioning of the ADI into tissue-specific safe concentrations is complete, derivation of the tolerance

TABLE 3.1 Assumptions for Calculation of Tolerance

Drug is intended to treat bovine respiratory disease
- Primary market—feedlot cattle
  - Long tissue withdrawal is OK
    - Small percentage of ADI is needed for tissues
- Secondary market—dairy cattle
  - Long tissue withdrawal is OK
    - Small percentage of ADI is needed for tissues
  - Needs short milk discard
    - Large percentage of ADI is needed for milk

Assume that the market development will *never* include laying hens

Assume an ADI = 10 μg/kg body weight (BW)

TABLE 3.2 Partitioning the ADI for Milk

To achieve the desired short milk discard time, partition 70% of the ADI to milk and 30% to tissues:

$$ADI_{milk} = 10\ \mu g/kg\ BW \times 70\% = 7\ \mu g/kg$$
$$ADI_{tissue} = 10\ \mu g/kg\ BW \times 30\% = 3\ \mu g/kg$$

TABLE 3.3 Calculating Safe Concentrations (SCs)

$$SC = \frac{(\text{partitioned ADI}) \times (60\text{ kg person})}{\text{food consumption value}}$$

$$SC_{milk} = \frac{(7\ \mu g/kg\text{ per day}) \times (60\text{-kg person})}{1.5\ l/\text{person per day}} = 280\ \mu g/l$$

$$SC_{muscle} = \frac{(3\ \mu g/kg\text{ per day}) \times (60\text{-kg person})}{0.3\ kg/\text{person per day}} = 600\ \mu g/kg$$

$$SC_{liver} = \frac{(3\ \mu g/kg\text{ per day}) \times (60\text{-kg person})}{0.1\ kg/\text{person per day}} = 1800\ \mu g/kg$$

TABLE 3.4 Calculating the Tolerance

Assume that the marker residue, using a specified HPLC/MS-MS method, represents 50% of the total residues as determined by combustion analysis of the total radiolabeled residues i.e., a marker-to-total ratio of 50%

Assume that this 50% relationship applies for all tissues and milk, at all timepoints

$$\text{Tolerance} = \text{tissue-specific SC} \times \left(\frac{\text{marker residue concentration}}{\text{total residue concentration}}\right)$$

$$\text{Tolerance}_{milk} = 280\ \mu g/l \times (0.5) = 140\ \mu g/l$$

$$\text{Tolerance}_{muscle} = 600\ \mu g/kg \times (0.5) = 300\ \mu g/kg$$

$$\text{Tolerance}_{liver} = 1800\ \mu g/kg \times (0.5) = 900\ \mu g/kg$$

will reflect the performance of the analytical method (i.e., the marker : total ratio) used in the selection of the marker residue, as shown in Table 3.4.

In the United States, the tolerance is linked directly to the analytical method by which it is determined. As part of the drug approval process, and prior to designating an official regulatory method for the residue monitoring program, this analytical method is assessed in a multi-laboratory trial to ensure its transferability. Thereafter, it is not possible to introduce a new method for residue monitoring until it has been bridged (i.e., directly compared) to the official analytical method.

It is interesting to note that extralabel use of a drug product in an approved species (e.g., at an unapproved dose or by unapproved routes of administration) may or may not result in residues that exceed the assigned tolerances. Specifically, when the ADI is large and incurred residues are low, the tolerances, calculated arithmetically by applying the marker : total ratio to the ADI-derived safe concentrations, will be larger than the incurred residues. As a consequence, while the tolerance approach very effectively allows for monitoring of the boundary between safe and unsafe residue concentrations, it is less effective in monitoring the boundary between on-label and extralabel uses. Although it is easy to conclude that residues in excess of the tolerance represent a public health concern and are deserving of prosecution, not all misuse situations will result in residues above tolerance, and these will escape regulatory action even though the use was illegal.

In summary, in the United States, the tolerance is derived arithmetically from the safe concentration, which is, in turn, linked directly back to the ADI. As such, the tolerance, determined using the specified analytical method, represents the maximum residue that, in routine monitoring for residues, can be considered safe for human consumption. The tolerance approach is extremely effective in ensuring that the consuming public will not be exposed to drug residues in excess of the ADI when residues in animal-derived food products are less than or equal to the assigned tolerance. The tolerance approach focuses limited regulatory resources on those residue cases that represent the greatest threat to the public health.

3.3.3.2 Maximum Residue Limits

Maximum residue limits (MRLs) also are linked directly to the ADI.[30] Significantly, however, the MRLs are not derived *from* the ADI and do not represent a direct partitioning of the ADI. Rather, they are reflective of the concentrations of residues incurred under the evaluated conditions of use, determined using appropriately validated analytical methods. Because the MRLs reflect only those uses available for evaluation at the time they are established, the MRLs may not be fully reflective of the eventual spectrum of product development and may require reassessment as new uses for a particular drug are realized. Inherent in relating MRLs back to the ADI is the assumption that all of the animal-derived edible products will be eaten at their maximum consumption values every day (i.e., no partitioning of the ADI), and quantifying human exposure to drug residues regulated through MRLs necessitates the assignment of MRLs for all appropriate edible products. Further, relating overall food safety regulated with MRLs to the ADI is often achieved using a theoretical maximum daily intake (TMDI) calculation: (tissue-specific MRL) × (marker : total ratio) × (tissue-specific consumption value) = tissue residue contribution to TMDI.

Examples of calculations of MRLs and the related TMDI are given in Tables 3.5 and 3.6.

The TMDI is compared to the ADI. The TMDI must not result in exposures in excess of the ADI. In the example above, the TMDI represents 95% of the ADI. While a specific study and its associated analytical method may provide the data used to establish the marker : total ratio, there is no unbreakable link between the MRLs and any specific analytical method for residue monitoring.

TABLE 3.5 Assumptions for Calculating an MRL

Assume an ADI of 0–0.8 μg/kg BW per day, corresponding to a total maximum acceptable exposure of 480 μg/person per day
Assume that incurred residues support MRLs as follows (tissue—MRL):
Milk—200 μg/l
Muscle—100 μg/kg
Liver—1000 μg/kg
Kidney—400 μg/kg
Fat—100 μg/kg
Assume that there is no additional correction of marker : total (M : T) based on microbiological activity: marker residue = 100% of total residue (M : T factor = 1)

TABLE 3.6 Calculating Theoretical Exposure as Determined by the TMDI

Tissue	MRL (μg/kg)	M : T	Food Basket (kg)	TMDI (μg)
Muscle	100	1	0.3	30
Liver	1000	1	0.1	100
Kidney	400	1	0.05	20
Fat	100	1	0.05	5
Milk	200	1	1.5	300
TMDI	—	—	—	455

Consequently, it is possible to support multiple methods of analysis for routine residue monitoring.

Because the MRLs are derived from incurred residue concentrations resulting from approved uses, residue concentrations in excess of the MRL represent, by definition, uses outside the conditions on the label. Multiple reassessments of the MRLs may result in revised MRLs that exceed the incurred residue concentrations associated with the original evaluation, thus weakening their direct application to prosecution of label violations. Nevertheless, MRLs provide for a more direct linkage between regulatory decision criteria and the labeled conditions of use than do tolerances. However, because MRLs may result in exposure estimates that are significantly below the ADI, it can be difficult to conclude that residues in excess of the MRLs represent a direct public health concern.

In summary, MRLs are derived following an assessment of incurred residues resulting from approved conditions of use and represent the maximum residue concentrations that are consistent with those label uses (e.g., dose and routes of administration). Residues in excess of the MRL are indicative of uses outside the approved conditions of use. Thus, the MRL approach is extremely effective in monitoring label compliance, focusing regulatory resources on those residue cases that represent deviations from the labeled conditions of use. However, because not all extralabel uses result in unsafe residues, the MRL approach may result in compliance cases that cannot claim to protect the public health.[2]

As noted above, MRLs can be monitored using all properly validated methods, whereas tolerances are tied inextricably to the method by which they are determined. Because MRL values may represent significantly less than the entire ADI, they are often lower than tolerance values elaborated for the same compound. Thus, the analytical methods used for the monitoring of MRLs may need to detect residue concentrations lower than those used for the monitoring of tolerances. In addition, because MRLs are needed for all edible tissues, where residue concentrations are very low, the MRL may be established at a concentration representing a multiple of the method's limit of quantification (LOQ). This option, not available for tolerances where the official analytical method must pass a multi-laboratory trial and demonstrate acceptable performance at one-half the tolerance, facilitates MRL assignments for tissues where residue concentrations are routinely low (i.e., muscle and fat). To compensate, in the United States, a tolerance is needed only for the marker residue in the target tissue, usually liver or kidney. This, in turn, lessens the need for extremely sensitive analytical methods. Also, because MRLs may be lower than tolerances, the withdrawal time assignments for MRLs may be extended relative to those for tolerances. This again speaks to the focus on label compliance (MRLs) versus a focus on the safe/unsafe bright line (tolerances).

3.3.4 International Harmonization

There have been considerable international efforts to harmonize evaluation of the safety of veterinary drug residues. The International Cooperation on Harmonization of Technical Requirements for Veterinary Products (VICH) was established in 1996 under the auspices of the World Organization for Animal Health, formally known as Office International des Epizooties, which has retained its historical acronym (OIE). VICH incorporates representatives of both government and industry of VICH member states (EU, Japan, and USA) and observers (Australia, Canada, and New Zealand). The program addresses a broad scope of requirements for the approval of veterinary drug products by the national authorities, including what would be needed to show safety for residues of the veterinary drug in food. Agreement has been reached for a number of requirements for the toxicology needed to establish the safety of veterinary drug products for the human consumer. The Codex Alimentarius, created by the United Nations Food and Agriculture Organization (FAO) and World Health Organization (WHO), establishes maximum residue limits (MRLs) for the residues of veterinary drugs in food; the MRLs serve as international standards for safety setting an upper concentration for residues of veterinary drugs in foods in international trade and can be accessed at `http://www.codexalimentarius.net/vetdrugs/data/index.html`.[31] The Joint FAO/WHO Expert Committee on Residues for Veterinary Drugs in Food (JECFA) performs independent expert toxicological evaluations used to establish an acceptable daily intake (ADI) of residues of the veterinary drug for the human consumer and residue evaluations that result in MRLs recommended to the CAC for their consideration. The Joint FAO/WHO Meeting on Pesticide Residues (JMPR) performs similar evaluations for pesticides, which may include antimicrobial compounds used in fruit and vegetable production, and also recommends MRLs to the CAC.

In the United States, the USDA Foreign Agriculture Service offers a consolidated database of international MRLs for pesticides and veterinary drugs.[32] In addition, a number of countries and regulatory authorities list their official MRLs online. Table 3.7 provides some online sources of information on regulatory requirements for the development of ADIs and MRLs/tolerances, while Table 3.8 provides the URLs for a selection of sites providing MRLs/tolerances established by national regulatory authorities.

3.4 INDIRECT CONSUMER EXPOSURE TO ANTIBIOTICS IN THE NATURAL ENVIRONMENT

Consumers also may be exposed to antibiotics that are released into the natural environment. Following administration to an animal, the antibiotic may be absorbed and metabolized to some extent. A mixture of the parent compound and any metabolites will then be excreted in the urine and feces. For animals on pasture, the compounds will be emitted directly to the pasture environment, whereas for intensively reared systems, the urine and feces will be collected and stored and may then be disposed of or applied to land as a fertilizer.[33,34] Consequently, antibiotics will enter the soil environment. A number of classes of antibiotics have been detected in soils, including sulfonamides, tetracyclines, macrolides, and 2,4-diaminopyrimidines.[35–37] Once released to the soil environment, selected antibiotics such as the tetracyclines have the potential to persist for months to years.[38] Antibiotics in soil may be transported to surface water and groundwater or be taken up into plants and ultimately may enter the food chain or contaminate drinking water supplies. This is partly recognized by existing risk assessment schemes for veterinary medicines. For example, VICH has developed a two-phase process for the environmental risk assessment of veterinary pharmaceuticals, and this process is employed in the marketing authorization process in Europe and the United States.[39,40] For antibiotics where significant environmental exposure is expected, the potential for contamination of aquifers needs to be assessed. If the predicted concentration in groundwater exceeds 100 ng/l (based on the action limit for pesticides in drinking water), then the risk assessment needs to be

TABLE 3.7 Online Sources[a] of Regulatory Requirements for ADI and MRL/Tolerance Development

Country/Regulatory Authority	Agency	URL for Requirements/Guidelines
Australia	Australian Pesticides and Veterinary Medicines Authority	http://www.apvma.gov.au/morag_vet/vol_4/index.php
Canada	Health Canada	http://www.hc-sc.gc.ca/dhp-mps/vet/legislation/guideld/vdd_nds_guide-eng.php
European Union	European Medicines Agency (EMA)	http://www.ema.europa.eu/ema/index.jsp?curl=pages/regulation/general/general_content_000384.jsp&murl=menus/regulations/regulations.jsp&mid=WC0b01ac058002dd37
Japan	Ministry of Agriculture, Forestry and Fisheries of Japan (JMAFF)	http://www.mhlw.go.jp/english/topics/foodsafety/residue/dl/03.pdf
Trilateral Agreement between EU, Japan, and United States of America	International Cooperation on Harmonization Requirements for Registration of Veterinary Products (VICH)	http://www.vichsec.org/en/guidelines.htm 1
United Nations	FAO/WHO Joint Expert Committee on Food Additives (JECFA)	http://www.who.int/ipcs/food/principles/en/index1.html
United States of America	US Food and Drug Administration (USFDA)	http://www.fda.gov/AnimalVeterinary/GuidanceComplianceEnforcement/GuidanceforIndustry/ucm123817.htm

[a]URLs accessed October 23, 2010.

TABLE 3.8 Online Sources[a] of National/Regulatory Authority MRL/Tolerance Information

Country/Regulatory Authority	Agency	URL for MRL/Tolerance Information
Australia	Dept. Agriculture, Fisheries, and Forestry (veterinary and pesticide MRLs)	http://www.daff.gov.au/agriculture-food/nrs/industry info/mrl/cattle-sheep-pigs
Canada	Health Canada (veterinary MRLs)	http://www.hc-sc.gc.ca/dhp-mps/vet/mrl-lmr/mrl-lmr_versus_new-nouveau-eng.php
Canada	Health Canada (pesticide MRLs)	http://www.hc-sc.gc.ca/cps-spc/pest/part/protect-proteger/food-nourriture/mrl-lmr-eng.php
European Union	European Medicines Agency	http://www.ema.europa.eu/ema/index.jsp?curl=pages/medicines/landing/vet_mrl_search.jsp&murl=menus/medicines/medicines.jsp&mid=WC0b01ac058008d7ad
European Union	Directorate General for Health and Consumers (pesticide MRLs)	http://www.efsa.europa.eu/en/praper/mrls.htm
New Zealand	NZ Food Safety Authority (veterinary and pesticide MRLs)	http://www.nzfsa.govt.nz/policy-law/legislation/food-standards/index.htm&mrl
United Nations	Codex Alimentarius (veterinary MRLs)	http://www.codexalimentarius.net/mrls/vetdrugs/jsp/vetd_q-e.jsp
United Nations	Codex Alimentarius (pesticide MRLs)	http://www.codexalimentarius.net/mrls/pestdes/jsp/pest_q-e.jsp
United States of America	US Food and Drug Administration	http://ecfr.gpoaccess.gov/cgi/t/text/text-idx?c=ecfr&sid=25ee42a2644a114d986ba66619dff1f0&rgn=div5&view=text&node=21:6.0.1.1.17&idno=21 or http://www.accessdata.fda.gov/scripts/animaldrugsatfda/
United States of America	US Environmental Protection Agency (pesticide MRLs)	http://www.access.gpo.gov/nara/cfr/waisidx_09/40cfr180_09.html

[a]URLs accessed October 23, 2010.

further refined, or risk management options must be considered.

An overview of our knowledge of the contamination of water bodies and crops by antibiotics is provided below.

3.4.1 Transport to and Occurrence in Surface Waters and Groundwaters

Contaminants applied to soil can be transported to surface waters in surface runoff, subsurface flow, and drainflow or to groundwaters via leaching. The extent of transport via any of these processes is determined by a range of factors, including: the solubility, sorption behavior, and persistence of the contaminant; the physical structure, pH, organic carbon content, and cation exchange capacity of the soil matrix; and climatic conditions such as temperature and rainfall volume and intensity. A number of studies have explored the fate and transport of veterinary antibiotics by these different pathways.[36,38,41–48] Field and semi-field studies have shown that sulfonamide, macrolide, and phenicol antibiotics have the potential to leach to groundwaters, probably because of their low sorption coefficients in soils, whereas the tetracyclines and fluoroquinolones do not leach.[47,49,50] Transport of veterinary medicines via runoff and drainflow has been observed for tetracycline antibiotics (i.e., oxytetracycline) and sulfonamide antibiotics (sulfadiazine, sulfamethazine, sulfathiazole, sulfachloropyridazine).[42,51] Just as with leaching, the transport of these substances is influenced by the sorption behavior of the compounds, the presence of manure in the soil matrix, and the nature of the land to which the manure is applied. Runoff of highly sorptive substances, such as tetracyclines, was observed to be significantly lower than that of the more mobile sulfonamides.[42] However, even for the relatively water-soluble sulfonamides, total mass losses to surface are small (between 0.04% and 0.6% of the mass applied) under actual field conditions.[52]

Once in the water column, substances may be degraded abiotically via photodegradation and/or hydrolysis or biotically by aerobic or anaerobic organisms. Highly sorptive substances may partition to the bed sediment. A significant amount of information is available on the fate and behavior of many veterinary antibiotics in sediment due to their use as aquaculture treatments.[34] While many compounds degrade very quickly (e.g., chloramphenicol, florfenicol, ormethoprim), others persist in the sediment for months to years (e.g., oxolinic acid, oxytetracycline, sarafloxacin, sulfadiazine, trimethoprim).

Alongside the fate experiments described above, a series of studies have monitored concentrations of veterinary antibiotics in surface waters and groundwaters (Table 3.9). In a national monitoring study in the United States, a wide range of medicines was monitored in watercourses.[53] A number of substances that are used as veterinary medicines, including sulfonamides, fluoroquinolones, tetracyclines, and macrolides, were detected in the ng/l range. Many of these substances also are used as human medicines, so the concentrations may result from a combination of inputs from both human and veterinary sources. Similar broad-scale monitoring studies have been done in other regions (including Europe and Asia) and show similar results. The majority of surface monitoring studies involve grab sampling on a number of occasions across a variety of sites. As inputs of many veterinary medicines are likely to be intermittent, concentrations reported in the studies will probably be significantly lower than peak concentrations. To address this, a recent UK study used continuous monitoring of water and sediment, at farms where veterinary medicines (including oxytetracycline, lincomycin, sulfadiazine, trimethoprim, ivermectin, and doramectin) were known to be in use, to determine typical exposure profiles for aquatic systems.[54] Maximum concentrations of antibacterials in streamwater ranged from 0.02 μg/l (trimethoprim) to 21.1 μg/l (lincomycin).

There are only a few reports of detection of veterinary medicines in groundwater.[49,55] In an extensive monitoring study conducted in Germany, a large number of groundwater samples were collected from agricultural areas in order to determine the extent of contamination by antibiotics.[49] The data show that in most areas with intensive livestock breeding, no antibiotics were present above the limit of detection (0.02–0.05 μg/l). Sulfonamide residues were, however, detected in four samples. While the source of contamination of two of these is considered attributable to irrigation with sewage, the authors concluded that sulfamethazine, detected at concentrations of 0.08 and 0.16 μg/l, could possibly have derived from veterinary applications, as it is not used in human medicine.

Veterinary medicines are also known to leach from landfill sites. In Denmark, high concentrations (mg/kg) of numerous sulfonamides were found in leachates close to a landfill site where a pharmaceutical manufacturer had previously disposed of large amounts of these drugs over a 45-year period.[56] Concentrations dropped off significantly tens of meters down gradient, most probably due to microbial attenuation. Although this is recognized as a specific problem, disposal of veterinary medicines to landfill should nevertheless be considered a potential route for environmental contamination.

3.4.2 Uptake of Antibiotics into Crops

Antibiotics may also be taken up from soil into crops.[54,64] The potential uptake of veterinary medicines into plants is receiving increasing attention. Studies with a wide range of veterinary medicines showed that a number of antibiotics are taken up by plants following exposure to soil at environmentally realistic concentrations of the

TABLE 3.9 Measured Concentrations of Veterinary Medicines in Surface Waters and Groundwaters

Compound	Location	Maximum Concentration (μg/l)	Reference
	Surface Waters		
Chloramphenicol	China, Canada	N/D–0.002	57, 58
Chlortetracycline	Canada, USA	0.192–0.21	58, 59
Doxycycline	Canada	0.073	58
Erythromycin	Canada, USA	0.051–0.45	58, 59
Erythromycin	USA	0.45	59
Lincomycin	Canada, UK	0.355–21.1	54, 58
Oxytetracycline	Japan, UK, Canada, USA	N/D–68	54, 58–60
Salinomycin	USA	0.007–0.04	61, 62
Sulfachloropyridazine	Canada, USA	0.007–0.03	58, 59
Sulfadiazine	UK	4.13	54
Sulfadimethoxine	Canada, USA	0.04–0.056	58, 59
Sulfamerazine	Canada	N/D–0.06	58, 59
Sulfamethazine	Canada, USA	0.02–0.408	58, 59
Sulfamethoxazole	Canada, USA	0.009–0.32	58, 59
Sulfathiazole	Canada, USA	0.016–0.03	58, 59
Tetracycline	Canada, USA	N/D–0.03	58, 59
Trimethoprim	Canada, UK	0.015–0.02	54, 58
Tylosin	Canada, USA	Trace–0.05	58, 59
	Groundwaters		
Lincomycin	USA	0.32	63
Sulfamethazine	Germany	0.16	55, 63
Sulfamethoxazole	Germany, USA	0.47–1.11	55

compounds, whereas other compounds were not observed to be accumulated.[54] The lack of uptake observed may be due to the underlying properties of the compound or other factors such as high limits of detection or significant degradation during the study. These studies looked at uptake into carrots and lettuce following exposure to antibiotics at concentrations that might be found in the natural environment. Florfenicol and trimethoprim were detected in lettuce leaves, and enrofloxacin, trimethoprim, and florfenicol were detected in carrot tubers.

3.4.3 Risks of Antibiotics in the Environment to Human Health

Measured concentrations of pharmaceuticals in water and crops in the studies described above, typically result in exposures that are well below human therapeutic dose levels or acceptable daily intakes (ADIs).[54,65] However, there is concern among the scientific and regulatory communities and the general public that exposure to pharmaceuticals, including antibiotics, in the environment may affect human health. These concerns arise from the following facts:

- Individual antibiotics do not occur in the environment on their own but occur as a mixture, which introduces the possibility of synergistic or additive interactions or environmental contraindications between an environmental residue and a medicine taken by a patient for an existing condition.
- Humans will be exposed to antibiotics via a number of routes, whereas most risk assessment studies have considered only one route of exposure.
- Degradation processes, particularly in drinking water treatment processes, may result in transformation products that may be of greater health concern than the parent compound. For example, some pharmaceuticals with amine functionality are possible precursors for nitrosamines—which can be mutagenic and carcinogenic.[66]
- Indirect effects of residues in the environment, such as the selection of antibiotic-resistant microorganisms, cannot currently be ruled out.[67–70]

In summary, the occurrence of antibiotics as mixtures in the environment can result in possibly complex interactions. In addition, humans have multiple routes of exposure to antibiotics. Finally, the potential impact of antibiotics includes effects from degradation products as well as indirect effects such as selection for antimicrobial resistance. There is therefore still much work to do to establish the degree of risk arising from indirect consumer exposure to antibiotics.

3.5 SUMMARY

Animals are routinely treated with antibiotics to prevent, treat, or control disease. There have been historic non-therapeutic uses of antibiotics in food-producing animals to improve production, but this practice is falling out of favor. A consequence of the use of veterinary drugs (including antibiotics) in food-producing animals is the production of residues of the drug in the edible tissues.

Regulatory agencies address the safety of antibiotic residues of veterinary drugs in food by evaluating the toxicity of the antibiotic and establishing an acceptable daily intake (ADI) or an acute reference dose (ARfD). Both ADI and ARfD represent the quantity of residue that may safely be consumed (daily or from a single exposure, respectively) in the human diet. Following the establishment of the ADI (or ARfD), the maximum concentration of residues permitted in edible tissues (meat, milk, eggs, etc.) is determined, following an evaluation of the nature and extent of the residues in the treated animal. The value is termed either the *maximum residue limit* (MRL) or *tolerance* (used in the United States). Whether an MRL or tolerance is used, either approach ensures that people consuming products derived from the animal treated with the antibiotic veterinary drug will not ingest quantities of residue that exceed the acceptable daily intake.

REFERENCES

1. World Health Organization, *The Medical Impact of Antimicrobial Use in Food Animals*, Report of a WHO meeting, Berlin, Oct. 13–17, 1997, WHO/EMC/ZOO/97.4, 1997 (available at `http://whqlibdoc.who.int/hq/1997/WHO_EMC_ZOO_97.4.pdf`; accessed 11/17/10).
2. World Health Organization and Food and Agriculture Organization of the United Nations, report of a JECFA/JMPR informal harmonization meeting, Rome, Feb. 1–2, 1999, Food and Agriculture Organization of the United Nations and World Health Organization, Rome 1999 (available at `http://www.fao.org/ag/agn/agns/jecfa_guidelines_1_en.asp`; accessed 11/17/10).
3. Codex Alimentarius Commission, report of 1st session of Codex ad hoc intergovernmental task force on antimicrobial resistance, Seoul, Oct. 23–26, 2007, in *Joint FAO/WHO Food Standards Program. Codex Alimentarius Commission*, 31st Session, Geneva, June 30–July 5, 2008, ALINORM 08/31/42 (available at `ftp://ftp.fao.org/codex/Alinorm08/al31_42e.pdf`; accessed 11/17/10).
4. Codex Alimentarius Commission, report of 2nd session of Codex ad hoc intergovernmental task force on antimicrobial resistance, Seoul, Oct. 20–24, 2008, in *Joint FAO/WHO Food Standards Program. Codex Alimentarius Commission*, 32nd Session, Geneva, June 30–July 5, 2009, ALINORM 09/32/42 (available at `ftp://ftp.fao.org/codex/Alinorm09/al32_42e.pdf`; accessed 11/17/10).
5. Codex Alimentarius Commission, report of 3rd session of Codex ad hoc intergovernmental task force on antimicrobial resistance, Seoul, Oct. 12–16, 2009, in *Joint FAO/WHO Food Standards Program. Codex Alimentarius Commission*, 33rd Session, Geneva, July 5–9, 2010, ALINORM 10/33/42 (available at `ftp://ftp.fao.org/codex/Alinorm10/al33_42e.pdf`; accessed 11/17/10).
6. Food and Drug Administration, NADA 140–973, *Ventipulmin Syrup—Original Approval*, Boehringer Ingelheim Vetmedica, Inc., Freedom of Information Summary, 1998 (available at `http://www.fda.gov/AnimalVeterinary/Products/ApprovedAnimalDrugProducts/FOIADrugSummaries/UCM054881`; accessed 11/17/10).
7. Committee for Veterinary Medicinal Products, *Clenbuterol Hydrochloride*, Summary Report (1), European Agency for the Evaluation of Medicinal Products, Veterinary Medicines and Information Technology Unit, 1995, EMEA/MRL/030/95-FINAL (available at `http://www.ema.europa.eu/docs/en_GB/document_library/Maximum_Residue_Limits_-_Report/2009/11/WC500012566.pdf`; accessed 11/17/10).
8. Dumestre-Toulet V, Cirimel V, Ludes B, Gromb S, Kintz P, Hair analysis of seven bodybuilders for anabolic steroids, ephedrine, and clenbuterol, *J. Forensic Sci*. 2002;47(1): 211–214.
9. Geyer H, Parr MK, Koehler K, Mareck U, Schänzer W, Thevis M, Nutritional supplements cross-contaminated and faked with doping substances, *J. Mass Spectrom*. 2008; 43:892–902.
10. Kierzkowska B, Stanczyk J, Kasprzak JD, Myocardial infarction in a 17-year-old body builder using clenbuterol, *Circ. J*. 2005;69(9):1144–1146.
11. *Clenbuterol Direct* (available at `http://www.clenbuteroldirect.com/index.html`; accessed 10/28/10).
12. Garssen GJ, Geesink GH, Hoving-Bolink AH, Verplanke JC, Effects of dietary clenbuterol and salbutamol on meat quality in veal calves, *Meat Sci*. 1995;40:337–350.
13. Salleras L, Dominguez A, Mata, E, Taberna, JL, Moro I, Salva, P, Epidemiologic study of an outbreak of clenbuterol poisoning in Catalonia, Spain, *Public Health Rep*. 1995;110(3):338–342.
14. Brambilla G, Cenci T, Franconi F, Galarini R, Macri A, Rondoni F, Strozzi M, Loizzo A, Clinical and pharmacological profile in a clenbuterol epidemic poisoning of contaminated beef meat in Italy, *Toxicol. Lett*. 2000;114(1–3):47–53.
15. Barbosa J, Cruz C, Martins J, Silva JM, Neves C, Alves C, Ramos F, Da Silviera MIN, Food poisoning by clenbuterol in Portugal, *Food Addit. Contam*. 2005;22(6):563–566.
16. Luk G, *Leanness-Enhancing Agents in Pork*, Centre for Food Safety, the Government of the Hong Kong Special Administrative Region (available at `http://www.cfs.gov.hk/english/multimedia/multimedia_pub/multimedia_pub_fsf_14_01.html`; accessed 10/31/10).
17. Rose MD, Shearer G, Farrington WHH, The effect of cooking on veterinary drug residues in food: 1. Clenbuterol, *Food Addit. Contam*. 1995;12(1):67–76.
18. Wallerstein, RO, PK Condit, CK Kasper, JW Brown, FR Morrison, Statewide study of chloramphenicol therapy and fatal aplastic anemia, *JAMA* 1969;208(11):2045–2050.

19. World Health Organization, Comments on chloramphenicol found at low levels in animal products, in *Evaluation of Certain Veterinary Drug Residues in Food*, 62nd report of Joint FAO/WHO Expert Committee on Food Additives, WHO Technical Report Series 925, Geneva, 2004 (available at http://whqlibdoc.who.int/trs/WHO_TRS_925.pdf; accessed 11/17/10).
20. Codex Alimentarius Commission, report of 19th Session of Codex Committee on *Residues of Veterinary Drugs in Foods*, Burlington, NC, Aug. 30–Sept. 3, 2010, REP11/RVDF (available at http://www.codexalimentarius.net/web/archives.jsp?year=11; accessed 11/17/10).
21. International Programme on Chemical Safety, Environmental Health Criteria 240, *Principles and Methods for the Risk Assessment of Chemicals in Food*, World Health Organization, Geneva, 2009 (available at http://www.who.int/ipcs/food/principles/en/index1.html; accessed 11/17/10).
22. International Programme on Chemical Safety, Environmental Health Criteria 239, *Principles for Modelling Dose-Response for the Risk Assessment of Chemicals*, World Health Organization, Geneva, 2009 (available at http://whqlibdoc.who.int/publications/2009/9789241572392_eng.pdf; accessed 11/17/10).
23. Dorne JL, Renwick AG, The refinement of uncertainty/safety factors in risk assessment by the incorporation of data on toxicokinetic variability in humans, *Toxicol. Sci*. 2005;86:20–26.
24. Dourson ML, Felter SP, Robinson D, Evolution of science-based uncertainty factors in non-cancer risk assessment, *Regul. Toxicol. Pharm*. 1996;24:108–120.
25. Cerniglia CE, Kotarski S, Approaches in the safety evaluations of veterinary antimicrobial agents in food to determine the effects on human intestinal microflora, *J. Vet. Pharmacol. Ther*. 2005;28(1):3–20.
26. Inter-Organization Programme for the Sound Management of Chemicals (IOMC), 2010, OECD Environment, Health, and Safety Publications Series on Testing and Assessment 124, *Guidance for the Derivation of an Acute Reference Dose*, Env/JM/MONO(2010)15 (available at http://www.oecd.org/officialdocuments/displaydocumentpdf?cote=env/jm/mono(2010)15&doclanguage=en; accessed 11/17/10).
27. Organisation for Economic Co-Operation and Development, *Chemicals Testing—Guidelines* (available at http://www.oecd.org/department/0,3355,en_2649_34377_1_1_1_1_1,00.html; accessed 10/27/10).
28. *Agreement on the Application of Sanitary and Phytosanitary Measures*, World Trade Organization, Geneva, 1995 (available at http://www.wto.org/english/docs_e/legal_e/15-sps.pdf; accessed 11/17/10).
29. Food and Drug Administration, Guidance for Industry 3, *General Principles for Evaluating the Safety of Compounds Used in Food-Producing Animals*, 2006 (available at http://www.fda.gov/downloads/AnimalVeterinary/GuidanceComplianceEnforcement/GuidanceforIndustry/UCM052180.pdf; accessed 11/17/10).
30. Joint FAO/WHO Expert Committee on Food Additives, *Procedures for Recommending Maximum Residue Limits-Residues of Veterinary Drugs in Food*, Food and Agriculture Organization of the United Nations and World Health Organization, Rome, 2000 (available at http://www.fda.gov/downloads/AnimalVeterinary/GuidanceComplianceEnforcement/GuidanceforIndustry/UCM052180.pdf; accessed 11/17/10).
31. Codex Alimentarius Commission, *Veterinary Drug Residues in Food*, Codex Veterinary Drug Residues in Food Online Database (available at http://www.codexalimentarius.net/vetdrugs/data/index.html; accessed 10/28/10).
32. United States Department of Agriculture, International Maximum Residue Level Database (available at http://www.mrldatabase.com/; accessed 10/31/10).
33. Boxall ABA, Kolpin DW, Halling-Sorensen B, Tolls J, Are veterinary medicines causing environmental risks? *Environ. Sci. Technol*. 2003;37:286A–294A.
34. Boxall ABA, Fogg LA, Kay P, Blackwell PA, Pemberton EJ, Croxford A, Veterinary medicines in the environment, *Rev. Environ. Contam. Toxicol*. 2004;180:1–91.
35. Boxall ABA, Fogg LA, Baird DJ, Lewis C, Telfer TC, Kolpin D, Gravell A, Pemberton E, Boucard T, *Targeted Monitoring Study for Veterinary Medicines in the Environment*, Environment Agency, Bristol, UK, 2006.
36. Hamscher G, Pawelzick HT, Hoper H, Nau H, Different behaviour of tetracyclines and sulfonamides in sandy soils after repeated fertilization with liquid manure, *Environ. Toxicol. Chem*. 2005;24:861–868.
37. Carlson JC, Mabury SA, Dissipation kinetics and mobility of chlortetracycline, tylosin, and monensin in an agricultural soil in Northumberland County, Ontario, Canada, *Environ. Toxicol. Chem*. 2006;25:1–10.
38. Kay P, Blackwell P, Boxall A, Fate of veterinary antibiotics in a macroporous drained clay soil, *Environ. Toxicol. Chem*. 2004;23:1136–1144.
39. Environmental Impact Assessment (EIAs) for Veterinary Medicinal Products (VMPs)—Phase I, VICH GL 6 (Ecotoxicity Phase I), CVMP/VICH/592/98—final, London, June 30, 2000 (available at http://www.vichsec.org/pdf/2000/Gl06_st7.pdf; accessed 11/17/10).
40. Environmental Impact Assessment for Veterinary Medicinal Products—Phase II, VICH GL 38 (Ecotoxicity Phase II), CVMP/VICH/790/03—final, London, Oct. 2004 (available at http://www.vichsec.org/pdf/10_2004/GL38_st7.pdf; accessed 11/17/10).
41. Aga DS, Goldfish R, Kulshrestha P, Application of ELISA in determining the fate of tetracyclines in land-applied livestock wastes, *Analyst* 2003;128:658–662.
42. Kay P, Blackwell PA, Boxall ABA, Column studies to investigate the fate of veterinary antibiotics in clay soils following slurry application to agricultural land, *Chemosphere* 2005;60:497–507.
43. Kay P, Blackwell PA, Boxall ABA, A lysimeter experiment to investigate the leaching of veterinary antibiotics through a clay soil and comparison with field data, *Environ. Pollut*. 2005;134:333–341.

44. Kay P, Blackwell PA, Boxall ABA, Transport of veterinary antibiotics in overland flow following the application of slurry to arable land, *Chemosphere* 2005;59:951–959.
45. Blackwell PA, Kay P, Ashauer R, Boxall ABA, Effects of agricultural conditions on the leaching behaviour of veterinary antibiotics in soils, *Chemosphere* 2009;75:13–19.
46. Burkhard M, Stamm S, Waul C, Singer H, Muller S, Surface runoff and transport of sulfonamide antibiotics on manured grassland, *J. Environ. Qual*. 2005;34:1363–1371.
47. Kreuzig R, Holtge S, Investigations on the fate of sulfadiazine in manured soil: Laboratory experiments and test plot studies, *Environ. Toxicol. Chem*. 2005;24:771–776.
48. Blackwell PA, Kay P, Boxall ABA, The dissipation and transport of veterinary antibiotics in a sandy loam soil, *Chemosphere* 2007;67:292–299.
49. Hamscher G, Abu-Quare A, Sczesny S, Höper H, Nau H, Determination of tetracyclines and tylosin in soil and water samples from agricultural areas in lower Saxony, in van Ginkel LA, Ruiter A, eds., *Proc. Euroresidue IV Conf*., Veldhoven, The Netherlands, May 8–10, 2000, National Institute of Public Health and the Environment (RIVM), Bilthoven.
50. Sinclair CJ et al., *Assessment and Management of Inputs of Veterinary Medicines from the Farmyard*, Final Report to DEFRA, CSL, York, UK, 2008.
51. Kreuzig R, Holtge S, Brunotte J, Berenzen N, Wogram J, Sculz R, Test-plot studies on runoff of sulfonamides from manured soils after sprinkler irrigation, *Environ. Toxicol. Chem*. 2005;24:777–781.
52. Stoob K, Singer, HP, Mueller SR, Schwarzenbach RP, Stamm CH, Dissipation and transport of veterinary sulphonamide antibiotics after manure application to grassland in a small catchment, *Environ. Sci. Technol*. 2007;41:7349–7355.
53. Kolpin DW, Furlong ET, Meyer MT, Thurman EM, Zaugg SD, Barber LB, Buxton HT, Pharmaceuticals, hormones, and other organic wastewater contaminants in US streams 1999–2000: A national reconnaissance, *Environ. Sci. Technol*. 2002;36:1202–1211.
54. Boxall ABA et al., Uptake of veterinary medicines from soils into plants, *J. Agric. Food Chem*. 2006;54(6):2288–2297.
55. Hirsch R, Ternes T, Haberer K, Kratz K-L, Occurrence of antibiotics in the aquatic environment, *Sci. Total Environ*. 1999;225:109–118.
56. Holm JV, Berg PL, Rugge K, Christensen TH, Occurrence and distribution of pharmaceutical organic-compounds in the groundwater down gradient of a landfill (Grinsted, Denmark), *Environ. Sci. Technol*. 1995;29:1415–1420.
57. Tong L, Li P, Wang YX, Zhu KZ, Analysis of veterinary antibiotic residues in swine wastewater and environmental water samples using optimized SPE-LC/MS/MS, *Chemosphere* 2009;74(8):1090–1097.
58. Lissemore L, Hao C, Yang P, Sibley PK, Mabury S, Solomon KR, An exposure assessment for selected pharmaceuticals within a watershed in Southern Ontario, *Chemosphere* 2006;64:717–7129.
59. Kim SC, Carlson K, Temporal and spatial trends in the occurrence of human and veterinary antibiotics in aqueous and river sediment matrices, *Environ. Sci. Technol*. 2007;41(1): 50–57.
60. Matsui Y, Ozu T, Inoue T, Matsushita T, Occurrence of veterinary antibiotic in streams in a small catchment area with livestock farms, *Desalination*. 2008;226(1–3):215–221.
61. Kim SC, Carlson K, Occurrence of ionophore antibiotics in water and sediments of a mixed-landscape watershed, *Water Res*. 2006;40(13):2549–2560.
62. Cha JM, Yang S, Carlson KH, Rapid analysis of trace levels of antibiotic polyether ionophores in surface water by solid-phase extraction and liquid chromatography with ion trap tandem mass spectrometric detection, *J. Chromatogr. A* 2005;1065(2):187–198.
63. Barnes KK, Kolpin DW, Furlong ET, Zaugg SD, Meyer MT, Barber LB, A national reconnaissance of pharmaceuticals and other organic wastewater contaminants in the United States—I) Groundwater, *Sci. Total Environ*. 2008;402: 192–200.
64. Kumar K, Gupta SC, Baidoo SK, Chander Y, Rosen CJ, Antibiotic uptake by plants from soil fertilized with animal manure, *J. Environ. Qual*. 2005;34(6):2082–2085.
65. Hughes J et al., *Evaluation of the Potential Risks to Consumers from Indirect Exposure to Veterinary Medicines*, IEH Final Report to DEFRA, 2006 (available at `http://randd.defra.gov.uk/Default.aspx?Menu=Menu&Module=More&Location=None&ProjectID=11778&FromSearch=Y&Publisher=1&SearchText=vm02130&SortString=ProjectCode&SortOrder=Asc&Paging=10#Description`; accessed 11/24/10).
66. Krasner SW, The formation and control of emerging disinfection by-products of health concern, *Phil. Trans. Royal Soc. A* 2009;367(104):4077–4095.
67. Witte W, Medical consequences of antibiotic use in agriculture, *Science* 1998;279(5353):996–997.
68. Boxall A, Blackwell P, Cavallo R, Kay P, Tolls J, The sorption and transport of a sulphonamide antibiotic in soil systems, *Toxicol. Lett*. 2002;131:19–28.
69. Heuer H, Smalla K, Manure and sulfadiazine synergistically increased bacterial antibiotic resistance in soil over at least two months, *Eviron. Microbiol*. 2007;9:657–666.
70. Byrne-Bailey KG, Gaze WH, Kay P, et al., Prevalence of sulfonamide resistance genes in bacterial isolates from manured agricultural soils and pig slurry in the United Kingdom, *Antimicrob. Agents Chem*. 2009;53(2): 696–702.

4

SAMPLE PREPARATION: EXTRACTION AND CLEAN-UP

Alida A. M. (Linda) Stolker and Martin Danaher

4.1 INTRODUCTION

Sample preparation affects all the later assay steps and therefore is critical for unequivocal identification, confirmation, and quantification of analytes. It includes both the isolation and/or pre-concentration of compounds of interest from various matrices and also makes the analytes more suitable for separation and detection. Sample preparation typically takes more than 70% of the total analysis time. Whereas chromatographic methods are preferred in the analysis of organic molecules following clean-up, sample preparation in bioanalytical methods regularly employs liquid–liquid extraction and solid phase extraction. In contrast with ultrarapid chromatographic analysis, conventional sample pre-treatment approaches are more labor-intensive and time-consuming, consisting of many steps. For this reason, many new sample preparation techniques have been developed and there is continuing interest in this aspect of analytical work.

An evaluation of the scientific literature reveals that over 500 papers on veterinary drug residue analysis were published in the 5-year period of 2005–2009.[1] Liquid extraction (LE) and liquid–solid extraction (LSE) were found to be very popular sample treatment techniques that were used in 30% and 60% of the reported studies, respectively. Here, LE includes all liquid-based approaches such as liquid–liquid extraction (LLE), extrelut liquid–liquid extraction, liquid–liquid micro-extraction, and pressurized liquid extraction (PLE). LSE includes solid phase extraction (SPE)[2] and all other sorbent-based extraction procedures, such as solid phase micro-extraction (SPME), stir bar sorptive extraction (SBSE), restricted-access materials (RAM), turbulent-flow chromatography (TFC), dispersive SPE (dSPE), and matrix solid phase dispersion (MSPD). Other techniques used for some specific applications are microwave-assisted and ultrasonically assisted extraction (MAE, UAE), immunoaffinity(-based) extraction (IAC),[3] and polymer imprinted types of extraction techniques [molecularly imprinted polymer (MIP)].[4]

There have been many changes in approach to sample preparation with the more recent widespread application of mass spectrometry. Previously, methods were capable of analyzing residues of only a limited number of compounds (usually a single class of drug);[5–9] but mass spectrometry now offers the possibility of analyzing residues of many compounds in a test sample.[10–13] As a result, there is now a trend to focus more on generic extraction and clean-up procedures to cover the wide range of antibiotics that can potentially occur in food of animal origin.[13–15] Although mass spectrometry permits the use of simpler generic clean-up methods, effective removal of matrix constituents is necessary, as these may affect the performance of the mass spectrometer, particularly through ion suppression and enhancement effects.[16]

There has also been a move from slow manual sample preparation techniques to faster automated techniques. Automated sample preparation can be carried out on-line (with sample preparation connected directly to the analysis system) or off-line (sample preparation is automated, but the sample has to be manually transferred to the analysis system). Automated sample preparation offers the potential of performing sample clean-up, concentration, and analyte separation in a closed system. This reduces the sample preparation time, and the whole sample becomes available for analysis, leading to improved limits of detection. It also removes some of the human element from a procedure, thereby improving precision and reproducibility. Furthermore, automated sample preparation reduces cost by using

Chemical Analysis of Antibiotic Residues in Food, First Edition. Edited by Jian Wang, James D. MacNeil, and Jack F. Kay.

less solvent and fewer personnel. Other advantages include reduced risk of sample contamination and elimination of analyte losses by evaporation or by degradation during sample pre-concentration.

Automated methods also have some disadvantages, including an increase in initial capital expenditure. There is a risk of increased downtime from equipment breakdowns, so parallel systems need to be operated to reduce downtime. In addition, there is also a potential for memory or carryover effects, although these can be eliminated using cartridge-based systems.

Two very comprehensive reviews of current trends in sample preparation have been published by Kinsella et al.[17] and by Noákavá and Vlcková.[18] This chapter includes topics discussed in these specific reviews. The different off- and on-line sample preparation procedures mentioned above are described. General items regarding extraction procedures are discussed, followed by a discussion of the current sample preparation techniques,[17,18] with some examples of applications.

4.2 SAMPLE SELECTION AND PRE-TREATMENT

The first selection that has to be made when setting up, for example, a monitoring program involves the type of sample material to be targeted for analysis. For monitoring drugs that have an established maximum residue limit (MRL),[19] matrices that may be selected include tissues, such as liver, kidney, muscle and fat, plus milk, eggs, or honey. Since the drug concentrations in the consumable parts or products of an animal must be below the MRL, these matrices are therefore of most interest. Kidney and liver are the target organs for most antibiotics because the drug concentrations in these organs are typically higher than in other edible tissues. One disadvantage of selecting animal organs, muscle or fat is that they can be analyzed only after slaughter.

Muscle can present analytical difficulties because of variability in residue distribution,[20–22] particularly in the area surrounding injection sites.[23–26] There is also the concern of a lower probability of finding non-compliant samples in muscle compared to matrices such as liver and kidney.[27] Schneider et al.[28] described the variability of residues from treatment with penicillin G (Pen G) in muscle. It was observed that Pen G residue concentrations varied between and within different muscles, although no reproducible pattern was identified between cows or related to withdrawal times. Because of the potential for variation of residue concentrations within muscles, all samples taken need to be large enough to be representative.[28]

Antibiotics are also frequently monitored in animal feed and drinking water. Feed is a difficult matrix; it is not easy to extract the drugs because of the large amounts of proteins and carbohydrates. However, the drug concentrations in feed are usually much higher (1–10 mg/kg) than in animal tissues (1–100 μg/kg); consequently drugs can be more easily detected. Analysis of residues can be complicated because of the high variability of matrix effects in different types of feed. Quantification can be improved through the introduction of internal standards or the use of the standard addition approach. However, when matrix effects result in the masking of antibiotic residues, there is no choice but to include a suitable clean-up step in the assay.

Be aware, however, that analysis of feed samples or injection sites should be kept separate from the analysis of tissue, milk, or other such materials for residues. The difference in concentrations of drugs present in feed samples or injection sites may be orders of magnitude greater than those found in typical tissue, milk, egg, or honey samples, and therefore the risk of contamination of such samples becomes high unless great care is taken to prevent such an occurrence. It is generally preferable to simply physically separate the analyses of those materials that potentially contain target analyte concentrations that are very different from the analysis of samples that contain residues at or below MRLs.

Manure and urine are a third group of matrices targeted in some monitoring programs. They are mostly used to monitor prohibited substances and can be taken at or prior to slaughter. This has the advantage that when "non-compliant" results are obtained, the animals can be prevented from reaching the market. Alternative matrices allow detection of residues for longer periods post-treatment (e.g., hair[29]) or may allow detection of residues using less sophisticated equipment (e.g., HPLC detection of semicarbazide in retina[30]).

Because of the usual lag time between sample collection and analysis at the laboratory, sample storage is an important step. The potential effects of physicochemical factors such as oxidation, proteolysis, and precipitation and biological factors that include microbiological and enzymatic reactions need to be considered when storing samples.[17] For example, the production of the enzyme penicillinase, which is capable of reducing the concentration of penicillin in kidney tissue stored at 4°C,[31] has been reported in some studies. Preservation can be achieved through the addition of enzyme inhibitors (e.g., piperonylbutoxide inhibits cytochrome P450).

A number of studies have highlighted the degradation of residues during frozen storage, including β-lactam antibiotics in milk;[32] ampicillin in pig muscle;[33] chlortetracycline in incurred pig muscle, liver, and kidney;[34] sulfamethazine in incurred pig muscle and bovine milk;[35] and gentamicin residues in egg.[36] The EU validation criteria[37] require that stability be determined for the analytes in matrix and in solution at various stages of the sample preparation process. Preferably, incurred tissue should be used; otherwise matrix fortified material is used.[38] From a practical perspective it is helpful to run a test to see how long a sample and/or analyte can be held without degradation. Analysis of samples should be completed within that time (see Chapter 8 for a discussion of typical analyte and sample stability experiments).

The variation of residues within a single organ or tissue is often ignored, but it is an important factor to consider prior to sample preparation. For example, residue variations may occur in the kidney between the medulla and the cortex.[39–41] It is very important to take a representative aliquot of the sample, which may require removal of several portions distributed throughout the composite sample to give a representative sub-sample from which test portions for analysis will be taken. Another critical point is the homogenization. Homogenization with the use of a blender is often advantageous for obtaining a homogenous sample but can result in the release of enzymes, which can degrade residues and provide inaccurate results. It is therefore easier to process liquid samples, such as blood, plasma, serum, milk, bile, or water samples. In comparison with solid samples, the residues are more uniformly distributed and homogeneity can be easily achieved by mixing or shaking the samples.

4.3 SAMPLE EXTRACTION

4.3.1 Target Marker Residue

Because of the extensive metabolism in animals that often occurs after drug administration, the residues present can vary greatly between target tissues. The target residue for analysis is sometimes only the parent drug but can also be a metabolite, the sum of the parent drug, and/or metabolites or a compound formed by chemical conversion of the parent compound and metabolites. The free parent and metabolite residues are readily extracted by organic solvents, H_2O, or aqueous buffers, depending on their solubilities and polarities. However, residues of some compounds may be present in the conjugated forms (glucuronides or sulfates) and require liberation through enzymatic or chemical hydrolysis prior to extraction. Hydrolysis conditions (viz., pH, temperature, time) have to be carefully optimized to ensure efficient deconjugation of residues. There are different procedures available for hydrolysis, but enzymatic hydrolysis generally ensures milder conditions than does acid or alkaline hydrolysis. Helix promatia juice (a mixture of β-glucuronidase and arylsulfatase) and *Escherichia coli* β-glucuronidase are commonly used for enzymatic hydrolysis.

Free residues and conjugates can be easily extracted after dialysis, proteolysis, or denaturation of proteins by heat or acid treatments. In general, the bound residues are hydrolyzed before analysis. Analysis of bound residues is required for very few antibiotics, namely, nitrofurans and florfenicol. Nitrofuran antibiotics are rapidly metabolized to form bound residues, which persist for many weeks after treatment.[42] These bound metabolites may pose a health risk and are used as marker residues to monitor for evidence of nitrofuran use.[43] It is proposed that binding of residues occurs through cleavage of the nitrofuran ring by stomach acid, releasing the side chains, which then become bound to protein in the tissue with which they come into contact.[44] In analytical applications, these side chains are cleaved from tissue samples under mildly acidic conditions before undergoing derivatization to increase the analytical response.[45] Metabolism studies of florfenicol depletion demonstrated that non-extractable residues of florfenicol were predominant in tissues.[46] Acid hydrolysis of non-extractable residues not only liberates bound residues but also converts them to florfenicolamine (FFA), which is the marker residue for florfenicol.[47,48]

4.3.2 Stability of Biological Samples

Because of the complex nature of biological matrices, sample preparation steps are the most important integral part of the bioanalytical method. One of the key problems of analysis of biological samples is the instability of drugs, metabolites, and pro-drugs in these kinds of samples.[18] The stability of drugs in biological materials may be affected by storage temperature, exposure to enzymes, the pH of the biological samples, anticoagulants, and freeze–thaw cycles. Moreover, instability may occur during any of the numerous steps of bioanalytical methods:

- In the biological matrix before taking aliquots of samples for the analysis
- During the extraction step
- During the evaporation to dryness or reconstitution
- In the solution inside injection vials
- In the case of mass spectrometry, in the ion source as well

Therefore, short-term, long-term, and freeze–thaw stability studies should be performed for standard solutions as well as for real samples (see Chapter 8 for additional details). The degradation of a drug during sample pretreatment can cause an underestimation of drug concentration. Generally, degradation occurs naturally and could be caused by exposure to light, or may be a result of a reaction with the biological fluid. The drug may also be adsorbed onto the surface of containers or synthetic barriers, such as polymers or separation gels.[49]

Some groups of compounds undergo interconversion reactions; thus special precautions must be taken when analyzing acylglucuronides, lactone, and open-hydroxy acid compounds, or samples that contain a thiol group and a corresponding disulfide. Minimizing interconversion depends on controlling conditions during the bio-analytical procedure; pH is one of the most important factors, together with temperature. Other reasons for sample instability include epimerization (such as with tetracyclines) and E→Z isomersization reactions,[50] influenced again by pH or light exposure. Highly unstable metabolites or parent drug residues may be stabilized by the addition of stabilizing agents such as citric acid to blood samples, or citrate or phosphate buffers to a plasma sample. These additives can be used to maintain the pH of plasma during storage or processing, as the pH of biological samples changes during storage, ultrafiltration, centrifugation, and extraction.[51]

The type of anticoagulant used during collection of the blood sample may also affect the stability of drugs tested for or their metabolites. Various chemical agents, such as EDTA, formic acid, acetic acid, sodium fluoride, lithium heparin, potassium oxalate, and methylacrylate have been used to stabilize analytes in biological matrices.[52]

4.4 EXTRACTION TECHNIQUES

The disruption or homogenization of samples is key to obtaining good extractability of residues from test samples. This has been highlighted by McCracken et al.[53] when comparing four different disruption techniques (probe blender, Stomacher, ultrasonic bath, and end-over-end mixer) for isolating chlortetracycline, sulfadiazine, and flumequine residues from incurred and spiked chicken muscle. An apparatus has been developed for tissue disruption, and several vendors have developed equipment to automate this technique, which was until recently a manual operation. A more detailed discussion on sample disruption can be found in the paper by Kinsella et al.[17]

4.4.1 Liquid–Liquid Extraction

Liquid–liquid extraction (LLE) or liquid–liquid partitioning (LLP) was one of the first sample preparation techniques and continues to be widely used for biological sample analysis. LLE is based on the transfer of analyte from the aqueous sample to a water-immiscible solvent based on the octanol–water partition coefficient. Nevertheless, some shortcomings, such as emulsion formation, the use of large sample volumes, and toxic organic solvents and above all, the production of a large volume of hazardous waste, make LLE expensive, time-consuming, and environmentally harmful. Another drawback of LLE is its unsuitability for hydrophilic compounds.[18]

Despite this, LLE is still widely used in the sample preparation of biological fluids. In general, the majority of LLE methods employ more efficient organic solvents as extracting agents. Acetonitrile is the preferred extraction solvent as it gives good yields of residues but low concentrations of matrix co-extractives and is effective at denaturing proteins and inactivating enzymes. Methanol and ethyl acetate are also widely used solvents but result in the extraction of additional matrix components.[13] However, in the area of multi-residue analysis there is always a compromise between recovery and the purity of sample extracts. LLE was the most widely applied extraction procedure in residue analysis because of its high selectivity compared to simple solvent extraction. LLE applications can also include polar ionizable compounds, which can be extracted by non-polar organic solvents using the ion-pair technique: transforming positively charged substances into non-polar neutral compounds in the presence of organic anions, or vice versa. Examples of the successful application of ion-pair extraction to antibiotic residues include aminoglycosides[54] and oxytetracycline analysis.[55] As an advance, LLE may be conducted using 96-well plates and a 96-channel robotic liquid-handling workstation to automate the process.

4.4.2 Dilute and Shoot

"Dilute and shoot" is perhaps the simplest sample preparation strategy, especially when designing multi-class methods, where the selectivity of other techniques such as SPE may become a disadvantage.[56] Dilution of the extracts can reduce matrix effects to a certain degree, but extensive maintenance of the LC-MS system is needed to ensure reproducible chromatograms and MS sensitivity. Typical maintenance includes thorough column cleaning and regeneration, MS ion source cleaning, and/or the use of a divert valve. As an example, Chico et al.[57] developed a simple method for the analysis of 39 antimicrobials (tetracyclines, quinolones, penicillins, sulfonamides, and macrolides) in animal muscle tissue. It consisted of an extraction with ethanol : water (70 : 30, v/v) containing EDTA to improve the extraction of tetracylines, followed by dilution of the extracts before injection into the chromatographic system. Matrix-matched standards were used for correct quantification of the samples. The simplicity of the sample preparation procedure along with the use of UPLC enabled a high sample throughput to be realized. The method was successfully applied in an Official Public Health Laboratory for the routine analysis of >1000 samples over a 6-month period.[57] Moreover, the method was tested in several inter-laboratory studies with good results.

In another application, Granelli et al.[58] developed a method for the extraction of a total of 19 antimicrobials, including tetracyclines, sulfonamides, quinolones, β-lactams, and macrolides, from muscle or kidney samples. Extraction was performed with 70% methanol, and the extracts were then diluted five-fold with water before LC-MS injection. The method was suitable only for screening purposes at the MRLs. Matrix effects were more pronounced in kidney samples than in muscle, especially for tetracyclines and macrolides. Signal suppression resulted in poor precision for these two antimicrobial classes. This can be attributed to the fact that, during extraction with 70% methanol, salts are co-extracted with the analytes, causing suppression. Moreover, lower recoveries were also observed for tetracyclines, macrolides, and quinolones in kidney samples (<66%) compared with results obtained for muscle.

A 2008 paper has described for the first time a "dilute and shoot" strategy for the simultaneous extraction of wide variety of residues and contaminants (pesticides, mycotoxins, plant toxins, and veterinary drugs) from different foods (meat, milk, honey, and eggs) and feed matrices.[59] Several antimicrobial classes were included (sulfonamides, quinolones, β-lactams, macrolides, ionophores, tetracyclines, and nitroimidazoles) in the analytical method. Sample extraction was performed with water/acetonitrile or acetone/1% formic acid, but instead of dilution of the extracts before analysis by UPLC-MS/MS, small extract volumes (typically 5 μl) were injected to minimize matrix effects. Despite the absence of clean-up steps and the inherent complexity of the different sample matrices, adequate recoveries were obtained for the majority of the analyte/matrix combinations (typical values for antimicrobials were in the range of 70–120%). In addition, the use of UPLC allows high-speed analysis, since all analytes eluted within 9 min.

4.4.3 Liquid–Liquid Based Extraction Procedures

4.4.3.1 QuEChERS

Anastassiades et al. developed a variation of LLE in the QuEChERS (quick, easy, cheap, effective, rugged, and safe) sample preparation procedure, which has been successfully applied to the analysis of hundreds of pesticide residues.[60] In QuEChERS, the high-moisture sample (H_2O is added to dry foods) is extracted with an organic solvent [acetonitrile (ACN), ethylacetate, or acetone]. The addition of salts (anhydrous $MgSO_4$, NaCl, and/or buffering agents) to the extraction medium induces separation of solvent and aqueous phases. The residues of interest partition into the organic phase, and matrix co-extractives go into the aqueous phase. On shaking and centrifugation, an aliquot of the organic phase is subjected to further purification using dispersive SPE (dSPE), which entails mixing sorbents such as $MgSO_4$, primary secondary amine (PSA), C18, and/or graphitized carbon black with the extract. The approach is very flexible because it uses very little labware and generates little waste. Several modifications of the technique have been reported.[61–64] The technique provides high recovery for many LC- and GC-amenable residues, gives high reproducibility, and costs less than many typical sample preparation approaches.

Several groups have adapted the method to analyze residues in a variety of matrices. Acetic acid (HOAc, 1%) and sodium acetate have been widely used to adjust and maintain pH and promote stability and recovery of base-sensitive residues.[64] HOAc was used to adjust pH by Stubbings and Bigwood to determine residues [sulfonamides, quinolones, (fluoro)quinolones, ionophores, and nitroimidazoles] in chicken muscle.[15] Buffering to acidic conditions improved the extraction efficiency of quinolones. Acetonitrile extracts were subsequently purified by dSPE (see also Section 4.4.6.1) over Bondesil NH_2 sorbent. An aliquot of the extracts was evaporated to dryness and re-dissolved in acetonitrile : water (90 : 10, v/v) before LC-MS/MS analysis. Validation was performed on chicken muscle samples, and matrix-matched standards were used because suppression of the MS response was observed for many of the target analytes.

In addition to the antimicrobial classes discussed above, recent research has demonstrated that the method is also applicable to other antimicrobial classes, namely, macrolides and lincosamides. Aguilera-Luiz et al. described[63] a simple and fast procedure for the extraction of sulfonamides, quinolones, macrolides, and tetracyclines from milk samples based on a buffered QuEChERS liquid extraction method. The target compounds were extracted from milk with acidified acetonitrile in the presence of EDTA to increase the recoveries of macrolides and tetracyclines. Co-extracted water and protein were subsequently removed by addition of magnesium sulfate and sodium acetate followed by centrifugation and filtration of the organic phase; the diluted extracts were analyzed directly without further clean-up. No denaturing of proteins or fat removal was required prior to extraction, which was performed in a single step. The proposed method was less time-consuming and easier to perform than other currently available procedures. Furthermore, extraction times were less than 10 min per sample with recovery values for the antimicrobials ranging between 73% and 108%.

4.4.3.2 Bipolarity Extraction

Kaufmann et al. developed a "bipolarity extraction" method based on principles similar to those of the QuEChERS technique.[13] With the use of this isolation technique, polar and non-polar residues remained in the aqueous phase and underwent clean-up by SPE on a mixed-mode Oasis HLB cartridge. The residues were subsequently analyzed by UPLC-MS/MS. Extracts isolated using the bipolarity approach required a lengthy SPE procedure prior to analysis. Kaufmann stated that extracts produced at the end of the procedure contained less matrix components compared to QuEChERS. However, it can be concluded that

QuEChERS, due to its low cost, coupled to its flexibility and ease of use, will be increasingly applied in residue analysis.[17]

4.4.4 Pressurized Liquid Extraction (Including Supercritical Fluid Extraction)

Instrument-based extraction techniques such as supercritical fluid extraction (SFE) and pressurized liquid extraction (PLE) offer advantages because of their potential for automation, more selective isolation of residues through tuning of parameters, and on-line clean-up of samples. Their applications have been slowed by the limited number of commercially available instruments, additional extraction costs, and instrumental downtime. Although several applications have been developed using SFE and PLE, these techniques are not widely used in routine laboratories.

A *supercritical fluid* is defined as any substance that is above its critical temperature and pressure.[66] Supercritical fluids have physical properties intermediate between liquid and gas phases; the solvating power (density) of a SF is similar to that of a liquid, and its diffusivity and viscosity are similar to that of a gas. Carbon dioxide (CO_2) is the most widely used SF because of its inertness, low cost, high purity, low toxicity, and low critical parameters (CO_2 : $T_c = 31.3°C$, $P_c = 72.9$ atm).[65] If extraction cannot be achieved using CO_2, a more polar SF (e.g., N_2O or CHF_3) can be used. Alternatively, a polar modifier (MeOH, EtOH, or H_2O) may be added to the SF in order to increase the solvating power. Several SFE applications have been reported in peer-reviewed literature for selective isolation of residues from food.[66–68] The number of published applications has decreased in recent years, which may be attributed to the lack of automated SFE systems and limited advances in the area.

Pressurized liquid extraction (PLE) has received numerous names, such as accelerated solvent extraction (ASE), pressurized fluid extraction (PFE), pressurized hot-solvent extraction (PHSE), subcritical solvent extraction (SSE) and hot-water (H_2O) extraction (HWE).[69] PLE is carried out at temperatures above the boiling point of the solvent and uses high pressure to maintain the solvent in the liquid phase and achieve fast and efficient extraction of analytes from the solid matrix. HWE is being increasingly used in residue analysis, due to low cost, low toxicity, and ease of disposal. At ambient temperature and pressure H_2O is a polar solvent, but if the temperature and pressure are increased, the polarity decreases considerably, and H_2O can be used to extract medium to low polarity analytes.[69]

A schematic representation of PLE system is shown in Figure 4.1.[163] At elevated temperature and pressure, the PLE extraction process proceeds faster but selectivity decreases[70] and results in the co-extraction of unwanted matrix components. As a result, post-extraction clean-up is frequently required. Off-line clean-up can be achieved post-extraction or during extraction through trapping on sorbent on-line (outside the vessel) or in-line (in the vessel). The latter two approaches help to reduce the number of steps in the analytical process and thus reduce transfer losses. Examples of sorbents employed include Florisil (synthetic magnesium silicate), alumina, or silica gel, which, when used prior to analysis, can prevent lipids and other interferents from being co-extracted. Alternatively, samples may be pre-extracted with a non-polar solvent (e.g., hexane) to eliminate the hydrophobic compounds present in the sample prior to the extraction of analytes of interest. The lengthy time required to pack extraction cells has been

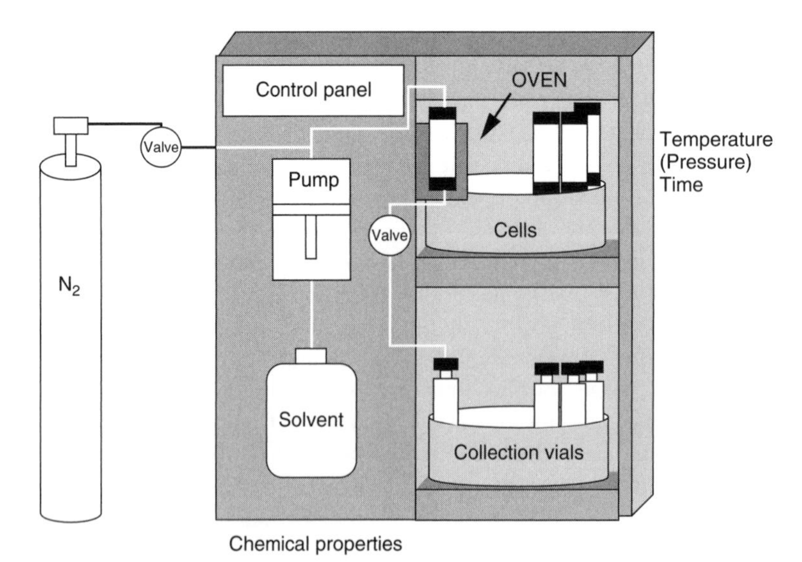

Figure 4.1 Principle of a pressurized fluid extraction (PFE) system and influencing parameters. (Adapted from Camel[163] with permission from Royal Society of Chemistry; copyright 2001.)

partially offset by short extraction times, low solvent usage, and the ability to use H_2O (cheap and environmentally friendly) as the extraction solvent. Hence, PLE has had more recent success that SFE in residue analysis.[17,18,71]

In 2008, Carretero et al. described a multi-class method for the analysis of 31 antibacterials (including β-lactams, macrolides, lincosamides, quinolones, sulfonamides, tetracyclines, nitroimidazoles, and trimethoprim) in meat samples by PLE-LC-MS/MS.[72] Meat samples were homogenized and blended with EDTA-washed sand, then extracted with water by applying 1500 psi (lb/in.^{2}), at 70°C. One extraction cycle was 10 min. A drawback of the method is the large volumes of extracts (40 ml) obtained, which required evaporation to concentrate the extract volume prior to final analysis. This evaporation step considerably increases the time required for sample preparation. The proposed method has been applied to the analysis of 152 samples of cattle and pig tissues, with the presence of quinolones, tetracyclines, and sulfonamides detected in 15% of the samples, although at concentrations below the MRLs.

Runnqvist et al.[73] reviewed the published PLE methods for antimicrobials, including multi-analyte methods. The methods are summarized and critically discussed. Regarding method optimization, the authors concluded that pressure may be less important while solvent composition and temperature are presumably the parameters that influence the extraction efficiencies to the greatest extent. Extraction temperature should be carefully determined since this parameter affects not only the absolute amounts extracted but also the degradation of analytes and the co-extraction of unwanted matrix components.[73]

4.4.5 Solid Phase Extraction (SPE)

4.4.5.1 Conventional SPE

Solid phase extraction (SPE) is the most important sample purification technique in residue analysis and has gradually replaced LLE. The objective of this section is to give a brief overview of SPE and sorbent materials. A number of books and review papers have already been written on this topic and can be consulted for more details.[74,75]

Solid phase extraction is used primarily to prepare liquid samples and extracts of semi-volatile or non-volatile analytes, but may also be used for solids pre-extracted into solvents. The choice of sorbent is the key factor in SPE, because this can control parameters such as selectivity, affinity, and capacity. This choice depends primarily on the analytes and their physicochemical properties, which should define the interactions with the chosen sorbent. However, results also depend on the kind of sample matrix and interactions with both the sorbent and the analyte. SPE sorbents range from chemically bonded silicas, such as with the C8 and C18 organic groups, to graphitized carbon, ion exchange materials, and polymeric materials (PS-DVB, cross-linked styrene–divinylbenzene, PMA, cross-linked methacrylate, MA-DVB, and many others). In addition, there are mixed-mode sorbents (containing both non-polar and strong cation or anion), immunosorbents, molecularly imprinted polymers, and also more recently developed monolith sorbents. Silica sorbents have several disadvantages when compared with polymeric sorbents. Silica sorbents are unstable in a broader pH range and contain silanols, which can cause the irreversible binding of some groups of compounds, such as tetracyclines.

Solid phase extraction method development has many similarities with HPLC. SPE may be performed either on-line or off-line. Conventional SPE cartridges are easy to handle by using vacuum or positive-pressure manifold. However, it is not always easy to control the flow rate, and in addition care should be taken to prevent the column from drying out prior to sample application. As it could be difficult to elute the analyte of interest from conventional SPE cartridges using minimal solvent volume unless organic solvent composition rises up to 100%, special SPE disks are typically used for these purposes. This approach is much quicker as evaporation to dryness and reconstitution are no longer necessary because elution can be performed directly by mobile phase. The on-line configuration of SPE utilizes a 96-well plate format for SPE automated with a robotic liquid-handling system, facilitating high-throughput analysis of biological samples (see Section 4.4.5.2). SPE is the most widely used sample preparation technique in bio-analytical laboratories.[18]

Given the diversity of chemistries involved, developing a viable method that could simultaneously extract different antibiotics (multi-analyte extraction) is challenging. It is essential to identify the right balance of analytical conditions through careful consideration of analyte physicochemical properties, such as solubility, pK_a, chemical and thermal stability, and polarity so as to maximize the analyte recoveries. Nevertheless, any such method may involve a compromise in optimal conditions for individual analytes.[76]

β-Lactams are sensitive to acids and bases, and this sensitivity varies with the nature of the sidechain. The maximum stability of monobasic compounds such as Pen G is exhibited in the pH range 6–7, whereas for ampicillin (an amphoteric compound), the maximum stability occurs at its isoelectric point of ~pH 5. The highly susceptible β-lactam nitrogen is prone to attack by nucleophiles such as methanol. Furthermore, this nucleophilic attack is accelerated by acid catalysis and application of heat. They are also readily isomerized in an acidic environment. β-lactams are typically extracted with water and/or polar organic solvents from solid matrices.

Tetracyclines (TCs) may degrade under extremes of pH with strong acids as well as alkali through epimerization, dehydration, isomerization, and other processes. A pH of 4 has been most commonly utilized for extraction of TCs from various matrices. There is a tendency of TCs to form chelation complexes with metal ions and bind with matrix constituents, which may cause problems in analysis.

The polypeptide antibiotic bacitracin C_l converts to epi-bacitracin in strong acid. It also undergoes thermal decomposition on prolonged heating. Dilute acid solutions, acid buffers, and polar organic solvents are typically used for extraction of this polar compound. Virginiamycin is a combination of two unrelated cyclic polypeptides that are non-polar and practically insoluble in water but soluble in organic solvents such as methanol.

Chloramphenicol is a highly polar, stable compound typically extracted under neutral conditions in aqueous buffers and/or polar organic solvents.

Macrolides are basic and lipophilic macrocyclic lactones that are slightly soluble in water but readily soluble in organic solvents. In general, neutral or slightly basic conditions are chosen to extract macrolides to avoid degradation of members of this group, such as erythromycin, which degrade in an acidic setting. For example, erythromycin degrades to anhydroerythromycin.

The polyether monensin is both acid- and base-labile because of its hemi-ketal group, and neutral conditions have commonly been employed in its extraction. It is non-polar, lipophilic, and only sparingly soluble in water.

Aminoglycosides, by contrast, are hydrophilic, highly polar, and resistant to acids, bases, and heat. They are typically extracted from food or biomatrices with acidic or basic aqueous solutions or aqueous/organic mixtures to facilitate their release from bound proteins.[76]

Typical SPE sorbents used for multi-class antibiotic analysis in food matrices include Oasis HLB (hydrophilic–lipophilic balanced) and Strata X. Oasis HLB cartridges have been preferred in many laboratories because of their good retention properties and highly reproducible recoveries of a wide range of compounds, whether polar or non-polar (due to their combined hydrophobic–hydrophiclic retention mechanism). Strata X cartridges, which are similar in functionality to Oasis HLB cartridges, provide comparable results.

Among the multi-class methods reported in the literature, a procedure that involves sample dissolution with EDTA under mildly acidic conditions (pH 4.0) followed by SPE with Oasis HLB cartridges has been applied for the simultaneous analysis of macrolides, tetracyclines, quinolones, and sulfonamides in honey samples.[77] Separation and determination by UPLC-MS/MS enabled the analysis of 17 compounds in <5 min. Mean recoveries ranged from 70% to 120%, except for three compounds (doxycycline, erythromycin, and tilmicosin), which had recoveries of >50%. Application of the method to the analysis of honey samples obtained from different beekeepers and local supermarkets revealed residues of erythromycin, sarafoxacin, and tylosin in three of the samples.

Turnipseed et al.[78] described a method for the multi-class residue determination of β-lactams, sulfonamides, tetracyclines, fluoroquinolones, and macrolides in milk and other dairy products. The sample preparation combines extraction with acetonitrile, clean-up with Oasis HLB cartridges, and ultrafiltration using molecular weight cut-off filters to improve the overall performance of the analysis. Acceptable recoveries were obtained for sulfanomides, macrolides, and quinolones (>70%); however, recoveries were rather low for tetracyclines (50–60%) and β-lactams (<50%). Despite the extensive clean-up procedure, significant matrix ion suppression was observed for many compounds, making it necessary to include matrix-matched calibration standards for quantification purposes.

Stolker et al.[11] developed a method that was suitable for screening more than 100 veterinary drugs in milk, including antibicaterials of different classes, namely, macrolides, penicillins, quinolones, sulfonamides, tetracyclines, nitroimidiazoles, ionophores, and phenicols. After protein precipitation with acetonitrile, followed by centrifugation and further clean-up with Strata X cartridges, the extracts were analyzed by UPLC-TOF-MS. The results were satisfactory in terms of repeatability [relative standard deviation (RSD) <20% for 86% of the compounds], reproducibility (RSD <40% for 96% of the compounds) and accuracy (80–120% for 88% of the compounds). However, identification criteria for TOF-MS detectors are not yet included in the EU 2002/657/EC guidelines,[37] so the method can be used only for screening purposes in EU laboratories, and those samples that are suspected to be positive must be confirmed by a tandem MS technique.

Heller et al. developed a method to screen antibacterial residues in eggs.[79] A total of 29 analytes belonging to four antibacterial classes (sulfonamides, tetracyclines, fluoroquinolones, β-lactams) were analyzed. The extraction of the antimicrobials from the matrix was achieved by adding succinate buffer followed by centrifugation; afterwards, the cloudy extract required a clean-up step with Oasis HLB cartridges. Recoveries for each drug class were as follows: sulfonamides 70–80%, tetracyclines 45–55%, fluoroquinolones 70–80%, and β-lactams 25–50%. The reproducibility of the method varied widely (from 10% to >30%). Therefore, the results provide an estimated concentration range, and the method could be useful for screening purposes, but not for quantification.

4.4.5.2 Automated SPE

The SPE process can be performed either on-line or off-line. The procedure using SPE cartridges with a vacum manifold is known as *off-line SPE*, and the eluate from the cartridge is introduced into the chromatographic instrument after concentration and reconstitution of sample extracts. In

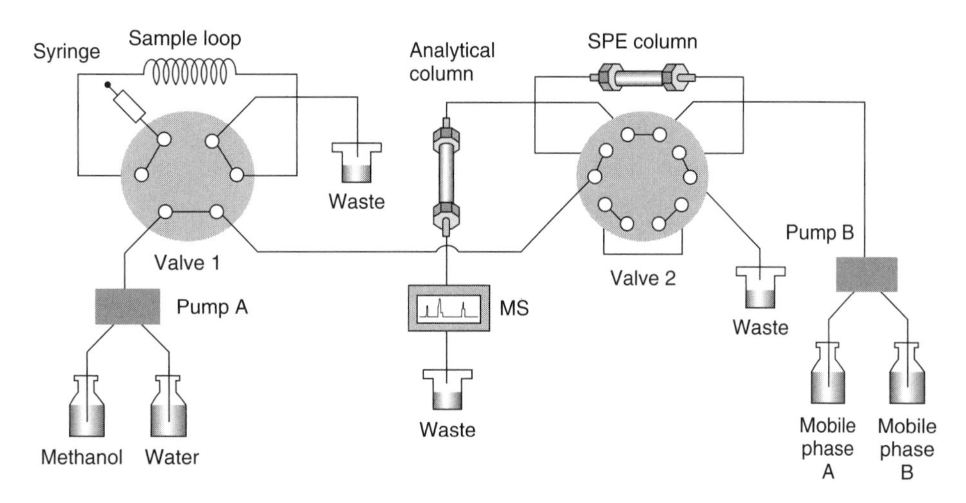

Figure 4.2 Schematic diagram of on-line coupling of SPE to LC–MS/MS. (Adapted from Ding et al.[123] with permission from Elsevier; copyright 2009.)

on-line configurations, the extraction cartridge is inserted as a part of the chromatographic equipment, frequently in the sample loop, and is directly connected to the stream of the mobile phase (Fig. 4.2). On-line SPE processes, also known as *precolumn concentration techniques*, may involve column switching or coupled column techniques.

A drawback with off-line SPE is the additional handling required and potential lower selectivity due to the requirement of eluting with 100% organic solvent. However, it is advantageous because of the absence of memory or sample carryover effects (observed with some on-line systems) through single-use SPE cartridges.[17] The Gilson ASPEC XLTM is a typical example of an automated off-line SPE system that can process four samples in parallel in cartridge and 96-well format. A number of applications have been developed using this platform, including quinolones in animal feed[80] and seafood[81] and sulfonamides in ovine plasma.[82] Alternatively, purification of samples can be achieved on-line using trace enrichment cartridges. On-line SPE offers better control of the sample preparation process and improves sensitivity through more selective isolation of target residues. This technique has been further refined by Spark Holland, through the development of an on-line system (Symbiosis) based on disposable single-use cartridges, which are automatically replaced for each sample to eliminate memory effects. The Symbiosis automated SPE unit has been successfully used in the analysis of β-lactams in bovine milk,[83] tetracyclines in milk,[84] and chloramphenicol in egg.[85]

A high-throughput method that combines on-line extraction and determination by LC-MS/MS has been developed for the screening of 13 multi-class antibacterials (macrolides, fluoroquinolones, lincosamides, and trimethoprim) in different animal muscle tissues.[86] After sample deproteinization with acetontirile, the extracts were directly loaded onto the SPE cartridge, packed with an Oasis HLB sorbent, and connected through a switching valve device to a short LC analytical column. In this way, a complete cycle of SPE clean-up and LC determination can be performed in only 6 min. The method has shown excellent selectivity, as no interfering peaks were observed in the retention windows of any of the target compounds. Furthermore, the performance of the extraction cartridge was found to remain consistent over 100 injections. The only precaution taken was to flush acetonitrile and methanol over the cartridge once the clean-up step was finished, in order to remove residual tissue matrix.

4.4.6 Solid Phase Extraction-Based Techniques

This section discusses dispersive SPE, matrix solid phase dispersion, solid phase micro-extraction, micro-extraction by packed sorbent, stir-bar sorbent extraction, and the restricted-access materials.[18]

4.4.6.1 Dispersive SPE

Dispersive SPE (dSPE) is a clean-up technique that involves mixing sorbent with a solvent extract. It is best known for its application in the QuEChERS method. The most widely adopted approach is to adsorb matrix co-extractives onto sorbents, while leaving analytes of interest in the solvent. Anhydrous $MgSO_4$ is added to provide additional clean-up by removing residual H_2O, enhance the elution strength of the solvent, and remove matrix components via chelation. Following centrifugation the supernatant can be analyzed directly or can be subjected to a concentration and/or solvent exchange step if necessary. It is an extremely effective technique and can be tailored to the different analytes and matrices by careful choice of sorbent.

Primary secondary amine (PSA) is the most common sorbent used in pesticide residue analysis because it can effectively retain organic acids present in food. C18 or a

PSA/C18 combination is more widely used in the analysis of food of animal origin because of the higher lipid content. In more recent research, it was found that the combination of PSA and C18 provided better clean-up than did PSA or C18 alone for 38 anthelmintics in liver and milk.[64] However, PSA/C18 gave a lower recovery for some analytes (due to PSA), compared with C18 alone, which gave sufficient clean-up and good recovery for all analytes and was therefore chosen as the preferred sorbent.

Graphitized carbon black (GCB) has been reported to be a highly effective sorbent for sample clean-up but also removes structurally planar analytes, limiting its applications.[62] Addition of HOAc to the extraction solvent may help to improve recovery of analytes, but it also inhibits the retention of acidic matrix compounds.[62] Several papers have reported the use of C18 for dSPE in veterinary residue analysis.[62,64,87] The use of PSA, NH_2, and silica has also been reported.[88] While dSPE does not provide the same degree of clean-up as SPE, it does provide good recovery and reproducibility, coupled with practical and cost advantages.[88]

Mastovska and Lightfield have improved a previous analytical method replacing conventional SPE by dSPE for the determination of 11 β-lactams in bovine kidney samples.[89] The method involves solvent extraction/deproteinization with H_2O/ACN (20 : 80, v/v), followed by dSPE clean-up with C18 sorbent and final LC-MS/MS determination. Taking into account the simplicity of this protocol, accuracy was satisfactory with recoveries ranging from 87% to 103% and RSD $<16\%$. Moreover, the use of dSPE increased the number of samples that can be prepared in a day by a factor of 3–4.

The same scheme has been adopted for the determination of several antibacterials (quinolones and erythromycin A), fungicides, and parasiticides in salmon tissue.[51,90] In this case, samples were extracted with acidic ACN, and then dSPE was carried out with a Bondesil NH_2 sorbent, followed by LC-TOF-MS determination. Excellent recoveries were obtained with the exception of enrofloxacin (40%), although matrix suppression effects were observed for most of target compounds.[90]

4.4.6.2 *Matrix Solid Phase Dispersion*

Matrix solid phase dispersion (MSPD) is an effective sample preparation technique that combines extraction and purification in one step. Barker et al.[91] defined MSPD procedures as those that use dispersing sorbents with chemical modification of the silica surface (e.g., C18, C8). Samples are blended and dispersed on particles (diameters of 40–100 μm) using a glass or agate mortar and pestle (Fig. 4.3).[92,93] The use of ceramic or clay mortars and pestles can result in loss of analytes. A disadvantage of the method is the traditionally high sorbent : sample ratios ultilized, with ratios of 1 : 1 to 4 : 1 typically used (2 g sorbent is the most common, with 0.5 g sample).

In early applications, the sample was first air-dried (for 5–15 min) prior to compression between two frits in a syringe barrel with a syringe plunger. The need for the air-drying step has since been eliminated through the application of non-bonded silica-based dispersion agents such as Na_2SO_4 or silica.[94,95] In contrast to typical reversed phase SPE applications, a range of novel solvent washes have been employed, including hexane, dichloromethane (DCM), alcohols, and hot H_2O. Hot H_2O has been used to extract several classes of drugs from various matrices, but care must be taken as some analytes can thermally degrade.[96] MSPD is beneficial because of the wide range of residues to which it can be applied as well as its potential to fractionate samples through sequential elution with solvents of increasing or decreasing polarity. It also eliminates protein precipitation and centrifugation steps. There has been a resurgence of use of the technique in recent years for the preparation of veterinary samples for drug residue analysis.

More recent reviews on this particular topic suggest that MSPD has attracted many researchers in environmental, clinical, and food analysis.[95,97,98] Different bulk materials have been used as matrix dispersing agents; the most popular is C18- and C8-bonded silica.

Zou et al. have described an MSPD procedure for the extraction of eight sulfonamides from honey samples using C18 as solid support.[99] After the MSPD, the sulfonamides were derivatized with 9-fluorenylmethylchloroformate (FMOC-Cl). The derivatives required further purification by silica gel SPE, prior to LC-UV determination. Average recoveries for most sulfonamides were $>70\%$. Other polar sorbents, such as silica gel, CN- or NH_2-bonded silica resulted in a strong absorption of the polar sulfonamides, providing very low recoveries.

As an alternative to the classical C18- or C8 apolar bonded phases, the use of normal phase supports has been proposed to improve the isolation of more polar compounds as well as to perform extraction and clean-up in a single step, prior to reversed phase LC determination. Following this trend, Kishida[100] has developed a simple method for the determination of six sulfonamides in meat samples, using normal phase MSPD with alumina N-S and 70% (v/v) ethanol solution as extraction solvent, followed by the evaporation of the extracts and LC-MS/diode-array detection (DAD) determination. Average recoveries were $>90\%$ in all cases, and the LOQs were well below the MRLs established by the EU.

Perhaps one of the most interesting MSPD-based techniques is MSPD with hot-water extraction. Bogialli et al. have published several papers using this sample preparation technique for the determination of different antibacterials in a great variety of foodstuffs, such as fluoroquinolones in milk.[101]

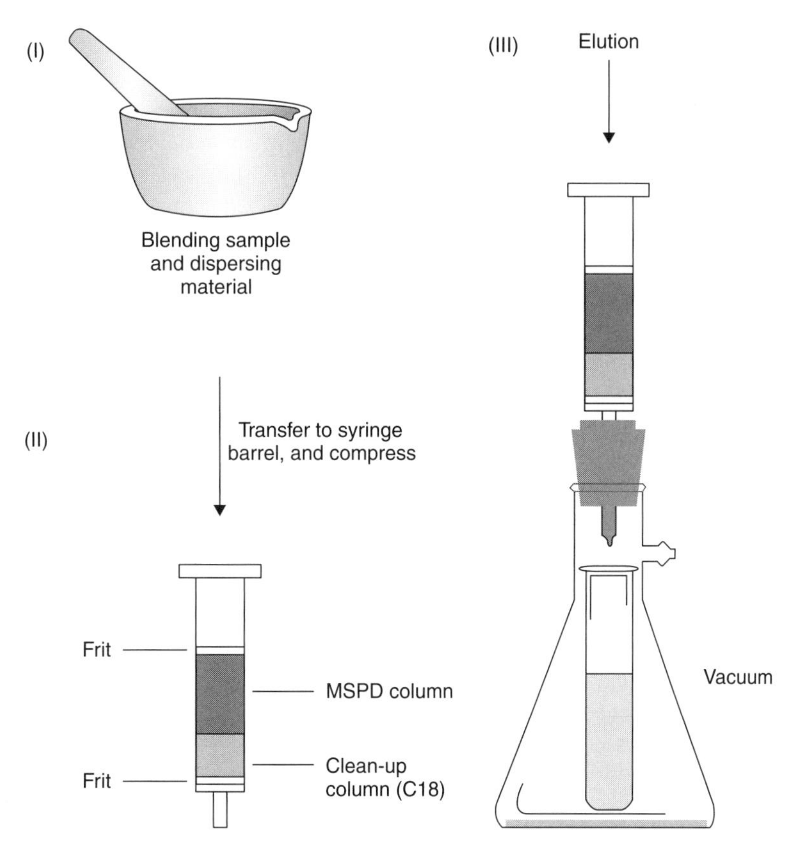

Figure 4.3 Matrix solid phase dispersion (MSPD) procedure. Main steps of MSPD: (I) the sample is blended with the dispersant material in a mortar with a pestle; (II) the homogenized powder is transferred in a solid phase extraction cartridge, and compressed; (III) elution with a suitable solvent or solvent mixture is performed by the aid of a vacuum pump. (Adapted from Capriotti et al.[93] and with permission from Elsevier; copyright 2010.)

Yan et al. have used molecularly imprinted polymers (MIPs) as selective dispersing media for sample clean-up in the determination of quinolones in eggs and swine muscle samples by liquid chromatography–fluorescence detection (LC-FLD).[102] The use of MIPs enhances the selectivity of the MSPD procedure, allowing recoveries of the target analytes above 85% without matrix interferences being observed in the final determination. Thus, extraction and clean-up can be performed in a single step, which simplifies sample preparation and reduces time and cost of the analysis. Moreover, the synthesized MIPs have shown good specific recognition of the quinolones in aqueous media, which is especially important for final application to complex matrices, such as food samples.[103]

4.4.6.3 Solid Phase Micro-extraction

Solid phase micro-extraction (SPME) was developed in 1989 by Pawliszyn and coworkers[104,105] as a simple and effective adsorption/adsorption and desorption technique that eliminates the need for solvents. Applications of SPME in the analysis of drugs have been reviewed,[106] as well as the possibilities of interfacing SPME with HPLC.[107] SPME may be performed in either of two formats: fiber SPME and in-tube SPME.

Fiber SPME is based on a modified syringe that contains stainless-steel microtubing within its needle. Inside there is a fused-silica fiber tip that is coated with organic polymer, typically polydimethylsiloxane (PDMS). This coated fiber can be moved inside and outside the needle by a plunger. As discussed by Gaurav et al.,[108] a significant effort was put into the development and optimization of these fibers over the years since the introduction of the technique. The extraction and pre-concentration of the analyte is completed with the coated fiber in the "outside" position. The penetration of the septum of a GC injection port is then performed with the fiber in the "inside" position. Once in the injection port, the desorption of analyte and transfer to a capillary column requires moving the fiber again to the outside position. Using such simple equipment, all steps, including extraction, pre-concentration, derivatization, and transfer to the chromatograph, are integrated in one device. Therefore, the main advantage of fiber SPME is the simplicity and automation of sample preparation procedures.[106]

In bio-analytical methods, both direct-immersion (DI-SPME) and head-space fiber (HS-SPME) have been applied with or without a derivatization step. Using the direct immersion approach, which means exposure of the fiber to the sample in solution, clenbuterol in urine and serum as well as citalopram, fluoxetine, and their main metabolites in urine, were determined without a derivatization step by HPLC analysis.[106]

The outstanding advantage of HS-SPME in bio-analytical methods is the prevention of direct contact of the fiber with the sample, and therefore the prevention of contamination of the surface of the fiber with organic polymers. However, the use of the fiber HS-SPME technique is limited to only analytes that have a suitably high vapor pressure. Furthermore, the transfer of fibers to the GC as well as desorption should be performed immediately after extraction because the high vapor pressure poses a risk of analyte loss during storage of the loaded fiber.

In summary, fiber SPME has several advantages, such as ease of use, non-usage of solvents, and minimal equipment requirements. It is fast and easy to automate and provides good linearity and high sensitivity. However, as was stated above, these advantages are of use only in some areas of bioanalysis. Thus, the matrix and volatility of the target analyte must be taken into consideration. The combination of the low volatility of analyte with a complex matrix (containing polymer components, including proteins in plasma or cell cultures) considerably limits the application of fiber SPME. Another drawback is the longer time needed for extraction, which is critical for some analytes and can become unacceptable. Generally, the recoveries are also considerably lower than those of LLE and SPE. Sample clean-up for fiber DI-SPME is not optimal; therefore interferences from endogenous trace substances could occur. On the other hand, the determination of volatile analytes that may be examined by HS-SPME is more favorable.

There are some principal disadvantages that limit the use of fiber SPME. These include the limited capacity of SPME fiber and a requirement for a very low initial temperature for GC temperature programs because of the necessity of cryofocusing of the analyte, thus prolonging GC analysis time. In addition, desorption takes more time than in the injection of LLE or SPE extracts, and carryover effects occur very easily. As SPME is by nature a "dirty" extraction, quantification is more prone to errors due to changes in the matrix even when internal standards are used. Finally, SPME fiber is quite fragile. Because of these restrictions and limitations, fiber SPME is not a universal sample preparation method, especially not in bio-analytical laboratories, and is unlikely to become so in the future.

However, the development and investigation of new sorbent materials for use in extraction and SPME techniques is both a growing and promising field.[109] The compilation presented in the Augusto et al.[109] paper provides a representative selection of materials and devices that are the main targets of research on novel sorbents and adsorbents: materials with improved (ad)sorptive capacity, capable of providing higher extraction efficiencies, and, therefore, better analytical detectabilities and sensitivities, as well as high stability and morphology compatible to fast mass transfer during extractions. Of course, it is virtually impossible to combine all those features in a single sorbent. Different classes of sorptive materials are designed to primarily address one of these goals (e.g., selectivity/specificity for MIP; fast mass transfer and chemical stability for SPME fiber coatings).

One point has to be carefully considered when one browses the literature dealing with some of the "novel" (ad)sorbents. The properties of several of the sorbents and adsorbents described in the more recent literature are literally the same as those of materials that are already well known or available commercially. For example, some of the solgel SPME fibers more recently reported, prepared from different organic modifiers and alkoxysilanes, have almost the same features as the original solgel coating described by Malik in 1997.[110] Therefore, any claims of improved selectivity, stability, or extraction efficiency of new sorbents and adsorbents should be carefully considered.

Finally, it should be mentioned that the number of new materials and devices described in the literature, which results in commercially available products, is limited. Several innovative sorbents and adsorbents have not progressed beyond the stage of academic research, despite their potential application to relevant analytical problems. The reasons for this range from the lack of experience (and even interest) of some research groups to enter partnerships with analytical instrumentation industries, to the perception that the excessive rigor required in the application of analytical validation and certification protocols may sometimes delay or even hinder the acceptance of new analytical techniques.[109]

McClure and Wong[111] described a SPME application for the analysis of macrolide and sulfonamide (including trimethoprim) antibiotics in wastewater. A SPME fiber of carbowax-templated resin (CW/TPR, 50 μm) was immersed into 1.5 ml of stirred aqueous sample in glass amber vials for 30 min, followed by desorption of analytes from the fiber by immersion into 1.5 ml of stirred methanol for 10 min. The resulting extracts were evaporated to dryness using a nitrogen evaporator, then reconstituted with 30% acetonitrile/70% water and internal standard solution to a final volume of 75 μl. Quantification of analytes was accomplished using standard addition curves generated by SPME extractions of wastewater spiked with increasing amounts of target analytes (0–5000 ng/l). The limits of quantification ranged from 16 to 1380 ng/l. The advantages of the SPME method over typical SPE methods for wastewater analysis include decreased sample volume (a few millilitres vs. 4 l), decreased cost, and ease of sample extraction. The major limitations of the method are its higher limits of quantification in comparison with conventional SPE.

Lu et al.[112] described the trace determination of sulfonamide residues in meat with a combination of SPME and LC-MS. Fiber coated with a 65 μm thickness of polydimethylsiloxane/divinylbenzene (PDMS/DVB) was used to extract sulfonamides at optimum conditions. Analytes were desorbed with static desorption in an SPME-HPLC desorption chamber for 15 min and then determined by LC-MS. The linear range was 50–2000 μg/kg, with RSD values below 15% (intra-day) and 19% (inter-day) and detection limits 16–39 μg/kg. Some meat samples collected from the local market contained residues of sulfonamides ranging from 66 to 157 μg/kg. The results demonstrated that the SPME-LC-MS system could effectively analyze residues of sulfonamides in meat products.

4.4.6.4 Micro-extraction by Packed Sorbent

Micro-extraction by packed sorbent (MEPS) is a new technique for sample preparation that can be connected on-line with LC or GC.[113] In MEPS approximately 1–2 mg of solid packing material is either inserted into the barrel of a syringe (100–250 μl) as a plug with polyethylene filters on both sides or inserted between the syringe barrel and the injection needle as a cartridge. Sample preparation takes place on the packed bed, which can be packed or coated to provide selective and suitable sampling conditions. The key factor in MEPS is that the volume of solvent used to elute the analytes from the extraction process is suitable for injection directly into an LC or GC system. MEPS can thus be described as a short LC column in a syringe. The bed dimensions are scaled from a conventional SPE cartridge bed, and in this way MEPS can be adapted to most existing SPE methods by simply adjusting the reagents and sample volumes from the conventional device to the MEPS.[113] MEPS can handle small sample volumes (10 μl of plasma, urine, or water) as well as larger volumes (1000 μl) and can be used for GC, LC, or CEC applications. Compared with LLE and SPE, MEPS reduces sample preparation time and organic solvent consumption. MEPS may be fully automated, and the extraction procedure takes only a few minutes for each sample. The MEPS technique is more robust than SPME, where the sampling fiber is quite sensitive to the nature of the sample matrix. MEPS can be used without major problems for complex matrices (such as plasma, urine, and organic solvents).

Drawbacks of the MEPS technique include the possibility of bubble formation and some difficulties connected to off-line arrangement as on-line coupling is not possible with every LC system. For off-line MEPS the speed of plunger movement is crucial for the recovery of analytes. An excessively high speed of movement does not allow adsorption of the analyte on the MEPS support and leads to misleading recovery results and poor repeatability.

4.4.6.5 Stir-bar Sorptive Extraction

This sorptive and solventless extraction (SBSE) technique, introduced in 1999, is based on the same principles as SPME. Instead of a polymer-coated fiber, a large amount of extracting phase is coated on a stir bar. The most widely used sorptive extraction phase is polydimethylsiloxane (PDMS). The extraction of an analyte from the aqueous phase into an extraction medium is controlled by the partitioning coefficient of the analyte between the silicone phase and the aqueous phase. For a PMDS coating and aqueous samples, this partitioning coefficient resembles the octanol–water partitioning coefficient. The amount of extraction phase in SBSE is 50–250 times greater than in SPME (typically 0.5 μl of extraction phase for 100 μl volume PDMS fiber). After extraction and thermal desorption the analyte can be introduced quantitatively into the analytical system. This process provides excellent concentration of analytes, since the complete extract can be analyzed. In contrast to SPME, the desorption process is slower because the extraction phase is extended; thus desorption needs to be combined with cold trapping and reconcentration.[114]

Since only PDMS coating is available as an extraction phase, SBSE has been used predominantly for low-polarity analytes, but the problems of extraction of polar compounds may be solved by *in situ* derivatiziation. It is clear that further developments in stir bar coatings and designs could extend the applicability of the method. The main drawback of this method is the duration of extraction, typically 30–150 min.[115] For this reason SBSE may be impractical for routine high-throughput laboratories.

Huang et al.[116] described a simple, rapid method for the quantitative monitoring of five sulfonamide antibacterial residues in milk. The analytes were concentrated by SBSE based on poly(vinylimidazole–divinylbenzene) monolithic material as coating, and analyzed by HPLC with diode-array detection. The extraction procedure was very simple. Milk was first diluted with water and then directly subjected to sorptive extraction without a requirement for additional steps to eliminate fats and protein in the samples. Under the optimized experimental conditions, low detection limits ($S/N = 3$) (where S/N = signal-to-noise ratio) and quantification limits ($S/N = 10$) were achieved for the target compounds within the range of 1.30–7.90 and 4.29–26.3 μg/l from spiked milk, respectively. Good linearities were obtained for the sulfonamides with correlation coefficients (R^2) above 0.996. Finally, the proposed method was successfully applied to the determination of sulfonamides in different milk samples.

Luo et al.[117] described a stirring rod (instead of stir bar) sorptive extraction (SRSE). A rod with monolithic polymer as coating was proposed to avoid the friction loss of coating during the stirring process. In this study, poly(2-acrylamide-2-methylpropanesulfonic acid-*co*-octadecyl methacrylate-*co*-ethylene glycol dimethacrylate) [poly(AMPS-*co*-OCMA-*co*-EDMA)] monolithic polymer was used as a coating for the rod. Four fluoroquinolones were selected as analytes to evaluate the extraction efficiency of SRSE. To achieve the optimum extraction conditions of SRSE toward fluorquinolones, various parameters, including extraction time, extraction temperature, stirring rate, sample solution pH, and contents of inorganic salt in the sample solution were investigated. Under the optimized conditions of SRSE, a method for the determination of fluorquinolones in honey samples was developed, based on the combination of SRSE and liquid chromatography electrospray ionization mass spectrometry (SRSE/LC/ESI-MS). The detection limits (LODs) of the proposed method ranged from 0.06 to 0.14 μg/kg and the recoveries were in the range 70.3–122.6% at different concentrations for honey samples. Good method performance was observed for intra- and inter-day precisions, yielding the relative standard deviations under 11.9% and 12.4%, respectively. The results demonstrated that SRSE with poly(AMPS-*co*-OCMA-*co*-EDMA) monolithic polymer as coating possessed good extraction capacity toward fluorquinolones in honey samples. In addition, the monolithic polymer-coated stirring rod was demonstrated to be sufficiently stable that it could be reused at least 60 times.

4.4.6.6 *Restricted-Access Materials*

Restricted-access materials (RAM) are biocompatible sample preparation supports that enable the direct injection of biological fluid into a chromatographic system. The technique was introduced in 1991 by Desilets et al.,[118] who also established the acronym RAM. Sorbents used in RAM represent a special class of materials that are able to fractionate a biological sample into a protein matrix and an analyte fraction, based on molecular weight cut-off. Macromolecules are excluded and interact only with the outer surface of the particle support, which is coated with hydrophilic groups. This minimizes the adsorption of matrix proteins. Applications of RAMs have been reviewed by several research groups.[119,120]

The basis of RAMs is the simultaneous size exclusion of macromolecules and extraction/enrichment of low-molecular-weight compounds into the interior phase via partition. The outer surface of the particles, which is in contact with biological matrix components such as proteins and nucleic acids, possesses a special chemistry to prevent adsorption of these molecules. Macromolecules can be excluded by a physical barrier, the pore diameter, or by a chemical diffusion barrier created by a protein (or polymer) network at the outer surface of the particle. RAMs can be classified according to the protein exclusion mechanism used into the following two groups: RAM with a physical barrier (reversed phase, alkyl-diol-silica material, porous silica with combined ligand) and RAM with a chemical barrier (semi-permeable surface, protein-coated silica, mixed-function materials, or shielded hydrophobic phase). Enantioselective RAMs using glycopeptides antibiotics as chiral selector or weak cation exchange RAM have also been developed.[121]

Oliveira and Cass[122] described a method for the analysis of cephalosporin antibiotics in bovine milk using RAM columns for on-line sample clean-up. The system was composed of a RAM bovine serum albumin (BSA) phenyl column coupled to a C18 analytical column. Milk samples were directly injected after addition of 0.8 mM solution of tetrabutylammonium phosphate. The standard curve was linear over the range 0.100–2.50 μg/ml for five cephalosporin antibiotics (cefoperazone, cephacetril, cephalexin, cephapirin, and ceftiofur). The limits of quantification and detection reported were 0.100 and 0.050 μg/ml, respectively. The method showed high intermediate precision [coefficient of variation percent (CV%) 2.37–2.63] and recovery (CV% 90.7–94.3) with adequate sensitivity for drug monitoring in bovine milk samples.

Ding et al.[123] described an automated on-line SPE-LC-MS/MS method for the determination of macrolide antibiotics, including erythromycin, roxithromycin, tylosin, and tilmicosin in environmental water samples. A Capcell Pak MF Ph-1 packed-column RAM was used as SPE column for the concentration of the analytes and clean-up of the sample. One millilitre of a water sample was injected into the conditioned SPE column, and the matrix was washed out with 3 ml high-purity water. By rotation of the switching valve (see Fig. 4.2), macrolides were eluted in the back-flush mode and transferred to the analytical column. The limits of detection and quantification obtained were 2–6 and 7–20 ng/l, respectively, which is suitable for trace analysis of macrolides. The intra- and inter-day precisions ranged within 2.9–7.2% and 3.3–8.9%, respectively. At the three fortification concentrations tested (20, 200, and 2000 ng/l), recoveries of macrolides ranged from 86.5% to 98.3%.

4.4.7 Solid Phase Extraction-Based Selective Approaches

This section discusses immunoaffinity chromatography, molecular imprinted polymers, and aptamers.

4.4.7.1 *Immunoaffinity Chromatography*

Immunoaffinity chromatography (IAC), in its various forms, is a rather specialized form of affinity chromatography wherein the separatory ligand is either an immobilized antibody or antigen. For antibiotic residue analysis, the antibody is the separatory ligand. The selective separation occurs through the classical antibody–antigen reaction $Ab+Ag \leftrightarrow AbAg$, where Ab is the antibody and Ag is the antigen and the complex formed is represented by AbAg. IAC for antigen isolation is completely dependent on the antibody to separate the target compound. The antibody ligand is immobilized on a support, and, as the target compound comes into contact with it, a complex is formed. The ligand–target compound complex is dissociated because of hydrophobic changes caused by the mobile phase, and the target compound is eluted from the column. The dissociation of the complex takes place after other materials have passed or been washed through the column. The specificity of the antibody leaves a minimum of interfering materials to be eluted from the column. Thus, the eluate can be relatively pure.[124]

Immunoaffinity chromatography IAC is particularly advantageous for the detection of banned substances, particularly when enhanced selectivity and detection at concentrations in the μg/kg–ng/kg range is required. The applications of the technique in the isolation of residues of licensed veterinary drug from food are limited because of the cost and specialized nature of columns. Despite this, some very good multi-residue applications of IAC have been reported.[125–127] Luo et al.[128] described the simultaneous determination of thiamphenciol, florfenicol, and florfenicol amine in swine muscle by LC-MS/MS with IAC clean-up. An IAC column based on polyclonal antibodies and protein A–sepharose CL 4B was employed. The dynamic column capacity was more than 512 ng/ml of gel after use for 15 cycles. Recovery of analytes from swine muscle fortified at concentrations of 0.4–50 μg/kg ranged from 85% to 99%. Limits of quantification ranged from 0.4 to 4 μg/kg.

4.4.7.2 Molecularly Imprinted Polymers

Molecularly imprinted polymers (MIPs) are selective materials used for solid phase extraction. MIPs can not only concentrate but also selectively separate the target analytes from matrices, which is crucial for the quantitative and selective determination of trace concentrations of analytes in complex matrices. The main benefit of MIPs is the possibility of preparing selective sorbents pre-determined for a particular substance or a group of structural analogs. MIPs are synthetic polymers with highly specific recognition ability for target molecules. Details on the classification and preparation of MIPs are described in a paper by He et al.[129] MIPs used for SPE may be synthesized by three imprinting techniques that enable the formation of complex template-functional monomer: (1) covalent imprinting, (2) non-covalent imprinting, and (3) hybridization of both covalent and non-covalent imprinting.

The imprinting molecule complexes one or several functional monomers. The next step is polymerization. As a result, MIPs possess a steric (size and shape) and chemical (special arrangements of complementary functionality) memory for a template.[130]

The first application of MIP for SPE was introduced by Sellergren in 1994,[131] in which pentamidine was determined in urine using a pentaimidine-imprinted dispersion polymer in an SPE column. More recently, a selective imprinted aminofunctionalized silica gel sorbent was prepared by combining a surface molecular imprinting technique with a solgel process for on-line solid phase extraction HPLC determination of three trace sulfonamides in pork and chicken muscle.[132] The imprinted functionalized silica gel sorbent exhibited selectivity and fast kinetics for the adsorption and desorption of sulfonamides. With a sample loading flow rate of 4 ml per minute for 12.5 min, enhancement factors and improved detection limits for the three sulfonamides ($S/N = 3$) were achieved. The precision (RSD) for nine replicate on-line sorbent extractions of 5 μg/l of the analytes was less than 4.5%. The sorbent also offered good linearity (R^2; 0.99) for on-line solid phase extraction of trace concentrations of sulfonamides.

Boyd et al.[133] described the analysis of chloramphenicol using MIP as the sample clean-up technique. The MIP was produced using an analog to chloramphenicol as the template molecule. Using an analog of the analyte as the template (Fig. 4.4) avoids a major traditional drawback associated with MIPs referred to as residual template leeching or bleeding.[109,134] The MIP was used as a SPE phase for the extraction of chloramphenicol from various sample matrices, including honey, urine, milk, and plasma.

Mohamed et al.[135] described advantages of MIP LC-ESI-MS/MS for the selective extraction and quantification of chloramphenicol in milk-based matrices and compared the MIP with the classical SPE type of sample clean-up. The method entails a single centrifugation step prior to loading the supernatant onto the MIP cartridge and subsequent elution with a mixture of solvents. The advantages of the MIP approach were assessed by comparing the data generated from classical SPE and LLE extraction procedures. A better recovery of chloramphenicol due to an enhanced selectivity and a faster turnaround time (18 samples processed within 3 h compared to 8 h with the classical approach) were evidenced when using the MIP clean-up. The analysis of chloramphenicol in raw milk was further validated according to the 2002/657/EC[37] criteria at the minimum required performance limit (MRPL) of 0.3 μg/kg, using a d_5-deuterated internal standard. Non-internal-standard-corrected recovery values ranged between 50% and 87% over the range of concentrations considered. The decision limit (CCα) and detection capability (CCβ) were calculated to be 0.06 and 0.10 μg/kg, respectively.

Other applications of MIPs have been described by Hung and Hwang[136] for the analysis of sulfonamides and by Turiel et al.[137] for the analysis of fluoroquinolones in soil.

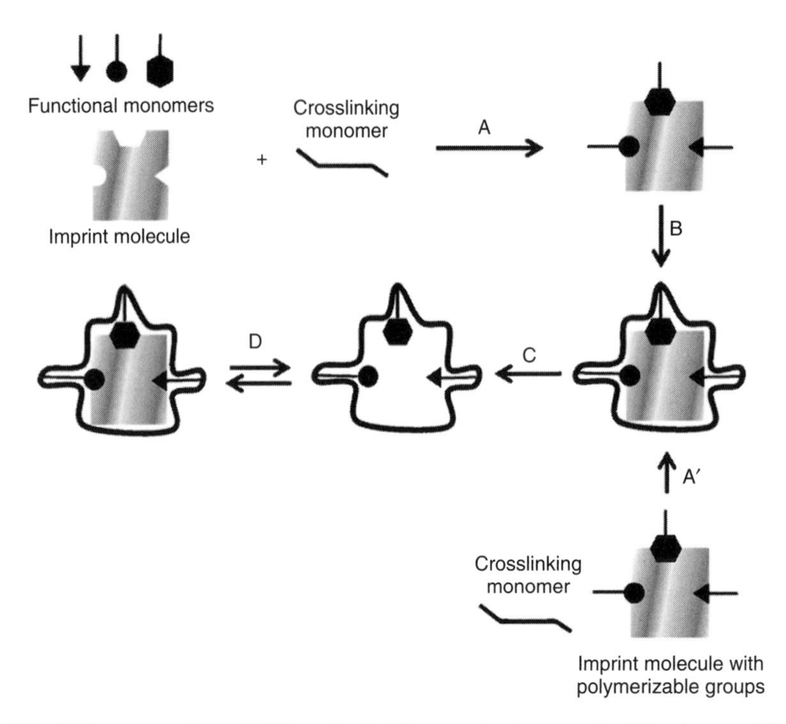

Figure 4.4 Main approaches to MIP synthesis. Non-covalent approach: (A) mixture of functional monomers, crosslinking agents, polymerisation initiator and templates dissolved on porogenic solvent to form template/functional monomer complex; (B) polymerization; (C) removal of template (by solvent extraction); (D) analyte binding (via non-covalent interactions) on the specific imprinted site. Covalent approach: (A′) template containing polymerizable groups mixed with crosslinking agent and initiator in proper solvent; (C) removal of template after polymerization (with breaking of covalent bonds between template and polymer); (D) analyte binding (via covalent bonds) on the specific imprinted site. (Adapted from Augusto et al.[109] with permission from Elsevier; copyright 2010.)

4.4.7.3 Aptamers

Aptamers are oligonucleotides (DNA or RNA) that bind with high affinity and specificity to a wide range of target molecules, such as drugs, proteins, and other organic or inorganic molecules.[18] Aptamers are generated using an *in vitro* selection process called *systematic evolution of ligands by exponential enrichment* (SELEX), which was first reported in 1990.[138,139] The SELEX method has permitted the identification of unique RNA/DNA molecules from very large populations of random sequence oligomers (DNA or RNA libraries). These molecules bind to the target molecule with very high affinity and specificity. Aptamers show a very high affinity for their targets, with dissociation constants typically ranging from micromolar to low picomolar, comparable to those of some monoclononal antibodies, sometimes even better.[140]

There is a limited number of studies on aptasensors (aptamer sensors), particularly for the detection of small organic molecules. An electrochemical aptasensor was developed by Kim et al.[141] for the detection of tetracycline using an ssDNA aptamer that selectively binds to tetracycline as the recognition element. The aptamer was highly selective for tetracycline and distinguishes minor structural changes on other tetracycline derivatives. The biotinylated ssDNA aptamer was immobilized on a streptavidin-modified screen-printed gold electrode, and the binding of tetracycline to the aptamer was analyzed by cyclic voltammetry and square-wave voltammetry. The results showed that the minimum detection limit of this sensor was in the 10 nM–micromolar range. The aptasensor showed high selectivity for tetracycline over the other structurally related tetracycline derivatives (oxytetracycline and doxycycline) in a mixture. This aptasensor can potentially be used for detection of tetracycline in pharmaceutical preparations, contaminated food products, and drinking water.[18]

4.4.8 Turbulent-Flow Chromatography

Turbulent-flow chromatography (TFC) is a high-throughput sample preparation technique that utilizes high flow rates (4–6 ml). The chromatographic efficiency of TFC is similar to that of laminar flow chromatography but at much lower flow rates. The columns used for TFC contain common LC sorbents but of larger particle sizes (30–60 μm). Because of the large pore size, there is only moderate back-pressure on the column, which serves as both extraction and analytical column. At the higher flow rate, solvent does not exhibit laminar flow but exhibits turbulent flow instead. This leads to the formation of eddies that promote cross-channel mass transfer and diffusion of the analytes into the particle pores. Samples are applied to the column using aqueous mobile phase (Fig. 4.5).[142] Small molecules diffuse more extensively than do macromolecules (e.g., proteins, lipids, sugars) and are driven into the pores of the sorbent. Because of the high flow rate, the larger molecules are flushed to waste and do not have an opportunity to diffuse into the particle pores. The trapped analytes are desorbed from the TFC column by back-flushing with a polar organic solvent, and the eluate can be transferred with a switching valve onto the HPLC system (normal low flow rate) for further separation and subsequent detection (usually by MS/MS). During LC-MS/MS analysis, the TFC column is reconditioned and primed for the next sample.

Liquid samples can be directly injected onto the system, while tissue samples require a crude extraction and sedimentation prior to analysis. TFC is also effective at separating residues that are bound to sample proteins.[143] The use of TFC eliminates time-consuming sample clean-up in the laboratory and results in a much shorter analysis time, higher productivity, and reduced solvent consumption without sacrificing sensitivity or reproducibility.

Despite these advantages, only a limited number of TFC applications for residue analysis have been reported in the literature. This is most likely due to the cost of the instrumentation and the limited number of vendors. Mottier et al. reported a TFC-LC-MS/MS method for 16 quinolones in honey.[144] The method is particularly advantageous because honey samples were simply diluted in water. Sample extraction time was 4.5 min, while the overall analysis took 18.5 min. Recovery of the analytes ranged from 85% to 127%, while the LOD of the method was 5 μg/kg. Krebber et al. used TFC-LC-MS/MS for the rapid determination of enrofloxacin and ciprofloxacin in edible tissues.[145] Tissue samples (bovine, porcine, turkey, rabbit) were extracted with ACN : H_2O : formic acid, filtered and an aliquot injected onto the TFC-LC-MS/MS system. The run time for the analysis was 4 min. The LOQ of the method was 25 μg/kg in all matrices. The recovery of the analytes ranged from 72% to 105%.

Stolker et al.[146] described an analytical method based on TFC-LC-MS/MS for the direct analysis of 11 veterinary drugs (belonging to seven different classes) in milk. The method was applied to a series of raw milk samples, and the analysis was carried out for albendazole, difloxacin, tetracycline, oxytetracycline, phenylbutazone, salinomycin-Na, spiramycin, and sulfamethazine in milk samples with various fat contents. Even without internal standards, results proved to be linear and quantitative in the concentration range of 50–500 μg/l, as well as repeatable (RSD<14%; sulfamethazine and difloxacin <20%). The limits of detection were between 0.1 and 5.2 μg/l, far below the maximum residue limits for milk set by the EU. While matrix effects, namely, ion suppression or enhancement, were observed for all the analytes, the method proved to be useful for screening purposes because of its detection limits, linearity, and repeatability. A set of blank and fortified raw milk samples was analyzed and no false-positive or false-negative results were obtained.

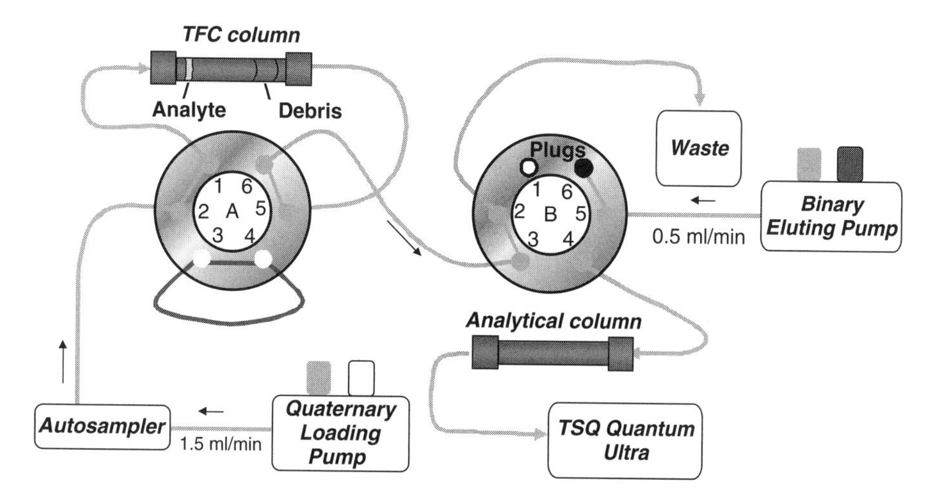

Figure 4.5 Schematic representation of the loading step in an Aria TLX-2 system. The sample is introduced into the system by an autosampler and a loading pump. The system also involves a solvent holding loop connected to valve A. This loop contains a solvent mixture strong enough to elute the analytes from the TurboFlow column (TFC) into the analytical column (transfer step). An eluting pump delivers a mixture of solvents and enables a normal chromatographic separation and detection. (Adapted from Stolker et al.[146] with permission from Springer; copyright 2010.)

4.4.9 Miscellaneous

This section discusses the ultrafiltration, microwave-assisted, and ultrasound-assisted extraction techniques.

4.4.9.1 Ultrafiltration

Ultrafiltration (UF) is used primarily to separate analytes of interest from macromolecules, such as proteins, peptides, lipids, and sugars, which may interfere with analyses, particularly affecting ionization in mass spectrometry. In residue analysis, molecular weight cut-off devices or spin filters coupled to microcentrifuge tubes are the most commonly used formats. Alternative formats such as the 96-well plate are also available, but require dedicated vacuum manifolds and pumps. All residue applications use centrifugal devices.

Goto et al.[147] described a simple, rapid, and simultaneous analysis method for oxytetracycline, tetracycline, chlortetracycline, penicillin G, ampicillin, and nafcillin in meat using electrospray ionization tandem mass spectrometry. The samples were homogenized with water followed by a centrifugal ultrafiltration after addition of internal standards (demeclocycline, penicillin G-d_5, ampicillin-d_5, and nafcillin-d_6). The MS/MS analysis involved the combined use of sample enrichment on the short column and a multiple reaction monitoring technique. The overall recoveries from bovine and swine muscle, kidney, and liver fortified at 50 and 100 μg/kg ranged from 70% to 115% with coefficients of variation ranging from 0.7% to 14.8% ($n = 5$). Analysis time, including sample preparation and determination, is only 3 h per eight samples, and detection limits for all antibiotics are 2 μg/kg. The method is considered to be satisfactory for rapid screening for tetracycline and penicillin residues in meat.

Other examples of applications include sulfonamides in milk;[148–150] eggs;[151,152] plasma;[153] edible tissues;[154,155] tetracyclines in egg;[156] penicillin G in muscle, kidney, and liver;[157] and spiramycin (a macrolide) in egg and chicken muscle.[158]

4.4.9.2 Microwave-Assisted Extraction

Microwave-assisted extraction (MAE) uses microwave energy to heat the solvent/sample mixture in order to partition analytes from the sample matrix into the solvent (see Fig. 4.6). Using microwave energy allows the solvent to be

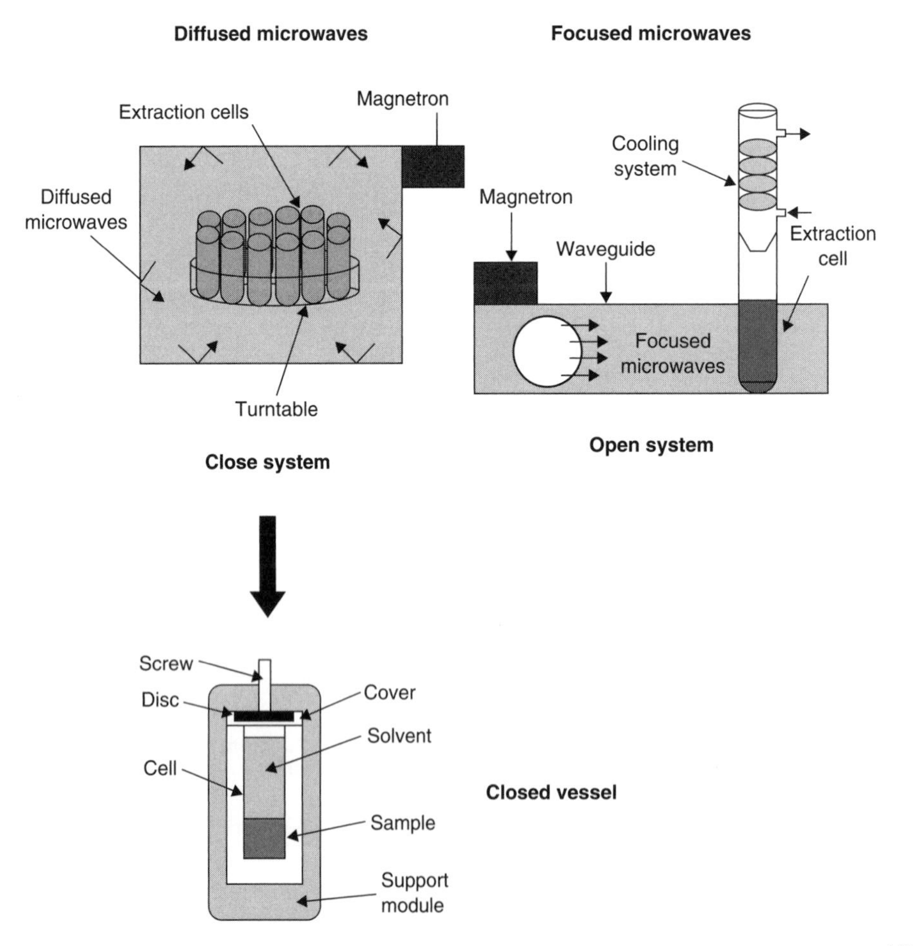

Figure 4.6 Principle of microwave-assisted extraction (MAE) systems. (Adapted from Camel[163] with permission from Royal Society of Chemistry; copyright 2001.)

heated rapidly; an average extraction takes 15–30 min.[159] MAE offers high sample throughput (several samples can be extracted simultaneously) with low solvent consumption (10–30 ml). A good review of MAE is available by Eskilsson and Bjorklund.[160]

Microwave-assisted extraction has received increasing attention as a potential alternative to traditional solid–liquid extraction methods.[161,162] Since the early 1990s, MAE has been used in an off-line mode for accelerating the sample preparation process. On-line MAE has also been used, allowing automation of the preliminary step of the analytical process. Even though microwave devices have been used for some years in analytical laboratories for sample digestion, their use to enhance extraction is more recent.[163] Preliminary studies performed in the late 1980s using domestic microwave ovens showed the great potential of microwaves for extraction. However, their extended use in laboratories began around 1997, with the commercialization of several instruments dedicated to extractions. MAE uses microwave radiation as the source of heating of the solvent–sample mixture. Because of the particular effects of microwaves on matter (viz., dipole rotation and ionic conductance), heating with microwaves is instantaneous and occurs in the heart of the sample, leading to very fast extractions. In most cases, the extraction solvent is chosen to absorb microwaves. As an alternative for thermolabile compounds, however, the microwaves may be absorbed only by the matrix, resulting in heating of the sample and release of the solutes into the cold solvent.

The results obtained to date have led to the conclusion that microwave radiation causes no degradation of the extracted compounds, unless an excessively high temperature arises in the vessel. However, a specific effect of microwaves on plant material has been found. Microwaves interact selectively with the free water molecules present in the gland and vascular systems, leading to rapid heating and temperature increase, followed by rupture of the walls and release of the essential oils into the solvent. Similar mechanisms are suspected in soils and sediments, where strong, localized heating should lead to an increase in pressure and subsequent destruction of the matrix macrostructure.

Microwave energy may be applied to samples using either of two technologies: closed vessels (under controlled pressure and temperature) or open vessels (at atmospheric pressure).[163] These two technologies are commonly termed *pressurized MAE* (PMAE) or *focused MAE* (FMAE), respectively. Whereas in open vessels the temperature is limited by the boiling point of the solvent at atmospheric pressure, in closed vessels the temperature may be elevated by simply applying the correct pressure. The latter system seems more suitable in the case of volatile compounds. However, with closed vessels, one needs to wait for the temperature to decrease after extraction before opening the vessel, increasing the overall extraction time (by ~20 min). Both systems have been shown to have similar extraction efficiencies for the recovery of polycyclic aromatic hydrocarbons (PAHs) from soils. The closed-vessel technology is quite similar to the pressurized fluid extraction (PFE) technology, as the solvent is heated and pressurized in both systems. The main difference is in the means of heating, either microwave energy or conventional oven heating. Consequently, unlike PFE, the number of parameters that influence method performance is reduced, thus making the application of this technique quite simple in practice.

The nature of the solvent is obviously of prime importance in MAE. As with other techniques, the solvent (or solvent mixture) should efficiently solubilize the analytes of interest without significantly extracting matrix material (i.e., the extraction should be as selective as possible to avoid further clean-up). In addition, it should be able to displace the solute molecules adsorbed onto matrix active sites in order to ensure efficient extractions. Finally, the microwave-absorbing properties of the solvent are of great importance as sufficient heating is required to allow efficient desorption and solubilization and thus efficient extraction. In general, the solvent chosen should absorb the microwaves without leading to strong heating so as to avoid degradation of the compounds. Thus, it is common practice to use a binary mixture (e.g., hexane–acetone, 1 + 1), with only one solvent absorbing microwaves. However, in some cases, polar solvents (such as water or alcohols) may provide efficient extractions. Alternatively, apolar solvents may be used if the matrix absorbs microwaves or if an additional microwave-absorbing material (e.g., Weflon) is added. Other important parameters are the power applied, the temperature, and the extraction time; the latter is dependent on the number of simultaneous extractions performed.

Sufficient heating is usually required to enable efficient extractions to be performed; however, an excessively high temperature may lead to solute degradation. As with other techniques, the nature of the matrix is also an important factor for the success of the extraction. In particular, the water content needs to be carefully controlled to avoid excessive heating and to allow reproducible results. Therefore, drying the matrix before the extraction or adding a drying agent, with subsequent addition of the required water content, may be advisable. Also, the strength of the analyte–matrix interactions may induce matrix effects and require a change in the extraction conditions from one matrix to another. MAE, especially with closed vessels, has been used successfully for several applications, most of them environmental.

Akhtar et al. developed a method for MAE extraction of fortified and incurred chloramphenicol residues in freeze-dried egg.[164] Sample extraction time was 10 s using a binary solvent mixture consisting of ACN and 2-propanol. Akhtar also compared MAE with conventional extraction (homogenization, vortexing) for the determination of incurred salinomycin in chicken eggs and tissues.[165] Raich-Montiu et al.[159] described the extraction of trace concentrations of sulfanomides from soil samples using MAE with acetonitrile, followed by further clean-up with SPE. The extraction efficiency was evaluated using three soil samples with different physicochemical characteristics. Recovery rates ranged from 60% to 98%, and detection limits were between 1 and 6 μg/kg.

4.4.9.3 Ultrasound-Assisted Extraction

More recent studies have shown that ultrasound-assisted extraction (UAE) enhances the extraction efficiency by increasing the yield and by shortening the time of extraction. Ultrasound is capable of accelerating the extraction of organic compounds contained within matrices such as plant tissues by disrupting the cell walls and enhancing mass transfer of cell contents.[166] The enhancement of extraction efficiency of organic compounds by ultrasound is attributed to a phenomenon called *acoustic cavitation*. As ultrasound passes through a liquid, the expansion cycles exert negative pressure on the liquid, pulling the molecules away from one another (Fig. 4.7[167]). If the ultrasound intensity is sufficient, the expansion cycle can create cavities or microbubbles in the liquid. This occurs when the negative pressure exceeds the local tensile strength of the liquid, which varies with the type and purity of the liquid. Once formed, these bubbles will absorb the energy from the sound waves and grow during the expansion cycles and recompress during the compression cycle. The increase in pressure and temperature caused by the compression leads to collapse of the bubbles, which causes a shockwave that passes through the solvent, enhancing the mass transfer within the system.

Ma et al. have described a method for the simultaneous determination of 22 sulfonamides in cosmetics.[168] Various cosmetic samples, including, cream, lotion, powder, shampoo, and lipstick, were extracted under ultrasonication. Qualitative and quantitative analysis was carried out by using UPLC-MS/MS. The limits of detection for the set of sulfonamides were 3.5–14.1 μg/kg. The mean recoveries at the three spiked concentrations were 80.3–103.6% with intra-day precision less than 12% and inter-day precision less than 15%.

4.5 FINAL REMARKS AND CONCLUSIONS

Sample preparation techniques must be chosen and optimized with careful regard to the method purpose. As stated in the ICH guidelines, the validated method must be appropriate for the intended purposes. Sample preparation procedures used in bioanalytical applications and their important features are described in Table 4.1. An appropriate technique should be chosen with regard to extraction time, selectivity, the number of steps, solvent consumption, and the possibility of using on-line techniques.

Current sample preparation techniques employ small amounts of sample as well as simpler methods that are "just enough" prior to analysis, as more steps may introduce more errors. New developments have attempted to enhance selectivity (immunoaffinity, MIP, and aptamers) as well as to reduce solvent consumption, thus making sample preparation more environmentally friendly (micro-extraction approaches). Finally, such methods also feature high-throughput automated techniques.

As shown in Table 4.2, the faster approaches are MEPS, on-line RAM, and TFC. The highest selectivity is obtained using MIP or aptamers. Among solventless techniques, micro-extractions such as SPME and SBSE or on-line RAM and TFC, which both use mobile phase for sample

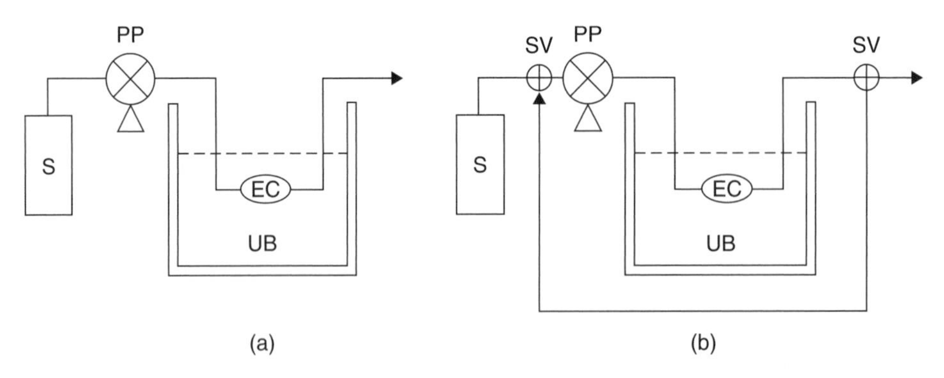

Figure 4.7 Schematic diagram of ultrasound-assisted extraction: (a) as an open system and (b) as a closed system (S—solvent; PP—peristaltic pump; EC—extraction chamber; UB—ultrasonic bath; SV—switching valve). (Adapted from Tadeo et al.[167] with permission from Elsevier; copyright 2010.)

TABLE 4.1 Summary of Sample Preparation Procedures for Determination of Multi-class Antibiotics in Food Samples

Compounds	Matrix	Extraction and Clean-up	Reference
S, Q, M, T, P, TMP	Muscle	Extraction with MeOH/H_2O (70 : 30, v/v) containing EDTA, followed by dilution with H_2O and filtration	167
S, Q, M, T, β-L	Muscle, kidney	Extraction with 70% MeOH and five-fold dilution with H_2O	58
S, Q, L, M, N, IP, T	Honey, milk, egg, muscle	Extraction with H_2O/ACN or acetone/1% formic acid (v/v)	59
S, Q, M, T	Milk	QuEChERS liquid extraction with acidified ACN-EDTA in presence of magnesium sulfate and sodium acetate for removal of water and proteins; centrifugation and filtration of organic phase and dilution of extracts	63
S, Q, IP, N	Chicken muscle	QuEChERS liquid extraction with 1% (v/v) acetic acid/ACN and sodium sulfate, followed by d-SPE clean-up with a Bondesil-NH_2 sorbent (plus additional strong cation exchange clean-up with Bond Elut SCX cartridges for NMZs)	15
S, Q, M, A, L, IP	Muscle	Extraction with ACN/MeOH (95 : 5, v/v) and defatting with n-hexane saturated in ACN, followed by evaporation and dissolution in MeOH	10
β-L, M, N, Q, S, T	Milk	Deproteinization with a solution of 0.1% (v/v) formic acid in ACN, followed by ultrafiltration for 1 h with cut-off membranes of 3 kDa; finally, the extracts were partially evaporated and centrifuged and the supernatants collected	12
S, M, T, β-L, A, AMP	Honey	Four sequential LLE steps: (1) ACN, (2) 10% (v/v) TCA + ACN, (3) NFPA + ACN, and (4) hydrolysis + ACN; the four extracts were individually resuspended in MeOH/H_2O (20/80, v/v), vortexed, sonicated, and filtered	169
M, T, Q, S	Honey	Extraction with aqueous EDTA under milk acidic conditions (pH 4.0) followed by clean-up with Oasis HLB cartridges	77
T, F, M, L, S, A,CAP	Honey	Dissolution of honey with water, centrifugation, and filtration; an aliquot of the supernatant was diluted with water for AG determination; the remaining portion was submitted to a clean-up with Strata X cartridges	170
β-L, S, T, FQ, N, IP, AMP	Milk	Extraction with ACN/0.1% formic acid (v/v) and clean-up with Oasis HLB cartridges; partial solvent evaporation for 20 min, followed by ultrafiltration for another 20 min with cut-off membranes	78
M, P, Q, S, T, N, IP, AMP	Milk	Deproteinization with ACN (shaking for 30 min) followed by centrifugation for a further 15 min; 10-fold dilution of supernatants with H_2O and clean-up with Strata X cartridges	11
S, Q, M, T, L, β-L	Muscle, kidney, liver	"Bipolarity" extraction with ACN-aqueous McIlvaine buffer/ammonium sulfate, followed by clean-up with Oasis HLB cartridges	13
S, T, FQ, β-L	Eggs	Extraction with sodium succinate buffer, centrifugation, and clean-up with Oasis HLB cartridges	79
β-L, M, L, Q, S, T, N, TMP	Muscle	PLE extraction: solvent H_2O, pressure 1500 psi, temperature 70°C, flush volume 60%, static time 10 min, 1 extraction cycle	171

Notation: A = aminoglycosides; AMP = amphenicols; β-L = β-lactams; d-SPE, dispersive SPE; FQ = fluoroquinolones; IP = ionophores; L = lincosamides; M = macrolides; NMZ = nitroimidazoles; P = penicillins; PLE = pressurized liquid extraction; Q = quinolones; S = sulfonamides; T = tetracyclines; TMP = trimethoprim. This table is adapted from QuEChERS = quick, easy, cheap, effective, rugged, and safe;

Source: Adapted from Moreno-Bondi et al.[56] with permission from Springer; copyright 2009.

elution, are the most environmentally friendly. However, most micro-extraction techniques, including LLE, SPME, and SBSE, may have to be excluded from use for high-throughput applications, due to the time required for the establishment of equilibrium. Therefore it is unlikely that these techniques will be widely used in modern residue laboratories. Overall, on-line RAM, TFC, and MEPS appear to be the most convenient techniques in terms of extraction time, ease of use, possibility of automation, and solvent consumption. However, more experimental work and development is necessary before their widespread implementation as sample preparation techniques in modern high-throughput laboratories.

Conventional sample preparation techniques such as SPE and LLE are still the most widely used in routine laboratories. However, their performance may be surpassed by new modern approaches, and their development should proceed further, following high-throughput, low volume, ease of

TABLE 4.2 Comparison of Sample Preparation Approaches

Sample Preparation Technique	Extraction Time (mins)	Selectivity	Multi-step	On-line Possibility	Solvent Consumption
LLE	15-25	Medium	Yes	+	High
PLE	10-15	Medium	No	+	Low
SPE	15-25	Medium	Yes	+	Relatively high
MEPS					
MSPD	10-20	Medium	No	−	Low
SPME	10-60	Medium	Adsorption/desorption	+	No
SBSE	30-240	Medium	Adsorption/desorption	−	No
MIP	15-20	High	Yes	+	Relatively high
RAM	<5	Medium	No (centrifugation)	+	No
TFC	<5	Low	No	+	No

Source: Adapted from Noákavá and Vlcková[18] with permission from Elsevier; copyright 2009.

use, and automated and environmental trends in order to reduce the contrast with fast LC approaches. Nevertheless, SPE and LLE continue to serve as effective and reliable sample preparation techniques in many analytical laboratories because of their automated (or semi-automated) and on-line connection with chromatographic techniques. This will probably not change in the near future.

REFERENCES

1. Scopus search veterinary and drugs and determination and chromatography and extraction and food and tissue and residues and cleanup; time period 2005–2010 (available at `http://www.scopus.com`).
2. Barker SA, Applications of matrix solid-phase dispersion in food analysis, *J. Chromatogr. A* 2000;880:63–68.
3. Stolker AAM, Schwillens PLWJ, van Ginkel LA, Brinkman UATh, Comparison of different liquid chromatography methods for the determination of corticosteroids in biological matrices, *J. Chromatogr. A* 2000;893:55–67.
4. Zheng MM, Gong R, Zhao X, Feng, YQ, Selective sample pretreatment by molecularly imprinted polymer monolith for the analysis of fluoroquinolones from milk samples, *J. Chromatogr. A* 2010;1217:2075–2081.
5. Hormazabal V, Yndestad M, Determination of amprolium, ethopabate, lasalocid, monensin, narasin, and salinomycin in chicken tissues, plasma, and egg using liquid chromatography-mass spectrometry, *J. Liq. Chromatogr. RT* 2000;23:1585–1598.
6. Anderson CR, Rupp HS, Wu WH, Complexities in tetracycline analysis—chemistry, matrix extraction, clean-up, and liquid chromatography, *J. Chromatogr. A* 2005;1075:23–32.
7. Danaher M, Howells LC, Crooks SRH, Cerkvenik-Flajs V, O'Keeffe M, Review of methodology for the determination of macrocyclic lactone residues in biological matrices, *J. Chromatogr. B* 2006;844:175–203.
8. Wang S, Zhang HY, Wang L, Duan ZJ, Kennedy I, Analysis of sulphonamide residues in edible animal products: A review, *Food Addit. Contam.* 2006;23:362–384.
9. Danaher M, DeRuyck H, Crooks SRH, Dowling G, O'Keeffe M, Review of methodology for the determination of benzimidazole residues in biological matrices, *J. Chromatogr. B* 2007;845:1–37.
10. Yamada R, Kozono M, Ohmori T, Morimatsu F, Kitayama M, Simultaneous determination of residual veterinary drugs in bovine, porcine, and chicken muscle using liquid chromatography coupled with electrospray ionization tandem mass spectrometry, *Biosci. Biotech. Biochem.* 2006;70:54–65.
11. Stolker AAM, Rutgers P, Oosterink E, Lasaroms JJP, Peters RJB, van Rhijn J. A, Nielen MWF, Comprehensive screening and quantification of veterinary drugs in milk using UPLC–ToF-MS, *Anal. Bioanal. Chem.* 2008;391:2309–2322.
12. Ortelli D, Cognard E, Jan P, Edder P, Comprehensive fast multiresidue screening of 150 veterinary drugs in milk by ultra-performance liquid chromatography coupled to time of flight mass spectrometry, *J. Chromatogr. B* 2009;877:2363–2374.
13. Kaufmann A, Butcher P, Maden K, Widmer M, Quantitative multiresidue method for about 100 veterinary drugs in different meat matrices by sub 2-μm particulate high-performance liquid chromatography coupled to time of flight mass spectrometry, *J. Chromatogr. A* 2008;1194: 66–74.
14. Kinsella B, Lehotay SJ, Mastovska K, Lightfield AR, Furey A, Danaher M, New method for the analysis of flukicide and other anthelmintic residues in bovine milk and liver using liquid chromatography–tandem mass spectrometry, *Anal. Chim. Acta* 2009;637:196–207.
15. Stubbings G, Bigwood T, The development and validation of a multi-class liquid chromatography tandem mass spectrometry (LC–MS/MS) procedure for the determination of veterinary drug residues in animal tissue using a QuEChERS (QUick, Easy, CHeap, Effective, Rugged and Safe) approach, *Anal. Chim. Acta* 2009;637:68–78.
16. Antignac JP, de Wasch K, Monteau F, De Brabander H, Andre F, Le Bizec B, *Proc. EuroResidue V Conf.* Noordwijkerhout, The Netherlands, 2004, p. 129.

17. Kinsella B, O'Mahony J, Malone E, Moloney M, Cantwell H, Furey A, Danaher M, Current trends in sample preparation for growth promoter and veterinary drug residue analysis, *J. Chromatogr. A* 2009;1216:7977–8015.
18. Noákavá L, Vlcková H, A review of current trends and advances in modern bio-analytical methods: Chromatography and sample preparation, *Anal. Chim. Acta* 2009;656:8–35.
19. Commission Regulations No 37/2010 of December 22 2009 on pharmacologically active substances and their classification regarding maximum residue limits in foodstuffs of animal origin, *Off. J. Eur. Commun*. 2010;L15:1–72.
20. De Ruyck H, Daeseleire E, Grijspeerdt K, De Ridder H, Van Renterghem R, Huyghebaert G, Determination of flubendazole and its metabolites in eggs and poultry muscle with liquid chromatography-tandem mass spectrometry, *J. Agric. Food Chem*. 2001;49:610–617.
21. De Ruyck H, Daeseleire E, Grijspeerdt K, De Ridder H, Van Renterghem R, Huyghebaert G, Distribution and depletion of flubendazole and its metabolites in edible tissues of guinea fowl, *Br. Poultry Sci*. 2004;45:540–549.
22. Reyes-Herrera I, Schneider MJ, Cole K, Farnell MB, Blore PJ, Donoghue DJ, Concentrations of antibiotic residues vary between different edible muscle tissues in poultry, *J. Food Prot*. 2005;68:2217–2219.
23. Delmas JM, Chapel AM, Gaudin V, Sanders P, Pharmacokinetics of flumequine in sheep after intravenous and intramuscular administration: bioavailability and tissue residue studies, *J. Vet. Pharmacol. Ther*. 1997;20:249–257.
24. Nappier JL, Hoffman GA, Arnold TS, Cox TD, Reeves DR, Hubbard VL, Determination of the tissue distribution and excretion of ^{14}C-fertirelin acetate in lactating goats and cows, *J. Agric. Food Chem*. 1998;46:4563–4567.
25. Lifschitz A, Imperiale F, Virkel G, Cobenas MM, Scherling N, DeLay R, Lanusse C, Depletion of moxidectin tissue residues in sheep, *J. Agric. Food Chem*. 2000;48:6011–6015.
26. Prats C, El Korchi G, Francesch R, Arboix M, Perez B, Tylosin depletion from edible pig tissues, *Res. Vet. Sci*. 2002;73:323–325.
27. Council Regulation 2377/90/EEC laying down a Community procedure for the establishment of maximum residue limits veterinary medicinal products in foodstuffs of animal origin, *Off. J. Eur. Commun*. 1990;L224:1–8.
28. Schneider MJ, Mastovska K, Solomon MB, Distribution of penicillin G residues in culled dairy cow muscles: Implications for residue monitoring, *J. Agric. Food Chem*. 2010;58:5408–5413.
29. Gratacós-Cubarsí M, Castellari M, Valero A, García-Regueiro JA, Hair analysis for veterinary drug monitoring in livestock production, *J. Chromatogr. B* 2006;834:14–25.
30. Cooper KM, Kennedy DG, Nitrofuran antibiotic metabolites detected at parts per million concentrations in retina of pigs—a new matrix for enhanced monitoring of nitrofuran abuse, *Analyst* 2005;130:466–468.
31. Rose MD, Bygrave J, Farrington WHH, Shearer G, The Effect of cooking on veterinary drug residues in food. Part 8, *Analyst* 1997;122:1095–1099.
32. Riediker S, Rytz A, Stadler RH, Cold-temperature stability of five β-lactam antibiotics in bovine milk and milk extracts prepared for liquid chromatography–electrospray ionization tandem mass spectrometry analysis, *J. Chromatogr. A* 2004;1054:359–363.
33. Verdon E, Fuselier R, Hurtaud-Pessel D, Couedor P, Cadieu N, Laurentie M, Cold-temperature stability of five β-lactam antibiotics in bovine milk and milk extracts prepared for liquid chromatography–electrospray ionization tandem mass spectrometry analysis, *J. Chromatogr. A* 2000;882:135–143.
34. McEvoy JDG, Ferguson JP, Crooks SRH, Kennedy DG, van Ginkel LA, Maghuin-Rogister G, Meyer HHD, Pfaff MW, Farrington WHH, Juhel-Gaugain M, *Proc. 3rd Intnatl. Symp. Hormone and Veterinary Drug Residue Analysis*, Oud St. Jan, Belgium, 1998, p. 2535.
35. Papapanagiotou EP, Fletouris DJ, Psomas EI, *Proc. EuroResidue V Conf*., Noordwijkerhout, The Netherlands, 2004, p. 305.
36. Sireli UT, Filazi A, Cadirci O, *Proc. Union of Scientists Natl. Conf*., Stara Zagora, Bulgaria, 2005, p. 441.
37. Commission Decision of 12 August 2002 implementing Council Directive 96/23/EC concerning the performance of analytical methods and the interpretation of results. 2002/657/EC, *Off. J. Eur. Commun*. 2002;L221:8–36.
38. Croubels S, De Baere S, De Backer P, Practical approach for the stability testing of veterinary drugs in solutions and in biological matrices during storage, *Anal. Chim. Acta* 2003;483:419–427.
39. McEvoy DG, Crooks SRH, Elliott CT, McCaughey WJ, Kennedy DG, *Proc. 2nd Intnatl. Symp. Hormone and Veterinary Drug Residue Analysis*, Oud St. Jan, Belgium, 1994, p. 2603.
40. Cooper AD, Tarbin JA, Farrington WHH, Shearer G, Aspects of extraction, spiking and distribution in the determination of incurred residues of chloramphenicol in animal tissues, *Food Addit. Contam*. 1998;15:637–644.
41. Heller DN, Peggins JO, Nochetto CB, Smith ML, Chiesa OA, Moulton K, LC/MS/MS measurement of gentamicin in bovine plasma, urine, milk, and biopsy samples taken from kidneys of standing animals, *J. Chromatogr. B* 2005;821:22–30.
42. Vass M, Hruska K, Franek M, Nitrofuran anti-biotics: A review on the application, prohibition and residual analysis, *Vet. Med. CzechPraha*. 2008;53:469–500.
43. EMEA Nitrofurans Summary Report (available at `http://www.emea.europa.eu/pdfs/vet/mrls/nitrofurans.pdf`; accessed 11/22/10).
44. Hoogenboom LAP, Berghmans MCJ, Polman THG, Parker R, Shaw IC, Depletion of protein-bound furazolidone metabolites containing the 3-amino-2-oxazolidinone side-chain from liver, kidney and muscle tissues from pigs, *Food Addit. Contam*. 1992;9:623–644.

45. Conneely A, Nugent A, O'Keeffe M, Mulder PPJ, van Rhijn JA, Kovacsics L, Fodor A, McCracken RJ, Kenned DG, Isolation of bound residues of nitrofuran drugs from tissue by solid-phase extraction with determination by liquid chromatography with UV and tandem mass spectrometric detection, *Anal. Chim. Acta* 2003;483:91–98.
46. FOI Summary, NADA 141-063 (original), NUFLOR (florfenicol), 1996 (available at `http://cpharm.vetmed.vt.edu/VM8784/ANTIMICROBIALS/FOI/141063.htm`; accessed 11/22/10).
47. Wrzesinski CL, Crouch LS, Endris R, Determination of florfenicol amine in channel catfish muscle by liquid chromatography, *J. AOAC Int*. 2003;86:515–520.
48. Wu JE, Chang C, Ding WP, He DP, Determination of florfenicol amine residues in animal edible tissues by an indirect competitive ELISA, *J. Agric. Food Chem*. 2008;56:8261–8267.
49. Xu Y, Du L, Rose MJ, Fu I, Wolf ED, Musson DG, Concerns in the development of an assay for determination of a highly conjugated adsorption-prone compound in human urine, *J. Chromatogr. B* 2005;818:241–248.
50. Wang CJ, Pao LH, Hsiong CH, Wu CY, Whang-Peng JJK, Hu OYP, Novel inhibition of cis/trans retinoic acid interconversion in biological fluids—an accurate method for determination of trans and 13-cis retinoic acid in biological fluids, *J. Chromatogr. B* 2003;796:283–291.
51. Murphy-Poulton SF, Boyle F, Gu XQ, Mater LE, Thalidomide enantiomers: Determination in biological samples by HPLC and vancomycin-CSP, *J. Chromatogr. B* 2006;831:48–56.
52. Evans MJ, Livesey JH, Ellis MJ, Yandle TG, Effect of anticoagulants and storage temperatures on stability of plasma and serum hormones, *Clin. Biochem*. 2001;34:107–112.
53. McCracken RJ, Spence DE, Kennedy DG, Comparison of extraction techniques for the recovery of veterinary drug residues from animal tissues, *Food Addit. Contam*. 2000;17:907–914.
54. Babin Y, Fortier S, A high-throughput analytical method for determination of aminoglycosides in veal tissues by liquid chromatography/tandem mass spectromety with automated clean-up, *J.AOAC Int*. 2007;90:1418–1426.
55. Fletouris DJ, Papapanagiotou EP, *Proc. 14th European Conf. Analytical Chemistry*, Antwerp, Belgium, 2007, p. 1189.
56. Moreno-Bondi MC, Marazuela MD, Herranz S, Rodriguez E, An overview of sample preparation procedures for LC-MS multi-class antibiotic determination in environmental and food samples, *Anal. Bioanal. Chem*. 2009;395:921–946.
57. Chico J, Rúbies A, Centrich F, Companyó R, Prat MD, Granados M, High-throughput multi-class method for antibiotic residue analysis by liquid chromatography–tandem mass spectrometry, *J. Chromatogr. A* 2008;1213:189–199.
58. Granelli K, Branzell C, Rapid multi-residue screening of antibiotics in muscle and kidney by liquid chromatography-electrospray ionization–tandem mass spectrometry, *Anal. Chim. Acta* 2007;586:289–295.
59. Mol HGJ, Plaza-Bolaños P, Zomer P, Rijk TC, Stolker AAM, Mulder PPJ, Toward a generic extraction method for simultaneous determination of pesticides, mycotoxins, plant toxins, and veterinary drugs in feed and food matrices, *Anal. Chem*. 2008;80:9450–9459.
60. Anastassiades M, Lehotay SJ, Stajnbaher D, Schenck FJ, *J. Assoc. Off. Anal. Chem. Intnatl*. 2003;86: 412.
61. Pinto CG, Laespada MEF, Martín SH, Ferreira AMC, Pavón JLP, Cordero BM, Simplified QuEChERS approach for the extraction of chlorinated compounds from soil samples, *Talanta* 2010;81:385–391.
62. Lehotay SJ, Determination of pesticide residues in foods by acetonitrile extraction and partitioning with magnesium sulfate: collaborative study, *J.AOAC Int*. 2007;90:485–520.
63. Aguilera-Luiz MM, Vidal JLM, Romero-González R, Frenich AG, Multi-residue determination of veterinary drugs in milk by ultra-high-pressure liquid chromatography–tandem mass spectrometry, *J. Chromatogr. A* 2008;1205:10–16.
64. Kinsella B, Lehotay SJ, Mastovska K, Lightfield AR, Furey A, Danaher M, New method for the analysis of flukicide and other anthelmintic residues in bovine milk and liver using liquid chromatography–tandem mass spectrometry, *Anal. Chim. Acta* 2009;37:196–207.
65. Taylor LT, *Supercritical Fluid Extraction*, Wiley-Interscience, New York, 1996.
66. Stolker AAM, van Ginkel LA, Stephany RW, Maxwell RJ, Parks OW, Lightfield AR, Supercritical fluid extraction of methyltestosterone, nortestosterone and testosterone at low ppb levels from fortified bovine urine, *J. Chromatogr. B*. 1999;726:121–131.
67. Stolker AAM, Zoontjes PW, Schwillens PLWJ, Kootstra P, van Ginkel LA, Stephany RW, Brinkman UATh, Determination of acetyl gestagenic steroids in kidney fat by automated supercritical fluid extraction and liquid chromatography ion-trap mass spectrometry, *Analyst* 2002;127:748–754.
68. Stolker AAM, van Tricht EF, Zoontjes PW, van Ginkel LA, Stephany RW, Rapid method for the determination of stanozolol in meat with supercritical fluid extraction and liquid chromatography–mass spectrometry, *Anal. Chim. Acta* 2003;483:1–9.
69. Carabias-Martinez R, Rodriguez-Gonzalo E, Revilla-Ruiz P, Hernandez-Mendez J, Pressurized liquid extraction in the analysis of food and biological samples, *J. Chromatogr. A* 2005;1089:1–17.
70. Giergielewicz-Mozajska H, Dabrowski L, Namiesnik J, Accelerated solvent extraction (ASE) in the analysis of environmental solid samples—some aspects of theory and practice, *Crit. Rev. Anal. Chem*. 2001;31:149–165.
71. Mendiola JA, Herrero M, Cifuentes A, Ibáñez E, Use of compressed fluids for sample preparation: Food applications, *J. Chromatogr. A* 2007;1152:234–246.
72. Carretero V, Blasco C, Picó Y, Multi-class determination of antimicrobials in meat by pressurized liquid extraction and liquid chromatography–tandem mass spectrometry, *J. Chromatogr. A* 2008;1209:162–173.

73. Runnqvist H, Bak SA, Hansen M, Styrishave B, Halling-Sørensen B, Björklund E, Determination of pharmaceuticals in environmental and biological matrices using pressurised liquid extraction—Are we developing sound extraction methods? *J. Chromatogr. A* 2010;1217:2447–2470.
74. Thurman EM, Mills MS, *Solid-Phase Extraction: Principles and Practice*, Wiley-Interscience, New York, 1998.
75. Fontanals N, Marcé RM, Borrull F, New hydrophilic materials for solid-phase extraction, *Trends Anal. Chem*. 2005;24:394–406.
76. De Alwis H, Heller DN, Multi-class, multi-residue method for the detection of antibiotic residues in distillers grains by liquid chromatography and ion trap tandem mass spectrometry, *J. Chromatogr. A* 2010;1217;3076–3084.
77. Martinex-Vidal JL, Aguilera-Luiz MM, Romero-Gonález R, Garido-Frenish A, Multi-class analysis of antibiotic residues in honey by ultraperformance liquid chromatography-tandem mass spectrometry, *J. Agric. Food Chem*. 2009;57:1760–1767.
78. Turnipseed SB, Andersen WC, Karbiwnyk CM, Madson MR, Miller KE, Multi-class, multi-residue liquid chromatography/tandem mass spectrometry screening and confirmation methods for drug residues in milk, *Rapid Commun. Mass Spectrom*. 20008;22:1467–1480.
79. Heller D, Nochetto CB, Rumel NG, Thomas MH, Development of multi-class methods for drug residues in eggs: hydrophilic solid-phase extraction clean-up and liquid chromatography/tandem mass spectrometry analysis of tetracycline, fluoroquinolone, sulfonamide, and β-lactam residues, *J. Agric. Food Chem*. 2006;54:5267–5278.
80. Pecorelli I, Galarini R, Bibi R, Floridi A, E. Casciarri E, Floridi A, Simultaneous determination of 13 quinolones from feeds using accelerated solvent extraction and liquid chromatography, *Anal. Chim. Acta* 2003;483:81–89.
81. Johnston L, Mackay L, Croft M, Determination of quinolones and fluoroquinolones in fish tissue and seafood by high-performance liquid chromatography with electrospray ionisation tandem mass spectrometric detection, *J. Chromatogr. A* 2002;982:97–109.
82. Hubert P, Chiap P, Evrard B, Delattre L, Crommen J, High-throughput screening for multi-class veterinary drug residues in animal muscle using liquid chromatography/tandem mass spectrometry with on-line solid-phase extraction, *J. Chromatogr. B* 1993;622:53–60.
83. Kantiani L, Farre M, Sibum M, Postigo C, Lopez de Alda M, Barcelo D, Fully automated analysis of β-lactams in bovine milk by online solid phase extraction-liquid chromatography-electrospray-tandem mass spectrometry, *Anal. Chem*. 2009;81:4285–4295.
84. Spark Holland Application Note 30 (2000) (available at `http://www.sparkholland.com/applications/application-search`; accessed 11/22/10).
85. Spark Holland Application Note 53074 (2007) (available at `http://www.sparkholland.com/applications/application-search`; accessed 11/22/10).
86. Tang HP, Ho C, Lai SS, High-throughput screening for multi-class veterinary drug residues in animal muscle using liquid chromatography/tandem mass spectrometry with on-line solid-phase extraction, *Rapid Commun. Mass Spectrom*. 2006;20:2565–2572.
87. Fagerquist CK, Lightfield AR, Lehotay SJ, Confirmatory and quantitative analysis of β-lactam antibiotics in bovine kidney tissue by dispersive solid-phase extraction and liquid chromatography-tandem mass spectrometry, *Anal. Chem*. 2005;77:1473–1482.
88. Anastassiades M, Lehotay SJ, Stajnbaher D, Schenck FJ, Fast and easy multiresidue method employing acetonitrile extraction/partitioning and "dispersive solid phase extraction" for the determination of pesticide residues in fruits and vegetables, *J.AOAC Int*. 2003;86: 412–431.
89. Mastovska K, Lightfield AR, Streamlining methodology for the multi-residue analysis of β-lactam antibiotics in bovine kidney using liquid chromatography–tandem mass spectrometry, *J. Chromatogr. A* 2008;1202:118–123.
90. Hernando MD, Mezcua M, Suarez-Barcena JM, Fernandez-Alba AR, Liquid chromatography with time-of-flight mass spectrometry for simultaneous determination of chemotherapeutant residues in salmon, *Anal. Chim. Acta* 2006;562:176–184.
91. Barker SA, Long AR, Short CR, Isolation of drug residues from tissues by solid phase dispersion, *J. Chromatogr*. 1989;475:353–361.
92. Kristenson EM, Ramos L, Brinkman UATh, Recent advances in matrix solid-phase dispersion, *Trends Anal. Chem*. 2006;25:96–111.
93. Capriotti AL, Cavaliere C, Giansanti P, Gubbiotti R, Samperia R, Lagana A, Recent developments in matrix solid-phase dispersion extraction, *J. Chromatogr. A* 2010;1217:2521–2532.
94. Liu Y, Zou QH, Xie MX, Han J, A novel approach for simultaneous determination of 2-mercaptobenzimidazole and derivatives of 2-thiouracil in animal tissue by gas chromatography/mass spectrometry, *Rapid Commun. Mass Spectrom*. 2007;21:1504–1510.
95. Bogialli S, Di Corcia A, Matrix solid-phase dispersion as a valuable tool for extracting contaminants from foodstuffs, *J. Biochem. Biophys. Meth*. 2007;70:163–179.
96. Bogialli S, D'Ascenzo G, Di Corcia A, Laganà A, Tramontana G, Simple assay for monitoring seven quinolone antibacterials in eggs: Extraction with hot water and liquid chromatography coupled to tandem mass spectrometry: Laboratory validation in line with the European Union Commission Decision 657/2002/EC, *J. Chromatogr. A* 2009;1216:794–800.
97. Barker SA, Matrix solid phase dispersion (MSPD), *Biochem. Biophys. Meth*. 2007;70:151–162.
98. Garcia-Lopez M, Canosa P, Rodriguez I, Trends and recent applications of matrix solid-phase dispersion, *Anal. Bioanal. Chem*. 2008;391:963–974.
99. Zou QH, Wang J, Wang XF, Liu Y, Han J, Hou F, Xie MX, Application of matrix solid-phase dispersion

and high-performance liquid chromatography for determination of sulfonamides in honey, *J. AOAC Int*. 2008;91: 252–258.

100. Kishida K, Quantitation and confirmation of six sulphonamides in meat by liquid chromatography–mass spectrometry with photodiode array detection, *Food Control* 2007;18:301–305.
101. Bogialli S, D'Ascenzo G, Di Corcia A, Lagana A, Nicolardi S, A simple and rapid assay based on hot water extraction and liquid chromatography–tandem mass spectrometry for monitoring quinolone residues in bovine milk, *Food Chem*. 2008;108:354–360.
102. Yan H, Qiao F, Row KH, Molecularly imprinted-matrix solid phase dispersion for selective extraction of five fluoroquinolones in eggs and tissue, *Anal. Chem*. 2007;79:8242–8248.
103. Marazuela MD, Bogialli S, A review of novel strategies of sample preparation for the determination of antibacterial residues in foodstuffs using liquid chromatography-based analytical methods, *Anal. Chim. Acta* 2009;645:5–17.
104. Beraldi RP, Pawliszyn J, The application of chemically modified fused silica fibers in the extraction of organics from water matrix samples and their rapid transfer to capillary columns, *Water Pollut. Res. J. Can*. 1989;24:179–191.
105. Artur CL, Pawliszyn J, Solid phase microextraction with thermal desorption using fused silica optical fibers, *Anal. Chem*. 1990;62:2145–2148.
106. Ullrich S, Solid-phase microextraction in biomedical analysis, *J. Chromatogr. A* 2000;902:167–194.
107. Lord HL, Strategies for interfacing solid-phase microextraction with liquid chromatography, *J. Chromatogr. A* 2007;1152:2–13.
108. Gaurav AK, Malik AK, Tewary DK, Singh B, A review on development of solid phase microextraction fibers by sol–gel methods and their applications, *Anal. Chim. Acta* 2008;610:1–14.
109. Augusto F, Carasek E, Costa Silva RG, Rivellino SR, Batista AD, Martendal E, New sorbents for extraction and microextraction techniques, *J. Chromatogr. A* 2010;1217:2533–2542.
110. Chong SL, Wang D, Hayes JD, Wilhite BW, Malik A, Sol-gel coating technology for the preparation of solid phase microextraction fibers of enhanced thermal stability, *Anal. Chem*. 1997;69:3889–3898.
111. McClure EL, Wong ChS, Solid phase microextraction of macrolide, trimethoprim, and sulfonamide antibiotics in wastewaters, *J. Chromatogr. A* 2007;1169:53–62.
112. Lu KH, Chen CY, Lee MR, Trace determination of sulfonamides residues in meat with a combination of solid-phase microextraction and liquid chromatography–mass spectrometry, *Talanta* 2007;72:1082–1087.
113. Altun Z, Abdel-Rehim M, Study of the factors affecting the performance of microextraction by packed sorbent (MEPS) using liquid scintillation counter and liquid chromatography-tandem mass spectrometry, *Anal. Chim. Acta* 2008;630:116–123.
114. Pavlovic DM, Babic S, Horvat ALM, Macan JK, Sample preparation in analysis of pharmaceuticals, *Trends Anal. Chem*. 2007;26:1062–1075.
115. David F, Sandra P, Stir bar sorptive extraction for trace analysis, *J. Chromatogr. A* 2007;1152:54–69.
116. Huang X, Qiu N, Yuan D, Simple and sensitive monitoring of sulfonamide veterinary residues in milk by stir bar sorptive extraction based on monolithic material and high performance liquid chromatography analysis, *J. Chromatogr. A* 2009;1216:8240–8245.
117. Luo Y-B, Ma Q, Feng Y-Q, Stir rod sorptive extraction with monolithic polymer as coating and its application to the analysis of fluoroquinolones in honey sample, *J. Chromatogr. A* 2010;1217:3583–3589.
118. Desilets CP, Rounds MA, Regnier FE, Semipermeable-surface reversed-phase media for high-performance liquid chromatography, *J. Chromatogr. A* 1991;544:25–39.
119. Cassiano NM, Lima VV, Oliveria RV, Pietro AC, Cass QB, Development of restricted-access media supports and their application to the direct analysis of biological fluid samples via high-performance liquid chromatography, *Anal. Bioanal. Chem*. 2006;384:1462–1469.
120. Sadílek P, Satínský D, Solich P, Using restricted-access materials and column switching in high-performance liquid chromatography for direct analysis of biologically-active compounds in complex matrices, *Trends Anal. Chem*. 2007;26:375–384.
121. Sato Y, Yamamoto E, Takaluwa S, Kato T, Asakawa N, Weak cation-exchange restricted-access material for on-line purification of basic drugs in plasma, *J. Chromatogr. A* 2008;1190:8–13.
122. Oliveira RV, Cass QB, Evaluation of liquid chromatographic behavior of cephalosporin antibiotics using restricted access medium columns for on-line sample clean-up of bovine milk, *J. Agric. Food Chem*. 2006;54:1180–1187.
123. Ding J, Ren N, Chen L, Ding L, On-line coupling of solid-phase extraction to liquid chromatography–tandem mass spectrometry for the determination of macrolide antibiotics in environmental water, *Anal. Chim. Acta* 2009;634:215–221.
124. Katz SE, Siewierski M, Drug residue analysis using immunoaffinity chromatography, *J. Chromatogr. A* 1992;624:403–409.
125. Li C, Wang ZH, Cao XY, Beier RC, Zhang SX, Ding SY, Li XW, Shen JZ, Development of an immunoaffinity column method using broad-specificity monoclonal antibodies for simultaneous extraction and cleanup of quinolone and sulfonamide antibiotics in animal muscle, *J. Chromatogr. A* 2008;1209:1–5.
126. Zhao SJ, Li XL, Ra YK, Li C, Jiang HY, Li JC, Qu ZN, Zhang SX, He FY, Wan YP, Feng CW, Zheng ZR, Shen JZ, Developing and optimizing an immunoaffinity cleanup technique for determination of quinolones from chicken muscle, *J. Agric. Food Chem*. 2009;57:365–371.
127. Li JS, Qian CF, Determination of avermectin B1 in biological samples by immunoaffinity column clean-up and

liquid chromatography with UV detection, *J.AOAC Int.* 1996;79:1062–1067.

128. Luo P, Chen X, Liang C, Kuang H, Lu L, Jiang Z, Wang Z, Li C, Zhang S, Shen J, Simultaneous determination of thiamphenicol, florfenicol and florfenicol amine in swine muscle by liquid chromatography–tandem mass spectrometry with immunoaffinity chromatography clean-up, *J. Chromatogr. B* 2010;878:207–212.
129. He C, Long Y, Pan J, Li K, Liu F, Application of molecularly imprinted polymers to solid-phase extraction of analytes from real samples, *J. Biochem. Biophys. Meth.* 2007;70:133–150.
130. Gupta R, Kumar A, Molecular imprinting in sol–gel matrix, *Biotechnol. Adv.* 2008;26:533–547.
131. Sellergren B, Direct drug determination by selective sample enrichment on an imprinted polymer, *Anal. Chem.* 1994;66:1578–1582.
132. He J, Wang S, Fang G, Zhu H, Zhang Y, Molecularly imprinted polymer online solid-phase extraction coupled with high-performance liquid chromatography-UV for the determination of three sulfonamides in pork and chicken, *J. Agric. Food Chem.* 2008;56:2919–2925.
133. Boyd B, Björk H, Billing J, Shimelis O, Axelsson S, Leonora M, Yilmaz E, Development of an improved method for trace analysis of chloramphenicol using molecularly imprinted polymers, *J. Chromatogr. A* 2007;1174:63–71.
134. Puoci F, Iemma F, Cirillo G, Curcio M, Parisi OI, Spizzirri UG, Picci N, New restricted access materials combined to molecularly imprinted polymers for selective recognition/release in water media, *Eur. Polym. J.* 2009;45:1634–1640.
135. Mohamed R, Richoz-Payot J, Gremaud E, Mottier P, Yilmaz E, Tabet JC, Gut PA, Advantages of molecularly imprinted polymers LC-ESI-MS/MS for the selective extraction and quantification of chloramphenicol in milk-based matrixes. Comparison with a classical sample preparation, *Anal. Chem.* 2007;79:9557–9565.
136. Hung ChY, Hwang ChCh, HPLC behaviour of sulfonamides on molecularly imprinted polymeric stationary phases, *Acta Chromatogr.* 2007;18:106–115.
137. Turiel E, Martín-Esteban A, Tadeo JL, Molecular imprinting-based separation methods for selective analysis of fluoroquinolones in soils, *J. Chromatogr. A* 2007;1172: 97–104.
138. Ellington AD, Szostak JW, *In vitro* selection of RNA molecules that bind specific ligands, *Nature* 1990;346: 818–822.
139. Tuerk C, Gold L, *Science* 1990;249:505–510.
140. Tombelli S, Minunni M, Mascini M, Analytical applications of aptamers, *Biosens Bioelectron.* 2005;20:2424–2434.
141. Kim YJ, Kim YS, Niazi JH, Gu MB, Electrochemical aptasensor for tetracycline detection, *Bioprocess. Biosyst. Eng.* 2010;33:31–37.
142. Xu Y, Willson KJ, Musson DG, Strategies on efficient method development of on-line extraction assays for determination of MK-0974 in human plasma and urine using turbulent-flow chromatography and tandem mass spectrometry, *J. Chromatogr. B* 2008;863:64–73.
143. Zimmer D, Pickard V, Czembor W, Muller C, *Proc. 15th Montreux Symp. LC–MS SFC–MS CE–MS and MS–MS*, Elsevier Science, Montreux, France, 1998, p. 23.
144. Mottier P, Hammel YA, Gremaud E, Philippe AG, Quantitative high-throughput analysis of 16 (fluoro)quinolones in honey using automated extraction by turbulent flow chromatography coupled to liquid chromatography-tandem mass spectrometry, *J. Agric. Food Chem.* 2008;56:35–43.
145. Krebber R, Hoffend FJ, Ruttman F, Simple and rapid determination of enrofloxacin and ciprofloxacin in edible tissues by turbulent flow chromatography/tandem mass spectrometry (TFC–MS/MS), *Anal. Chim. Acta* 2009;637:208–213.
146. Stolker AAM, Peters RJB, Zuiderent R, Bussolo JD, Martins C, Fully automated screening of veterinary drugs in milk by turbulent flow chromatography and tandem mass spectrometry, *Anal. Bioanal. Chem.* 2010;397:2841–2849.
147. Goto T, Ito Y, Yamada S, Matsumoto H, Oka H, High-throughput analysis of tetracycline and penicillin antibiotics in animal tissues using electrospray tandem mass spectrometry with selected reaction monitoring transition, *J. Chromatogr. A* 2005;1100:193–199.
148. Furusawa N, Liquid-chromatographic determination of sulfadimidine in milk and eggs, *Fresenius J. Anal. Chem.* 1999;364:270–272.
149. Furusawa N, Simplified determining procedure for routine residue monitoring of sulphamethazine and sulphadimethoxine in milk, *J. Chromatogr. A* 2000;898:185–191.
150. Furusawa N, Kishida K, High-performance liquid chromatographic procedure for routine residue monitoring of seven sulfonamides in milk, *Fresenius J. Anal. Chem.* 2001;371:1031–1033.
151. Furusawa N, Determination of sulfonamide residues in eggs by liquid chromatography, *J. AOAC Int.* 2002;85: 848–851.
152. Furusawa N, Rapid high-performance liquid chromatographic determining technique of sulfamonomethoxine, sulfadimethoxine, and sulfaquinoxaline in eggs without use of organic solvents, *Anal. Chim. Acta* 2003;481:255–259.
153. Furusawa N, Simultaneous high-performace liquid chromatographic determination of sulfamonomethoxine and its hydroxy/N4-acetyl metabolites following centrifugal ultra-filtration in animal blood plasma, *Chromatographia* 2000;52:653–656.
154. Muldoon MT, Buckley SA, Deshpande SS, Holtzapple CK, Beier RC, Stanker LH, Development of a monoclonal antibody-based cELISA for the analysis of sulfadimethoxine. 2. Evaluation of rapid extraction methods and implications for the analysis of incurred residues in chicken liver tissue, *J. Agric. Food Chem.* 2000;48:545–550.
155. Furusawa N, Determining the procedure for routine residue monitoring of sulfamethazine in edible animal tissues, *Biomed. Chromatogr.* 2001;15:235–239.

156. Furusawa N, Simplified liquid-chromatographic determination of residues of tetracycline antibiotics in eggs, *Chromatographia* 2001;53:47–50.
157. Furusawa N, Liquid chromatographic determination/identification of residueal penicillin G in food producing animal tissues, *J. Liq. Chromatogr. RT* 2001;24:161–172.
158. Furusawa N, Normal-phase high-performance liquid chromatographic determination of spiramycin in eggs and chicken, *Talanta* 1999;49:461–465.
159. Raich-Montiu J, Folch J, Compañó R, Granados M, Prat MD, Analysis of trace levels of sulfonamides in surface water and soil samples by liquid chromatography-fluorescence, *J. Chromatogr. A* 2007;1172:186–193.
160. Eskilsson CS, Bjorklund E, Analytical-scale microwave-assisted extraction, *J. Chromatogr. A* 2000;902:227–250.
161. Sharma U, Sharma K, Sharma N, Sharma S, Singh HP, Sinha AK, Microwave-assisted efficient extraction of different parts of hippophae rhamnoides for the comparative evaluation of antioxidant activity and quantification of its phenolic constituents by reverse-phase high-performance liquid chromatography (RPHPLC), *J. Agric. Food Chem.* 2008;56:374–379.
162. Vryzas Z, Papadopoulou-Mourkidou E, Determination of triazine and chloroacetanilide herbicides in soils by microwave-assisted extraction (MAE) coupled to gas chromatographic analysis with either GC-NPD or GC-MS, *J. Agric. Food Chem.* 2002;50:5026–5033.
163. Camel V, Recent extraction techniques for solid matrices—supercritical fluid extraction, pressurized fluid extraction and microwave-assisted extraction: Their potential and pitfalls, *Analyst* 2001;126:1182–1193.
164. Akhtar MH, Croteau LG, Dani C, AbouElSooud K, *Proc. 109th AOAC Int. Meeting*, IOS Press, Nashville, TN, 1995, p. 33.
165. Akhtar MH, Comparison of microwave assisted extraction with conventional (homogenization, vortexing) for the determination of incurred salinomycin in chicken eggs and tissues, *J. Environ. Sci. Health B* 2004;39:835–844.
166. Hemwimol S, Pavasant P, Shotipruk A, Ultrasound-assisted extraction of anthraquinones from roots of Morinda citrifolia, *Ultrason, Sonochem.* 2006;13:543–548.
167. Tadeo JL, Sánchez-Brunete C, Albero B, García-Valcárcel AI, Application of ultrasound-assisted extraction to the determination of contaminants in food and soil samples, *J. Chromatogr. A* 2010;1217:2415–2440.
168. Ma Q, Wang C, Wang X, Bai H, Dong YY, Wu T, Zhang Q, Wang JB, Tang YZ, Simultaneous determination of 22 sulfonamides in cosmetics by ultra performance liquid chromatography tandem mass spectrometry, *Fenxi Huaxue/Chin. J Anal.* 2008;12:1683–1689.
169. Hammel YA, Mohamed R, Gremaud E, Le Breton MH, Guy PA, Multi-screening approach to monitor and quantify 42 antibiotic residues in honey by liquid chromatography–tandem mass spectrometry, *J. Chromatogr. A* 2008;1177:58–76.
170. Lopez MI, Barton JS, Chu PS, Multi-class determination and confirmation of antibiotic residues in honey using LC-MS/MS, *J. Agric. Food Chem.* 2008;56:1553–1559.
171. Carretero V, Blasco C, Picó Y, Multi-class determination of antimicrobials in meat by pressurized liquid extraction and liquid chromatography–tandem mass spectrometry, *J. Chromatogr. A.* 2008;1209:162–173.

5

BIOANALYTICAL SCREENING METHODS

Sara Stead and Jacques Stark

5.1 INTRODUCTION

The use of veterinary medicines in food-producing animals raises concerns regarding the potential for the occurrence of residues in milk or in carcasses at slaughter.[1] A 2007 poll of European consumers on food safety issues indicated that concern about the presence of such residues received a high ranking.[2] Such concerns include residues of antimicrobial compounds, which constitute the largest class of approved veterinary compounds administered to farmed livestock globally.

Antimicrobial agents are administered to livestock for different purposes: (1) prevention and control of infections and (2) growth promotion.[3] The use of antibiotic growth promoters (AGPs) [also termed *antibiotic growth-promoting agents* (AGPAs)] has been forbidden in the European Union (EU) since 2006.[4] There is concern regarding the spread of resistant microorganisms in the human population, such as meticillin-resistant *Staphylococcus aureus* (MRSA). Antimicrobial resistance is a complex process arising through a variety of mechanisms including point mutations, gene conversions, rearrangements (translocations), and deletions or insertions of foreign DNA.[5] Resistant bacteria from animals can infect the human population via contact, the food chain, or occupational exposure. Residues of antimicrobial compounds in foods can cause allergic reactions (e.g., penicillin) or toxicity (e.g., chloramphenicol) and may affect the composition of the human intestinal flora. Finally, residues present in milk or meat at too high a concentration will inhibit the development of starter cultures in cheese, yogurt, and fermented meat production and thus may result in economic losses. For example, available studies on the effect of pirlimycin on starter cultures used in the production of cheeses, buttermilk, sour cream, and yogurt were considered by the Joint FAO/WHO Expert Committee on Food Additives (JECFA)[6] in recommending the MRL for pirlimycin residues in milk adopted by the Codex Alimentarius Commission (CAC) in 2006.[7]

In the EU the safe use of veterinary medicinal products (VMPs) is monitored, and member states are routinely audited to ensure compliance.[8] Countries exporting food of animal origin into the EU (termed "third countries") must also demonstrate equivalence within their legislative framework.[9] Veterinary medicine use is regulated *via* Council Regulations 470/2009[10] [repealing Council Regulation (EEC) 2377/90][11] and 37/2010.[12] Council Directive 96/23/EC[13] regulates the residue control and surveillance monitoring of pharmacologically active compounds.

Within the EU, in contrast with other areas of food control, there is no obligation to use standardized methods in the surveillance of veterinary medicine residues. Instead, a criterion-based approach applies that defines the performance characteristics that the methods used must meet.[14] However, within the United States and some other countries methods are statutorily prescribed. The method must be able to detect the marker residue, specifically, metabolite, sum of metabolites, or parent compound at/or below the appropriate regulatory limit (RL), as available. In the EU, the RL for authorized veterinary medicinal products is the maximum residue limit (MRL). The RL for prohibited and unauthorized substances is the minimum required performance limit (MRPL) or the reference point for action (RPA).[10] In other cases, especially for unauthorized substances, the Community Reference Laboratory (CRL) Recommended Concentration (RC)[15] can be applied, although this has no legal standing and is not a "limit" per se.

Chemical Analysis of Antibiotic Residues in Food, First Edition. Edited by Jian Wang, James D. MacNeil, and Jack F. Kay.

An internationally recognized method performance scheme is the AOAC Research Institute Performance Tested Method (PTM) program.[16] The AOAC-RI PTM program provides an independent third-party review of proprietary test method performance. Test methods demonstrated to meet acceptable performance criteria are granted PTM status. Kit manufacturers of approved PTM test methods are subsequently licensed to use the PTM certification mark. The PTM certification mark assures the end kit user that an independent assessment has found that the test method performance meets an appropriate standard for the claimed intended use. The PTM program is designed to be complementary to the Official Methods of Analysis (OMA) program. The PTM evaluation can serve as the OMA "precollaborative" study for a microbiology method or as the single laboratory validation for a chemical method. The PTM program has six distinct phases including, consulting, PTM application, method developer validation study, independent validation study, validation study report, and PTM review, which must be satisfactorily completed before the test kit can receive AOAC accreditation status. The list of test kits that have received AOAC accreditation status, including the information concerning the range of target analyte–matrix combination, is publically available via the AOAC website.[17]

A two-tier approach is often utilized by residue control laboratories whereby samples are first screened to identify the suspected positive (non-compliant) samples, which are subject to further quantitative and confirmatory analysis. Screening methods should be inexpensive and rapid and permit a high sample throughput. The basic criteria that should be met are a detection capability below the RL, a low incidence of false-negative (compliant) results ($\leq 5\%$), and a high degree of repeatability, reproducibility, and robustness.[18] A low incidence of false-positive results is also important to reduce the costs incurred by additional confirmatory analysis. False-positive results in screening analysis can occur for a number of reasons, such as if the test is sensitive to other structurally related compounds naturally present in the matrix or to co-contaminants.

Because of the diverse physicochemical properties of antimicrobial compounds, a variety of analytical techniques are commonly employed to screen for their residues in food of animal origin.[19] Chromatographic techniques coupled to ultraviolet (UV) or fluorescence detection have traditionally been used for this purpose, although they are no longer appropriate for the confirmatory analysis of all substance groups within the EU because of the criteria established in Commission Decision 2002/657/EC,[20] namely, the identification point (IP) system. The more recent trend is to replace these complicated LC methods, no longer sufficient for confirmation, with rapid bioanalytical screening assays, complemented by confirmatory mass spectrometric (MS) methods.[21] Screening techniques can generate qualitative (positive or negative), semi-quantitative (high, medium/low, or negative), or quantitative results. Screening assays may be sensitive to more than one compound in a related group or class, for example, sulfonamides or tetracyclines, giving either separate indications for each residue or a total for that group.

The modes of action of antimicrobial drugs are diverse and are broadly classified as bacteriostatic (inhibit cellular growth and/or reproduction) or bactericidal (directly kill the cells). Bacteria are classified as either Gram-positive or Gram-negative according to the original staining technique devised by Gram in 1884.[22] To exert their activity, compounds must permeate the microbial cell wall. Because of differences in cellular structure, antimicrobial agents have differing potencies against Gram-positive/negative bacteria. By exploiting the mode action of antimicrobial compounds several biologically based screening assays have been developed, such as microbial inhibition, applicable for the detection of residues in animal tissues, milk, and other food products.

The widespread use of antimicrobial compounds in animal husbandry and the stringent food safety legislation demand the availability of rapid and sensitive screening techniques for residue detection. For these reasons there is an increased need for rapid, easy-to-use, reliable, cost-effective, and broad-spectrum screening methods, which can be readily implemented in survey, surveillance, and compliance monitoring schemes. Another important aspect to consider when choosing to introduce a screening method is the extent of assay compliance with internationally recognized validation criteria. The focus of this chapter is on the different types, applications, and performance of commercially available bioanalytical screening test kits for the detection of antimicrobial residues in foods of animal origin. The advantages and disadvantages of the various screening assay formats and the general aspects of validation are also discussed.

5.2 MICROBIAL INHIBITION ASSAYS

5.2.1 The History and Basic Principles of Microbial Inhibition Assays

Microbial inhibition assays (MIAs) are routinely used screening techniques offering the advantage of detecting the total biological activity associated with unknown residues (non-targeted analysis). Microbial inhibition assays are the methods of choice when an estimate of antimicrobial potency is required.[23] The MIAs are sensitive to compounds that inhibit or disturb the growth of a test microorganism. For these reasons microbial inhibition assays are a widely employed screening method for determination of the presence or absence of antimicrobial residues in milk, animal tissues, and food products.

Shortly after their development in the 1940s, antibiotics were used in veterinary medicine, first to prevent or treat mastitis in cows and later for the treatment of other diseases. Initial concern about antibiotic residues in milk was not a public health issue but came from dairy processors who noticed inhibition of starter cultures used in the production of cheese and yogurt, thus generating a need for screening tests to examine milk for antibiotic residues.[24] Since inhibition of starter cultures by penicillin in milk was the main problem, the earliest microbial inhibition assays were based on growth inhibition of lactic acid bacteria. Spores of *Bacillus* species were also utilized; spores are easier to handle and far more stable than the vegetative cells.

The ability of bacteria to produce acid led to one of the earliest microbiological methods for the detection of antimicrobial compounds. A culture of *Streptococcus agalactiae* was used for the detection of penicillin in bovine milk by incubating the samples for several days at 37°C using litmus as a color indicator.[25] In a method developed by Berridge,[26,27] a culture of *Streptococcus thermophilus* and the acid–base color indicator bromocresol purple was added to a milk sample. During the incubation period the color was determined every 30 min, and the time it took for the indicator to change color was considered as an indication for the concentration of antimicrobial compounds present in the milk. Galesloot and Hassing[28] optimized the acid–base method with *S. thermophilis* and using methylene blue as the indicator and were able to detect 1 μg/l of benzylpenicillin.

Another method to determine the presence of antibiotics in milk consisted in measuring coagulation. One of the earliest coagulation tests was described in 1952 by Lemoigne et al.,[29] who used four different strains of *Streptococcus lactis*. A defined number of cells of each culture were added to different milk samples and incubated overnight. If antimicrobial compounds were present in the milk at a certain concentration, the milk would not coagulate. This method was further optimized by Pien and co-workers[30] using *S. lactis* strains with a defined resistance for certain concentrations of penicillin that enabled a semi-quantitative estimation of the antimicrobial concentration within the milk sample to be obtained.

Agar diffusion methods based on determining inhibition zones of a standard test organism seeded in agar plates is perhaps the most widely used screening technology. During the incubation period the liquid sample diffuses from a carrier into the agar medium. After an incubation of several hours (between 8 and 36 h depending on the particular test) the size of the inhibition zones can be measured. If antimicrobial compounds are present above a certain concentration, the microorganism will be inhibited (as a result of microbial death and/or inhibition of growth), and clear zones are visible on the agar plate (Fig. 5.1).

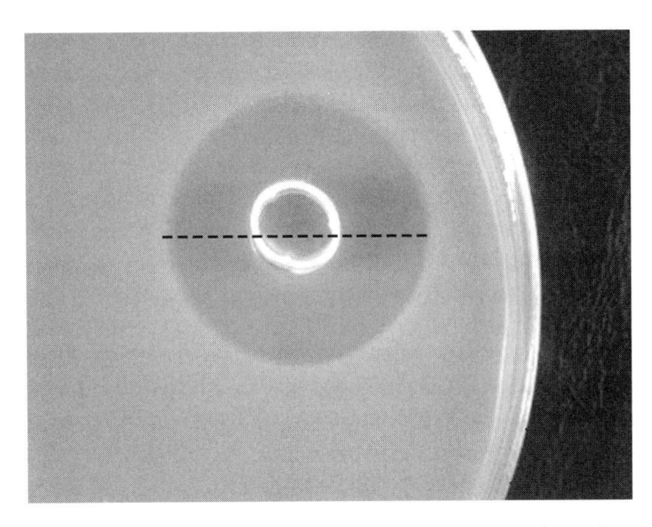

Figure 5.1 A photograph of the area on an agar plate where the growth of the test microorganism (*Bacillus subtilis*) has been prevented by the presence of an antibiotic-impregnated disk (white circular spot), termed the zone of inhibition. The dashed line indicates the diameter of the zone of inhibition.

In 1944, Foster et al.[31] were one of the first groups to develop an agar diffusion test for the determination of penicillin in liquid samples. Agar seeded with *Bacillus subtilis* spores was added to a petri dish. Sterile glass cups were then filled with liquid samples containing different concentrations of penicillin and placed on the agar. Following an incubation period of 12–16 h at 30°C, inhibition zones of different sizes were formed indicating the approximate concentration of penicillin in the test solution. As an alternative to glass cups, filter-paper disks saturated with milk or meat juice can also be applied. In some methods small pieces of muscle or kidney tissue are placed directly on the agar surface.

In 1948, Welsh et al.[32] developed a disk plate method using paper disks soaked in milk, which were placed on agar seeded with *B. subtilis* spores. Whey agar was first seeded with *B. subtilis* spores, and paper disks (7 mm) were soaked in the milk samples. The impregnated disks were placed on the agar surface, and the plates were incubated at 37°C for 4 h, after which time the zones of inhibition were determined. The sensitivity of this test for benzylpenicillin was found to be 5 μg/l.

When using agar-plate based methods, multiple plates (typically between 2 and 12) with different indicator microorganisms are employed. This approach allows the laboratory to determine different antimicrobial compounds. The test organisms commonly used are *Geotrichum stearothermophilis* (also known as *Bacillus stearothermophilis*), *Bacillus subtilis, Bacillus cereus, Escherichia coli*, and lactic acid bacteria such as *Streptococcus thermophilus*. The advantages of these multi-plate tests are that the sampling is very easy (direct application), and, due to the use of more test organisms, a broader spectrum of antimicrobial compounds can be detected (e.g., Gram-positive microbial inhibitors by *Bacillus* strains and Gram-negative inhibitors such as quinolones by an *E. coli* strain).

The limited potential for automation and the necessity to grow fresh bacterial cultures and prepare fresh agar make the use of plate tests labor-intensive and limited to containment level II (CL2) laboratories requiring operation by skilled technicians. Other disadvantages include long incubation times, large variations in test performance, and often the lack of meaningful quality assurance data. Variations in test performance can be caused by the size of the inhibition zones, which may vary with the quality of the agar, the depth of the agar layer, the age of the plates, the heat of the agar during plate preparation, the duration of incubation, and the incubation temperature.

5.2.2 The Four-Plate Test and the New Dutch Kidney Test

Despite the limitations associated with plate-based tests, the majority of routine antimicrobial screening is carried out using in-house prepared plate-based assays. Commonly used examples of such plate-based tests are the four-plate test (FPT) and the modified four-plate test (mFPT) and are widely used as official screening tests within the EU.[33]

The basic FPT method comprises three plates impregnated with *Bacillus subtilis* (BGA) endospores maintained at different pH values and a fourth plate containing *Kocuria rhizophila*. The mode of operation is as follows. Tissue ($\sim$1 cm^2) is placed directly on the agar plate surface, then the plates are incubated at the optimum temperature and conditions required for cellular growth. Following the incubation period, the zone of inhibition is measured and compared against the control sample. The method of sample preparation allows many samples to be processed rapidly and cost-effectively. However, the direct introduction of tissue samples onto the agar plate may result in poor test repeatability as a result of the reliance on passive diffusion of liquid from the tissue sample into the agar, which at worst could result in an unacceptably high falsely compliant (negative) rate. Without an adequate sample extraction step the test may be limited to use only in samples of the same type with specified moisture content.

In 2002, the European CRL for antibiotics, Agence Nationale de Sécurité Sanitaire ANSES (Fougeres, France), organized a collaborative study of the screening test for antibiotic residues (STAR) protocol, a five-plate microbial inhibition test designed for meat and milk. A large variation between the participant's data was recorded. The organizers concluded that the "STAR protocol" was too sensitive to the macrolide, quinolone, and tetracycline classes with a false-positive rate of 8.3%.[34] As an alternative to the EU four-plate method, the new Dutch kidney test (NDKT) was developed.[35] The NDKT consists of a single plate and uses *Bacillus subtilis* (BGA) as the test strain. It is based on the analysis of paper disks impregnated with renal pelvis fluid, a fluid reported to contain the highest residue concentrations.[36] However, it is questionable whether renal pelvis fluid from kidneys is a representative matrix for the determination of antimicrobial residues in muscle and other edible tissues. The NDKT, and more recently the improved five-plate Nouws antibiotic test (NAT) screening assay, has been the statutorily prescribed residue test for the monitoring of slaughtered animals in The Netherlands since 1988. The method has proved effective in obtaining lower detection limits than the EU four-plate test. Notably, however, field evaluation studies indicate that a negative NDKT does not ensure MRL compliance.[37]

The United States Department of Agriculture (USDA), Food Safety and Inspection Service (FSIS) developed a bioassay system which incorporates a seven-plate agar diffusion assay that can detect and quantify a range of antibiotic residues found in meat and poultry products.[38] The USDA/FSIS system utilizes bacteria that are relatively sensitive or resistant to a particular class of antibiotic. The bacteria are used in combination with specific antibiotic test agars and four pH-specific, buffered sample extracts. If a detectable antibiotic residue is present in a sample, it produces a zone of clearing (inhibition) on one or more of the test plates. Certain antibiotic residues can be identified according to their characteristic patterns of inhibition. A chart called an *antibiogram* was developed that depicts expected patterns of inhibition that specific antibiotics are anticipated to produce on the seven plates.

5.2.3 Commercial Microbial Inhibition Assays for Milk

The main focus of this chapter is on the commercially available screening tests for antimicrobial compounds, and the current state-of-the-art technology will be presented in the following sections.

In 1974, at the former Gist-brocades (now DSM) laboratories (Delft, The Netherlands), the first tube diffusion microbial inhibition assay for the rapid detection of antimicrobial residues in milk was developed, called the DelvoTest.[39] After decades, the DelvoTest remains the "gold standard" for screening antimicrobial residues in milk and the most widely used commercially available microbial inhibition assay on the market at the present time. DelvoTest is based on the inhibition of growth of *Bacillus stearothermophilis*, a thermophilic bacterium with high sensitivity for many antimicrobial compounds, especially the β-lactam class. The test is presented in a single ampoule or as a 96-well plate in which a standardized number of bacterial endospores are embedded in an agar medium containing selected nutrients and a pH-sensitive dye. Since the spores of the thermophilic bacterium are not able to germinate at room temperature, the test is robust and has a shelf life of about one year when stored at room temperature.

After the addition of a fixed volume of a milk sample (100 μl) to the ampoule, the test is incubated for ~2 h at 64°C in a water bath, block heater, or reader-incubator. In the absence of an inhibitory substance present in the sample (at or above the threshold concentration), the bacterial spores will germinate and multiply. Cellular respiratory activity increases the generation of carbonic acid, causing a decrease in pH that can be measured as a purple-yellow color change *via* the acid–base indicator dye, bromocresol purple. If the sample contains inhibitory substances above the detection concentration of the test, no growth will occur and the agar color will remain purple. The test end-point can be determined by visual interpretation of the ampoule color change following the defined incubation period. The ampoule agar bed is divided into three theoretical zones. A visual score is assigned to the sample on the basis of the coloration pattern, specifically, three zones are yellow = Y, $\frac{2}{3}$ yellow = YYP, $\frac{1}{2}$ yellow = YP, $\frac{2}{3}$ purple = PPY; three zones are purple = P. A sample scored as Y is reported as a negative result and a sample scored as P is reported as a positive result. A sample having a score of $\geq$ YP is classified as a suspected positive. In the case of a weak purple (YYP), visual discrimination between the colored interferences present in certain matrices and a low-concentration positive sample is difficult and subjective.

Since its original development in the 1970s the DelvoTest has been further refined to meet the customers' requirements in terms of both detection capabilities and presentation of the test format. The current commercially available versions of the test are described below:

- *DelvoTest*. This is a broad-spectrum screening test for the detection of an array of different antibiotic residues in milk. The manufacturer claimed sensitivity for the β-lactam class is at or below the EU MRL and US Food and Drug Administration (FDA)-permitted limits. The test is applicable for use on the farm site or in a laboratory. The end results are indicated by a color change and available within 120–150 min. The test kit includes individual plastic ampoules containing *Bacillus stearothermophilus* in a solid medium, nutrient tablets, dosing syringe, disposable pipettes, tweezers, and instructions. The only additional equipment required is an incubator device capable of maintaining 64°C. The test offers a simple, reliable, and economical test for antibiotic residues in milk.
- *DelvoTest SP*. This test has enhanced sensitivity to the sulfonamide class, as well as the ability to detect the same broad spectrum of antibiotics as the original DelvoTest. The incubation time is longer than the DelvoTest, at around 180 min.
- *DelvoTest SP-NT*. This test has performance characteristics similar to those of DelvoTest SP; however, there is no requirement to add a nutrient tablet to the ampoule. A longer kit shelf life is also reported. The DelvoTest SP-NT is available in either single-ampoule or 96-well multi-plate formats. The ampoule version was designed for the small-scale screening of individual milk samples applicable for use on the farm, while the 96-well plate version allows mass testing by dairy industry and control laboratories. When applied for mass testing the test can provide results for hundreds of samples within a few hours. Both the ampoule and plate versions of the DelvoTest SP have received AOAC approval[17] (see Section 5.1)
- *BR-Test AS Brilliant*. This test, developed specifically for the German–Austrian market, is a brilliant black reduction test to detect antibiotic and sulfonamide residues in milk samples. The test is suitable for various dairy product applications such as the analysis of cow, goat, and sheep milk. The test device consists of wells combined into multiple plates with 96 wells containing a solid and buffered agar medium with nutrients, a standardized number of spores of the test organism, an antifolate, and a blue-colored redox indicator (brilliant black). The BR-Test AS Brilliant is based on the sensitivity of a selected strain of *Bacillus stearothermophilus* to a large number of inhibitory substances such as antibiotic residues that may be present in milk. The BR-Test AS Brilliant conforms to the article 35 of the German (Foods and Other Commodities Act) LMBG (Lebensmittel und bedarfsgegenstände-gesetz).

In addition to the DelvoTest range, other diffusion tube format MIAs are currently available from different manufacturers. Such tests include the range of inhibition tests from Charm Sciences Inc., the Blue Yellow II, and the Cowside II (Massachusetts, USA), from Zeu-Immunotec SL in the Eclipse range (Zaragosa, Spain) and from Copan, the Innovation Copan Milk Test (CMT) (DSM Food Sciences Ltd Delft, The Netherlands). These tests also employ strains of *Bacillus stearothermophilis* and function in a similar mode to that described for the DelvoTest with similar operational characteristics. The performance of some these commercially available tests is compared in terms of the manufacturers' quoted detection limit in bovine milk (Table 5.1). A disk diffusion format assay, *Bacillus stearothermophilis* disk assay (BsDA), is also available from Charm Sciences Inc. This format employs *B. stearothermophilus* var. *calidolactis*–inoculated agar plates (Section 5.2.2) onto which milk-sample soaked disks are

TABLE 5.1 Manufacturers' Reported Detection Limits[a] for Some of the Most Widely Used Commercially Available Tube Diffusion Microbial Inhibition Assays in Fortified Bovine Milk

		Testkit Manufacturer					
		Zeu Immunotec (www.zeu-immunotec.com)	DSM Food Specialities Ltd. (www.dsm.com)			Charm Sciences Inc. (www.charm.com)	
		Manufacturer's Reported Detection Limits in Bovine Milk (μg/kg).					
Antibiotic (Marker Residue)	EU MRL[b] in Bovine Milk (μg/kg)	Eclipse 100	DelvoTest SP	DelvoTest SP-NT	CMT	Blue Yellow II	Cowside II
Benzylpenicillin	4	4	2	1–2	2	2–3	2–3
Amoxicillin	4	5	2	2–3	2	2–3	3–4
Ampicillin	4	4	2–3	4	2	2–3	3–4
Cloxacillin	30	N/A	15	20	12	10–20	10–25
Dicloxacillin	30	N/A	10	10	5	10–20	5–10
Oxacillin	30	25	5	10	5	8–10	5–10
Naficillin	30	N/A	5	5	4	N/A	NA
Ceftiofur[c]	100	N/A	<50	25–50	25	50–100	50–100
Cefacetril	125	N/A	20	N/A	25	N/A	N/A
Cefalexin	100	75	40–60	50	>45	N/A	N/A
Cefapirin[d]	60	8	5	4–6	4	4–6	8–10
Cefalonium	20	N/A	5–10	N/A	12–15	N/A	N/A
Cefoperazone	50	N/A	40	N/A	30	N/A	N/A
Cefazolin	50	N/A	NA	N/A	6	N/A	N/A
Cefquinome	20	N/A	NA	N/A	80	N/A	N/A
Sulfadiazine[e]	100	N/A	50	25–50	50	80–100	40–60
Sulfamethazine[e]	100	150	25	25–100	150	75–125	75–125
Sulfadimethoxine[e]	100	N/A	50	100	50	50–75	25–50
Sulfathiazole[e]	100	75	50	50	50	N/A	N/A
Sulfadioxine[e]	100	N/A	N/A	N/A	150	N/A	N/A
Sulfamethoxazole[e]	100	N/A	N/A	N/A	50	N/A	N/A
Sulfamerazine[e]	100	N/A	N/A	N/A	60	N/A	N/A
Tetracycline[f]	100	150	100	250–500	450	75–100	50–100
Oxytetracycline[f]	100	150	100	250–500	450	75–100	75–100
Chlortetracycline[f]	100	N/A	100–150	200	450	N/A	N/A
Doxcycline	N/A	N/A	N/A	N/A	150	N/A	N/A
Spiramycin[g]	200	N/A	200	400–600	5000	400–500	300–400
Erythromycin A	40	500	50	40–80	600	100–150	75–100
Tylosin A	50	N/A	10–20	30	100	20–30	20–30
Tilmicosin	50	N/A	N/A	N/A	75–100	N/A	N/A
Streptomycin	200	N/A	N/A	N/A	1750	N/A	N/A
Dihydrostreptomycin	200	N/A	300–500	>1000	1750	N/A	N/A
Gentamicin	100	300	100–300	50	400	75–100	75–150
Neomycin B (including framycetin)	1500	500	100–200	100–200	5000–10,000	75–150	100–150
Kanamycin A	150	N/A	2500	5000	4000–5000	N/A	N/A
Dapsone[h]	0	N/A	1	0.5–1	2–4	1–2	1–2
Trimethoprim	50	N/A	50	50–100	135	200–300	200–300
Lincomycin	150	300	100	200	500–700	N/A	N/A
Chloramphenicol[h]	0	N/A	2500	2500	5000–7500	N/A	N/A
Pirlimycin	100	N/A	N/A	N/A	N/A	N/A	25–50

[a]Detection limits quoted are claimed by the kit manufacturers (March 2010) for reference purposes only and collated from manufacturers websites and brochures (N/A = not available).
[b]Commission Regulation (EU) 37/2010.[10]
[c]Sum of the parent drug and the metabolites.
[d]Sum of cephapirin and desacetylcephapirin.
[e]Combined total residues of all substances within the sulfonamide group should not exceed 100 μg/l.
[f]Sum of parent drug and its 4-epimer.
[g]Sum of spiramycin and neospiramycin.
[h]Not authorized for use in food-producing species within the EU.

placed. Following the incubation period the zones of inhibition can be measured and recorded. The assay is marketed for the detection of penicillin G, ampicillin, amoxicillin, and cephapirin at the USFDA tolerance levels and is an official standard test for regulatory use in reference laboratories.[40]

Milk production in small ruminants has increased and the use of antimicrobial substances in dairy ewes has become common practice for the treatment of mastitis and other diseases. It is, therefore, important to consider the applicability of the commercial MIAs in milk from various species due to the natural variation in milk composition. The performance of some of the MIA assays in ovine and caprine milk has been investigated, including the Blue-Yellow II Test (BY test),[41] the DelvoTest, and the BsDA.[42] Linage et al.[41] reported that only 5 of the 25 antimicrobial compounds studied were detected by the BY test at concentrations similar to the EU MRLs and concluded that although the BY test showed improved sensitivity compared with other screening tests studied in ovine milk, an improvement in the sensitivity of screening tests to detect a greater number of residues of antimicrobial agents in ovine milk is warranted. The findings of the study conducted by Zeng et al.[42] showed the DelvoTest and BsDA to have high sensitivity to the two compounds (pencillin G and cephapirin) included in the study with false non-compliant rates of 0% for the BsDA and 7% for the DelvoTest in caprine milk.

5.2.4 Commercial Microbial Inhibition Assays for Meat-, Egg-, and Honey-Based Foods

A limited range of microbial inhibition assays has also been developed and is commercially available for the analysis of tissue samples (muscle, kidney, liver), egg, fish, and honey. These assays include the PremiTest, produced by DSM Food Specialities Ltd. (Delft, The Netherlands); the Explorer test, developed by Zeu Inmunotec (Zaragoza, Spain); and the kidney inhibition swab (KIS test), produced by Charm Sciences Inc. (Massachusetts, USA).

The PremiTest is a rapid broad-spectrum screening test that detects antimicrobial substances in various applications, such as fresh meat, fish, eggs, honey, urine, and feed. This test is designed to facilitate onsite screening for antibiotic residues in the food chain by taking a sample of the fluid from the meat or honey. The PremiTest is also based on the inhibition of growth of *Bacillus stearothermophilus*. It is presented in a single ampoule in which a standardized number of bacterial endospores are embedded in an agar medium containing selected growth nutrients, diffusion salts, and a pH-sensitive dye directly comparable to the DelvoTest (Section 5.2.3). The incubation time for the PremiTest ranges between 180 and 240 min at 64°C.

In the basic PremiTest method recommended by the test manufacturer, the sample extract is prepared via fluid expression from the "wet" tissue samples. A small aliquot of the expressed fluid (100 μl) is added to the test ampoule. The fluid is collected following extrusion of a piece of tissue using a garlic press device or a multi-press to allow simultaneous processing of up to 12 samples. Physical fluid extraction is simple to perform and is appropriate for on-site testing, such as at an abattoir. From some tissues with lower moisture content such as poultry, meat, fish, and shrimps, it is difficult to obtain sufficient fluid by squeezing alone. For drier tissues the test manufacturer recommends a mild heating step performed (5 min at 60°C) prior to the fluid expression step. Despite the simplicity of the fluid expression procedure, there are some claims that this sample procedure does not generate a representative analytical sample and there are further difficulties relating the results obtained from tissue fluid to regulatory control limits, which are expressed in μg/kg.

More recently, the Explorer test has been developed. The Explorer test principle is the same as the PremiTest and also uses the same microorganism species (*Geobacillus stearothermophilus*). However, the Explorer test is presented in a microplate plate format (96-well) and a different indicator (sensitive to changes in the redox potential of the medium) is used to visualize the test end-point. The incubation time for the test is around 210 min at 65°C. Following the incubation step, the color change (from blue to yellow) can be measured photometrically using a standard plate reader device at two wavelengths: 590 and 650 nm. The sample pre-treatment recommended is based on a simple fluid extraction protocol similar to that described for the PremiTest. The manufacturer has optimized the test for muscle tissue from different animals, including porcine, poultry, bovine, and ovine species. However, limited information concerning the detection capabilities of the test is available from the manufacturer. In a study conducted by Gaudin,[43] the specificity and detection capabilities for five compounds from the major antimicrobial classes were studied in muscle from porcine, ovine, bovine, and poultry species. The group reported detection limits for amoxicillin (10 μg/kg), tylosin (100 μg/kg), doxcycline (200 μg/kg), sulfathiazole (200 μg/kg), and cefalexin (500 μg/kg). These findings show that only amoxicillin and tylosin can be detected at concentrations at or below the current EU MRLs.

The KIS test is marketed for use at either abattoirs or laboratories for the analysis of kidney samples to identify animal carcasses with non-compliant concentrations of antimicrobial residues present in the edible tissues. The KIS reagents are supplied as self-contained and pre-measured for a single use presented in a disposable swab format. The swab is directly inserted into an incision made in the kidney test sample to collect a sample of the serum for analysis via microbial inhibition. The impregnated swap is then inserted into a device containing a tube diffusion MIA

TABLE 5.2 Manufacturer's Quoted Detection Capability for KIS Test in Fortified Kidney Serum in Relation to Current EU MRLs and US Tolerance Limits

Antimicrobial Compound	Detection Capability in Fortified Kidney Serum (μg/kg)	EU MRL[a] in Kidney (μg/kg)	USFDA Tolerance in Kidney (μg/kg)
Benzylpenicillin	30	50	50
Oxytetracycline	3000	600[b]	12,000[c]
Tylosin	400	100	200
Gentamicin	750	750[d]	400
Sulfadimethoxine	250	100[e]	100
Sulfamethazine	500	100[e]	100
Neomycin B (including framycetin)	4000	5000	7,200

[a]Commission Regulation (EU) 37/2010.[10]
[b]Sum of parent drug and its epimer.
[c]Sum of residues of the tetracyclines: chlortetracycline, oxytetracycline, and tetracycline.
[d]Sum of gentamicin C_1, gentamicin C_{1a}, gentamicin C_2, and gentamicin C_{2a}.
[e]Combined total residues of all substances within the sulfonamide group should not exceed 100 μg/kg.

and incubated for about 180 min at 64°C until the color change end-point is observed. The test microorganism used is *Bacillus subtilus*. The manufacturer's reported detection capabilities in kidney in relation to current EU MRLs and USFDA permitted tolerances are shown in Table 5.2. The test can detect a wide range of antimicrobial compounds at concentrations equivalent to or greater than the RL and as such is not considered reliable as a broad-spectrum screening assay.

Other tests based on the use of swabs and microbial inhibition assays were developed at the US Department of Agriculture and Food Safety and are widely used in the United States of America and Canada primarily by meat inspection agencies, including the swab test on premises (STOP), calf antibiotic screen test (CAST), and the fast antibiotic screen test (FAST).[44,45] The STOP assay employs *Bacillus subtilis*, and the CAST and FAST assays use *Bacillus megaterium*. However, these tests will not be discussed further in the context of commercially available assays.

Notably, the tube diffusion based MIAs that are commercially available are applicable only for the detection of Gram-positive inhibitory antimicrobial compounds at concentrations relevant to RLs, due to the type of microorganism employed. Certain antimicrobial compounds belonging to the Gram-negative inhibitory class are also authorized for use in various countries, are subject to RLs, and are used globally in animal husbandry practice. For example, the 4-quinolone and fluoroquinolone compounds, namely, the (fluoro)quinolones, are a class of semi-synthetic antimicrobial agents. EU MRLs currently exist for eight (fluoro)quinolone compounds and range from 10 to 1900 μg/kg depending on the species and tissue type.[12] The UK Department of Food and Rural Affairs (Defra) has funded research in this area, and a prototype MIA in ampoule incorporating an *E.coli* species selective for the (fluoro)quinolone class has been reported.[46] The prototype *E. coli* MIA has detection capabilities close to the MRL concentrations for the target (fluoro)quinolones and is applicable for a variety of food matrices. However, at the present time a commercially available MIA for this important group of compounds is not available.

As a general observation, all the tube-diffusion based MIAs currently available show broadly similar detection capability profiles and comparable performance characteristics. However, the applicability of such tests can be more readily determined on the basis of other parameters such as the robustness and reliability (false compliant/noncompliant rates). Other features, such as the nature of the sample preparation (especially important for the analysis of solid samples, e.g., tissue and feeds) and the potential for automation, are important considerations when choosing to implement a commercial MIA for multi-residue screening.

5.2.5 Further Developments of Microbial Inhibition Assays and Future Prospects

Although the basic principles of commercial tube diffusion or multi-plate screening assays for the detection of antimicrobial residues are still the same as the original concept in 1974, substantial improvements to the technology have been made. To complement these developments, specific tests and sampling procedures for other food products, including animal tissues, eggs, honey, and fish, have been developed. The most recent developments involve the automation of MIAs using incubators, readers, and dedicated software programs. Improvements to the basic MIAs can be made in the following aspects: test sensitivity, duration, ease of use, automation, sampling protocols, and post-screening confirmation tests.

5.2.5.1 Sensitivity

For broad-spectrum screening tests, the most important antimicrobial compounds should be detected at or very close to the appropriate regulatory concentrations. Most commonly, MIAs are used in relation to local MRL legislation, which varies per country and sample type. In reality, screening tests per definition will never be able to detect all antimicrobial compounds at the required concentrations.

The starting point in the technical development of a microbial inhibition assay is the sensitivity of the test microorganism to different antimicrobial compounds under different media conditions. Factors such as the pH and nutrient profile of the media can be varied to obtain the optimum performance. The sensitivity of MIAs may also be modulated by the addition of specific compounds to the test composition. For certain antimicrobial compounds an enhancement in sensitivity may be advantageous, whereas for others, a sensitivity decrease is required to avoid the number of false non-compliant results. For example, cysteine can reduce the overall sensitivity to the β-lactam class. Antifolates, such as ormethoprim or trimethoprim, are known to improve the sensitivity of the test organism to sulfonamide compounds. Sensitivity to sulfonamides can also be modulated by using the enzyme dihydropteroate synthetase that selectively inhibits the response of sulfonamides.[47] Antibiotic receptors such as antibodies can also be employed to decrease the sensitivity of the test to specific compounds, as required.

5.2.5.2 *Test Duration*

Most commercial tube diffusion or multi-plate screening tests have a duration of approximately 120–240 min. In principle, a faster screening test of 60 min for the detection of β-lactam compounds is feasible. Although rapid determination might be an advantage, a faster test would simply lose too many beneficial characteristics of a microbial screening assay, such as the ability to detect a broad spectrum of antimicrobial classes. The duration of the test is limited by the time needed for spores to germinate plus the time required to obtain sufficient acid production or reduced metabolite profile. Making use of vegetative cells is not a viable option for a commercial test, since the shelf life would be limited to only a few days. As an alternative to pH-sensitive acid–base or redox indicators for the end-point determination, more sensitive analytical methods are available. However, at the present time such methods are considered too expensive and/or cannot be readily implemented in this type of "low technology" diagnostic kit.

5.2.5.3 *Ease of Use*

Commercial tube diffusion or multi-plate assays are by nature simple to use and do not require the use of expensive instrumental equipment. Over the years some improvements have been realized. For example, the incubation step is typically performed using a water bath. However, the use of water baths has some disadvantages. Temperature gradients can occur within the unit (unless the water is adequately circulated), which can give rise to falsely compliant or non-compliant results. If not disinfected properly, a water bath may be a source of contamination as the water can enter the test medium contained within the ampoule/plate well. Finally, there are health and safety concerns relating to the use of water baths, and a laboratory environment is required to perform the test. To overcome the problems associated with water baths, specific dry block heaters have been developed by some kit manufacturers. The tests (ampoules or multi-plates) are placed directly into the heater device, and a constant temperature is maintained throughout the duration of the incubation phase. The temperature profile of the heater block can even be monitored and recorded for quality control assurance purposes throughout the incubation, if required.

5.2.5.4 *Automation*

The classical approach to reading the results of tube diffusion MIAs is by visual end-point determination and a scoring system (Section 5.2.3). Visual evaluation is subjective by nature, and is susceptible to contradictory results being recorded for the same sample by different trained analysts.[48,49] Different types of sample can lead to different end-point colors; for example, blood present in meat fluid can disturb the indicator dye color intensity. To manage and archive sample results effectively, laboratory information systems (LIMSs) require a numerical test output. For this reason, subjective screening assays are rarely selected for implementation in official residue monitoring laboratories. To overcome the limitations associated with visual interpretation, scanner-based techniques have been developed by DSM Food Specialties (Delft, The Netherlands) to complement their product range (DelvoScan and PremiScan). The scanner systems offer the advantage of removing the subjectivity associated with visual assessment and provide a much greater discriminatory power than is possible by the naked eye.

More recently, the DelvoTest Accelerator system for the combined incubation and measurement of the DelvoTest SP-NT ampoules and multi-plates (Section 5.2.3) has been launched by DSM Food Specialties. The test ampoules or multi-plates are placed directly onto the scanner glass top plate; the glass plate is heated to 63°C to perform the incubation step. The Accelerator system can run up to 100 individual ampoules or four 96-well multi-plates (potential for 384 samples) in parallel. A scanner measurement is taken every minute until the test end-point is reached, which is determined automatically via the software control system. The incubation time is shorter than the conventional system, at around 100 min. Once the end-point is reached, the test is automatically terminated and the results stored for subsequent assessment, meaning that the tests can be performed out of the normal working hours.

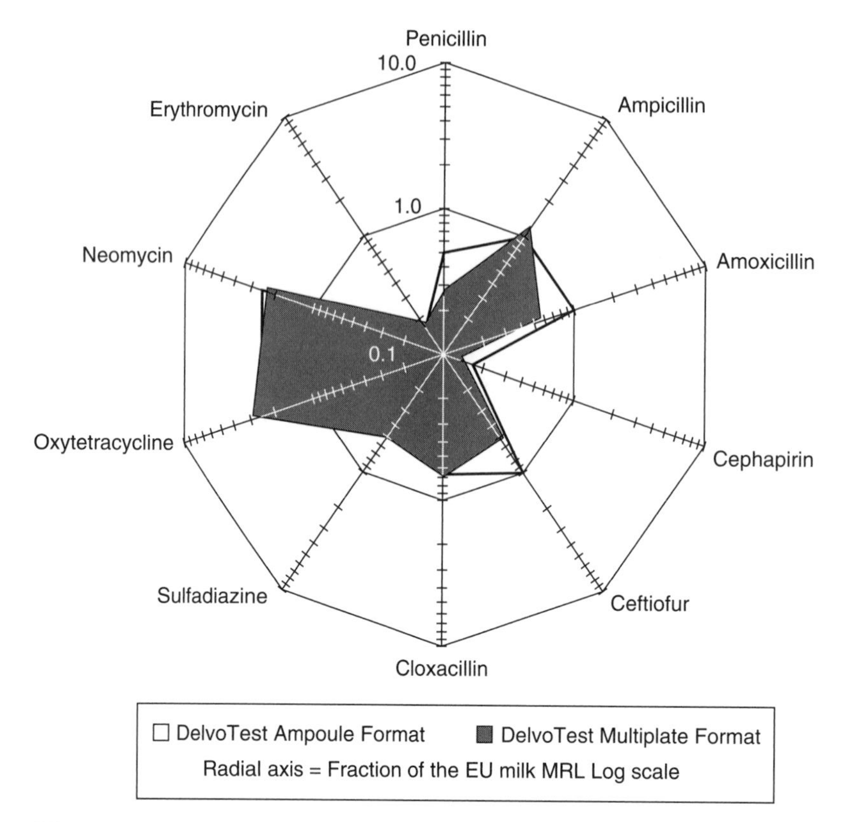

Figure 5.2 Radar plot showing the DelvoTest (SP-NT ampoules and multi-plate formats) lowest limit of detection for 10 antimicrobial compounds in bovine milk expressed as a fraction of the current EU MRLs.

The DelvoTest Accelerator system in combination with the DelvoTest SP-NT ampoules and multi-plates were validated in milk in accordance to the requirements of the International Dairy Federation.[50] The estimated detection limits of the DelvoTest SP-NT ampoules and multi-plates are expressed as a fraction of the current EU MRLs (Fig. 5.2). With the exceptions of oxytetracycline and neomycin, all the analytes included in the study were detected at or below the EU MRL by both test formats. This finding reflects the degree of susceptibility of the microorganism employed in the test to the range of antimicrobial compounds tested.

5.2.5.5 Pre-treatment of Samples

Another important consideration when selecting an MIA is the method of sample pre-treatment available. Certain matrices, especially eggs, kidney, liver, and honey, may contain high concentrations of naturally occurring inhibitory compounds such as lysozyme and/or other biocidal agents that are known to inhibit the test organism, leading to a high rate of falsely non-compliant results. To overcome the problem posed by inhibitory proteins and biocides, a simple step whereby the sample in contact with the test ampoule/multi-plate is exposed to a brief heat-shock (10 min at 80°C) has been developed.[51,52] The heat-shock technique has been found to degrade lysozyme and not to affect the test response of the antimicrobial compounds.

Numerous challenges are presented in the analysis of honey samples; for example, because of its high viscosity, honey is often difficult to handle. Furthermore, certain honey samples might have low pH values of 3–4, which can interfere with the acid–base indicator dye and give a high rate of false-compliant results. Both sample viscosity and any pH effects can be reduced by simply diluting the sample, for instance, in a 1 : 1 ratio (by weight) with phosphate buffer solution pH 5.6. (*Note*: The detection capability of the test will be altered following any additional dilution of the sample, and the quality control samples should be diluted in a similar manner.)

For the analysis of solid tissue samples such as muscle, kidney, or liver, a representative liquid sample first has to be obtained. Fluids can be simply collected by the fluid expression technique (Section 5.2.4). To improve detection limits for certain compounds and provide an alternative method for samples where the fluid cannot be obtained by expression alone, Stead et al.[53] described a solvent extraction method applicable for the generic recovery of antimicrobial compounds from a wide variety of matrices. A comparison of the fluid expression and solvent extraction techniques in combination with the

TABLE 5.3 Comparison of Fluid Expression (Garlic Press) versus Generic Solvent Extraction in Combination with PremiTest Using Tissue Fortified with Various Antimicrobial Compounds

Antibiotic Compound	EU MRL[a] (μg/kg) in Porcine Muscle	PremiTest Limit Detection in Porcine Muscle (μg/kg)	
		Fluid Expression	Solvent Extraction
β-Lactams			
Benzylpenicillin	50	<2.5	<2.5
Amoxicillin	50	5	5
Ampicillin	50	5–8	5
Cloxacillin	300	100	100
Oxacillin	300	100–200	100
Sulfonamides			
Sulfaguanidine	100[b]	150	150
Sulfadimethoxine	100[b]	25–50	<25
Sulfapyridine	100[b]	50	25–50
Sulfamethizole	100[b]	50–100	50
Sulfamethoxypyridine	100[b]	25	<25
Sulfisoxazole	100[b]	25	<25
Sulfathiazole	100[b]	25	<25
Sulfadiazine	100[b]	50	25
Sulfachloropyridazine	100[b]	25	<25
Sulfamerazine	100[b]	25	<25
Sulfanilamide	100[b]	150	100
Sulfaquinoxaline	100[b]	50	<50
Sulfamethazine	100[b]	50–100	75
Tetracyclines			
Tetracycline	100[c]	100	50
Oxytetracycline	100[c]	100–200	50–75
Chlortetracycline	100[c]	>200	50–75
Doxcycline	100[c]	75–100	50
Others			
Tylosin A	100	25	12.5

[a]Commission Regulation (EU) 37/2010.[12]
[b]Combined total residues of all substances within the sulfonamide group should not exceed 100 μg/kg.
[c]Sum of parent drug and its 4-epimer.
Notation: "Less than" sign < indicates that a strong positive response was recorded at the denoted concentration. The actual limit of detection will be, by definition, lower than this value.

PremiTest was performed in porcine tissue fortified with a wide range of antimicrobial compounds. The use of a solvent extraction was found to generate lower detection capabilities, especially for the weaker inhibitors (e.g., the tetracycline class) compared to the fluid expression approach (Table 5.3).

Chemically and mechanically denaturing the tissue facilitates the release of drug residues from tissue. The solvent extraction method also provides a known sample weight that can be taken through the sample preparation to a known volume, enabling the result to be more easily related to the appropriate RLs. Use of a solvent extraction step enables effective concentration of the extract to provide an increased sensitivity profile that otherwise would not be achievable. In the laboratory, screening of samples is generally performed to half the regulatory limit, and samples showing a positive response are often subject to further investigations based on chromatographic methods with specific detectors.

5.2.5.6 *Confirmation/Class-Specific Identification*

A general limitation associated with broad-spectrum MIAs is the inconclusive nature of a positive test result. Without the use of a secondary diagnostic assay, it is difficult to attribute a positive response as being elicited by a specific antimicrobial class. More recently, post-screen classification methods selective for the most commonly detected antimicrobial classes (β-lactam, sulfonamide, and tetracyclines) have been described[54,55] in direct combination with the PremiTest assay applicable for a range of tissues, milk, egg, and honey.

According to this approach, following a positive result being obtained during the primary screen, aliquots of the sample (or stored sample extract) are further treated with agents selective for the β-lactam, sulfonamide, and tetracycline classes, and the MIA is repeated. A negative secondary screening result indicates the presence of one (or more) classes of antimicrobial compound in the sample. Positive samples resulting from the secondary screening assays may be rapidly directed to the appropriate quantitative/confirmatory chemical analysis, such as liquid chromatography coupled to a tandem mass spectrometer (LC-MS/MS) (see Chapters 6 and 7).

The mechanisms of action for the three post-screening assays are described below.

1. *Identification of β-Lactam Compounds*. The enzyme β-lactamase is used to identify penicillin and cephalosporin compounds *via* targeted cleavage of the β-lactam ring system. It should be noted, however, that certain β-lactam compounds, including cloxacillin, are resistant to β-lactamase inhibition.[56]
2. *Identification of Sulfonamide Compounds*. Sulfonamide compounds are antagonists of the prokaryotic enzyme tetrahydropteroic synthetase; hence, they inhibit the production of folic acid. By establishing a competition for the occupancy of the enzyme active site in the presence of the natural agonist, *para*-aminobenzoic acid (*p*-ABA), the antagonistic effect of the sulfonamides can be selectively reversed.[57]
3. *Identification of Tetracycline Compounds*. Tetracyclines are a class of naturally occurring and semi-synthetic compounds that inhibit protein synthesis by preventing the attachment of aminoacyl-tRNA to the ribosomal acceptor (A).[58] The ability of the tetracyclines to chelate polyvalent metal cations is an established property. Metal ion chelation has been exploited as a mechanism for the selective identification of tetracycline compounds. Both the antimicrobial and pharmacokinetic characteristics of the tetracyclines are influenced by the chelation status of the molecule.[59]

5.2.6 Conclusions Regarding Microbial Inhibition Assays

In summary, MIAs are the methods of choice when a cost-effective qualitative measure of the biological activity associated with unknown antimicrobial residues is required. When samples are received for analysis at a control laboratory or at a farm/abattoir site, there may be very limited (if any) information concerning what antimicrobial compounds the animal was treated with; therefore, the use of multi-residue broad-spectrum MIAs can rapidly and efficiently identify the presence of many of the commonly used compounds. At the present time, a selection of commercial kits is available in the form of tube diffusion assays, disk diffusion assays, and swab devices.

Validation studies have demonstrated that the majority of commercial MIA kits can detect the important antimicrobial compounds around the current EU MRLs or USFDA tolerances,[60–63] although some have enhanced performance in terms of specific analytes and important operational considerations, such as automated test measurement. Typically, few (<5%) false non-compliant or false compliant results are reported. However, the operator should be aware that the assay might be sensitive to inhibitory compounds other than the target antimicrobial compounds and should take effective steps to eliminate or reduce the presence of naturally occurring inhibitors known to be present in certain sample types (e.g., lysozyme in egg and kidney). Depending on the sensitivity profile of the specific MIA used (Table 5.1), the assay may be over-sensitive to certain antimicrobial compounds below the RLs, resulting in the requirement for a higher rate of confirmatory analysis. Another important consideration is inclusion of the appropriate quality control (QC) check samples within each analytical batch. The majority of commercial MIA kits do not supply QC reagents as part of the kit, leaving the performance of the assay open to interpretation.

While the MIAs are easy to perform and valuable tools for primary screening purposes, none of the current commercial assays offers a perfect match for the needs of a broad antibiotic testing program.

5.3 RAPID TEST KITS

Different types of rapid tests are currently available for antibiotic residue screening, and they can be broadly classified as either immunoassay or enzymatic tests according to the principle of operation. Rapid tests generally provide qualitative or semi-quantitative results and are often portable tests applicable for *in situ* operation. In a production line, the term *rapid* might mean having analytical results in sufficient time to remove non-compliant material prior to processing. In a dairy plant, it might mean having results for a tanker or silo of milk prior to the product manufacture, while only a few minutes' lag time would be admissible for removal of non-compliant carcasses to meet veterinary inspection requirements in an abattoir.[64] The rapid assay should, therefore, provide real-time results, that is, results obtained in a timescale that allows for a response matched to the analytical needs. Ideally, test results should be available within a 30-min period and the requirement for sample preparation should be minimal.

5.3.1 Basic Principles of Immunoassay Format Rapid Tests

The term *immunoassay* describes methods that detect the specific interaction between an antibody and an antigenic analyte.[65] The modern-day immunoassays originate from the work of Yalow and Berson[66] using anti-insulin antibodies to measure hormones circulating in human blood plasma. Immunoassays are divided into two basic categories, direct or indirect, depending on whether the primary antibody–antigen reaction or a secondary reaction is measured, respectively. The advantages generally associated with immunoassays include low limits of detection, specificity, simplicity, reduced analysis time, and the ease of automation. A plethora of formats has been developed, each with particular advantages and limitations. There are many reported applications of immunoassay for veterinary medicine residue analysis ranging from simple techniques, including lateral-flow devices (LFDs) and dipsticks,[67,68] enzyme-linked immunosorbent assay (ELISA),[69] radioimmunoassays (RIA) to sophisticated instrumental formats, such as surface plasmon resonance (SPR)-based optical biosensors.[70] The focus of this section is on the rapid format immunoassays, for example, the commercial LFDs, dipsticks and other tests having test times of around 30 min.

The design of new assays is often limited by the availability of antibodies with the desired characteristics. In conventional immunoassays, antibodies are used, however, other recognition elements, including binding proteins,[71] receptors,[72] molecular imprinted polymers[73] (MIPs), and more recently aptamers[74–76] have also been used for the recognition of antimicrobial compounds. Binding proteins and receptors offer the advantage of wide range specificity and are often capable of detecting a class of compounds. Aptamers are short, single-stranded oligonucleotide sequences derived from either DNA or RNA, which are generated against specific targets. These three-dimensional (3D) shapes possess specific structural, ligand-binding, and catalytic properties.[77] Comparisons can be drawn between aptamers and antibodies and many advantages have been attributed to aptamers, including *in vitro* synthesis (no requirement for the use of animals), short timescale for production, robustness, batch-to-batch consistency, and the ease of modification and manipulation essential for subsequent assay development.

Antibodies are glycoproteins, termed *immunoglobulins* (Ig). They are composed of four polypeptides, two identical copies of a heavy (circa ~55-kDa) and light (circa ~25-kDa) chain held together by disulfide and non-covalent bonds. Depending on the Ig class, up to five structural molecules may be combined to form one antibody.[78] Antibodies are secreted by specialized B-lymphocytes (termed *plasma cells*) of higher vertebrates in response to challenge by an immunogenic antigen. They have the ability to recognise and bind to a defined molecular structure associated with that antigen, termed the *epitope*. Most antigens are complex and possess numerous epitopes that are recognized by different lymphocytes. Each lymphocyte is activated to proliferate, differentiating into plasma cells. The resulting antibody response is polyclonal (heterogeneous). By contrast, monoclonal antibodies (MAbs) are produced by a single B-lymphocyte clone. Thus, they are homogeneous, exhibiting monospecificity and high affinity for the epitope.[71] Monoclonal antibodies can be generated in limitless supply via the fusion of splenic and myeloma cells creating immortal hybridomas, each producing a unique MAb.

A good immunogen should have a molecular mass of at least 3–5 kDa.[79] Antibodies to smaller molecular mass compounds (<1 kDa) can be produced by linking the small target molecule (hapten) to a larger molecule (carrier) to produce an immunogen. High-affinity antibodies have been raised to a variety of low-molecular-mass compounds, including a wide range of veterinary medicines[80] using hapten carrier conjugates.

5.3.2 Lateral-Flow Immunoassays

Lateral-flow immunoassays (LFIAs), also known as *immunochromatographic test strips* or *dipsticks*, are a form of rapid and portable immunoassay in which the test sample extract flows along a solid substrate *via* capillary action. The dipstick system is simple operationally, whereas the technology behind it is advanced and subtle.

Lateral-flow devices are generally composed of multiple component parts, each serving one or more functions. The component parts overlap and are mounted on a solid backing surface using pressure-sensitive adhesive. The assay consists of several zones constructed of segments of different materials. The assay principle is briefly explained below. The test sample is firstly applied to the proximal end of the strip (sample pad). The sample migrates through this region to the conjugate pad, where a particle conjugate has been pre-immobilized. The particle is conjugated to one of the specific biological components of the assay (either the antigen or antibody and other recognition molecules) depending on the assay format. The particle component is required for visualization of the interaction between the immunoreactive components. Colored particles such as blue latex beads or colloidal gold nanoparticles (red in color) are typically used for this purpose. The gold particles are red in color because of the localized surface plasmon resonance effect. Fluorescent or magnetic labeled particles have also been used; however, the use of an electronic reader to assess the test result is required.

The liquid sample re-mobilizes the dried conjugate component, and the analyte (if present) in the sample interacts with the conjugate as both migrate into the next section of the strip, termed the *reaction matrix*. The reaction matrix is a porous nitrocellulose membrane onto which the other specific biological component has been immobilized as bands in specific areas of the membrane. These bands (termed *test* and *control lines*) serve to capture the analyte and/or the conjugate as they migrate past these zones. Excess reagents flow past the capture lines and are entrapped in the absorbent pad. The absorbent pad, also called a *wick* or *wicking pad*, draws fluid from the membrane to maintain the correct directionality of the capillary flow at the appropriate rate. The results are interpreted on the reaction matrix as the presence or absence of lines of captured conjugate and can be read by eye or using a reader. Lateral-flow tests can operate as either direct (sandwich) or competitive (inhibition) format assays. The basic principles are described below. However, many variations are possible, and different test kit manufacturers favor different formats.

5.3.2.1 Sandwich Format

Direct sandwich assays are generally used when testing for higher-molecular-weight analytes, which possess multiple epitope regions, such as macromolecules. The sample first encounters colored particles, conjugated to antibodies (or other recognition element) raised against the target analyte(s). Analyte-specific antibodies are also immobilized at the test line, although these antibodies may bind to a different epitope region on the analyte. A non-compliant (positive) result is indicated by the presence of a colored band at the test line position. The excess of sample mixture continues to flow to the second line of immobilized antibodies present at the control line. The control line typically contains a species-specific anti-immunoglobulin antibody, which captures the antibody in the particular conjugate and serves to indicate the validity of the test (Fig. 5.3).

5.3.2.2 Competitive Format

Competitive formats, direct (Fig. 5.4a and 5.4b) or indirect (Fig. 5.4c and 5.4d) are commonly used when testing for low-molecular-weight compounds, which often possess a single-epitope region. In both competitive formats, the

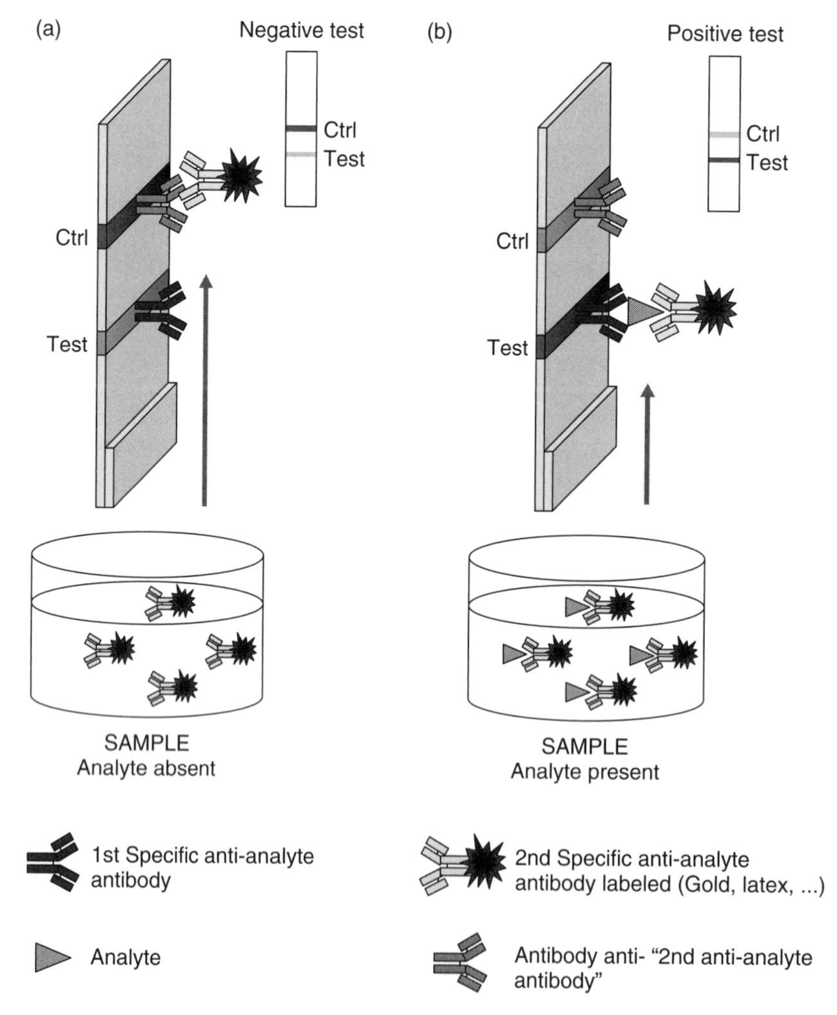

Figure 5.3 Schematic representation of the mode of operation of a direct sandwich format lateral flow immunoassay showing test results for a compliant (negative) sample (a) and a non-compliant (positive) sample (b).

sample is first pre-mixed with analyte-labeled conjugate (direct) or antibody-labeled conjugate (indirect) particles. In competitive formats, the conjugated reagent is seldom incorporated in the LFD because the binding competition reaction may require a specific incubation step, which is often more successful when performed in the liquid phase, where critical parameters such as temperature can be more readily controlled. The test line can contain antibodies or other recognition elements specific to the target analyte(s) (direct) or a protein conjugate linked to the analyte (indirect).

If the sample mixture contains the unlabeled analyte, it will preferentially occupy the target-specific binding sites at the test line in competition with the analyte conjugated with colored particles (direct) or the immobilized carrier protein (indirect). In both formats, a non-compliant (positive) sample is indicated by the absence of the test line or a test line appearing with a weaker color intensity in comparison to the control line. Generally, the control line is composed of a secondary antibody that recognizes the labeled molecules in either orientation. The control line serves as validation of the test and is also used as a threshold indicator for interpretation of the result.

The majority of LFD kits are intended to operate on a qualitative basis. However, it is possible to measure the intensity of the test line to determine the quantity of analyte in the sample. Diagnostic devices known as *dipstick readers* are produced by several test manufacturers to provide a semi-quantitative assay result, such as the Readsensor produced by Unisensor SA (Liege, Belgium). By utilizing unique wavelengths of light for illumination in conjunction with either complementary metal oxide semiconductor (CMOS) or charge-coupled device (CCD) detection technology, a signal-rich image can be produced

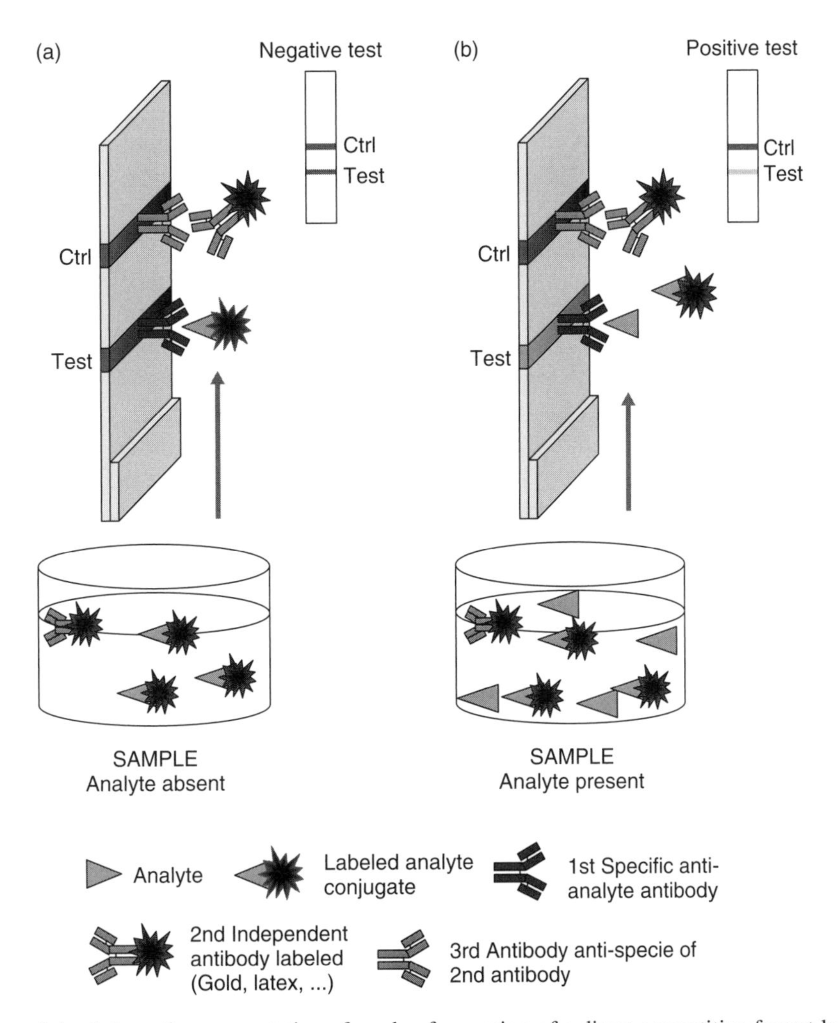

Figure 5.4 Schematic representation of mode of operation of a direct competitive format lateral-flow immunoassay showing test results for a compliant (negative) sample (a) and a non-compliant (positive) sample (b); schematic representation of mode of operation of an indirect competitive format lateral-flow immunoassay showing test results for a compliant (negative) sample (c) and a non-compliant (positive) sample (d).

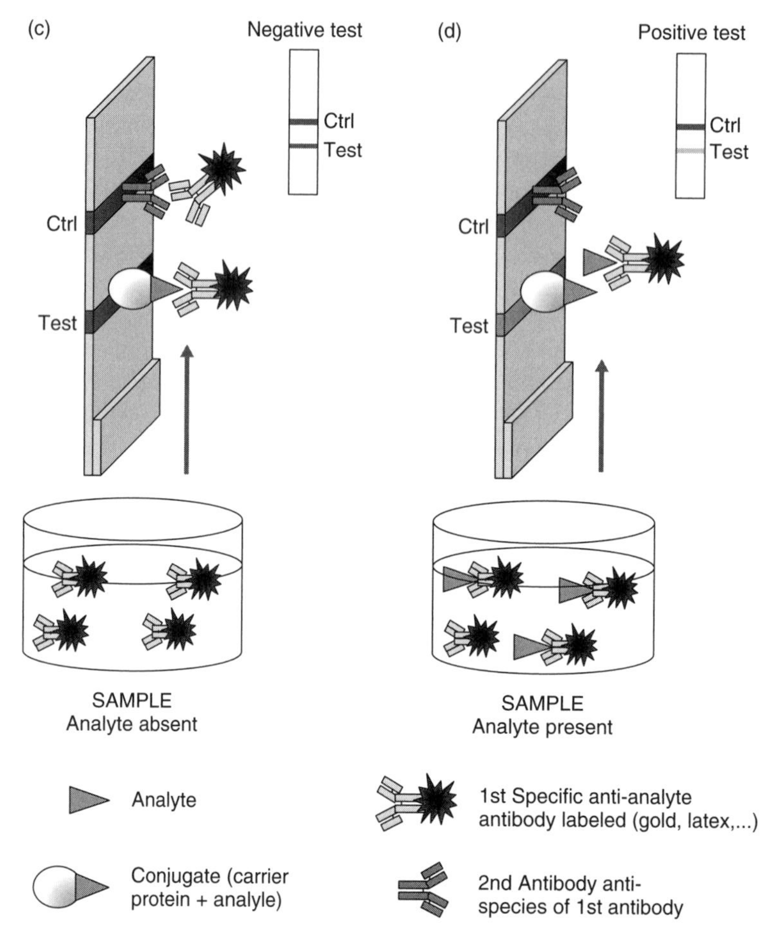

Figure 5.4 (*Continued*)

of the actual test and control lines. Using image-processing algorithms specifically designed for a particular test type and medium, one can then correlate line intensities with analyte concentrations.

The advantages of lateral-flow immunoassays are numerous; it is a well-established and mature technology, and the test devices are easy to manufacture and amenable to high-volume production. The devices are stable and have typical shelf lives of between 12 and 24 months, often under room temperature storage conditions. Small volumes of multiple sample types can be handled. The assays can be highly sensitive for the target analyte and exhibit good specificity. LFIAs have become the format of choice for many diagnostic applications, including medical point-of-care (POC) tests (e.g., pregnancy, infectious disease, and drug use indicators), environmental monitoring, food safety, and chemical biological radioactive and nuclear (CBRN) analysis. The key feature of a LFD is that it is very simple for the end user and can be used *in situ*. Generally, the user has simply to apply the liquid sample, and the test will run within a few minutes, giving a qualitative or semi-quantitative response.

5.3.3 Commercial Lateral-Flow Immunoassays for Milk, Animal Tissues, and Honey

A wide variety of commercial LFIA kits for the detection of antibiotic residues is available and the most well-characterised of these are summarized in Table 5.4, including the rapid one-step assay (ROSA) range from Charm Sciences Inc., and the Tetrasensor, Twinsensor, Trisensor, and Sulfasensor from Unisensor SA and the Betastar from Neogen Corporation. Other LFIA assays have been reported in the scientific literature for the detection of antimicrobial residues, including a lateral-flow device for nicarbazin detection in animal feedstuffs.[81] However, at present these are not commercially available. These LFIA tests incorporate either a receptor protein or an antibody as the specific capture molecule and operate in the competitive assay format (most applicable for small-molecule detection). The sample preparation protocols are based on either direct analysis of the liquid sample (e.g., milk) or a simple extraction step for solid or complex matrices using buffer(s) supplied in the test kit. In general, the time required to perform these tests is less than 30 min with only basic laboratory equipment, if any, required.

TABLE 5.4 Summary of Commercially Available Lateral-Flow Immunoassays for Detection of Antibiotic Residues in Foods of Animal Origin

Testkit Manufacturer and Website	Test Name	Class and Number of Target Analyte(s)	Application(s)	Test Time (mins)
Charm Sciences Inc. www.charm.com	ROSASLBL	β-Lactams[a] (5)	Milk	8
	ROSASL6	β-Lactams[a] (6)	Milk	8
	ROSAMRL	β-Lactams[b] (14)	Milk	8
	ROSAMRL-3	β-Lactams[b] (14)	Milk	3
	ROSASulfa/Tetra	Sulfonamides[a] (2) and tetracyclines (3)	Milk	8
	ROSAMRLBL/Tet	β-Lactams (10) and tetracyclines (3)	Milk	8
	ROSASulfa	Sulfonamides[c] (14)	Milk	8
	ROSASMZ	Sulfamethazine[c]	Milk and urine	8
	ROSAStrep	Streptomycin[b]	Milk	8
	ROSACAP	Chloramphenicol[d]	Milk	8
	ROSAEnroflox	Enrofloxacin[c]	Milk, tissue, egg, urine, and fish	8
Unisensor SA www.tetrasensor.com www.twinsensor.com	Tetrasensor—animal tissue	Tetracyclines[b,e] (5)	Tissues, egg, fish, urine, and feed	<15
	Tetrasensor—milk		Milk	10
	Tetrasensor—honey		Honey	30
	Twinsensor	β-Lactams[b] (9) and tetracyclines[b] (4)	Milk	6
	Twinsensor Express	β-Lactams[b] (9) and tetracyclines[b] (4)	Milk	3
	Trisensor	β-Lactams[b] (9), tetracyclines[b] (4), and sulfonamides[b] (10)	Milk	6
	Sulfasensor	Sulfonamides (10) ≤25 μg/kg	Honey	6
Neogen Corporation www.neogen.com	Betastar	β-Lactams[a] (5)	Milk	5
IDEXX Laboratories www.idexx.com	SNAPBeta-lactam	β-Lactams[a] (5)	Milk	<10
	SNAPMRL Beta-lactam	β-Lactams[b] (5)	Milk	<10
	SNAPTetracycline	Tetracyclines[a,b] (3)	Milk	<10
	SNAPGentamicin	Gentamicin[a,b]	Milk	<10
	SNAPSulfamethazine	Sulfamethazine[a,b]	Milk	<10

[a] Manufacturer quotes limits of detection at or below USFDA-established tolerance/safe levels (μg/l) in raw and commingled bovine milk.
[b] Manufacturer quotes limits of detection at or below EU MRLs (μg/l) in raw and commingled bovine milk.
[c] Manufacturer quotes limits of detection at or below EU MRLs and USFDA established tolerance/safe levels (μg/l) in raw and commingled bovine milk.
[d] Manufacturer quotes limits of detection at or below EU MRPL (μg/l) in raw and commingled bovine milk.
[e] Manufacturer quotes limits of detection at or below EU MRLs (μg/kg) (as applicable).

Advances in the technology allow multiple test lines to be incorporated within a single device and hence, the possibility of multi-residue screening. The new Trisensor test (Unisensor SA) allows the simultaneous screening of the tetracycline, sulfonamide, and β-lactam classes in bovine milk. The device incorporates three test lines and one control line, thus allowing the analyst to determine the class identity of a residue in a non-compliant sample in one single dipstick. The SNAP test (IDEXX Laboratories, Maine, USA) is an enzyme-linked receptor-binding assay presented in a multi-compartment lateral-flow format. An enzyme-labeled conjugate is employed to visualize the binding interaction. The sample is first pre-mixed and incubated with the enzyme-labeled binding reagents. After the sample mixture is applied to the device and the liquid is observed at the activator circle position, the activator button is "snapped" down, serving to release the enzyme substrate and indicator dye into the reaction chamber. Enzymatic cleavage of the substrate results in the production of a blue-colored product. The test result is determined by the comparison of the blue color intensities at the sample and control spot positions on the test device. The higher the concentration of the target antibiotic(s) in the milk sample, the lighter blue the color is in the sample spot compared to the control spot and vice versa. The results can be obtained visually or using a SNAP image reader (SNAPShot DSR Reader).

More recently, a range of LDF-based kits produced by Hangzhou Nankai Biotech Co. Ltd (Zhejiang, China; `http:www//chinatestkit.en.ec21.com/`) for the detection of antibiotic residues in foods has been introduced. The product range includes rapid inspection kits for the following antibiotic substances, β-lactams, sulfonamides, streptomycin, gentamicin, and chloramphenicol.

In general, the detection limits quoted by the test kit manufacturers are either at or below the current EU MRLs/MRPLs and USFDA tolerances, as appropriate for the various matrices. To verify the manufacturers' claims and assess the applicability of the tests using the variety of analytical samples likely to be encountered, validation and comparative studies should be conducted by independent laboratories. Reybroeck et al.[82] conducted a study to investigate the performance of the Tetrasensor honey kit (Unisensor SA, Liege, Belgium) when challenged with 100 different honey samples. Depending on the particular tetracycline compound, a detection range of 6–12 μg/kg was reported. No falsely compliant or non-compliant results were recorded throughout the study, and the group concluded that the test is very simple and rugged, and the detection capability was not influenced with regard to the geographic or botanical origin or by other physical parameters of the honey. Similar evaluations have also been reported[83,84] where the Tetrasensor assay performance in honey, egg, and raw bovine milk was compared to performances with liquid chromatography tandem mass spectrometry (LC-MS/MS) and high-performance liquid chromatography (HPLC) methods. Alfredsson reported the Tetrasensor assay to be rugged and applicable for the detection of tetracyclines at concentrations relevant to the EU RLs in these matrices.

Okerman et al. compared the performance of a Tetrasensor with three microbial inhibition assays for the analysis of tetracycline antibiotics in tissue.[85] The group concluded that when large numbers of samples have to be analyzed without the requirement for immediate results, classical agar diffusion tests with thin plates and performed as prescribed for the FPT still seem the most economical choice. However, the receptor-based test Tetrasensor was recommended for use in official surveys and also in cases when immediate results are required. Unlike the inhibition tests, this receptor test does not require a well-equipped laboratory for use and is more suited for the meat industry.[77]

5.3.4 Receptor-Based Radioimmunoassay: Charm II System

The radioimmunoassay (RIA) technique is long established as a highly sensitive and specific assay method that uses the competition between radiolabeled and unlabeled substances in an antigen–antibody (or other binding molecule) reaction to determine the concentration of the unlabeled substance. RIA can be used to determine antibody concentrations or the concentration of any substance against which specific binding molecule can be produced. Although the technique requires the use of specialized laboratory equipment and certain safety precautions due to the use of radiolabeled tracers, modern assay formats offer the possibility of screening for multiple antibiotic residues in around 30 min and therefore, can be considered as a laboratory-based, multi-residue rapid screening technique.

The Charm II (Charm Sciences Inc., Massachusetts, USA) is a commercial scintillation-based detection system for chemical families of drug residues utilizing class-specific receptors or antibodies in an immunobinding assay format (Fig. 5.5). The Charm II uses ^{3}H and ^{14}C tagged drug tracers with broadly specific binding agents in a receptor assay format. The tracer molecules and any analyte(s) present in the analytical sample compete for the binding sites. Following the binding interaction the reaction is stopped and unbound tracer is separated from the tracer–binder complex via a centrifugation step. Following the centrifugation step, the pellet (containing the tracer–binder complex) is analyzed in a scintillation counter for one minute to give a counting result expressed as counts per minute (cpm).

The assay operates in a competitive inhibition format. The higher the count, the less drug contamination in the

Figure 5.5 The Charm II detection system for receptor-based radioimmunoassays. (Schematic diagram provided courtesy of Charm Sciences Inc.)

sample and conversely, the lower the count, the more drug contamination present in the sample. The result can be simplified to a present/absent result using a control point. The *control points* are numbers determined from the negative reference (two standard deviations less than the average negative count) or the positive spiked sample (two standard deviations greater than the positive count). Samples with counts greater than the negative control point are considered to be compliant (negative) for the target analyte(s), while samples with counts equal to greater than the negative control point are considered presumptive non-compliant (positive) for the target analyte(s). There are many references in the scientific literature concerning the validation of the Charm II system for antibiotic residue analysis in a variety of matrices, including comparisons of performance to chemical analytical methods.[86–88]

The range of kits currently available for the Charm II system and their applications are given in Table 5.5. The test manufacturer claims that the Charm II tests are capable of detecting compounds belonging to the antimicrobial class at or below their defined MRLs (or USFDA tolerances, as indicated) within the relevant matrix, including, milk, urine, serum, animal tissue, honey, and other substances at concentrations of interest to regulatory agencies.

5.3.5 Basic Principles of Enzymatic Tests

Rapid test kits that monitor enzymatic activity for detection of the β-lactam class of antibiotics are also commercially available and now constitute well-established technology. Enzymatic tests are generally considered qualitative technologies that detect the presence of specific chemical residues in analytical samples and are normally based on a color change reaction signaling the test end-point. All enzyme assays measure either the consumption of substrate or the production of a specific product over time. A large number of different methods of measuring the concentrations of substrates and products exist, and many enzymes can be assayed in several different ways. Four main types of experiment are usually used to study enzyme-catalyzed reactions, termed *initial rate*, *progress curve*, *transient kinetics*, and *relaxation experiments*. However, the details of the different permutations will not be discussed in further detail here.

The β-lactam class (penicillins, cephalosporins, monobactams, and carbapenems), exert their activity via inhibition of membrane-bound enzymes, resulting in interference with bacterial cell wall peptidoglycan synthesis.[89] These enzymes are termed *penicillin-binding proteins* (PBPs) because of their ability to covalently bind β-lactam antibiotics.[90]

The PBPs are classified into two major groups based on molecular mass: high-molecular-mass PBPs (~50–100 kDa) and the low-molecular-mass PBPs (~30–40 kDa). High-molecular-mass PBPs can be sub-divided into two classes: class A forms, which are bifunctional enzymes that catalyze both transpeptidation and transglycosylation during cell wall synthesis; and class B forms, which exert only transpeptidation activity. Low-molecular-mass PBPs are D-alanyl D-alanine carboxypeptidases that control peptidoglycan cross-linking.[91] Following attachment to the PBPs, β-lactams inhibit the transpeptidation enzyme responsible for cross-linking peptide chains attached to the backbone of peptidoglycan. Penicillins are structural analogs of D-Ala-D-Ala, the transpeptidase substrate. The final bactericidal event is the inactivation of an inhibitor of the autolytic enzymes in the cell wall, leading to cell lysis.

5.3.5.1 The Penzyme Milk Test

The enzyme exploited in the Penzyme milk test (Neogen Corporation, Michigan, USA) for β-lactam residues is D,D-carboxypeptidase. The assay is based on the following two properties of D, D-carboxypeptidase:

1. It is specifically and quantitatively inhibited by β-lactam antibiotics; thus, the more the sample is contaminated with these antibiotics, the further the residual enzyme activity will be reduced.
2. It specifically hydrolyzes substrates of the *R*-D-Ala-D-Ala type with the liberation of D-alanine.

In order to measure the activity of the enzyme, the liberated D-alanine is transformed by a stereospecific oxidase into pyruvic acid with the liberation of hydrogen peroxide. The peroxide oxidizes (under the action of peroxidase) an organic dye with a resulting change in color. The assay protocol involves the direct addition of the enzyme to the milk sample, a mixing step followed by a short incubation step at 47°C, during which time any β-lactams present in the sample will bind to the enzyme and specifically inhibit its activity. A tablet containing the color reagents is then added to the milk, and further incubation is performed. During this step an orange color develops that is proportional in intensity to the amount of active D, D-carboxypeptidase remaining. The results are interpreted by comparison of the color observed of the assay tube with the color chart provided in the test kit. The total test time is quoted as 15 min.

In general, the Penzyme test is applicable for screening milk samples for the presence of β-lactam compounds with detection limits close to the USFDA tolerances (Table 5.6) and is, therefore, applicable for the regulatory monitoring in accordance with the USA system. However, in all cases the detection limits quoted are greater than the current EU MRLs, and for that reason the test is not considered as a useful monitoring tool for the European market.

TABLE 5.5 Range of Charm II Test Kits and Their Applications Currently Available for Antibiotic Residue Screening from Charm Sciences Inc.

Charm II Tests for Antibiotics	Test Kits Available (number of individual compounds for which detection concentration are quoted)	Application(s)	Sample Preparation Time (minutes)	Assay Time (minutes)
Dairy products[a]	β-Lactams (12)	Raw, commingled, and pasteurized milk	None	10
	Sulfonamides (4)			
	Tetracyclines (3)			
	Aminoglycosides (4)			
	Macrolides (5)			
	Lincosamides (1)			
	Novobiocin			
	Spectinomycin			
	Chloramphenicol			
Seafood	β-Lactams (8)	Aquaculture products (shrimp and fish)	75	12–22[b]
	Sulfonamides (6)			
	Tetracyclines (3)			
	Aminoglycosides (4)			
	Macrolides (6)			
	Amphenicols (2)			
	Chloramphenicol			
	Nitrofurans[c] (AMOZ/AOZ)			~120
Grain	β-Lactams (19)	Grain-based animal feedstuffs	10–15	10
	Sulfonamides (15)			
	Tetracyclines (3)			
	Aminoglycosides (4)			
	Macrolides (6)			
	Chloramphenicol			
Honey	β-Lactams (19)	Raw and processed honey	<5	10 or 40 for the nitrofuran tests
	Sulfonamides (12)			
	Tetracyclines (3)			
	Aminoglycosides (2)			
	Macrolides (6)			
	Amphenicols (4)			
	Chloramphenicol			
	Nitrofurans (AMOZ/AOZ)			
Tissue	β-Lactams (8)	Edible tissue including muscle, liver and kidney	12	30
	Sulfonamides (5)			
	Tetracyclines (3)			
	Aminoglycosides (4)			
	Macrolides (6)			
	Amphenicols (2)			
	Chloramphenicol			

[a]Two dairy product versions are available: US-level and MRL-level.
[b]ELISA-based test.
[c]The exact assay time depends on the particular antibiotic test
Notation: Numbers in parentheses in column 2 represent the number of individual compounds for which detection concentration values are quoted.
Source: Information obtained from the manufacturer's website (www.charm.com/content/view/61/104/lang,en/;accessed during April 2010).

5.3.5.2 *The Delvo-X-PRESS*

The Delvo-X-PRESS (DSM Food Specialties, Delft, The Netherlands) is an enzyme-based rapid test, which operates in a similar mode to that of the Penzyme. The Delvo-X-PRESS is specially designed to control milk from bulk tank and silos within minutes for the presence of β-lactam residues. The Delvo-X-PRESS is a qualitative, competitive, receptor–enzyme assay. A specific receptor isolated from *Bacillus stearothermophilus* is employed that recognizes and binds a wide spectrum of β-lactam antibiotics with detection capabilities around the appropriate RLs (Table 5.6). The enzyme, horseradish peroxidase is employed to mediate a reaction signaled via the formation of a blue coloration specifically indicating the absence or

TABLE 5.6 Manufacturer's Reported Detection Limits for the Penzyme Milk Test and the Delvo-X-PRESS for Various β-Lactam Compounds in Spiked Bovine Milk Compared to USFDA Tolerances and EU MRLs

	Manufacturer's Reported Detection Limit in Bovine Milk (μg/l)			
β-lactam Compound	Penzyme Neogen Corporation	Delvo-X-PRESS DSM Ltd	USFDA Tolerances in Milk (μg/l)	EU MRLs[a] in Milk (μg/l)
Benzylpencillin	4.3	2–4	5	4
Ampicillin	5.6	4–8	10	4
Amoxicillin	5.3	4–8	10	4
Cloxacillin	N/A	30–60	10	30
Ceftiofur	N/A	4–8	100	100[b]
Cefapirin	14.3	4–8	20	60[c]
Procaine–penicillin	N/A	3–5	—	4
Hetacillin	N/A	6–10	—	—
Penicillin-V	N/A	3–5	—	4
Piperacillin	N/A	5–10	—	—
Cephalonium	N/A	3–4	—	20
Meticillin	N/A	10–20	—	—
Ticarcillin	N/A	30–100	—	—
Cefadroxil	N/A	5–25	—	—
Cefotaxime	N/A	4–5	—	—
Cefaperazone	N/A	5–20	—	50
Cephalexin	N/A	25–50	—	100
Oxacillin	N/A	25–50	—	30
Dicloxacillin	N/A	25–50	—	30
Cephradine	N/A	25–50	—	—
Cefuroxime	N/A	4–20	—	—
Cephoxazole	N/A	75–100	—	—

[a]Commission Regulation (EU) 37/2010.[10]
[b]Sum of the parent drug and the metabolites.
[c]Sum of cephapirin+descetylcephapirin.
Notation: N/A = not available; N/A& (does not meet FDA safety level).
Source: Information (in columns 2 and 3 above) obtained from the manufacturer's website (www.neogen.com/FoodSafety/pdf/ProdInfo/Page_94.pdf; accessed 8/04/10); product brochure available at www.dsm.com/en_US/html/dfsd/tests.htm.

presence of β-lactam residues. The test result can be determined visually or using an automated reader system. The total test time is quoted as under 10 min.

An important consideration is the applicability of the rapid test to the range of sample matrices, and its susceptibility to naturally occurring interference. In a study conducted by Andrew et al.,[92] the performance of rapid antibiotic residue screening tests including the Penzyme test were evaluated in presence of different milk compositions and qualities. Metabolic changes that occur because of disease affect the composition and quality of the milk produced. Mastitis is the major disease for which antibiotic treatment is used and after which antibiotic residue screening assays are employed.

During intramammary infection, the somatic cell count (SCC) in mastitic milk increases, and the concentrations of plasma components, including bovine serum albumin (BSA) and immunoglobin, are higher in comparison to normal milk. Several components in mastitic milk are known to interfere with various antibiotic residue screening tests. These components include somatic cells, lactoferrin, lysozyme, microbes, and free fatty acids. Depending on the analytical principle of the screening test, these milk components may have a major impact on the outcome of the test, especially if these screening tests are used when the animal is recovering from a health-related condition and the milk may still contain these natural components in relatively high concentrations.[93] Consequently, the evaluation of the screening assays should account for variable SCC and bacteria counts in milk from individual cows, which is representative of the variation in a typical dairy herd. Andrew et al.[92] reported that a failure of the Penzyme test resulted in a positive and they concluded that there may be factors in milk that inhibit the enzymatic reaction of the test that could give rise to higher rates of false non-compliant results.

Gibbons-Burgener et al.[94] evaluated the performance of the rapid SNAPBL, DelvoTest SP, and Penzyme milk test-kits for testing raw milk samples from individual cows.

All three tests are indicated by the manufacturers for use in commingled milk. The sample population consisted of milk from 111 cows diagnosed with mild clinical mastitis. Approximately half of these cows were treated with USFDA-approved intramammary antimicrobial therapy (pirlimycin, hetacillin, or cephapirin), and the other half received no treatment (control group). Post-treatment milk samples were collected at the first milking after the recommended withholding period and then analyzed in duplicate on each rapid screening test and once by HPLC. The sensitivity, specificity, and false-compliant/non-compliant predictive rates were determined for each assay. The findings showed that the sensitivities for the DelvoTest SP and the SNAPBL were >90%, whereas the sensitivity of the Penzyme milk test was 60%. The Penzyme milk test was also found to have a higher rate of false-compliant results. The specificities were comparable between the three assays.

The *positive predictive value* (PPV) is a measure of the likelihood that a positive screening assay truly identifies a sample with an antimicrobial residue concentration equal to or greater than the appropriate RL. The PPVs were calculated for the three assays and were considered to be poor, ranging from 39% to 74%. The group concluded that because of the low PPVs, these three assays may not be useful for detecting antimicrobial residues in individual milk samples from cows treated for mild clinical mastitis. However, the repeatability of each assay was considered to be excellent.[94]

5.3.6 Conclusions Regarding Rapid Test Kits

In summary, rapid tests, either immunoassay or enzymatic formats, are the methods of choice when qualitative or semi-quantitative results are required within a short timescale, specifically, around 30 min for targeted residue screening. In general, these assay formats are portable (suitable for *in situ* testing) and simple to both operate and interpret. A wide variety of test formats are commercially available, many of which are applicable for the detection of classes of antibiotics, such as β-lactams and tetracyclines with detection capabilities at or below the appropriate RLs. As with other test kits, it is important to determine the applicability of the assay for the specific matrix type prior to use as certain matrices are known to contain interference that causes elevated false non-compliant/compliant rates.

5.4 SURFACE PLASMON RESONANCE (SPR) BIOSENSOR TECHNOLOGY

5.4.1 Basic Principles of SPR Biosensor

The term *biosensor* describes a device that responds to analyte(s), and can interpret concentration as an electrical signal via a combination of a biological recognition element (BRE) and an electrochemical transducer.[95] The earliest biosensors were catalytic systems that integrated bioreceptors, that is, enzymes, cellular organelles, or microorganisms with transducers to convert the biological response into digital electronic signals.[96] When biological interactions take place, changes in other physiochemical parameters, including enthalpy, ionic conductance, and mass, also occur. Such effects can be exploited by coupling the biocatalytic reaction with a transducer.[97] Optical, electrochemical, thermometric, piezoelectric, and magnetic transducers are all commonly used transduction mechanisms.

The focus of this section is on the application of optical, surface plasmon resonance (SPR) biosensors for the detection of antibiotic residues in foods of animal origin. SPR offers the advantage of detecting the specific binding event between a target and a recognition element without the use of the enzyme labels or fluorescent tags required by the majority of immunochemical techniques. SPR systems have been combined with miniaturized flow systems to permit continuous, real-time monitoring of the binding complex.[98] SPR can provide information about the concentration, binding specificity, binding affinity, kinetics, and cooperativity of a target molecule.

Optical excitation of surface plasmons by the method of attenuated total reflection (ATR) was demonstrated in 1968 by Kretschmann and Raether[99] and Otto.[100] SPR is a charge density oscillation that exists at the interface of two media with dielectric constants of opposite signs. The charge density wave is associated with an electromagnetic wave, the field vectors of which reach their maxima at the interface and decay evanescently into both media. The surface plasmon wave (SPW) is a transverse magnetic (TM) polarized wave; the magnetic vector is perpendicular to the direction of propagation of the SPW and parallel to the plane of interface. At optical wavelengths this condition is fulfilled by several metals, of which gold and silver are commonly used.[101] The energy carried by photons can be coupled, or transferred to electrons in the metal.

Coupling results in the creation of a group of excited electrons (plasmons) at the metal surface. The intensity of the plasmons is influenced by the type of metal and the environment at the metal surface. Variations in the optical parameters of the transduction medium in the plasmon field range (e.g., antibody–antigen binding event) are detected by monitoring the interaction between the SPW and the optical wave. The propagation length of the SPW is limited. Thus, sensing is performed directly in the area where the SPW is excited by an optical wave. The optical system used to excite the SPW is simultaneously used for the interrogation of SPR. Surface plasmon resonance manifests itself by the resonant absorption of the energy of the optical wave.[102] The propagation constant of the SPW is always higher than that of the optical wave. The SPW cannot be excited directly by an incident optical wave at a planar metal–dielectric interface. The momentum of the incident optical wave is enhanced to match that of the SPW. The momentum change is achieved using ATR in prism couplers, optical waveguides, and diffraction at the surface of diffraction gratings.[103] Prism-based SPR sensors using angular interrogation form the basis of the Biacore technology.[97]

To monitor the binding event using SPR, either the biological recognition element or target ligand is immobilized (chemically tethered) on the surface of the sensor chip. The binding event between the BRE and target(s) occurs on the sensor chip surface. The basic requirement is that the binding event generate a measurable mass change. Mass changes influence the refractive index of the solution at the sensing surface, which, in turn, alters the angle at which reduced-intensity polarized light is reflected from a supporting glass plane. The change in resonance angle caused by the binding or dissociation of molecules from the sensing surface is proportional to the mass of bound material. Inhibition (indirect) assays, based on a binding competition for the recognition element between immobilized and free analyte are applicable for lower molecular mass target compounds (<1 kDa), which includes the majority of antibiotic residues. To generate a strong response, the smaller molecular mass partner is immobilized to the sensor surface. In this orientation the binding event results in a larger mass change. The inhibition format gives results that are inversely proportional to the concentration of analyte present in the extract. In the direct assay format, the recognition element is immobilized on the sensor surface. The direct format is most applicable for macromolecular targets (>5 kDa). The measured response derives directly from the binding of the target analyte to the detecting molecule at the sensing surface.

Instrumentation based on SPR detection consists of the following components (Fig. 5.6): (1) a sensing surface containing a coupling matrix bearing one of the interacting pair; (2) SPR optics based on convergent incident light and position-sensitive detection of the reflected light; (3) an on-line computer for determination of the location of the resonance angle and for data handling; and (4) a microfluidic system for injection of the sample, rinsing, and regeneration solutions.

5.4.2 Commercially Available SPR Biosensor Applications for Milk, Animal Tissues, Feed, and Honey

At present a limited number of commercial SPR-based biosensor systems are available. Further information and the technical specifications relating to the different biosensor instruments can be obtained from the following manufacturers' websites: GE Healthcare (Biacore),[104] Sensate Technologies Inc. (Spreeta),[105] and Reichert (SPR) Inc.[106]

Of the commercial SPR biosensors listed above, the Biacore brand (GE Healthcare) is the most well established technology and has the largest market share in terms of food analysis. A range of test kits (QFlex kits) are produced specifically for use with the Biacore Q SPR

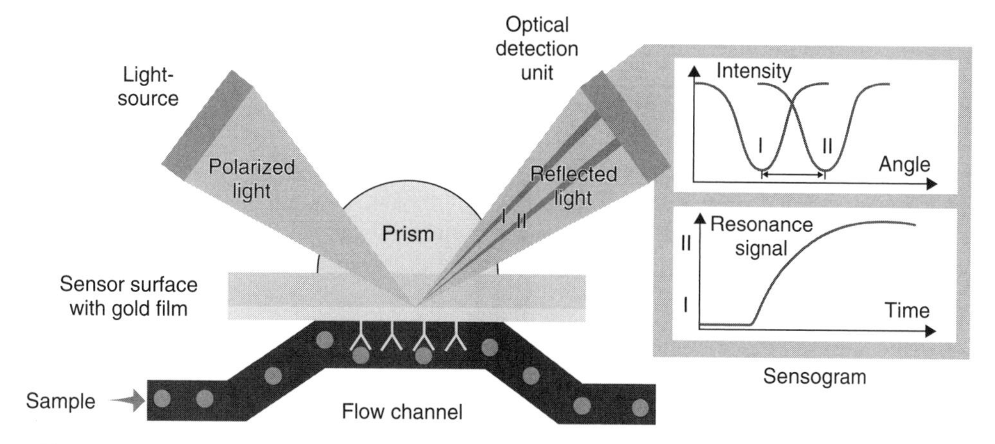

Figure 5.6 A prism-based surface plasmon resonance (SPR) biosensor. The microfluidic flow channel interfaces with the sensor chip (glass support and thin gold layer), the optical unit contains a light source, prism, and detection unit. The device detects changes in the refractive index in the immediate vicinity of the chip surface layer. SPR is observed as a sharp shadow in the reflected light from the surface at an angle dependent on the mass of the material at the surface. The SPR angle shifts from I to II (lower right) when biomolecules bind to the surface and change the mass of the surface layer. The change in resonant angle is monitored non-invasively in real time as a plot of resonance signal (proportional to mass change) versus time. (Schematic diagram provided courtesy of GE Healthcare Bio-Sciences AB.)

biosensor instrument for the detection of veterinary drug residues (and other contaminants and residues) in foods of animal origin. The Biacore concentration test kits for low molecular weight analytes are designed as indirect (inhibition) assays, as discussed previously. A known concentration of a relevant binding protein is mixed with the sample and injected over a sensor surface on which a corresponding target analyte or derivative is immobilized. Any target molecules present in the sample bind to the binding protein and thus inhibit it from binding to the sensor surface. The higher the concentration of the target molecule in the sample, the higher the degree of inhibition and hence the lower the SPR response. Concentrations are calculated by interpolation of the binding responses on a calibration curve.

Biacore systems monitor protein interactions in real time using a label-free detection method. Sample in solution is injected over a sensor surface on which potential interacting partners are immobilized. As the injected sample interacts with the immobilized partners, the refractive index at the interface between the sensor surface and the solution alters to a degree proportional to the change in mass at the surface. The SPR phenomenon is exploited to detect these changes in real time, and data are presented in a sensorgram (SPR response plotted against time). The key events in the sensorgram are (1) a flow of buffer is injected over the sensor surface to establish the baseline response, (2) the sample extract is injected, and (3) the response signal generated relates to the magnitude of the complex formation (association phase). Following the injection, the bound analyte gradually dissociates in the buffer flow (dissociation phase). A regeneration solution can be used to facilitate the dissociation of the remaining analyte, thus returning the signal to the baseline response.

The range of test kit applications, detection capabilities, and cross-reactivity profiles reported by the Biacore are shown in Table 5.7. The detection limits reported show the combination of the Biacore Q instrument and the QFlex kits to be very sensitive and capable of detecting the target analytes at concentrations below the current EU RLs or recommended concentrations (RCs) proposed by the EU CRLs.[15]

The capability of SPR biosensor technology to perform rapid and reliable analysis for food samples has been clearly demonstrated using Biacore technology. A long list of applications also appears in the literature demonstrating its suitability in food analysis, including its ease of use, sensitivity, selectivity, flexibility, and reliability. Assays, among others, are reported for the detection of chloramphenicol[107,108] and other fenicols,[109,110] fluoroquinolones,[111] β-lactams,[112] streptomycin,[113,114] gentamicin,[115] and sulfonamides.[116,117] More recent developments have focused on multi-residue analysis, array platforms, miniaturization, and portable SPR-based detection systems.[118,119]

The use of SPR biosensor technology for residue monitoring as part of the analytical screening strategy can offer a number of advantages over conventional immunoassay formats. For example, semi-automated high-throughput screening, shorter analysis times, lower detection capabilities, less susceptibility to non-specific binding effects, and lower false compliant/non-compliant rates have been reported. SPR biosensors can generate semi-quantitative results and are particularly well suited to the demands of laboratory-based screening. Many samples can be rapidly analyzed for known (targeted) residues with only those samples indicated as presumptive non-compliant requiring further confirmatory analysis. However, these advantages should be offset against the high cost of the biosensor instrumentation, which can be comparable to the cost of certain mass spectrometry instruments and the ongoing requirement for relatively large quantities of immunoreagents per assay and/or the availability of kits.

Sample preparation is a major challenge for any rapid screening assay procedure. It has been estimated that sample preparation can claim about 50–75% of the total analysis time, especially for tissues and other solid matrices, which require the liberation of plasma protein-bound residues.[120] Despite this limitation, most of the technical innovations and improvements continue to be made in the detection procedures.

For certain matrices most notably, honey and animal feed stuffs, extensive sample preparation and clean-up (incorporating liquid–liquid extraction and/or solid-phase extraction) prior to the biosensor analysis may be necessary to achieve the desired detection concentrations. Honey is a complex matrix containing high concentrations of natural sugars and other carbohydrates. The typical composition is fructose (38.5%) and glucose (31.0%). The remaining 30% or so includes maltose, melezitose, sucrose, and other complex carbohydrates, vitamins, minerals, and certain enzymes (e.g., catalase). Depending on the type of honey, natural components such as waxes, royal jelly, propolis, and bee carcasses may also be present. Some of these components, especially the waxes, can foul the sensor interface if not adequately removed during the extraction step, resulting in spurious response signals.

5.4.3 Conclusions Regarding Surface Plasmon Resonance (SPR) Technology

In summary, the SPR biosensor is a sophisticated screening tool applicable for the analysis of residues (at μg/kg concentrations) in complex food matrices. The results can be generated within the work day, providing information on the presence and presumptive concentration of targeted residues (and their metabolites for some applications). The

TABLE 5.7 Current Range of QFlex Kits for Use with Biacore Q Surface Plasmon Resonance Biosensor Instrument for Antibiotic Detection in Foods of Animal Origin (at time of publication)

Antibiotic	Currently Validated Application(s) and Detection Limits (μg/kg or μg/l)	Cross-Reactivity (%) Profile
Chloramphenicol	Poultry muscle (0.02) Bovine milk (0.03) Shellfish (0.07) Honey (0.07) Values quoted for chloramphenicol	Chloramphenicol—100 Chloramphenicol glucuronide—73.8
Streptomycin	Porcine muscle (69) Porcine kidney (50) Bovine milk (28) Honey (15) Values quoted for streptomycin	Streptomycin—100 Dihydrostreptomycin—97.3
Sulfadiazine	Porcine muscle (6) Value quoted for sulfadiazine	Sulfadiazine—100 *N*-(4)-Acetylsulfadiazine—230 Sulfamerazine—11 Sulfathiazole—9 Sulfamethoxypyridazine—8
Sulfamethazine	Porcine muscle (8) Value quoted for sulfamethazine	Sulfamethazine—100 *N*-(4)-Acetylsulfamethazine—160
Multiple sulfonamides	Porcine muscle (18.2) Value quoted for sulfamethazine	Sulfamethazine—100 Sulfadiazine—123.8 Sulfathiazole—191.6 Sulfaquinoxaline—127.3 Sulfamethoxazole—106.4 Sulfachloropyridazine—185.7 Sulfamerazine—123.0 Sulfadimethoxine—185.7 Sulfaguanidine—143.3 Sulfamethoxypyridazine—171.3 Sufamonomethoxine—178.4 Sulfamethizole—209.2 Sulfapyridine—204.5 Sulfaphenazole—193.6 Sulfapyrazine—125.5 Sulfadoxine—87.5 Sulfisoxazole—63.9 Sulfatroxazole—30.2 Sulfanitran—17.9 Dapsone—214.1
Tylosin	Poultry muscle (5.6) Honey rapid assay (5.7) Honey (2.8) Feedstuffs (195) Value quoted for tylosin A	Tylosin A—100 Tylosin B—141.9 Tylosin C—71.3 Tylosin D—323

Source: Information (in columns 2 and 3 above) collated from `http://www.biacore.com/food/food_analysis/index.html`; accessed during April 2010.

Biacore Q is a fully automated system allowing off-hours operation. The data analysis is also automated and does not require specialist interpretation. At present, the technology is expensive and not readily amenable to field test operation. Cost and portability may be important considerations when deciding whether to implement screening assays. Nevertheless, the use of SPR biosensors for screening prior to confirmatory analysis allows the rapid identification of the non-compliant samples. This approach can offer both time- and cost-saving advantages by limiting the number of samples requiring confirmatory (e.g., LC-MS/MS) analysis, especially when employed in large-scale surveillance monitoring schemes. More recent advancements have facilitated the coupling of biosensors to LC-MS^n instrumentation to allow screened non-compliant sample extract to be recovered and diverted for in-line confirmatory analysis.

5.5 ENZYME-LINKED IMMUNOSORBENT ASSAY (ELISA)

5.5.1 Basic Principles of ELISA

To date, the commonest type of immunoassay is the enzyme-linked immunosorbent assay (ELISA). The ELISA technique was developed through the pioneering work of Engvall and Perlmann[121] in the 1970s. By immobilizing the reagents to a surface, the facile separation of bound and unbound material is achieved, making ELISA a powerful tool for the measurement of analytes in crude sample preparations.[121] ELISA has become the basic immunoassay on which many of the modern assay formats are based.

In its simplest form, ELISA is the direct competitive assay in which the antigen is bound to the solid surface. The measurement step is achieved via the use of an enzyme label. The enzyme label can be attached to either the antigen or the antibody in such a way that the binding reaction is not impaired. Horseradish peroxidase (HRP), alkaline phosphatase (AP), and β-D-galactosidase are commonly used for this purpose.[122] When the appropriate enzymic substrate is supplied, the labeled component can be detected and even quantified. The conversion of the substrate to product either generates a signal or releases an ion that reacts with a secondary compound, resulting in a change that can be measured spectrophotometrically or electrochemically. ELISAs offer numerous advantages over other immunoassay techniques because the end signal is amplified by the formation of a large number of product molecules. ELISAs are typically performed in 96-well (or 384-well) polystyrene plates, and are applicable for high-throughput screening (HTS) analysis with the possibility of automation.

The sandwich assay (competitive or non-competitive) is the most commonly used ELISA format in commercial test kits, although a variety of other configurations exist. In the sandwich configuration the target analyte is bound between two antibodies, termed the *capture and detection antibodies*. An important consideration in designing a sandwich ELISA is that the capture and detection antibodies recognize two non-overlapping structural regions on the target molecule. When the antigen binds to the capture antibody, the region recognized by the detection antibody must not be obscured or altered in any way. Monoclonal antibodies have an inherent monospecificity toward a single region providing exquisite detection limits and reliable quantification (Section 5.3.1). Thus, they are often used as the detection antibody. A complementary polyclonal antibody may be used as the capture antibody to recover as much of the antigen as possible.

Calibration standards can be included in the ELISA experiment to produce a sigmoidal standard curve, and can provide the means to quantify the extent of the binding event and hence, the amount of target analyte present in an unknown sample. ELISAs can be employed as quantitative screening assays, depending on the particular performance characteristics of the assay and assuming the use of adequate control samples.

Traditional ELISA typically involves the use chromogenic reporters and substrates, which generate an observable color change, and the majority of ELISA applications reported in the literature or commercially available for antibiotic residue analysis are based on such reporters.[123,124] The most recent advances in ELISA technology utilize fluorgenic, electrochemiluminescent, and real-time polymerase chain reaction (rtPCR) reporters to create quantifiable signals. These non-enzyme reporters can have various advantages, including higher sensitivities (via cascade amplified systems) and greater multiplexing possibilities, such as microarray configurations. The successful application of such techniques for the detection of antibiotics in foods has been reported.[125,126] While many ELISA methods have been reported using both polyclonal and monoclonal antibodies (Section 5.3.1) as capture molecule, there is an increase in the development of methods based on monoclonals, due to the immortal nature of the hybridoma providing a constant source of pure antibodies.

5.5.2 Automated ELISA Systems

The ELISA experiment can be broken down into a series of four basic steps: dispensing, incubation, washing and reading. The number of individual steps, the order in which they are performed, and the number of repetitions will vary depending on the specific protocol. As such, ELISA is readily amenable to automation. Commercial microplate-based devices have been developed to accomplish the following basic steps: (1) automatic plate washers, (2) liquid handlers or microplate dispensers, (3) plate stacks (ambient) or automated incubators, and (4) microplate readers.

Robotic components, such as robotic arms, have been developed that specifically facilitate the flow of process between the different component devices. Scheduling software allows the analyst to program for the specific requirements of the method to make the system compatible with virtually any ELISA. Using robotic workstations, it is possible to screen hundreds of sample extracts during off-hours operation. High-throughput screening of this nature is extensively used in clinical diagnostics where large numbers of blood, serum, and urine samples require analysis for the presence of marker compounds, such as metabolites or hormones. Food samples represent much more complicated systems because of the various consistencies and compositions and requirements for extract processing, as discussed previously. However, the use of ELISA in the automated HTS mode is particularly applicable for use in "outbreak" situations where a laboratory is required to screen many samples of a similar type for the presence of suspected residue. Besides such incidents, abbatoirs, milk processors, and residue analysis laboratories also have a need for high-volume screening analysis.

5.5.3 Alternative Immunoassay Formats

In addition to ELISA, numerous alternative format immunoassays for the determination of antibiotic residues in a range of foods are also available, including time-resolved fluorescence immunoassay (TRFIA) and flow cytometry technologies. These immunoassay techniques are applicable for high-throughput multi-residue analysis providing semi-quantitative screening results.[127] Screening assays for the detection of antibiotics have been reported in the scientific literature on the basis of these technologies; however, at the present time there are no commercial kits available, and thus a laboratory wishing to implement this technology would have to develop assays using in-house or commercially available immunoreagents.

Luminex Corporation (Austin, TX, USA) produces commercial systems applicable for the development of multiplex immunoassays. The Luminex technology is based on the principles of flow cytometry and the use of color-coded polystyrene microspheres (termed xMAP technology) At present, 100 distinct microsphere sets are available. Each microsphere set can be coated with a specific immunoreagent, allowing the capture and detection of target analytes in the sample. Within the Luminex analyzer dual lasers are used to excite the microsphere internal dyes (to identify each individual microsphere) and then the reporter dye captured during the immunoassay reaction. The technology allows multiplexing of up to 100 unique assays within a single sample. The development of a multiplex method based on the Luminex xMAP technology has been reported for the analysis of the sulfonamide class in milk.[128] The method was validated for the simultaneous detection of 11 sulfonamides at 100 μg/kg in milk with the potential for further multiplexing to allow the detection of analytes.

5.5.4 Commercially Available ELISA Kits for Antibiotic Residues

Various ELISA kits are commercially available for the detection of antibiotics in foods of animal origin. There at least four kits on the market for the detection of chloramphenicol (which is banned for use in food-producing species in the EU and many other countries) in tissue, honey, milk, and aquaculture products and five kits claiming multi-residue detection for the fluoroquinolone class. A survey of the range of kits currently available for different antibiotic residues from the major manufacturers is presented in Table 5.8. In general, all the commercial kits are broadly capable of meeting the EU RLs or CRL recommended concentrations (RCs)[15] within the wide variety of matrices routinely analyzed, including muscle, kidney, liver, milk, eggs, honey, feeds, fish, and aquaculture products. The main condition for the wide use of immunoassay in residue analysis is the availability of commercial kits. It should be noted that the range of kits currently available does not provide full coverage of all the antibiotic residues of interest. While the kits generally demonstrate compliance with the appropriate RLs, the performance in terms of detection capability (CCβ), cross-reactivity profile, susceptibility to matrix interference, and false compliant/non-compliant rates varies and should be taken into consideration. As with the other immunoassay formats, the success of a particular ELISA kit is dependent largely on the quality of the antibody in relation to its ability to specifically recognize the target analyte(s).

The commercially available ELISA kits are applicable for the analysis of a wide variety of matrices. Most kit manufacturers supply basic protocols for the preparation of sample extracts compatible with the ELISA; for instance, the final extract is largely free of organic solvent and closely matched to the optimum operating conditions for the immunoreagents. Some laboratories prefer to use in-house extraction procedures, which may be very well characterized in terms of analyte recovery profile and designed to minimize the presence of interfering components originating from the matrix. Some kits, however, are specifically designed for the analysis of one matrix type where that analyte poses a particular problem, such as the Eurodiagnostica multi-Antimicrobial Growth Promoter (AGP) kit for animal feedingstuffs from ELISA Technologies (Florida, USA).

A common problem associated with ELISA is non-specific binding (NSB) interactions. There are two main kinds of NSB event: (1) binding events to exposed surfaces, such as microplate wells and (2) binding to interfering substances naturally present in certain matrices, such as albumins. These NSB events can result in high background readings and higher rates of falsely non-compliant and compliant results being observed. However, NSB can be controlled by using wash solutions incorporating surfactants and choatrophic agents to disrupt these weak binding interactions.

Another important consideration when using the ELISA (and other immunoassay formats) is the cross-reactivity of the method. *Cross-reactivity* (CR) can be described as the interaction between an antigen and an antibody, which was generated against a different but similar antigen. The cross-reactivity profile should be characterized during the validation of a new immunoassay. The cross-reactivity,

TABLE 5.8 Survey of Commercially Produced ELISA Kits for Detection of Antibiotic Residues in Various Foods of Animal Origin Available from Major European and US Manufacturers

Antibiotic	Manufacturer				
	Randox	Eurodiagnostica	Abraxis	R-Biopharm AG	Biooscientific
Bacitracin	N	Y[a]	—	—	—
β-Lactams (multi)	Y (11)	—	—	—	Y
Combined chloramphenicol and chloramphenicol glucuronide	Y	Y	Y	—	Y
Chloramphenicol	—	—	—	Y	—
Chloramphenicol glucuronide	—	—	—	Y	—
Crystal violet	—	Y	Y	—	Y
Enrofloxacin	—	Y	Y	Y	Y
Flumequine	Y	Y	—	—	Y
Fluoroquinolone (multi)	Y (17)	Y (20)	Y (10)	Y (10)	Y
Furaltadone (AMOZ)	Y	—	Y	Y	Y
Furazolidone (AOZ)	Y	—	Y	Y	Y
Gentamicin	—	—	Y	—	Y
Ionophore (multi)	—	Y (2)	—	—	—
Leucomalachite green	Y	—	—	—	—
Malachite green	—	Y	Y	—	Y
Malachite green and leucomalachite green	—	Y	—	—	Y
Nitrofurantoin (AHD)	Y	—	—	—	—
Nitrofurazone (SEM)	Y	—	—	—	—
Neomycin	—	—	Y	—	—
Norfloxacin	—	—	—	—	Y
Olaquindox	—	Y[a]	—	—	—
Spiramycin	—	Y[a]	—	—	—
Streptomycin	Y	—	—	Y	Y
Sulfadiazine	Y	—	Y	Y	Y
Sulfamethazine	Y	—	Y	—	Y
Sulfamethoxazole	—	—	Y	—	Y
Sulfaquinoxaline	Y	—	—	—	Y
Sulfonamide (multi)	—	Y (9)	Y (9)	Y (19)	Y
Tetracycline (multi)	Y	—	Y (5)	Y (6)	Y
Tylosin	—	Y[a]	Y	—	—
Timicosin	—	—	Y	—	—
Virginiamycin	—	Y[a]	—	—	—

[a]Available as part of the "multi-AGP EIA" kit.

Notation: Manufacturer data (columns 2–6 above) indicate relative ELISA kit availability (Y = yes, kit available; N = no, kit not available; Multi = multi-residue kit) and number of analytes detected with a cross-reactivity of $\geq$1%.

Source: Information obtained from manufacturers' websites: Randox, www.randox.com; ELISA Technologies Inc. Eurodiagnostica products, www.elisa-tek.com/eurodiagnostica product list.htm; Abraxis,/www.abraxiskits.com/product_veterinary.htm; R-Biopharm AG, www.r-biopharm.com/product_site.php?; and Biooscientific, www.biooscientific.com/.

expressed as a percentage (CR%), is normally quantified by comparing the assay response to a range of structurally related compounds to that of the primary antigen (the compound against which the antibody was raised). In practice, calibration curves are produced using fixed concentration ranges for a selection of structurally related compounds. The mid-points (expressed as the concentration of analyte eliciting half the maximal response) of the calibration curves are calculated and compared. The CR% provides an estimate of the response of the assay to possible interfering compounds relative to the target analyte. Depending on the ultimate objective of the analysis, a wide cross-reactivity profile may be considered advantageous, for instance, multi-residue screening for a group analytes. However, in such a situation the analyst must be aware that a non-compliant screening result will require further investigation to determine both the identity of the residue and the concentration in relation to the appropriate RLs.

5.5.5 Conclusions Regarding ELISA

In summary, ELISA has been routinely used as a qualitative, semi-quantitative, or even quantitative screening technique in official residue control and quality control laboratories since the late 1970s and hence, is a well-established bioanalytical technique. ELISA is generally accepted as a cost-effective and highly specific laboratory-based method that does not require the need for expensive equipment in its most basic operation.

5.6 GENERAL CONSIDERATIONS CONCERNING THE PERFORMANCE CRITERIA FOR SCREENING ASSAYS

When choosing to implement a particular screening assay for antibiotic residue analysis, the laboratory must ensure that the performance criterion is fit for the intended purpose and in line with the appropriate legislative requirements within that country or the country for which the testing is undertaken, for example, the country to which the produce is to be exported.

By way of example, within the EU, Commission Decision 2002/657/EC[20] defines the performance criteria for analytical residue methods including decision limit, detection capability, reproducibility, selectivity, specificity, applicability, and robustness. Screening assays are defined by Commission Decision 2002/657/EC as "methods that are used to detect the presence of a substance or class of substances at the level of interest. These methods have the capability for a high sample throughput and are used to sift large numbers of samples for potential non-compliant results. They are specifically designed to avoid false compliant results." To aid laboratories wishing to implement screening assays a guidance document regarding validation of screening tests has been published by the European CRLs.[129] The level of interest, termed the *screening target concentration* (STC), should be established for each analyte–matrix combination based on any RLs. The STC is the concentration at which a screening test aims to categorize the sample as screen-positive (potentially non-compliant) and trigger a confirmatory test. For the regulatory monitoring of authorized substances, wherever possible, the STC is usually set at one-half the RL (MRL). For prohibited and unauthorized substances, the STC must be less than or equal to the MRPL or RPA except for substances with only a RC, where the STC should be below the RC, if possible. The farther the STC is below the RL, the lower the probability of obtaining a false-compliant result in samples containing the drug at the RL.

Two statistical limits (CCα and CCβ) have been established for determination of the concentrations above which a method reliably distinguishes and quantifies residues, on the basis of method variability and the risk of an incorrect decision.[20] The decision limit (CCα) is the concentration of a residue in a sample at which it is decided that the sample is non-compliant with a pre-defined statistical certainty (α error). The detection capability (CCβ) is the residue limit that can be detected, identified, and/or quantified with an error probability of β. The statistics are based on the α error, considering the false-positive (non-compliant) rate and β error considering the false-negative (compliant) rate. For screening methods only the detection capability (CCβ) is required to be determined during the method validation. In the EU, the β error is set at 5% for both substances with established permitted limits and substances for which no permitted limits have been established. The CCβ can be determined via a number of different methods according to the type of screening and whether it is a qualitative or quantitative assay (see also Chapter 10).

A key performance factor for the evaluation of a screening method is demonstration that the chosen STC can be achieved. The STC should be low enough to provide confidence that the antibiotic residue can be detected at the appropriate RL in the sample, that is, that there is sufficient margin of difference between the STC and the RL. This implies that the CCβ is equal to or less than the RL. It does not necessarily require an estimation of the numeric value of CCβ depending on the capability of the screening assay.

5.7 OVERALL CONCLUSIONS ON BIOANALYTICAL SCREENING ASSAYS

Bioanalytical screening assays have made enormous progress since the 1980s, which has directly contributed to the improved effectiveness of residue control programs. This progress is evident in a number of aspects: (1) the variety of assay formats commercially available ranging from "low technology" field test kits to sophisticated laboratory-based instrumental systems, (2) the scope of antibiotics that can be detected, and (3) the flexibility of the kits to the ever-increasing types of matrices requiring analysis. There is a growing demand for screening for an increasing number of possible analytes, and proportionally there will be an increasing demand for sensitive, rapid, and reliable formats. There is also an increasing trend that the application of rapid assays is performed not only by trained residue analysts in a laboratory environment but also by non-technical staff often *in situ*, for example, at food-processing plants. For these reasons, the majority of commercially available tests are presented as robust and technically simple formats.

Despite the advances in detection technology and the availability of an increasing number of rapid diagnostic assays, the bottleneck remains sample preparation, especially for the complex matrices, consuming in general the major part of the total analysis time. Substantial progress in this area is necessary to circumvent the need for time-consuming sample clean-up procedures and to keep pace with the advances in test kit design. Irrespective of these advances, screening assays should always be applied with care, and the user should be aware of the capabilities and limitations associated with the particular assay format. Ultimately, a screening assay should enable the user to make technically and administratively correct decisions about the regulatory compliance of samples in a rapid, cost-effective, and reliable manner.

ABBREVIATIONS

AGP	Antimicrobial growth promoter
AP	Alkaline phosphatase
ATR	Attenuated total reflection
BRE	Biological recognition element
BSA	Bovine serum albumin
CAST	Calf antibiotic screening test
CCα	Decision limit
CCβ	Detection capability
cpm	Counts per minute
CRL	Community reference laboratory
CR%	Cross-reactivity percentage
CV%	Coefficient of variation percentage
EIA	Enzyme immunoassay
ELISA	Enzyme-linked immunosorbent assay
FAST	Fast antibiotic screening test
FDA	Food and Drug Administration (USA)
FPT	Four-plate test
GC	Gas chromatography
HPLC	High-performance liquid chromatography
HRP	Horseradish peroxidase
HTS	High-throughput screening
Ig	Immunoglobin
IP	Identification point
KIS	Kidney inhibition swab
LC	Liquid chromatography
LC-MS	Liquid chromatography–mass spectrometry
LFD	Lateral-flow device
LFIA	Lateral-flow immunoassay
LIMS	Laboratory information management system
MAb	Monoclonal antibody
MIA	Microbial inhibition assay
MRL	Maximum residue limit
MRM	Multiple-reaction monitoring
MRPL	Minimum required performance limit
MRSA	Meticillin-resistant *Staphyloccocus aureus*
MS	Mass spectrometer
MS^n	Mass spectrometer in tandem
NDKT	New Dutch kidney test
NSB	Non-specific binding
PBP	Pencillin-binding protein
PCR	Polymerase chain reaction
POC	Point of care
PPV	Positive predictive value
RC	Recommended concentration
RIA	Radioimmunoassay
RL	Regulatory limit
ROSA	Rapid one-step assay
RPA	Reference point for action
SCC	Somatic cell count
SIM	Selected ion monitoring
SPR	Surface plasmon resonance
TRFIA	Time-resolved fluorescence immunoassay
SPW	Surface plasmon wave
STAR	Screening test for antimicrobial residues
STC	Screening target concentration
STOP	Swab on the premises
TM	Transverse magnetic
VMP	Veterinary medicinal product

REFERENCES

1. Schwarz S, Chaslus-Dancla E, Use of antimicrobials in veterinary medicine and mechanisms of resistance, *Vet. Res.* 2001;32:201–225.
2. Verbeke W, Frewer LJ, Scholderer J, De Brabander HF, Why consumers behave as they do with respect to food safety and risk information, *Anal. Chim. Acta* 2007;586:2–7.
3. Schwarz S, Kehrenberg C, Walsh TR, Use of antimicrobial agents in veterinary medicine and food animal production, *Intnatl. J. Antimicrob. Agents* 2001;17:431–437.
4. EC Commission Regulation (EC) No 1831/2003, of the European Parliament and of the Council of 22 September 2003 on additives for use in animal nutrition, *Off. J. Eur. Commun*. 2003;L268:29–43.
5. Levy SB, Active efflux, a common mechanism for biocide and antibiotic resistance, *J. Appl. Microbiol. Symp. Supp*. 2002;92:65–71.
6. World Health Organization, *Evaluation of Certain Veterinary Drug Residues in Food*, 62nd report Joint FAO/WHO Expert Committee on Food Additives, WHO Technical Report Series 925, 2004, pp. 26–37 (available at `http://whqlibdoc.who.int/trs/WHO_TRS_925.pdf`; accessed 11/24/10).
7. *Codex Veterinary Drug Residues in Food Online* (available at Database `http://www.codexalimentarius.net/vetdrugs/data/index.html`; accessed 11/11/10).
8. EC Regulation (EC) No 178/2002 of the European Parliament and of the council of 28 January 2002 laying down the general principles and requirements of food law, establishing the European Food Safety Authority and laying down procedures in matters of food safety, *Off. J. Eur. Commun*. 2002;L31:1–24.
9. EC Commission Decision 2005/34/EC of 11 January 2005 laying down harmonised standards for the testing for certain residues in products of animal origin imported from third countries, *Off. J. Eur. Commun*. 2005;L16:61–63.
10. EC Regulation (EC) No. 470/2009 of the European Parliament and of the Council of 6 May 2009 laying down

community procedures for the establishment of residue limits of pharmacologically active substances in foodstuffs of animal origin, repealing Council Regulation (EEC) No 2377/90 and amending Directive 2001/82/EC of the European Parliament and of the Council and Regulation (EC) No 726/2004 of the European Parliament and of the Council, *Off. J. Eur. Commun*. 2009;L152:11–22.

11. EC Council Regulation (EEC) No 2377/90 of 26 June 1990 laying down a Community procedure for the establishment of maximum residue limits of veterinary medicinal products in foodstuffs of animal origin, *Off. J. Eur. Commun*. 1990; L224:1–8.
12. EC Commission Regulation (EU) No. 37/2010 of 22 December 2009 on pharmacologically active substances and their classification regarding maximum residue limits in foodstuffs of animal origin, *Off. J. Eur. Commun*. 2010;L15: 1–69.
13. EC Council Directive 96/23/EC of 29 April 1996 on measures to monitor certain substances and residues thereof in live animals and animal products and repealing Directives 85/358/EEC and 86/469/EEC and Decisions 89/187/EEC and 91/664/EEC, *Off. J. Eur. Commun*. 1996;L125:10–32.
14. Gowik P, Criteria and requirements of commission decision 2002/657/EC, *Bull. Intnatl. Dairy Fed*. 2003;383:52–56.
15. CRL Guidance Paper (Dec. 7, 2007), *CRLs View on the State of the Art of Analytical Methods for National Residue Control Plans*.
16. *AOAC Research Institute Performance Tested Methods*^{SM} *Program Policies and Procedures*, AOAC Research Institute, Gaithersburg, MD, 2009 (available at `http://www.aoac.org/testkits/Policies%20&%20Procedures.pdf`; accessed 11/01/10).
17. *Performance Tested*^{SM} *Methods Validated Methods*, AOAC Research Institute, Gaithersburg, MD, 2010 (available at `http://www.aoac.org/testkits/testedmethods.html`; accessed 11/01/10).
18. Watson DH, *Food Chemical Safety*, Woodhead Publishing, Cambridge, UK, 2001.
19. Turnipseed SB, Analysis of drug residues in food, in Hui YH, Bruinsma BL, Richard Gorham J, et al., eds., *Food Plant Sanitation. Food Science and Technology*, CRC Press, Boca Raton, FL, 2003.
20. EC Commission Decision 2002/657/EC of 12 August 2002 implementing Council Directive 96/23/EC concerning the performance of analytical methods and the interpretation of results, *Off. J. Eur. Commun*. 2002;L221:8–36.
21. Alfredsson G, Branzell C, Granelli K, Lundstrom A, Simple and rapid screening and confirmation of tetracyclines in honey and egg by a dipstick test and LC-MS/MS, *Anal. Chim. Acta* 2005;529:47–51.
22. Gram C, Ueber die isolirte Farbung der Schizomyceten in Schitt-Und Trockenpreparaten, Fortschritte der Medicin 2, in Brock TB, ed, *Milestones in Microbiology: 1556–1940*, American Society for Microbiology Press, 1998, pp. 1556–1940.
23. Stead DA, Current methodologies for the analysis of aminoglycosides, *J. Chromatogr. B* 2000;747:69–93.
24. Mittchell JM, Griffiths MW, McEwen SA, McNab WB, Yee AJ, Antimicrobial drug residues in milk and meat: Causes, concerns, prevalence, regulations, tests and test performance, *J. Food Prot*. 1998;61:742–756.
25. Watts PS, McLeod D, The estimation of penicillin in blood serum and milk of bovines after intramuscular injection, *J. Comp Pathol. Therap*. 1946;56:170–176.
26. Berridge NJ, Testing for penicillin in milk, *Dairy Ind*. 1953;18:586.
27. Berridge NJ, Penicillin in milk. 1. The rapid routine assay of low concentrations of penicillin in milk, *J Dairy Res*. 1956;23:336–341.
28. Galesloot, TE, Hassing, F, A rapid and sensitive paper disc method for the detection of penicillin in milk, *Neth. Milk Dairy J*. 1962;6:89–95.
29. Lemoigne M, Sanchez G, Girard H, Caracterisation de la penicilline et de la streptomycine dans le lait, *C. R. Acad Agric. Fr*. 1952;38:608–609.
30. Pien J, Lignac J, Claude P, Detection biologilue des antiseptiques et des antibiotiques dans le lait, *Ann. Falsi fraudes* 1953;46:258–270.
31. Foster JW, Woodruff HB, Microbiological aspects of penicillin, *J. Bacteriol*. 1944;47:43–58.
32. Welsh M, Langer PH, Burkhardt RL, Schroeder CR, Penicillin blood and milk concentrations in the normal cow following parenteral administration, *Science* 1948;108:185–187.
33. Chang CS, Tai TF, Li HP, Evaluating the applicability of the modified four-plate test on the determination of antimicrobial agent residues in pork, *J. Food Drug Anal*. 2000;8:25–34.
34. Rault A, Gaudin V, Maris P, Fuselier R, Ribouchon JL, Cadieu N, Validation of a microbiological method: The STAR protocol, a five-plate test, for the screening of antibiotic residues in milk, *Food Addit. Contam*. 2004;21: 422–433.
35. Nouws JFM, Broex NJG, Den Hartog JMP, Diserens F, The new Dutch kidney test, *Arch. Lebensmittelhyg*. 1988;39:135–138.
36. Nouws JFM, Tolerances and detection of antimicrobial residues in slaughtered animals, *Arch. Lebensmittelhyg*. 1981;32:103–110.
37. Pikkemaat MG, Oostra-van Dijk S, Schouten J, Rapallini M, van Egmond HJ, A new microbial screening method for the detection of antimicrobial residues in slaughter animals: The Nouws antibiotic test (NAT-screening), *Food Control* 2007;19:781–789.
38. Microbiology Laboratory Guidebook, Method 34.02, *Bioassay for the Detection, Identification and Quantitation of Antimicrobial Residues in Meat and Poultry Tissue*, US Dept. Agriculture, Food Safety and Inspection Service (available at `http://www.fsis.usda.gov/PDF/MLG_34_02.pdf`; accessed 11/01/10).
39. Laméris SA, van Os JL, Oostendorp JG, Method for Determination of the Presence of Antibiotics, US Patent 3,941,658, 1976.

40. Kantiani L, Farre M, Barcelo D, Analytical methodologies for the detection of β-lactam antibiotics in milk and feed samples, *Trends Anal. Chem*. 2009;28(6):729–733.
41. Linage B, Gonzalo C, Carriedo JA, Asensio JA, Blanco MA, De La Fuente LF, Performance of blue-yellow screening test for antimicrobial detection in ovine milk, *J. Dairy Sci*. 2007;90:5374–5379.
42. Zeng SS, Escobar EN, Brown-Crowder I, Evaluation of screening tests for detection of antimicrobial residues in goat milk, *Small Rumin. Res*. 1996;21:155–160.
43. Gaudin V, Hedou C, Verdon E, Validation of a wide-spectrum microbiological tube test, the EXPLORER_ test, for the detection of antimicrobials in muscle from different animal species, *Food Addit. Contam*. 2009;26:1162–1171.
44. Korsrud GO, Boison JO, Nouws JF, MacNeil JD. Bacterial inhibition tests used to screen for antimicrobial veterinary drug residues in slaughtered animals, *J. AOAC Int*. 1998;81:21–24.
45. Dey BP, Reamer RP, Thaker NH, Thaker AM, Calf antibiotic and sulfonamide test (CAST) for screening antibiotic and sulfonamide residues in calf carcasses, *J. AOAC Int*. 2005;88(2):440–446.
46. Ashwin H, Stead S, Caldow M, Sharman M, Stark J, de Rijk A, Keely BJ, A rapid microbial inhibition-based screening strategy for fluoroquinolone and quinolone residues in foods of animal origin, *Anal. Chim. Acta* 2009;637(1–2):241–246.
47. Braham R, Black WD, Claxon J, Yee AJ, A rapid assay for detecting sulfonamides in tissues of slaughtered animals, *J. Food Prot*. 2001;64:1565–1573.
48. Suhren G, Luitz M, Evaluation of microbial inhibitor tests with indicator in microtitre plates by photometric measurements, *Milchwissenschaften* 1995;50:467–470.
49. Lotto RB, Purves D, The effects of color on brightness, *Nature Neurosci*. 1999;2:1019–1014.
50. Stead SL, Ashwin H, Richmond SF, Sharman M, Langeveld PC, Barendse JP, Stark J, Keely BJ, Evaluation and validation according to international standards of the Delvotest SP-NT screening assay for antimicrobial drugs in milk, *Intnatl. Dairy J*. 2008;18:3–11.
51. Langeveld PC, Stark J, *Detection of Antimicrobial Residues in Eggs*, WO Patent 01/25795, 2001.
52. Langeveld PC, Stark J, Van Paridon PA, *Method for the Detection of Antimicrobial Residues in Food and Bodily Fluid Samples*, US Patent 7,462,464, 2008.
53. Stead S, Sharman M, Tarbin JA, Gibson E, Richmond S, Stark J, Geijp E, Meeting maximum residue limits: An improved screening technique for the rapid detection of antimicrobial residues in animal food products, *Food Addit. Contam*. 2004;21:216–221.
54. Stead S, Richmond S, Sharman M, Stark J, Geijp E, A new approach for detection of antimicrobial drugs in food; PremiTest coupled to scanner technology, *Anal. Chim. Acta* 2005;529:83–88.
55. Stead SL, Caldow M, Sharma A, Ashwin HM, Sharman M, De-Rijk A, Stark J, New method for the rapid identification of tetracycline residues in foods of animal origin—using the PremiTest in combination with a metal ion chelation assay, *Food Addit. Contam*. 2007;24:583–589.
56. Vermunt AEM, Stadhouders J, Loeffen GJM, Bakker R, Improvements of the tube diffusion method for detection of antibiotics and sulfonamides in raw milk, *Neth. Milk Dairy J*. 1993;47:31–40.
57. Woods DD, The relation of p-aminobenzoic acid to the mechanism of the action of sulphanilamide, *Br. J. Exp. Pathol*. 1940;21:74–90.
58. Mitscher LA, *The Chemistry of the Tetracycline Antibiotics*, Marcel Dekker, New York, 1978.
59. Blackwood RK, Structure determination and total synthesis of the tetracyclines, in Hlavka JJ, Boothe JH, eds., *Handbook of Experimental Pharmacology*, Vol. 78, Springer-Verlag, Berlin, 1985, pp. 59–136.
60. Le Breton MH, Savoy-Perroud MC, Diserens JM, Validation and comparison of the Copan milk test and Delvotest SP-NT for the detection of antimicrobials in milk, *Anal. Chim. Acta* 2007;586(1–2):280–283.
61. Sierra D, Sanchez A, Contreras A et al., Detection limits of four antimicrobial residue screening tests for beta-lactams in goat's milk, *J. Dairy Sci*. 2009;92(8):3585–3591.
62. Stead SL, Ashwin H, Richmond SF, et al., Evaluation and validation according to international standards of the Delvotest (R) SP-NT screening assay for antimicrobial drugs in milk, *Intnatl. Dairy J*. 2008;18(1):3–11.
63. Schneider MJ, Mastovska K, Lehotay SJ, Lightfield AR, Kinella B, Shultz CE, Comparison of screening methods for antibiotics in beef kidney juice and serum, *Anal. Chim. Acta* 2009;37:290–297.
64. Bergwerff AA, Rapid assays for detection of residues of veterinary drugs, in van Amerongen A, Barug D, Lauwars M, eds., *Rapid Methods for Biological and Chemical Contaminants in Food and Feed*, Wageningen Academic Publishers, 2005, pp. 259–292.
65. Rosner MH, Grassman JA, Haas RA, Immunochemical techniques in biological monitoring, *Environ. Health Perspect*. 1991;94:131–134.
66. Yallow RS, Berson A, Immunological specificity of human insulin: Application to immunoassay of insulin, *J. Clin. Invest*. 1961;40:2190–2198.
67. O'Keeffe M, Crabbe P, Salden M, Wichers J, van Peteghem C, Kohen F, Pieraccini G, Moneti G, Preliminary evaluation of a lateral flow immunoassay device for screening urine samples for the presence of sulphamethazine, *J. Immun. Meth*. 2003;278:117–126.
68. Campbell K, Fodey T, Flint J, Danks C, Danaher M, O'Keeffe M, Kennedy DG, Elliott C, Development and validation of a lateral flow device for the detection of nicarbazin contamination in poultry feeds, *J. Agric. Food Chem*. 2007;55:2497–2503.
69. Huet AC, Charlier C, Tittlemier SA, Benrejeb S, Delahaut P, Simultaneous determination of (fluoro)quinolone antibiotics in kidney, marine products, eggs, and muscle by enzyme-linked immunosorbent assay (ELISA), *J. Agric. Food Chem*. 2006;54:2822–2827.

70. Haughey S, Baxter CA, Biosensor screening for veterinary drug residues in foodstuffs, *J. AOAC Int*. 2006;89: 862–867.
71. Lamar J, Petz M, Development of a receptor-based microplate assay for the detection of beta-lactam antibiotics in different food matrices, *Anal. Chim. Acta* 2007;586:296–303.
72. Rake JB, Gerber R, Metha RJ, Newman DJ, OH YK, Phelen C, Shearer MC, Sitrin RD, Nisbet LJ, Glycopeptide antibiotics: A mechanism-based screen employing a bacterial cell wall receptor mimetic, *J. Antibiot*. 1986;39(1):58–67.
73. Levi R, McNiven S, Piletsky SA, Cheong SH, Yano K, Karube I, Optical detection of chloramphenicol using molecularly imprinted polymers, *Anal. Chem*. 1997;69(11): 2017–2021.
74. Tombelli S, Minunni M, Mascini M, Aptamers-based assays for diagnostics, environmental and food analysis, *Biomol. Eng*. 2007;24:191–200.
75. Niazi JH, Lee SJ, Gu MB, Single-stranded DNA aptamers specific for antibiotics tetracyclines, *Bioorg. Med. Chem*. 2008;16:1254–1261.
76. Stead SL, Ashwin H, Johnston B, Tarbin JA, Sharman M, Kay J, Keely BJ, An RNA-aptamer-based assay for the detection and analysis of malachite green and leucomalachite green residues in fish tissue, *Anal. Chem*. 2010;82(7):2652–2660.
77. Huang Z, Szostak JW, Evolution of aptamers with a new specificity and new secondary structures from an ATP aptamer, *RNA* 2003;9:1456–1463.
78. Lipman NS, Jackson LR, Trudel LJ, Weis-Garcia F, Laboratory animals and immunization procedures: Challenges and opportunities, *Inst. Lab. Anim. Res. J*. 2005:46:258–268.
79. Chaudhry MQ, Immunogens and standards, in Gosling JP, ed., *Immunoassays: A Practical Approach*, Oxford Univ. Press, Oxford, UK, 2000, Chap. 6.
80. Franek M, Hruska K, Antibody based methods for environmental and food analysis: A review, *Vet. Med. Czech*. 2005;50:1–10.
81. Campbell K, Fodey T, Flint J, Danks C, Danaher M, O'Keeffe M, Kennedy DG, Elliott C, Development and validation of a lateral flow device for the detection of nicarbazin contamination in poultry feeds, *J. Agric. Food Chem*. 2007;55(6):2497–2503.
82. Reybroeck W, Ooghe S, De Brabander H, Daeseleire E, Validation of the tetrasensor honey test kit for the screening of tetracyclines in honey, *J. Agric. Food Chem*. 2007;55:8359–8366.
83. Alfredsson G, Branzell C, Granelli K, Lundstrom A, Simple and rapid screening and confirmation of tetracyclines in honey and egg by a dipstick test and LC-MS/MS, *Anal. Chim. Acta* 2005;529(1–2):47–51.
84. Navratilova P, Borkovcova I, Drackova M, Janstova B, Vorlova L, Occurrence of tetracycline, chlortetracyclin, and oxytetracycline residues in raw cow's milk, *Czech. J. Food Sci*. 2009;27(5):379–385.
85. Okerman L, Croubels S, Cherlet M, De Wasch K, De Backer P, Van Hoof J, Evaluation and establishing the performance of different screening tests for tetracycline residues in animal tissues, *Food Addit. Contam*. 2004;21(2):145–153.
86. Al-Mazeedi HM, Abbas AB, Alomirah HF, et al., Screening for tetracycline residues in food products of animal origin in the State of Kuwait using Charm II radioimmunoassay and LC/MS/MS methods, *Food Addit. Contam*. 2010;27(3):291–301.
87. Bonvehi IS, Gutierrez AL, Residues of antibiotics and sulfonamides in honeys from Basque Country (NE Spain), *J. Sci. Food Agric*. 2009;89(1):63–72.
88. Zomer E, Quintana J, Scheemaker J, Saul S, Charm SE, High-performance liquid chromatography-receptorgram: A comprehensive method for identification of veterinary drugs and their active metabolites, in Moats WA, Medina MB, eds., *Veterinary Drug Residues*, American Chemical Society, 1996, Vol. 636, pp. 149–160.
89. Holtje JV, Growth of the stress-bearing and shape-maintaining murein Sacculus of Escherichia coli, *Microbiol. Mol. Biol. Rev*. 1998;62:181–203.
90. Grebe T, Hakenbeck R, Penicillin-binding proteins 2b and 2x of Streptococcus pneumoniae are primary resistance determinants for different classes of beta-lactam antibiotics, *Antimicrob. Agents Chem*. 1996;40:829–834.
91. Goffin C, Ghuysen JM, Multimodular penicillin-binding proteins: An enigmatic family of orthologs and paralogs, *Microbiol. Mol. Biol. Rev*. 1998;64:1079–1093.
92. Andrew SM, Frobish RA, Paape MJ, Maturin LJ, Evaluation of selected antibiotic residue screening tests for milk from individual cows and examination of factors that affect the probability of false-positive outcomes, *J. Dairy Sci*. 1997;80(11):3050–3057.
93. Andrew SM, Effect of fat and protein content of milk from individual cows on the specificity rates of antibiotic residue screening tests, *J. Dairy Sci*. 2000;83(12):2992–2997.
94. Gibbons-Burgener SN, Kaneene JB, Lloyd JW, Leykam JF, Erskine RJ, Reliability of three bulk-tank antimicrobial residue detection assays used to test individual milk samples from cows with mild clinical mastitis, *Am. J. Vet. Res*. 2001;62(11):1716–1720.
95. Lowe CR, Biosensors, *Phil. Trans. Royal Soc. Lond. B* 1989;324:487–496.
96. Turner APF, Biosensors-sense and sensitivity, *Science* 2000;290:1315–1317.
97. Higgins IJ, Lowe CR, Introduction to the principles and applications of biosensors, *Phil. Trans. Royal Soc. Lond. B* 1987;316:3–11.
98. Karlsson R, SPR for molecular interaction analysis: A review of emerging application areas, *J. Mol. Recognit*. 2004; 17:151–161.
99. Kretschmann E, Raether H, Radiative decay of non-radiative surface plasmons excited by light, *Z. Naturforschung*. 1968;23A:2135–2136.
100. Otto A, Excitation of surface plasma waves in silver by the method of frustrated total reflection, *Z. Phys*. 1968;216:398–410.

101. Homola J, Yee SS, Gauglitz G, Surface plasmon resonance sensors: Review, *Sensor Actuat. B Chem*. 1999;54:3–5.

102. Leidberg B, Lundström I, Stenberg E, Principles of biosensing with an extended coupling matrix and surface plasmon resonance, *Sensor Actuat. B Chem*. 1993;11:63–72.

103. Homola J, Koudela I, Yee SS, Surface plasmon resonance sensors based on diffraction gratings and prism couplers: Sensitivity comparison, *Sensor Actuat. B Chem*. 1999;54:16–24.

104. GE Healthcare Biacore Life Sciences (available at `http://www.biacore.com/lifesciences/products/systems_overview/index.html`; accessed 11/03/10).

105. Sensata Technologies (available at `http://www.sensata.com/sensors/spreeta-analytical-sensor-highlights.htm`; accessed 11/03/10).

106. Reichert Life Sciences (available at `http://www.reichertspr.com/?gclid=CLLzprini6ACFcJd4wodG3CRdg`; accessed 11/03/10).

107. Ferguson J, Baxter A, Young P, Kennedy G, Elliott C, Weigel S, Gatermann R, Ashwin H, Stead S, Sharman M, Detection of chloramphenicol and chloramphenicol glucuronide residues in poultry muscle, honey, prawn and milk using a surface plasmon resonance biosensor and Qflex kit chloramphenicol, *Anal. Chim. Acta* 2005;529:109–113.

108. Ashwin HM, Stead SL, Taylor JC, Startin JR, Richmond SF, Homer V, Bigwood T, Sharman M, Development and validation of screening and confirmatory methods for the detection of chloramphenicol and chloramphenicol glucuronide using SPR biosensor and liquid chromatography-tandem mass spectrometry, *Anal. Chim. Acta* 2005;529:103–108.

109. Gaudin V, Maris P, Development of a biosensor-based immunoassay for screening of chloramphenicol residues in milk, *Food Agric. Immunol*. 2001;13:77–86.

110. Dumont V, Huet AC, Traynor I, Elliott C, Delahaut P, A surface plasmon resonance biosensor assay for the simultaneous determination of thiamphenicol, florefenicol, florefenicol amine and chloramphenicol residues in shrimps, *Anal. Chim. Acta* 2006;567:179–183.

111. Huet AC, Charlier C, Singh G, et al., Development of an optical surface plasmon resonance biosensor assay for (fluoro)quinolones in egg, fish, and poultry meat, *Anal. Chim. Acta* 2008;623(2):195–203.

112. Cacciatore G, Bergwerff AA, Petz M, Development of screening assays for veterinary drug residues utilizing surface plasmon resonance-based biosensor, in Stephany R, Bergwerff A, eds., *Proc. EuroResidue V*, Natl. Inst. Public Health and the Environment and Utrecht Univ., Utrecht, The Netherlands, 2004, pp. 143–150.

113. Baxter GA, Ferguson JA, O'Connor MC, Elliott CT, Detection of streptomycin residues in whole milk using an optical immunobiosensor, *J. Agric. Food Chem*. 2001;49(7):3204–3207.

114. Ferguson JP, Baxter GA, McEvoy JDG, Stead S, Rawlings E, Sharman M, Detection of streptomycin and dihydrostreptomycin residues in milk, honey and meat samples using an optical biosensor, *Analyst* 2002;127:951–956.

115. Haasnoot W, Verheijen R, A direct (non-competitive) immunoassay for gentamicin residues with an optical biosensor, *Food Agric. Immunol*. 2001;13:131–134.

116. McGrath T, Baxter A, Ferguson J, Haughey S, Bjurling P, Multi-residue screening in porcine muscle using a surface Plasmon resonance biosensor, *Anal. Chim. Acta* 2004;529(1–2):123–127.

117. Situ C, Crooks SRH, Baxter AG, Ferguson J, Elliott CT, On-line detection of sulfamethazine and sulfadiazine in porcine bile using a multi-channel high throughput SPR biosensor, *Anal. Chim. Acta* 2002;473(1–2):143–149.

118. Connolly L, Thompson CS, Haughey SA, Traynor IM, Tittlemeiser S, Elliott CT, The development of a multi-nitroimidazole residue analysis assay by optical biosensor via a proof of concept project to develop and assess a prototype test kit, *Anal. Chim. Acta* 2007;598(1):155–161.

119. Petz M, Recent applications of surface plasmon resonance biosensors for analyzing residues and contaminants in food, *Monatsh. Chem*. 2009;140(8):953–964.

120. Stolker AAM, Current trends and developments in sample preparation, in Van Ginkel L, Ruiter A, eds., *Proc. EuroResidue IV*, National Institute for Public Health and the Environment, Bilthoven, The Netherlands, 2000, pp. 148–158.

121. Engvall E, Perlmann P, Enzyme-linked immunosorbent assay (ELISA) quantitative assay of immunoglobulin G, *Immunochemistry* 1971;8:871–874.

122. Bonwick GA, Smith CJ, Immunoassays: Their history, development and current place in food science and technology, *Intnatl. J. Food Sci. Technol*. 2004;39:817–827.

123. Franek M, Hruska K, Antibody based methods for environmental and food analysis: A review, *Vet. Med. Czech*. 2005;50(1):1–10.

124. Adrian J, Pinacho DG, Granier B, Diserens JM, Sanchez-Baeza F, Marco MP, A multianalyte ELISA for immunochemical screening of sulfonamide, fluoroquinolone and ß-lactam antibiotics in milk samples using class-selective bioreceptors, *Anal. Bioanal. Chem*. 2008;391:1703–1712.

125. Lin S, Han SQ, Xu WG, Guan GY, Chemiluminescence immunoassay for chloramphenicol, *Anal. Bioanal. Chem*. 2005;382:1250–1255.

126. Knecht BG, Strasser A, Dietrich R, Martlbauer E, Niessner R, Weller MG, Automated microarray system for the simultaneous detection of antibiotics in milk, *Anal. Chem*. 2004;76(3):646–654.

127. Zhang Z, Liu JF, Shao B, Jiang GB, Time-resolved fluoroimmunoassay as an advantageous approach for highly ffficient determination of sulfonamides in environmental waters, *Environ. Sci. Technol*. 2010;44(3):1030–1035.

128. de Keizer W, Bienenmann-Ploum ME, Bergwerff AA, Haasnoot W, Flow cytometric immunoassay for sulfonamides in raw milk, *Anal. Chim. Acta* 2008;620(1–2):142–149.

129. Community Reference Laboratories (CRLs)O, Eurpoean Union (EU) (available at `http://ec.europa.eu/food/food/chemicalsafety/residues/GuidelineValidationScreening_en.pdf`; accessed 10/10/2010).

6

CHEMICAL ANALYSIS: QUANTITATIVE AND CONFIRMATORY METHODS

JIAN WANG AND SHERRI B. TURNIPSEED

6.1 INTRODUCTION

Analytical methods for the detection and/or determination of antibiotic residues in food fall into two categories: (1) screening methods such as microbial inhibition tests and rapid test kits, as discussed in Chapter 4 and (2) quantitative and/or confirmatory methods, including gas chromatography with electron capture, flame ionization, or mass spectrometry detection, as well as liquid chromatography (LC) with ultraviolet (UV), fluorometric or electrochemical detection, or mass spectrometry (MS). Almost all antibiotics are LC-amenable compounds and can be analyzed by LC techniques, although some compounds such as chloramphenicol, florfenicol, and thiamphenicol were historically determined by gas chromatography with mass spectrometry or electron capture detection.[1] According to the European Commission Decision 2002/657/EC,[2] "confirmatory methods for organic residues or contaminants shall provide information on the chemical structure of the analyte. Consequently, methods based only on chromatographic analysis without the use of spectrometric detection are not suitable on their own for use as confirmatory methods. However, if a single technique lacks sufficient specificity, the desired specificity shall be achieved by analytical procedures consisting of suitable combinations of clean-up, chromatographic separation(s) and spectrometric detection."

With the development and/or implementation of new LC-MS interfaces, modern column chemistries and advanced mass analyzers, LC-MS has largely superseded other detection techniques for quantification, confirmation, and identification of antibiotic residues in food because of its sensitivity and specificity.

6.2 SINGLE-CLASS AND MULTI-CLASS METHODS

Historically, grouping for the analysis of antibiotic residues in food has been based on a single class or related families, and the number of compounds in one analysis was typically less than 20. A single-class method is relatively easy to optimize for both extraction and instrumental parameters because of the similar physical and chemical properties of antibiotics from the same group. However, a multi-class approach is always employed when analyzing pharmaceuticals, antibiotics, and pesticides in environmental samples, as well as pesticides in fruits and vegetables. A multi-class method typically covers as many analytes as possible, up to a few hundred, regardless of the type of analytes and the nature of samples; the method may or may not be optimized for individual classes or compounds.[3–5] The main advantage of a multi-class approach is cost-effectiveness, especially for screening purposes. There has been an increased number of publications on multi-class methods (Table 6.1) for analysis of antibiotics in food.[6–12] Good examples include the use of UHPLC TOF-MS to analyze 100 veterinary drugs in egg, fish, and meat,[10] and to screen 150 veterinary drugs from different classes in milk.[11]

Chemical Analysis of Antibiotic Residues in Food, First Edition. Edited by Jian Wang, James D. MacNeil, and Jack F. Kay.

TABLE 6.1 Summary of LC-MS Analysis of Antibiotic Residues in Food Matrices and Other Substances

	Single-Class				Exact Mass[a,b]					LC		MS[a]			
Example	Class	Matrix	Compound	Molecular Formula	Molecular Weight	$[M+H]^+$	$[M-H]^-$	$[M+NH_4]^+$	$[M+Na]^+$	Column	Mobile Phase	Type	Ionization	Mass or Transitions	Reference
1	Aminoglycosides	Animal tissues, honey, and milk	Amikacin	$C_{22}H_{43}N_5O_{13}$	585.285740	**586.293016**	584.278464	603.319565	608.274961	Capcell Pak C18 UG120, 150 × 2.0 mm, 5 μm	Gradient profile: mobile phases A—acetonitrile/water (5 : 95, v/v) containing 20 mM HFBA; B—acetonitrile/water (50 : 50, v/v) containing 20 mM HFBA, IPC application	QqQ	ESI^+	586 > 425, 163	41
			Apramycin	$C_{21}H_{41}N_5O_{11}$	539.280260	**540.287536**	538.272984	557.314085	562.269481				ESI^+	540 > 217, 378	
			Dihydrostreptomycin	$C_{21}H_{41}N_7O_{12}$	583.281323	**584.288599**	582.274047	601.315148	606.270544				ESI^+	584 > 263, 246	
			Gentamicin C_1	$C_{21}H_{43}N_5O_7$	477.316250	**478.323526**	476.308974	495.350075	500.305471				ESI^+	478 > 157, 322	
			Gentamicin C_2	$C_{20}H_{41}N_5O_7$	463.300600	**464.307876**	462.293324	481.334425	486.289821				ESI^+	464 > 322, 160	
			Gentamicin C_{1a}	$C_{19}H_{39}N_5O_7$	449.284950	**450.292226**	448.277674	467.318775	472.274171				ESI^+	450 > 160, 322	
			Hygromycin B	$C_{20}H_{37}N_3O_{13}$	527.232641	**528.239917**	526.225365	545.266466	550.221862				ESI^+	528 > 177, 352	
			Kanamycin A	$C_{18}H_{36}N_4O_{11}$	484.238061	**485.245337**	483.230785	502.271886	507.227282				ESI^+	485 > 163, 324	
			Neomycin B	$C_{23}H_{46}N_6O_{13}$	614.312289	**615.319565**	613.305013	632.346114	637.301510				ESI^+	615 > 161, 293	
			Paromomycin	$C_{23}H_{45}N_5O_{14}$	615.296305	**616.303581**	614.289029	633.330130	638.285526				ESI^+	616 > 163, 293	
			Spectinomycin	$C_{14}H_{24}N_2O_7$	332.158353	333.165629	331.151077	350.192178	355.147574				ESI^+	**$[M+H_2O+H]^+$** : 351 > 333, 207	
			Streptomycin	$C_{21}H_{39}N_7O_{12}$	581.265673	**582.272949**	580.258397	599.299498	604.254894				ESI^+	582 > 263, 246	
			Tobramycin	$C_{18}H_{37}N_5O_9$	467.259130	**468.266406**	466.251854	485.292955	490.248351				ESI^+	468 > 163, 324	
2	Aminoglycosides	Animal tissues	Apramycin	$C_{21}H_{41}N_5O_{11}$	539.280260	**540.287536**	538.272984	557.314085	562.269481	ZIC-HILIC, 100 × 2.1 mm, 5 μm	Gradient profile: mobile phases A—1% formic acid and 150 mM ammonium acetate in water; B—acetonitrile, HILIC application	QqQ	ESI^+	540 > 378, 217	57
			Dihydrostreptomycin	$C_{21}H_{41}N_7O_{12}$	583.281323	**584.288599**	582.274047	601.315148	606.270544				ESI^+	584 > 263, 246	
			Gentamicin C_1	$C_{21}H_{43}N_5O_7$	477.316250	**478.323526**	476.308974	495.350075	500.305471				ESI^+	478 > 322, 160	
			Gentamicin C_2, C_{2a}	$C_{20}H_{41}N_5O_7$	463.300600	**464.307876**	462.293324	481.334425	486.289821				ESI^+	464 > 322, 160	
			Kanamycin A	$C_{18}H_{36}N_4O_{11}$	484.238061	**485.245337**	483.230785	502.271886	507.227282				ESI^+	485 > 163, 324	
			Neomycin B	$C_{23}H_{46}N_6O_{13}$	614.312289	**615.319565**	613.305013	632.346114	637.301510				ESI^+	615 > 161, 455	
			Spectinomycin	$C_{14}H_{24}N_2O_7$	332.158353	333.165629	331.151077	350.192178	355.147574				ESI^+	**$[M+H_2O+H]^+$** : 351 > 333, 98	
			Streptomycin	$C_{21}H_{39}N_7O_{12}$	581.265673	**582.272949**	580.258397	599.299498	604.254894				ESI^+	582 > 263, 246	
3	Aminoglycosides	Bovine milk	Aminosidine	$C_{23}H_{45}N_5O_{14}$	615.296305	616.303581	614.289029	633.330130	638.285526	Alltima C18, 250 × 4.6 mm, 5 μm	Gradient profile: mobile phases A—water with 1 mM HFBA; B—methanol with 1 mM HFBA, IPC application	QqQ	ESI^+	**$[M+2H]^{2+}$** : 309 > 161, 455	42
			Apramycin	$C_{21}H_{41}N_5O_{11}$	539.280260	540.287536	538.272984	557.314085	562.269481				ESI^+	**$[M+2H]^{2+}$** : 271 > 163, 217	
			Dihydrostreptomycin	$C_{21}H_{41}N_7O_{12}$	583.281323	584.288599	582.274047	601.315148	606.270544				ESI^+	**$[M+2H]^{2+}$** : 293 > 176, 409	
			Gentamicin C_1	$C_{21}H_{43}N_5O_7$	477.316250	478.323526	476.308974	495.350075	500.305471				ESI^+	**$[M+2H]^{2+}$** : 240 > 139, 157, 322	
			Gentamicin C_{1a}	$C_{19}H_{39}N_5O_7$	449.284950	450.292226	448.277674	467.318775	472.274171				ESI^+	**$[M+2H]^{2+}$** : 226 > 129, 322	
			Gentamicin C_2, C_{2a}	$C_{20}H_{41}N_5O_7$	463.300600	464.307876	462.293324	481.334425	486.289821				ESI^+	**$[M+2H]^{2+}$** : 233 > 126, 143, 322	
			Neomycin B	$C_{23}H_{46}N_6O_{13}$	614.312289	615.319565	613.305013	632.346114	637.301510				ESI^+	**$[M+2H]^{2+}$** : 308 > 161, 455	
			Spectinomycin	$C_{14}H_{24}N_2O_7$	332.158353	333.165629	331.151077	350.192178	355.147574				ESI^+	**$[M+H_2O+H]^+$** : 351 > 315, 333	
			Streptomycin	$C_{21}H_{39}N_7O_{12}$	581.265673	582.272949	580.258397	599.299498	604.254894				ESI^+	**$[M+CH_3OH+2H]^{2+}$** : 308 > 176, 263	
4	β-Lactam	Bovine kidney	Amoxicillin	$C_{16}H_{19}N_3O_5S$	365.104544	**366.111820**	364.097268	383.138369	388.093765	Prodigy ODS3, 150 × 3 mm, 5 μm	Gradient profile: mobile phases A—0.1% formic acid in water; B—0.1% formic acid in acetonitrile.	QqQ	ESI^+	366 > 349, 208	26
			Ampicillin	$C_{16}H_{19}N_3O_4S$	349.109629	**350.116905**	348.102353	367.143454	372.098850				ESI^+	350 > 106, 192	
			Cefazolin	$C_{14}H_{14}N_8O_4S_3$	454.030018	**455.037294**	453.022742	472.063843	477.019239				ESI^+	455 > 323, 156	
			Cephalexin	$C_{16}H_{17}N_3O_4S$	347.093979	**348.101255**	346.086703	365.127804	370.083200				ESI^+	348 > 158, 174	
			Cloxacillin	$C_{19}H_{18}ClN_3O_5S$	435.065572	**436.072848**	434.058296	453.099397	458.054793				ESI^+	436 > 277, 160	
			Deacetylcephapirin	$C_{15}H_{15}N_3O_5S_2$	381.045316	**382.052592**	380.038040	399.079141	404.034537				ESI^+	382 > 152, 226	
			Desfuroylceftiofur cysteine disulfide	$C_{17}H_{20}N_6O_7S_4$	548.027636	**549.034912**	547.020360	566.061461	571.016857				ESI^+	549 > 183, 241	
			Dicloxacillin	$C_{19}H_{17}Cl_2N_3O_5S$	469.026600	**470.033876**	468.019324	487.060425	492.015821				ESI^+	470 > 160, 311	
			Nafcillin	$C_{21}H_{22}N_2O_5S$	414.124945	**415.132221**	413.117669	432.158770	437.114166				ESI^+	415 > 199, 171	
			Oxacillin	$C_{19}H_{19}N_3O_5S$	401.104544	**402.111820**	400.097268	419.138369	424.093765				ESI^+	402 > 160, 243	
			Penicillin G	$C_{16}H_{18}N_2O_4S$	334.098730	**335.106006**	333.091454	352.132555	357.087951				ESI^+	335 > 160, 176	

5	β-Lactams	Milk	Amoxicillin	$C_{16}H_{19}N_3O_5S$	365.104544	**366.111820**	364.097268	383.138369	388.093765	YMC ODS-AQ, 50 × 2 mm, 3 µm	Gradient profile: mobile phases A—0.1% formic acid in water; B—0.1% formic acid in water/ acetonitrile (35+65, v/v)		ESI$^+$	366 > 114, 208, 349	20
			Ampicillin	$C_{16}H_{19}N_3O_4S$	349.109629	**350.116905**	348.102353	367.143454	372.098850				ESI$^+$	350 > 106, 160, 192	
			Cloxacillin	$C_{19}H_{18}ClN_3O_5S$	435.065572	**436.072848**	434.058296	453.099397	458.054793				ESI$^+$	436 > 160, 277	
													ESI$^+$	438 > 279	
			Oxacillin	$C_{19}H_{19}N_3O_5S$	401.104544	**402.111820**	400.097268	419.138369	424.093765				ESI$^+$	402 > 160, 243	
			Penicillin G	$C_{16}H_{18}N_2O_4S$	334.098730	**335.106006**	333.091454	352.132555	357.087951				ESI$^+$	335 > 160, 176	
			Penicillin-d_7 G (IS)	$C_{16}H_{11}D_7N_2O_4S$	341.142669	**342.149945**	340.135393	359.176494	364.131890				ESI$^+$	342 > 160	
6	Macrolides	Eggs, raw milk, and honey	Spiramycin I	$C_{43}H_{74}N_2O_{14}$	842.514008	**843.521284**	841.506732	860.547833	865.503229	UHPLC: Acquity UPLC BEH C18, 100 × 2.1 mm, 1.7 µm	Gradient profile: UHPLC mobile phases A—10 mM ammonium acetate; B—acetonitrile, column temperature 45°C; HPLC mobile phases A—0.1% formic acid; B—acetonitrile	QqTOF and QqQ	ESI$^+$	QTOF: 843.5218 MS/MS: 843 > 174, 142	110
			Erythromycin A	$C_{37}H_{67}NO_{13}$	733.461244	**734.468520**	732.453968	751.495069	756.450465	HPLC: YMC ODS-AQ S-3, 50 × 2 mm			ESI$^+$	QTOF: 734.4690 MS/MS: 734 > 158, 576	
			Neospiramycin I	$C_{36}H_{62}N_2O_{11}$	698.435363	**699.442639**	697.428087	716.469188	721.424584				ESI$^+$	QTOF: 699.4432 MS/MS: 699 > 174, 142	
			Oleandomycin	$C_{35}H_{61}NO_{12}$	687.419378	**688.426654**	686.412102	705.453203	710.408599					QTOF: 688.4272 MS/MS: 688 > 158, 544	
			Tilmicosin	$C_{46}H_{80}N_2O_{13}$	868.566043	**869.573319**	867.558767	886.599868	891.555264					QTOF: 869.5738 MS/MS: 869 > 174, 132	
			Tylosin A	$C_{46}H_{77}NO_{17}$	915.519154	**916.526430**	914.511878	933.552979	938.508375					QTOF: 916.5270 MS/MS: 916 > 174, 145	
			Tylosin B (desmycosin)	$C_{39}H_{65}NO_{14}$	771.440509	**772.447785**	770.433233	789.474334	794.429730					QTOF: 772.4483	
7	Macrolides	Liver and kidney	Erythromycin A	$C_{37}H_{67}NO_{13}$	733.461244	**734.468520**	732.453968	751.495069	756.450465	Kromasil 100 C18, 250 × 4.6 mm, 5 µm	Gradient profile: mobile phases A—1% acetic acid; B—acetonitrile	Q	ESI$^+$	734, 576, 158	138
			Josamycin	$C_{42}H_{69}NO_{15}$	827.466724	**828.474000**	826.459448	845.500549	850.455945				ESI$^+$	829, 174	
			Roxithromycin	$C_{41}H_{76}N_2O_{15}$	836.524573	**837.531849**	835.517297	854.558398	859.513794				ESI$^+$	838, 414, 679	
			Spiramycin I	$C_{43}H_{74}N_2O_{14}$	842.514008	**843.521284**	841.506732	860.547833	865.503229				ESI$^+$	422 (**[M+2H]$^{2+}$**), 174, 843	
			Tilmicosin	$C_{46}H_{80}N_2O_{13}$	868.566043	**869.573319**	867.558767	886.599868	891.555264				ESI$^+$	435 (**[M+2H]$^{2+}$**), 869, 174	
			Troleandomycin	$C_{41}H_{67}NO_{15}$	813.451074	**814.458350**	812.443798	831.484899	836.440295				ESI$^+$	722, 814, 435	
			Tylosin A	$C_{46}H_{77}NO_{17}$	915.519154	**916.526430**	914.511878	933.552979	938.508375				ESI$^+$	916, 174, 772	
8	Macrolides Lincosamides	Honey	Tylosin A	$C_{46}H_{77}NO_{17}$	915.519154	**916.526430**	914.511878	933.552979	938.508375	Luna C8, 150 × 2 mm, 5µm	Gradient profile: mobile phases A—0.04% HFBA; B—acetonitrile+ethyl acetate; IPC application	QqQ	ESI$^+$	916 > 174, 772	37
			Lincomycin	$C_{18}H_{34}N_2O_6S$	406.213760	**407.221036**	405.206484	424.247585	429.202981				ESI$^+$	407 > 126, 359	
9	Polyether antibiotics	Eggs	Diclazuril	$C_{17}H_9Cl_3N_4O_2$	405.979110	406.986386	**404.971834**	424.012935	428.968331	Zorbax Eclipse XDB C8, 150 × 3 mm, 5 µm	Gradient profile: mobile phases A—0.1% formic acid in water; B—methanol with 0.1% formic acid. C—acetonitrile with 0.1% formic acid		ESI$^-$	405 > 334, 336	139
			Halofuginone	$\mathbf{C_{16}H_{17}BrClN_3O_3}$	413.014181	**414.021457**	412.006905	431.048006	436.003402				ESI$^+$	416 > 120, 138	
			Lasalocid	$C_{34}H_{54}O_8$	590.381869	591.389145	589.374593	608.415694	**613.371090**				ESI$^+$	613 > 377, 577	
			Maduramicin	$C_{47}H_{80}O_{17}$	916.539555	917.546831	915.532279	934.573380	**939.528776**				ESI$^+$	940 > 878, 720	
			Monensin	$C_{36}H_{62}O_{11}$	670.429215	671.436491	669.421939	688.463040	**693.418436**				ESI$^+$	693 > 675, 462	
			Narasin A	$C_{43}H_{72}O_{11}$	764.507465	765.514741	763.500189	782.541290	**787.496686**				ESI$^+$	787 > 431, 532	
			Nicarbazin	$C_{13}H_{10}N_4O_5$	302.065121	303.072397	**301.057845**	320.098946	325.054342				ESI$^-$	301 > 137, 107	
			Nicarbazin-d_8 (IS)	$C_{13}H_2D_8N_4O_5$	310.115337	311.122613	**309.108061**	328.149162	333.104558				ESI$^-$	309 > 141	
			Robenidine	$C_{15}H_{13}Cl_2N_5$	333.054801	**334.062077**	332.047525	351.088626	356.044022				ESI$^+$	334 > 138, 155	
			Salinomycin	$C_{42}H_{70}O_{11}$	750.491815	751.499091	749.484539	768.525640	**773.481036**				ESI$^+$	774 > 431, 531	
			Semduramicin	$C_{45}H_{76}O_{16}$	872.513340	873.520616	871.506064	890.547165	**895.502561**				ESI$^+$	895 > 833, 851	

(*continued*)

TABLE 6.1 (Continued)

Single-Class					Exact Mass[a,b]					LC		MS[a]			
Example	Class	Matrix	Compound	Molecular Formula	Molecular Weight	$[M+H]^+$	$[M-H]^-$	$[M+NH_4]^+$	$[M+Na]^+$	Column	Mobile Phase	Type	Ionization	Mass or Transitions	Reference
10	Polyether antibiotics	Eggs and chicken	Amprolium	$C_{14}H_{18}N_4$	242.153146	**243.160422**	241.145870	260.186971	265.142367	Acquity UPLC BEH C18, 100 × 2.1 mm, 1.7 µm	Gradient profile: mobile phases A—0.1% formic acid in water; B—methanol	QqQ	ESI+	243 > 150, 94	140
			Diaveridine	$C_{13}H_{16}N_4O_2$	260.127326	**261.134602**	259.120050	278.161151	283.116547				ESI+	261 > 123, 245	
			Diclazuril	$C_{17}H_9Cl_3N_4O_2$	405.979110	406.986386	**404.971834**	424.012935	428.968331				ESI−	405 > 334, 299	
			Dimetridazole	$C_5H_7N_3O_2$	141.053827	**142.061103**	140.046551	159.087652	164.043048				ESI+	142 > 96, 81	
			Ethopabate	$C_{12}H_{15}NO_4$	237.100109	**238.107385**	236.092833	255.133934	260.089330				ESI+	238 > 206, 164	
			Halofuginone	$C_{16}H_{17}BrClN_3O_3$	413.014181	**414.021457**	412.006905	431.048006	436.003402				ESI+	416 > 138, 398	
			Lasalocid	$C_{34}H_{54}O_8$	590.381869	591.389145	589.374593	608.415694	**613.371090**				ESI+	613 > 595, 377	
			Maduramicin	$C_{47}H_{80}O_{17}$	916.539555	917.546831	915.532279	934.573380	**939.528776**				ESI+	940 > 877, 895	
			Metronidazole	$C_6H_9N_3O_3$	171.064392	**172.071668**	170.057116	189.098217	194.053613				ESI+	172 > 128, 82	
			Monensin	$C_{36}H_{62}O_{11}$	670.429215	671.436491	669.421939	688.463040	**693.418436**				ESI+	693 > 675, 461	
			Nicarbazin	$C_{13}H_{10}N_4O_5$	302.065121	303.072397	**301.057845**	320.098946	325.054342				ESI−	301 > 137, 107	
			Robenidine	$C_{15}H_{13}Cl_2N_5$	333.054801	**334.062077**	332.047525	351.088626	356.044022				ESI+	334 > 138, 155	
			Ronidazole	$C_6H_8N_4O_4$	200.054556	**201.061832**	199.047280	218.088381	223.043777				ESI+	201 > 140, 55	
			Salinomycin	$C_{42}H_{70}O_{11}$	750.491815	751.499091	749.484539	768.525640	**773.481036**				ESI+	774 > 431, 531	
11	Polypeptides	Milk and animal tissues	Bacitracin A	$C_{66}H_{103}N_{17}O_{16}S$	1421.748945	1422.756221	1420.741669	1439.782770	1444.738166	Luna C18, 150 × 2.1 mm, 5 µm	Gradient profile: mobile phases A—0.1% formic acid in water; B—0.1% formic acid in acetonitrile	QqQ	ESI+	**$[M+3H]^{3+}$** : 475 > 199, 670	83
			Colistin A (polymyxin E_1)	$C_{53}H_{100}N_{16}O_{13}$	1168.765579	1169.772855	1167.758303	1186.799404	1191.754800				ESI+	**$[M+3H]^{3+}$** : 391 > 385, 379	
			Colistin B (polymyxin E_2)	$C_{52}H_{98}N_{16}O_{13}$	1154.749929	1155.757205	1153.742653	1172.783754	1177.739150				ESI+	**$[M+3H]^{3+}$** : 386 > 380, 374	
12	Polypeptides	Muscle, kidney or liver	Virginiamycin S_1	$C_{43}H_{49}N_7O_{10}$	823.354092	**824.361368**	822.346816	841.387917	846.343313	Inertsil OSD-2, 150 × 2.0 mm, 5 µm	Mobile phases A—3 mM ammonium formate; B—methanol/ acetonitrile (50+50, v/v)	Q or QqQ	ESI+	824	141
			Virginiamycin M_1	$C_{28}H_{35}N_3O_7$	525.247501	**526.254777**	524.240225	543.281326	548.236722				ESI+	526 or 526 > 508, 355, 377	
13	Quinolones or fluoroquinolones	Fish	Ciprofloxacin	$C_{17}H_{18}FN_3O_3$	331.133220	**332.140496**	330.125944	349.167045	354.122441	Perfectsil ODS-2, 250 × 4 mm, 5 µm	Gradients A—acetonitrile; B—methanol and C—water; all mobile phases contained 0.2% formic acid	QqQ	ESI+	332 > 314, 231, 294	142
			Danofloxacin	$C_{19}H_{20}FN_3O_3$	357.148870	**358.156146**	356.141594	375.182695	380.138091				ESI+	358 > 340, 255, 82	
			Enrofloxacin	$C_{19}H_{22}FN_3O_3$	359.164520	**360.171796**	358.157244	377.198345	382.153741				ESI+	360 > 342, 316, 286, 245	
			Flumequine	$C_{14}H_{12}FNO_3$	261.080122	**262.087398**	260.072846	279.113947	284.069343				ESI+	262 > 202, 244, 174	
			Nalidixic acid	$C_{12}H_{12}N_2O_3$	232.084793	**233.092069**	231.077517	250.118618	255.074014				ESI+	233 > 215, 187, 158	
			Oxolinic acid	$C_{13}H_{11}NO_5$	261.063724	**262.071000**	260.056448	279.097549	284.052945				ESI+	262 > 244, 216	
			Sarafloxacin	$C_{20}H_{17}F_2N_3O_3$	385.123797	**386.131073**	384.116521	403.157622	408.113018				ESI+	386 > 368, 348, 299	
14	Quinolones or fluoroquinolones	Honey	Cinoxacin	$C_{12}H_{10}N_2O_5$	262.058973	**263.066249**	261.051697	280.092798	285.048194	Zorbax SB C18, 50 × 2.1 mm, 1.8 µm	Gradient profile: mobile phases A—0.5% formic acid+1 mM nonafluoropentanoic acid (NFPA) in water; B—0.5% formic acid in methanol/acetonitrile (50+50, v/v)	QqLIT	ESI+	263 > 245, 217, 189	143
			Ciprofloxacin	$C_{17}H_{18}FN_3O_3$	331.133220	**332.140496**	330.125944	349.167045	354.122441				ESI+	332 > 314, 228, 245	
			Danofloxacin	$C_{19}H_{20}FN_3O_3$	357.148870	**358.156146**	356.141594	375.182695	380.138091				ESI+	358 > 340, 314, 283	
			Difloxacin	$C_{21}H_{19}F_2N_3O_3$	399.139448	**400.146724**	398.132172	417.173273	422.128669				ESI+	400 > 382, 356, 299	
			Enoxacin	$C_{15}H_{17}FN_4O_3$	320.128469	**321.135745**	319.121193	338.162294	343.117690				ESI+	321 > 303, 277, 257	
			Enrofloxacin	$C_{19}H_{22}FN_3O_3$	359.164520	**360.171796**	358.157244	377.198345	382.153741				ESI+	360 > 342, 316, 245	
			Fleroxacin	$C_{17}H_{18}F_3N_3O_3$	369.130026	**370.137302**	368.122750	387.163851	392.119247				ESI+	370 > 326, 352, 269	
			Flumequine	$C_{14}H_{12}FNO_3$	261.080122	**262.087398**	260.072846	279.113947	284.069343				ESI+	262 > 244, 202, 220	
			Lomefloxacin	$C_{17}H_{19}F_2N_3O_3$	351.139448	**352.146724**	350.132172	369.173273	374.128669				ESI+	352 > 334, 308, 251	

			Marbofloxacin	$C_{17}H_{19}FN_4O_4$	362.139034	**363.146310**	361.131758	380.172859	385.128255				ESI$^+$	363 > 345, 320, 277	
			Nalidixic acid	$C_{12}H_{12}N_2O_3$	232.084793	**233.092069**	231.077517	250.118618	255.074014				ESI$^+$	233 > 215, 187, 159	
			Norfloxacin	$C_{16}H_{18}FN_3O_3$	319.133220	**320.140496**	318.125944	337.167045	342.122441				ESI$^+$	320 > 302, 276, 233	
			Ofloxacin	$C_{18}H_{20}FN_3O_4$	361.143785	**362.151061**	360.136509	379.177610	384.133006				ESI$^+$	362 > 344, 318, 261	
			Oxolinic acid	$C_{13}H_{11}NO_5$	261.063724	**262.071000**	260.056448	279.097549	284.052945				ESI$^+$	262 > 244, 216, 160	
			Pipemidic acid	$C_{14}H_{17}N_5O_3$	303.133140	**304.140416**	302.125864	321.166965	326.122361				ESI$^+$	304 > 286, 217, 189	
			Sarafloxacin	$C_{20}H_{17}F_2N_3O_3$	385.123797	**386.131073**	384.116521	403.157622	408.113018				ESI$^+$	386 > 368, 342, 299	
15	Quinolones or fluoroquinolones	Milk	Cinoxacin	$C_{12}H_{10}N_2O_5$	262.058973	**263.066249**	261.051697	280.092798	285.048194	Acquity UPLC Shield RP18, 100 × 2.1 mm, 1.7μm	Gradient profile: mobile phases A—0.2% formic acid in water (pH 3.0); B—methanol/ acetonitrile (40+60, v/v), column temperature 40°C	QqQ	ESI$^+$	263 > 217, 245	144
			Ciprofloxacin	$C_{17}H_{18}FN_3O_3$	331.133220	**332.140496**	330.125944	349.167045	354.122441				ESI$^+$	332 > 288, 314	
			Danofloxacin	$C_{19}H_{20}FN_3O_3$	357.148870	**358.156146**	356.141594	375.182695	380.138091				ESI$^+$	358 > 82, 340	
			Difloxacin	$C_{21}H_{19}F_2N_3O_3$	399.139448	**400.146724**	398.132172	417.173273	422.128669				ESI$^+$	400 > 299, 357	
			Enoxacin	$C_{15}H_{17}FN_4O_3$	320.128469	**321.135745**	319.121193	338.162294	343.117690				ESI$^+$	321 > 232, 303	
			Enrofloxacin	$C_{19}H_{22}FN_3O_3$	359.164520	**360.171796**	358.157244	377.198345	382.153741				ESI$^+$	360 > 316, 342	
			Fleroxacin	$C_{17}H_{18}F_3N_3O_3$	369.130026	**370.137302**	368.122750	387.163851	392.119247				ESI$^+$	370 > 269, 326	
			Flumequine	$C_{14}H_{12}FNO_3$	261.080122	**262.087398**	260.072846	279.113947	284.069343				ESI$^+$	262 > 202, 244	
			Gatifloxacin	$C_{19}H_{22}FN_3O_4$	375.159435	**376.166711**	374.152159	393.193260	398.148656				ESI$^+$	376 > 261, 332	
			Lomefloxacin	$C_{17}H_{19}F_2N_3O_3$	351.139448	**352.146724**	350.132172	369.173273	374.128669				ESI$^+$	352 > 265, 308	
			Marbofloxacin	$C_{17}H_{19}FN_4O_4$	362.139034	**363.146310**	361.131758	380.172859	385.128255				ESI$^+$	363 > 320, 346	
			Moxifloxacin	$C_{21}H_{24}FN_3O_4$	401.175084	**402.182360**	400.167808	419.208909	424.164305				ESI$^+$	402 > 364, 385	
			Nadifloxacin	$C_{19}H_{21}FN_2O_4$	360.148536	**361.155812**	359.141260	378.182361	383.137757				ESI$^+$	361 > 283, 343	
			Nalidixic acid	$C_{12}H_{12}N_2O_3$	232.084793	**233.092069**	231.077517	250.118618	255.074014				ESI$^+$	233 > 187, 215	
			Norfloxacin	$C_{16}H_{18}FN_3O_3$	319.133220	**320.140496**	318.125944	337.167045	342.122441				ESI$^+$	320 > 276, 302	
			Ofloxacin	$C_{18}H_{20}FN_3O_4$	361.143785	**362.151061**	360.136509	379.177610	384.133006				ESI$^+$	362 > 318, 344	
			Oxolinic acid	$C_{13}H_{11}NO_5$	261.063724	**262.071000**	260.056448	279.097549	284.052945				ESI$^+$	262 > 216, 244	
			Pazufloxacin	$C_{16}H_{15}FN_2O_4$	318.101586	**319.108862**	317.094310	336.135411	341.090807				ESI$^+$	319 > 281, 301	
			Pefloxacin	$C_{17}H_{20}FN_3O_3$	333.148870	**334.156146**	332.141594	351.182695	356.138091				ESI$^+$	334 > 290, 316	
			Pipemidic acid	$C_{14}H_{17}N_5O_3$	303.133140	**304.140416**	302.125864	321.166965	326.122361				ESI$^+$	304 > 217, 287	
			Sarafloxacin	$C_{20}H_{17}F_2N_3O_3$	385.123797	**386.131073**	384.116521	403.157622	408.113018				ESI$^+$	386 > 342, 368	
			Sparfloxacin	$C_{19}H_{22}F_2N_4O_3$	392.165996	**393.173272**	391.158720	410.199821	415.155217				ESI$^+$	393 > 292, 349	
16	Sulfonamides	Honey	Sulfachloropyridazine	$C_{10}H_9ClN_4O_2S$	284.013475	285.020751	**283.006199**	302.047300	307.002696	Xterra MS C18, 150 × 2.1 mm, 3.5 μm	Gradient profile: mobile phases A—0.15% acetic acid; B—0.15% acetic acid in methanol	QqQ	ESI$^-$	283 > 156, 92	81
			Sulfadiazine	$C_{10}H_{10}N_4O_2S$	250.052448	251.059724	**249.045172**	268.086273	273.041669				ESI$^-$	249 > 185, 92	
			Sulfadimethoxine	$C_{12}H_{14}N_4O_4S$	310.073578	311.080854	**309.066302**	328.107403	333.062799				ESI$^-$	309 > 66, 122	
			Sulfadoxine	$C_{12}H_{14}N_4O_4S$	310.073578	311.080854	**309.066302**	328.107403	333.062799				ESI$^-$	309 > 156, 251	
			Sulfamerazine	$C_{11}H_{12}N_4O_2S$	264.068097	265.075373	**263.060821**	282.101922	287.057318				ESI$^-$	263 > 199, 108	
			Sulfameter	$C_{11}H_{12}N_4O_3S$	280.063012	281.070288	**279.055736**	298.096837	303.052233				ESI$^-$	279 > 264, 196	
			Sulfamethazine	$C_{12}H_{14}N_4O_2S$	278.083748	279.091024	**277.076472**	296.117573	301.072969				ESI$^-$	277 > 106, 122	
			Sulfamethoxazole	$C_{10}H_{11}N_3O_3S$	253.052114	254.059390	**252.044838**	271.085939	276.041335				ESI$^-$	252 > 156, 92	
			Sulfamethoxypyridazine	$C_{11}H_{12}N_4O_3S$	280.063012	281.070288	**279.055736**	298.096837	303.052233				ESI$^-$	279 > 156, 264	
			Sulfamonomethoxine	$C_{11}H_{12}N_4O_3S$	280.063012	281.070288	**279.055736**	298.096837	303.052233				ESI$^-$	279 > 132, 66	
			Sulfapyridine	$C_{11}H_{11}N_3O_2S$	249.057199	250.064475	**248.049923**	267.091024	272.046420				ESI$^-$	248 > 184, 93	
			Sulfaquinoxaline	$C_{14}H_{12}N_4O_2S$	300.068097	301.075373	**299.060821**	318.101922	323.057318				ESI$^-$	299 > 144, 117	
			Sulfathiazole	$C_9H_9N_3O_2S_2$	255.013621	256.020897	**254.006345**	273.047446	278.002842				ESI$^-$	254 > 156, 98	
			Sulfisoxazole	$C_{11}H_{13}N_3O_3S$	267.067764	268.075040	**266.060488**	285.101589	290.056985				ESI$^-$	266 > 171, 239	
17	Sulfonamides	Milk and eggs	Sulfachloropyridazine	$C_{10}H_9ClN_4O_2S$	284.013475	**285.020751**	283.006199	302.047300	307.002696	Luna NH_2 (HILIC column), 150 × 2.0 mm, 3 μm	Gradient profile: mobile phases A—0.05% formic acid in water; B—0.05% formic acid in acetonitrile, HILIC application	Q	ESI$^+$	285	58
			Sulfadiazine	$C_{10}H_{10}N_4O_2S$	250.052448	**251.059724**	249.045172	268.086273	273.041669				ESI$^+$	251	
			Sulfadimethoxine	$C_{12}H_{14}N_4O_4S$	310.073578	**311.080854**	309.066302	328.107403	333.062799				ESI$^+$	311	
			Sulfadoxine	$C_{12}H_{14}N_4O_4S$	310.073578	**311.080854**	309.066302	328.107403	333.062799				ESI$^+$	311	
			Sulfamerazine	$C_{11}H_{12}N_4O_2S$	264.068097	**265.075373**	263.060821	282.101922	287.057318				ESI$^+$	265	
			Sulfameter (IS)	$C_{11}H_{12}N_4O_3S$	280.063012	**281.070288**	279.055736	298.096837	303.052233				ESI$^+$	281	
			Sulfamethazine	$C_{12}H_{14}N_4O_2S$	278.083748	**279.091024**	277.076472	296.117573	301.072969				ESI$^+$	279	
			Sulfamethizole	$C_9H_{10}N_4O_2S_2$	270.024520	**271.031796**	269.017244	288.058345	293.013741				ESI$^+$	271	
			Sulfamethoxazole	$C_{10}H_{11}N_3O_3S$	253.052114	**254.059390**	252.044838	271.085939	276.041335				ESI$^+$	254	
			Sulfamethoxypyridazine	$C_{11}H_{12}N_4O_3S$	280.063012	**281.070288**	279.055736	298.096837	303.052233				ESI$^+$	281	
			Sulfamonomethoxine	$C_{11}H_{12}N_4O_3S$	280.063012	**281.070288**	279.055736	298.096837	303.052233				ESI$^+$	281	
			Sulfapyridine	$C_{11}H_{11}N_3O_2S$	249.057199	**250.064475**	248.049923	267.091024	272.046420				ESI$^+$	250	

(continued)

TABLE 6.1 (*Continued*)

Example	Single-Class Class	Matrix	Compound	Molecular Formula	Exact Mass[a,b] Molecular Weight	$[M+H]^+$	$[M-H]^-$	$[M+NH_4]^+$	$[M+Na]^+$	LC Column	LC Mobile Phase	MS[a] Type	MS Ionization	MS Mass or Transitions	Reference
			Sulfathiazole	$C_9H_9N_3O_2S_2$	255.013621	**256.020897**	254.006345	273.047446	278.002842				ESI^+	256	
			Sulfisoxazole	$C_{11}H_{13}N_3O_3S$	267.067764	**268.075040**	266.060488	285.101589	290.056985				ESI^+	268	
18	Sulfonamides		Dapsone	$C_{12}H_{12}N_2O_2S$	248.061950	**249.069226**	247.054674	266.095775	271.051171	Zorbax SB C18, 50 × 2.1 mm, 1.8 μm	Gradient profile: mobile phases A—0.5% formic acid +1 mM NFPA; B—methanol/acetonitrile (50 : 50, v/v) containing 0.5% formic acid	QqLIT	$APPI^+$	249 > 156, 92	80
			Sulfabenzamide	$C_{13}H_{12}N_2O_3S$	276.056865	**277.064141**	275.049589	294.090690	299.046086				$APPI^+$	277 > 156, 92	
			Sulfachloropyridazine	$C_{10}H_9ClN_4O_2S$	284.013475	**285.020751**	283.006199	302.047300	307.002696				$APPI^+$	285 > 156, 92	
			Sulfadiazine	$C_{10}H_{10}N_4O_2S$	250.052448	**251.059724**	249.045172	268.086273	273.041669				$APPI^+$	251 > 156, 92	
			Sulfadiazine-d_4 (IS)	$C_{10}H_6D_4N_4O_2S$	254.077560	**255.084836**	253.070284	272.111385	277.066781				$APPI^+$	255 > 160, 96	
			Sulfadimethoxine	$C_{12}H_{14}N_4O_4S$	310.073578	**311.080854**	309.066302	328.107403	333.062799				$APPI^+$	311 > 156, 92	
			Sulfadimethoxine-d_4 (IS)	$C_{12}H_{10}D_4N_4O_4S$	314.098686	**315.105962**	313.091410	332.132511	337.087907				$APPI^+$	315 > 160, 96	
			Sulfadoxine	$C_{12}H_{14}N_4O_4S$	310.073578	**311.080854**	309.066302	328.107403	333.062799				$APPI^+$	311 > 156, 108	
			Sulfamerazine	$C_{11}H_{12}N_4O_2S$	264.068097	**265.075373**	263.060821	282.101922	287.057318				$APPI^+$	265 > 156, 92	
			Sulfamerazine-d_4 (IS)	$C_{11}H_8D_4N_4O_2S$	268.093205	**269.100481**	267.085929	286.127030	291.082426				$APPI^+$	269 > 160, 96	
			Sulfameter	$C_{11}H_{12}N_4O_3S$	280.063012	**281.070288**	279.055736	298.096837	303.052233				$APPI^+$	281 > 156, 92	
			Sulfamethazine	$C_{12}H_{14}N_4O_2S$	278.083748	**279.091024**	277.076472	296.117573	301.072969				$APPI^+$	279 > 186, 156	
			Sulfamethazine-d_4 (IS)	$C_{12}H_{10}D_4N_4O_2S$	282.108856	**283.116132**	281.101580	300.142681	305.098077				$APPI^+$	283 > 186, 160	
			Sulfamethoxazole	$C_{10}H_{11}N_3O_3S$	253.052114	**254.059390**	252.044838	271.085939	276.041335				$APPI^+$	254 > 156, 92	
			Sulfamethoxazole-d_4 (IS)	$C_{10}H_7D_4N_3O_3S$	257.077222	**258.084498**	256.069946	275.111047	280.066443				$APPI^+$	258 > 160, 96	
			Sulfamethoxypyridazine	$C_{11}H_{12}N_4O_3S$	280.063012	**281.070288**	279.055736	298.096837	303.052233				$APPI^+$	281 > 156, 92	
			Sulfamoxole	$C_{11}H_{13}N_3O_3S$	267.067764	**268.075040**	266.060488	285.101589	290.056985				$APPI^+$	268 > 156, 92	
			Sulfapyridine	$C_{11}H_{11}N_3O_2S$	249.057199	**250.064475**	248.049923	267.091024	272.046420				$APPI^+$	250 > 156, 108	
			Sulfaquinoxaline	$C_{14}H_{12}N_4O_2S$	300.068097	**301.075373**	299.060821	318.101922	323.057318				$APPI^+$	301 > 156, 92	
			Sulfathiazole	$C_9H_9N_3O_2S_2$	255.013621	**256.020897**	254.006345	273.047446	278.002842				$APPI^+$	256 > 156, 92	
			Sulfathiazole-d_4 (IS)	$C_9H_5D_4N_3O_2S_2$	259.038729	**260.046005**	258.031453	277.072554	282.027950				$APPI^+$	260 > 160, 96	
			Sulfisomidine	$C_{12}H_{14}N_4O_2S$	278.083748	**279.091024**	277.076472	296.117573	301.072969				$APPI^+$	279 > 156, 124	
							MULTI-CLASS								
19	Aminoglycosides	Honey	Streptomycin	$C_{21}H_{39}N_7O_{12}$	581.265673	**582.272949**	580.258397	599.299498	604.254894	Polar-RP Synergi, 50 × 2.0 mm, 4 μm	Gradient profile: mobile phases A—0.1% formic acid in water; B—0.1 % formic acid in acetonitrile	QqQ	ESI^+	582 > 263, 246	12
	Lincosamides		Lincomycin	$C_{18}H_{34}N_2O_6S$	406.213760	**407.221036**	405.206484	424.247585	429.202981				ESI^+	407 > 126, 359	
	Macrolides		Tylosin A	$C_{46}H_{77}NO_{17}$	915.519154	**916.526430**	914.511878	933.552979	938.508375				ESI^+	916 > 772, 174	
			Erythromycin A	$C_{37}H_{67}NO_{13}$	733.461244	**734.468520**	732.453968	751.495069	756.450465				ESI^+	734 > 576, 522	
	Phenicols		Chloramphenicol	$C_{11}H_{12}Cl_2N_2O_5$	322.012329	323.019605	**321.005053**	340.046154	345.001550				ESI^-	321 > 152, 194	
	Polyether antibiotics		Monensin	$C_{36}H_{62}O_{11}$	670.429215	671.436491	669.421939	688.463040	**693.418436**				ESI^+	693 > 461, 479	
	Quinolones or fluoroquinolones		Ciprofloxacin	$C_{17}H_{18}FN_3O_3$	331.133220	**332.140496**	330.125944	349.167045	354.122441				ESI^+	332 > 288, 254	
			Danofloxacin	$C_{19}H_{20}FN_3O_3$	357.148870	**358.156146**	356.141594	375.182695	380.138091				ESI^+	358 > 314, 283	
			Difloxacin	$C_{21}H_{19}F_2N_3O_3$	399.139448	**400.146724**	398.132172	417.173273	422.128669				ESI^+	400 > 356, 299	
			Enrofloxacin	$C_{19}H_{22}FN_3O_3$	359.164520	**360.171796**	358.157244	377.198345	382.153741				ESI^+	360 > 316, 245	
			Sarafloxacin	$C_{20}H_{17}F_2N_3O_3$	385.123797	**386.131073**	384.116521	403.157622	408.113018				ESI^+	386 > 342, 299	
	Sulfonamides		Sulfathiazole	$C_9H_9N_3O_2S_2$	255.013621	**256.020897**	254.006345	273.047446	278.002842				ESI^+	256 > 156, 92	
	Tetracyclines		Chlortetracycline	$C_{22}H_{23}ClN_2O_8$	478.114296	**479.121572**	477.107020	496.148121	501.103517				ESI^+	479 > 444, 462	
			Doxycycline	$C_{22}H_{24}N_2O_8$	444.153268	**445.160544**	443.145992	462.187093	467.142489				ESI^+	445 > 321, 410	
			Oxytetracycline	$C_{22}H_{24}N_2O_9$	460.148183	**461.155459**	459.140907	478.182008	483.137404				ESI^+	461 > 426, 444	
			Tetracycline	$C_{22}H_{24}N_2O_8$	444.153268	**445.160544**	443.145992	462.187093	467.142489				ESI^+	445 > 410, 154	
	Others		Fumagillin	$C_{26}H_{34}O_7$	458.230455	**459.237731**	457.223179	476.264280	481.219676				ESI^+	459 > 233, 215	
20	Macrolides	Honey	Erythromycin A	$C_{37}H_{67}NO_{13}$	733.461244	**734.468520**	732.453968	751.495069	756.450465	Acquity UPLC BEH C18, 100 × 2.1 mm, 1.7 μm	Gradient profile: mobile phases A—0.05% formic acid in water; B—methanol, runtime 7.5 min, column temperature 30°C.	QqQ	ESI^+	**$[M-H_2O+H]^+$**: 716 > 158, 116	9
			Josamycin	$C_{42}H_{69}NO_{15}$	827.466724	**828.474000**	826.459448	845.500549	850.455945				ESI^+	829 > 174, 109	
			Tilmicosin	$C_{46}H_{80}N_2O_{13}$	868.566043	**869.573319**	867.558767	886.599868	891.555264				ESI^+	870 > 174, 697	
			Tylosin A	$C_{46}H_{77}NO_{17}$	915.519154	**916.526430**	914.511878	933.552979	938.508375				ESI^+	916 > 174, 101	
	Quinolones or fluoroquinolones		Danofloxacin	$C_{19}H_{20}FN_3O_3$	357.148870	**358.156146**	356.141594	375.182695	380.138091				ESI^+	358 > 340, 255	
			Difloxacin	$C_{21}H_{19}F_2N_3O_3$	399.139448	**400.146724**	398.132172	417.173273	422.128669				ESI^+	400 > 382, 356	
			Enrofloxacin	$C_{19}H_{22}FN_3O_3$	359.164520	**360.171796**	358.157244	377.198345	382.153741				ESI^+	360 > 342, 316	

			Marbofloxacin	$C_{17}H_{19}FN_4O_4$	362.139034	**363.146310**	361.131758	380.172859	385.128255				ESI+	363 > 320, 345	
			Sarafloxacin	$C_{20}H_{17}F_2N_3O_3$	385.123797	**386.131073**	384.116521	403.157622	408.113018				ESI+	386 > 368, 348	
	Sulfonamides		Sulfachloropyridazine	$C_{10}H_9ClN_4O_2S$	284.013475	**285.020751**	283.006199	302.047300	307.002696				ESI+	285 > 156, 80	
			Sulfadimethoxine	$C_{12}H_{14}N_4O_4S$	310.073578	**311.080854**	309.066302	328.107403	333.062799				ESI+	311 > 156, 245	
			Sulfadimidine	$C_{12}H_{14}N_4O_2S$	278.083748	**279.091024**	277.076472	296.117573	301.072969				ESI+	279 > 92, 124	
			Sulfaquinoxaline	$C_{14}H_{12}N_4O_2S$	300.068097	**301.075373**	299.060821	318.101922	323.057318				ESI+	301 > 156, 108	
	Tetracyclines		Chlortetracycline	$C_{22}H_{23}ClN_2O_8$	478.114296	**479.121572**	477.107020	496.148121	501.103517				ESI+	479 > 444, 462	
			Doxycycline	$C_{22}H_{24}N_2O_8$	444.153268	**445.160544**	443.145992	462.187093	467.142489				ESI+	445 > 428, 154	
			Oxytetracycline	$C_{22}H_{24}N_2O_9$	460.148183	**461.155459**	459.140907	478.182008	483.137404				ESI+	461 > 443, 426	
			Tetracycline	$C_{22}H_{24}N_2O_8$	444.153268	**445.160544**	443.145992	462.187093	467.142489				ESI+	445 > 410, 427	
21	Aminoglycosides	Honey	Dihydrostreptomycin	$C_{21}H_{41}N_7O_{12}$	583.281323	**584.288599**	582.274047	601.315148	606.270544	Zorbax SB C18, 50 × 2.1 mm, 1.8 μm	Gradient profile: mobile phases A—1 mM, nonafluoropentanoic acid+0.5% formic acid in water; B—acetonitrile/ methanol (50+50, v/v) with 0.5% formic acid	QqLIT	ESI+	584 > 263, 246	7
			Neomycin B	$C_{23}H_{46}N_6O_{13}$	614.312289	**615.319565**	613.305013	632.346114	637.301510				ESI+	615 > 455, 161	
			Streptomycin	$C_{21}H_{39}N_7O_{12}$	581.265673	**582.272949**	580.258397	599.299498	604.254894				ESI+	582 > 263, 407	
	β-Lactams		Amoxicillin	$C_{16}H_{19}N_3O_5S$	365.104544	**366.111820**	364.097268	383.138369	388.093765				ESI+	366 > 208, 349	
			Ampicillin	$C_{16}H_{19}N_3O_4S$	349.109629	**350.116905**	348.102353	367.143454	372.098850				ESI+	350 > 192, 174	
			Cloxacillin	$C_{19}H_{18}ClN_3O_5S$	435.065572	**436.072848**	434.058296	453.099397	458.054793				ESI+	436 > 277, 160	
			Dicloxacillin	$C_{19}H_{17}Cl_2N_3O_5S$	469.026600	**470.033876**	468.019324	487.060425	492.015821				ESI+	470 > 160, 311	
			Nafcillin	$C_{21}H_{22}N_2O_5S$	414.124945	**415.132221**	413.117669	432.158770	437.114166				ESI+	415 > 199, 181	
			Oxacillin	$C_{19}H_{19}N_3O_5S$	401.104544	**402.111820**	400.097268	419.138369	424.093765				ESI+	402 > 243, 160	
			Penethamate	$C_{22}H_{31}N_3O_4S$	433.203529	**434.210805**	432.196253	451.237354	456.192750				ESI+	434 > 259, 100	
			Penicillin G	$C_{16}H_{18}N_2O_4S$	334.098730	**335.106006**	333.091454	352.132555	357.087951				ESI+	335 > 160, 176	
	Macrolides		Erythromycin A	$C_{37}H_{67}NO_{13}$	733.461244	**734.468520**	732.453968	751.495069	756.450465				ESI+	734 > 576, 158	
			Oleandomycin	$C_{35}H_{61}NO_{12}$	687.419378	**688.426654**	686.412102	705.453203	710.408599				ESI+	688 > 158, 544	
			Roxithromycin	$C_{41}H_{76}N_2O_{15}$	836.524573	**837.531849**	835.517297	854.558398	859.513794				ESI+	837 > 158, 679	
			Spiramycin I	$C_{43}H_{74}N_2O_{14}$	842.514008	**843.521284**	841.506732	860.547833	865.503229				ESI+	843 > 174, 540	
			Tilmicosin	$C_{46}H_{80}N_2O_{13}$	868.566043	**869.573319**	867.558767	886.599868	891.555264				ESI+	869 > 174, 156	
			Tylosin A	$C_{46}H_{77}NO_{17}$	915.519154	**916.526430**	914.511878	933.552979	938.508375				ESI+	916 > 174, 156	
			Tylosin B (desmycosin)	$C_{39}H_{65}NO_{14}$	771.440509	**772.447785**	770.433233	789.474334	794.429730				ESI+	772 > 174, 156	
	Phenicols		Chloramphenicol	$C_{11}H_{12}Cl_2N_2O_5$	322.012329	323.019605	321.005053	340.046154	345.001550				ESI+	**[M−H$_2$O+H]$^+$**: 305 > 275, 165	
			Thiamphenicol	$C_{12}H_{15}Cl_2NO_5S$	355.004802	356.012078	353.997526	373.038627	377.994023				ESI+	**[M−H$_2$O+H]$^+$**: 338 > 308, 229	
	Sulfonamides		Dapsone	$C_{12}H_{12}N_2O_2S$	248.061950	**249.069226**	247.054674	266.095775	271.051171				ESI+	249 > 156, 108	
			Sulfabenzamide	$C_{13}H_{12}N_2O_3S$	276.056865	**277.064141**	275.049589	294.090690	299.046086				ESI+	277 > 156, 108	
			Sulfachloropyridazine	$C_{10}H_9ClN_4O_2S$	284.013475	**285.020751**	283.006199	302.047300	307.002696				ESI+	285 > 156, 92	
			Sulfadiazine	$C_{10}H_{10}N_4O_2S$	250.052448	**251.059724**	249.045172	268.086273	273.041669				ESI+	251 > 156, 92	
			Sulfadimethoxine	$C_{12}H_{14}N_4O_4S$	310.073578	**311.080854**	309.066302	328.107403	333.062799				ESI+	311 > 156, 245	
			Sulfadoxine	$C_{12}H_{14}N_4O_4S$	310.073578	**311.080854**	309.066302	328.107403	333.062799				ESI+	311 > 156, 108	
			Sulfamerazine	$C_{11}H_{12}N_4O_2S$	264.068097	**265.075373**	263.060821	282.101922	287.057318				ESI+	265 > 156, 172	
			Sulfameter	$C_{11}H_{12}N_4O_3S$	280.063012	**281.070288**	279.055736	298.096837	303.052233				ESI+	281 > 156, 215	
			Sulfamethazine	$C_{12}H_{14}N_4O_2S$	278.083748	**279.091024**	277.076472	296.117573	301.072969				ESI+	279 > 124, 108	
			Sulfamethoxazole	$C_{10}H_{11}N_3O_3S$	253.052114	**254.059390**	252.044838	271.085939	276.041335				ESI+	254 > 156, 92	
			Sulfamethoxypyridazine	$C_{11}H_{12}N_4O_3S$	280.063012	**281.070288**	279.055736	298.096837	303.052233				ESI+	281 > 156, 126	
			Sulfamoxole	$C_{11}H_{13}N_3O_3S$	267.067764	**268.075040**	266.060488	285.101589	290.056985				ESI+	268 > 156, 108	
			Sulfanilamide	$C_6H_8N_2O_2S$	172.030650	**173.037926**	171.023374	190.064475	195.019871				ESI+	173 > 156, 108	
			Sulfapyridine	$C_{11}H_{11}N_3O_2S$	249.057199	**250.064475**	248.049923	267.091024	272.046420				ESI+	250 > 156, 108	
			Sulfaquinoxaline	$C_{14}H_{12}N_4O_2S$	300.068097	**301.075373**	299.060821	318.101922	323.057318				ESI+	301 > 156, 92	
			Sulfathiazole	$C_9H_9N_3O_2S_2$	255.013621	**256.020897**	254.006345	273.047446	278.002842				ESI+	256 > 156, 92	
			Sulfisomidine	$C_{12}H_{14}N_4O_2S$	278.083748	**279.091024**	277.076472	296.117573	301.072969				ESI+	279 > 124, 156	
	Tetracyclines		Chlortetracycline	$C_{22}H_{23}ClN_2O_8$	478.114296	**479.121572**	477.107020	496.148121	501.103517				ESI+	479 > 444, 462	
			Demeclocycline	$C_{21}H_{21}ClN_2O_8$	464.098646	**465.105922**	463.091370	482.132471	487.087867				ESI+	465 > 448, 430	
			Doxycycline	$C_{22}H_{24}N_2O_8$	444.153268	**445.160544**	443.145992	462.187093	467.142489				ESI+	445 > 428, 410	
			Oxytetracycline	$C_{22}H_{24}N_2O_9$	460.148183	**461.155459**	459.140907	478.182008	483.137404				ESI+	461 > 426, 443	
			Tetracycline	$C_{22}H_{24}N_2O_8$	444.153268	**445.160544**	443.145992	462.187093	467.142489				ESI+	445 > 410, 427	

[a] Number or text in bold font style indicates ionization form or charge state.

[b] The electron mass (0.000549 amu) is substracted or added depending on the charge state when calculating exact mass. Exact mass is calculated by the authors of this chapter.

There are a few aspects that need to be considered when developing an LC-MS multi-class method:

1. The differences in physical and chemical properties make it a challenge to extract and analyze all antibiotics from a matrix using a generic procedure. Under certain circumstances, it is necessary to adjust pH and/or to add chelating agents during the extraction to enhance extraction efficiency, especially when solid phase extraction (SPE) is used. Antibiotics have a wide range of pK_a values, which are approximately 2–2.5 and 5–7.5 for sulfonamides, 7.5–9 for macrolides, 3–4, 7–8, and 9–10 for tetracyclines and 3–4, 6, 7.5–9, and 10–11 for fluoroquinolones.[13,14] The pK_a values of individual antibiotics can be found in Chapter 1. The pH of a sample or an extraction solvent determines the charge state of antibiotics and eventually affects the extraction efficiency from SPE cartridges. For example, sulfonamides are amphoteric, and react as either an acid or a base. They have values of $pK_{a,1}$ 2–2.5 and $pK_{a,2}$ 5–7.5 for arylamin and sulfonamino groups, respectively. The sulfonamides are positively charged at pH values ≤ 2 and negatively charged when pH ≥ 5. The recovery enrichment of sulfonamides observed between pH 2 and 6 is a result of their increased hydrophobicity, and highest recoveries are achieved for sulfonamides in their uncharged forms (pH 4) when using reversed phase SPE cartridges.[15,16] At pH 4, tetracyclines form a zwitterionic state that favors hydrophobic interactions, and they have high molar solubility. In addition, tetracyclines tend to form strong complex with cations or metals (Ca^{2+} and Mg^{2+} ions) and bind to protein and silanol groups. Therefore, chelating agents such as McIlvain buffer, EDTA, Na_2EDTA, citric acid, and oxalic acid are used to prevent chelation between tetracyclines and metals to improve extraction efficiency. McIlvaine/EDTA at pH 4 is the most prevalent buffer for extracting tetracyclines. In addition, tetracyclines are susceptible to conformational degradation to their 4-epimers in aqueous solution and during sample preparation depending on the pH and temperature. Therefore, chromatographic separation and quantification of tetracyclines and their 4-epimers residues remain a challenge, which is discussed further in Chapter 7.[14,17,18]

2. The stability of various classes at different pH values makes it less than ideal to extract some antibiotics under the same conditions. Tetracyclines are more stable at pH 3–4 than at pH > 5, and changes in pH can lead to the formation of their epimers and even further degradation.[19] Macrolides and β-lactams are stable at neutral or slightly basic conditions (i.e., pH ≥ 8 or 8.5).[20–23] Erythromycin degrades to erythromycin-H_2O with loss of one H_2O molecule at low pH[24] and is often measured as erythromycin-H_2O at *m/z* 716 in environmental samples but not in food.[18] The degradation of erythromycin to erythromycin-H_2O was even observed in a mixture of 0.1% formic acid and acetonitrile (50+50, v/v).[22,23] Tylosin A degrades gradually to tylosin B (desmycosin) under acidic conditions such as in honey.[25] Some β-lactams, including amoxicillin, ampicillin, cefazolin, cephapirin, cloxacillin, dicloxacillin, nafcillin, oxacillin, penicillin G, and penicillin V, degrade rapidly in methanol (somewhat more slowly for cephalosporins) but slowly in a mixture of methanol/water (50+50, v/v); they are stable in water, acetonitrile, and acetonitrile/water solutions. Monobasic penicillins, especially penicillin G and nafcillin, degrade rapidly in 0.1% formic acid, whereas the degradation of amoxicillin, ampicillin, and cephalosporins is less pronounced. The presence of methanol and/or 0.1% formic acid in the extraction solvent or final extracts can cause significant degradation of the β-lactams.[26]

3. Under certain circumstances, respective parent antibiotics cannot be targeted as marker residues, and therefore, derivatization and chemical conversion are required. Nitrofurans including furazolidone, furaltadone, nitrofurazone, and nitrofurantoin are metabolized rapidly to 3-amino-2-oxazolidinone (AOZ), 3-amino-5-morpholinomethyl-2-oxazolidinone (AMOZ), semicarbazide (SC), and 1-aminohydantoin (AH) respectively. These metabolites bind strongly to proteins in animals. Therefore, analytical methods for the determination of nitrofurans often focus on the detection of protein-bound residues or active side-chains as discussed in Chapter 7. The extraction procedure involves an overnight acid hydrolysis and simultaneous derivatization of the released side-chains with 2-nitrobenzaldehyde (2-NBA) to form their nitrophenyl derivatives that can be analyzed by LC-MS.[27,28] Ceftiofur is administered parenterally to lactating dairy cattle, and it is metabolized quickly to desfuroylceftiofur, which then forms a variety of metabolites and conjugates or is bound to proteins. Its maximum residue limit (MRL) is defined as the sum of all residues retaining the β-lactam structure expressed as desfuroylceftiofur. Therefore the extraction involves the release of desfuroylceftiofur from the various conjugated forms with a reducing agent such as dithioerythritol followed by derivatization with iodoacetamide to form an acetamide derivative, namely, desfuroylceftiofur acetamide, which is stable and suitable for LC-MS analysis.[29,30]

4. MRLs or required method validation concentrations can be different for various residues. It may not be practical to quantify all antibiotics of interest in the same analytical range, if it is required. The analytical ranges for quantification are usually based on MRLs, the minimum required performance limit (MRPL) concentration, or a required method validation concentration. LC-MS typically has a linear or quadratic dynamic range of 2–3 orders of magnitude, and some antibiotics demonstrate better linear response than others. In general, it may be possible

to quantify antibiotics in the same range only if the concentrations of interest are within 2 orders of magnitude. For example, chloramphenicol has a MRPL of 0.3 μg/kg, whereas lincomycin has a MRL of 150 μg/kg in milk. The difference in concentration is 2.7 orders of magnitude, and it would be a challenge to analyze both compounds with a single calibration curve. In addition, a high concentration can cause cross-contamination or carryover that potentially leads to a false positive.

5. Not all antibiotics are retained and separated well on the same type of analytical columns. Columns of different retention mechanism and/or mobile-phase additives are required for improved chromatographic performance, which is discussed further in Section 6.3.3.

6.3 CHROMATOGRAPHIC SEPARATION

6.3.1 Chromatographic Parameters

Several chromatographic parameters or characteristics are important when choosing a column or selecting an LC elution profile for residue method development and validation. These characteristics include the following:[31,32]

Retention factor (k) or *capacity factor*, which is a measure of the relative speed of an analyte through a column and is calculated as:

$$k = \frac{t_R - t_0}{t_0}$$

where t_R is the retention time of an analyte and t_0 is the retention time of an unretained peak. Ideally, k should be greater than 2 so that the analyte peak is definitely separated from the solvent peak.

Separation factor (α), also termed *selectivity* or *relative retention*, which is a measure of the chromatographic separation of two analytes and is calculated as:

$$\alpha = \frac{k_2}{k_1} = \frac{t_{R2} - t_0}{t_{R1} - t_0}$$

with $k_2 > k_1$. When $\alpha = 1$, the two peaks elute at the same time.

Resolution (R_s), which describes the degree of separation between two analytes, is based on the distance between two peaks and their dispersion. Resolution is calculated as a ratio of the distance between peaks to their width.

$$R_s = \frac{\Delta t_R}{\frac{1}{2}(w_{t,1} + w_{t,2})} = 1.18\frac{\Delta t_R}{w_{1/2,1} + w_{1/2,2}}$$

where Δt_R is the retention time difference betweens peaks, w_t is peak width with $w = 4\sigma$, and $w_{1/2}$ is the peak width at half-height. For two truly Gaussian peaks with equal peak size, when $R_s = 1$, two peaks can be identified but not completely separated; whereas when $R_s = 1.5$, two peaks are separated with baseline resolution.

Column efficiency (*plate count* N) or *height equivalent to a theoretical plate* (HETP), which is a measure of the quality of separation. Column efficiency takes into account peak dispersion. The efficiency is expressed as the number of theoretical plates:

$$N = 5.54\left(\frac{t_R}{w_{1/2}}\right)^2$$

where t_R is the retention time of analyte and $w_{1/2}$ is the peak width at half peak height. An efficient column minimizes dispersion. High efficiency results in effective separation and improved sensitivity. HETP is calculated as column length (L) divided by plate count (N).

Column void volume (V_c, μl), which is the volume between and within the pores of the individual particles. The column void volume is calculated as:

$$V_c = \frac{d^2 \times \pi \times L \times V_p}{4}$$

where d is column internal or inner diameter (mm), L is column length (mm), and V_p is pore volume or constant (no unit) and is between 0.6 and 0.7. With respect to column void volume and flow rate, the *dead* or *breakthrough time*, which is the time required by the mobile phase to pass through a column, can be roughly estimated and then used to calculate the retention factor of an analyte. Extra void volumes, such as from injector and interconnecting tubing, should be considered.

6.3.2 Mobile Phase

The composition of the LC mobile phase, including buffer or additive concentrations and pH, is critical for the optimal ionization efficiency, ion spray stability, and chromatographic separation of antibiotics. The selection of proper solvents depends on the compounds being analyzed, as well as the ionization techniques and columns that are utilized. Solvents that are most suitable for LC-MS mobile phases are water, methanol, and acetonitrile. Methanol (viscosity, $\eta = 0.60$ mPa·s) produces higher column back-pressure than does acetonitrile ($\eta = 0.37$ mPa·s). The column back-pressure changes during the course of an LC gradient and is proportional to the mobile-phase viscosity. A mixture of methanol and water generates even higher viscosity than pure methanol or water ($\eta = 1.00$ mPa·s). The viscosity reaches a maximum of 1.62 mPa·s when 40% methanol is mixed in water.[31] However, the mixture of acetonitrile and water leads to a decrease in viscosity with an increase of acetonitrile. Consequently, acetonitrile and water have become the most commonly used solvents for LC-MS. Under certain circumstances such as in reversed-phase LC with APCI or APPI, methanol is preferred for improved ionization efficiency because it has lower proton affinity (PA) than acetonitrile, especially for analysis of the analytes with relatively low PAs. For example, methanol is a better choice than acetonitrile for LC/APCI-MS analysis of steroids, whose PAs are low. In APPI, acetonitrile absorbs photons more efficiently than methanol, which results in a decreased number of photons available for ionization reactions and therefore in decreased sensitivity.[33] It is worth mentioning that methanol or a mixture of methanol and acetonitrile can improve chromatographic peak shapes, resulting in relatively narrow peaks.

Liquid chromatography–mass spectrometry mobile-phase additives or modifiers are utilized to enhance ion abundance, to suppress sodium adducts, and to improve chromatographic peak shape. Common acidic additives used for positive ionization include formic acid (pH = 2.6–2.8), acetic acid (pH = 3.2–3.4), or trifluoroacetic acid (pH = 1.8–2.0) prepared at a concentration of 0.1% (v/v). Formic acid is the most commonly used acid. Ammonium formate ($pK_{a,1} = 3.7$, $pK_{a,2} = 9.2$) and ammonium acetate ($pK_{a,1} = 4.8$, $pK_{a,2} = 9.2$), prepared at a concentration of 5–20 mM, are volatile salts that can be used for either positive or negative ionization under neutral conditions. Basic additives, commonly prepared as 10 mM solutions, include triethylamine, ammonium hydroxide and other compounds. These are applicable to negative ionization under basic conditions. Occasionally, hydrophobic ion-pairing reagents such as heptafluorobutyric acid (HFBA, prepared at 0.04%), trifluoroacetic acid (TFA, prepared at 0.02–0.1%), and nonafluoropentanoic acid (NFPA, prepared at 1 mM) are used to improve chromatographic peak shape[34,35] and to extend the retention of polar analytes such as lincomycin or aminoglycosides.[7,36,37] Ion-pairing reagents are further discussed in Section 6.3.3.2.

Generally, LC gradient profiles are applied for better retention and separation, especially for analysis of multiple antibiotics. An LC runtime that uses a gradient is relatively long, and about one-third or one-half of the total runtime is required to regenerate the column for the next injection. An LC runtime that utilizes an isocratic condition can be fairly short, and there is no need to regenerate the column between runs. However, an isocratic condition is practical only when a small number of analytes are intended to be analyzed.

6.3.3 Conventional Liquid Chromatography

6.3.3.1 Reversed Phase Chromatography

The most common liquid chromatography employed for analysis of antibiotics is reversed phase liquid chromatography (RPLC). RPLC stationary phases (3.5- or 5.0-μm particles) consist of non-polar or hydrophobic organic species (e.g., octyl, octadecyl, phenyl, or cyanopropyl groups) attached by siloxane (or silyl ether) bonds (—Si—O—Si—) to the surface of a silica support. Chemically bonded octadecylsilane (ODS, C18) or an 18-carbon atom *n*-alkane is the most frequently used stationary phase. Shorter alkyl chain columns such as C8, phenyl- or cyanopropyl bonded phases are less non-polar and occasionally serve as alternatives. The surfaces of reversed phase packing are hydrophobic, and their retention mechanisms are functions of hydrophobicity, partitioning, and/or adsorption.[32] Most antibiotics are well retained and chromatographically separated on reversed phase columns (Table 6.1) except for aminoglycosides and lincomycin, which require the use of an ion-pairing reagent or hydrophilic interaction liquid chromatographic (HILIC) columns. RPLC mobile phases consist of acetonitrile or methanol and water with additives. The initial RPLC profile usually starts with >90% water followed by a gradient for which acetonitrile or methanol is used as the elution solvent. The degree of retention and the selectivity of RPLC are largely dependent on the nature and composition of the mobile phases, pH, modifier, and/or the ionic strength of the mobile phases.

6.3.3.2 Ion-Pairing Chromatography

The words "ion-pairing chromatography" (IPC) refers to the chromatographic application that uses lipophilic or hydrophobic ions, namely, an ion-pairing agent (IPA), to perform the separation of organic and inorganic ionic solutes with adequate retention, good column efficiency, and resolution on traditional reversed phase columns.[38] IPAs are ionic compounds that contain a relatively hydrophobic *n*-alkyl chain. When paired with polar analytes as counter ions, they reduce analyte hydrophilicity, and thus increase the retention on RPLC columns as well as reduce peak tailing. IPAs for acidic compounds include dipropylamine acetate and other dialkylamines such as dibutylamine acetate and dipentylamine acetate. These can be utilized for negatively charged acidic compounds, but their applications are rare in antibiotic analysis. IPAs for basic compounds, which include trifluoroacetic acid (C_1, TFA) and other perfluorinated carboxylic acids such as pentafluoropropionic acid (C_2, PFPA), heptafluorobutyric acid (C_3, HFBA, the most commonly used), and nonafluoropentanoic acid (C_4, NFPA), are used for positively charged basic analytes. In general, IPA retention ability increases with increasing alkyl chain length and increasing IPA concentration.[39,40] The IPA concentration or pH in the mobile phase is a key factor used to manipulate analyte charge status and thus affect retention and selectivity, but the final concentration in the mobile phase should be less than 20 mM. High acid concentration results in hydrolysis of octyl- or octadecyl-bonded silica and can damage RPLC columns. Ideally, the concentration should be less than 5 mM as long as significant retention, for example, retention factor $k > 2$ and/or selectivity for analytes are achieved. IPAs for basic analytes have been widely used for LC-MS analysis of aminoglycosides (up to 13 analytes) and lincosamides (lincomycin) (Table 6.1),[7,18,37,41,42] especially for dihydrostreptomycin and streptomycin,[43–45] on RPLC columns in various food matrices. In this case, IPAs form ion pairs with the amine functional groups on the aminoglycosides that are positively charged under acidic conditions, thereby increasing hydrophobicity for improved retention. However, it is known that acidic IPAs such as HFBA, TFA, and NFPA can lead to ESI ion suppression and produce significant loss in signal, especially for nitrogen-atom-containing compounds, due to ion-pairing effects in the ESI process.[46] Therefore, ion-pairing reagents should be avoided, if possible, to achieve a low limit of detection (LOD) for trace antibiotic detection. Occasionally IPAs have been added to the mobile phases simply because aminoglycosides and/or lincomycin were included in the list of antibiotics that are analyzed.

The decreased sensitivity seen with the use of IPAs can be regained to a certain degree through post-column addition of a volatile organic acid such as propionic acid (75% in isopropanol) that minimizes ion-pairing effects in the ion source. For example, the ESI-MS response of amphetamine (counter ion : TFA)[47] or ceftiofur (counter ion : HFBA)[48] was enhanced more than six-fold by the post-column addition of propionic acid. The enhancement effect is attributed to the fact that a less volatile organic acid (e.g., propionic acid) at a relatively high concentration would displace a strong ion-pair (TFA or HFBA-analyte) and form a weak ion-pair (propionic acid–analyte) driven by mass action, which subsequently generates the protonated analyte ion and thus increases sensitivity.[47,49]

6.3.3.3 Hydrophilic Interaction Liquid Chromatography

Hydrophilic interaction liquid chromatography (HILIC) was first introduced in the early 1990s by Alpert.[50] More recently it has become an emerging chromatographic technique for retention and separation of polar and hydrophilic compounds, especially for small molecules that are basic or contain nitrogen atoms. HILIC stationary phases, whose primary function is to bind water at the surface, are made from bare silica or derivatized silica bonded with different polar functional groups such as amine, amide, cyano, diol, and sulfoalkylbetaine.[51–53] Acetonitrile is the most frequently used organic solvent in HILIC, and methanol is utilized when analytes are not soluble in acetonitrile.[51,53] The solvent strength for HILIC applications is roughly inverted to what is observed for RPLC separations, and the relative solvent strength may be outlined as tetrahydrofuran (polarity index 4.0) < acetone (5.1) < acetonitrile (5.8) < isopropanol (3.9) < ethanol (5.2) < methanol (5.1) < water (9.0).[54] In general, HILIC mobile phases consist of 40–95% acetonitrile, and the water content in mobile phases is maintained at 5–60%. The initial HILIC profile starts with 95% acetonitrile followed by a gradient where water is used as the elution solvent. Like RPLC, the degree of retention and the selectivity of HILIC are largely dependent on the mobile-phase composition, including parameters such as pH, modifier choice, and its concentration. The HILIC retention mechanism involves a combination of partitioning, electrostatic interaction, and/or hydrogen bonding. Partitioning has been considered the key retention mechanism; analytes partition between a water-enriched layer of stagnant eluent on the hydrophilic stationary phase and a relatively hydrophobic bulk eluent (40–95% acetonitrile in water). The more hydrophilic an analyte is, the longer its retention time on a HILIC column. To keep the water-enriched layer, HILIC columns should not run under either 100% organic or 100% aqueous. However, HILIC separation is not explained solely by the partitioning mechanism and might involve some kinds of dipole–dipole interactions and hydrogen bonding as well.[50,51] Electrostatic interactions, which add additional selectivity to the separation, result from the coulombic attraction between charged analytes and a deprotonated silanol group or a sulfoalkylbetaine zwitterionic ion on the stationary phases. Ammonium formate or acetate in low concentration (5–20 mM, with or without 0.1% (v/v) formic or acetic acid), is often used to disrupt the electrostatic interactions, eluting the analytes and reducing peak tailing. Depending on the application, a high concentration of buffer (>100 mM) may be used to improve chromatographic peak shapes, but this could lead to a decrease in ESI sensitivity.[55] The elution order in HILIC is generally opposite that from RPLC, giving useful alternative selectivity. Analytes that have little or no retention to RPLC columns may be well retained on HILIC. Mass spectrometer sensitivity is enhanced by the high organic content in the mobile phase and the high desolvation efficiency of organic solvent.[56] In addition, extracts eluted from reversed phase SPE cartridges, that is, with solvents of high organic content, can be injected directly to HILIC, avoiding solvent evaporation and reconstitution steps that could cause loss of sample.

The applications of HILIC for antibiotics have focused mainly on aminoglycosides and lincomycin due to their poor RPLC retention. One example was the use of HILIC/ESI-MS/MS for the analysis of seven aminoglycoside antibiotics, including spectinomycin, dihydrostreptomycin, streptomycin, kanamycin, apramycin, gentamicin, and neomycin in kidney and muscle tissues (Table 6.1).[57] The method could achieve limits of quantification as low as 25 μg/kg for gentamicin; 50 μg/kg for spectinomycin, dihydrostreptomycin, kanamycin, and apramycin; and 100 μg/kg for streptomycin and neomycin. A relatively high concentration of ammonium acetate (150 mM) along with 1% formic acid was used in a mobile phase in order to obtain sharp chromatographic peaks for all compounds as

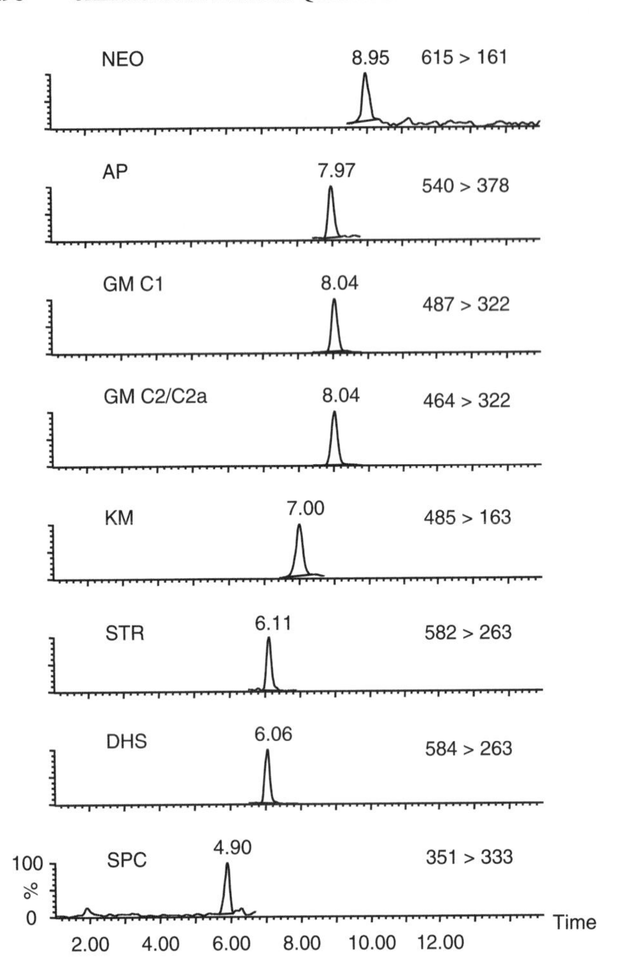

Figure 6.1 HILIC ESI/MS/MS chromatograms of a swine kidney sample spiked with gentamicin (GM) at 25 ng/g, spectinomycin (SPC), dihydrostreptomycin (DHS), kanamycin (KM), and apramycin (AP) at 50 ng/g, and streptomycin (STR) and neomycin (NEO) at 100 ng/g. (Reproduced from Ishii et al.[57] with permission from Taylor & Francis; copyright 2008.)

shown in Figure 6.1. Other studies that explored the potential uses of HILIC for antibiotics included analyses of sulfonamide residues in milk and eggs[58] and tetracyclines.[59] However, in those applications, HILIC was not truly employed for the purpose of improved retention but to avoid steps or problems associated with RPLC such as sample extract evaporation and reconstitution.

6.3.4 Ultra-High-Performance or Ultra-High-Pressure Liquid Chromatography

Conventional LC with C18-modified silica stationary-phase or HILIC columns has been widely used in antibiotic residue analysis as discussed above. Since the 1980s, the need for faster and more powerful separations for the analyses of constituents, such as protein digests, in complex matrices led to the development of highly mechanically stable (to high back-pressures) columns and ultra-high-performance or ultra-high-pressure (>10,000-psi) liquid chromatography (UHPLC).[60] UHPLC is an emerging technique in separation science with wide applications, including the analysis of antibiotic residues. It takes advantage of <2-μm-particle columns, which results in significant improvements in analytical speed, chromatographic resolution, and detection sensitivity, particularly when coupled with mass spectrometers capable of performing high-speed data acquisition. The <2-μm columns available for UHPLC include those packed with BEH C18, Shield RP18, C8, and phenyl-bonded packing materials. The small particles used in these columns typically result in high system back-pressure (>10,000 psi). UHPLC makes it possible to achieve a fewfold faster separations compared to conventional LC, while maintaining or providing even higher column efficiency. This can be explained by the van Deemter equation, an empirical formula that describes the relationship between linear velocity and HETP.[61,62] The van Deemter equation is defined as:

$$H = A + \frac{B}{u} + Cu \approx 2\lambda d_p + \frac{2\gamma D_m}{u} + f(k)\frac{d_p^2 u}{D_m}$$

where A, B, and C are constants that account for contributions to band broadening from eddy diffusion (A), longitudinal diffusion (B), and mass transfer resistance (C). Other terms in the expanded equation include: u is the linear velocity, γ is a constant called tortuosity or obstruction factor, D_m is the diffusion coefficient of an analyte in the mobile phase, d_p is the diameter of the packing material, and k is the retention factor for an analyte. The particle size of column packing material plays a major role in affecting the terms A and C. Columns that are packed with small particles provide enhanced efficiency by virtue of the relatively small intra-particulate mass transfer resistance due to short diffusion distances and, to a lesser extent, the small contribution of eddy diffusion to plate height.[62] As illustrated in the van Deemter plot (Fig. 6.2), 1.7-μm particles demonstrate improved performance over the commonly used 3.5 and 5.0 μm sizes. The 1.7-μm particles give 2–3 times lower plate height values, which are translated into better resolution, improved sensitivity, faster separation, or reduced analysis time. To make it simple, the benefits of UHPLC are explained for isocratic conditions, but they apply to gradient separations as well. The resolution is proportional to the square root of the column efficiency, and the efficiency at the optimum linear velocity is inversely proportional to the particle size. Therefore, the resolution is inversely proportional to the square root of the particle size. As an example, 1.7-μm particles offer $\sqrt{\frac{3.5}{1.7}} =$ 1.4 times and $\sqrt{\frac{5}{1.7}} = 1.7$ times greater resolution than do 3.5- and 5.0-μm particles, respectively, at equal column lengths. Furthermore, column efficiency at the optimum

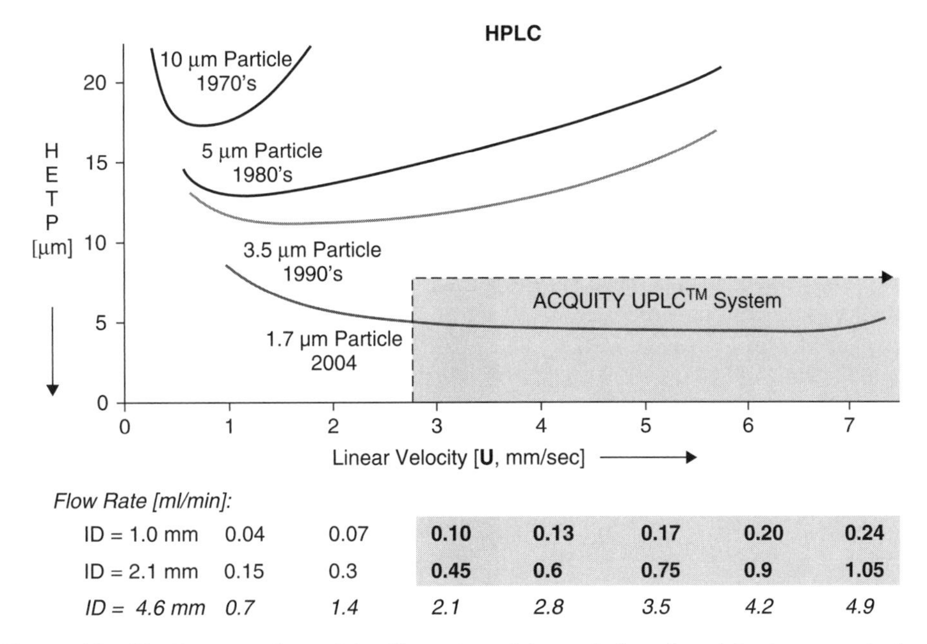

Figure 6.2 Van Deemter plot and the illustration of the evolution of particle sizes over the last three decades. (Graph provided courtesy of Waters Corp.)

linear velocity is proportional to column length L divided by particle size d_p. With constant L/d_p, the sensitivity gain is inversely proportional to the column length that is used to generate the same efficiency because shorter and smaller particle columns generate narrower and taller peaks. For example, a 34-mm column with 1.7-μm particles is 2.06 ≈ 2 times shorter than a 70 mm length column with 3.5-μm particles and 2.94 ≈ 3 times shorter than a 100-mm column with 5.0-μm particles. All of these columns provide the same column efficiency and resolution as a result of a constant $L/d_p = 20{,}000$. Therefore, sensitivity gain for the column with the smaller particles is $\frac{70}{34} = 2.06 \approx 2$ times over the 3.5-μm column or $\frac{100}{34} = 2.94 \approx 3$ times over the one with 5-μm particles. Finally, because the optimum flow rate is also inversely proportional to particle size and analysis time was inversely proportional to flow rate, the optimum flow rate of 1.7-μm particles is $\frac{3.5}{1.7} = 2.06 \approx 2$ times or $\frac{5.0}{1.7} = 2.94 \approx 3$ times higher than that of 3.5 or 5 μm sizes, respectively, which translates to 4 or 9 times faster analysis time with the same resolution.[60]

The fact that the van Deemter plots of small particles are generally flat beyond the optimum linear velocity means that higher linear velocity would not compromise column efficiency significantly. Therefore, in practice, rapid gradients and high flow rates are usually applied in small particle columns to shorten the runtime and increase sample throughput. The column temperature also affects separation speed and column efficiency, and is used to control the selectivity of the separation as well. An increase in temperature leads to a decrease in retention in reversed phase UHPLC. It is generally considered that the A term does not depend on temperature, but the B term and C term are temperature-dependent. The B term is directly proportional to the diffusion coefficient, whereas the C term is inversely proportional to the diffusion coefficient. Thus, with increasing temperature, the diffusion of analytes in the mobile and stationary phase is increased. The viscosity of the mobile phase decreases with an increase in temperature, and thereby also enhances the diffusion of analytes and leads to an increase of the absolute plate number for a given column. A lower viscosity and higher diffusivity of a mobile phase at a higher temperature result in a flatter van Deemter curve. Therefore, a higher column temperature speeds up the separation process and allows for fast LC without sacrificing column efficiency.[62,63] Current instruments can easily produce 60–90°C eluent, and mobile-phase pre-heating and column heating are two common approaches used to generate high eluent temperatures (> 60°C). Temperature programming, which includes heating and then cooling the mobile phase before and after the column, is still in the early development stage for applications. Additionally, studies are required to investigate the benefits of high eluent temperature for antibiotic analysis, especially their stability under a relatively high temperature on a column.

Ultra-high-performance liquid chromatography has been increasingly used for the determination of multi-residue or multi-class antibiotics in food (Table 6.1). For example, an UHPLC/ESI-MS/MS method was reported to simultaneously analyze 17 different veterinary drug residues belonging to several classes of antibiotics such as macrolides, tetracyclines, quinolones, and sulfonamides in honey in 7.5 min.[9] The method was able to quantify and confirm the residues at low concentrations (<4 μg/kg). In another example, an UHPLC TOF MS was employed to determine 150 veterinary drugs and metabolite residues, which included avermectins, benzimidazoles, β-agonists, β-lactams, corticoids, macrolides, nitroimidazoles, quinolones, sulfonamides, tetracyclines, and some others, in raw milk in 9 min.[11] The method achieved LODs in a range of 0.5–25 μg/l. Except for compounds that have very low MRLs such as clenbuterol (0.05 μg/l) or corticosteroids (0.3 μg/l), the LODs were generally lower than the MRLs. Other good examples included UHPLC TOF-MS analyses of 100 veterinary drugs in various animal tissues in 14.6 min[64] and 100 veterinary drugs in egg, fish, and meat in 12 min.[10]

It should be mentioned that there is another type of relatively new column that is made from the 2.7-μm fused-core silica particles, bonded with C18 alkyl chains, by fusing a 0.5-μm porous silica layer onto 1.7-μm non-porous silica cores. The selectivity of the fused-core particle columns is very similar to that of certain <2-μm C18 columns and has the advantage of a substantially lower back-pressure at much higher flow rates, which allows rapid separations to be performed even routinely on a conventional LC system without significant loss in efficiency or resolution.[65] The fused-core columns are new to antibiotic analysis and may serve as good alternatives to <2-μm columns in the field.

6.4 MASS SPECTROMETRY

Mass spectrometry is the most widely used analytical technique available to scientists for quantification, confirmation, identification, and chemical structural elucidation. It is based on the *in vacuo* separation of ions, in the gas phase, according to their mass-to-charge (*m/z*) ratio. Current LC-MS instruments allow for the determination of almost all antibiotics in food, which have molecular weights between 100 and 1200 Da, with the majority in the range of 200–500 Da.

6.4.1 Ionization and Interfaces

Analytes are introduced into mass spectrometers in gas, liquid, or solid states. In the latter two cases, volatilization must be accomplished either prior to or accompanying ionization. Many ionization techniques are available to produce charged species from analytes; the most common ones are electron ionization (electron impact ionization), chemical ionization, matrix-assisted laser desoprtion ionization and atmospheric-pressure ionization (electrospray, atmospheric-pressure chemical ionization, and atmospheric-pressure photoionization). Electron ionization utilizes accelerated electrons (70 eV) colliding with gaseous analyte molecules to knock off free electrons and produce positively charged ions. Electron ionization induces extensive fragmentation, and therefore, both molecular ions and fragment ions are produced. These ions result in reproducible mass spectra, which can be searched against detailed spectral libraries to determine the structure or identity of analytes of interest. Chemical ionization relies on the interaction or collision of analytes of interest with primary ions, typically ionized reagent gas, present in the MS source to produce analyte ions in either positive or negative mode. Generally chemical ionization produces less fragmentation than electron impact, and consequently, is complementary to electron ionization in GC-MS. Molecular ions can be more easily identified with chemical ionization. Both electron impact and chemical ionization are applied to thermally stable lower mass (<1000) volatile compounds and are commonly used for GC-MS applications. Chemical ionization may be preferable for quantification, but is generally less useful for confirmation.

Matrix-assisted laser desoprtion ionization (MALDI) and electrospray (ESI) were revolutionary ionization techniques that marked a breakthrough in analytical chemistry in the late 1980s. These ionization methods allow for the determination of a broad range of large and small, non-volatile, and thermally labile compounds. They are known as *soft ionization* because they produce intact molecular species, and these techniques have been widely used for analysis of molecules in proteomics, as well as for pharmaceuticals and/or chemical residues or contaminants. MALDI was introduced and developed in the middle to late 1980s by Karas et al.[66,67] and Tanaka et al.[68] It has been utilized mainly for the determination of intact molecules of proteins, oligonucleotides, polysaccharides, and synthetic polymers, with minimal fragmentation. MALDI is usually coupled to time-of-flight (TOF) mass analyzers but occasionally to other mass analyzers such as the triple-quadrupole, Fourier transform ion cyclotron resonance, or quadrupole ion trap. In MALDI MS, sample preparation is very simple, and the sample size is very small 1 or 2 μl. Samples or sample extracts are mixed and cocrystallized with an organic acid, the so-called matrix, such as 2,5-dihydroxybenzoic acid, 3-aminoquinoline, 2′,4′,6′-trihydroxyacetophenone, and α-cyano-4-hydroxycinnamic acid on a probe. The sample–matrix crystals are bombarded with UV laser beam (a nitrogen laser at a wavelength of 337 nm), leading to absorption of the laser energy by the matrix, and subsequent desorption and ionization of the analytes of interest, mainly as singly charged species.[69] MALDI is tolerant toward alkali metals in buffer. The time used to analyze a sample is very short and could be <1 min. Although it is used mainly for large molecules, it has been employed to characterize and quantify small molecules such as anthocyanins and flavonoids in food.[70] Its applications for antibiotics are scarce. This may be because the "matrix" produces many low-mass fragments (*m/z* < 500) that can interfere with many antibiotics in that mass range. However, if the matrix background can be subtracted from the mass spectrum or if any deconvolution software is able to identify analytes in the spectrum, MALDI could be a very promising technique for antibiotic screening because of its speed of analysis.

Electrospray was first described as an ionization technique to produce macroions (macromolecules) in the gas phase by Dole et al. in 1968.[71] It was in 1984 that Yamashita and Fenn[72] developed electrospray as a true interface for mass spectrometry. Later on, Fenn and coworkers found its applications for mass spectrometric analysis of biological macromolecules, and this work was awarded the Nobel Prize in Chemistry in 2002.[73] ESI, which is applicable to both large and small molecules, is a process by which analytes in solution are brought into the gas phase through a mechanism of evaporation and desorption. Analytes are ionized mainly through protonation ($[M+H]^+$) and occasionally by formation of adduct ions ($[M+NH_4]^+$ or $[M+Na]^+$) in the positive mode. Some are negatively charged by deprotonation ($[M-H]^-$) in the negative mode. ESI produces singly charged ions and multiply charged ions, and the number of charges tends to increase as the molecular weight increases. It is applicable to polar and medium-polarity analytes, and has become the "gold standard" of LC-MS interfaces (Fig. 6.3).[74] ESI dominates 95% of LC-MS applications for antibiotic analysis, and therefore it is the main focus of discussion in this chapter.

Atmospheric-pressure chemical ionization (APCI), which was introduced in the early 1970s,[33] is an ionization technique in which analytes are ionized through gas-phase reactions between analytes and primary reactant ions produced by corona discharges that take place at atmospheric pressure. The primary reactant ions originate from mobile phases and desolvation gas. The ionization process includes proton transfer, charge exchange, and/or adduct formation, which are the same as those in classical chemical ionization. APCI often generates single-charged ion species and is applied mainly to less polar and relatively non-polar analytes that are small molecules (mass <2000 Da) (Fig. 6.3). APCI tolerates higher salt and additive concentrations and therefore, shows less matrix effects than ESI. Its applications for antibiotic analysis are not as common. Examples of its applications include the determination of tetracyclines in edible swine tissues[75] and fluoroquinolones in chicken tissues.[76]

Atmospheric-pressure photoionization (APPI), which was first applied as an LC-MS interface in 2000,[77] is a relatively new ionization technique. The ionization process is initiated by an ultraviolet lamp (krypton discharge lamp), which emits 10.0 and 10.6 eV photons. Any compounds that have ionization energies below 10 or 10.6 eV can be ionized directly by the photons. The ionization often takes place with the use of a dopant such as toluene, acetone, or benzene. These solvents possess ionization energies below 10.0 or 10.6 eV and can be ionized by the photons to enhance or initiate the ionization of analytes. Thus, the formation of analyte ions is directly through photoionization, or proton transfer and/or charge exchange.[33,78] APPI has an ionization range similar to that of APCI and is expected to broaden the range of ionizable compounds toward the low-polarity side (Fig. 6.3). It can play a complementary role to ESI for analysis of relatively non-polar compounds and has demonstrated some applications for the determination of drugs, lipids, pesticides, synthetic organics, petroleum

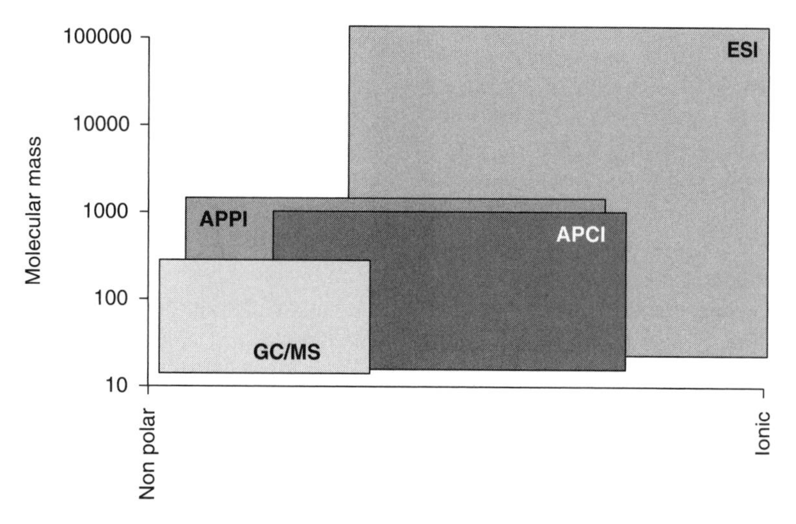

Figure 6.3 Application range of different LC-MS interfaces as a function of compound polarity and molecular weight. (Reproduced from Hernandez et al.[74] with permission from Springer; copyright 2005.)

derivatives, and other compounds. Examples include the LC/APPI-MS/MS determination of chloramphenicol ($[M-H]^-$) in fish tissues with LODs between 0.1 and 0.27 μg/kg[79] and the analysis of 16 sulfonamides in honey (Table 6.1) with a LOD in a range from 0.4 to 4.5 μg/kg for a targeted concentration of 50 μg/kg.[80] Although ESI dominates ~95% of LC-MS applications, APPI and APCI tend to improve sensitivity, provide relatively wide linear dynamic range, and reduce matrix effects under certain circumstances.

Table 6.1 lists the ionized forms, MRM transitions, and exact masses of antibiotics that are commonly reported in the literature or are monitored by the international bodies related to food safety. Most of the antibiotics are ionized by protonation ($[M+H]^+$), forming a singly charged ion. Polyether antibiotics or ionophore coccidiostats form sodium adducts ($[M+Na]^+$) in positive mode. Sulfonamides have been reported to ionize in both positive and negative electrospray modes.[81] In general, the positive mode provides better sensitivity for most antibiotics as compared to negative mode, except for chloramphenicol, diclazuril, and nicarbazin. When an antibiotic contains more than one nitrogen atom, it can be ionized to form singly, doubly, or triply charged molecular ions in the positive ESI mode, depending on the number of nitrogens it contains.[18,82] For example, erythromycin, tylosin, and oleadomycin, each of which contains one nitrogen, form only singly charged ($[M+H]^+$) ions. Spiramycin, neospiramycin, tilmicosin, and roxithromycin, each of which contains two nitrogen atoms, form singly ($[M+H]^+$) and doubly charged ($[M+2H]^{2+}$) ions (Table 6.1). The same ionization phenomena (Table 6.1) were observed for aminoglycosides[42] and polypeptides.[83]

6.4.2 Matrix Effects

Matrix effects, which result from the interference of LC co-eluting compounds on the ionization of analytes during the ESI process, induce either ion suppression or enhancement. The effects are matrix-dependent, and ultimately affect the LC-MS quantitative results. Several measures, which include sample extraction, clean-up, dilution, and chromatography, are mandatory and effective to reduce matrix effects. Sample extraction and/or clean-up as discussed in Chapter 4 remove the majority of endogenous compounds present in samples, but a small amount often remains in the final sample extracts. Dilution is the simplest "clean-up" approach and should be considered first as long as the required detection concentrations are achieved. LC or UHPLC separates analytes from some matrices, which definitely helps to reduce matrix effects. However, no matter what procedures are adopted, matrix effects may not be completely eliminated. Consequently, matrix effects need to be evaluated and compensation is made to achieve the best accuracy for an LC-MS quantitative method. Matrix effects are commonly estimated through either post-column infusion[84] or on-column injection.[85] The post-column infusion (Fig. 6.4a) entails the injection of a blank sample extract onto an LC-MS system while continuously teeing in a standard solution after the column but prior to the LC-MS interface. An observed valley (ion suppression) or peak (ion enhancement) at the corresponding retention time of an analyte indicates the existence of matrix effects. Matrix effects can be better quantified with the on-column injection method. The on-column injection technique (Fig. 6.4b) compares the response of a standard prepared in solvent to that of a standard prepared in a matrix at the same concentration. Three aspects of quantitative information, which include matrix effects (MEs), recovery (RE, extraction efficiency), and "process efficiency" (PE, a combination of both matrix effects and extraction efficiency), can be obtained by this process.[85] The ME, RE, and PE values are calculated as follows:

$$\mathrm{ME}(\%) = \frac{B}{A} \times 100$$

$$\mathrm{RE}(\%) = \frac{C}{B} \times 100$$

$$\mathrm{PE}(\%) = \frac{C}{A} \times 100 = \frac{\mathrm{ME} \times \mathrm{RE}}{100}$$

where A is the peak response (area or height) of a standard in solvent, B is the corresponding peak response for a standard spiked into a blank matrix extract after extraction, and C is the peak response for a standard spiked into a sample before extraction. In general, when ME (%) is between 70% and 110%, the extraction or clean-up is considered to be adequate or sufficient to provide reproducible LC chromatography and consistent mass spectrometric responses. When RE (%) is between 70% and 110% and a matrix-matched standard calibration curve is used, a method provides acceptable accuracy for quantification. When PE (%) is between 70% and 110%, a solvent standard calibration curve may be acceptable for quantification.

To improve the accuracy of LC-MS quantitative results, matrix effects can be compensated for by means of isotopically labeled internal standards, matrix-matched standard calibration curves, standard addition, echo-peak technique, post-column infusion, extrapolative dilution, and so on. Isotopically labeled internal standards and/or matrix-matched standard calibration curves are two common approaches that have been widely used. Table 6.1 lists some commercially available isotopically labeled internal standards. Although this method provides the most accurate result, sometimes it is not realistic to have isotopically labeled standards for each individual analyte. Therefore, matrix-matched standard calibration curves, with or without

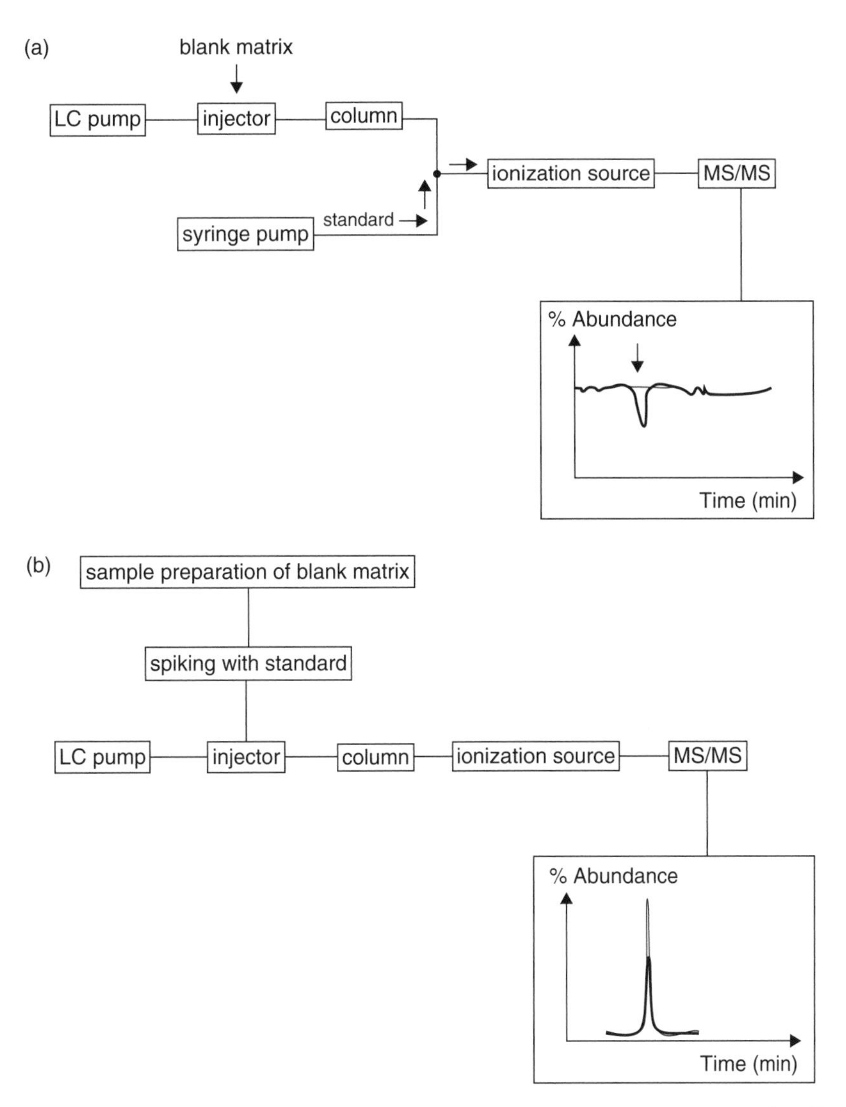

Figure 6.4 Schematic overview of the two commonly used methods to assess matrix effects in LC/ESI–MS/MS. (a) The post-column infusion method. The dashed line represents the signal of the analyte. The full line is obtained when injecting blank matrix. The arrow indicates the region of ion suppression. (b) The post-extraction spike method. The dashed peak represents the standard in neat solution. The full-line peak is obtained with standard spiked in matrix post-extraction. A clear reduction of the peak area is observed, which indicates ion suppression. (Reproduced from Van Eeckhaut et al.[135] with permission from Elsevier; copyright 2009.)

the use of a chemical analog as an internal standard, serve as a practical alternative approach. Matrix-matched standard calibration curves can be constructed through either post-extraction spike or pre-extraction fortification. The former is called a *matrix-matched standard calibration curve* (MSCC), in which the standards are added to sample extracts after extraction (post-extraction spike), and the latter is referred to as a *method matrix-matched standard calibration curve* (MMSCC), in which the standards are added to samples prior to extraction (pre-extraction fortification).[86] For antibiotic analysis, method matrix-matched standard calibration curves (MMSCCs) are often used, especially for calculations of CCα (decision limit) and CCβ (detection capability).

Under certain circumstances, matrix-matched standard curves may not be fit for the intended purpose, for example, when ME is $<70\%$ or $>110\%$, due to the complex nature of samples. This approach is also not applicable when there is no blank matrix available. The other techniques mentioned above may provide possible solutions, although these procedures are tedious and require additional sample preparation and calculations. Standard addition[86] is a technique that introduces the standard of a target analyte directly into samples—in other words, a procedure in which known amounts of an analyte are added to aliquots of sample extracts containing the analyte to produce new concentrations. The analyte responses generated by the spiked sample and original extracts are measured, and the analyte concentration in the original sample is determined from the slope and intercept of the response curve. A linear response in that narrow range of the curve is a prerequisite. The amount of analyte present in the sample can be determined in as little as three injections, specifically, the original sample (extract) and two spiked samples (extracts), as shown in Figure 6.5.

The echo-peak technique,[87,88] which simulates the use of an internal standard, is based on two consecutive injections of a reference standard and a sample extract within the same LC analysis. The first and second injections are made to either pass through or bypass a pre-column, respectively. This results in a short difference in retention times between a known amount of reference standard, which is also the analyte of interest, and an analyte from a sample. As a result, the reference elutes close to the analyte. The peak from a reference is referred to as an *echo-peak*, and the peak from a sample is called a *sample peak* (Fig. 6.6). Provided that the retention times of these two peaks are close enough, both the reference and the analyte encounter the same matrix effects. Concentration of the analyte in a sample is calculated from the peak area ratio of the analyte and the reference standard, and consequently it is possible to obtain an accurate quantitative result.

Post-column infusion[89] is a technique that utilizes continuous post-column infusion of a representative reference standard, and the response from this standard is used to compensate for matrix effects. This method is based on an assumption that most analytes, even those with diverse physical and chemical properties, will behave similarly in terms of ion suppression or enhancement during the entire chromatographic run in the presence of matrices. Therefore, the response of this representative reference standard at any given retention times of other analytes is possibly utilized to compensate for the matrix effects for most of analytes.

The extrapolative dilution approach[90] is based on the fact that matrix effects are a function of dilution. Consecutive dilutions of a sample extract are made, and the concentration of an analyte in a sample is calculated from the most diluted solution. Matrix effects should be the least intense in that solution and can be compared to a standard prepared in solvent. This technique takes more instrument time than others because of the series of sample dilutions.

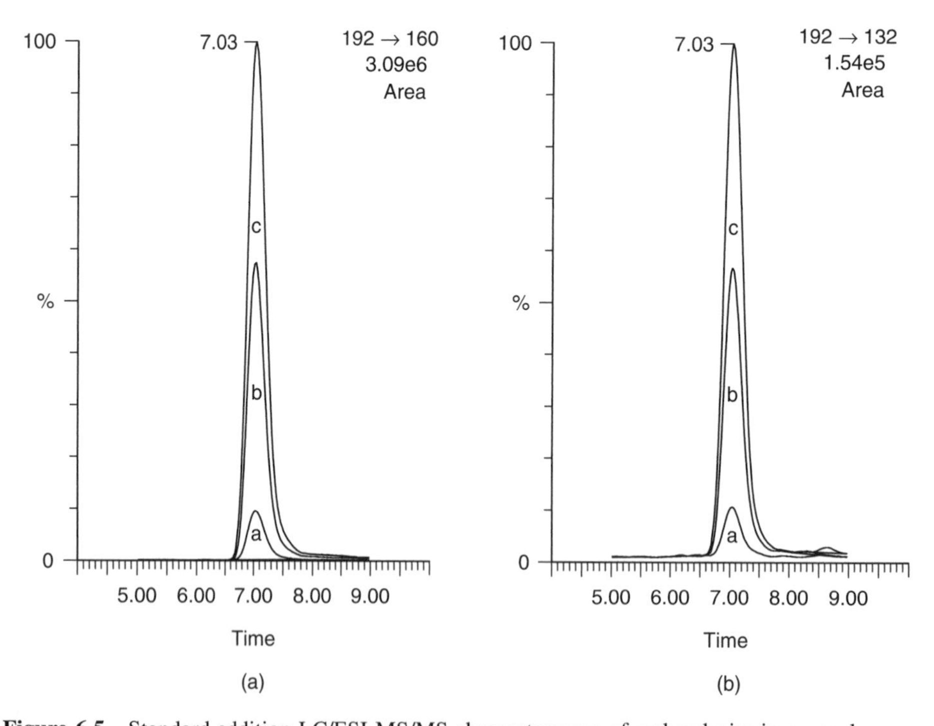

Figure 6.5 Standard addition LC/ESI-MS/MS chromatograms of carbendazim in an apple sauce sample [peak *a*—unknown concentration (fortified at 5.4 μg/kg equivalent in sample); peak *b*—unknown + 28.8 μg/kg of carbendazim (in sample); peak *c*—unknown + 57.6 μg/kg of carbendazim (in sample)]: (a) transition at 190 > 160 is used for quantification; (b) transition at 190 > 132 is used for confirmation. (Reproduced from Wang et al.[86] with permission from the Canadian Food Inspection Agency; published by American Chemical Society; copyright 2005. Crown in the right of Canada.)

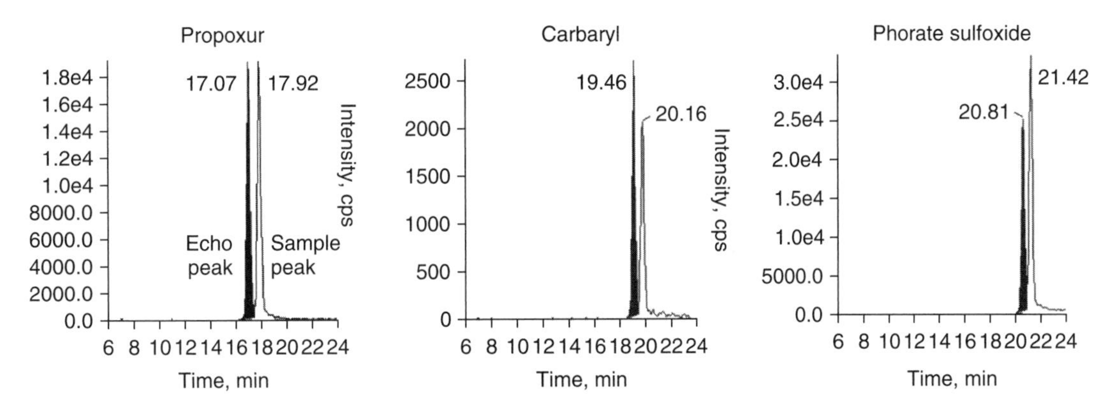

Figure 6.6 An example of the echo-peak technique. First peak (echo-peak): pesticides in solvent (0.1 μg/ml, solid peak). Second peak (sample peak): pesticides spiked in lemon extracts (0.1 μg/ml). (Reproduced from Alder et al.[88] with permission from Elsevier; copyright 2004.)

6.4.3 Mass Spectrometers

Mass spectrometers or analyzers that have been used for antibiotic residue analysis include single-quadrupole (Q), quadrupole ion trap (QIT), linear ion trap (LIT), time-of-flight (TOF), and Orbitrap configurations.[18,69,91–94] Applications of magnetic sector and ion cyclotron resonance (ICR) instruments are rarely seen in this field. Mass spectrometers can be configured or used with either stand-alone mass analyzers or mass analyzers in tandem, namely, triple-quadrupole (QqQ), quadrupole linear ion trap (QqLIT), quadrupole time-of-flight (QqTOF), linear ion trap Orbitrap (LIT Orbitrap), and so forth. Each system has unique design, funtionality, and performance characteristics with corresponding advantages and disadvantages with regard to its applications for antibiotic analysis. Quadrupole and ion trap are low-resolution (unit mass) mass analyzers, which typically have 0.7±0.1 amu widths at half peak height at any given *m/z* in its operating range. TOF and Orbitrap are medium-range to high-resolution mass spectrometers, which have the ability to discriminate between ions that differ by only a few millidaltons (mDa) in *m/z* with accurate mass measurement capability. Low-resolution mass spectrometers are often operated in multiple- or selected-reaction monitoring (MRM or SRM) mode, and this type of analysis is called *pre-target analysis*. TOF and Orbitrap instruments, because of their high mass resolving power and accurate mass measurement, allow for post-target analysis or unknown identification.

6.4.3.1 Single Quadrupole

The single quadrupole is one of the earliest mass analyzers to become widely available. A quadrupole separates ions according to their *m/z* ratio as a function of their trajectory through an oscillating electric field. With a transmission quadrupole mass analyzer, direct-current (DC) and radiofrequency (RF) voltages are applied to four parallel rods (Fig. 6.7a). As ions are accelerated through this dynamic electric field, they oscillate in such a way that only ions of specific *m/z* ratios have a stable trajectory to the detector. Lighter and heavier ions crash into the walls of the rods and are not detected; therefore, the quadrupole acts as a mass filter. The voltages applied can be ramped to let ions of different *m/z* ratios pass through the quadrupole in a sequential manner, thereby scanning over a range of ions. Scan data collected over a wide *m/z* range have the potential to yield the most structurally significant information about a compound. In addition to operating in this scanning mode, single quadrupoles can also be operated in the selected-ion monitoring (SIM) mode. With SIM, the voltages are set so that only ions within *m/z* values of interest are constantly allowed to pass through the quadrupole to the detector. Using SIM significantly decreases LOD. When used alone, quadrupoles are considered low-resolution instruments in that their ability to discriminate between ions that have similar *m/z* values is limited.[95] Generally, separation of singly charged ions that differ by one dalton is possible.

Single quadrupoles have traditionally been part of GC-MS instruments that employ electron ionization. This is because the spectra obtained from the high-energy ionization process of electron impact contain many fragment ions; further dissociation in a second stage of MS is not necessary. Quadrupole instruments are small, relatively inexpensive, and easy to maintain. Single quadrupoles can also be utilized in combination with an LC. However, because the ions formed by ESI or APCI tend to be protonated (or deprotonated) molecules, little structural information is obtained by scanning with a single quadrupole. It is possible to break down the protonated molecules in the ion source by adding energy and allowing collisions to occur in this high-pressure area. This "in-source" collision-induced dissociation (CID) does not first select for the ions of interest, so the resulting spectra can contain large amounts of chemical noise. The applications of a single quadrupole coupled with LC are very limited because of its lack of selectivity,

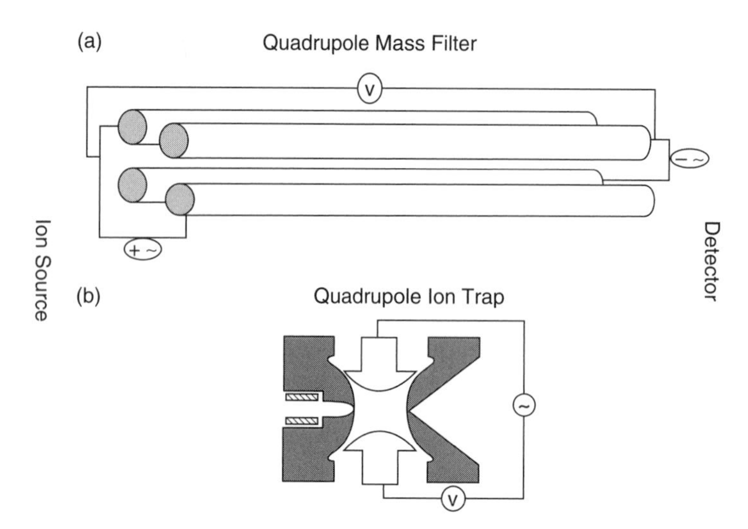

Figure 6.7 Schematics of single-quadrupole (a) and quadrupole ion trap (b) mass analyzers. (Reproduced from Jonscher and Yates[136] with permission from ABRF News the Association of Biomolecular Resource Facilities; copyright 1996.)

and it has been replaced with tandem or high-resolution mass spectrometers. Table 6.1 lists some examples of single quadrupoles utilized for analysis of macrolides in liver and kidney, polypeptides in animal tissues, and sulfonamides in milk and eggs.

6.4.3.2 *Triple Quadrupole*

A triple-quadrupole (QqQ) mass spectrometer is the most commonly used tandem mass spectrometer for antibiotic analysis because of its ability to generate quantitative data with outstanding sensitivity, repeatability, and dynamic range. It can perform tandem mass spectrometric experiments (MS/MS) using the same principles of ion separation as the single quadrupole described above. In a triple quadrupole MS, there are three sets of quadrupole rods in series. The first and last quadrupoles in a QqQ instrument act as ion filters using varying DC and RF voltages like a single quadrupole. The middle quadrupole (q) is a reaction cell where CID can occur. Like other tandem ion beam mass spectrometers ("tandem in space"), a QqQ mass spectrometer can be operated in four different functions with CID: multiple-reaction monitoring (MRM) or selected-reaction monitoring (SRM), product ion scan, precursor ion scan, and constant neutral loss scan (Fig. 6.8).

Multiple-reaction monitoring is the most commonly used function for quantification and confirmation. The first quadrupole is set to only filter the selected precursor ion into the collision cell. This ion is then collisionally dissociated to form product ions. A small number of these product ions, which are chosen by the analyst for their structural significance, are then allowed to pass through the third quadrupole to the detector. MRM filters out chemical matrix noise, which results in superior selectivity and great sensitivity.[93] Table 6.1 lists the MRM transitions of common antibiotics, which are typically pre-determined by infusion of individual standards to a QqQ. As an example, the product ion spectrum of oleandomycin (Fig. 6.9a) is first acquired using a reference standard, where MRM transitions are defined and selected to perform LC/ESI-MS/MS analysis (Fig. 6.9b). The instrument parameters, especially collision energy, are optimized to obtain the best sensitivity for a method. This process can be achieved automatically by software. The MS fragmentation patterns, or MRM transitions, for an antibiotic may vary somewhat depending on the instrumentation, especially in terms of relative ion abundances. LC/ESI-MS/MS data acquisition occurs in single or multiple retention time windows that contain a large number of MRM transitions. The dwell time for each transition and interscan delay or pause time can be as short as 2 ms. Duty cycle and cycle time are two parameters that affect the LC/ESI-MS/MS data quality and should be considered during method development. Duty cycle is the amount of time spent on monitoring an analyte, while cycle time is a total of MRM dwell time and interscan delay or pause time added together. Duty cycle is inversely proportional to the number of concurrent MRMs monitored, but the total cycle time is proportional to the number of MRMs in the same retention period. A high duty cycle provides good sensitivity, and a short cycle time increases the sampling rate across an LC peak, which results in a more reproducible quantitative result. The total number of MRM transitions or analytes that can be monitored in each retention time window is limited, but can be practically extended to 200–300 in one period with some newer QqQ instruments without suffering a significant loss in sensitivity and repeatability. The latest generations of triple-quadrupole systems feature the so-called scheduled MRM, where individual transitions are scheduled to be

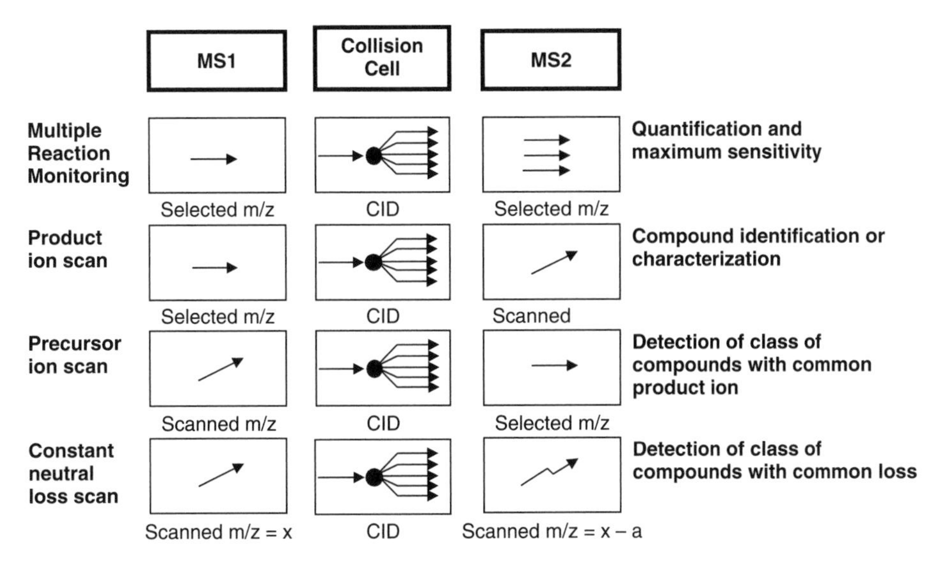

Figure 6.8 MS/MS experiments based on analysis using two spatially separate mass analyzers or tandem mass spectrometers ("tandem in space"). A collision cell is set between two mass analyzers. (Reproduced from de Hoffmann[137] with permission from John Wiley and Sons; copyright 1996.)

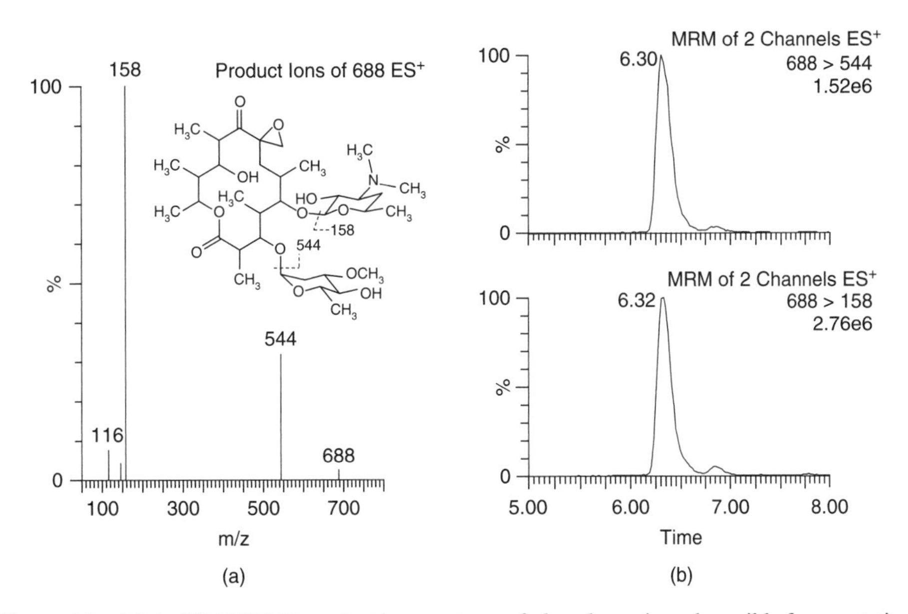

Figure 6.9 (a) An ESI-MS/MS product ion spectrum of oleandomycin and possible fragmentation; (b) LC/ESI-MS/MS chromatograms of a blank raw milk sample fortified with oleandomycin (5 μg/kg). (The mass spectrum and chromatograms are unpublished data and are obtained from the author's previous research project, Calgary Laboratory, Canadian Food Inspection Agency. Instrumental parameters are described in a paper by Wang and Leung.[110])

monitored in narrowly designated retention windows at the time when analytes are eluted. Therefore, the number of concurrent MRM transitions is significantly reduced at any acquisition time, resulting in much a higher duty cycle for each analyte and a shorter cycle time overall. The scheduled MRM allows the instrument to maintain maximized dwell times for improved sensitivity and optimized cycle times for better repeatability. It may be possible to monitor over 1000 transitions in a single LC analysis using scheduled MRM.

Tandem LC-MS/MS is the method of choice for both quantification and confirmation. In routine practice, two or more transitions are monitored (Table 6.1). The most intense transition is used for a quantitative work, whereas the second or third transition is utilized for confirmatory purposes. An LC-MS/MS instrument is able to provide a dynamic range of 2–3 orders of magnitude for quantification with LODs at sub-μg/kg concentration and RSD <15%. LC retention times and ion ratios serve as two important characteristics for confirmation. LC-MS confirmatory criteria are well defined in many international method performance guidelines as discussed in Chapter 8. FDA Guidance for Industry #118[96] and EC Decision 2002/657/EC[2] are taken as examples to explain how the criteria are applied for confirmation. There are two basic guidelines: (1) the LC retention time of an analyte should match that of the calibration standard within a specified relative retention time window ±2.5%, and (2) the relative ion ratios of an analyte should fall within the maximum permitted tolerances of the comparison standard (Table 6.2). The FDA guidance recommends that the relative ion ratio be within ±10% absolute when two transitions are monitored. This acceptable range is calculated by addition and subtraction. For example, at 50% relative abundance, the acceptable range would be 40–60%, not 45–55%. If three or more transitions are monitored, then the relative ion ratios should match the comparison standard within ±20% absolute. Those tolerances are applied regardless of the range of ion ratios.

Decision 2002/657/EC has set relative abundance criteria that are dependent on the relative intensities of two transitions (Table 6.2), and has also established an identification points (IPs) system in order to confirm organic residues and contaminants in live animals and animal products (Table 6.3). For instance, where the relative intensity of the confirmation MRM is >50%, the maximum permitted tolerance is 20% relative. Therefore, at 60% relative abundance, the acceptable range would be 48–72%. With regard to the assignment of IPs, one precursor ion and one transition product are assigned 1 IP and 1.5 IPs, respectively, from a low-resolution (unit mass) mass spectrometer. Therefore, one precursor and two transitions from a QqQ mass spectrometer earn a total of 4 IPs. For the confirmation of banned substances listed in group A in Council Directive 96/23/EC,[97] a minimum of 4 IPs are required. For substances listed in the group B in the directive, a minimum of 3 IPs are mandatory.

TABLE 6.2 Maximum Permitted Tolerances for MS Relative Ion Intensity

Relative Intensity (Percent of Base Peak) (%)	EU[a] (Relative) (%)	FDA[b] (Absolute) (%)
>50	±20	
>20–50	±25	2 transitions: ±10
>10–20	±30	>2 transitions: ±20
≤10	±50	

[a]Criterion set by the Commission Decision 2002/657/EC.[2]
[b]Criterion set by US FDA.[96]

TABLE 6.3 Relationships between MS Data, Mass Accuracy, and Identification Points (IPs)

MS Technique	IPs Obtained for Each Ion
Low-resolution mass spectrometry (LR)	1[a]
LR-MS^n precursor ion	1[a]
LR-MS^n product ion or transition products	1.5[a]
HRMS	2[a]
HR- MS^n precursor ion	2[a]
HR-MS^n product ion or transition products	2.5[a]
Mass accuracy	
Error >10 mDa[b] or ppm[c]	
Single ion	1[b]
Precursor ion	1[b]
Product ion or transition products	1.5[b]
Error 2–10 mDa or ppm	
Single ion	1.5[b]
Precursor ion	1.5[b]
Product ion or transition products	2[b]
Error<2 mDa or ppm	
Single ion	2[b]
Precursor ion	2[b]
Product ion or transition products	2.5[b]

[a]Criterion proposed by Commission Decision 2002/657/EC.[2]
[b]Criterion proposed by Hernandez et al.[111] and mass error in mDa.
[c]Criterion proposed by Wang and Leung[110] and mass error in ppm.

In addition to MRM, the other scan modes available on a QqQ have occasionally been used for residue analysis as well. A precursor ion scan can be used to identify precursor ions from a product ion, and therefore to identify analytes and metabolites or impurities, which generate the same product ion, in complex matrices. For example, erythromycin B was identified in yogurt using this function.[98] In this application, Q3 was held constant to measure a fragment ion at *m/z* 158, which is a typical product ion of compounds or impurities related to erythromycin A with a desosamine residue. Q1 was then scanned over an appropriate range, from which a precursor ion at *m/z* 718 was detected. The latter was identified as erythromycin B, which was an impurity in the erythromycin fermentation product. Constant neutral loss scan, which has rare applications for antibiotic analysis, records spectra that show all the precursor ions that have fragmented by the loss of a specific neutral mass. In this instance, both Q1 and Q3 scan together with a constant mass offset between the two quadrupoles. Both precursor ion and constant neutral loss scans can be performed only with ion beam tandem in-space mass spectrometers.

6.4.3.3 *Quadrupole Ion Trap*

Quadrupole ion trap (QIT) mass analyzers are similar to the quadrupoles already discussed in that trajectories within DC and RF fields are used to differentiate ions according to their *m/z* ratios. A commonly used QIT is called the *three-dimensional trap* (e.g., 3D ion trap; also known as *Paul ion trap*). This device consists of a metal ring electrode with two endcap electrodes on either side, creating an internal space where ions can be trapped (Fig. 6.7b). DC and RF voltages are applied to the ring and endcap electrodes so that the trajectory of ions with specified *m/z* values are stabilized inside the trap. When the RF voltage is ramped, the oscillation of the ions becomes unstable and the ions are ejected from the trap. Ions can exit the trap to the vacuum system or to the detector to be counted as signal. An advantage of a QIT is that ions across a relatively large *m/z* range can be simultaneously trapped and then detected, so there is less overall loss than in the sequential scanning process that occurs in a transmission quadrupole. As a result, with an ion trap, good-quality full-scan spectra can be obtained with relatively low amounts of analyte.

A QIT mass analyzer is unique in its ability to conduct multiple-stage mass fragmentation. It performs MS/MS and MS^n experiments in a "tandem in time" fashion, but it cannot accomplish the precursor ion scan and neutral loss scan that are available scan functions with a QqQ mass spectrometer.[99,100] Product ions are generated by CID, which occurs by a process somewhat different from that described for a triple quadrupole. When introduced into a 3D ion trap, the precursor ion is first trapped while all other ions are scanned out. Additional potential corresponding to the resonance frequency of the molecule is then applied to the trap, causing the ions to dissociate as they collide with the helium gas that is present in the trap. The product ions generated are then scanned into the detector and evaluated. This is considered a tandem MS experiment "in time" as opposed to the tandem MS "in space" that occurs with ion beam instruments such as a triple-quadrupole MS. The CID can be performed multiple times to give several generations of product ions (MS^n). For example, a precursor ion can be trapped and dissociated to give product ions. One or more of those product ions can then be trapped, while allowing other product ions to exit, and be dissociated to generate MS^3 spectra. This can be done as long as there are sufficient ions present to detect signal. For each MS^n experiment, the overall signal is decreased, but the noise is also less. The MS^n capability of a QIT renders it a powerful tool for characterizing antibiotics and their impurities, in addition to its MRM ("tandem in time") for quantification and confirmation. The step-by-step MS^n clearly indicates the sequential dissociation of several parallel product ion paths. In residue analysis this may translate to obtaining information rich spectra. The MS^n spectra along with fragmentation patterns associated with the nature of functional groups in a molecule allow for the identification and structural elucidation of antibiotics and their impurities or metabolites,[101] especially when searching against an established mass spectral library.

One concern with a QIT mass analyzer is that the number of ions allowed in the trap at any given time must be controlled. When too many ions are present, they can interact with each other, causing space-charge effects that lead to distortions in the electric field and ion trajectories. This can result in loss of mass resolution and accuracy as well as overall sensitivity. Another limitation to be aware of is the so-called "one-third rule," where, because of limitations in the range of ions that can be stabilized in a trap at any given time, a product ion scan has a lower *m/z* limit equal to approximately one-third that of the precursor ion. For example, if the precursor ion has an *m/z* ratio of 300, the mass range of the product ion scan would be limited to approximately *m/z* 100 and higher. It is also generally thought that an ion trap gives less precise quantitative results at low concentrations of analyte when compared to a triple-quadrupole instrument. This is due to the fact that the number of data points across a chromatographic peak can vary depending on the number of ions needed to optimally fill the trap. In general, longer times are required at the start and end of an eluting peak. In addition, it is not an advantage to focus the duty cycle of the instrument to monitor just a few ions. For qualitative work, however, an ion trap is ideal because full scan spectra can be collected without a loss in sensitivity.

6.4.3.4 Linear Ion Trap

Linear ion trap (LIT), also known as the *2D ion trap*, is a mass analyser based on the four-rod quadrupole and end electrodes that confine ions radially by a two-dimensional (2D) RF field and axially by stopping potentials or DC voltages applied to end electrodes. The trapped ions are ejected either axially to a detector[102] or radially to one or dual detectors through slots cut along the middle of the center section of two opposite rods.[103] The advantages of a LIT versus a 3D ion trap include enhanced ion-trapping capacity and reduced space charge effects due to the increased ion storage volume. More ions can be introduced into the LIT, resulting in increased sensitivity and dynamic range as compared to a 3D ion trap. A LIT is often built as part of a hybrid instrument such as QqLIT, LIT TOF, or LIT Orbitrap, with only a few instruments using the 2D ion traps as standalone mass spectrometers.

A triple-quadrupole linear ion trap (QqLIT), which is the most widely used hybrid linear ion trap, is based on the ion path of a triple-quadrupole mass spectrometer with Q3 operated as either a conventional RF/DC quadrupole mass filter or a linear ion trap mass spectrometer.[99,102,104] A QqLIT combines the advantages of a QqQ and a QIT within the same platform without compromising the performance of either mass spectrometer. It retains classical QqQ functions such as MRM, product ion scan, precursor ion scan, and constant neutral loss scan for quantitative and qualitative analysis, and possesses MS^n ion accumulation and fragmentation capacity for structural elucidation. When in its MRM mode, a QqLIT performs the same as a QqQ does, and is used as a conventional triple-quadrupole for quantification. Similar to other tandem mass analyzers, a QqLIT can perform data-dependent analysis based on a survey scan or the previous data acquisition such as full-scan, MRM,[105,106] or precursor ion scan[80] to yield valuable information in real time. For example, when a MRM transition goes above a pre-set intensity threshold, a product ion scan is triggered by software to obtain an enhanced product ion (EPI) spectrum. Thus, the MRM quantitative information and a qualitative EPI spectrum are recorded in the same chromatographic analysis. Collecting EPI near the MRM limits of quantification is possible because the instrument accumulates ions with an enhanced instrument duty cycle which increases the product ion scan sensitivity.[105] The EPI spectra of incurred antibiotics can be compared to a mass spectral library for confirmation. Therefore, a QqLIT along with EPI can serve as a practical tool for antibiotic screening and confirmation in food and represents an emerging technique in the residue analysis. In addition, precursor ion survey scans potentially identify additional analytes belonging to the same class of molecules.[80] As an example, an LC/QqLIT precursor ion scan of *m/z* 92 (i.e., a common product ion of many sulfonamides) was carried out as a survey experiment. This led to the identification of sulfamethoxazole in honey through an EPI spectrum matching via a library search.[80]

6.4.3.5 *Time-of-Flight*

Time-of-flight (TOF) mass analyzers are the simplest mass spectrometers, where ions that have the same kinetic energy but different *m/z* values are separated in a field-free flight tube (typically 1–2 m) and reach the detector at different times. The smallest ions are faster and will be detected first, while the larger, slower ions will reach the detector at a delayed time. This is a fairly simple relationship described by the following equation: $m/z = 2eVt^2/L^2$. In a TOF mass analyzer, the flight path (L) is fixed and the potential (V) at which ions are accelerated is also kept constant. Therefore the time it takes an ion to reach the detector (t) is proportional to its *m/z* value. Ions entering the device are accelerated to a constant velocity via an ion pulse. Modern instruments also employ a reflectron, or electrostatic mirror, reversing the direction of the ions in the middle of the flight tube. Not only does this provide a longer overall flight tube without requiring additional laboratory space, but the physics of the electrostatic reflectron also corrects for small differences in velocities of ions with the same *m/z*. This results in excellent mass resolution. Some instrument designs employ a second reflectron (*W* mode), which can improve resolution even more, although decreases in sensitivity are observed.

Time-of-flight instruments are configured as either a stand-alone TOF mass analyzer (TOF MS) or as a hybrid quadrupole time-of-flight (QqTOF) mass spectrometer; the latter consists of a quadrupole front-end and an orthogonal acceleration TOF back-end for MS/MS experiments (Fig. 6.10).[107,108] The orthogonal design minimizes the ions' initial velocity spread as they are accelerated into the TOF by a pulsed potential. A QqTOF can be operated as a TOF mass analyzer (QqTOF MS, full-scan) or a quadrupole TOF tandem mass spectrometer (QqTOF MS/MS, product ion scan). Compared to a TOF MS, the

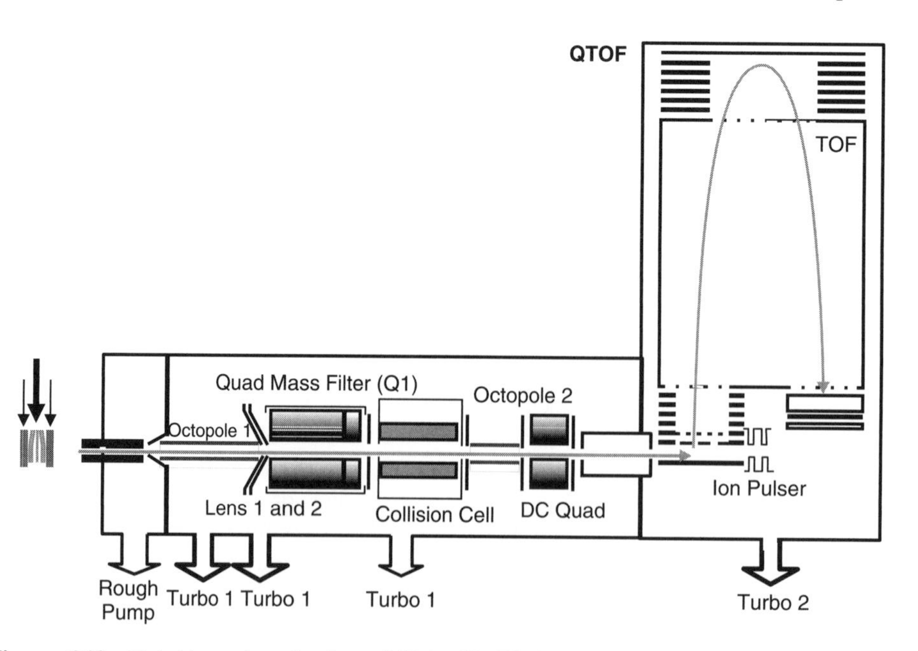

Figure 6.10 Hybrid quadrupole time-of-flight (QqTOF) mass analyzer. (Schematic diagram provided courtesy of Agilent Technologies.)

main advantage of QqTOF MS/MS is the additional information or confirmation achieved by the acquisition of a full accurate mass product ion spectrum through collision-induced dissociation after the precursor ion is selected in the first quadrupole, which increases the confidence about the product ion origin and decreases the chemical noise.[109] TOF mass spectrometers offer medium-range to high resolution ($\leq$40,000 FWHM), accurate mass measurement ($<$5 ppm), excellent full-scan sensitivity, and complete mass spectral information. They are complementary to low-resolution (unit mass) mass spectrometers. TOF data allow for screening target analytes; quantifying selected compounds; identifying unknowns, degradation products, or metabolites; and confirming positive findings on the basis of accurate mass measurement. When acquired in the full-scan mode, the data can be processed in a retrospective manner. This is in contrast to the pre-target analysis that is based on the predetermined MRM transitions collected with a QqQ and a QqLIT. Although a QqLIT and a QIT can perform full-scan or multiple-stage fragmentation for structural elucidation and searching for unknowns from a mass spectral library, their MS data are in nominal mass. This does not permit the assignment of possible elemental compositions of any compounds of interest. In addition to providing high resolution and accurate mass measurement MS data, other advantages of a TOF instrument include virtually unlimited *m/z* scan range as well as speed because all ions can reach the detector eventually and be scanned in the order of few milliseconds.

Time-of-flight instruments have shown great promise for antibiotic screening or high sample throughput because the instrument data acquisition is inherently fast and can be coupled with UHPLC. For example, UHPLC TOF-MS has been used to screen or detect about 100 veterinary drugs in meat, fish, and egg[10] and 150 veterinary drugs in milk[11] in a single analysis. In these applications, most of veterinary drugs could be determined below regulatory limits (e.g., EU MRLs). A TOF is also applicable for quantitative work with the acceptable performance characteristics that include linearity, accuracy, precision, and LOD. In one study, both UPLC/QqTOF-MS and LC/ESI-MS/MS were applied for the determination of six macrolide residues in eggs, raw milk, and honey, and the performance characteristics were compared.[110] LC/ESI-MS/MS demonstrated better repeatability and wider dynamic range for quantification than did UPLC/QqTOF-MS, but both provided essentially the same accuracy. A UPLC/QqTOF MS was able to achieve LODs between 0.1 and 1.0 μg/kg, whereas LC/ESI-MS/MS LODs were in a range of 0.01–0.2 μg/kg. In general, it has been found that the minimum concentrations detected with a TOF instrument may be 1–2 orders of magnitude poorer than those attained using a QqQ in the MRM mode. Therefore, it could be a challenge to use a TOF for analysis of zero-tolerance antibiotics or banned substances in food as a result of their extremely low concentrations present in the samples.

Time-of-flight is a powerful tool for the identification of unknowns, degradation products or metabolites, as well as confirmation of incurred antibiotic residues, because accurate mass measurement facilitates the identification of compounds by providing possible molecular formulas or elemental compositions. As an example, tylosin A (Fig. 6.11) is a macrocyclic lactone (tylonolide) with three sugar moieties (shown from the left in the tylosin A structure in Fig. 6.11: mycinose, mycaminose, and mycarose). Hydrolysis of tylosin A at pH $<$4 removes the mycarose to form tylosin B, that is desmycosin (Fig. 6.11). Tylosin A hydrolyzes slowly to tylosin B in honey, which typically has a pH value in the range from 3.4 to 6.1. An UPLC/QqTOF MS was able to acquire accurate masses and use these data to identify the degradation product (tylosin B) in honey. As shown in Figure 6.11, the compound, with *m/z* 772.4446 (Fig. 6.11 plot B2), or with *m/z* 772.4448 and *m/z* 174.1126 (Fig. 6.11 plot D2), eluted at 3.88 (Fig. 6.11 plot B1) or 3.87 min (Fig. 6.11 plot D1), was identified as tylosin B with a mass error of $<$5 ppm.[110]

In the Decision 2002/657/EC document, high-resolution mass spectrometry (HRMS) is defined as the resolution that shall typically be greater than 10,000 for the entire mass range at 10% valley. Traditionally, TOF and Orbitrap instruments use a FWHM definition for resolution, whereas magnetic sector and ion cyclotron resonance (ICR) instruments use a 10% peak valley definition. The FWHM definition of the resolving power gives values twice those obtained by the 10% peak valley definition, when a peak is in a Gaussian distribution. Therefore, a TOF with $>$ 20,000 FWHM at the mass range of interest meets the Decision's high-resolution mass spectrometry criterion. According to the 2002/657/EC identification point (IP) system, when high-resolution measurements are performed at a resolving power $\geq$ 10,000 (10% valley over the complete mass range), one ion gives 2 IP and one transition provides 2.5 IP. However, the Decision does not take into account mass accuracy, which can be used for IP assignment, as a confirmatory parameter. The criterion, which could use either absolute (in mDa)[111] or relative [in parts per million (ppm)] mass errors[110] for IPs assignment (Table 6.3), based on mass accuracy, has been reported. The ppm concept has an advantage that the IP rating criterion is consistent across a mass range or independent to *m/z* values. Thus, for substances with established MRLs, at least two ions must be detected to achieve a minimum of 3 IPs for satisfactory confirmation of the compound's identity with mass errors between 2 and 10 mDa or ppm. Figure 6.11 shows an example that a total of 4.5 IPs could be assigned to tylosin A in a sample using an UPLC/QqTOF MS; the 4.5 IPs include 1.5 IPs from the precursor ion at *m/z* 916.5273 or 916.5306 (theoretical mass

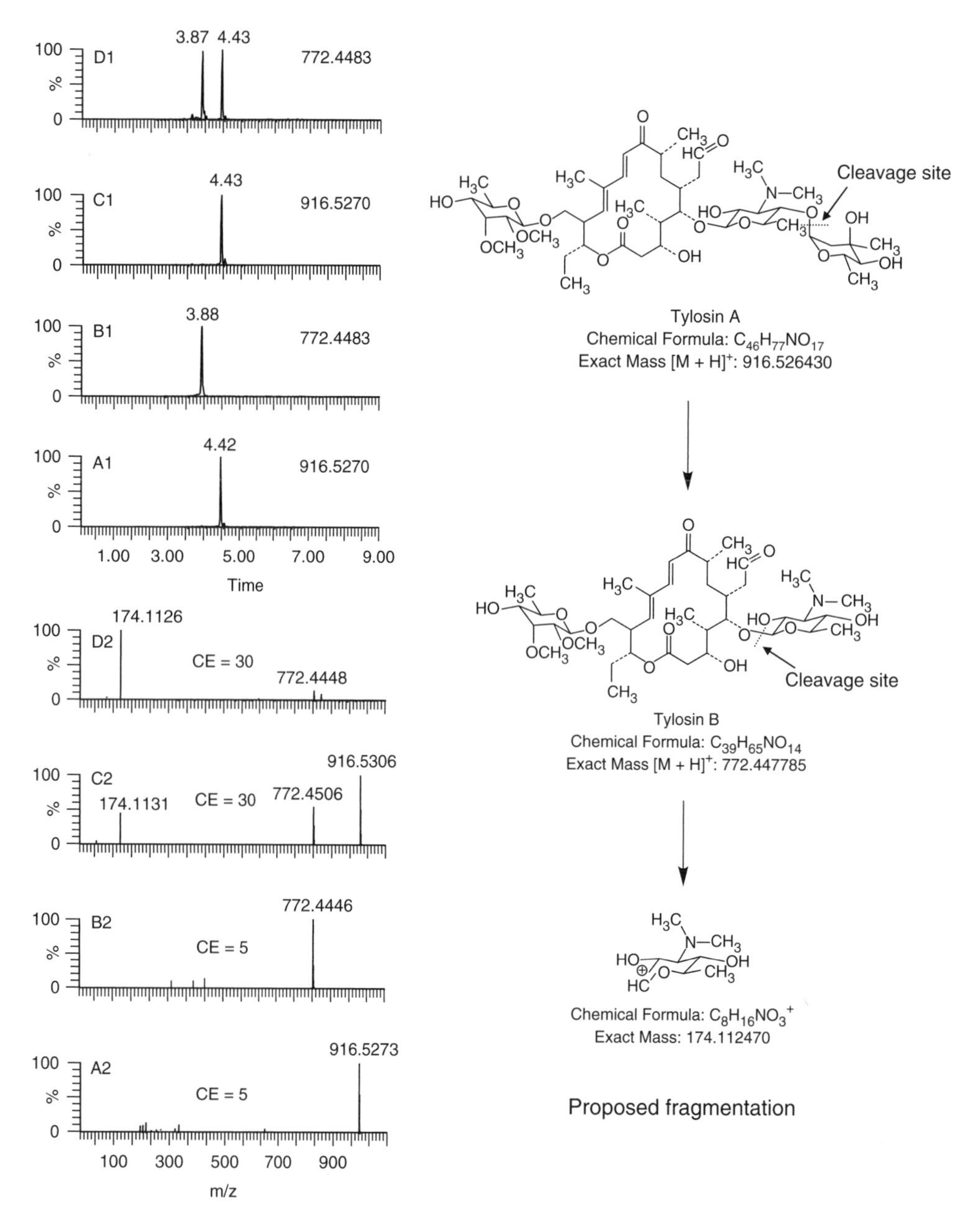

Figure 6.11 UPLC QqTOF MS chromatograms of a tylosin (4.1 μg/kg, RSD = 1.5%, $n = 3$) incurred honey sample (CE—collision energy). Plots A and C: traces of tylosin A. Plots B and D: traces of tylosin B. Plots A2—spectrum at 4.42 min from plot A1; Plot B2—spectrum at 3.88 min from B1; plot C2—spectrum at 4.43 min from plot C1. Plot D2—spectrum at 3.87 min from plot D1. Proposed fragmentation is based on the nitrogen rule and accurate mass measurement. (Reproduced and modified from Wang and Leung[110] with permission from the Canadian Food Inspection Agency; published by John Wiley & Sons; copyright 2007. Crown in the right of Canada.)

916.5270) (Fig. 6.11 plots A2 and C2) and 3 IPs from two fragments at *m/z* 772.4506 (theoretical mass 772.4483) and 174.1131 (theoretical mass 174.1130) (Fig. 6.11 plot C2). In this example, CID with low and high collision energy was applied to acquire fragment-rich spectra, and additional IPs were assigned for confirmation.

6.4.3.6 *Orbitrap*

An Orbitrap mass spectrometer, as a result of its outstanding mass resolving power, high sensitivity, accurate mass measurement (<5 ppm), full-scan, and/or MS^n capabilities, is an attractive alternative to a TOF instrument, especially for analysis of small molecules (m/z < 1000). The Orbitrap is analogous to a quadrupole ion trap device, but it operates by radially trapping ions that orbit around a central spindle-like electrode. An outer barrel-like electrode is coaxial with the inner electrode and *m/z* values are measured from the frequency of harmonic ion oscillations, along the axis of the electric field, undergone by the orbitally trapped ions. The characteristic ion frequencies are taken as an "ion image current" or transient that is Fourier-transformed to create a mass spectrum. A more detailed explanation of the Orbitrap can be found elsewhere.[112,113] This device performs as a high resolution (>100,000) mass analyzer because the important axial frequencies are independent of energy fluctuations. An Orbitrap instrument is formatted as either a stand-alone (a single-stage) mass analyzer (Exactive) or a tandem mass spectrometer such as a linear ion trap Orbitrap (LIT Orbitrap) with either atmospheric-pressure or matrix-assisted laser desorption as an ion source. The former (Exactive) is well suited for chemical contaminant residue screening, and the latter (LIT Orbitrap) is a perfect means for structural elucidation. Figure 6.12 shows the instrumentation of the standalone Orbitrap (Exactive). This design provides very high-resolution mass spectra at a relatively low cost. Because the Orbitrap analyzer operates as a pulsed device and the ion source is continuous, a storage region (or the "C-Trap" shown in Fig. 6.12) is needed to provide the mass analyzer with a pulse of ions. Product ion spectra can be obtained by diverting ions to a collision cell prior to analysis. There is no filtering of ions prior to entering the collision cell, however, so all ions present are fragmented and the entire mixture is introduced into the Orbitrap. In comparison, the LIT Orbitrap provides true product ion spectra with accurate mass measurement because the LIT selects ions of interest and fragments them prior to Orbitrap analysis.

The resolving power of a current Orbitrap instrument is defined at either *m/z* 400 for LIT Orbitrap (LTQ Orbitrap XL or LTQ Orbitrap Velos) or *m/z* 200 for a stand-alone Orbitrap (Exactive) to achieve various mass resolutions. The LTQ Orbitrap XL has a set of resolutions available at 7,500, 15,000, 30,000, 60,000, or 100,000, whereas the Exactive has a selection at 10,000, 25,000, 50,000, or 100,000. The resolving power is inversely proportional to the square root of *m/z*. A resolving power of 100,000 defined at *m/z* 200 results in an effective resolving power of 81,650 at *m/z* 300. The scan speed is a function of the mass resolution setting of the Orbitrap instrument. The higher the resolution is, the longer it takes to acquire data. For example, the scan rate is 10 Hz (10 scans per second) at a resolution of 10,000, whereas the scan rate is 1 Hz (1 scan per second) at a resolution of 100,000 at *m/z* 200. A high resolution is essential to resolve two peaks that share the same nominal mass. This can also serve to reduce interference in complex matrices and therefore, to assign the accurate masses of analytes even at a low concentration. The application of an Orbitap instrument for chemical contaminant residue analysis is in its early stage and represents a frontier technique in the field for simple and complex sample screening. As an example, an UHPLC Orbitrap (Exactive) has demonstrated its power to screen as many as 151 compounds, including

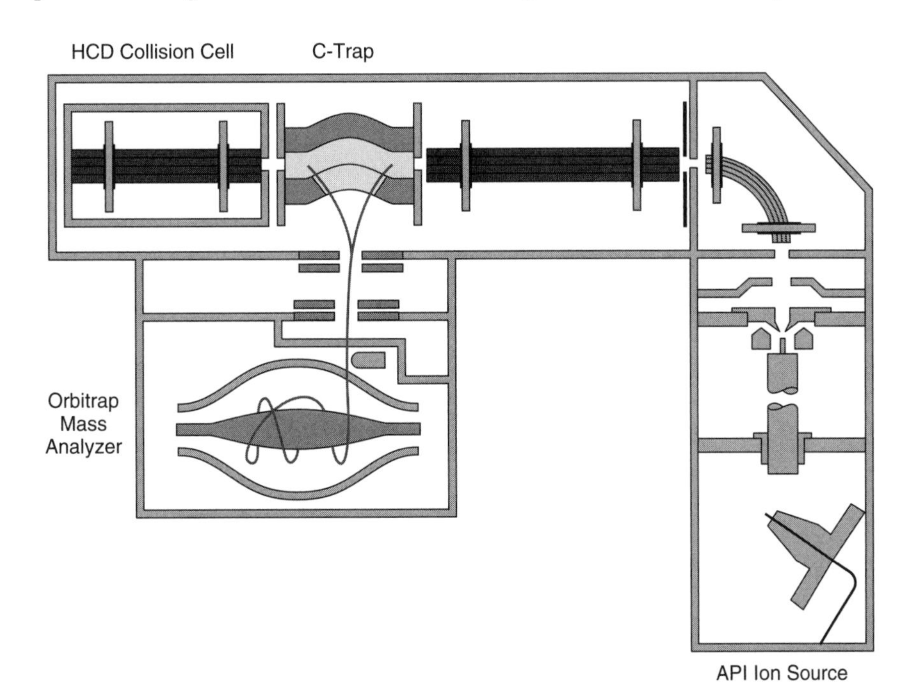

Figure 6.12 Standalone Orbitrap (Exactive) mass spectrometer. (Schematic diagram provided courtesy of Thermo Fisher Scientific.)

pesticides, veterinary drugs, mycotoxins, and plant toxins in honey and animal feed in the range 10–250 μg/kg in a single analysis.[114] The superior resolving power enabled the Exactive to differentiate two isobaric compounds, imazalil (*m/z* 297.05560) and flunixin (*m/z* 297.08454), from which the mass difference is 29 mDa (Fig. 6.13). Imazalil and flunixin were not mass-separated at the resolution of 10,000 but were resolved at 100,000 with accurate mass assignment (<2 ppm).[114]

The selection of resolving power in an Orbitrap must be "fit for purpose" and is associated with both the analyte concentration and the complexity of matrices. A study[114] indicated that a resolving power of 7,000–10,000 could be sufficient for detection of analytes in samples of intermediate complexity at concentrations down to 25 μg/kg with accurate mass measurement (mass errors <5 ppm); for lower concentration levels and/or more accurate mass assignment, a higher resolving power (18,000–25,000) was needed; and in highly complex extracts, a resolving power of 35,000–50,000 or even 70,000–100,000 was required. Although a higher resolution is attributed to a better selectivity, there is a need to find a balance between a resolving power and a scan speed to achieve sufficient data points across an LC peak for quantification. For example, a resolving power of 10,000 was not sufficient to ensure the reliable screening of banned steroid esters in hair by an UHPLC LTQ Orbitrap XL; therefore, a resolving power of 60,000 had to be applied to avoid false-positive findings.[115] When a resolving power was set at 60,000, the number of spectra acquired across an UHPLC peak could be as few as five, which may not be enough to obtain reproducible quantitative results.

6.4.4 Other Advanced Mass Spectrometric Techniques

Innovations in the field of mass spectrometry are constantly being developed. Often they are initially investigated for basic research and then eventually transferred to more applied areas such as the analyses of contaminants in food. Several advanced mass spectrometric techniques are described in this section, along with possible applications to residue analysis.

6.4.4.1 Ion Mobility Spectrometry

Ion mobility can add an extra dimension of separation when coupled to a mass analyzer. Ion mobility spectrometry (IMS) separates ions according to their interactions with a buffer gas, in addition to differences in their *m/z* ratios.[116,117] This can provide separation of ions (i.e., isobaric or conformational isomers), which cannot be accomplished using traditional mass analyzers. It can also be used to reduce interfering chemical noise. Ion mobility separates ions based on how long they take to migrate through a buffer gas with an electric field applied. Factors that affect this migration can include ion size, shape, and structure. Traditional IMS instruments measure the time required for ions to drift through the device, similar to a TOF mass analyzer. In a newer application of ion mobility, high-field asymmetric waveform ion mobility spectrometry (FAIMS)[117] is used between an electrospray source and a mass analyzer to improve the signal-to-noise ratio of low concentrations of analytes in complex matrices. With FAIMS, a buffer gas (usually nitrogen or a nitrogen/helium mixture) and analyte ions flow between two parallel plates with voltages applied. An asymmetric waveform is applied to the upper plate (whereas the lower plate is kept at ground) that causes ions to adopt up-and-down (vertical) trajectories. The ions will eventually collide with the lower plate unless an additional "compensation voltage" is applied. This additional voltage can be varied to discriminate for ions on the basis of their relative mobilities. Species of interest are allowed to pass through, while other ions are filtered out. The pressure, flow, and composition of the buffer gas used in the FAIMS device can also impact ion separation. Another type of IMS used in commercial mass spectrometers is a traveling-wave IMS. In this case, an electric field is applied sequentially along sections of the IMS region to separate ions on the basis of their mobility. In an instrument developed by Waters Corporation, additional ion specificity is achieved by utilizing ion mobility prior to a powerful time-of-flight instrument.[118] Ions are trapped, separated by ion mobility using traveling-wave IMS, and transferred between the quadrupole and TOF of a QqTOF instrument. This instrument is used to characterize the conformation of biomolecules in the gas phase.

6.4.4.2 Ambient Mass Spectrometry

A new generation of mass spectrometer inlets allow for direct sampling of a substrate under ambient conditions.[119] Theoretically, this eliminates the need for any sample preparation. Examples include direct analysis in real time (DART) and desorption electrospray ionization (DESI), as well as desorption atmospheric-pressure chemical ionization (DAPCI) and atmospheric solids analysis probe (ASAP). These techniques utilize a source of energy interacting directly with a sample surface at ambient pressure, causing molecules of interest to desorb, ionize, and be sampled by a mass spectrometer.

With DART, an electric potential is applied to a gas forming a high-energy plasma containing ions, electrons, and excited-state (metastable) atoms and molecules.[119] This plasma interacts with the sample, causing desorption and ionization of compounds. Some ionization of analytes may occur via proton transfer as the plasma produces ionized water species. In DESI, a charged solvent spray hits the sample surface.[119] Large molecules are desorbed

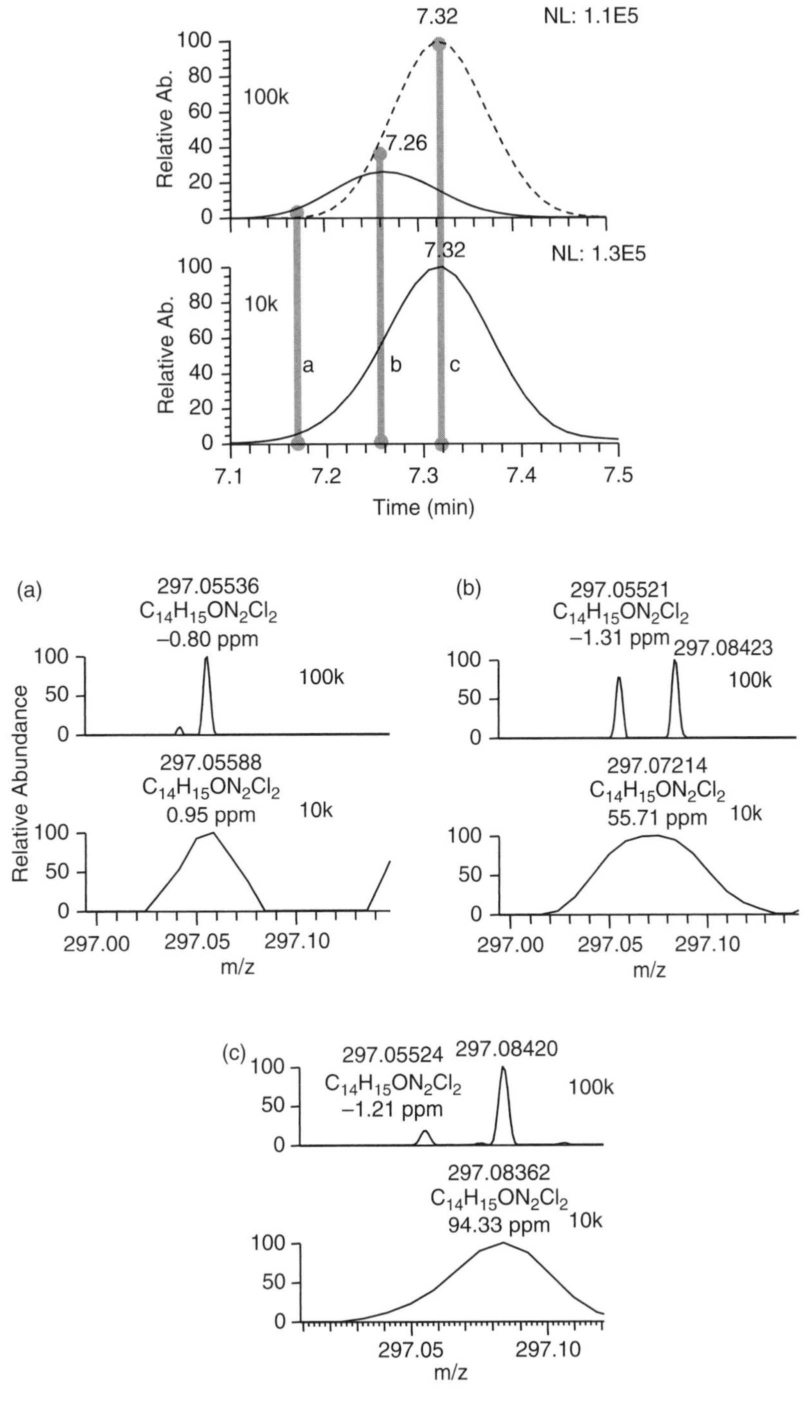

Figure 6.13 Effect of resolving power on assigned mass accuracy of two co-eluting analytes: imazalil (solid line) and flunixin (dashed line). Imazalil ($C_{14}H_{14}Cl_2N_2O$, $[M+H]^+$, 297.05560, RT = 7.26 min) and flunixin ($C_{14}H_{11}F_3N_2O_2$, $[M+H]^+$, 297.08454, RT = 7.32 min). Upper plots: extracted ion chromatograms (±5 ppm and ±100 ppm, respectively). Bottom plots: mass profiles at two resolving power settings: 10,000 and 100,000 FWHM. Letters *a*, *b*, and *c* denote three data acquisition points across the imazalil elution profile. (Reproduced from Kellmann et al.[114] with permission from Elsevier; copyright 2009.)

and ionized with charged droplets, and lower-molecular-weight compounds interact with gas-phase solvent ions. The operating conditions for DESI can be optimized to favor detection of either small or large molecules. DAPCI[120] is a variant of DESI that uses gas-phase ions generated by a corona discharge. Like APCI, DAPCI may be a more effective ionization technique for compounds of moderate to low polarity. ASAP is similar in that ions are formed by a corona discharge after being desorbed by a heated gas jet.[119] It works well for the analysis of volatile and semi-volatile compounds from a solid sample surface. These MS inlets are often coupled to a high-resolution mass analyzer such as time-of-flight or Orbitrap instruments that facilitate identification of any unknown compounds that may be desorbed from the sample. Many applications of these techniques have been demonstrated for forensic work, including the analysis of cocaine on dollar bills and counterfeit drug formulations.

6.4.4.3 Other Recently Developed Desorption Ionization Techniques

Several modifications of MALDI have been developed to couple additional sampling and reaction capabilities to this technique. Surface-enhanced laser desorption ionization (SELDI)[121] is one type of modified MALDI and describes an ionization process that involves reacting a sample with an enhanced surface. With SELDI, the sample interacts with a surface modified with some chemical functionality prior to laser desorption ionization and mass analysis. For example, an analyte could bind with receptors or affinity media on the surface, and be selectively captured and sampled by laser desorption. A SELDI surface can be modified for chemical (hydrophobic, ionic, immunoaffinity) or biochemical (antibody, DNA, enzyme, receptor) interactions with the sample. This technique can act as another dimension of separation or sample clean-up for analytes in complex matrices. As discussed before, one disadvantage of MALDI is that the matrix (usually a substituted cinnamic acid) that is mixed with the sample can directly interfere with the analysis of small molecules. There have been several areas of research to overcome this issue.[122] Direct ionization on silicon (DIOS) is an example of a modification of MADLI that eliminates the matrix.[123] In this case, analytes are captured on a silicon surface prior to laser desorption and ionization. Other examples of "matrix-free" laser desorption techniques include the use of siloxane or carbon-based polymers.

Another emerging technique in this area is *laser diode thermal desorption* (LDTD), which is used along with a commercially available APCI source to rapidly introduce a large number of samples into the MS without LC separation. With LDTD, samples are adsorbed onto the surface of a well in a $\geq$96-well plate. An IR laser beam is then used to vaporize the material in each well into the APCI source.[124] This technique has been successfully demonstrated for the analysis of sulfonamide residues in milk[125] and steroid hormones in wastewater.[126]

6.4.5 Fragmentation

It is evident from this chapter that there are many examples of methods for the analysis of antibiotic residues in food that utilize mass spectrometry. As a result, the fragmentation patterns for different classes of antibiotics have been proposed and described in several multi-residue methods,[7,127] as well as in procedures for specific groups of compounds. Table 6.4 and Figure 6.14 provide examples of the common product ions and expected neutral losses seen in MS/MS spectra for major classes of antibiotics. Specific examples, along with relevant citations, are also provided. As MS methods begin to search for and identify more non-targeted analytes, it will become more important to be familiar with the fragmentation patterns of common analytes.

6.4.6 Mass Spectral Library

Mass spectral data containing fragment or product ions can provide fingerprint like information for compound characterization. Therefore, the importance of creating and maintaining searchable libraries of mass spectra to assist in unknown identification and confirm the identity of target analytes was realized early on. Libraries for electron ionization mass spectrometry, generally obtained by GC-MS, are well established. Historically these spectra collections were distributed as large multi-volume texts; the databases are now available electronically. For example, the NIST/EPA/NIH database and the Wiley Registry of Mass Spectral Data each contain several hundred thousand entries. In addition, the NIST program includes the AMDIS (automated mass spectral deconvolution and identification system) program, which assists in searching and matching spectra obtained by GC/MS to those in the reference libraries.

Compound libraries for LC-MS data have been slower to establish and disseminate. This is due partly to the fact that product ions formed by CID can vary significantly depending on the instrument used to obtain the data. With a triple quadrupole instrument, dissociation occurs with relatively high energy as precursor ions are accelerated through space and collide with a relatively heavy neutral gas (argon or nitrogen). Sequential dissociation, where product ions break up into smaller pieces as the ions travel down the collision cell, can occur. CID in an ion trap is a more subtle process with the occurrence of multiple low-energy collisions. Only the weaker bonds that can be cleaved with low activation energy break to form product ions.[128] As a result, ions formed by the neutral loss of small groups (such as water) are more common.

TABLE 6.4 Fragmentation Patterns of Antibiotics Using tandem LC-MS/MS

Drug Class	Common Losses	Common Ions (m/z)	Specific Example of Fragmentation (m/z)	Reference
Aminoglycosides	Aminosugars such as amino-α-D-glucopyranose or −2-deoxy-D-glucopyranose	Aminosugars	Gentamicin C1: 478 $[M+H]^+$; 322 $[M+H-C_8H_{16}N_2O]^+$; 157 $[C_8H_{17}N_2O]^+$	41,145
β-Lactams	$-H_2O$; $-CO$; $-CO_2$; $-NH_3$; cleavage of C4 ring; $-COOH$, −R group	160 $[C_7H_{12}SO_2]^+$; 114 $[C_6H_{10}S]^+$	Ampicillin: 350 $[M+H]^+$; 333$[M+H-NH_3]^+$; 192$[M+H-158]^+$; 160; 114; 106 $[C_7H_6NH_2]^+$	146,147
Lincosamides	$-SHCH_3-H_2O$	126 $[C_8H_{16}N]^+$	Lincomycin: 407 $[M+H]^+$; 359$[MH-SHCH_3]^+$; 126	148
Macrolides	$-H_2O$; −methanol; −propionaldehyde; −sugars (cladinose, desosamine)	158 $[C_8H_{15}NO_2]^+$ (desosamine fragment)	Erythromycin: 734 $[M+H]^+$; 716+$[M+H-H_2O]^+$; 576[M+H−158$^+$; 158	18
Phenicols	$-H_2O-$, COHCl; −HF (florfenicol)		Chloramphenicol: 321 $[M-H]^-$; 257 $[M-H-COHCl]^-$; 194 $[M-H-C_2H_3ONCl_2]^-$; 152 $[C_7H_6NO_3]^-$	149–151
Polyether antibiotics	$-H_2O$; cleavage of C−C bonds on either side of carbonyl for narasin and salinomycin; ring cleavage of monensin	Narasin/salinomycin: 531 $[C_{29}H_{48}O_7Na]^+$; 431 $[C_{23}H_{36}O_6Na]^+$	Monensin A: 693 $[M+Na]^+$; 675 $[M+Na-H_2O]^+$; 479 $[C_{25}H_{44}O_7Na]^+$; 461[m/z 479 $H_2O]^+$	152
Quinolones	$-CO_2$; $-CO$; $-H_2O$; −alkyl sidechains; −HF (for fluoroquinolones)		Sarafloxacin: 386$[M+H]^+$; 342 $[M+H-CO_2]^+$; 322 $[M+H-CO_2-HF]^+$; 299 $[M+H-CO_2-C_2H_5N]^+$	127,153
Sulfonamides	$-RNH_2$; $-C_6H_6NH$	156 $[C_6H_6NSO_2]^+$; 108$[156\text{-}SO]^+$; 92$[156\text{-}SO_2]^+$	Sulfamethazine: 279 $[M+H]^+$; 186$[SO_2NHC_6H_7N_2]^+$; 156; 108; 92	154
Tetracyclines	$-xH_2O$; $-xNH_3$; −NR		Oxytetracycline: 461$[M+H]^+$; 426$[M+H-H_2O-NH_3]^+$; 408$[M+H-2H_2O-NH_3]^+$	154

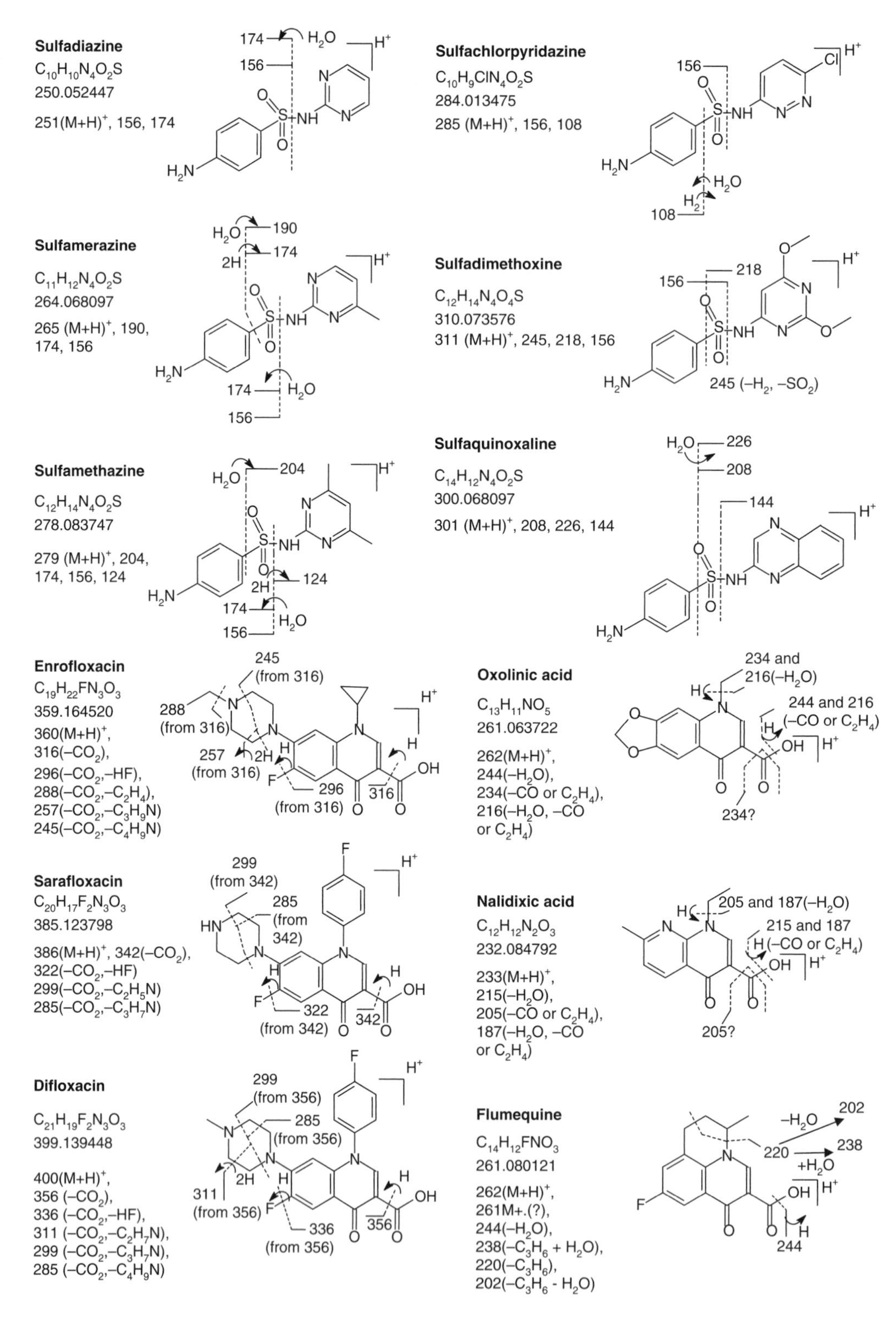

Figure 6.14 Proposed fragmentation pathways for selected veterinary drugs by ion trap CID. (Reproduced and modified from Li et al.[127] with permission from Elsevier; copyright 2006.)

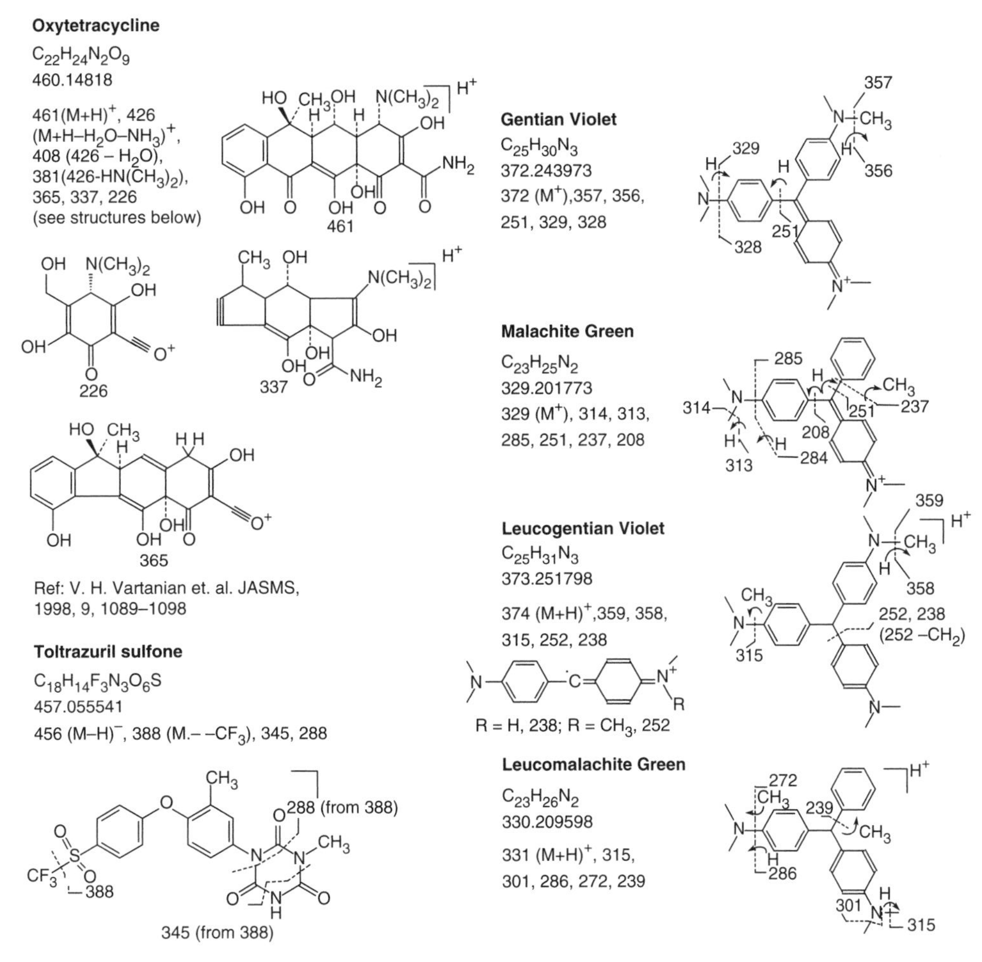

Figure 6.14 *(Continued)*

User libraries and databases that are specific to a particular instrument have been generated by individual users, as well as by instrument vendors. For example, an electrospray LC-MS/MS database of over 1200 compounds was generated with a QqLIT instrument for compounds of interest in forensics and toxicology.[129] It has been difficult, however, to meld these efforts together to make a more universal searching tool. One study describes using a tuning point technique to try to standardize product ion spectra obtained with different types of mass analyzers.[130] In this example, MS conditions where the product ion *m/z* 397 of reserpine was 80%±10% of the precursor ion (*m/z* 609) were defined as the calibration point. These settings were then used in 11 different instruments to obtain product ion spectra for 48 compounds. When the resulting spectra were compared, there was some reproducibility between all instruments for about 30% of the compounds. When similar type instruments were compared (beam to beam and trap to trap), this number improved to approximately 60%. Other researchers have been working toward the goal of standardizing LC-MS libraries for small molecules by utilizing sophisticated searching algorithms.[131,132]

With the advent of accurate mass instruments, access to a reference database or library of compounds with the exact masses of precursor and product ions is becoming increasingly important. Many instrument manufacturers, as well as independent companies or institutions, are developing software that allows chemists to manage this type of information for specific applications such as proteomics or chemical contaminants. Examples of on-line resources that may be useful for searching for unknown residues based on molecular formulas obtained from accurate mass data include "Metlin" from the Scripps Center for Mass Spectrometry[133] and "ChemSpider" from the Royal Society of Chemistry.[134]

ACKNOWLEDGMENT

We hereby indicate that any of the columns, UHPLC, and mass spectrometers described or mentioned in this chapter is by no means exhaustive and that any commercially available items cited above do not in any way constitute an endorsement by the authors.

ABBREVIATIONS

APCI	Atmospheric-pressure chemical ionization
APPI	Atmospheric-pressure photoionization
ASAP	Atmospheric solids analysis probe
CID	Collision-induced dissociation
DAPCI	Desorption atmospheric-pressure chemical ionization
DART	Direct analysis in real time
DC	Direct current
DESI	Desorption electrospray ionization
DIOS	Direct ionization on silicon
EPI	Enhanced product ion
ESI	Electrospray ionization
EU	European Union
FAIMS	High-field asymmetric waveform ion mobility spectrometry
FDA	Food and Drug Administration (US)
FWHM	Full-width at half-height maximum
FWHM	Full-width at half-height maximum
GC-MS	Gas chromatography mass spectrometry
HETP	Height equivalent to a theoretical plate
HILIC	Hydrophilic interaction liquid chromatography
IMS	Ion mobility spectrometry
IPA	Ion-pairing agent
IPC	Ion-pairing chromatography
IPs	Identification points
LC	Liquid chromatography
LC-MS	Liquid chromatography mass spectrometry
LC/MS/MS	Liquid chromatography coupled with a triple-quadrupole mass spectrometer operated in MRM mode
LDTD	Laser diode thermal desorption
LIT Orbitrap	Linear ion trap Orbitrap
LOD	Limit of detection
MRL	Maximum residue limit
MRM	Multiple-reaction monitoring
MS	Mass spectrometer or mass spectrometry
MS^n	Multiple-stages fragmentation
MS/MS	Tandem mass spectrometry
m/z	Mass-to-charge ratio
PA	Proton affinity
QIT	Quadrupole ion trap
QqLIT	Triple-quadrupole linear ion trap
QqQ	Triple quadrupole
QqTOF	Quadrupole time-of-flight
QqTOF MS	QqTOF operated as a straight TOF mass analyzer in full-scan mode
QqTOF MS/MS	QqTOF operated in MS/MS mode with Q1 enabled as a mass filter
Q1	First quadrupole
Q2 (q)	Collision cell
Q3	The third quadrupole
RF	Radiofrequency
SELDI	Surface-enhanced laser desorption ionization
SIM	Selected-ion monitoring
SPE	Solid-phase extraction
TOF	Time-of-flight
UHPLC	Ultra high-performance or ultra-high pressure liquid chromatography
UPLC/QqTOF MS	Ultra performance liquid chromatography coupled with a quadrupole time-of-flight mass spectrometer operated as a straight TOF mass analyzer in full-scan mode
UV	Ultraviolet

REFERENCES

1. Le Bizec B, Pinel G, Antignac J-P, Options for veterinary drug analysis using mass spectrometry, *J. Chromatogr. A* 2009;1216(46):8016–8034.
2. Commission Decision of 12 August 2002 implementing Council Directive 96/23/EC concerning the performance of analytical methods and the interpretation of results. 2002/657/EC, *Off. J. Eur. Commun*. 2002;L221:8–36.
3. Alder L, Greulich K, Kempe G, Vieth B, Residue analysis of 500 high priority pesticides: Better by GC-MS or LC-MS/MS? *Mass Spectrom. Rev*. 2006;25(6):838–865.
4. Gros M, Petrovic M, Barcelo D. Multi-residue analytical methods using LC-tandem MS for the determination of pharmaceuticals in environmental and wastewater samples: a review, *Anal. Bioanal. Chem*. 2006;386(4):941–952.
5. Kuster M, Lopez de Alda M, Barcelo D, Liquid chromatography-tandem mass spectrometric analysis and regulatory issues of polar pesticides in natural and treated waters, *J. Chromatogr. A* 2008;1216(3):520–5229.
6. Turnipseed SB, Andersen WC, Karbiwnyk CM, Madson MR, Miller KE, Multi-class, multi-residue liquid chromatography/tandem mass spectrometry screening and confirmation methods for drug residues in milk, *Rapid Commun. Mass Spectrom*. 2008;22(10):1467–1480.
7. Hammel YA, Mohamed R, Gremaud E, Lebreton MH, Guy PA, Multi-screening approach to monitor and quantify 42 antibiotic residues in honey by liquid chromatography-tandem mass spectrometry, *J. Chromatogr. A* 2008;1177(1):58–76.
8. Stolker AA, Rutgers P, Oosterink E, et al., Comprehensive screening and quantification of veterinary drugs in milk using UPLC-ToF-MS, *Anal. Bioanal. Chem*. 2008;391(6):2309–2322.
9. Vidal JLM, Aguilera-Luiz MdM, Romero-Gonzalez R, Frenich AG, Multiclass analysis of antibiotic residues in honey by ultraperformance liquid chromatography-tandem mass spectrometry, *J. Agric. Food Chem*. 2009;57(5):1760–1767.

10. Peters RJB, Bolck YJC, Rutgers P, Stolker AAM, Nielen MWF, Multi-residue screening of veterinary drugs in egg, fish and meat using high-resolution liquid chromatography accurate mass time-of-flight mass spectrometry, *J. Chromatogr. A* 2009;1216(46):8206–8216.
11. Ortelli D, Cognard E, Jan P, Edder P, Comprehensive fast multiresidue screening of 150 veterinary drugs in milk by ultra-performance liquid chromatography coupled to time of flight mass spectrometry, *J. Chromatogr. B* 2009;877(23):2363–2374.
12. Lopez MI, Pettis JS, Smith IB, Chu PS, Multiclass determination and confirmation of antibiotic residues in honey using LC-MS/MS, *J. Agric. Food Chem*. 2008;56(5):1553–1559.
13. Qiang Z, Adams C, Potentiometric determination of acid dissociation constants (pKa) for human and veterinary antibiotics, *Water Res*. 2004;38(12):2874–2890.
14. Anderson CR, Rupp HS, Wu WH, Complexities in tetracycline analysis—chemistry, matrix extraction, cleanup, and liquid chromatography, *J. Chromatogr. A* 2005;1075(1–2):23–32.
15. Lin CE, Chang CC, Lin WC, Migration behavior and separation of sulfonamides in capillary zone electrophoresis. III. Citrate buffer as a background electolyte, *J. Chromatogr. A* 1997;768(1):105–112.
16. Gobel A, McArdell CS, Suter MJ, Giger W, Trace determination of macrolide and sulfonamide antimicrobials, a human sulfonamide metabolite, and trimethoprim in wastewater using liquid chromatography coupled to electrospray tandem mass spectrometry, *Anal. Chem*. 2004;76(16):4756–4764.
17. Khong SP, Hammel YA, Guy PA, Analysis of tetracyclines in honey by high-performance liquid chromatography/tandem mass spectrometry, *Rapid Commun. Mass Spectrom*. 2005;19(4):493–502.
18. Wang J, Analysis of macrolide antibiotics, using liquid chromatography-mass spectrometry, in food, biological and environmental matrices, *Mass Spectrom Rev*. 2009;28(1):50–92.
19. Soeborg T, Ingerslev F, Halling-Sorensen B, Chemical stability of chlortetracycline and chlortetracycline degradation products and epimers in soil interstitial water, *Chemosphere* 2004;57(10):1515–1524.
20. Riediker S, Stadler RH, Simultaneous determination of five beta-lactam antibiotics in bovine milk using liquid chromatography coupled with electrospray ionization tandem mass spectrometry, *Anal. Chem*. 2001;73(7):1614–1621.
21. Holstege DM, Puschner B, Whitehead G, Galey FD, Screening and mass spectral confirmation of beta-lactam antibiotic residues in milk using LC-MS/MS, *J. Agric. Food Chem*. 2002;50(2):406–411.
22. Wang J, Leung D, Butterworth F, Determination of five macrolide antibiotic residues in eggs using liquid chromatography/electrospray ionization tandem mass spectrometry, *J. Agric. Food Chem*. 2005;53(6):1857–1865.
23. Wang J, Determination of five macrolide antibiotic residues in honey by LC-ESI-MS and LC-ESI-MS/MS, *J. Agric. Food Chem*. 2004;52(2):171–181.
24. Yang S, Carlson KH, Solid-phase extraction-high-performance liquid chromatography-ion trap mass spectrometry for analysis of trace concentrations of macrolide antibiotics in natural and waste water matrices, *J. Chromatogr. A* 2004; 1038(1–2):141–155.
25. Kochansky J, Degradation of tylosin residues in honey, *J. Apicult. Res*. 2004;43(2):65–68.
26. Mastovska K, Lightfield AR. Streamlining methodology for the multiresidue analysis of beta-lactam antibiotics in bovine kidney using liquid chromatography-tandem mass spectrometry, *J. Chromatogr. A* 2008;1202(2):118–123.
27. Lopez MI, Feldlaufer MF, Williams AD, Chu PS, Determination and confirmation of nitrofuran residues in honey using LC-MS/MS, *J. Agric. Food Chem*. 2007;55(4):1103–1108.
28. Chu PS, Lopez MI, Determination of nitrofuran residues in milk of dairy cows using liquid chromatography-tandem mass spectrometry, *J. Agric. Food Chem*. 2007;55(6):2129–2135.
29. Becker M, Zittlau E, Petz M, Quantitative determination of ceftiofur-related residues in bovine raw milk by LC-MS/MS with electrospray ionization, *Eur. Food Res. Technol*. 2003;217(5):449–456.
30. Makeswaran S, Patterson I, Points J, An analytical method to determine conjugated residues of ceftiofur in milk using liquid chromatography with tandem mass spectrometry, *Anal. Chim. Acta* 2005;529(1–2):151–157.
31. Meyer VR, *Practical High-Performance Liquid Chromatography*, 4th ed., Wiley, Chichester, UK, 2004.
32. Neue UD, *HPLC Columns: Theory, Technology, and Practice*, Wiley-VCH, New York, 1997.
33. Kostiainen R, Kauppila TJ, Effect of eluent on the ionization process in liquid chromatography-mass spectrometry, *J. Chromatogr. A* 2009;1216(4):685–699.
34. Thompson TS, Noot DK, Calvert J, Pernal SF, Determination of lincomycin and tylosin residues in honey using solid-phase extraction and liquid chromatography-atmospheric pressure chemical ionization mass spectrometry, *J. Chromatogr. A* 2003;1020(2):241–250.
35. Codony R, Compano R, Granados M, Garcia-Regueiro JA, Prat MD, Residue analysis of macrolides in poultry muscle by liquid chromatography-electrospray mass spectrometry, *J. Chromatogr. A* 2002;959(1–2):131–141.
36. Thompson TS, Pernal SF, Noot DK, Melathopoulos AP, van den Heever JP, Degradation of incurred tylosin to desmycosin-implications for residue analysis of honey, *Anal. Chim. Acta* 2007;586(1–2):304–311.
37. Thompson TS, Noot DK, Calvert J, Pernal SF, Determination of lincomycin and tylosin residues in honey by liquid chromatography/tandem mass spectrometry, *Rapid Commun. Mass Spectrom*. 2005;19(3):309–316.
38. Cecchi T, Ion pairing chromatography, *Crit. Rev. Anal. Chem*. 2008;38(3):161–213.
39. Wybraniec S, Mizrahi Y, Influence of perfluorinated carboxylic acids on ion-pair reversed-phase high-performance liquid chromatographic separation of betacyanins and

17-decarboxy-betacyanins, *J. Chromatogr. A* 2004; 1029(1–2):97–101.

40. Petritis KN, Chaimbault P, Elfakir C, Dreux M, Ion-pair reversed-phase liquid chromatography for determination of polar underivatized amino acids using perfluorinated carboxylic acids as ion pairing agent, *J. Chromatogr. A* 1999;833(2):147–155.
41. Zhu WX, Yang JZ, Wei W, Liu YF, Zhang SS, Simultaneous determination of 13 aminoglycoside residues in foods of animal origin by liquid chromatography-electrospray ionization tandem mass spectrometry with two consecutive solid-phase extraction steps, *J. Chromatogr. A* 2008;1207(1–2):29–37.
42. Bogialli S, Curini R, Di Corcia A, Lagana A, Mele M, Nazzari M, Simple confirmatory assay for analyzing residues of aminoglycoside antibiotics in bovine milk: Hot water extraction followed by liquid chromatography-tandem mass spectrometry, *J. Chromatogr. A* 2005;1067(1–2):93–100.
43. Kaufmann A, Butcher P, Kolbener P, Trace level quantification of streptomycin in honey with liquid chromatography/tandem mass spectrometry, *Rapid Commun. Mass Spectrom*. 2003;17(22):2575–2577.
44. van Bruijnsvoort M, Ottink SJ, Jonker KM, de Boer E, Determination of streptomycin and dihydrostreptomycin in milk and honey by liquid chromatography with tandem mass spectrometry, *J. Chromatogr. A* 2004;1058(1–2):137–142.
45. Cherlet M, De Baere S, De Backer P, Quantitative determination of dihydrostreptomycin in bovine tissues and milk by liquid chromatography-electrospray ionization-tandem mass spectrometry, *J. Mass Spectrom*. 2007;42(5):647–656.
46. Gustavsson SA, Samskog J, Markides KE, Langstrom B, Studies of signal suppression in liquid chromatography–electrospray ionization mass spectrometry using volatile ion-pairing reagents, *J. Chromatogr. A* 2001; 937(1–2):41–47.
47. Fuh M-R, Haung C-H, Lin S-L, Pan WHT, Determination of free-form amphetamine in rat brain by ion-pair liquid chromatography-electrospray mass spectrometry with *in vivo* microdialysis, *J. Chromatogr. A* 2004;1031(1–2):197–201.
48. Keever J, Voyksner RD, Tyczkowska KL, Quantitative determination of ceftiofur in milk by liquid chromatography-electrospray mass spectrometry, *J. Chromatogr. A* 1998;794(1–2):57–62.
49. Kuhlmann FE, Apffel A, Fischer SM, Goldberg G, Goodley PC, Signal enhancement for gradient reverse-phase high-performance liquid chromatography-electrospray ionization mass spectrometry analysis with trifluoroacetic and other strong acid modifiers by postcolumn addition of propionic acid and isopropanol, *J. Am. Soc. Mass Spectrom*. 1995;6(12):1221–1225.
50. Alpert AJ, Hydrophilic-interaction chromatography for the separation of peptides, nucleic acids and other polar compounds, *J. Chromatogr. A* 1990;499:177–196.
51. Hemstrom P, Irgum K, Hydrophilic interaction chromatography, *J. Sep. Sci*. 2006;29(12):1784–1821.
52. Hao Z, Xiao B, Weng N, Impact of column temperature and mobile phase components on selectivity of hydrophilic interaction chromatography (HILIC), *J. Sep. Sci*. 2008;31(9):1449–1464.
53. Nguyen HP, Schug KA, The advantages of ESI-MS detection in conjunction with HILIC mode separations: Fundamentals and applications, *J. Sep. Sci*. 2008;31(9):1465–1480.
54. *A Practical Guide to HILIC. A Tutorial and Application Book*, SeQuant AB 907 19 Umea, Sweden, 2008.
55. Shin-ichi K, Analysis of impurities in streptomycin and dihydrostreptomycin by hydrophilic interaction chromatography/electrospray ionization quadrupole ion trap/time-of-flight mass spectrometry, *Rapid Commun. Mass Spectrom*. 2009;23(6):907–914.
56. McCalley DV, Is hydrophilic interaction chromatography with silica columns a viable alternative to reversed-phase liquid chromatography for the analysis of ionisable compounds? *J. Chromatogr. A* 2007;1171(1–2):46–55.
57. Ishii R, Horie M, Chan W, MacNeil J, Multi-residue quantitation of aminoglycoside antibiotics in kidney and meat by liquid chromatography with tandem mass spectrometry, *Food Addit. Contam*. 2008;25(12):1509–1519.
58. Zheng MM, Zhang MY, Peng GY, Feng YQ, Monitoring of sulfonamide antibacterial residues in milk and egg by polymer monolith microextraction coupled to hydrophilic interaction chromatography/mass spectrometry, *Anal. Chim. Acta* 2008;625(2):160–172.
59. Valette JC, Demesmay C, Rocca JL, Verdon E, Separation of tetracycline antibiotics by hydrophilic interaction chromatography using an amino-propyl stationary phase, *Chromatographia* 2004;59(1–2):55–60.
60. Mazzeo JR, Neue UD, Kele M, Plumb RS, Advancing LC performance with smaller particles and higher pressure, *Anal. Chem*. 2005;77(23):460A–467A.
61. van Deemter JJ, Zuiderweg FJ, Klinkenberg A, Longitudinal diffusion and resistance to mass transfer as causes of nonideality in chromatography, *Chem. Eng. Sci*. 1956;5(6):271–289.
62. Wu N, Thompson R, Fast and efficient separations using reversed phase liquid chromatography, *J. Liq. Chromatogr. RT* 2006;29(7):949–988.
63. Teutenberg T, Potential of high temperature liquid chromatography for the improvement of separation efficiency—a review, *Anal. Chim. Acta* 2009;643(1–2):1–12.
64. Kaufmann A, Butcher P, Maden K, Widmer M, Quantitative multiresidue method for about 100 veterinary drugs in different meat matrices by sub 2-μm particulate high-performance liquid chromatography coupled to time of flight mass spectrometry, *J. Chromatogr. A* 2008;1194(1):66–79.
65. Abrahim A, Al-Sayah M, Skrdla P, Bereznitski Y, Chen Y, Wu N, Practical comparison of 2.7 μm fused-core silica particles and porous sub-2 μm particles for fast separations in pharmaceutical process development, *J. Pharmaceut. Biomed. Anal*. 2010;51(1):131–137.
66. Karas M, Bachmann D, Bahr U, Hillenkamp F, Matrix-assisted ultraviolet laser desorption of non-volatile

compounds, *Int. J. Mass Spectrom. Ion Process*. 1987;78(C):53–68.

67. Karas M, Hillenkamp F, Laser desorption ionization of proteins with molecular masses exceeding 10 000 daltons, *Anal. Chem*. 1988;60(20):2299–2301.
68. Tanaka K, Waki H, Ido Y, et al., Protein and polymer analyses up to m/z 100,000 by laser ionization time-of-flight mass spectrometry, *Rapid Commun. Mass Spectrom*. 1988;2(8):151–153.
69. El-Aneed A, Cohen A, Banoub J, Mass spectrometry, review of the basics: Electrospray, MALDI, and commonly used mass analyzers, *Appl. Spectrosc. Rev*. 2009;44(3):210–230.
70. Careri M, Bianchi F, Corradini C, Recent advances in the application of mass spectrometry in food-related analysis, *J. Chromatogr. A* 2002;970(1–2):3–64.
71. Dole M, Mack LL, Hines RL, Mobley RC, Ferguson LD, Alice MB, Molecular beams of macroions, *J. Chem. Phys*. 1968;49(5):2240–2249.
72. Yamashita M, Fenn JB, Electrospray ion source. Another variation on the free-jet theme, *J. Phys. Chem*. 1984;88(20): 4451–4459.
73. Fenn JB, Tanaka K, Wuthrich K, The Nobel Prize in Chemistry 2002 (available at `http://nobelprize.org/nobel_prizes/chemistry/laureates/2002/index.html`; accessed 6/16/09).
74. Hernandez F, Sancho JV, Pozo OJ, Critical review of the application of liquid chromatography/mass spectrometry to the determination of pesticide residues in biological samples, *Anal. Bioanal. Chem*. 2005;382(4):934–946.
75. Pena A, Lino CM, Alonso R, Barcelo D, Determination of tetracycline antibiotic residues in edible swine tissues by liquid chromatography with spectrofluorometric detection and confirmation by mass spectrometry, *J. Agric. Food Chem*. 2007;55(13):4973–4979.
76. Bailac S, Barron D, Sanz-Nebot V, Barbosa J, Determination of fluoroquinolones in chicken tissues by LC-coupled electrospray ionisation and atmospheric pressure chemical ionisation, *J. Sep. Sci*. 2006;29(1):131–136.
77. Robb DB, Covey TR, Bruins AP, Atmospheric pressure photoionization: An ionization method for liquid chromatography–mass spectrometry, *Anal. Chem*. 2000; 72(15):3653–3659.
78. Marchi I, Rudaz S, Veuthey JL, Atmospheric pressure photoionization for coupling liquid-chromatography to mass spectrometry: A review, *Talanta* 2009;78(1):1–18.
79. Takino M, Daishima S, Nakahara T, Determination of chloramphenicol residues in fish meats by liquid chromatography-atmospheric pressure photoionization mass spectrometry, *J. Chromatogr. A* 2003;1011(1–2):67–75.
80. Mohamed R, Hammel YA, LeBreton MH, Tabet JC, Jullien L, Guy PA, Evaluation of atmospheric pressure ionization interfaces for quantitative measurement of sulfonamides in honey using isotope dilution liquid chromatography coupled with tandem mass spectrometry techniques, *J. Chromatogr. A* 2007;1160(1–2):194–205.
81. Sheridan R, Policastro B, Thomas S, Rice D, Analysis and occurrence of 14 sulfonamide antibacterials and chloramphenicol in honey by solid-phase extraction followed by LC/MS/MS analysis, *J. Agric. Food Chem*. 2008;56(10):3509–3516.
82. Wang J, Leung D, Lenz SP, Determination of five macrolide antibiotic residues in raw milk using liquid chromatography-electrospray ionization tandem mass spectrometry, *J. Agric. Food Chem*. 2006;54(8):2873–2880.
83. Wan EC, Ho C, Sin DW, Wong YC, Detection of residual bacitracin A, colistin A, and colistin B in milk and animal tissues by liquid chromatography tandem mass spectrometry, *Anal. Bioanal. Chem*. 2006;385(1):181–188.
84. Bonfiglio R, King RC, Olah TV, Merkle K, The effects of sample preparation methods on the variability of the electrospray ionization response for model drug compounds, *Rapid Commun. Mass Spectrom*. 1999;13(12):1175–1185.
85. Matuszewski BK, Constanzer ML, Chavez-Eng CM, Strategies for the assessment of matrix effect in quantitative bioanalytical methods based on HPLC-MS/MS, *Anal Chem*. 2003;75(13):3019–3030.
86. Wang J, Cheung W, Grant D, Determination of pesticides in apple-based infant foods using liquid chromatography electrospray ionization tandem mass spectrometry, *J. Agric. Food Chem*. 2005;53(3):528–537.
87. Zrostlikova J, Hajslova J, Poustka J, Begany P, Alternative calibration approaches to compensate the effect of co-extracted matrix components in liquid chromatography-electrospray ionisation tandem mass spectrometry analysis of pesticide residues in plant materials, *J. Chromatogr. A* 2002;973(1–2):13–26.
88. Alder L, Luderitz S, Lindtner K, Stan HJ, The ECHO technique—the more effective way of data evaluation in liquid chromatography-tandem mass spectrometry analysis, *J. Chromatogr. A* 2004;1058(1–2):67–79.
89. Stahnke H, Reemtsma T, Alder L, Compensation of matrix effects by postcolumn infusion of a monitor substance in multiresidue analysis with LC-MS/MS, *Anal. Chem*. 2009;81(6):2185–2192.
90. Kruve A, Leito I, Herodes K, Combating matrix effects in LC/ESI/MS: The extrapolative dilution approach, *Anal. Chim. Acta* 2009;651(1):75–80.
91. Zwiener C, Frimmel FH, LC-MS analysis in the aquatic environment and in water treatment—a critical review. Part I: Instrumentation and general aspects of analysis and detection, *Anal. Bioanal. Chem*. 2004;378(4):851–861.
92. Volmer DA, Sieno L, Tutorial—mass analyzers: An overview of several designs and their applications, Part II, *Spectroscopy*. 2005;20(12):90–95.
93. Volmer DA, Sieno L, Tutorial—mass analyzers: An overview of several designs and their applications, Part I, *Spectroscopy*. 2005;20(111):20–26.
94. Boyd RK, Basic C, Bethem RA, *Trace Quantitative Analysis by Mass Spectrometry*, Wiley, Chichester, UK, 2008.
95. Balogh M, Debating resolution and mass accuracy in mass spectrometry, *Spectroscopy* 2004;19:34–40.
96. FDA, Guidance for Industry #118, *Mass Spectrometry for Confirmation of the Identity of Animal Drug*

Residues, Final Guidance, US Dept. Health and Human Services Food and Drug Administration Center for Veterinary Medicine, May 1, 2003 (available at `http://www.fda.gov/cvm/guidance/guide118.pdf`; accessed 4/07/10).

97. Council Directive 96/23/EC of 29 April 1996 on meaures to monitor certain substances and residues thereof in live animals and animal products and repealing Directives 85/358/EEC and 86/469/EEC and Decisions 89/187/EEC and 91/664/EEC, *Off. J. Eur. Commun*. 1996;L125:10–32
98. Bogialli S, Di Corcia A, Lagana A, Mastrantoni V, Sergi M, A simple and rapid confirmatory assay for analyzing antibiotic residues of the macrolide class and lincomycin in bovine milk and yoghurt: Hot water extraction followed by liquid chromatography/tandem mass spectrometry, *Rapid Commun. Mass Spectrom*. 2007;21(2):237–246.
99. Hager JW, Le Blanc JC, High-performance liquid chromatography-tandem mass spectrometry with a new quadrupole/linear ion trap instrument, *J. Chromatogr. A* 2003;1020(1):3–9.
100. Heller DN, Nochetto CB, Development of multiclass methods for drug residues in eggs: Silica SPE cleanup and LC-MS/MS analysis of ionophore and macrolide residues, *J. Agric. Food Chem*. 2004;52(23):6848–6856.
101. Leonard S, Ferraro M, Adams E, Hoogmartens J, Van Schepdael A, Application of liquid chromatography/ion trap mass spectrometry to the characterization of the related substances of clarithromycin, *Rapid Commun. Mass Spectrom*. 2006;20(20):3101–3110.
102. Hager JW, A new linear ion trap mass spectrometer, *Rapid Commun. Mass Spectrom*. 2002;16(6):512–526.
103. Schwartz JC, Senko MW, Syka JE, A two-dimensional quadrupole ion trap mass spectrometer, *J. Am. Soc. Mass Spectrom*. 2002;13(6):659–669.
104. Hager JW, Yves Le Blanc JC, Product ion scanning using a Q-q-Q linear ion trap (Q TRAP) mass spectrometer, *Rapid Commun. Mass Spectrom*. 2003;17(10):1056–1064.
105. Bueno MJM, Aguera A, Hernando MD, Gomez MJ, Fernandez-Alba AR, Evaluation of various liquid chromatography-quadrupole-linear ion trap-mass spectrometry operation modes applied to the analysis of organic pollutants in wastewaters, *J. Chromatogr. A* 2009;1216(32):5995–6002.
106. Bueno MJM, Aguera A, Gomez MJ, Hernando MD, Garcia-Reyes JF, Fernandez-Alba AR, Application of liquid chromatography/quadrupole-linear ion trap mass spectrometry and time-of-flight mass spectrometry to the determination of pharmaceuticals and related contaminants in wastewater, *Anal. Chem*. 2007;79(24):9372–9384.
107. Guilhaus M, Selby D, Mlynski V, Orthogonal acceleration time-of-flight mass spectrometry, *Mass Spectrom. Rev*. 2000;19(2):65–107.
108. Chernushevich IV, Loboda AV, Thomson BA, An introduction to quadrupole-time-of-flight mass spectrometry, *J. Mass Spectrom*. 2001;36(8):849–865.
109. Ibanez M, Guerrero C, Sancho JV, Hernandez F, Screening of antibiotics in surface and wastewater samples by ultra-high-pressure liquid chromatography coupled to hybrid quadrupole time-of-flight mass spectrometry, *J. Chromatogr. A* 2009;1216(12):2529–2539.
110. Wang J, Leung D, Analyses of macrolide antibiotic residues in eggs, raw milk, and honey using both ultra-performance liquid chromatography/quadrupole time-of-flight mass spectrometry and high-performance liquid chromatography/tandem mass spectrometry, *Rapid Commun. Mass Spectrom*. 2007;21(19):3213–3222.
111. Hernandez F, Ibanez M, Sancho JV, Pozo OJ, Comparison of different mass spectrometric techniques combined with liquid chromatography for confirmation of pesticides in environmental water based on the use of identification points, *Anal. Chem*. 2004;76(15):4349–4357.
112. Makarov A, Denisov E, Lange O, Performance evaluation of a high-field orbitrap mass analyzer, *J. Am. Soc. Mass Spectrom*. 2009;20(8):1391–1396.
113. Hu Q, Noll RJ, Li H, Makarov A, Hardman M, Graham Cooks R, The Orbitrap: A new mass spectrometer, *J. Mass Spectrom*. 2005;40(4):430–443.
114. Kellmann M, Muenster H, Zomer P, Mol H, Full scan MS in comprehensive qualitative and quantitative residue analysis in food and feed matrices: How much resolving power is required? *J. Am. Soc. Mass Spectrom*. 2009;20(8):1464–1476.
115. Van der Heeft E, Bolck YJC, Beumer B, Nijrolder AWJM, Stolker AAM, Nielen MWF, Full-scan accurate mass selectivity of ultra performance liquid chromatography combined with time-of-flight and orbitrap mass spectrometry in hormone and veterinary drug residue analysis, *J. Am. Soc. Mass Spectrom*. 2009;20(3):451–463.
116. Kanu AB, Dwivedi P, Tam M, Matz L, Hill HH, Ion mobility-mass spectrometry, *J. Mass Spectrom*. 2008;43(1):1–22.
117. Kolakowski BM, Mester Z, Review of applications of high-field asymmetric waveform ion mobility spectrometry (FAIMS) and differential mobility spectrometry (DMS), *Analyst* 2007;132(9):842–864.
118. Pringle SD, Giles K, Wildgoose JL, et al., An investigation of the mobility separation of some peptide and protein ions using a new hybrid quadrupole/travelling wave IMS/oa-ToF instrument, *Int. J. Mass Spectrom*. 2007;261(1):1–12.
119. Cooks RG, Ouyang Z, Takats Z, Wiseman JM, Ambient mass spectrometry. *Science* 2006;311(5767):1566–1570.
120. Williams JP, Patel VJ, Holland R, Scrivens JH, The use of recently described ionisation techniques for the rapid analysis of some common drugs and samples of biological origin, *Rapid Commun. Mass Spectrom*. 2006;20(9):1447–1456.
121. Merchant M, Weinberger SR, Recent advancements in surface-enhanced laser desorption/ionization-time of flight-mass spectrometry, *Electrophoresis* 2000;21(6):1164–1177.
122. Peterson DS, Matrix-free methods for laser desorption/ionization mass spectrometry, *Mass Spectrom Rev*. 2007;26(1):19–34.

123. Deng G, Sanyal G, Applications of mass spectrometry in early stages of target based drug discovery, *J. Pharm. Biomed. Anal*. 2006;40(3):528–538.

124. Wu J, Hughes CS, Picard P, et al., High-throughput cytochrome P450 inhibition assays using laser diode thermal desorption-atmospheric pressure chemical ionization-tandem mass spectrometry, *Anal. Chem*. 2007;79(12):4657–4665.

125. Segura PA, Tremblay P, Picard P, Gagnon C, Sauve S, High-throughput quantitation of seven sulfonamide residues in dairy milk using laser diode thermal desorption-negative mode atmospheric pressure chemical ionization tandem mass spectrometry, *J. Agric. Food Chem*. 2010;58(3):1442–1446.

126. Fayad PB, Prevost M, Sauve S, Laser diode thermal desorption/atmospheric pressure chemical ionization tandem mass spectrometry analysis of selected steroid hormones in wastewater: Method optimization and application, *Anal. Chem*. 2010;82(2):639–645.

127. Li H, Kijak PJ, Turnipseed SB, Cui W, Analysis of veterinary drug residues in shrimp: a multi-class method by liquid chromatography-quadrupole ion trap mass spectrometry, *J. Chromatogr. B* 2006;836(1–2):22–38.

128. Niessen WMA, *Liquid Chromatography–Mass Spectrometry*, 3rd ed., Taylor & Francis, Boca Raton, FL, 2006.

129. Dresen S, Gergov M, Politi L, Halter C, Weinmann W, ESI-MS/MS library of 1,253 compounds for application in forensic and clinical toxicology, *Anal. Bioanal. Chem*. 2009;395(8):2521–2526.

130. Hopley C, Bristow T, Lubben A, et al., Towards a universal product ion mass spectral library—reproducibility of product ion spectra across eleven different mass spectrometers, *Rapid Commun. Mass Spectrom*. 2008;22(12):1779–1786.

131. Oberacher H, Pavlic M, Libiseller K, et al., On the inter-instrument and the inter-laboratory transferability of a tandem mass spectral reference library: 2. Optimization and characterization of the search algorithm, *J. Mass Spectrom*. 2009;44(4):494–502.

132. Mylonas R, Mauron Y, Masselot A, et al., X-Rank: A robust algorithm for small molecule identification using tandem mass spectrometry, *Anal. Chem*. 2009;81(18):7604–7610.

133. Metlin. (available at `http://metlin.scripps.edu/index.php`; accessed 4/07/10).

134. ChemSpider (available at `http://www.chemspider.com/Search.aspx`; accessed 4/07/10).

135. Van Eeckhaut A, Lanckmans K, Sarre S, Smolders I, Michotte Y, Validation of bioanalytical LC-MS/MS assays: Evaluation of matrix effects, *J. Chromatogr. B* 2009; 877(23):2198–2207.

136. Jonscher KR, Yates JR, The whys and wherefores of quadrupole ion trap mass spectrometry, *ABRF News* (Vol. 7), Sept. 1996 (available at `http://www.abrf.org/abrfnews/1996/september1996/sep96iontrap.html`; accessed 4/07/10).

137. de Hoffmann E, Tandem mass spectrometry: A primer, *J. Mass Spectrom*. 1996;31(2):129–137.

138. Berrada H, Borrull F, Font G, Molto JC, Marce RM, Validation of a confirmatory method for the determination of macrolides in liver and kidney animal tissues in accordance with the European Union regulation 2002/657/EC, *J. Chromatogr. A* 2007;1157(1–2):281–288.

139. Dubreil-Chéneau E, Bessiral M, Roudaut B, Verdon E, Sanders P, Validation of a multi-residue liquid chromatography-tandem mass spectrometry confirmatory method for 10 anticoccidials in eggs according to Commission Decision 2002/657/EC, *J. Chromatogr. A* 2009; 1216(46):8149–8157.

140. Shao B, Wu X, Zhang J, Duan H, Chu X, Wu Y, Development of a rapid LC-MS-MS method for multi-class determination of 14 coccidiostat residues in eggs and chicken, *Chromatographia* 2009;69(9–10):1083–1088.

141. Boison J, Lee S, Gedir R, Analytical determination of virginiamycin drug residues in edible porcine tissues by LC-MS with confirmation by LC-MS/MS, *J. AOAC. Int*. 2009;92(1):329–339.

142. Samanidou V, Evaggelopoulou E, Trotzmuller M, Guo X, Lankmayr E, Multi-residue determination of seven quinolones antibiotics in gilthead seabream using liquid chromatography-tandem mass spectrometry, *J. Chromatogr. A* 2008;1203(2):115–123.

143. Mottier P, Hammel YA, Gremaud E, Guy PA, Quantitative high-throughput analysis of 16 (fluoro)quinolones in honey using automated extraction by turbulent flow chromatography coupled to liquid chromatography-tandem mass spectrometry, *J. Agric. Food Chem*. 2008;56(1):35–43.

144. Zhang H, Ren Y, Bao X, Simultaneous determination of (fluoro)quinolones antibacterials residues in bovine milk using ultra performance liquid chromatography-tandem mass spectrometry, *J Pharm. Biomed. Anal*. 2009;49(2):367–374.

145. Heller DN, Clark SB, Righter HF, Confirmation of gentamicin and neomycin in milk by weak cation-exchange extraction and electrospray ionization/ion trap tandem mass spectrometry, *J. Mass Spectrom*. 2000;35(1):39–49.

146. Daeseleire E, De Ruyck H, Van Renterghem R, Confirmatory assay for the simultaneous detection of penicillins and cephalosporins in milk using liquid chromatography/tandem mass spectrometry, *Rapid Commun. Mass Spectrom*. 2000;14(15):1404–1409.

147. Heller DN, Ngoh MA, Electrospray ionization and tandem ion trap mass spectrometry for the confirmation of seven beta-lactam antibiotics in bovine milk, *Rapid Commun. Mass Spectrom*. 1998;12(24):2031–2040.

148. Sin DW, Wong YC, Ip AC, Quantitative analysis of lincomycin in animal tissues and bovine milk by liquid chromatography electrospray ionization tandem mass spectrometry, *J. Pharm. Biomed. Anal*. 2004;34(3):651–659.

149. Hammack W, Carson MC, Neuhaus BK, et al., Multilaboratory validation of a method to confirm chloramphenicol in shrimp and crabmeat by liquid chromatography-tandem mass spectrometry, *J. AOAC. Int*. 2003;86(6):1135–1143.

150. Mottier P, Parisod V, Gremaud E, Guy PA, Stadler RH, Determination of the antibiotic chloramphenicol in meat and seafood products by liquid chromatography-electrospray ionization tandem mass spectrometry, *J. Chromatogr. A* 2003;994(1–2):75–84.

151. Turnipseed SB, Roybal JE, Pfennig AP, Kijak PJ, Use of ion-trap liquid chromatography-mass spectrometry to screen and confirm drug residues in aquacultured products, *Anal. Chim. Acta* 2003;483(1–2):373–386.
152. Volmer DA, Lock CM, Electrospray ionization and collision-induced dissociation of antibiotic polyether ionophores, *Rapid Commun. Mass Spectrom*. 1998;12(4):157–164.
153. Van Hoof N, De Wasch K, Okerman L, et al., Validation of a liquid chromatography–tandem mass spectrometric method for the quantification of eight quinolones in bovine muscle, milk and aquacultured products, *Anal. Chim. Acta* 2005;529(1–2):265–272.
154. Samanidou VF, Tolika EP, Papadoyannis IN, Chromatographic residue analysis of sulfonamides in foodstuffs of animal origin, *Sep. Purif. Rev*. 2008;37(4): 327–373.
155. Kamel AM, Fouda HG, Brown PR, Munson B, Mass spectral characterization of tetracyclines by electrospray ionization, H/D exchange, and multiple stage mass spectrometry, *J. Am. Soc. Mass Spectrom*. 2002;13(5):543–557.

7

SINGLE-RESIDUE QUANTITATIVE AND CONFIRMATORY METHODS

JONATHAN A. TARBIN, ROSS A. POTTER, ALIDA A. M. (LINDA) STOLKER, AND BJORN BERENDSEN

7.1 INTRODUCTION

A number of analytes or analyte classes require special treatment during extraction and clean-up. This may be because maximum residue limit (MRL) definitions are formulated in terms of a marker residue that requires either chemical conversion before detection and quantification or monitoring for multiple components followed by a mathematical treatment of the data to factor the concentrations back to the required marker. For example, within the European Union (EU), MRLs for tulathromycin are expressed as a marker residue (back calculated to be expressed as equivalents of the parent compound) generated as part of the extraction procedure. For others, extensive *in vivo* metabolism means that a conversion step is required to convert compounds into a form that may be detected, such as the nitrofurans, which can undergo significant protein binding. Sometimes this requirement may be important only in specific matrices, as, for example, a result of glucuronidation of chloramphenicol in kidney tissue. More recently, while there has been heavy emphasis on the incorporation of a wider range of analytes to create true multi-residue methods, it is insufficient for quantification and confirmation purposes to monitor for unconverted analytes, whether parent compound or major metabolite. For this reason, there still is a need for stand-alone methods capable of meeting the MRL definitions for these drugs. The aim of this chapter is to summarize the current position for analytes in this category, give some guidance for method requirements, and indicate where gaps still remain, particularly in the open literature. Other members of the various classes discussed (such as the majority of macrolide antibiotics) have been dealt with elsewhere in this book (Chapters 4 and 6), as have approaches to screening analysis.

7.2 CARBADOX AND OLAQUINDOX

7.2.1 Background

Carbadox and olaquindox are quinoxaline *N*-oxide antimicrobial drugs (Fig. 7.1) that were employed in the past as growth promoters in pigs, to increase feed efficiency and rate of weight gain and also to prevent swine dysentery and bacterial enteritis in young pigs.[1] Carbadox is known to metabolize rapidly *in vivo* to the desoxy compounds and eventually through to quinoxaline-2-carboxylic acid (QCA). Similarly, 3-methylquinoxaline-2-carboxylic acid (mQCA) is the last major detectable metabolite of olaquindox (Fig. 7.1). Carbadox and the didesoxy metabolite are suspected carcinogens and mutagens.[2] The non-carcinogenic metabolites QCA and mQCA have long been designated as suitable markers of use of carbadox and olaquindox, respectively, as they were thought to be the most persistent metabolites. The Joint FAO/WHO Expert Committee on Food Additives (JECFA) proposed in 1991 that MRLs of 30 and 5 μg/kg QCA be used in liver and muscle, respectively.[2]

Carbadox and olaquindox were banned for use as growth promoters in the EU in 1999 because of concerns about their suspected carcinogenicity and mutagenicity.[3] The responsible EU Community Reference Laboratory (CRL)

Chemical Analysis of Antibiotic Residues in Food, First Edition. Edited by Jian Wang, James D. MacNeil, and Jack F. Kay.

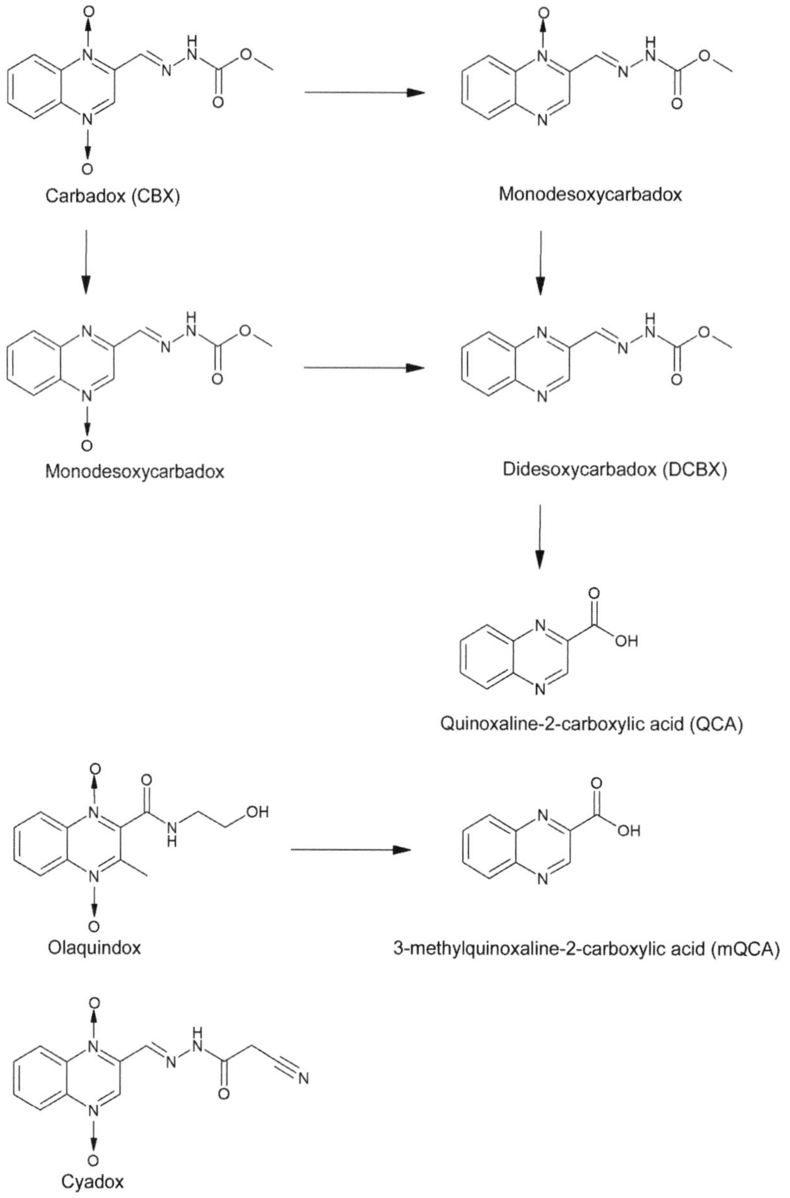

Carbadox (CBX)

Monodesoxycarbadox

Monodesoxycarbadox

Didesoxycarbadox (DCBX)

Quinoxaline-2-carboxylic acid (QCA)

Olaquindox

3-methylquinoxaline-2-carboxylic acid (mQCA)

Cyadox

Figure 7.1 Quinoxaline antimicrobial agents and their marker compounds.

has recommended that analysis methods have sensitivities of at least 10 μg/kg for both QCA and mQCA.[4] The use of QCA as a marker for carbadox is complicated as it is also a known metabolite of cyadox, a structurally related quinoxaline antimicrobial agent (Fig. 7.1).[5] Similarly, mQCA is a possible metabolite of a number of compounds.

The accepted position for monitoring carbadox use by the determination of QCA has been challenged in more recent studies, and is summarized in the report from the 60th meeting of JECFA in 2003.[6] The introduction of a new analytical procedure capable of detecting both carbadox and didesoxycarbadox (DCBX), the compounds of concern toxicologically, at <1 μg/kg meant that new depletion data were made available. In porcine liver tissue, both DCBX and QCA were detectable 15 days after dosing (LOQs 0.030 μg/kg for DCBX and 15 μg/kg for QCA), while carbadox was detectable for only 48 h after dosing (LOQ 0.050 μg/kg). In muscle, QCA was not consistently detected, whereas parent carbadox could be detected 6 h and DCBX 10 days after dosing. From the data generated, a relationship between QCA and DCBX concentrations could be established in liver but not in muscle. Because of these new data, JECFA recommended withdrawal of the MRLs proposed at the 36th meeting.[2] The position at present still needs clarification, although in the EU, at least, QCA and mQCA remain the markers of choice. However, with the introduction of newer, more sensitive methods, the markers of choice may be changed to the didesoxy compounds. This would also allow distinction between metabolites arising from dosing with carbadox and olaquindox and related compounds such as cyadox.

There is also the possibility that residues might arise from environmental contamination rather than deliberate dosing. This is of concern as exposure even at low concentrations to the excreta of previously medicated animals may result in detectable residue concentrations. This has legal implications whereby an innocent farmer might be penalized for illegal use by another farmer. A study by Hutchinson et al.[7] tentatively concluded that the ratio of QCA concentration in urine to that in liver could be used to provide evidence of the origin of residues. A ratio of <0.8 was deemed to have resulted from dosing; >4.5, from environmental contamination; and results in the range 0.9–4.4 were considered inconclusive, warranting further investigation.

7.2.2 Analysis

More recent methods of analysis for the markers and metabolites of carbadox and olaquindox, including QCA, mQCA, and DCBX, are summarized in Table 7.1. LC-MS/MS transitions are summarized in Table 7.6 at the end of the chapter. Most methods are complex multi-step procedures. Analysis of QCA and mQCA in animal tissues first requires release of any tissue-bound residues. Originally this was achieved by alkaline hydrolysis using sodium hydroxide at elevated temperatures (~100°C).[9] As an alternative, an enzymatic digestion using a protease in alkaline-buffered solution has also been used.[10] Elevated temperatures (~55°C) were again required. A mixed acidic extraction solvent consisting of 0.3% metaphosphoric acid and methanol in the ratio 8 : 2 (v/v) was used at room temperature.[12] It was reported subsequently that attempts to duplicate this extraction method resulted in low recoveries, and it was suggested that incomplete deproteination was to blame.[11] Increasing the metaphosphoric acid concentration to 2% and the acid : methanol ratio to 7 : 3 (v/v) resulted in improved recoveries from spiked samples.

A variety of protocols utilising combinations of liquid–liquid and solid-phase extractions (LLE and SPE) have been used to clean-up tissue extracts. Alkaline extracts are commonly made acidic, extracted into ethyl acetate and then back-extracted into aqueous buffer at alkaline pH.[9,10] Acidic extracts have been extracted directly into ethyl acetate and then back-extracted into buffer.[11] QCA and mQCA may act as acids or bases, and both of these properties have been utilized in the SPE clean-up of the buffered extracts from the initial liquid–liquid partitions. Extracts were acidified prior to clean-up on non-endcapped SCX (strong cation exchange) SPE columns.[9,10] The analytes of interest were eluted from the columns using a mixture of sodium hydroxide and methanol. Further clean-up and transfer to an appropriate solution for instrumental analysis was achieved by re-acidification

TABLE 7.1 Summary of Methods of Analysis for Metabolites of Carbadox and Olaquindox

Analyte	Matrix	Extraction and Clean-up	Instrumentation	Detection (μg/kg)	Reference
QCA	Porcine liver	Sodium hydroxide/heat	LC-MS/MS (+ve)	CCα 0.16	9
		LLE	Columbus C_{18} 5 μm	CCβ 0.27	
		SPE (SCX)	MeOH:Water:AcOH		
		LLE	Isocratic		
QCA	Porcine liver	Tris/HCl/protease/heat	LC-MS/MS (+ve)	CCα 0.4	10
		LLE	Luna C_{18} 3 μm	CCβ 1.2	
mQCA		SPE (SCX)	MeOH:MeCN:water:	CCα 0.7	
		LLE	AcOH	CCβ 3.6	
			Gradient		
	Porcine liver	Metaphosphoric acid/MeOH	GC-ECNI-MS (−ve)	LOQ 0.7	11
		LLE	DB-5 MS	CCα 32	
		SPE (Oasis MAX)	30 m × 0.25 mm, 0.25 μm	CCβ 34	
		Derivatization	80–300°C, 10°C/min		
QCA (−ve)	Porcine liver	Metaphosphoric acid/MeOH	LC-MS/MS (±ve)	LOD 1	12
DCBX (+ve)	Porcine muscle	SPE (Oasis HLB)	Cadenza CD C_{18}		
		LLE	100 × 2 mm		
			Aq AcOH–MeCN gradient		
QCA	Bovine muscle	0.6% formic acid	LC-MS/MS (+ve)	LOD 0.5	13
DCBX	Porcine liver	Tris/protease	Nova-Pak C_{18} 4 μm	LOD 0.05	
mQCA	Porcine muscle	SPE (Oasis MAX)	150 × 2.1 mm	LOD 0.5	
			Formic acid–MeOH		
			gradient		

and back-extraction into ethyl acetate before evaporation under nitrogen and redissolution. The use of reversed phase (RP)[12] and RP-anion exchange mixed-mode SPE cartridges[11] simplifies the clean-up procedure. QCA and mQCA are retained on HLB (a polymeric reversed phase sorbent) and MAX (a mixed-mode reversed phase anion exchange sorbent) from aqueous buffer at neutral pH. Use of the mixed-mode SPE cartridge is particularly efficacious as the retention mechanism allows for additional basic wash steps prior to elution using 2% trifluoroacetic acid in methanol. Trifluoroacetic acid offers the dual advantages of being a stronger acid than acetic acid, commonly used in similar protocols, and more volatile, facilitating any subsequent solvent removal steps.

Final determination is most commonly via liquid chromatography–mass spectrometry (LC-MS). LC-ESI-MS has been used in both positive and negative modes for the quantification of QCA and in positive mode for DCBX.[12] Separation was achieved on a reversed phase column (Cadenza CD C18) with an acetic acid–acetonitrile gradient. Molecular ions [$[M+H]^+$ 175 (QCA), 231 (DCBX), and $[M-H]^-$ 173 (QCA)] were monitored in a selected-ion mode. For QCA, negative-ion mode typically gave a cleaner ion trace. Under EU legislation[8] this method does not provide sufficient identification points to be regarded as a confirmatory method. More typically, liquid chromatography–tandem mass spectrometry (LC-MS/MS) is used for confirmatory purposes where monitoring at least two transitions is considered to be sufficient for confirmatory purposes. For example, Hutchinson et al.[10] separated QCA and mQCA on a Luna C18 column using a four-component mobile-phase (methanol, acetonitrile, water, acetic acid) gradient prior to ESI-positive MS/MS, monitoring the 175 > 102 and 175 > 75 or 175 > 129 transitions for QCA and 189 > 145 and 189 > 102 transitions for mQCA. Deuterated analogs of both QCA and mQCA are available and are commonly used for quantification purposes.

Gas chromatography–mass spectrometry (GC-MS) has also been used, but an additional derivatization step is required to render the analytes suitable for analysis. Sin et al.[11] proposed two alternative derivatizations. First, silylation using *N-tert*-butyldimethylsilyltrifluoroacetamide containing 1% *tert*-butyldimethylchlorosilane was employed. Then, to provide some measure of confirmation, an alternative methylation, using trimethylsilyldiazomethane, was also used. These derivatives were separated on a DB-5 MS column (30 m × 0.25 mm, 0.25 μm film thickness) using a temperature gradient of 80–300°C at 10°C/min. Mass spectrometric detection was in negative chemical ionization mode, monitoring ions at 292 and 293 for the silyl ester and 188 and 189 for the methyl ester of QCA. Deuterated QCA, derivatized as appropriate, was used as internal standard.

As discussed above, the validity of all these procedures is now under scrutiny. The original residue data submitted to JECFA[2] were obtained via a method utilizing high-temperature alkaline digestion. It has now been demonstrated that DCBX is not stable under these conditions, and a new method was required to establish the correct depletion profile for this metabolite.[13] In order to release the residues, samples were first digested with 0.6% formic acid at 47°C, a temperature at which DCBX is stable, to deactivate any native enzymes present. The digest was then buffered and subjected to a protease digest, again at 47°C. After acidification, the extracts were cleaned up using mixed-mode anion exchange. DCBX was eluted under neutral conditions in dichloromethane. QCA and mQCA were eluted in acidic ethyl acetate. Following evaporation and reconstitution, final determination was by LC-MS/MS using a reversed phase column (Nova-Pak C18) for separation. The transitions monitored for QCA and mQCA are the same as those used by other workers. In addition, the 231>143 and 231>102 transitions were monitored for DCBX. The LOQs of 0.5 μg/kg for QCA and mQCA are more than sufficient to confirm the presence of these residues at the previous JECFA-recommended concentrations and the current EU recommendations. An LOQ of 0.05 μg/kg for DCBX meant that low concentrations present in tissues even after 15 days withdrawal could now be detected.

7.2.3 Conclusions

The analytical position for both carbadox and olaquindox is somewhat obscure at present. As the previous discussion has shown, there are analytical methods available for the analysis of the designated markers of carbadox and olaquindox, namely, QCA and mQCA, and also for the carbadox metabolite DCBX. As yet there is no consensus on whether DCBX is a better marker for carbadox in some tissue types. Similarly, there is no knowledge at present as to whether desoxyolaquindox might also prove to be a better marker for olaquindox.

7.3 CEFTIOFUR AND DESFUROYLCEFTIOFUR

7.3.1 Background

Ceftiofur is a semi-synthetic antibiotic of the cephalosporin class. Cephalosporins, like penicillins, belong to the group of β-lactam antibiotics. β-Lactams are probably the class of antibiotics most widely used in veterinary medicine for the treatment of bacterial infections of animals used in livestock farming and bovine milk production. There are EU MRLs for all food-producing species ranging from 4 μg/l for ampicillin in milk to 300 μg/kg for oxacillin, cloxacillin, and dicloxacillin in bovine tissues such as

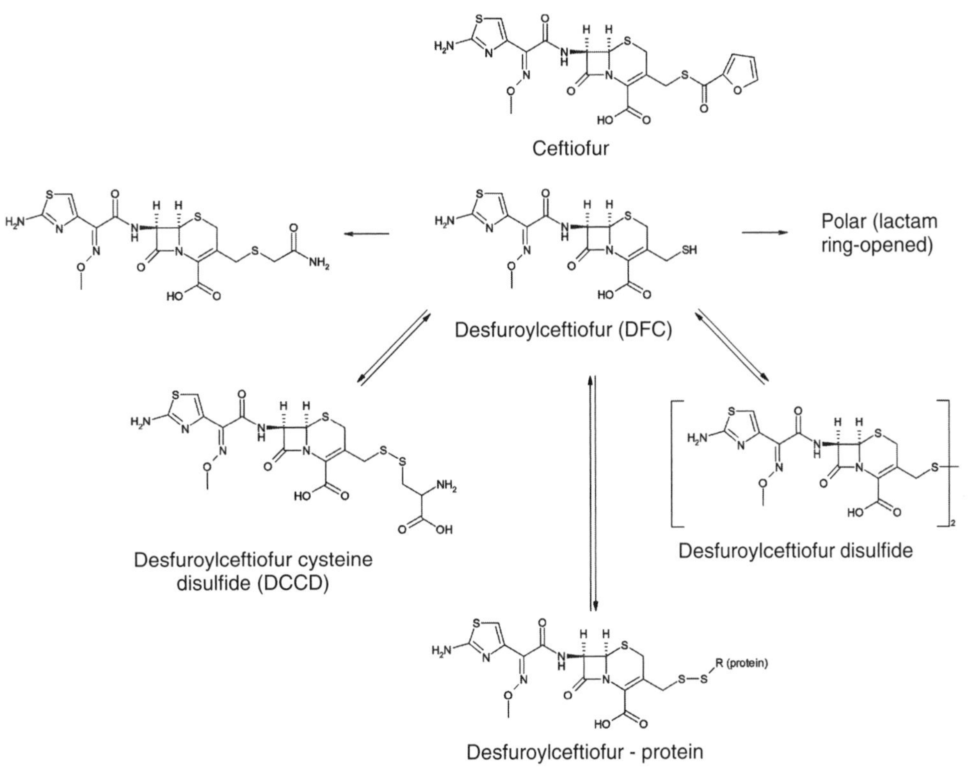

Ceftiofur

Polar (lactam ring-opened)

Desfuroylceftiofur (DFC)

Desfuroylceftiofur cysteine disulfide (DCCD)

Desfuroylceftiofur disulfide

Desfuroylceftiofur - protein

Figure 7.2 Structure of ceftiofur and marker compounds.

muscle, fat, liver, and kidney.[14] Both classes contain bulky side-chains attached to 6-aminopenicillanic acid and 7-aminocephalosporanic acid nuclei, respectively, as is shown in Figure 7.2. The presence of an unstable four-member ring in the β-lactam structures renders these compounds prone to degradation by heat and in the presence of alcohols.

Ceftiofur is based on 7-aminocephalosporanic acid, which is responsible for the biological activity of the compounds. Only a few methods involving the chemical analysis of ceftiofur are reported in literature. This might be caused by the difficulty of analyzing this compound, which is a result of its metabolism and instability combined with the rather complex EU MRL definition.

Ceftiofur is known to rapidly metabolize after intramuscular administration, resulting in metabolite residues found in milk[15] and tissue.[16] Reported metabolites include desfuroylceftiofur (DFC), desfuroylceftiofur cysteine disulfide (DCCD), protein-bound DFC, and ceftiofur thiolactone.[15–18] Because these metabolites are all microbiologically active, the EU MRL was defined as the sum of all residues retaining the β-lactam structure, expressed as DFC,[14] whereas the Codex Alimentarius Commission (CAC) defined DFC as the only marker residue, simplifying the analysis.[19] This approach is also used in the United States of America.

Three main approaches for the analysis of ceftiofur are reported. The first approach, which is reported for milk[20–24] and plasma,[22] does not take any of the metabolites into account. It will result in an underestimation of the total amount of active ceftiofur-related metabolites and will not be discussed here. The second approach includes a deconjugation of all protein-bound DFC in slightly alkaline buffer ($pH = 9$) using dithioerythritol followed by a derivatization of the resulting free DFC at $pH = 2.5$ using iodoacetamide. The third approach focuses on the analysis of ceftiofur and/or one or a few of its metabolites.[25,26] A fourth approach for the analysis of ceftiofur, which is still under investigation, is the derivatization of ceftiofur metabolites in alkaline solution at elevated temperature to produce (2*E*)-2-(2-amino-1,3-thiazol-4-yl)-2-(methoxyimino)acetamide (ATMA) as a marker compound.[18] This approach does not fit with existing MRL definitions but may provide a more accurate assessment of ceftiofur-related residue concentrations.

7.3.2 Analysis Using Deconjugation

The analysis of ceftiofur by applying a deconjugation using dithioerythritol is reported for plasma,[22,27,28] milk,[29] and tissue samples.[30] Dithioerythritol, a reducing agent that causes cleavage of disulfide bonds, is added to the sample extract to release bound ceftiofur and its metabolites from proteins. Next, iodoacetamide is added to derivatize the free ceftiofur and its metabolites to desfuroylceftiofur acetamide, which is analyzed (Fig. 7.2) by LC-UV[30] or LC-MS/MS.[29]

After the deconjugation–derivatization procedure, a sample clean-up involving two consecutive SPE steps is needed to remove excess reagents and to concentrate ceftiofur. The main disadvantages of this extensive procedure are the low sample throughput,[27] its poor robustness,[30] and the limitation that this approach is applicable only for single-analyte methods.[31] Furthermore, it is questioned whether this approach takes all relevant metabolites of ceftiofur into account.[18] DFC-thiolactone was found as one of the main metabolites of ceftiofur in plasma,[17] urine,[17] and kidney extract,[18] and may not be factored in when applying a deconjugation procedure using dithioerythritol and iodoacetamide, because it does not contain the complete DFC structure. Therefore, it is thought that the deconjugation procedure can result in an underestimation of the total amount of ceftiofur and microbiologically active ceftiofur metabolites present in the sample.[18]

7.3.3 Analysis of Individual Metabolites

The analysis of ceftiofur and DFC in milk and serum is reported by Tyczkowska et al.[32] After extraction of the sample using a mixture of acetonitrile and water (1 : 1, v/v) the sample is filtered through a 10-kDa molecular mass cutoff filter. The ultrafiltrate was injected onto an LC system equipped with a phenyl analytical column to establish a separation on the basis of the ion-pair mechanism using octanesulfonate and dodecanesulfonate. Detection was carried out using a quadrupole mass spectrometer (MS) operating in full-scan mode equipped with a thermospray interface. Using this method, ceftiofur and DFC can be detected with a limit of detection (LOD) of 50 μg/kg in plasma and milk. However, protein-bound residues and DCCD, reported to constitute 65% of the total ceftiofur residues,[15] are not included using this method, resulting in a severe underestimation of the total ceftiofur-related residues.

The analysis of DCCD, the main metabolite of ceftiofur in kidney extract,[18] has been reported by researchers of the US Western and Eastern Regional Research Center of the Agricultural Research Service.[31,33,34] The initial method[33] consisted of an extensive SPE procedure and made use of a quadrupole ion trap MS. Because the sample clean-up was time-consuming, and because the instrumentation resulted in poor linearity and reproducibility, an improved method applying dispersive SPE and using a triple-quadrupole MS was developed.[31] In 2008, a further optimized method was reported.[34] As degradation of β-lactams[32] in methanol resulting in β-lactam methyl esters[35] was reported, all methanol used (e.g., stock solution solvent and constituent of the mobile phase) was replaced by non-proton-donating solvents.

In the optimized method, kidney tissue was extracted using a mixture of water and acetonitrile (1 : 4, v/v). After mixing and centrifugation, the supernatant was isolated and 500 mg C18 SPE sorbent was added. The acetonitrile in the extract was removed, and the extract was concentrated and filtered before injection onto a reversed phase HPLC system. A separation was established with gradient elution using 0.1% formic acid in water and in acetonitrile as the mobile phases. Detection was carried out using LC-MS/MS equipped with an electrospray interface operating in positive mode. Using this method DCCD was detected at concentrations of 10 μg/kg showing an average recovery of 60%. Although DCCD is the most abundant ceftiofur-related metabolite, other metabolites, such as DFC, protein-bound metabolites, and DFC-thiolactone, significantly contribute to the total concentration of active ceftiofur metabolites in kidney.[15,18] This approach will therefore also result in an underestimation of the total amount of ceftiofur and active ceftiofur-related metabolites.

A method combining the analysis of ceftiofur, DCCD, and the dimer of DFC in kidney and liver tissue has been reported.[26] In this case, a portion of sample was extracted using a mixture of water and acetonitrile containing tetraethylammonium chloride and potassium dihydrogen phosphate. The organic solvent was evaporated to 1–2 ml and made up to 4 ml with water. After filtration the extract was fractioned on an LC fractioning system and the fractions were tested for microbial activity using the DelvoTest P-mini™. Ceftiofur and DCCD suspect fractions were reduced to 1 ml by evaporation under reduced pressure and injected onto the HPLC system equipped with a C18 analytical column. Separation was established via an ion-pair mechanism using sodium dodecylsulfate as the ion-pair reagent. Detection was carried out using a UV detector. This method proved to be able to detect concentrations as low as 0.31 μg/kg DCCD in kidney tissue. Unfortunately this sample clean-up procedure is very time-consuming and the method is not able to identify the detected compounds according to EU legislation.[8] Furthermore, DFC and protein-bound metabolites, other than DCCD, are not included in this method, and therefore this approach will also result in an underestimation of the total amount of ceftiofur and active ceftiofur-related metabolites.

7.3.4 Analysis after Alkaline Hydrolysis

A study of the degradation of ceftiofur at elevated temperature in kidney extract and in alkaline environment has been reported.[18] Degradation products in kidney extract and in kidney extract after addition of ammonia were identified using a combination of triple-quadrupole MS, LC-time-of-Flight MS (TOF/MS), nuclear magnetic resonance, and microbial techniques. After addition of ammonia to kidney extract at elevated temperature, ATMA is produced. This could be a suitable marker for the sum of ceftiofur and its active metabolites because it is the non-reactive component of ceftiofur. Therefore, it is expected that ATMA can be produced from ceftiofur and all its main metabolites (DFC, DCCD, protein-bound DFC, and DFC-thiolactone).

A clean-up procedure for the analysis of ATMA as a marker compound for ceftiofur and ceftiofur-related metabolites, based on ion exchange SPE, has been developed. Separation was carried out using reversed phase HPLC and detection by LC-MS/MS. At the time when this chapter went to press (December 2010), this method had not yet been tested using ceftiofur incurred tissue material, and therefore this new approach has not yet been compared to current approaches.

7.3.5 Conclusions

For ceftiofur, the EU MRL includes ceftiofur and all active ceftiofur metabolites. To be able to comply with this MRL definition, all active metabolites of ceftiofur have to be included in the analysis method.

Four approaches for the analysis of ceftiofur residues are reported here. The first approach involves the analysis of ceftiofur residues only and therefore is not suited for control purposes. The second approach involves a deconjugation of protein-bound DFC followed by a derivatization. The main disadvantages of this approach are the low sample throughput[27] and its low robustness,[30] and it is questioned whether it takes all relevant metabolites of ceftiofur into account.[18] The third approach involves the analysis of ceftiofur and/or one or more metabolites. However, none of the reported methods include all relevant metabolites, and therefore it is suggested that for quantification and confirmation purposes, these methods result in an underestimation of the amount of ceftiofur residues in the sample. Simple single-residue methods for DFC are suitable to meet the requirements of the CAC MRL and US tolerance definitions. A fourth approach is under development and involves the hydrolysis of ceftiofur under alkaline conditions to produce ATMA.[18] Although this procedure may give a more complete summation of ceftiofur-related residues, within the EU, at least, a mathematical correction factor will need to be applied to the data to fulfill the MRL definition in terms of desfuroylceftiofur. This procedure has not yet been thoroughly tested at the time of going to press.

7.4 CHLORAMPHENICOL

7.4.1 Background

Chloramphenciol (CAP) is a broad-spectrum antibiotic with historical veterinary uses in all major food-producing animals. CAP is biosynthesized by the soil organism *Streptomyces venezuelae* and several other actinomycetes, but is produced for commercial use by chemical synthesis (see Chapter 1).[36] The drug has been evaluated by a number of agencies, including the International Agency for Research on Cancer (1990),[37] the European Committee for Veterinary Medicinal Products (1996),[38] the US Food and Drug Administration (1985),[39] and more recently (in 2005) by JECFA at its 62nd meeting.[40] CAP is a suspected carcinogen and is known to cause aplastic anemia as a rare but potentially fatal side effect, and for this reason the drug is banned for use in food-producing animals in the EU and in many other countries, including the United States of America, Canada, Australia, Japan, and China. A series of EU Commission Decisions describe the required testing for animal-derived food products entering the European market.[41–43]

Thiamphenicol and florfenicol, which have structures similar to that of CAP (Fig. 7.3), are permitted as substitutes within the EU.[44,45] In the EU, MRLs for thiamphenicol are 50 μg/kg for bovine and chicken tissues, and for florfenicol, 100 μg/kg for muscle to 3000 μg/kg for bovine liver. Within the EU florfenicol also has a complex MRL definition, and this will be discussed in Section 7.9.2.1. Because of the ban on CAP, methods with very low detection limits have been developed. A minimum required performance limit (MRPL) of 0.3 μg/kg was assigned by the European Commission for the analytical methods testing for CAP in products of animal origin.[46]

7.4.2 Analysis by GC-MS and LC-MS

With regard to analysis, an organic solvent, predominantly ethyl acetate, or an aqueous phosphate buffer, is used as extraction solvent for CAP from biological matrices.[47–50,53] Next, the primary extract is cleaned by a variety of LLE and/or SPE steps. Gas chromatography (GC), in combination with chemical ionization (CI)-MS, provides excellent analyte detectability down to 0.1 μg/kg in muscle tissues; the results for urine analysis are often less reliable because of matrix interference. GC-MS in the electron impact (EI; now referred as *electron ionization*) mode is slightly less sensitive but has the distinct advantage of yielding spectra that can be searched in electronic libraries. The main drawback of using GC-MS for CAP analysis is the need for derivatization in order to improve its chromatographic properties. Gantverg et al.[47] described a GC-EI-MS method for CAP in urine. After hydrolysis, washing with ethyl acetate and clean-up by C18-SPE, the analyte was derivatized with a mixture of *N,O*-bis[trimethylsilyl]trifluoroacetamide and 10 vol% trimethylchlorosilane. An HP-5MS column with 30 m × 0.25 mm i.d. and 0.25 μm film thickness was used. The LOD in "dirty" urine was 2 μg/l. One alternative[48] used

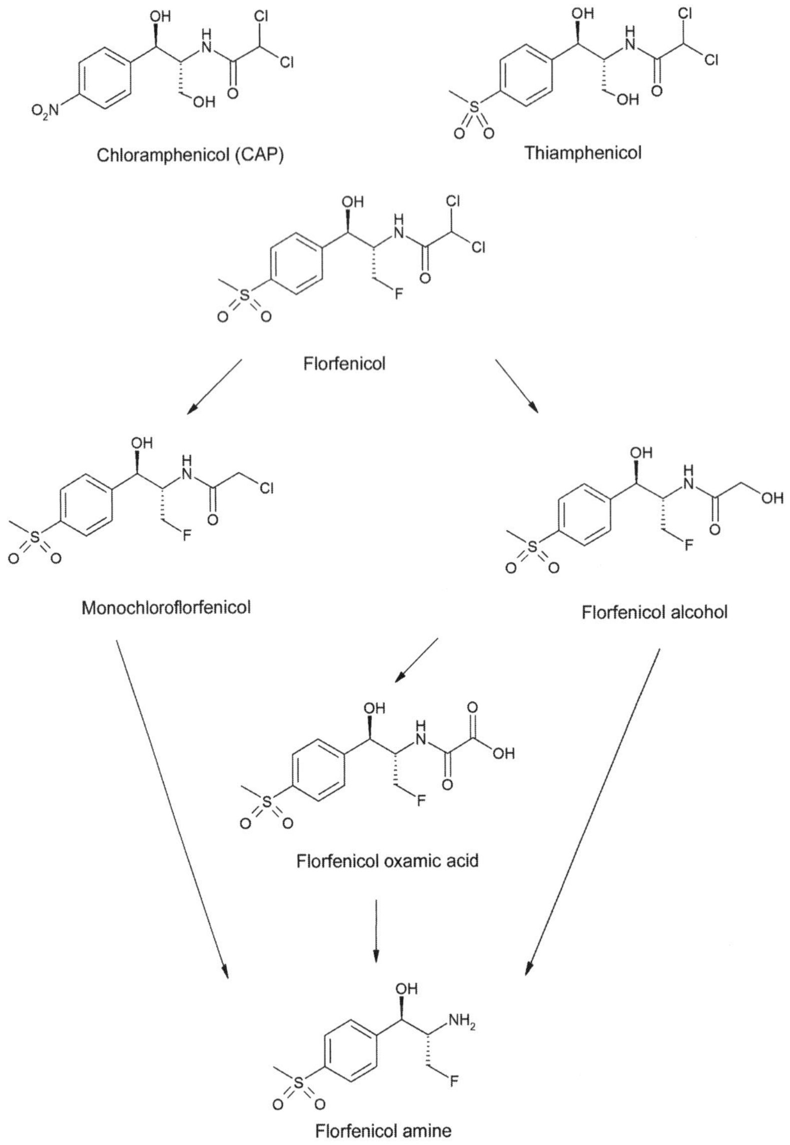

Chloramphenicol (CAP)

Thiamphenicol

Florfenicol

Monochloroflorfenicol

Florfenicol alcohol

Florfenicol oxamic acid

Florfenicol amine

Figure 7.3 Amphenicols and metabolites.

GC-ECD after selective extraction of CAP from muscle by means of matrix solid-phase dispersion (MSPD) and subsequent conversion into the trimethylsilyl derivative. Although the method is rapid and uses only a few millilitres of organic solvent, the LODs of 2–4 μg/kg found for cattle, pig, and horse muscle tissue do not permit CAP monitoring at the MRPL of 0.3 μg/kg.

Originally, the interest in LC-MS/MS as a confirmatory method for CAP was limited because of the availability of GC-MS procedures. However, it is often used today because LC-MS does not require derivatization and CAP detectability in sophisticated LC-MS procedures approaches that of GC-MS. In 2003, interest in the determination of CAP in shrimps suddenly increased because a number of non-compliant results were reported in The Netherlands and Germany. As a consequence, several new LC-MS procedures were developed. Gantverg et al.[47] suggested that LC-APCI(−)-MS/MS offered sensitivity and selectivity superior to that of GC-MS. Even in urine, the LOD was 0.02 μg/kg, in contrast to 2 μg/kg for GC-MS. Mottier et al.[49] also reported an LC-MS/MS method for CAP in meat and seafood. After ethyl acetate extraction and clean-up on silica-SPE, the analysis was on a C18 column with a water–acetonitrile eluent. The use of ESI(−)-MS/MS enabled highly precise quantification of CAP down to 0.05 μg/kg in fish and shrimps. The overall absolute recovery of CAP spiked at 2.5 μg/kg into a blank chicken meat was 60 ± 5% (n = 4). Ramos et al.[50] used LC-ESI(−)-MS for the determination of CAP in shrimps. After phosphate extraction and C18-SPE clean-up, an additional LLE with ethyl acetate was performed followed by a conventional RP-LC separation; the LOQ was 0.2 μg/kg.

Van de Riet et al.[51] used LC-ESI(−)-MS to determine chloramphenicol, thiamphenicol, and florfenicol in farmed aquatic species. After pressurized liquid extraction (PLE) with acetone, the extracts were partitioned with dichloromethane, the aqueous layer removed, and the organic layer evaporated to dryness. The residue was dissolved in dilute acid and defatted with hexane, and the aqueous layer was prepared for LC analysis on a C18 column with a water–acetonitrile gradient. Recoveries were 71–107%; LODs were 0.1 μg/kg for florfenicol and chloramphenicol, and 0.3 μg/kg for thiamphenicol.

Kaufmann et al.[52] also used the LC-ESI(−)-MS/MS technique. However, it was claimed that by using <2-μm particulate high-performance LC (UPLC) columns, the detection limits and speed were improved significantly. The proposed analytical method included an enzymatic digestion, which liberates glucuronide-bound CAP from kidney tissue. The extracts obtained after an Extrelut clean-up were sufficiently pure to permit routine injection of biological samples into the <2-μm UPLC column, without observing rapid deterioration of peak shape or column clogging problems. The time for one chromatographic run was 4.2 min. CCα concentrations were 0.007 μg/kg (honey) and 0.011 μg/kg (kidney). These concentrations are significantly below the EU MRPL of 0.3 μg/kg.

Table 7.2 lists some other published LC-MS techniques required to reach the very low EU MRPL of 0.3 μg/kg. LC-MS/MS transitions are summarized in Table 7.6 at the end of the chapter.

7.4.3 An Investigation into the Possible Natural Occurrence of CAP

In recent years (as of 2011) the detection of residues of CAP in food such as poultry and honey has had a major impact on international trade. Follow-up investigations in Thailand relating to non-compliant findings in poultry by European laboratories have, in some cases, been unable to establish the cause of the residues, since there was no recent history of the use of the drug. A possible source of contamination is the biosynthesis of CAP by *Streptomyces venezuelae* or other Actinomycetes. A study was instigated[54] to investigate the possibility in a typical poultry production environment. *Streptomyces venezuelae* in CAP-producing phase was spiked into poultry litter under various conditions, and the litter was tested for growth of the organism and for CAP concentration. Results showed that *S. venezuelae* was not viable after 3–4 weeks and initial concentrations of CAP from the *S. venezuelae*/normal saline solution added (maximum 0.6 μg/kg) decreased rapidly and were below the LOQ of the analytical method (0.04 μg/kg) by week 3. Litter samples collected from five poultry farms with a history of poultry contamination with CAP were tested and found to be negative for both CAP and *S. venezuelae*. The results suggest that residues of chloramphenicol on the farms tested were extremely unlikely to have been caused by natural biosynthesis of the drug in the production environment.

Several hypotheses for the contamination of food products by possible naturally occurring CAP are described by JECFA. The possibility of contamination due to ingestion of natural or externally contaminated soil was evaluated. The final conclusion from the evaluation was that the Committee could not completely rule out the possibility that foods are occasionally contaminated from environmental sources. However, due to lack of analytical methods to detect the relevant concentrations of CAP in soil, there are no analytical data available to support this suggestion.

Another hypothesis is the possibility that grass and herbs (plant materials) absorb and accumulate CAP from the soil. The CAP-containing grass and herbs are used as pasture or harvested as animal feed or forage, and consequently products of animal origin are contaminated with residues of CAP. It has been shown that plants are able to absorb veterinary drugs such as tetracyclines from soil.[55] To test this hypothesis, samples of grass and herbs were analyzed for CAP content.

TABLE 7.2 Selected Methods for Chloramphenicol (CAP)

Matrix	Extraction and Clean-up	Instrumentation	Detection (μg/kg)	Reference
Meat, seafood	LPE/LLE/SiOH SPE	LC-MS/MS [ESI(−)]	0.01	49
Shrimp	LPE/LLE/C18 SPE	LC-MS [ESI(−)]	0.02	50
	LLE			
Muscle, urine	LPE/C18 SPE + derivatization	LC- MS/MS [APCI(−)]	0.02	47
		GC-E-MS	2	
Muscle	MSPD+derivatization	GC-ECD	2–4	48
Meat, seafood	LLE	LC- MS/MS [ESI(−)]	<0.01	53
Eggs, honey, milk				
Urine, plasma	LLE/SPE	LC- MS/MS [ESI(−)]	<0.01	

7.4.4 Analysis of CAP in Herbs and Grass (Feed) Using LC-MS

Details of the analytical method used to analyze feed products (grass and herbs) for research on the natural occurrence of CAP have been published.[56] The separation and detection of CAP from the sample components was carried out by LC-MS/MS using a Waters Quattro Ultima mass spectrometer with ESI operating in negative ionization mode. CAP was fragmented using collision-induced dissociation, and selected-reaction monitoring transitions at 321 > 152, 321 > 194, and 321 > 257 were monitored. $^{37}Cl_2$-CAP was detected by monitoring the transition $m/z = 324.8 > 152.0$. Approximately 110 samples of herbs and grasses were analyzed using this method. CAP was detected in 26 samples with concentrations up to 450 μg/kg.

Chromatograms of a blank herb mixture sample, a blank herb mixture sample fortified with 2 μg/kg CAP, a non-compliant herb mixture sample (4 μg/kg), and the same herb mixture sample with the addition of 2 μg/kg CAP are presented in Figure 7.4.

7.4.5 Conclusions

For the monitoring of CAP in products of animal origin, the most widely used approach is a liquid–liquid extraction with a relatively polar solvent sometimes—depending on the matrix—followed by a solid-phase extraction clean-up/concentration step. The final analysis applies the very selective and sensitive LC-ESI(−)-MS/MS system using UPLC. Detection concentrations are mostly far below the EU MRPL of 0.3 μg/kg.

The LC-MS/MS analysis of plant materials has demonstrated that it is possible that plant materials can contain CAP. CAP was detected in plants of the families Artemisa or Thalictrum but was also detected in grass. It is known that the soil organism *Streptomyces venezuelae* and related organisms can biosynthesize CAP. Therefore it is suggested, from the results obtained, that CAP is produced in the soil and that the plants absorb CAP through their root systems. Further research is required to confirm this supposition and to elaborate the environmental parameters affecting CAP occurrence in plants.

These findings make it a much more realistic prospect that products of animal origin can contain residues of CAP that are not due to (illegal) use of the drug, but rather due the natural occurrence of CAP. These results also have significant implications for the application of legislation with respect to the detection of CAP in food products and may need, if not a change in the legislation, at least a change in the interpretation of analytical results and in follow-up actions and penalties to producers for the suspected illegal use of CAP. Furthermore, the finding of CAP in samples of herbal products purchased at retail outlets must be a cause for concern in relation to human exposure to this suspected carcinogen. As CAP is still used in human medication, for example, in eye drops, and has been used as a preservative in domestic products, an additional concern is the real potential for cross-contamination of samples. Laboratory procedures should take this aspect into consideration.

7.5 NITROFURANS

7.5.1 Background

Furazolidone, furaltadone, nitrofurazone, and nitrofurantoin (Fig. 7.5) are nitrofuran antibacterial agents that have been widely used as food additives for the treatment of gastronintestinal infections (bacterial enteritis caused by *Escherichia coli* and *Salmonella*) in cattle, pigs, and poultry. After research showed that furazolidone is a mutagenic and genotoxic drug, legislation was enforced to remove this and similar compounds from the market. The use of furaltadone has been prohibited by the US Food and Drug Administration (FDA) since 1985 and the other nitrofuran drugs (except some topical uses) since 1992. The topical use of furazolidone and nitrofurazone in food-producing animals was prohibited in 2002. The use of nitrofuran antimicrobials in food-producing animals has been prohibited within the EU since 1997.

In 2002 residues of nitrofuran drugs were frequently detected in poultry and shellfish imported into the EU. Action was taken, and the MRPLs for nitrofuran metabolites in poultry meat and aquaculture products were set at 1 μg/kg in 2003.[46] Nitrofuran metabolites are still found primarily in aquaculture products originating from Southeast Asia, with semicarbazide (SEM, the metabolite/marker of nitrofurazone) having the highest incidence.[57]

7.5.2 Analysis of Nitrofurans

Methods for detecting residues of nitrofurans should not aim for the parent drugs because these are rapidly metabolized and do not persist in edible tissues. Nitrofurans form protein-bound metabolites that may persist in tissues for considerable periods after treatment.

A common procedure for the analysis of nitrofuran metabolites involves hydrolysis of the protein-bound metabolites under acidic conditions followed by derivatization with 2-nitrobenzaldehyde (Fig. 7.5). After neutralization of the digest, solvent extraction is carried out with ethyl acetate. Residues are detected by LC-UV or LC-MS/MS.[58,59] In some cases an additional liquid–liquid extraction[60] or solid-phase extraction[61] step is applied to remove excessive matrix compounds. A broad overview of applied methods was published by Vass et al. in 2008.[57]

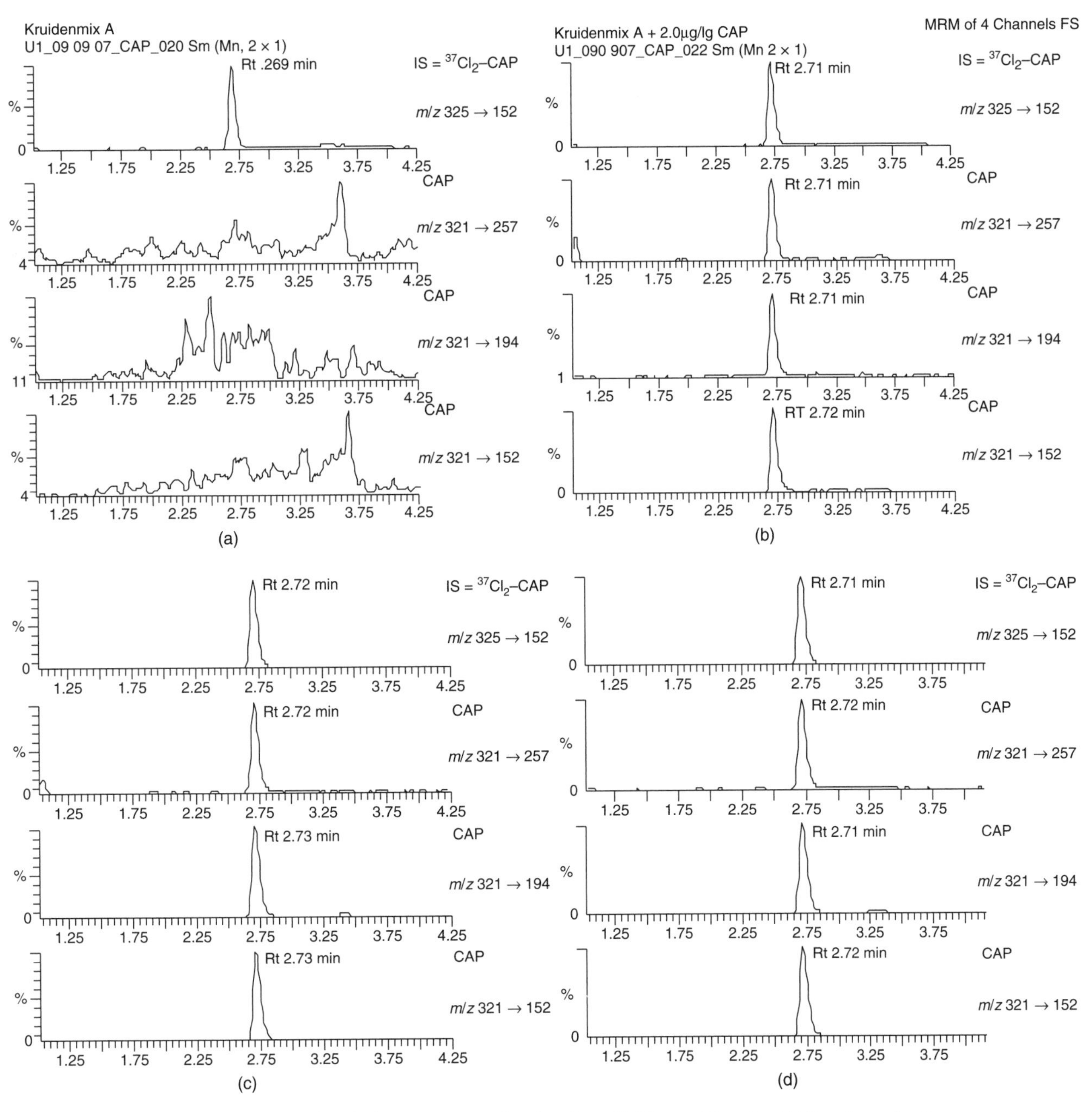

Figure 7.4 LC-MS/MS chromatograms showing three SRM transitions for CAP and one for the internal standard of (a) a blank herb sample, (b) a blank herb sample with addition of 2 μg/kg CAP, (c) a herb mixture form a local shop, and (d) the same herb sample with addition of 2 μg/kg CAP.

7.5.3 Identification of Nitrofuran Metabolites

Because nitrofuran metabolites are very small molecules, the marker metabolites are not highly specific. This is especially the case for SEM. However, the presence of tissue-bound metabolites is more specific, because this indicates the administration of nitrofurans. Therefore, analytical methods are described focusing on bound residues.[61,62] Prior to acidic hydrolysis, the samples are extracted several times with water, methanol, and/or ethyl acetate to remove unbound residues. After removal of excessive organic solvent, samples are hydrolyzed and treated according to standard procedures, resulting in the detection of bound nitrofuran metabolites only.

An example of the low selectivity of SEM as a marker metabolite for nitrofurazone is the false non-compliant findings caused by the use of azodicarbonamide (ADC,

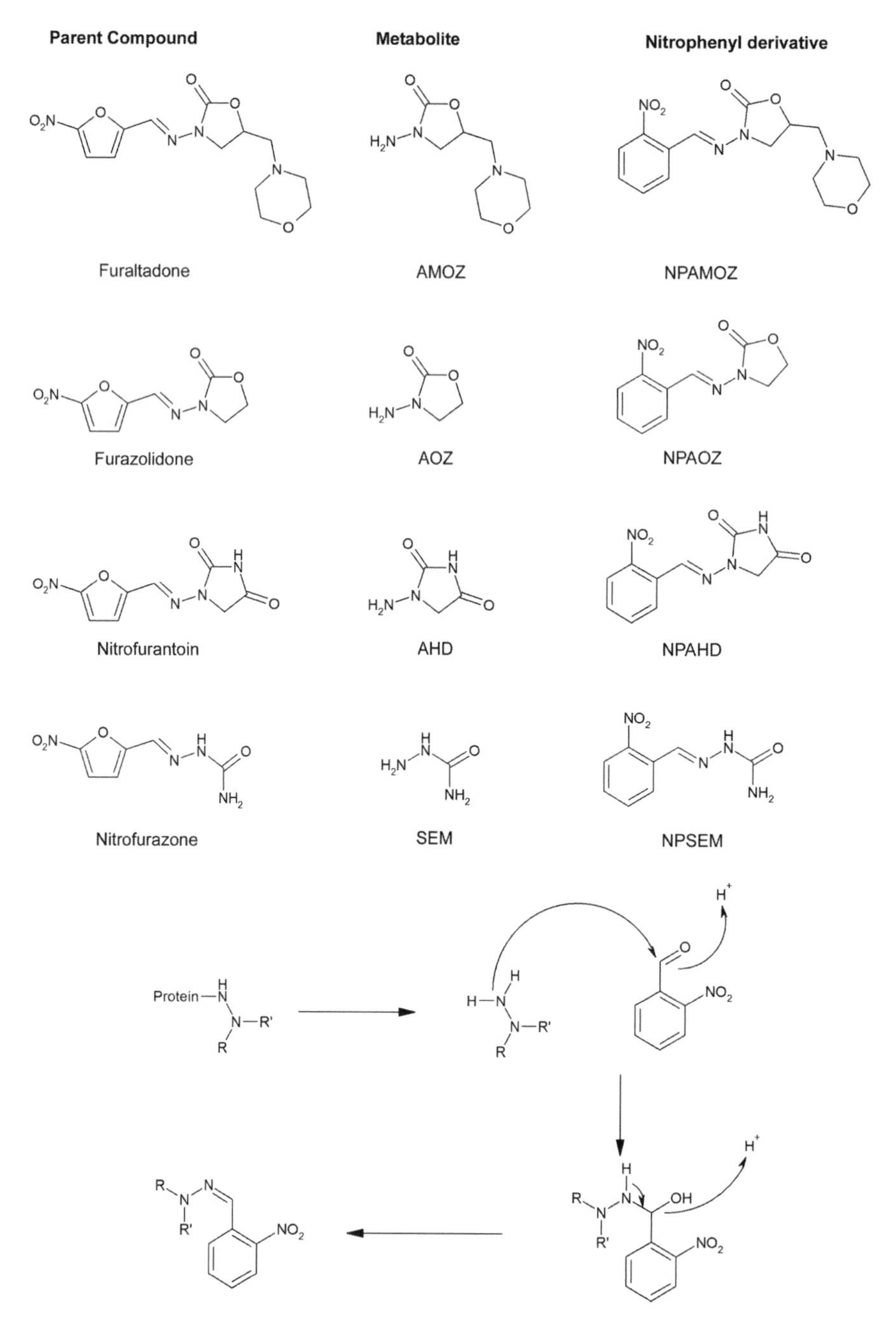

Figure 7.5 Molecular structures of the nitrofuran antibiotics, their free metabolites (AMOZ—3-amino-5-morpholinomethyl-2-oxazolidinone, AOZ—3-amino-2 oxazolidinone; AHD—1-aminohydantoin; SEM—semicarbazide) and their nitrophenyl derivatives.

Fig. 7.6) and its migration into food products.[63,64] ADC has been widely used as a blowing agent for the production of gasket seals in food jars, and it has been connected to the elevated concentrations of SEM encountered in baby food products.[63,64] It was shown that SEM is a minor decomposition product formed from ADC during heat treatment needed for the sealing of the jars.[65] In 2005 the EU prohibited the use of ADC as a blowing agent for cap seals that may come into contact with foodstuffs.[66]

Azodicarbonamide is also used as a flour-improving agent and as a dough conditioner at concentrations of $\leq$45 mg/kg.[67] During dough preparation ADC is almost quantitatively converted into biurea by the reaction with the wet flour.[68,69] It has been postulated that acid hydrolysis of biurea can produce SEM with an efficiency of ~0.1% (Fig. 7.6).[68] The use of ADC-treated flour or dough in coated or breaded food products may result in the migration of SEM residues and generate false non-compliant results

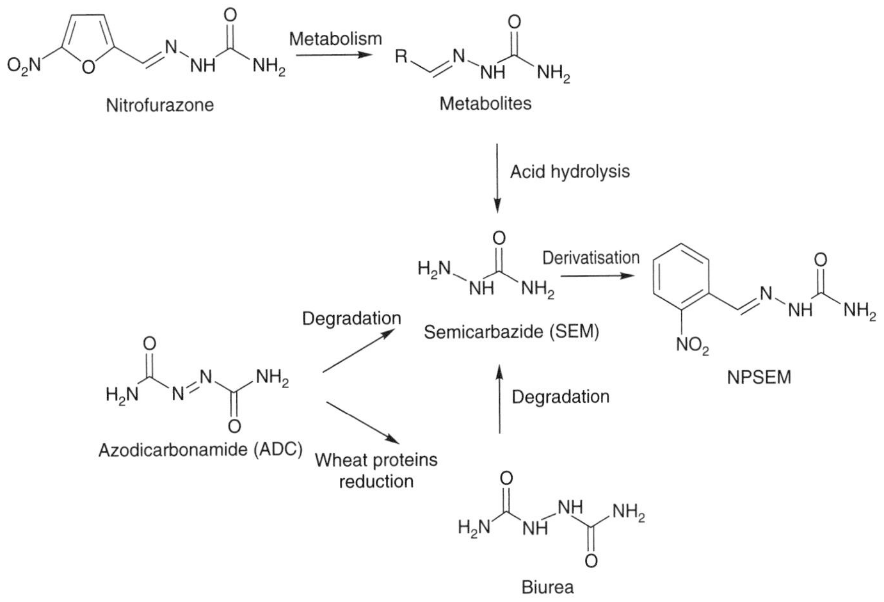

Figure 7.6 The formation of SEM by acid hydrolysis from nitrofurazone metabolites and azodicarbonamide.

in the analysis for nitrofurazone metabolites. In Europe, ADC was removed from the list of permitted dough treatment agents in the mid 1990s, but in the United States of America, Brazil, and Canada, the use of ADC is still permitted.[70] A method was developed to analyze for biurea, a suitable marker for ADC, in non-compliant SEM containing samples.[70] This method can be used in nitrofurazone analysis to effectively eliminate the risk of false non-compliant results for SEM due to the presence of ADC-treated components in the food product. The European Community Reference Laboratory has also recommended that coatings be removed and samples washed to remove unbound residues. This approach is also used in the laboratories of the CFIA in Canada.

7.5.4 Conclusions

Nitrofurans are banned substances within the EU and in some other countries because of their mutagenic and genotoxic characteristics. Nitrofuran metabolites are still found, primarily in aquaculture products originating from Southeast Asia, with SEM (the metabolite of nitrofurazone) having the highest incidence. Methods for detecting residues of nitrofurans aim for protein-bound metabolites that may persist in tissues for considerable periods after treatment. Methods are reported for the detection and identification of nitrofuran metabolites in many different food products. The main difficulty in nitrofuran metabolite analysis is the low selectivity of SEM as a marker metabolite of nitrofurazone. Several other possible sources of SEM have been identified and investigated, the most important of which is the use of ADC as a blowing agent or flour-improving agent. Analysis for biurea may be used as a means of identifying the source of SEM residues. Additionally, removal of coatings such as breadcrumbs and washing of samples to remove unbound residues are also recommended. Both approaches are useful in nitrofurazone analysis to effectively eliminate the risk of false non-compliant results for SEM due to the presence of ADC.

7.6 NITROIMIDAZOLES AND THEIR METABOLITES

7.6.1 Background

The nitroimidazoles are a class of veterinary drugs characterized by their heterocyclic imidazole ring structure. Commonly referred to as 5-nitroimidazoles, due to the presence of a NO_2 group in the fifth position of their ring structure,[71] they were permitted in the EU until 1990. The nitroimidazoles were used for both prophylactic and therapeutic treatment of diseases such as hemorrhagic enteritis in swine,[72] histominasis and coccidiosis in poultry, genital tricchoniasis in beef,[71] and aquatic parasites in fish.[73] Because of their suspected carcinogenicity and mutagenicity, the nitroimidazoles were banned for use in food-producing animals by the EU in 1990, under EU Regulation 2377/90.[14,74] In addition, the Codex Committee on Residues of Veterinary Drugs in Foods (CCRVDF) has not been able to set MRLs for the nitroimidazoles, as JECFA was unable to establish ADIs for these compounds.[75] Subsequently, they were banned for use in the United States of America (1994),[72] China (1999),[72,76]

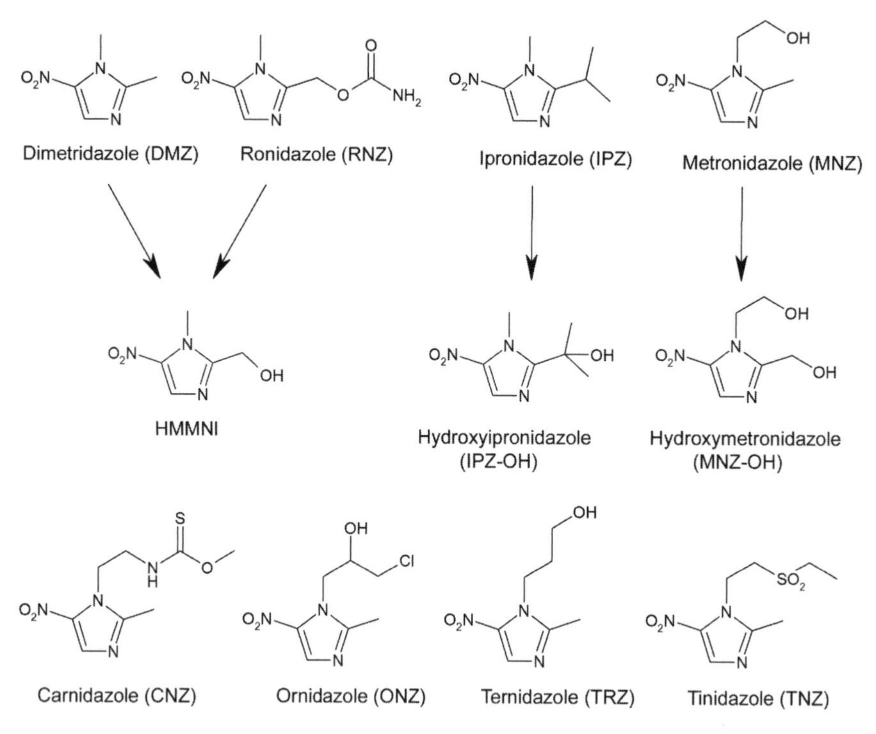

Figure 7.7 Common nitroimidazoles and metabolites.

and Canada (2003).[77] In addition, as a result of Council Directive 96/23/EC, their monitoring is now required within the EU as well as by countries exporting to the EU.[78]

The most commonly reported nitroimidazoles in the literature include dimetridazole (DMZ), metronidazole (MNZ), ronidazole (RNZ), and ipronidazole (IPZ) (Fig. 7.7). More recently, additional members of this class such as carnidazole (CNZ), ornidazole (ONZ), tinidazole (TNZ), and ternidazole (TRZ) have also been incorporated into analytical methods. Extensive studies on the metabolism of nitroimidazoles in poultry, beef, swine, and fish[79] have shown that the hydroxy metabolites of the parent compounds, formed by oxidation of the side-chain in the C2 position of the imidazole ring, are of additional concern. These compounds have been shown to have toxicity equal to that of their parent forms and are often rapidly formed by metabolism within the animal. Studies in both poultry[80] and rainbow trout[73] have shown that the distribution of parents to metabolites is analyte- and species-specific. As a result, it has been suggested that it is imperative that a residue control program considers both the parent and the metabolites.[79] The metabolites indicated in the literature include MNZ-OH, IPZ-OH, and HMMNI (2-hydroxymethyl-1-methyl-5-nitroimidazole) as the hydroxy metabolites of MNZ, IPZ, and DMZ, respectively. HMMNI has also been identified as a metabolite of RNZ, but through a different pathway.[71] HMMNI is not a major metabolite of RNZ, however, and is therefore not suitable on its own for use as a marker of RNZ use.

7.6.2 Analysis

Analytical methods exist for swine, beef, poultry, fish, and honey, with the recommended target tissue varying depending on commodity. In poultry, for example, plasma, retina, and eggs have been suggested as target matrices,[71,80] while plasma, liver, and kidney are recommended for swine and beef.[71,81] There is evidence suggesting that at least in poultry, nitroimidazoles have limited stability in muscle tissue and are far more persistent in plasma and retina, when stored at 4°C.[80] In addition, the same study reported homogeneity issues with regard to the analytes in muscle tissue. Although the recommended target matrix may not always be available, it is a factor that should be considered for regulatory testing.

In terms of instrumentation, earlier quantitative approaches concentrated on HPLC with UV detection.[73,82] Early confirmatory approaches utilizing both GC and GC-MS[83,84] for quantification and confirmation were abandoned because of the need for a derivatization step that made data interpretation challenging because the same end product was created for RNZ and the metabolite of DMZ. Most current methods reported in the literature are LC-MS/MS-based, due to its selectivity and confirmatory capability in the <1 μg/kg range. A summary of current quantitative and confirmatory methodology has been provided in Table 7.3. In addition, Table 7.6 at the end of this chapter outlines the common MS/MS transitions currently reported in the literature.

TABLE 7.3 A Summary of Current Quantitative and Confirmatory Methods for Analysis of Nitroimidazoles and Their Related Hydroxy Metabolites

Analyte[a]	Matrix	Instrumentation	Detection	Reference
DMZ, IPZ, MNZ, RNZ, ONZ, TRZ, CNZ, MNZ-OH, HMMNI, IPZ-OH	Animal plasma	LC-MS/MS	0.5–1.6 μg/kg	71
DMZ, MNZ, IPZ	Swine liver	LC-MS	0.5–1.0 μg/kg	72
DMZ, MNZ, RNZ, HMMNI	Poultry, swine muscle, eggs	LC-MS/MS	0.5–0.27 μg/kg	76
DMZ, IPZ, MNZ, RNZ, MNZ-OH, HMMNI, IPZ-OH	Poultry muscle, fish, eggs	LC-MS/MS	0.07–0.36 μg/kg	79
DMZ, IPZ, MNZ, RNZ, MNZ-OH, HMMNI, IPZ-OH	Swine plasma	LC-MS/MS	0.25–1.0 ng/ml	81
DMZ, IPZ, MNZ, RNZ, MNZ-OH, HMMNI, IPZ-OH	Swine kidney	UPLC-MS/MS	0.05–0.5 μg/kg	88
DMZ, IPZ, MNZ, RNZ, ONZ, TRZ, CNZ, TNZ, MNZ-OH, HMMNI, IPZ-OH	Eggs	LC-MS/MS	0.3–1.26 μg/kg	89
DMZ, MNZ, RNZ	Eggs	LC-MS/MS	0.5 μg/kg	90
DMZ, IPZ, MNZ, RNZ, MNZ-OH, HMMNI, IPZ-OH	Eggs	LC-MS/MS	0.3 μg/kg	91
DMZ, IPZ, MNZ, RNZ, HMMNI	Water	LC-MS	0.2 ng/ml	92
DMZ, MNZ, RNZ, HMMNI	Swine urine	LC-MS/MS	0.03–0.05 ng/ml	93
DMZ, MNZ, RNZ, HMMNI	Swine liver	LC-MS/MS	0.1–0.5 μg/kg	94
DMZ, MNZ, RNZ, ONZ, TNZ, MNZ-OH, HMMNI	Natural casings	LC-MS/MS	0.03–0.05 μg/kg	95
DMZ, IPZ, MNZ, RNZ, DMZ-OH	Poultry muscle	LC-MS	5.0 μg/kg	85
DMZ	Poultry tissue, eggs	LC-MS	<1.0 μg/kg	86
DMZ, RNZ, HMMNI	Poultry muscle, eggs	LC-MS	0.1–0.5 μg/kg	87

[a] Abbreviations are defined in list at end of this chapter.

With regard to extraction techniques, acetonitrile[71,76] and ethyl acetate[85,88] are the organic solvents most commonly used for deproteinization. Other approaches reported in the literature include extracting with dichloromethane or toluene[86] and the use of sodium chloride/potassium dihydrogen phosphate buffers and acidic protease digestion overnight.[81] The inclusion of a protease step is based on the premise that residues may be tissue-bound, but has not been widely adopted. An additional extraction technique used by several authors is the addition of salt to the organic acetonitrile extracts,[71,79] which is incorporated to remove impurities from the extracts[71] and any co-extracted water.[79] Deuterated internal standards have been incorporated into several modern methods, reducing the effects of matrix interferences.[71,88]

Most nitroimidazole methods also incorporate further clean-up steps such as solid phase extraction (SPE). Some common phases used include Oasis HLB,[79,92] Oasis MCX,[88] Chromabond XTR,[81] SCX,[87,95] Bakerbond Silica,[86] and MIP-SPE (molecular imprinted polymer).[91] In addition, most methods incorporate a solvent washing step with solvents such as hexane prior to instrumental analysis for removal of non-polar material. More recent work[71,76] bypassed extensive clean-up techniques such as SPE as it was felt that recent advances in MS/MS technologies and the use of internal standards limited matrix interferences and warranted such a simple clean-up.

7.6.3 Conclusions

On the basis of existing knowledge, it is recommended that novel quantitative methods use organic solvents such as acetonitrile or ethyl acetate for extraction. Methods should include at least seven nitroimidazoles (the parent compounds DMZ, IPZ, MNZ, and RNZ, and the metabolites HMMNI, IPZ-OH, and MNZ-OH), target tissues for the species under test, incorporate LC-MS/MS technologies, and have confirmatory capabilities to concentrations below 3 μg/kg. Inclusion of a deconjugation step is generally accepted as being unnecessary, although there is room for a study to definitively establish this.

7.7 SULFONAMIDES AND THEIR N^4-ACETYL METABOLITES

7.7.1 Background

The sulfonamides are a class of broad-spectrum antibiotics that have been widely used in aquaculture and animal husbandry on a global scale for several decades,[96,97] with usage dating to the 1960s or earlier.[98] Their popularity is largely due to their low cost and effectiveness in preventing and treating diseases and infections. Sulfonamides have been used to treat foot rot, acute mastitis, respiratory infections, and coccidiosis in a variety of livestock species[99] and American and European foulbrood in beekeeping[100,101] and to combat diseases caused by *Leukocytoza* in the poultry industry.[102] Characterized by their *p*-aminophenyl ring structure with an aromatic amino group at the N^4 position, the sulfonamides are all derivatives of sulfanilamide, differing from one another by substitutions in the N^1 position.[96]

Currently, the use of sulfonamides is permitted in many jurisdictions. Regulations vary between those based on single residues[103,104] and those based on the sum of the class.[105,106] For example, 15 sulfonamides are approved in Canada for use in various species, including cattle, sheep, swine, horses, chickens, turkeys, rabbits, and salmonids.[103] Generally, the Canadian MRLs are set at 100 μg/kg in edible tissues and 10 μg/kg in milk for each individual compound. Specific regulations are sulfonamide- and species-dependant. For example, in salmonids, only sulfadiazine and sulfadimethoxine are permitted for use. In addition the potentiators, ormetoprim and trimethoprim are permitted for use with an MRL of 100 μg/kg. Guidelines within the United States of America are also based on the individual sulfonamides.[104] Most established tolerances are at 100 μg/kg in edible tissues and 10 μg/kg in milk. As in Canada, the residues permitted are species- and sulfonamide-specific. Currently eight sulfonamides are permitted for use in the United States of America. Interestingly, the USA tolerance (set at zero) for sulfanitran also takes into account its metabolites.[107] Within the EU, the regulations are based on the sum of all sulfonamides present as parent compounds only. In all food-producing species, a total residue concentration of 100 μg/kg has been set for edible tissues as well as in milk.[14] Similarly to the EU, in China the summation is used to establish regulations,[99,106] which are set at 100 μg/kg. The only MRLs currently set by the CAC for the sulfonamides are for sulfadimidine (sulfamethazine) at 25 μg/kg in milk and 100 μg/kg in muscle, fat, kidney, and liver of unspecified animal species.[19] Also of note is that sulfonamides are not permitted in honey or eggs in any jurisdiction. Currently only two JECFA evaluations of sulfonamides exist, addressing sulfadimidine (sulfamethazine) and sulfathiazole.[75] A minimum of 16 sulfonamides have been identified as of concern by regulatory authorities on a global scale.[14,103] These include sulfabenzamide, sulfacetamide, sulfachlorpyridazine, sulfadiazine, sulfadimethoxine, sulfadoxine, sulfaethoxypyridazine, sulfaguanidine, sulfamerazine, sulfamethazine (sulfadimidine), sulfanilamide, sulfanitran, sulfapyridine, sulfaquinoxaline, and sulfathiazole. In addition, methods reported in the literature have included sulfamethoxypyridazine, sulfamethoxazole, sulfamonomethoxine,[102] sulfamethizole,[108] sulfisomidine, sulfamoxole, sulfameter, sulfisoxazole, and sulfaphenazole.[109] Current analytical methods in the literature are applicable to 10–24 sulfonamide residues, depending on the matrix and the intended use of the method.[102,109]

7.7.2 N^4-Acetyl Metabolites

In addition to the parent form, the metabolites of the sulfonamides are of additional concern. Although metabolites are not addressed in the current regulations, studies have shown that sulfonamides can undergo extensive metabolism, and in some cases it has been suggested that the metabolites should be used as the marker residue for regulatory purposes.[110,111] Although species-dependent, they are metabolized mainly in liver tissues, where the major metabolites are products of oxidation or acetylation.[112] The primary metabolic pathway for sulfonamides in animals and humans has been reported to be N^4-acetylation and is the primary focus of the existing literature.[112,113] Although the ratio of parent to metabolite is often low,[111,114] the metabolites should be considered with new methods, as it is possible that they may bind more tightly to proteins such as those found in plasma or be deacetylated to the parent compound *in vivo* or *in vitro*,[115,116] both factors that need to be considered when developing methods.

The N^4-acetyl metabolites pose a unique analytical challenge because of their relative instability and possible deacetylation back to their parent form when in the presence of acidic conditions.[117] As the current regulations are based on the incurred parent compounds, conversion due to extraction procedures may lead to false positives or inaccurate estimates. Although the detection of the N^4-acetyl metabolites is important, it is imperative that analytical methods be able to detect them separately from the parent compound. In cases where analytical standards are not available, it is also important to be aware of the relative response factors as they may differ between parent and metabolite.[117] Heller et al. studied 15 sulfonamide residues individually dosed in laying hens.[111] The formation of several N^4-acetyl metabolites was observed with N^4-acetylsulfamethazine, sulfadimethoxine, and sulfaquinoxaline all being present at low concentrations. The authors concluded that N^4-acetylsulfanilamide was the appropriate marker residue for monitoring sulfacetamide and sulfanilamide residues in laying hens, whereas the parent compounds were appropriate for the other sulfonamides. Several methods addressing the metabolites in eggs,[111,115] milk,[118] animal tissue,[119,120] and animal plasma[113,116] have been reported. Although the current EU regulations address the sulfonamides only in their parent form, future studies should consider the formation and presence of the metabolites, as, for example, hydrolysis of metabolites back to the parent compounds could generate false non-compliant results under current EU MRL definitions.

7.7.3 Analysis

Methods currently exist in the literature for the analysis of sulfonamides in various tissues of pork, beef, and poultry as well as fish, honey, and milk with different modes of detection. Quantitative approaches have used HPLC-UV,[121,122] CE-FLD,[123] CE-MS,[124,125] and LC-MS[126] as well as GC-MS,[127,128] with the latter often incorporating cumbersome derivatization steps. By far the most frequent quantitative approach is the use of LC-MS/MS for its selectivity and superior confirmatory capability.[97,102] A review of more recent mass spectrometric approaches is presented in Table 7.4. In addition to the existing methods outlined in Table 7.4, several comprehensive review articles have been published that can be used as a valuable resource with regard to the analysis of sulfonamides.[96,129] A summary of the MS/MS transitions currently utilized in the literature for the confirmation of sulfonamide residues can be found in Table 7.6; the most common product ions are m/z 156, 108, and 92, as all sulfonamides have a common base structure. With regard to the N^4-acetyl metabolites, the parent compound +42 Da is monitored (e.g., SDM at m/z 251 and N^4-acetyl SDM at m/z 292).[119] Product ions monitored for the N^4-acetyl metabolites are the same as those used for the parent compounds.[111]

Several extraction techniques have been reported in the literature for the analysis of sulfonamides. Because of their polar nature, sulfonamides are readily extracted by organic solvents;[109,110] the most commonly used are acetonitrile.[109,133] Other organic solvents used for analyte extraction and protein precipitation include dichloromethane,[102,134] acetone,[102] ethanol,[118] chloroform,[119] and ethyl acetate,[142] which are often used either alone or in conjunction with one another. Other techniques used for protein precipitation include the use of acids such as perchloric[120] or formic[108] and the use of basic buffers such as potassium hydrogen phosphate[139] and ammonium sulfate.[113,115] In the case of honey, the use of acids such as trichloroacetic,[134,140] hydrochloric,[100,136] and phosphoric[137] is necessary for hydrolysis, releasing carbohydrate-bound sulfonamide residues. Other extraction techniques reported in the literature include the use of pressurized liquid extractions,[124,138] matrix solid-phase dispersion,[126,130] and magnetic molecularly imprinted polymers.[101] Of additional note, several authors have observed that analyte recoveries were largely

TABLE 7.4 A Summary of Current Quantitative and Confirmatory MS-Based Methods for Analysis of Sulfonamides and Their Related Degradative Metabolites

Analytes	Matrix	Instrumentation	Detection	Reference
17 sulfonamides	Salmon	LC-MS/MS	0.1–0.9 μg/kg	97
14 sulfonamides	Honey	LC-MS/MS	0.5–5.0 μg/kg	100
7 sulfonamides	Honey	LC-MS/MS	1.0–4.0 μg/kg	101
10 sulfonamides	Eggs	LC-MS/MS	16–20 μg/kg	102
17 sulfonamides	Swine tissues	LC-MS/MS	0.01–1.0 μg/kg	106
9 sulfonamides	Milk	LC-MS/MS	0.2–2.0 ng/ml	108
24 sulfonamides	Meat	UPLC-MS/MS	0.04–0.37 μg/kg	109
15 sulfonamide+metabolites	**Eggs**[a]	**LC-UV/LC-MS/MS**	**5.0–10 μg/kg**	111
4 sulfonamides+metabolites	**Swine muscle**[a]	**LC-UV/LC-MS**	**25 μg/kg**	119
12 sulfonamides	Swine muscle	CE-MS/MS	<12.5 μg/kg	124
10 sulfonamides	Meat	CE-MS/MS	5-80 μg/kg	125
15 sulfonamides	Eggs	GC-MS/LC-MS	<25 μg/kg	128
6 sulfonamides	Beef, pork, chicken	LC-MS	2.0–12.0 μg/kg	130
25 sulfonamides	Livestock, seafood	LC-MS/MS	2.5-5.0 μg/kg	131
5 sulfonamides	Milk	LC-MS/MS	<3.0 μg/kg	132
10 sulfonamides	Eggs, honey	UPLC-MS/MS	7.0–25 μg/kg	133
16 sulfonamides	Honey	LC-MS/MS	0.4–4.5 μg/kg	134
12 sulfonamides	Cheese	LC-MS/MS	<0.2 μg/kg	135
7 sulfonamides	Honey	LC-MS/MS	<2.0 μg/kg	136
16 sulfonamides	Honey	LC-MS/MS	0.5–6.0 μg/kg	137
13 sulfonamides	Meat, infant foods	LC-MS/MS	<2.6 μg/kg	138
3 sulfonamides	Milk, cheese	LC-MS/MS	1.2 μg/kg	139
10 sulfonamides	Honey	LC-MS/MS	<10 μg/kg	140
7 sulfonamides	Milk	LDTD-MS/MS	<5.0 μg/kg	141

[a]This method includes consideration of the metabolites.

pH-dependent, with the observation attributed to the wide range of pK_a values of the various sulfonamides.[110,137]

Several clean-up methods have incorporated the use of additional liquid–liquid steps with solvents such as chloroform[97,106] and ethyl acetate.[109] Alternatively, some authors have turned to ultrafiltration for further purification.[115,116] Additionally, hexane[97,133] is extensively used in several methods to further remove non-polar matrix components such as fats and lipids. By far, the most popular clean-up technique reported in the literature is the use of solid phase extraction (SPE). Some common phases reported include C18 cartridges such as SepPak C18,[111] Strata X,[133] Chromobond C18,[143] and Mega Bond Elut C18.[131] The use of silica,[105,119] alumina,[131] and Extralut[144] has also been reported, as well as polymeric cartridges such as Oasis HLB.[100,139] The latter phase has been reported as the basis of an on-line clean-up procedure.[136] Because of their amphoteric nature, as both weak acids and weak bases, the sulfonamides also lend themselves to analysis by ion exchange. Cationic SPE has been reported by several authors.[101,128] On the basis of the same principles, anionic exchange has also been used.[128,142]

7.7.4 Conclusions

It is recommended that novel methods used for the analysis of sulfonamides in foods consider at a minimum the 16 sulfonamides that are regulated on an international scale. Of additional concern are metabolites such as the N^4-acetyl forms, which in some cases have been recommended as the marker residues. Hydrolysis of these metabolites back to the parent compounds within the analytical method could lead to false non-compliant results under current regulations. In contrast, hydrolysis is recommended for the detection of sulfonamide use in honey. In terms of technology, liquid chromatography with tandem mass spectrometry detection has become a popular technique for the quantification and confirmation of sulfonamides. Although current regulations are set at the 10–100 μg/kg range, lower limits of detection may be desirable as not all sulfonamides are regulated in all commodities.

On a cautionary note, it has been shown that residues of sulfanilamide may occur in honey as a result of environmental contamination from the herbicide asulam, of which it is a breakdown product.[145]

7.8 TETRACYCLINES AND THEIR 4-EPIMERS

7.8.1 Background

The tetracyclines are a class of broad-spectrum antibiotics, characterized by their hydronapthacene structure, which contains four fused rings.[146] They are widely used in animal husbandry for the therapeutic and prophylactic treatment of diseases as well as for growth promotion.[147] In aquaculture they are used to combat bacterial hemorrhagic septicemia in catfish as well as diseases caused by *Pseudomonas liquifaciens*.[148] In dairy farming they are used to combat mastitis,[149] and in beekeeping they are used to combat the devastating effects of American and European foulbrood.[150]

Currently, nine tetracyclines are commercially available: tetracycline (TC), chlortetracycline (CTC), doxycycline (DC), oxytetracycline (OTC), minocycline (MC), demeclocycline (DMC), methacycline (MTC), rolitetracycline (RTC, pro-drug for tetracycline), and tigecycline (a new drug of importance in human medicine).[150,151] Globally, regulations are limited to the use of OTC, TC, CTC, and DC with respect to food-producing animals.[14,103] As such, residues of these four compounds should be the primary focus of new methods being developed. MC, DMC, MTC, and RTC are used primarily in human medicine, but there is a trend to incorporate them into modern methods as they could possibly be used illegally.[150,152] Along with the parent forms, the degradative metabolites of the abovementioned tetracyclines are also of interest. Tetracyclines are susceptible to chemical transformations such as isomerization and epimerization when exposed to conditions of weak acid, strong acid, strong base, or heat.[151,153] To this end the 4-epimers of CTC, TC, and OTC are also included in the MRLs established by the EU.[14]

The regulatory approach varies depending on jurisdiction. The CAC has set limits for the tetracyclines OTC, TC, and CTC used singly or in combination for cattle, pig, sheep, poultry, fish (OTC only), and prawn (OTC only).[19] The current CAC MRLs are as follows: muscle 200 μg/kg, liver 600 μg/kg, kidney 1200 μg/kg, milk 100 μg/kg, and eggs 400 μg/kg, and are based on a JECFA evaluation, where an ADI of 0–30 μg/kg body weight was established.[75] In Canada, OTC, TC, and CTC are permitted for use in cattle, swine, sheep, chickens, and turkey. However, only OTC is permitted in salmonids and lobster, and TC is not permitted in eggs.[103] Canadian MRLs for specific matrices are as follows: muscle 200 μg/kg, liver 600 μg/kg, kidney 1200 μg/kg, milk 100 μg/kg, and eggs 400 μg/kg. The US Food and Drug Administration (FDA) guidelines are based on the sum of the tetracycline residues present. Currently, OTC, CTC, and TC are permitted in beef cattle, non-lactating dairy cows, calves, swine, sheep, chickens, turkeys, and ducks; only OTC is permitted in lobster and fin fish, and CTC is permitted in eggs.[104] The following USA tolerances have been established: muscle 2000 μg/kg, liver 6000 μg/kg, fat and kidney 12,000 μg/kg, eggs 400 μg/kg, and milk 300 μg/kg. EU regulations are based on the sum of the parent and the 4-epimer for OTC, TC, and CTC.[14] As such, the EU MRLs have been set as follows: muscle 100 μg/kg, liver 300 μg/kg, kidney 600 μg/kg, milk 100 μg/kg, and egg 200 μg/kg for all food-producing species.

In addition, parent-only MRLs have been established for DC in beef, swine, and poultry as 100 μg/kg in muscle, 300 μg/kg in liver, 600 μg/kg in kidney, and 300 μg/kg in skin/fat (swine and poultry only).

The selection of an appropriate method (marker residues, analytical range) therefore depends in part on the regulatory application, given the different regulatory limits applied in different jurisdictions. Since the epimerization of CTC, OTC, and TC is an equilibrium reaction, the epimers are often present to a greater or lesser extent in analytical standards and can occur throughout the extraction and clean-up procedure. Analytical method development is further complicated by the need to separate and quantify parent and epimer, even when the local legislation does not demand it.

7.8.2 Analysis

Currently, analytical methods exist for beef, swine, poultry, sheep, and fish tissues as well as for eggs, honey, and milk; using various modes of detection. Quantitative methods have used both CE[148,159] and more frequently HPLC with UV,[160,161] DAD,[152] or FLD.[162,163] More recent approaches make use of LC-MS[164,165] and LC-MS/MS[149,153] for their specificity, low detection limits, and superior confirmatory capability. Because of the concentrations of tetracyclines permitted, some authors feel that HPLC-DAD is sufficient for quantification; however, if confirmation is required, LC-MS/MS is recommended.[8] A summary of current mass spectrometry-based methods in various matrices is presented in Table 7.5. In addition, Table 7.6 provides the MS/MS transitions currently used for monitoring tetracyclines and their respective 4-epimers. Also available for use as reference methods are AOAC *Official Methods of Analysis* for the detection of tetracyclines in animal tissues[161] as well as milk.[166]

Several challenges have been highlighted for consideration related to the chemical characteristics of tetracyclines. The formation of degradative metabolites due to extraction conditions has been discussed as a source of concern. The most significant of these are the 4-epimers of OTC, TC, and CTC, possibly formed during extraction as a result of mildly acidic conditions or excessive heat and/or during long-term storage. The challenge for regulatory scientists is that the approach to the analysis of the tetracyclines and their corresponding epimers differs depending on the jurisdiction. In the EU, as the regulations are based on the sum of the individual tetracycline parent and epimer (with the exception of DC), it is important to measure each in its natural proportions and not as an artifact of the method. In the cases where the MRLs or tolerances are based only on the parent, care must be taken to reduce epimerization as a

TABLE 7.5 A Summary of Current Quantitative and Confirmatory MS Methods for Analysis of Tetracyclines and Their Related Degradative Metabolites

Analyte	Matrix	Instrumentation	Detection (μg/kg)	Reference
OTC,TC,CTC,DC,4-epi-TC, 4-epi-CTC, 4-epi-OTC	Animal muscle	LC-MS/MS	0.3–3.0	146
OTC,TC,CTC,DC,DMC,MTC, 4-epi TC, 4-epi OTC, 4-epi CTC, other degradation products	Milk	LC-MS/MS	0.3–3.7	149
OTC,TC,CTC,DC,MC,MTC, DMC,RTC	Honey	LC-TOF-MS	0.02–1.0	150
OTC,TC,CTC,DC,DMC,MC,MTC, 4-epi-TC, 4-epi-CTC, 4-epi OTC	Milk, animal tissues	LC-MS/MS	0.5–10.0	151
OTC,TC,CTC,DC,DMC,4-epi-TC, 4-epi-CTC, 4-epi OTC	Beef, chicken, pork, lamb	LC-MS/MS	0.1–0.3	153
OTC,TC,CTC,DC,4-epi-TC, 4-epi-CTC, 4-epi-OTC	Swine tissues	LC-MS/MS	0.5–4.5	154
OTC, 4-epi OTC	Calf muscle, liver,kidney	LC-MS/MS	0.8–48.2	155
OTC,TC,CTC	Honey and royal jelly	LC-MS/MS	1.0–3.0	156
OTC,TC, 4-epi TC	Calf hair	UPLC-MS/MS	6.0–10.0	157
OTC,TC,CTC,4-epi-TC, 4-epi-CTC, 4-epi OTC	Eggs	Bioassay/LC-MS/MS	4.0	170
OTC,TC,CTC,DC, MTC	Honey and eggs	Biosensor/LC-MS/MS	5.0–25.0	158
OTC,TC,CTC	Swine kidney, muscle	LC-FLD/LC-MS	15.0–30.0	164
OTC, TC, CTC	Shrimp and milk	LC-UV/LC-MS	15.0	165
OTC,TC,CTC,DC	Royal jelly	LC-MS/MS	<1.0	173
OTC,TC,CTC,DC	Lobster, duck, honey, eggs	LC-MS/MS	0.1–0.3	175
OTC,TC,CTC,DC	Animal tissues, milk	LC-MS/MS	1.0–4.0	176
OTC,TC,CTC,DC	Animal muscle	LC-MS/MS	5.0–30.0	178

function of the process as it would lead to underestimation of the analytes of interest. For quantification purposes it is important to be aware of the response factors of the tetracyclines to their individual epimers. Correction factors of 0.9, 1.2, 0.84, 1.26, and 1.32 have been reported for 4-epi OTC, 4-epi TC, 4-epi DMC, 4-epi CTC, and 6-epi DC, respectively, in honey.[167] Also of concern is the formation of isomers in alkaline conditions and keto-enol tautomerism in aqueous solutions.[149] Because tetracyclines and these degradation products are diastereomeric (with only slight conformational changes), they have the same chemical formula and fragment in a similar fashion in the MS detector, making mass spectral distinction difficult and chromatographic separation imperative to distinguish between the forms.[151]

Because of their polar nature, tetracyclines also have the ability to strongly bind to proteins as well as to chelate with divalent metal ions.[168,169] For this reason most extractions incorporate acidic solvents with the addition of metal-chelating agents[146] and include further de-proteinization steps. The most commonly used extraction approach uses Na_2EDTA-McIlvaine buffer (pH 4).[161] Known as the *universal tetracycline extractant*, McIlvaine buffer consists of citric acid and disodium hydrogen phosphate.[163] Other common buffers used for extraction include oxalic acid,[151] succinic acid,[165] and citric acid.[170] They are often used in conjunction with organic solvents such as ethyl acetate[171] and acetonitrile.[149] Stronger acids such as trifluoroacetic acid,[152] trichloroacetic acid,[172,173] and sulfuric acid[169] may possibly be utilized to further deproteinize in the extract. Most reported extraction techniques operate under mildly acidic (pH 2–6) conditions, because stronger acidic conditions cause degradative changes.[149,153] The use of PLE has also been reported in the literature.[153]

Most of the current methods reported in the literature incorporate solid phase extraction (SPE) as an additional clean-up step. However, it must be noted that tetracyclines have the ability to bind to free silanol groups of silica-based materials, causing losses during SPE, low column efficiency, poor recoveries, and peak tailing during liquid chromatography.[150,168] Some attempts to resolve this issue have included using non-silica-based SPE and HPLC columns as well as adding constituents such as oxalic acid to the mobile phase.[174] Because of the factors described earlier, most methods are tending toward the use of polymeric sorbents such as Oasis HLB,[149,153] which eliminates silanol group interactions or other non-silica-based C18 products. Some common cartridges include Agilent Sampli Q OPT,[151] MIP,[175] Amberlite resins,[159] Bond Elut LRC-PRS,[163] Bond Elut ENV,[176] LiChrolut,[152] Nexus,[152,174] Discovery DSC phenyl,[177] MCX,[178] carboxylic acid,[171] and chelating resins (using a technique called *metal chelate affinity chromatography*, in which the tetracyclines' ability to chelate to metal ions is put to good use).[168] When silica-based sorbents are used, metal-chelating materials such as EDTA are often incorporated.[147,150] Several review articles have been published that discuss the analysis of tetracyclines in foods and can be referenced for further details.[179–182]

7.8.3 Conclusions

On the basis of existing knowledge, novel quantitative approaches for the analysis of tetracyclines should involve extraction under mild acidic conditions, incorporate metal-chelating agents, and investigate polymeric SPE sorbents. In addition, they should be able to detect and quantify OTC, TC, CTC, DC, and their associated 4-epimers at concentrations that are fit for their intended purpose. If confirmation is required, LC-MS/MS is recommended for its selectivity.

7.9 MISCELLANEOUS

7.9.1 Aminoglycosides

The quantification and confirmation of aminoglycoside antibiotics (Fig. 7.8) at trace residue concentrations in animal tissues has long been a challenge because of the high polarity of the analytes and lack of a suitable chromophore for detection. MRLs are often relatively high; for example, for neomycin, MRLs range from 500 to 10000 μg/kg in Australia in line with the revised MRLs established by CAC, from 500 to 5000 μg/kg in the EU and from 150 to 7200 μg/kg as tolerances in the United States of America.[183]

Application of LC-MS to the analysis of aminoglycosides has also been problematic, due to the low retention of the parent compounds on the reversed phase columns commonly used.[184] To try to resolve this problem, ion-pairing reagents, such as fluorinated acids, have been used.[185] The introduction of hydrophilic interaction liquid chromatography (HILIC) has provided an alternative means for the retention and separation of aminoglycosides prior to mass spectrometric determination (see Chapter 6).[186] However, high buffer concentrations (>0.1 M) are often required before satisfactory chromatography can be achieved.

One innovation has been to derivatize the aminoglycosides by reaction with phenyl isocyanate (Fig. 7.8).[187] Each aminoglycoside has multiple sites that can be derivatized. The reaction is facile, going rapidly to completion. The derivatives are better retained on reversed phase columns than the parent compounds and respond well to LC-MS analysis. This procedure has been used for the determination of a range of aminoglycosides in bovine milk down to concentrations of 10–12 μg/l.[188] Weak cation exchange

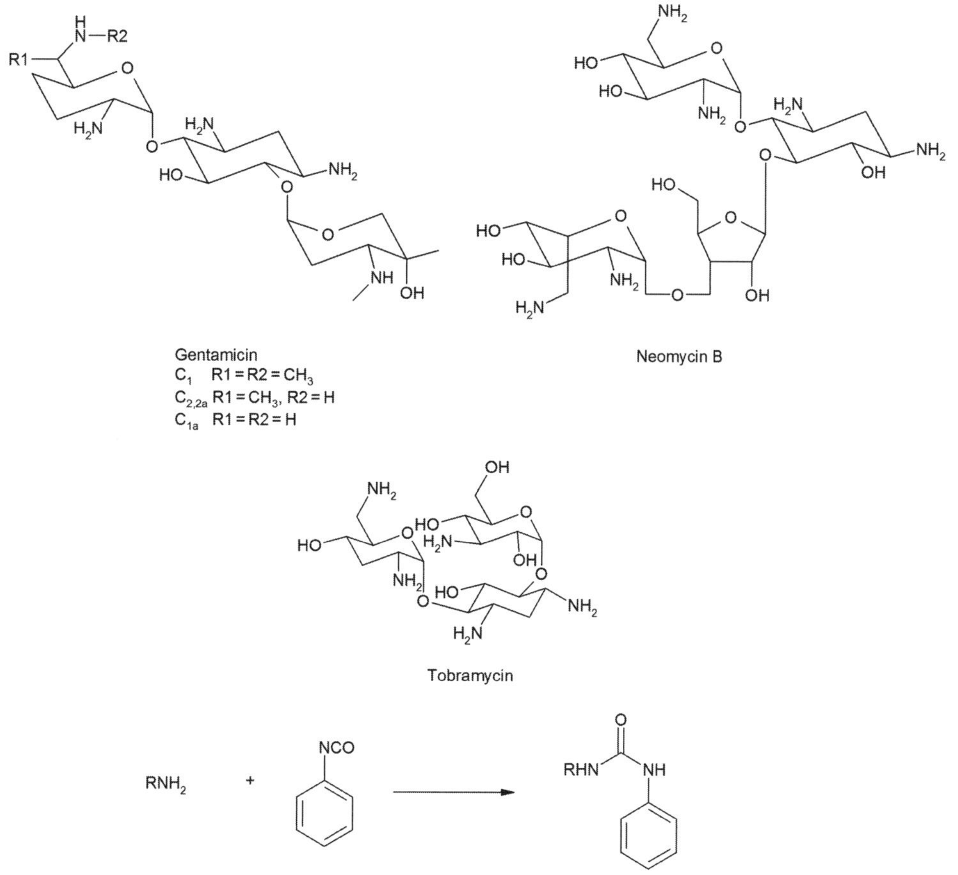

Gentamicin
C_1 R1 = R2 = CH_3
$C_{2,2a}$ R1 = CH_3, R2 = H
C_{1a} R1 = R2 = H

Neomycin B

Tobramycin

Figure 7.8 Structures of anminoglycoside antibiotics and derivatives.

solid-phase extraction (SPE) was used to clean-up defatted and buffered milk. The aminoglycosides were eluted from the cartridge using acidic methanol prior to evaporation to dryness. The derivatization was carried out by redissolving the residue in water, triethylamine in acetonitrile, and phenyl isocyanate in acetonitrile. The reaction is virtually instantaneous, requiring no incubation period. The derivatives were separated on a YMC-AQ reversed phase column using an aqueous formic acid–acetonitrile mobile-phase gradient. MS^2 analysis was carried out using a Thermo DECA ion trap mass spectrometer. Precursor and product ions are summarized in Table 7.6. The recoveries quoted for this method are in the range 80–120% with RSDs (relative standard deviations) of <25%. Illustrative ion traces are shown in Figure 7.9. This procedure represents an effective means of determining aminoglycoside residues without the use of speciality columns and mobile-phase additives.

7.9.2 Compounds with Marker Residues Requiring Chemical Conversion

7.9.2.1 Florfenicol

The position with regard to the analysis of this antibiotic is complicated by variations in the identification of the most appropriate marker of use. In many countries, such as Japan and the United States of America, the marker residue is identified as either the parent compound or the amine metabolite (Fig. 7.3) depending on the species and tissue to be tested. As such, confirmatory analysis of florfenicol may be undertaken in the context of more generic multi-residue procedures. However, in the EU and Canada, the MRL for florfenicol is defined in terms of the sum of the parent compound and its metabolites expressed as florfenicol amine. There are a number of possible metabolites (Fig. 7.3).[189] Typically, a hydrolysis step would be required to carry out the conversion.[190] Within the EU, MRLs range from 100 μg/kg in poultry muscle to 3000 μg/kg in bovine liver.[14]

Published procedures originating from outside the EU usually test for either the parent or the amine metabolite or both, along with other members of the phenicol group of antibiotics. Both GC-MS after derivatization and LC-MS have been used for quantification and confirmation. For example, Nagata and Oka used GC-MS following derivatization with *N,O*-bis(trimethylsilyl)acetamide for the quantitative analysis of three phenicols, including florfenicol.[191] Residues were extracted from fish muscle

TABLE 7.6 Summary of LC-MS/MS Transitions for Confirmation of Antibiotics

Analyte	Abbreviation (Where Used)	Precursor Ion (*m/z*)	Product Ion(s) (*m/z*)
Quinoxalines			
Quinoxaline-2-carboxylic acid	QCA	175	129/102/75
3-Methylquinoxaline-2-carboxylic acid	mQCA	189	145/134/102
Didesoxycarbadox	DCBX	231	143/102
Phenicols			
Chloramphenicol	CAP	321	257/194/152
Florfenicol		356	336/185
Florfenicol amine		248	230/130
Nitrofuran-2-nitrobenzaldehyde derivatives			
1-Amino-2,4-imidazolidinedione	NPAHD	249	134/178
3-Amino-5-morpholinomethyl-2-oxazolidinone	NPAMOZ	335	262/291
3-Amino-2-oxazolidinone	NPAOZ	236	104/134
Semicarbazide	MPSEM	209	166/192/134
Nitroimidazoles			
Dimetridazole	DMZ	142	96/81
Metronidazole	MNZ	172	128/82
Ronidazole	RNZ	201	140/110/55
Ipronidazole	IPZ	170	124/109
Carnidazole	CNZ	245	118/75
Ornidazole	ONZ	220	128/82
Tinidazole	TNZ	248	202/121
Ternidazole	TRZ	186	128/111/82
Hydroxymetronidazole	MNZ-OH	188	129/126/123
Hydroxyipronidazole	IPZ-OH	186	168/122
Hydroxydimetridazole	HMMNI	158	140/110/55
Sulfonamides			
Sulfabenzamide		277	156/108/92
Sulfacetamide		215	156/108
Sulfachloropyridazine		285	201/156/108/92
Sulfadiazine		251	174/156/108/92
Sulfadimethoxine		311	245/218/156/108/92
Sulfadoxine		311	218/156/108
Sulfaethoxypyridazine		295	156
Sulfaguanidine		215	156/108/92
Sulfamerazine		265	172/156/108/92
Sulfamethazine (sulfadimidine)		279	204/186/156/124/108/92
Sulfanilamide		173	156/132/108/92
Sulfanitran		336	294/198/156
Sulfapyridine		250	184/156/108/92
Sulfaquinoxaline		301	156/108/92
Sulfathiazole		256	156/108/92
Sulfamethoxypyridazine		281	156/126/108/92
Sulfamethoxazole		254	188/156/147/108/92
Sulfamonomethoxine		281	215/156/126/108/92
Sulfamethizole		271	156/108/92
Sulfisomidine		279	186/156/124
Sulfamoxole		268	156/113/108/92
Sulfameter		281	215/156/126/108/92
Sulfisoxazole		268	156/113/108
Sulfaphenazole		315	222/156

TABLE 7.6 *(Continued)*

Analyte	Abbreviation (Where Used)	Precursor Ion (m/z)	Product Ion(s) (m/z)
Tetracyclines			
Oxytetracycline	OTC	461	426/444/443/337/127
4-Epioxytetracycline	epi-OTC	461	444/443/426/201
Tetracycline	TC	445	428/427/410/337/154
4-Epitetracycline	epi-TC	445	428/427/410
Chlortetracycline	CTC	479	462/461/444/401/402/154
4-Epichlortetracycline	epi-CTC	479	462/444
Doxycycline	DC	445	428/410/321/154
Minocycline	MC	458	441/352
Demeclocycline	DMC	465	448/430
Aminoglycosidephenylisocyanate derivatives			
Gentamicin C_{1a} $(PhIC)_5$		1045	767/679/401/367
Gentamicin $C_{2,2a}$ $(PhIC)_5$		1059	781/679/401/381
Gentamicin C_1 $(PhIC)_{5)}$		1073	795/679/401/395
Neomycin $(PhIC)_6$		1329	799/706/531/399
Tobramycin $(PhIC)_5$		1063	681/588/401/383

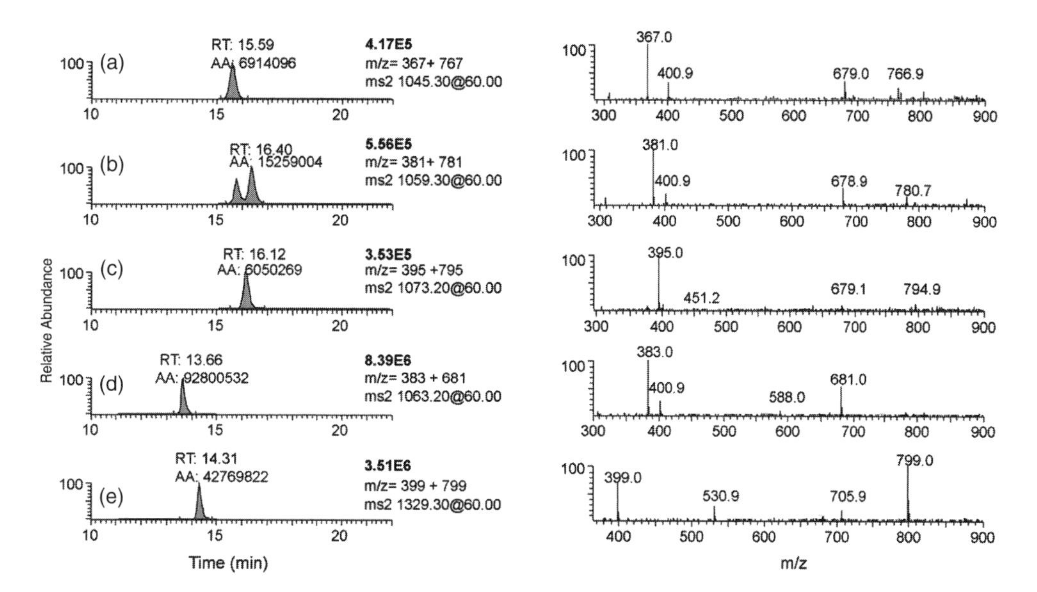

Figure 7.9 Determination of aminoglycosides in bovine milk fortified at 0.06 μg/kg as their phenylisocyanate derivatives, by LC-MS^2 analysis: (a) [gentamicin C1a(PhIC)5H]$^+$; (b) [gentamicin C2, 2a(PhIC)5H]$^+$; (c) [gentamicin C1(PhIC)5H]$^+$; (d) [tobramycin(PhIC)5H]$^+$; (e) [neomycin B(PhIC)6H]$^+$.

samples using ethyl acetate prior to blowdown and dissolution in aqueous sodium chloride. The extracts were further cleaned up using a series of partition steps and a Florisil solid-phase extractor. After derivatization, analysis was carried out on a 5% phenylmethylsilicone column, monitoring a single ion at m/z 257 for florfenicol, sufficient for quantification with a detection limit of 5 μg/kg, but not for confirmation.

A combination of liquid–liquid partitioning steps sometimes coupled to SPE is common to most recent methods of analysis of florfenicol. Basic ethyl acetate has been used as extraction solvent from chicken muscle and from shrimp.[192,193] Following a defatting partitioning step, cleanup was by C18 or mixed-mode (reversed phase cation exchange) SPE. In both cases analytical separation was achieved using an Xterra C18 column with detection by MS/MS. In both methods, two transitions (356 > 336, 356 > 185) were monitored, fulfilling the identification point criteria used in the EU for confirmation.[8] The method of Zhang and colleagues was also capable of detecting florfenicol amine.[192] Detection limits were <1 μg/kg for both methods. A somewhat simpler protocol for the determination of florfenicol and its amine, eliminating the need for solid phase extraction, has been developed at the laboratory of one of the authors.[51] Samples were extracted with acetone and then water was removed by addition of dichloromethane. After blowdown and reconstitution in aqueous acid, the extracts were defatted with hexane prior to analysis by LC-MS on a Hypersil C18-BD column using an acetonitrile–aqueous acetic acid gradient. Florfenicol was detected in negative-mode monitoring a single ion at m/z 356, whereas florfenicol amine was detected in positive-mode monitoring at mass m/z 248. Limits of detection were <1 μg/kg.

These methods are sufficient to satisfy the requirements in their country of origin. However, they do not strictly meet the current EU requirements, although the proportions of the other metabolites are such that the underestimation of concentrations according to the MRL definition is relatively small. However, since florfenicol is often included in an analytical method with chloramphenicol and thiamphenicol and therefore is detected at concentrations 50–100 times below existing MRLs or tolerances, a practical approach is to include florfenicol (or florfenicol and florfenicol amine) in such methods for screening purposes, and then confirm any residue findings approaching MRLs or US tolerances with a method that addresses the specific requirements of the marker residue definition (e.g., residues converted as florfenicol amine).

7.9.3 Miscellaneous Analytical Issues

7.9.3.1 Lincosamides

The lincosamides are a group of antibiotics linked by the commonality of a thioether-substituted sugar. Lincomycin, the first lincosamide (isolated from *Streptomyces lincolnensis*) and pirlimycin have a veterinary application, while clindamycin, produced by chemical modification of lincomycin, is used in human medicine. A fourth member of the class, mirincamycin, has been investigated as a means of treating *Plasmodium* spp. infections (malaria). Residue limits, where set, are based on the parent compound as marker residue. For example, in the EU, lincomycin has MRLs in all food-producing species ranging from 50 μg/kg in fat and eggs to 1500 μg/kg in kidney. Pirlimycin has MRLs ranging from 100 to 1000 μg/kg in bovine tissues in the EU and tolerances ranging from 300 to 400 μg/kg in the United States of America.

The lincosamides can undergo extensive transformation by metabolism. For example, in pigs the presence of 26 metabolites was indicated in liver.[194] Lincomycin sulfoxide, N-demethyllincomycin, and N-demethyllincomycin sulfoxide were all identified in poultry liver. In bovines, pirlimycin sulfoxide and sulfone have been identified as metabolites.[195] In liver and kidney the major residue is the sulfoxide for both lincomycin and pirlimycin, accounting for approximately 40–65% of the radioactivity observed following dosing with ^{14}C-labeled drug. The presence of higher concentrations of the sulfoxide metabolite is of significance from a regulatory perspective because of the phenomenon of "reverse metabolism" post-mortem, which has been observed for both pirlimycin[196] and lincomycin.[194] Extractable residue concentrations of lincomycin increased significantly following overnight incubation of liver samples at room temperature. Similarly, pirlimycin concentrations increased by up to 345% in bovine liver following incubation at temperatures ranging from 4°C to 37°C. This effect was not observed in kidney or muscle tissue. A concomitant reduction in pirlimycin sulfoxide residue concentration was also observed, suggesting that this effect is caused by residual reductase enzyme activity in liver.

As a result of these observations, recommendations have been made to include an incubation step in methods of analysis for lincosamides in liver.[194,197] If the aim is to detect lincosamide use, then inclusion of an incubation step is necessary to maximize the possibility of detection. However, as indicated earlier, MRLs, where set, are based on the parent compound only with no reference to the metabolites, either by summation of separate measurements or by conversion to a suitable marker. Thus, inclusion of an incubation step could lead to falsely high residue concentrations and the possibility of samples that fall within the legal limits being reported as non-compliant. As a result, analytically, steps should be taken to minimize enzymatic activity to prevent this reverse metabolism. A long-term solution to this issue might be to include the sulfoxides in the legislation as additional markers in some form.

Lincomycin (and to some extent the other lincosamides) is commonly mislabeled as a macrolide[198] and as such is often included in multi-residue macrolide procedures.[199,200] Honey has been the most common target matrix, because of interest in the use of lincomycin for control of foulbrood disease in bees.[199,201] Extraction from honey has principally been by dissolution/dilution with aqueous solvents. Acetonitrile has been used as extractant for tissues and milk.[202] Dispersion onto sand prior to hot-water extraction has also been used for the analysis of milk.[203] Solid-phase[199,204] and liquid–liquid[202] extraction have been used for clean-up. Separation is customarily achieved on reversed phase (C18) columns prior to determination by MS or MS/MS with APCI[205] or more commonly ESI.[199,201]

There are a few analytical methods for pirlimycin. The method of Hornish et al.[206] uses acidified acetonitrile as extractant for both tissues and milk, followed by a solvent partition and C18 SPE. Final determination is by RP-LC-thermospray MS. The method of Martos et al.[198] uses acetonitrile for extraction followed by a simple defatting step with hexane prior to determination by LC-ESI-MS/MS.

7.9.3.2 *Enrofloxacin*

As discussed in Chapters 1 and 2, enrofloxacin (ENR) is a second-generation fluoroquinolone antibiotic with a wide range of activity against both Gram-negative and Gram-positive bacterial infections. ENR is readily metabolized by deethylation to ciprofloxacin (CIP), a compound that also has extensive antimicrobial activity and is of importance in the area of human medicine as a frontline treatment for a number of infections, including anthrax. CIP is also metabolized further. Residue limits have been set in a number of countries, which differ in both the permitted species and the marker residue definition. For example, in the United States of America, ENR is not permitted for use in poultry but has tolerances of 100 μg/kg in cattle liver, based on desethylene CIP as marker and 500 μg/kg in porcine liver based on ENR as marker. In Japan, the parent compound is also used as the marker residue with MRLs of 50–100 μg/kg. In the EU the position is slightly more complicated, as the marker residue definition is expressed as the sum of the parent compound and the major metabolite CIP. MRLs range from 100 to 300 μg/kg and cover all food-producing species.

There are two issues, then, to be faced when setting up a method of analysis for ENR:

1. It is necessary to ensure that all the components within the definition are included in the procedure. For ENR this is relatively simple, as a maximum of two components are required—ENR+CIP or ENR or desethylene CIP, depending on country and species. The main concern within the EU, since the MRL is set on a summation, is to ensure that the method has sufficient sensitivity for the individual components. A value of 10–25% of the MRL is usually regarded as adequate. There is a vast literature available on the determination of fluoroquinolones in animal tissues, and most multi-residue methods include both ENR and CIP. It is beyond the scope of this chapter to give a comprehensive breakdown of all the available procedures. The following text gives a small indication of the techniques that have been applied to the analysis of fluoroquinolones in general and ENR + CIP in particular. There is some evidence of protein binding,[207] so extractions should be capable of breaking this interaction. Acidic extraction solvents are commonly used.[208,209] Liquid–liquid extraction[209] and SPE in both cartridge[210] and dispersive[211] formats have been used for clean-up. Chromatographic separation is usually achieved under reversed phase conditions (C8 or C18) using acidic mobile phases before determination by ESI-MS/MS[209] or TOF-MS.[212] Other techniques such as CE-MS/MS[213] have also been used for the end determination.
2. It is also important, in any ENR analysis, to determine whether a (legal) use of ENR can be distinguished from a (illegal) use of CIP. The ratio of ENR to CIP is dependent on species and tissue type and time of depletion. For example, 2 days after dosing, the ENR/CIP ratio in bovine kidney drops from approximately 8 : 1 to 1 : 1 after 4 days, with a concomitant drop in residue concentration to close to the detection limit of the method. In contrast the ratio in bovine liver remains roughly constant at 1 : 1 from 2 days post-dosing to 4 days.[214] It has also been noted that at 7 days post-dosing the main residue in bovine liver is desethylene CIP.[215] In chicken, ENR accounted for 61–85% of the total residues at 6 h post-dosing, depending on the matrix type. In muscle, the proportion of ENR went up to 98% at 15 h post-dosing.[215] Pharmacokinetic studies on ENR and CIP in bovines have shown that the plasma elimination half-life of both are comparable.[207] In freshwater prawn, CIP was not detected (<25 μg/kg) 2 days post-dosing, whereas ENR could be detected (>15 μg/kg) up to 15 days post-dosing.[216] Pharmacokinetic data in chickens also indicated that CIP has a shorter residence time than ENR.[217] However, another study was able to detect CIP in chicken muscle kidney and liver 12 days after a 4-day oral dosing regime, whereas ENR was detected only in the liver.[218]

From these studies it may be concluded that the presence of parent ENR, even in the absence of CIP, is indicative of ENR dosing. The absence of ENR and presence of CIP is not necessarily indicative of CIP dosing, particularly at low residue concentrations. Similarly, the presence of desethylene ciprofloxacin may be an indicator of either ENR or CIP dosing. Further studies are required to establish the criteria by which CIP dosing may be distinguished from ENR dosing.

7.9.4 Gaps in Analytical Coverage

A number of veterinary medicines licensed worldwide have residue limits set on the basis of marker residues that are formed by chemical conversion, such as hydrolysis and/or oxidation. For many of these, methods suitable for the quantification and confirmation of residues are not available in the open literature. The availability of analytical methods has been discussed by the CCRVDF, and it is considering how this important area can be addressed (see minutes of the 19th session of the CCRVDF meeting in September 2010).[219]

Tiamulin, a pleuromutilin antimicrobial agent, is extensively metabolized (Fig. 7.10). With the exception of eggs, EU MRLs for tiamulin are defined as the sum of those metabolites that may be hydrolyzed to 8α-hydroxymutilin—this does not include the parent compound. Parent tiamulin forms a significant part of the residue profile only in eggs.

The macrolide antibiotic tulathromycin also has residue limits defined in terms of a marker residue obtained by the hydrolysis of a range of metabolites. Again, analysis of incurred tissues indicates that the marker residue is not a major component of the residues present.[220]

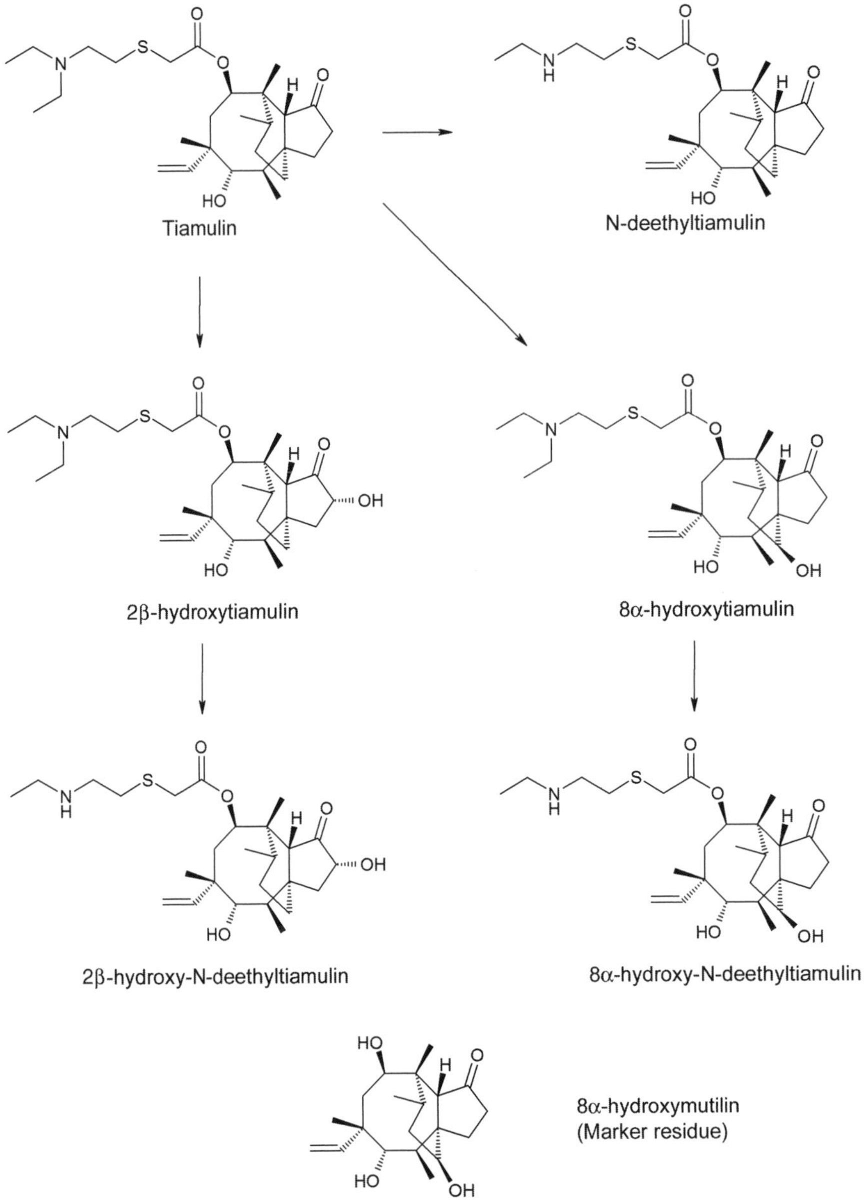

Tiamulin

N-deethyltiamulin

2β-hydroxytiamulin

8α-hydroxytiamulin

2β-hydroxy-N-deethyltiamulin

8α-hydroxy-N-deethyltiamulin

8α-hydroxymutilin
(Marker residue)

Figure 7.10 Tiamulin metabolism.

7.10 SUMMARY

In summary, it has been demonstrated that a number of compounds have specific analytical requirements due either to extensive metabolism or the legal definitions of the marker residues to be used for detection of use. For the quinoxaline-type antibiotics, the question of the most appropriate marker of use remains unanswered at present. For ceftiofur, differences in legislation worldwide mean that analytical requirements differ from region to region. As with the quinoxalines, a question has been raised as to whether the generally accepted marker residue is the optimum.

The challenges for chloramphenicol analysis lie in the low LODs required, taking into account the potential for formation of the glucuronide in some matrices. Analysis is further complicated by the presence of naturally occurring chloramphenicol in herbs and grasses that might be considered a feed source, its use in human medicine, and the potential for cross-contamination. Analysis of the nitrofurans is governed by their extensive metabolism and protein binding. For this reason, marker compounds have been designated. However, as these are small molecules, they are not very specific. Semicarbazide, the designated marker for nitrofurazone, can occur from multiple sources, not simply from nitrofurazone dosing. Protocols for nitrofuran analysis therefore have to be designed to eliminate the possibility of incorrect identification of residues from nitrofuran dosing.

The nitroimidazoles, sulfonamides, and tetracyclines all present analytical challenges because of metabolism and/or chemical degradation. In the case of the nitroimidazoles, this is further complicated by the relatively low requirements for detection. Method development therefore has to take into account both metabolites as additional target compounds and low detection limits. Sulfonamide analysis has to take into account the potential for conversion of N^4-acetyl metabolites back to the parent compound. In contrast, in the analysis of honey, deconjugation is regarded as necessary to accurately determine sulfonamide concentrations. The facile, reversible formation of epimers is of particular concern in the analysis of those tetracyclines that can epimerize in the 4 position. Protein and metal binding are other issues that have to be overcome for successful tetracycline residue determination.

Finally, gaps in the analytical coverage still remain, due to the lack of suitable procedures in the open literature for compounds with defined marker residues. These include florfenicol, tiamulin, and tulathromycin. Where any method exists, it has tended to focus on the parent drug or a single metabolite, which may not be appropriate for either screening or confirmatory purposes.

ABBREVIATIONS

AcOH	Acetic acid
ADC	Azodicarbonamide
ADI	Acceptable daily intake
AHD	1-Amino-2,4-imidazolidinedione
AMOZ	3-Amino-5-morpholinomethyl-2-oxazolidinone
AOZ	3-Amino-2-oxazolidinone
APCI	Atmospheric-pressure chemical ionization
ATMA	(2*E*)-2-(2-Amino-1,3-thiazol-4-yl)-2-(methoxyimino)acetamide
CAC	Codex Alimentarius Commission
CAP	Chloramphenicol
CCRVDF	Codex Committee on Residues of Veterinary Drugs in Foods
CCα	Decision limit[8]
CCβ	Detection capability[8]
CE	Capillary electrophoresis
CFIA	Canadian Food Inspection Agency
CI	Chemical ionization
CNZ	Carnidazole
CRL	Community Reference Laboratory
CTC	Chlortetracycline
DAD	Diode-array detection
DC	Doxycycline
DCBX	Desoxycarbadox
DCCD	Desfuroylceftiofur disulfide
DFC	Desfuroylceftiofur
DMC	Demeclocycline
DMZ	Dimetridazole
ECD	Electron capture detection
EDTA	Ethylenediaminetetraacetic acid
EI	Electron impact (ionization)
ELISA	Enzyme-linked immunosorbent assay
ESI	Electrospray ionization
EU	European Union
FLD	Fluorescence detection
GC	Gas chromatography
HCl	Hydrochloric acid
HMMNI	2-Hydroxymethyl-1-methyl-5-nitroimidazole
HPLC	High-performance liquid chromatography
IPZ	Ipronidazole
IPZ-OH	Hydroxyipronidazole
JECFA	Joint FAO/WHO Expert Committee on Food Additives
LC	Liquid chromaotgraphy
LLE	Liquid–liquid extraction
LPE	Liquid-phase extraction
LOD	Limit of detection
LOQ	Limit of quantification
MC	Minocycline
MeCN	Acetonitrile
MeOH	Methanol
MIP	Molecularly imprinted polymer
MNZ	Metronidazole
MNZ-OH	Hydroxymetronidazole
mQCA	3-Methylquinoxaline-2-carboxylic acid
MRL	Maximum residue limit
MRPL	Minimum required performance limit
MS	Mass spectrometry(er)
MS/MS	Tandem mass spectrometry
MSPD	Matrix solid-phase dispersion
MTC	Methacycline
NPAHD	2-Nitrophenyl 1-amino-2,4-imidazolidinedione
NPAMOZ	2-Nitrophenyl 3-amino-5-morpholinomethyl-2-oxazolidinone
NPAOZ	2-Nitrophenyl 3-amino-2-oxazolidinone
NPSEM	2-Nitrophenyl semicarbazide
ONZ	Ornidazole
OTC	Oxytetracycline
PLE	Pressurized liquid extraction
QCA	Quinoxaline-2-carboxylic acid
QqQ	Triple quadrupole
RNZ	Ronidazole
RP	Reversed phase
RTC	Rolitetracycline
SCX	Strong cation exchange
SEM	Semicarbazide
SEP	Sulfaethoxypyridazine

SPE	Solid-phase extraction
TC	Tetracycline
TNZ	Tinidazole
TOF/MS	Time-of-flight mass spectrometry
TRZ	Ternidazole
UPLC	Ultraperformance liquid chromatography
UV	Ultraviolet

REFERENCES

1. Yen JT, Nienbaber JA, Pond WG, Varel VN, Effect of carbadox on growth, fasting metabolism, thyroid function and gastrointestinal tract in young pigs, *J. Nutr.* 1985; 115(8):970–979.
2. World Health Organization, *Evaluation of Certain Food Additives and Contaminants*, 35th report of Joint FAO/WHO Expert Committee on Food Additives, WHO Food Additives Series 27, 1991, pp. 45–54 (available at `http://whqlibdoc.who.int/trs/WHO_TRS_799.pdf`; accessed 11/26/10).
3. Commission Regulation No. 2788/98 of 22 December 1998 amending Council Directive 70/524/EEC concerning additives in feedingstuffs as regards the withdrawal of authorisation for certain growth promoters, *Off. J. Eur. Commun.* 1998;L347:31–32.
4. CRL guidance paper (Dec. 7, 2007), *CRLs View on State of the Art Analytical Methods for National Residue Control Plans*, 2007. (available at `http://www.bvl.bund.de/SharedDocs/Downloads/09_Untersuchungen/EURL_Empfehlungen_Konzentrationsauswahl_Methodenvalierungen.pdf?_blob=publicationFile&v=2`; accessed 08/03/11)
5. Huang L, Wang Y, Tao Y, et al., Development of high performance liquid chromatographic methods for the determination of cyadox and its metabolites in plasma and tissues of chicken, *J. Chromatogr. B* 2008;874:7–14.
6. Fernández Suárez A, Arnold D, *"Carbadox" in Residues of Some Veterinary Drugs in Animals and Foods*, FAO Food and Nutrition Paper 41/15, 2003, pp. 1–19 (available at `ftp://ftp.fao.org/ag/agn/jecfa/vetdrug/41-15-carbadox.pdf`; accessed 11/26/10).
7. Hutchinson MJ, Young PB, Kennedy DG, Quinoxaline-2-carboxylic acid in pigs: Criteria to distinguish between the illegal use of carbadox and environmental contamination, *Food Addit. Contam.* 2004;21(6):538–544.
8. Commission Decision 2002/657/EC implementing Council Directive 96/23/EC concerning the performance of analytical methods and the interpretation of results, *Off. J. Eur. Commun.* 2002;L221/ 8–36.
9. Hutchinson MJ, Young PY, Hewitt SA, et al., Confirmation of the carbadox metabolite, quinoxaline-2-carboxylic acid, in porcine liver using LC-electrospray MS-MS according to revised EU criteria for veterinary drug residue analysis, *Analyst* 2002;127:342–346.
10. Hutchinson MJ, Young PB, Kennedy DG, Confirmation of carbadox and olaquindox metabolites in porcine liver using liquid chromatography-electrospray, tandem mass spectrometry, *J. Chromatogr. B* 2005;816:15–20.
11. Sin DWM, Chung LPK, Lai MMC, et al., Determination of quinoxaline-2-carboxylic acid, the major metabolite of carbadox, in porcine liver by isotope dilution gas chromatography-electron capture negative ionisation mass spectrometry, *Anal. Chim. Acta* 2004;508:147–158.
12. Horie M, Murayama M, Determination of carbadox metabolites, quinoxaline-2-carboxylic acid and desoxycarbadox, in swine muscle and liver by liquid chromatography/mass spectrometry, *J. Food Hyg. Soc. Japan* 2004;45(3):135–140.
13. Boison JO, Lee SC, Gedir RG, A determinative and confirmatory method for residues of the metabolites of carbadox and olaquindox in porcine tissues, *Anal. Chim. Acta* 2009;637:128–134.
14. Commission Regulation 37/2010 of 22 December 2009 on pharmacologically active substances and their classification regarding maximum residue limits in foodstuffs of animal origin, *Off. J. Eur. Commun.* 2010;L15/ 1–71.
15. Jaglan PS, Yein FS, Hornish RE, et al., Depletion of intramuscularly injected ceftiofur from the milk of dairy cattle, *J. Dairy Sci.* 1992;75:1870–1876.
16. Gilbertson TJ, Roof RD, Nappier JL, et al., Disposition of ceftiofur sodium in swine following intramuscular treatment, *J. Agric. Food Chem.* 1995;43(1):229–234.
17. Jaglan PS, Kubicek MF, Arnold TS, et al., Metabolism of ceftiofur. Nature of urinary and plasma metabolites in rats and cattle, *J. Agric. Food Chem.* 1989;37:1112–1118.
18. Berendsen B, Essers M, Mulder P, et al., Newly identified degradation products of ceftiofur and cephapirin impact the analytical approach for quantitative analysis of kidney, *J. Chromatogr. A* 2006;1216:8177–8186.
19. *Codex Veterinary Drug Residues in Food Online Database* (available at `http://www.codexalimentarius.net/vetdrugs/data/index.html`; accessed 11/11/10).
20. Sørensen LK, Snor, LK, Determination of cephalosporins in raw bovine milk by high-performance liquid chromatography, *J. Chromatogr. A* 2000;882:145–151.
21. Keever J, Voyksner RD, Tyczkowska KL, Quantitative determination of ceftiofur in milk by liquid chromatography-electrospray mass spectrometry, *J. Chromatogr. A* 1998;794: 57–62.
22. Navarre CB, Zhang L, Sunkara G, et al., Ceftiofur distribution in plasma and joint fluid following regional limb injection in cattle, *J. Vet. Pharmacol. Ther.* 1999;22:13–19
23. Bruno F, Curini R, Corcia AD, et al., Solid-phase extraction followed by liquid chromatography-mass spectrometry for trace determination of β-lactam antibiotics in bovine milk, *J. Agric. Food Chem.* 2001;49:3463–3470.
24. Kantiani L, Farre M, Sibum M, et al., Fully automated analysis of beta-lactams in bovine milk by online solid phase extraction-liquid chromatography-electrospray-tandem mass spectrometry, *Anal. Chem.* 2009;81:4285–4295.

25. Tyczkowska K, Voyksner R, Straub R, Aronson A, Simultaneous multiresidue analysis of beta-lactam antibiotics in bovine milk by liquid chromatography with utraviolet detection and confirmation by electrospray mass spectrometry, *J. AOAC Int*. 1994;77:1122–1131.
26. Moats WA, Romanowski RD, Medina MB, Identification of β-lactam antibiotics in tissue samples containing unknown microbial inhibitors, *J. AOAC Int*. 1998;81: 1135–1140.
27. Baere SD, Pille P, Croubels S, et al. High-performance liquid chromatographic-UV detection analysis of ceftiofur and its active metabolite desfuroylceftiofur in horse plasma and synovial fluid after regional intravenous perfusion and systemic intravenous injection of ceftiofur sodium. *Anal Chim Acta*. 2004;512:75–84.
28. Jaglan PS, Cox BL, Arnold TS, et al. Liquid chromatographic determination of desfuroylceftiofur metabolite of ceftiofur as residue in cattle plasma. *J. Assoc. Off. Anal. Chem*. 1990;73:26–30.
29. Makeswaran S, Patterson I, Points J, An analytical method to determine conjugated residues of ceftiofur in milk using liquid chromatography with tandem mass spectrometry, *Anal. Chim. Acta* 2005;529:151–157.
30. Beconi-Barker MG, Roof RDM, Kausche LFM, et al., Determination of ceftiofur and its desfuroylceftiofur-related metabolites in swine tissues by high-performance liquid chromatography, *J. Chromatogr. B* 1995;673:231–244.
31. Fagerquist CK, Lightfield AR, Lehotay SJ, Confirmatory and quantitative analysis of β-lactam antibiotics in bovine kidney tissue by dispersive solid-phase extraction and liquid chromatography-tandem mass spectrometry, *Anal. Chem*. 2005;77:1473–1482.
32. Tyczkowska KL, Voyksner RD, Anderson KL, Aronson AL, Determination of ceftiofur and its metabolite desfuroylceftiofur in bovine serum and milk by ion-paired liquid chromatography, *J. Chromatogr. B Biomed. Sci. Appl*. 1993;614:23–134.
33. Fagerquist CK, Lightfield AR, Confirmatory analysis of beta-lactam antibiotics in kidney tissue by liquid chromatography/electrospray ionization selective reaction monitoring ion trap tandem mass spectrometry, *Rapid Commun. Mass Spectrom*. 2003;17:660–671.
34. Mastovska K, Lightfield A, Streamlining methodology for the multiresidue analysis of β-lactam antibiotics in bovine kidney using liquid chromatography-tandem mass spectrometry, *J. Chromatogr. A* 2008;1202:118–123.
35. Page MI, *The Mechanisms of Reactions of β-Lactam Antibiotics*, Advances in Physical Organic Chemistry Series, Vol. 23, Academic Press, San Diego, CA, 1987, pp. 165–270.
36. Ehrlich J, Bartz QR, Smith RM, et al. Chloromycetin, a new antibiotic from a soil actinomycete, *Science* 1947; 106:417.
37. IARC (International Agency for Research on Cancer), *Chloramphenicol*, Vol. 50, Monographs on the Evaluation of Carcinogenic Risk of Chemicals to Humans, IARC Press, Lyon, France, 1990, pp. 169–193.
38. *EMEA Chloramphenicol Summary Report*, 1996 (available at `http://www.ema.europa.eu/docs/en_GB/document_library/Maximum_Residue_Limits_-_Report/2009/11/WC500012060.pdf`; accessed 11/29/10).
39. US Food and Drug Administration, Chloramphenicol oral solution; opportunity for hearing, *Federal Register* 1985;50:27059–27064.
40. Wongtavatchai J, McLean JG, Ramos F, Arnold D, *Chloramphenicol*, WHO Food Additives Series, Vol. 53, JECFA (WHO: Joint FAO/WHO Expert Committee on Food Additives), IPCS (International Programme on Chemical Safety) INCHEM., 2004, pp. 7–85 (available at `http://whqlibdoc.who.int/publications/2004/9241660538_chloramphenicol.pdf`; accessed 08/02/11).
41. Commission Decision 2002/994/EC Concerning certain protective measures with regard to the products of animal origin imported from China, *Off. J. Eur. Commun*. 2002;L348/ 154–156.
42. Commission Decision 2001/705/EC Concerning certain protective measures with regard to certain fishery and aquaculture products intended for human consumption and originating in Indonesia, *Off. J. Eur. Commun*. 2002;L260/ 35–36.
43. Commission Decision 2002/251/EC Concerning certain protective measures with regard to poultry, meat and certain fishery and aquaculture products intended for human consumption and imported from Thailand, *Off. J. Eur. Commun*. 2002;L84/ 77–78.
44. Di Corcia A, Nazzari M, Liquid chromatographic-mass spectrometric methods for analyzing antibiotic and antibacterial agents in animal food products, *J. Chromatogr. A* 2002;974:53–89.
45. Stolker AAM, Brinkman UATh, Analytical strategies for residue analysis of veterinary drugs and growth-promoting agents in food-producing animals—a review, *J. Chromatogr. A* 2005;1067:15–53.
46. Commission Decision 2003/181/EC amending Decision 2002/657/EC as regards the setting of minimum required performance limits (MRPLs) for certain residues in food of animal origin, *Off. J. Eur. Commun*. 2003;L71/ 17–18.
47. Gantverg A, Shishani I, Hoffman M, Determination of chloramphenicol in animal tissues and urine—liquid chromatography-tandem mass spectrometry versus gas chromatography-mass spectrometry, *Anal. Chim. Acta* 2003;483(1–2):125–135.
48. Kubala-Drincic H, Bazulic D, Sapunar-Postruznik J, et al., Matrix solid-phase dispersion extraction and gas chromatographic determination of chloramphenicol in muscle tissue, *J. Agric. Food Chem*. 2003;51(4):871–875.
49. Mottier P, Parisod V, Gremaud E, et al., Determination of the antibiotic chloramphenicol in meat and seafood products by liquid chromatography–electrospray ionization tandem mass spectrometry, *J. Chromatogr. A* 2003;994(1–2):75–84.
50. Ramos M, Muñoz P, Aranda A, et al., Determination of chloramphenicol residues in shrimps by liquid

chromatography-mass spectrometry, *J. Chromatogr. B* 2003; 791(1–2):31–38.

51. Van de Riet JM, Potter RA, Christie-Fougere M, Burns BG, Simultaneous determination of residues of chloramphenicol, thiamphenicol, florfenicol and florfenicol amine in farmed aquatic species by liquid chromatography/mass spectrometry, *J. AOAC Int*. 2003;86(3):510–514.
52. Kaufmann A, Butcher P, Quantitative liquid chromatography/tandem mass spectrometry determination of chloramphenicol residues in food using sub-2 μm particulate high-performance liquid chromatography columns for sensitivity and speed, *Rapid Commun. Mass Spectrom*. 2005;19:3694–3700.
53. Rønning HT, Einarsen K, Asp TN, Determination of chloramphenicol residues in meat, seafood, egg, honey, milk, plasma and urine with liquid chromatography-tandem mass spectrometry, and the validation of the method based on 2002/657/EC, *J. Chromatogr. A* 2006;1118:226–233.
54. Kanarat S, Tangsirisup N, Nijthavorn N, et al., An investigation into the possible occurrence of chloramphenicol in poultry litter, *Proc. EuroResidue VI Conf. Residues of Veterinary Drugs in Food*, Egmond aan Zee, The Netherlands, May 19–21, 2008.
55. Boxall ABA, Johnson P, Smith EJ, et al., Uptake of veterinary medicines from soils into plants, *J. Agric. Food Chem*. 2006;54:2288–2297.
56. Berendsen B, Stolker AAM, Jong de J, et al., Evidence of natural occurrence of the banned antibiotic chloramphenicol in herbs and grass, *Anal. Bioanal. Chem*. 2010;397:1955–1963.
57. Vass M, Hruska K, Franek M, Nitrofuran antibiotics: A review on the application, prohibition and residual analysis, *Vet. Med*. 2008;53(9):469–500.
58. Finzi JK, Donato JL, Sucupira M, et al., Determination of nitrofuran metabolites in poultry muscle and eggs by liquid chromatography-tandem mass spectrometry, *J. Chromatogr. B* 2005;824(1–2):30–35.
59. Verdon E, Couedor P, Sanders P, Multi-residue monitoring for the simultaneous determination of five nitrofurans (furazolidone, furaltadone, nitrofurazone, nitrofurantoine, nifursol) in poultry muscle tissue through the detection of their five major metabolites (AOZ, AMOZ, SEM, AHD, DNSAH) by liquid chromatography coupled to electrospray tandem mass spectrometry—in-house validation in line with Commission Decision 657/2002/EC, *Anal. Chim. Acta* 2007;586(1–2):336–347.
60. Bock C, Gowik P, Stachel C, Matrix-comprehensive in-house validation and robustness check of a confirmatory method for the determination of four nitrofuran metabolites in poultry muscle and shrimp by LC-MS/MS, *J. Chromatogr. B* 2007;856:178–189.
61. Conneely A, Nugent A, O'Keeffe M, et al., Isolation of bound residues of nitrofuran drugs from tissue by solid-phase extraction with determination by liquid chromatography with UV and tandem mass spectrometric detection, *Anal. Chim. Acta* 2003;483(1–2):91–98.
62. Cooper KM, Kennedy DG, Nitrofuran antibiotic metabolites detected at parts per million concentrations in retina of pigs—a new matrix for enhanced monitoring of nitrofuran abuse, *Analyst* 2005;130(4):466–468.
63. European Food Safety Authority, *Advice of the Ad Hoc Expert Group Set up to Advise the European Food Safety Authority (EFSA) on the Possible Occurrence of Semicarbazide in Packaged Foods*, July 28, 2003, AFC/ad hoc SEM/1, Brussels, 2003.
64. European Food Safety Authority, *Statement of the Scientific Panel on Food Additives, Flavouring, Processing Aids and Materials in Contact with Food*, updating the advice available on semicarbazide in packaged foods, adopted Oct. 1, 2003, EFSA/AFA/FCM/17-final, Brussels, 2003.
65. Stadler RH, Mottier P, Guy P, et al., Semicarbazide is a minor thermal decomposition product of azodicarbonamide used in the gaskets of certain food jars, *Analyst* 2004;129(3):276–281.
66. Commission Directive 2004/1/EC of 6 January 2004 amending Directive 2002/72/EC as regards the suspension of the use of azodicarbonamide as blowing agent, *Off. J. Eur. Commun*. 2004;L7:45–46.
67. *Code of Federal Regulations*, Requirements for specific standardized bakery products, Title 21, Parts 136.110, 137.105, and 137.200, 2004.
68. Pereira AS, Donato JA, De Nucci G, Implications of the use of semicarbazide as a metabolic target of nitrofurazone contamination in coated products, *Food Addit. Contam*. 2004;21(1):63–69.
69. Becalski A, Lau BPY, Lewis D, Seaman SW, Semicarbazide formation in azodicarbonamide-treated flour: A model study, *J. Agric. Food Chem*. 2004;52(18):5730–5734.
70. Mulder PPJ, Beumer B, Van Rhijn JA, The determination of biurea: A novel method to discriminate between nitrofurazone and azodicarbonamide use in food products, *Anal. Chim. Acta* 2007;586(1–2):366–373.
71. Cronly M, Behan P, Foley B, et al., Development and validation of a rapid method for the determination and confirmation of 10 nitroimidazoles in animal plasma using liquid chromatography tandem mass spectrometry, *J. Chromatogr. B* 2009;877:1494–1500.
72. Wang H, Wang Z, Liu S, et al., Quantification of nitroimidazoles residues in swine liver by liquid chromatography-mass spectrometry with atmospheric pressure chemical ionization, *Bull. Environ. Contam. Toxicol*. 2009;82(4):411–414.
73. Sørensen LK, Hansen H, Determination of metronidazole and hydroxymetronidazole in trout by a high-performance liquid chromatographic method, *Food Addit. Contam*. 2000;17(3):197–203.
74. *European Union Maximum Residue Limits Reports*. (available at `http://www.ema.europa.eu/ema/index.jsp?curl=pages/medicines/landing/vet_mrl_search.jsp&murl=menus/regulations/regulations.jsp&mid=WC0b01ac058006488e`; accessed 11/29/10).
75. JECFA-Joint FAO/WHO Expert Committee on Food Additives (available at `www.fao.org/ag/agn/jecfa-vetdrugs/search.html`; accessed 12/01/10).

76. Xia X, Li X, Shen J, Zhang S, et al., Determination of four nitroimidazoles in poultry and swine muscle and eggs by liquid chromatography/tandem mass spectrometry, *J. AOAC Int*. 2006;89(1):94–99.
77. Canada Gazette Part II, *Food and Drug Regulations (1277—Prohibition of Certain Veterinary, Drugs)*, 2003; 137(18):2306–2312 available at http://gazette.gc.ca/archives/p2/2003/2003-08-27/pdf/g2-13718.pdf; accessed 12/02/10).
78. Commission Directive No 96/23/EC of 29 April 1996 on measures to monitor certain substances and residues thereof in live animals and animal products and repealing Directives 85/358/EEC and 86/469/EEC and Decisions 89/187/EEC and 91/664/EEC, *Off. J. Eur. Commun*. 1996; L125:10–32.
79. Mottier P, Hurè I, Gremaud E, Guy P, Analysis of four 5-nitroimidazoles and their corresponding hydroxylated metabolites in egg, processed egg, and chicken meat by isotope dilution liquid chromatography tandem mass spectrometry, *J. Agric. Food Chem*. 2006;54:2018–2026.
80. Polzer J, Gowik P, Treatment of turkeys with nitroimidazoles—impact of the selection of target analytes and matrices on an effective residue control, *Anal. Chim. Acta* 2004;521;189–200.
81. Fraselle S, Derop V, Degroodt JM, Loco JV, Validation of a method for the detection and confirmation of nitroimidazoles and the corresponding hydroxyl metabolites in pig plasma by high performance liquid chromatography-tandem mass spectrometry, *Anal. Chim. Acta* 2007;586:383–393.
82. Zhou J, Shen J, Xue X, et al., Simultaneous determination of nitroimidazole residues in honey samples by high-performance liquid chromatography with ultraviolet detection, *J. AOAC Int*. 2007;90(3):872–878.
83. Wang JH, Determination of three nitroimidazoles residues in poultry meat by gas chromatography with nitrogen-phosphorus detection, *J. Chromatogr. A* 2001;918:435–438.
84. Polzer J, Gowik P, Validation of a method for the detection and confirmation of nitroimidazoles and corresponding hydroxyl metabolites in turkey and swine muscle by means of gas chromatography-negative ion chemical ionization mass spectrometry, *J. Chromatogr. B* 2001;761: 47–60.
85. Hurtud-Pessel D, Delèpine B, Laurentie M, Determination of four nitroimidazole residues in poultry meat by liquid chromatography-mass spectrometry, *J. Chromatogr. A* 2000;882:89–98.
86. Cannavan A, Kennedy G, Determination of dimetridazole in poultry tissues and eggs using liquid chromatography-thermospray mass spectrometry, *Analyst*. 1997;122: 963–966.
87. Sams M, Strutt P, Barnes K, et al., Determination of dimetridazole, ronidazole and their common metabolite in poultry muscle and eggs by high performance liquid chromatography with UV detection and confirmatory analysis by atmospheric pressure chemical ionisation mass spectrometry, *Analyst* 1998;123:2545–2549.
88. Xia X, Li X, Ding S, et al., Determination of 5-nitroimidazoles and corresponding hydroxyl metabolites in swine kidney by ultra-performance liquid chromatography coupled to electrospray tandem mass spectrometry, *Anal. Chim. Acta* 2009;637:79–86.
89. Cronly M, Behan P, Foley B, et al., Rapid confirmatory method for the determination of 11 nitroimidazoles in egg using liquid chromatography tandem mass spectrometry, *J. Chromatogr. A* 2009;1216:8101–8109.
90. Daeseleire E, De Ruyck H, Van Renterghem R, Rapid confirmatory assay for the simultaneous detection of ronidazole, metronidazole and dimetridazole in eggs using liquid chromatography-tandem mass spectrometry, *Analyst* 2000;125:1533–1535.
91. Mohamed R, Mottier P, Treguier L, et al., Use of molecularly imprinted solid-phase extraction sorbent for the determination of four 5-nitroimidazoles and three of their metabolites from egg-based samples before tandem LC-ESIMS/MS analysis, *J. Agric. Food Chem*. 2008;56:3500–3508.
92. Captitan-Valvey LF, Ariza A, Checa R, Nava, N, Determination of five nitroimidazoles in water by liquid chromatography-mass spectrometry, *J. Chromatogr. A* 2002;978(1–2):243–248.
93. Xia X, Li X, Shen J, et al., Determination of nitroimidazole residues in porcine urine by liquid chromatography/tandem mass spectrometry, *J. AOAC Int*. 2006;89(4):1116–1119.
94. Xia X, Li X, Zhang S, et al., Confirmation of four nitroimidazoles in porcine liver by liquid chromatography-tandem mass spectrometry, *Anal. Chim. Acta* 2007;586:394–398.
95. Sun H, Wang F, Ai L, et al., Validated method for determination of eight banned nitroimidazole residues in natural casings by LC/MS/MS with solid-phase extraction, *J. AOAC Int*. 2009;92(2):612–621.
96. Wang S, Zhang H, Wang L, et al., Analysis of sulphonamide residues in edible animal products: A review, *Food Addit. Contam*. 2006;23(4):362–384.
97. Potter R, Burns B, van de Riet J, et al., Simultaneous determination of 17 sulfonamides and the potentiators ormetoprim and trimethoprim in salmon muscle by liquid chromatography with tandem mass spectrometry detection, *J. AOAC Int*. 2007;90(1):343–348.
98. Tunnicliff E, Swingle K, Sulfonamide concentrations in milk and plasma from normal and mastitic ewes treated with sulfamethazine, *Am. J. Vet. Res*. 1965;26(113):920–927.
99. He J, Shen J, Suo X, et al., Development of a monoclonal antibody based ELISA for detection of sulfamethazine and N4-acetyl sulfamethazine in chicken breast tissue, *J. Food Sci*. 2005;70(1):113–117.
100. Sheridan R, Policasreo B, Thomas S, Rice D, Analysis and occurrence of 14 sulfonamide antibacterials and chloramphenicol in honey by solid phase extraction followed by LC-MS/MS analysis, *J. Agric. Food Chem*. 2008;56:3509–3516.
101. Chen L, Zhang X, Sun L, et al., Fast selective extraction of sulfonamides from honey based on magnetic molecularly imprinted polymer, *J. Agric. Food Chem*. 2009; 57:10073–10080

102. Forti A, Scortichini, G, Determination of ten sulfonamides in egg by liquid chromatography-tandem mass spectrometry, *Anal. Chim. Acta* 2009;637:214–219.

103. Health Canada, Veterinary Drugs Directorate, *Administrative Maximum Residue Limits (AMRLS) and Maximum Residue Limits (MRLS) Set by Canada*, 2010 (available at www.hc-sc.gc.ca/dhp-mps/vet/mrl-lmr/mrl-lmr_versus_new-nouveau-eng.php; accessed 12/01/10).

104. US Food and Drug Administration, *Code of Federal Regulations*, Tolerances for residues of new animal drugs in foods, Title 21, Part 556, 2010 (available at www.accessdata.fda.gov/scripts/cdrh/cfdocs/cfcfr/CFRSearch.cfm?CFRPart=556; accessed 12/01/10).

105. Council Regulation (ECC) No 508/1999. Amending Annexes I to IV to Council Regulation (EEC) No 2377/90 laying down a Community procedure for the establishment of MRLs of veterinary medical products in food stuffs of animal origin, *Off. J. Eur. Commun*. 1999;L60:9.3:16–52.

106. Shao B, Dong D, Wu Y, et al., Simultaneous determination of 17 sulfonamide residues in porcine meat, kidney and liver by solid phase extraction and liquid chromatography-tandem mass spectrometry, *Anal. Chim. Acta* 2005;546:174–181.

107. *Code of Federal Regulations*, Specific tolerances for residues of new animal drugs, Title 21, Food and Drugs, Chap. I, Subchap. E, Part 556, Subpart B, Sec. 556.680, Sulfanitran, 2009.

108. Gonzalez C, Usher K, Brooks A, Majors R, *Determination of Sulfonamides in Milk Using Solid-Phase Extraction and Liquid Chromatography-Tandem Mass Spectrometry*, Agilent Technologies Application Note 2009, pp. 5990–3713.

109. Cai Z, Zhang Y, Pan H, et al., Simultaneous determination of 24 sulfonamide residues in meat by ultra-performance liquid chromatography tandem mass spectrometry, *J. Chromatogr. A* 2008;1200:144–185.

110. Gentili A, Perret D, Marchese S, Liquid chromatography tandem mass spectrometry for performing confirmatory analysis of veterinary drugs in animal food products, *Trends Anal. Chem*. 2005;24(7); 704–733.

111. Heller D, Ngoh M, Donoghue, D, et al., Identification of incurred sulfonamide residues in eggs: Methods for confirmation by liquid chromatography-tandem mass spectrometry and quantitation by liquid chromatography with ultraviolet detection, *J. Chromatgr. B* 2002;774:39–52.

112. García-Galán M, Díaz-Cruz M, Barceló D, Identification and determination of metabolites and degradation products of sulfonamide antibiotics, *Trends Anal. Chem*. 2008;27(11):1008–1022.

113. Kishida K, Nishinari K, Furusawa N, Liquid chromatographic determination of sulfamonmethoxine, sulfadimethoxine and their N^4-acetyl metabolites in chicken plasma, *Chromatographia* 2005;61:81–84.

114. Shaikh B, Rummel N, Donoghue D, Determination of sulfamethazine and its major metabolites in egg albumin and egg yolk by high performance liquid chromatography, *J. Liq. Chromatogr. Rel. Technol*. 1999;22(17):2651–2662.

115. Kishida, K, Restricted access media liquid chromatography for the determination of sulfamonomethoxine, sulfadimethoxine and their N^4-acetyl metabolites in eggs, *Food Chem*. 2007;101:281–285.

116. Furusawa N, Simultaneous high performance liquid chromatographic determination of sulfamonomethoxine and its hydroxy/N^4-acetyl metabolites following centrifugal ultra-filtration in animal blood plasma, *Chromatographia* 2000;52:653–656.

117. Grant G, Frison S, Sporns P, A sensitive method for the detection of sulfamethazine and N^4-acetylsulfamethazine residues in environmental samples using solid phase immunoextraction coupled with MALDI-TOF MS, *J. Agric. Food Chem*. 2003;51:5367–5375.

118. Kishida K, Furusawa N, Application of shielded column liquid chromatography for determination of sulfamonomethaoxine, sulfadimethoxine and their N^4-acetyl metabolites in milk, *J. Chromatogr. A* 2004;1028:175–177.

119. Balizs G, Benesch-Girke L, Börner S, Hewitt S, Comparison of the determination of four sulphonamides and their N^4-acetyl metabolites in swine muscle tissue using liquid chromatography with ultraviolet and mass spectral detection, *J. Chromatogr. B* 1994;661:75–84.

120. Furusawa N, Organic solvents free technique for determining sulfadimethoxine and its metabolites in chicken meat, *J. Chromatogr. A* 2007;1172:92–95.

121. AOAC, Sulfonamide residues in raw bovine milk- liquid chromatographic method, 993.32, in *Official Methods of Analysis*, 18th ed. (revised), AOAC International, Gaithersburg, MD (available at http://www.eoma.aoac.org/gateway/readFile.asp?id=993_32.pdf; accessed 3/24/10).

122. AOAC, Sulfamethazine residues raw bovine milk-liquid chromatographic method, 992.21, in *Official Methods of Analysis*, 18th ed. (revised), AOAC International, Gaithersburg, MD (available at http://www.eoma.aoac.org/gateway/readFile.asp?id=992_21.pdf; accessed 3/24/10).

123. Hoff R, Barreto F, Kist T, Use of capillary electrophoresis with laser-induced fluorescence detection to screen and liquid chromatography-tandem mass spectrometry to confirm sulphonamide residues: Validation according to European Union 2002/657/EC, *J. Chromatogr. A* 2009;1216:8254–8261.

124. Font G, Juan-Garcí A, Picó Y, Pressurized liquid extraction combined with capillary electrophoresis-mass spectrometry as an improved methodology for the determination of sulfonamide residues in meat, *J. Chromatogr. A* 2007;1159:233–241.

125. Soto-Chonchilla J, Garcia-Campaña A, Gámiz-Gracia L, Analytical methods for multiresidue determination of sulfonamides and trimethoprim in meat and ground water samples by CE-MS and CE-MS/MS, *Electrophoresis* 2007;28:4164–4172.

126. Bogialli S, Curini R, Di Corcia A, et al., Rapid confirmatory assay for determining 12 sulfonamide antimicrobials in milk and eggs by matrix solid phase dispersion and liquid

chromatography-mass spectrometry, *J. Agric. Food Chem.* 2003;51:4225–4232.

127. AOAC, Sulfamethazine residues in swine tissues-gas chromatographic mass spectrometric method, 982.40, in *Official Methods of Analysis*, 18th ed. (revised), AOAC International, Gaithersburg, MD (available at `http://www.eoma.aoac.org/gateway/readFile.asp?id=982_40.pdf`; accessed 3/24/10).
128. Tarbin J, Clarke P, Shearer G, Screening of sulfonamides in egg using gas chromatography-mass selective detection and liquid chromatography-mass spectrometry, *J. Chromatogr. A* 1999;729:127–138.
129. Hoff R, Kist T, Analysis of sulfonamides by capillary electrphoresis, *J. Sep. Sci*. 2009;32:854–866.
130. Kishida K, Quantitation and confirmation of six sulphonamides in meat by liquid chromatography-mass spectrometry with photodiode array detection, *Food Control* 2007;18:301–305.
131. Fujita M, Taguchi S, Obana H, Determination of sulfonamides in livestock products and seafoods by liquid chromatography-tandem mass spectrometry using glass bead homogenization, *J. Food Hyg. Soc. Japan*. 2008;49(6):411–415.
132. van Rhijn J, Lasaroms J, Berendsen B, Brinkman U, Liquid chromatographic—tandem mass spectrometric determination of selected sulphonamides in milk, *J. Chromatogr. A* 2002;960:121–133.
133. Tamošiūnas V, Padarauskas A, Comparison of LC and UPLC coupled to MS-MS for the determination of sulfonamides in egg and honey, *Chromatographia* 2008:67: 783–788.
134. Mohamed R, Hammel Y, LeBreton M, et al., Evaluation of atmospheric pressure ionization interferences for the quantitative measurement of sulfonamides in honey using isotope dilution liquid chromatography coupled with tandem mass spectrometry techniques, *J. Chromatogr. A* 2007;1160:194–205.
135. Berardi G, Bogialli S, Curini R, et al., Evaluation of a method for assaying sulfonamide antimicrobial residues in cheese: Hot water extraction and liquid chromatography-tandem mass spectrometry, *J. Agric. Food Chem*. 2006;54:4537–4543.
136. Thompson T, Noot D, Determination of sulfonamides in honey by liquid chromatography-tandem mass spectrometry, *Anal. Chim. Acta* 2005;551:168–176.
137. Pang G, Cao Y, Zhang J, et al., Simultaneous determination of 16 sulfonamides in honey by liquid chromatography-tandem mass spectrometry, *J. AOAC Int*. 2005;88(5):1304–1311.
138. Gentili A, Perret D, Marchese S, et al., Accelerated solvent extraction and confirmatory analysis of sulfonamide residues in raw meat and infant foods by liquid chromatography electrospray tandem mass spectrometry, *J. Agric. Food Chem*. 2004;52:4614–4624.
139. Clark S, Turnipseed S, Madson M, et al., Confirmation of sulfamethazine, sulfathiazole and sulfadimethoxine residues in condensed milk and soft-cheese products by liquid chromatography-tandem mass spectrometry, *J. AOAC Int*. 2005;88(3):736–743.
140. Verzegnassi L, Savoy-Perroud M, Stadler R, Application of liquid chromatography-electrospray ionization tandem mass spectrometry to the detection of 10 sulfonamides in honey, *J. Chromatogr. A* 2002;997:77–87
141. Segura P, Tremblay P, Picard P, et al., High throughput quantitation of seven sulfonamide residues in dairy milk using laser diode thermal desorption-negative mode atmospheric pressure chemical ionization tandem mass spectrometry, *J. Agric. Food Chem*. 2010;58:1442–1446.
142. Ito Y, Oka H, Ikai Y, et al., Application of ion-exchange cleanup in food analysis V. Simultaneous determination of sulphonamide antibacterials in animal liver and kidney using high performance liquid chromatography with ultra violet and mass spectrometric detection, *J. Chromatogr. A* 2000;898:95–102.
143. Krivohlavek A, Šmit Z, Baštinac M, et al., The determination of sulfonamides in honey by high performance liquid chromatography-mass spectrometry, *J. Sep. Sci*. 2005;28:1434–1439.
144. Mengelers M, Oorsprong M, Kuiper H, et al., Determination of sulfadimethoxine, sulfamethoxazole, trimethoprim and their main metabolites in porcine plasma by column switching HPLC, *J. Pharm. Biomed. Anal*. 1989; 7(12):1765–1776.
145. Kaufmann A, Kaenzig A, Contamination of honey by the herbicide asulam and its antibacterial active metabolite sulfanilamide, *Food Addit. Contam*. 2004;21(6):564–571.
146. Bogialli S, Curini R, Di Corcia A, et al., A rapid confirmatory method for analysing tetracycline antibiotics in bovine, swine and poultry muscle tissues: Matrix solid phase dispersion with heated water as extractant followed by liquid chromatography tandem mass spectrometry, *J. Agric. Food Chem*. 2006;54:1564–1570.
147. Fritz J, Zuo Y, Simultaneous determination of tetracycline, oxytetracycline, and 4-epitetracycline in milk by high performance liquid chromatography, *Food Chem*. 2007;107: 1297–1301.
148. Huang T, Du W, Marshall M, Wei C, Determination of oxytetracycline in raw and cooked channel catfish by capillary electrophoresis, *J. Agric. Food Chem*. 1997;45: 2602–2605.
149. Spisso B, de Araújo M, Monteiro M, et al., A liquid chromatography-tandem mass spectrometry confirmatory assay for the simultaneous determination of several tetracyclines in milk considering keto-enol tautomerism and epimerization phenomena, *Anal. Chim. Acta* 2009;656:72–84.
150. Carrasco-Pancorbo A, Casado-Terrones S, Segura-Carretero A, Fernández-Gutiérrez A, reversed phase high performance liquid chromatography coupled to ultraviolet and electrospray time of flight mass spectrometry on-line detection for the separation of eight tetracyclines in honey samples, *J. Chromatogr. A* 2008;1195: 107–116.

151. Fang Y, Zhai H, Zho Y, Determination of Multi-Residue Tetracyclines and Their Metabolites in Milk by High Performance Liquid Chromatography-Tandem Mass Spectrometry, Agilent Technologies Application Note, Wilmington, DE, 2009.

152. Nikolaidou K, Samanidou V, Papadoyannis I, Development and validation of an HPLC method for the determination of seven tetracycline antibiotics residues in chicken muscle and egg yolk according to 2002/657/EC, *J. Liq. Chromatogr. Rel. Technol*. 2008:31:2141–2158.

153. Blasco C, Corcia A, Picó Y, Determination of tetracyclines in multi-specie animal tissues by pressurized liquid extraction and liquid chromatography-tandem mass spectrometry, *Food Chem*. 2009;116:1005–1012.

154. Cherlet M, Schelkens M, Croubels S, De Backer P, Quantitative multi-residue analysis of tetracyclines and their 4-epimers in pig tissues by high-performance liquid chromatography combined with positive ion electrospray ionization mass spectrometry, *Anal. Chim. Acta* 2003;492: 199–213.

155. Cherlet M, De Baere S, De Backer P, Quantitative analysis of oxytetracycline and its 4-epimer in calf tissues by liquid chromatography combined with positive electrospray ionization mass spectrometry, *Analyst* 2003;128:871–878.

156. Ishii R, Horie M, Murayama M, Maitani T, Analysis of tetracyclines in honey and royal jelly by LC-MS/MS, *J. Food Hyg. Soc. Japan* 2006;47(6):277–283.

157. Castellari M, Gratacós-Cubarsí M, Garćia-Regueiro J, Detection of tetracycline and oxytetracycline residues in pig and calf hair by ultra-high performance liquid chromatography tandem mass spectrometry, *J. Chromatogr. A* 2009;1216:8096–8100.

158. Alfredsson G, Branzell C. Granelli K, Lundström Ä, Simple and rapid screening and confirmation of tetracyclines in honey and egg by a dipstick and LC-MS/MS, *Anal. Chim. Acta* 2005;529:47–51.

159. Miranda J, Rodríguez J, Galán-Vidal C, Simultaneous determination of tetracyclines in poultry muscle by capillary zone electrophoresis, *J. Chromatogr. A* 2009;1216:3366–3371.

160. Moats W, Determination of tetracycline antibiotics in beef and pork tissues using ion-paired liquid chromatography, *J. Agric. Food Chem*. 2000;48:2244–2248.

161. AOAC, Chlortetracycline, oxytetracycline and tetracycline in edible animal tissues, 995.09, in *Official Methods of Analysis*, 18th ed. (revised), AOAC International, Gaithersburg, MD (available at `http://www.eoma.aoac.org/gateway/readFile.asp?id=995_09.pdf`; accessed 3/24/10).

162. Spisso B, de Oliveira e Jesus A, et al., Validation of a high performance liquid chromatographic method with fluorescence detection for the simultaneous determination of tetracycline residues in bovine milk, *Anal. Chim. Acta* 2007;581:108–117.

163. Pena A, Lino C, Silveira M, Determination of tetracycline antibiotics in salmon muscle by liquid chromatography using post column derivatization with fluorescence detection, *J. AOAC Int*. 2003;86(5):925–929.

164. Pena A, Lino C, Alonso R, Barceló D, Determination of tetracycline antibiotic residues in edible tissues by liquid chromatography with spectrofluorometric detection and confirmation by mass spectrometry, *J. Agric. Food Chem*. 2007;55:4973–4979.

165. Anderson W, Roybal J, Gonzales S, et al., Determination of tetracycline residues in shrimp and whole milk using liquid chromatography with ultraviolet detection and residue confirmation by mass spectrometry, *Anal. Chim. Acta* 2005;529- 145–150.

166. AOAC, Multiple tetracycline residues in milk, 995.04, in *Official Methods of Analysis*, 18th ed. (revised), AOAC International, Gaithersburg, MD (available at `http://www.eoma.aoac.org/gateway/readFile.asp?id=995_04.pdf`; accessed 3/24/10).

167. Khong S-P, Hammel Y-A, Guy P, Analysis of tetracyclines in honey by high-performance liquid chromatography-tandem mass spectrometry, *Rapid Commun. Mass Spectrom*. 2005;19:493–502.

168. Cristofani E, Antonini C, Tovo G, et al., A confirmatory method for the determination of tetracyclines in muscle using high performance liquid chromatography with diode-array detection, *Anal. Chim. Acta* 2009;637:40–46.

169. Fletouris D, Papapanagiotou E, A new liquid chromatographic method for routine determination of oxytetracycline marker residue in the edible tissues of farm animals, *Anal. Bioanal. Chem*. 2008;391:1189–1198.

170. Sczesny S, Nau H, Hamscher G, Residue analysis of tetracyclines and their metabolites in eggs and in the environment by HPLC coupled with microbiological assay and tandem mass spectrometry, *J. Agric. Food Chem*. 2003;51:697–703.

171. Gajda A, Posyniak A. Pietruszka K, Analytical procedure for the determination of doxycycline residues in animal tissues by liquid chromatography, *Bull. Vet. Inst. Pulawy* 2008;52:417–420.

172. Cinquina A, Longo F, Anastasi G, et al., Validation of a high performance liquid chromatography method for the determination of oxytetracycline, tetracycline, chlortetracycline and doxycycline in bovine milk and muscle, *J. Chromatogr. A* 2003;987:227–233.

173. Xu J, Ding T, Wu B, et al., Analysis if tetracycline residues in royal jelly by liquid chromatography-tandem mass spectrometry, *J. Chromatogr. B* 2008;868:42–48.

174. Nikolaidou K, Samanidou V, Papadoyannis I, Development and validation of an HPLC confirmatory method for the determination of seven tetracycline antibiotics residues in bovine and porcine muscle tissues according to 2002/657/EC, *J. Liq. Chromatogr. Rel. Technol*. 2008;31:3032–3054.

175. Jing T, Gao X, Wang P, et al., Determination of trace tetracycline antibiotics in foodstuffs by liquid chromatography-tandem mass spectrometry coupled with selective molecular-imprinted solid-phase extraction, *Anal. Bioanal. Chem*. 2009;393:2009–2018.

176. Nakazawa H, Ino S, Kato K, et al., Simultaneous determination of residual tetracyclines in foods by high performance

liquid chromatography with atmospheric pressure chemical ionization tandem mass spectrometry, *J. Chromatogr. B* 1999;732:55–64.

177. Viñas P, Balsalobre N, López-Erroz C, Hernández-Córdoba M, Liquid chromatography with ultraviolet absorbance detection for the analysis of tetracycline residues in honey, *J. Chromatogr. A* 2004;1022:125–129.
178. Kanda M, Kusano T, Osanai T, et al., Rapid determination of residues of 4 tetracyclines in meat by a microbiological screening, HPLC and LC-MS/MS, *J. Food Hyg. Soc Japan* 2008;43(1):37–44.
179. Oka H, Ito Y. Ikai Y, et al., Mass spectrometric analysis of tetracycline antibiotics in foods, *J. Chromatogr. A.* 1998;812:309–319.
180. Oka H, Ito Y, Matsumoto H, Chromatographic analysis of tetracycline antibiotics in foods, *J Chromatogr A* 2000;882:109–133.
181. Gentili A, Perret D, Marchese S, Liquid chromatography-tandem mass spectrometry for performing confirmatory analysis of veterinary drugs in animal-food products, *Trends Anal. Chem.* 2005;24(7):704–733.
182. Stolker A, Brinkman U, Analytical strategies for residue analysis of veterinary drugs and growth-promoting agents in food-producing animals-a review, *J. Chromatogr. A* 2005;1067:15–53.
183. *Survey on Use of Veterinary Medicinal Products in Third Countries* [available (login required) at `https://secure.fera.defra.gov.uk/vetdrugscan/index.cfm`; accessed 4/15/10)].
184. Isoherranen N, Soback S, Chromatographic methods for analysis of aminoglycoside antibiotics, *J. AOAC Int.* 1999;82(5):1017–1045.
185. Heller DN, Clark SB, Righter HF, Confirmation of gentamicin and neomycin in milk by weak cation-exchange extraction and electrospray ionisation/ion trap tandem mass spectrometry, *J. Mass Spectrom.* 2000;35(1):39–40.
186. Ishii R, Horie M, Chan W, MacNeil J, Multi-residue quantitation of aminoglycoside antibiotics in kidney and meat by liquid chromatography with tandem mass spectrometry, *Food Addit. Contamin.* (Part A) 2008;25(12):1509–1519.
187. Kim B-H, Lee SC, Lee HJ, Ok JH, reversed phase liquid chromatographic method for the analysis of aminoglycoside antibiotics using pre-column derivatization with phenylisocyanate, *Biomed. Chromatogr.* 2003;17:396–403.
188. Turnipseed SB, Clark SB, Karbiwnyk CM, et al., Analysis of aminoglycoside residues in bovine milk by liquid chromatography electrospray ion trap mass spectrometry after derivatization with phenyl isocyanate, *J. Chromatogr. B* 2009;877:1487–1493.
189. Anadon A, Martinez MA, Martinez M, et al., Plasma and tissue depletion of florfenicol and florfenicol amine in chickens, *J. Agric. Food Chem.* 2008;56(22):11049–11056.
190. *EMEA Florfenicol* [extension to all food producing animals] *Summary Report* (6), 2002, EMEA/MRL/822/02-final (available at `http://www.ema.europa.eu/docs/en_GB/document_library/Maximum_Residue_Limits_-_Report/2009/11/WC500014282.pdf`; accessed 11/29/10).
191. Nagata T, Oka H, Detection of residual chloramphenicol, florfenicol and thiamphenicol in yellowtail fish muscles by capillary gas chromatography—mass spectrometry, *J. Agric. Food Chem.* 1996;44:1280–1284.
192. Zhang S, Liu Z, Guo X, et al., Simultaneous determination of chloramphenicol, thiamphenicol, florfenicol and florfenicol amine in chicken muscle by liquid chromatography-tandem mass spectrometry, *J. Chromatogr. B* 2008;875:399–404.
193. Peng T, Li S, Chu X, et al., Simultaneous determination of residues of chloramphenicol, thiamphenicol and florfenicol in shrimp by high performance liquid chromatography-tandem mass spectrometry, *Chin. J. Anal. Chem.* 2005;33(4):463–466.
194. Rostel B, Zmudzki J, MacNeil JD, Lincomycin, in *Residues of Some Veterinary Drugs in Foods*, FAO Food and Nutrition Paper 41/13, 2000, pp. 59–74 (available at `ftp://ftp.fao.org/ag/agn/jecfa/vetdrug/41-13-lincomycin.pdf`; accessed 11/26/10).
195. *EMEA Pirlimycin Summary Report* (1), 1998, EMEA/MRL/460/98-final (available at `http://www.ema.europa.eu/docs/en_GB/document_library/Maximum_Residue_Limits_-_Report/2009/11/WC500015685.pdf`; accessed 11/29/10).
196. Hornish RE, Roof RD, Wiest JR, Pirlimycin residue in bovine liver—a case of reverse metabolism, *Analyst* 1998;123:2463–2467.
197. Friedlander, LG, Moulin G, Pirlimycin, in *Residues of Some Veterinary Drugs in Foods*, FAO Food and Nutrition Paper 41/16, Food and Agriculture Organization of the United Nations, Rome, 2004, pp. 55–73.
198. Martos PA, Lahotay SJ, Shurner B, Ultratrace analysis of nine macrolides, including tulathromycin A (Draxxin), in edible animal tissues with minicolumn liquid chromatography tandem mass spectrometry, *J. Agric. Food Chem.* 2008;56(19):8844–8850.
199. Lopez MI, Pettis JS, Smith IB, Chu PS, Multiclass determination and confirmation of antibiotic residues in honey using LC-MS/MS, *J. Agric. Food Chem.* 2008;56(5):1553–1559.
200. Tang HPO, Ho C, Lai SSL, High throughput screening for multi-class veterinary drug residues in animal muscle using liquid chromatography/tandem mass spectrometry with on-line solid-phase extraction, *Rapid Commun. Mass Spectrom.* 2006;20(17):2565–2572.
201. Adams S, Fussell RJ, Dickinson M, et al., Study of the depletion of lincomycin residues in honey extracted from treated honeybee (Apis mellifera L) colonies and the effect of the shook swarm procedure, *Anal. Chim. Acta* 2009;637(1–2):315–320.
202. Sin DWM, Wong YC, Ip ACB, Quantitative analysis of lincomycin in animal tissues and bovine milk by liquid chromatography electrospray ionisation tandem mass spectrometry, *J. Pharm. Biomed. Anal.* 2004;34(3):651–659.
203. Bogiall S, Di Corcia A, Lagana A, et al., A simple and rapid confirmatory assay for analysing antibiotic residues of the macrolide class and lincomycin in bovine milk

and yoghurt: Hot water extraction followed by liquid chromatography/tandem mass spectrometry, *Rapid Commun. Mass Spectrom*. 2007;21(2):237–246.

204. Benetti C, Piro R, Binato G, et al., Simultaneous determination of lincomycin and five macrolide antibiotic residues in honey by liquid chromatography coupled to electrospray ionisation mass spectrometry (LC-MS/MS), *Food Addit. Contam*. 2006;23(11):1009–1108.
205. Thompson TS, Noot DK, Calvert J, Pernal SF, Determination of lincomycin and tylosin residues in honey using solid-phase extraction and liquid chromatography-atmospheric pressure chemical ionisation mass spectrometry, *J. Chromatogr. A* 2003;1020(2):241–250.
206. Hornish RE, Cazers AR, Chester ST Jr, Rool RD, Identification and determination of pirlimycin residue in bovine milk and liver by high-performance liquid chromatography-thermospray mass spectrometry, *J. Chromatogr. B* 1995; 674:219–235.
207. Idowu OR, Peggins JO, Cullison R, von Bredow J, Comparative pharmacokinetics of enrofloxacin and ciprofloxacin in lactating dairy cows and beef steers following intravenous administration of enrofloxacin, *Res. Vet. Sci*. 2010;89(2):230–235.
208. Chang CS, Wang WH, Tsai CE, Simultaneous determination of 18 quinolone residues in marine and livestock products by liquid chromatography/tandem mass spectrometry, *J. Food Drug Anal*. 2010;18(2):87–97.
209. Pearce JN, Burns BG, van de Riet JM, et al., Determination of fluoroquinolones in aquaculture products by ultra-performance liquid chromatography-tandem mass spectrometry (UPLC-MS/MS), *Food Addit. Contam*. (Part A) 2009;26(10):39–46.
210. Tang QF, Yang TT, Tan XM, Luo JB, Simultaneous determination of fluoroquinolone antibiotic residues in milk sample by solid-phase extraction-liquid chromatography-tandem mass spectrometry, *J. Agric. Food Chem*. 2009; 57(11):4535–4539.
211. McMullen SE, Schenck FJ, Vega VA, Rapid method for the determination and confirmation of fluoroquinolones residues in catfish using liquid chromatography/fluorescence detection and liquid chromatography-tandem mass spectrometry, *J. AOAC Int*. 2009;92(4):1233–1240.
212. Hernando MD, Mezcua M, Suarez-Barcena JM, Fernandez-Alba AR, Liquid chromatography with time-of-flight mass spectrometry for simultaneous determination of chemotherapeutant residues in salmon, *Anal. Chim. Acta* 2006;562(2):176–184.
213. Lara FJ, Garcia-Campana AM, Ales-Barrero F, Bosque-Sendra JM, In-line solid-phase extraction preconcentration in capillary electrophoresis tandem mass spectrometry for the multiresidue detection of quinolones in meat by pressurized liquid extraction, *Eletrophoresis I* 2008;29(10): 2117–2125.
214. *EMEA Enrofloxacin* [extension to sheep, rabbits, and lactating cows] *Summary Report* (3), 1998; EMEA/MRL/389/98-final (available at `http://www.ema.europa.eu/docs/en_GB/document_library/Maximum_Residue_Limits_-_Report/2009/11/WC500014144.pdf`; accessed 11/29/10).
215. *EMEA Enrofloxacin* [modification for bovine, porcine and poultry] *Summary Report* (2), 1998, EMEA/MRL/388/98-final (available at `http://www.ema.europa.eu/docs/en_GB/document_library/Maximum_Residue_Limits_-_Report/2009/11/WC500014142.pdf`; accessed 11/29/10).
216. Poapolathep A, Jermnak U, Chareonsan A et al., Disposition and residue depletion of enrofloxacin and its metabolite ciprofloxacin in muscle tissue of giant freshwater prawns (Macrobrachium rosenbergii), *J. Vet. Pharmacol. Ther*. 2009;32(3):229–234.
217. Ovando HG, Gorla N, Luders C, et al., Comparative pharmacokinetics of enrofloxacin and ciprofloxacin in chickens, *J. Vet. Pharmacol Ther*. 1999;22(3):209–212.
218. Anadon A, Martinez-Larranaga MR, Diaz MJ, et al., Pharmacokinetics and residues of enrofloxacin in chickens, *Am. J. Vet. Res*. 1995;56(4):501–506.
219. Codex Alimentarius Commission, REP11/RVDF, *Report of the 19th Session of the Codex Committee on Residues of Veterinary Drugs in Foods*, Burlington, United States, Aug. 30–Sept. 3, 2010 (available at `http://www.codexalimentarius.net/web/archives.jsp?year=11`; accessed 11/26/10).
220. Tarbin JA, unpublished data.

8

METHOD DEVELOPMENT AND METHOD VALIDATION

JACK F. KAY AND JAMES D. MACNEIL

8.1 INTRODUCTION

The origins of the concept of *analytical method validation*, as we use the term today, can be traced to the early days of AOAC International, when the association was initially formed as the Convention of Agricultural Chemists, then later as the Association of Agricultural Chemists.[1] The initial aim of the Association was to standardize analytical methods so that the same method could be used successfully in different laboratories to achieve comparable results. Statistical and experimental requirements for collaborative studies have evolved over the years, as have concepts of method validation. In the current environment, it is expected in most jurisdictions that methods used in regulatory analysis will be conducted in laboratories that have been accredited to a recognized standard, such as ISO/IEC-17025 (or equivalent), which requires that laboratories demonstrate the "fitness of purpose" of test methods in routine use under the scope of the accreditation.[2] To be fit for purpose, the laboratory should have sufficient data to support a claim that in their hands the analytical method can produce robust and defensible data and that the results can be used to take the action for which the analysis was undertaken, such as controlling food production, regulatory enforcement, and associated legal proceedings.

The Codex Alimentarius Commission (CAC) issued a guideline in 1997 for laboratories involved in the import/export testing of foods recommending that such laboratories meet the following four criteria:[3]

- Accreditation under a recognized system for laboratory accreditation[2]
- Participation in appropriate proficiency testing (PT) programs[4]
- Demonstration of an effective quality assurance (QA) system[5]
- Use of methods validated according to criteria specified by the Codex Alimentarius Commission

8.2 SOURCES OF GUIDANCE ON METHOD VALIDATION

There are many sources of information on method validation, including papers in the scientific literature; guidance issued by scientific bodies such as the International Union of Pure and Applied Chemistry (IUPAC) and Eurachem; guidance from international organizations involved in the establishment of harmonized standards, such as the CAC; as well as guidance from national and regional regulatory authorities, such as the US Food and Drug Administration (USFDA) and the European Commission (EC). In some cases, the requirements for validation may differ, so it is then the responsibility of the individual scientist and their organization to assess the purpose for which the method will be used and to then choose the appropriate validation criteria, based on the guidance documents that are available from authoritative bodies. For example, a laboratory developing and validating a method of analysis to support a new animal drug application to the USFDA should obviously ascertain that their validation work meets the standards required by that organization, while a method being developed for use in a laboratory involved in the import/export testing of foods should ensure that the validation work is in accordance with the guidelines published by the CAC and, if the trade involves the European Union (EU) or the laboratory is in a member state of the EU, then the validation design should also incorporate requirements laid down by the EC. This is part of establishing the "fitness for purpose."

Chemical Analysis of Antibiotic Residues in Food, First Edition. Edited by Jian Wang, James D. MacNeil, and Jack F. Kay.

8.2.1 Organizations that Are Sources of Guidance on Method Validation

Numerous organizations are sources of information pertaining to the validation of analytical methods and may at times seem to present different requirements, so a choice of an appropriate source of guidance to follow is important. Basically, the sources of guidance may be grouped into those from independent scientific sources, such as IUPAC, Eurachem, and AOAC International, and those elaborated by national/regional regulatory authorities and international organizations with an interest in the establishment of international regulatory standards and practices. Such international organizations include the CAC and the International Cooperation on Harmonisation of Technical Requirements for Registration of Veterinary Medicinal Products (VICH), while national/regional organizations providing regulatory guidance on methods include the EC and the USFDA. While the guidance provided by independent scientific organizations is generally applicable to a wide range of analytical methods and procedures, guidance issued by regulatory authorities is more directed to specific applications of methods or even specific types of analytes and it may be mandatory to follow the guidance issues.

8.2.1.1 International Union of Pure and Applied Chemistry (IUPAC)

IUPAC is an independent, non-governmental scientific organization with a network of national adhering organizations that represent the interests of chemists in their country. A "national adhering organization" may be a national society (such as the Royal Society of Chemistry in the UK), a national academy of science (e.g., the Australian Academy of Science, US National Academy of Science) or "any other institution or association of institutions representative of national chemical interests" (e.g., Hungarian National Committee for IUPAC). While perhaps best known for its role in chemical nomenclature, IUPAC's role is the development of scientific standards and guidance, including terminology[6–8] and guidance on important issues in chemical analysis such as the conduct of collaborative studies,[9] laboratory quality assurance,[5] proficiency testing,[4] recovery correction,[10] and validation of methods within a single laboratory.[11] Information on IUPAC projects and access to the journal *Pure and Applied Chemistry* are available via the IUPAC website (`http://www.iupac.org/`).

8.2.1.2 AOAC International

AOAC International was formed in the United States of America in 1880 at a meeting attended by government and university scientists and officials concerned with the analysis of fertilizers.[1,12] Initially, the association was formed as a section of the chemistry subdivision of the American Academy of Sciences, adopting a committee system to deal with specific analytical issues and limiting voting privileges to practicing analytical chemists. The first meeting of the Association of Official Agricultural Chemists (AOAC) took place in 1884, with support from the US Department of Agriculture. The connection with USA government agencies continued with the creation of the Food and Drug Administration and the transfer of staff involved with the management of AOAC to the new organization. Although the original structure of AOAC limited membership to North America, cooperation with international regulatory groups began at early stages of the organization, leading to the establishment of a committee on international cooperation in 1968 and more formalized international cooperation, including cooperation with other international scientific organizations. Reflecting the broadening interests of the work of the association, which had expanded to include food and drugs, the Association changed its name to the Association of Official Analytical Chemists in 1965, still maintaining the acronym AOAC. In 1979, AOAC became an independent scientific association, with membership open to interested scientists in any country. Reflecting the expansion of membership to include greater international membership and other disciplines, such as microbiology, important in regulatory analysis, the association changed its name again in 1991.

Over the years, AOAC International (and its previous incarnations) has cooperated with joint projects with IUPAC and the International Standards Organization (ISO) to develop harmonized protocols related to the performance of collaborative studies,[9] proficiency testing,[4] quality assurance in analytical laboratories,[5] and recovery correction.[10] However, the primary focus of the association has been on the validation of analytical methods used in regulatory laboratories and laboratories working in the regulated industry sector. The *Journal of AOAC International* is a source of information on methods developed in regulatory laboratories, as well as information on analytical issues, while the committees within the organization are a source of information on current activities within the association and the analytical community on work on method validation. As with IUPAC and ISO, AOAC International (`http://www.aoac.org/`) is a source of the independent scientific advice and standards referenced by national and international regulatory organizations. Methods published in the *Official Methods of Analysis*[13] by AOAC International have been validated through collaborative study and are suitable for use as reference methods when validating a method within a single laboratory.

8.2.1.3 International Standards Organization (ISO)

The International Standards Organization (`http://www.iso.org/`) is a broad-based network of national standards institutes with membership from over 160 countries. It is a non-governmental organization, although many of the members are government organizations within their own countries, and develops consensus-based standards covering a broad field, including analytical chemistry. ISO cooperated with IUPAC and AOAC International in the development of various harmonized protocols, including those dealing with collaborative studies,[9] laboratory quality assurance,[5] proficiency testing,[4] and recovery correction in analytical measurement.[10] Standards developed by ISO are the primary international standards for laboratory accreditation,[2] provide guidance on analytical issues such as the use of control charts,[14] and are a primary source for definitions of analytical terminology.[15]

8.2.1.4 Eurachem

Eurachem was established in 1989 as a network of European organizations interested in analytical chemistry, with the stated objectives of supporting international traceability of measurement and the promotion of good-quality practices. Membership includes representatives of member states of the EU and the European Free Trade Association (EFTA), the EC and is also open to European countries recognized as accession states by the EU and EFTA. Other European countries and international organizations that share the same interests may be granted observer status. Work within Eurachem includes efforts to promote collaboration, organization of technical workshops, and the production of guides to facilitate the quality of analytical work, including such topics as fitness for purpose,[16] measurement uncertainty (MU),[17] and quality of measurement.[18] For those new to the world of laboratory accreditation and method validation, these guidance documents, which may be downloaded from the Eurachem website (`http://www.eurachem.org/`) can be excellent information sources.

8.2.1.5 VICH

The International Cooperation on Harmonisation of Technical Requirements for Registration of Veterinary Medicinal Products (VICH) was formed in 1996 as a trilateral organization by the United States of America, the European Union, and Japan to develop a harmonized approach to the technical requirements for registration of veterinary drugs in these respective jurisdictions. Currently, Australia, Canada, and New Zealand have observer status. VICH has issued a number of harmonized guidelines, including two dealing with the validation of analytical methods for residues of veterinary drugs.[19,20] These guidelines are intended to apply to the analytical methods that are provided by drug sponsors as part of a new analytical drug application to ensure that a suitable method of analysis will be available for regulatory use once an approved use has been granted. Information on the activities of VICH and access to VICH guidelines may be obtained from the VICH website (`http://www.vichsec.org/`).

8.2.1.6 Codex Alimentarius Commission (CAC)

The Codex Alimentarius Commission is an international organization open to all member states of the United Nations. It was established in 1963 by two bodies of the United Nations, the Food and Agriculture Organization (FAO) and the World Health Organization (WHO), with a mandate under the Joint FAO/WHO Food Standards Programme that includes the development of food standards [including maximum residue limits for veterinary drug residues (MRLVDs) in foods], guidelines (including those dealing with laboratory practices and methods of analysis), and other related texts such as codes of practice for industry. The primary objectives stated for the CAC are to protect the health of consumers and to ensure fair trade practices, which includes providing guidance on technical matters, such as regulatory programmes in member states.

For example, a guideline approved in 2009 by the CAC defines a validated method as an "accepted test method for which validation studies have been completed to determine the accuracy and reliability of this method for a specific purpose,"[21] referencing a definition provided by the Interagency Coordinating Committee on the Validation of Alternative Methods (ICCVAM).[22] This definition is similar in content to definitions used by Eurachem,[16] AOAC International,[23] the International Conference on Harmonisation of Technical Requirements for Registration of Pharmaceuticals for Human Use (ICH),[24] and the VICH.[19]

There are two sources of information from the CAC that provide guidance on the validation of analytical methods. General guidance on method validation, including "single-laboratory method validation," may be found in the CAC *Procedural Manual*.[25] This is supplemented by Codex Alimentarius Commission guidelines, a number of which have been adopted from harmonized guidelines previously developed by independent international scientific organizations.[26–31] In addition, a 2009 CAC guideline has been adopted that deals specifically with methods used for analysis of veterinary drug residues in foods and residue control laboratories using these methods.[32]

8.2.1.7 Joint FAO/WHO Expert Committee on Food Additives (JECFA)

JECFA is an independent expert committee established (and jointly administered) by the FAO and the WHO in 1956 to evaluate the safety of food additives. The work has since expanded to include the evaluation of the safety of "contaminants, naturally occurring toxicants and residues of veterinary drugs in food."[33] JECFA serves as the risk assessor for the Codex Committee on Residues of Veterinary Drugs in Foods (CCRVDF), establishing an acceptable daily intake (ADI) for a veterinary drug when sufficient information is available, recommending maximum residue limits (MRLs) for consideration by

CCRVDF, and also evaluating the suitability of analytical methods for these residues. Two aspects are considered by JECFA with respect to the performance of analytical methods: (1) JECFA must ensure that methods used in the pharmacokinetic and residue depletion studies considered in the establishment of an ADI or recommendation of MRLs are suitably validated to support the quality of the data reported in these studies, and (2) JECFA has been asked by CCRVDF to recommend when a suitably validated analytical method is available for regulatory use to support the recommended MRLs. JECFA has published a guidance document that states the requirements for validation of analytical methods submitted for JECFA consideration.[34]

8.2.1.8 *European Commission*

The European Commission Decision 2002/657/EC, which is legally binding and applies to laboratories within member states engaged in the analysis of foods for residues of veterinary drugs, specifies requirements for performance of the analytical methods used in those laboratories.[35] Laboratories engaged in the testing of food for export to the European Union must also demonstrate that they meet these requirements when audited by inspectors from the Food and Veterinary Office of the EU. Decision 2002/657/EC incorporates the "criteria approach," in which the use of specific methods is not required. Instead, each laboratory must demonstrate that the methods it uses for the analysis of residues of veterinary drugs in foods are fit for purpose in that they have been validated to meet the performance criteria specified in the Decision. Guidance on interpretation and implementation of the criteria contained in the Decision 2002/657/EC is provided in an additional document issued by the Health and Consumer Protection Directorate of the European Commission.[36]

8.2.1.9 *US Food and Drug Administration (USFDA)*

The USFDA requires that a new submission for use of a veterinary drug in a food animal includes a suitably validated analytical method for marker residues in target tissues except when use of the drug has been demonstrated to result in no detectable residues. As discussed in Chapter 3, this method is directly related to the tolerances established for residues of that drug and is intended to be used as the reference method for regulatory action. Two guidance documents related to requirements for residue methods submitted to the USFDA are available from their website (`http://www.fda.gov/`). The first, *General Principles for Evaluating the Safety of Compounds Used in Food-Producing Animals*, deals with general requirements for the validation of analytical methods for the determination of veterinary drug residues, with a focus on method specificity, precision, and recovery.[37] A second document provides guidance on methods based on mass spectrometry.[38] The two VICH guidelines on validation of analytical methods[19,20] are also included in the list of guidance documents available from the USFDA website.

8.3 THE EVOLUTION OF APPROACHES TO METHOD VALIDATION FOR VETERINARY DRUG RESIDUES IN FOODS

The inclusion of a validated analytical method for the control of residues in veterinary drugs in foods in a submission prepared for regulatory approval of a drug for use in food-producing animals is a requirement for product registrations in many jurisdictions. Typically, such methods are developed and validated for only a single compound by (or on behalf of) pharmaceutical companies, the marker residue for the drug, the nature of which is discussed in more detail in Section 8.5.1 and that is the subject of the submission, rather than for residues of multiple drugs. For authorities responsible for the delivery of national residue control programs, single-residue methods are of limited value, particularly when there are multiple drugs from the same class available for use. Thus, the regulatory model that required submission of a validated method as a requirement for registration of a veterinary drug did not readily translate into methods in routine use in many regulatory control programmes.

8.3.1 Evolution of "Single-Laboratory Validation" and the "Criteria Approach"

During the early 1990s, the conventional approach to validation of a method for control of veterinary drug residues in foods in some jurisdictions such as the United States required a multi-laboratory method trial, involving a minimum of three analysts and preferably three independent laboratories.[39] However, by the late 1990s, measures such as the routine accreditation of laboratories meant that not only did demonstrable quality procedures have to be in place in the accredited laboratory but also that there had been independent expert review of methods in use (this is discussed in more detail in Chapter 10). Rapidly changing technologies and the expansion of the range of analytes covered in typical regulatory control programs made inter-laboratory trials of methods a less viable approach to method validation, as this added significantly to the time required to complete a "method validation." This led to alternative approaches usually referred to as "fitness for purpose" or "the criteria approach," with more emphasis on internal quality control (QC) measures during method development and within-laboratory method validation and less reliance on inter-laboratory studies.

Within the EU, there was an almost complete abandonment of inter-laboratory method trials to characterize performance of methods used in veterinary drug residue testing. Instead, each official residue control laboratory in an EU member state was required to demonstrate fitness for purpose of the methods it uses according to the criteria contained in the Decision 2002/657/EC.[33]

8.3.2 The Vienna Consultation

Beginning in 1997, a series of international consultations and workshops were held, which led to the adoption of a single-laboratory validation approach by the CAC[25,31] and the publication of guidance by the International Union of Pure and Applied Chemistry.[11] In 1996, the CCRVDF, recognizing that information was not available on interlaboratory method trials for many of the drugs for which Codex standards (maximum residue limits; more often referred to as MRLs) were under development, requested that FAO organize an expert consultation on method validation. This consultation, held in Vienna in 1997 and hosted by the International Atomic Energy Agency (IAEA), produced agreement on the following principles:[40]

- Laboratories carrying out validation studies should operate under a suitable quality system based on internationally recognized principles.
- Method validation requires a third-party review (ISO/IEC 17025 : 2005, GLP, etc.).
- Methods validated to support Codex standards should be assessed against Codex criteria, with an emphasis on the assessment of the limit of quantification (LOQ), not the limit of detection (LOD).
- The validation work must be fully documented in a validation report that clearly identifies the analytes and matrices included in the validation work.
- There should be evidence of transferability for methods intended for Codex use (multi-laboratory trial or, at a minimum, ruggedness testing).
- Member countries should make such methods available to Codex.

8.3.3 The Budapest Workshop and the Miskolc Consultation

The subsequent IUPAC/ISO/AOAC Symposium on Single Laboratory Validation, held in Budapest, November 4–5, 1999, resulted in the draft working paper that, after international review and comment, was adopted as a harmonized guideline and published by IUPAC in 2002.[11] This IUPAC guideline has also been adopted as a guideline by the Codex Alimentarius Commission.[31]

A Joint FAO/IAEA Expert Consultation on Practical Procedures to Validate Method Performance of Analysis of Pesticide and Veterinary Drug Residues, and Trace Organic Contaminants in Food was held in Miskolc, Hungary, immediately following the IUPAC/ISO/AOAC Symposium on Single Laboratory Validation. The invited consultants were participants in the preceding IUPAC symposium who also served as key members of working groups associated with the Codex committees dealing with analytical methods.[41] The outcome of this consultation was subsequently used in training courses for scientists from developing countries offered by the IAEA Training Centre located at its laboratory facilities in Seibersdorf, Austria, and was also used by the Codex Committee on Pesticide Residues (CCPR) and CCRVDF working groups on methods of analysis to update and revise Codex guidance for methods for pesticide and veterinary drug residue analysis (see also Chapter 10), in conjunction with the concurrent work undertaken by the Codex Committee on Methods of Analysis and Sampling (CCMAS) to develop guidelines on single-laboratory validation of analytical methods now included in the Codex *Procedural Manual*.

8.3.4 Codex Alimentarius Commission Guidelines

In 2009, the CAC approved a new guideline for the validation of methods used for the determination of veterinary drug residues in foods (CAC/GL 71–2009),[32] which replaced the previous guidance.[39] The new guidance states that methods may be validated in a single laboratory, provided that the validation meets "the General Criteria for the Selection of Methods of Analysis" contained in the CAC *Procedural Manual*.[25] In addition:

- The validation should be conducted according to the requirements of an internationally recognized protocol (such as the IUPAC harmonized guideline).[11]
- The method should be used in a laboratory with a quality management system that is in compliance with "the ISO/IEC 17025 : 2005 Standard or with the Principles of Good Laboratory Practice."
- Method accuracy should be demonstrated through regular participation in suitable proficiency testing programs (when available) and use of certified reference materials in calibration (when available).
- Within-laboratory recovery studies for the analyte should be conducted at appropriate concentrations, with comparison of results to those obtained with another validated method.

These recommendations are consistent with those contained in the IUPAC harmonized guidelines.[11] Additional guidance on the specifics of single-laboratory method validation is the topic of a current IUPAC project.[42] In 2009 the CCRVDF also agreed to establish an electronic working group to draft performance criteria for multi-residue analytical methods, as these are becoming increasingly important to laboratories juggling the conflicting demands of increased analytical throughput with reduced resources. On completion, the performance criteria for multi-residue analytical methods will appear as an annex to the more recently published CAC/GL 71–2009.

8.4 METHOD PERFORMANCE CHARACTERISTICS

The recommendations on design of method validation experiments contained in this chapter represent the authors' assessment of current best practices, and these should be considered in the context of any subsequent guidance that may be provided by IUPAC or other authoritative scientific bodies. As guidance is constantly evolving, it is important that a check be undertaken to ensure that the most recent relevant guidance is consulted before planning any method validation studies.

The current list of performance characteristics or related factors (discussed in greater detail in Section 8.7) to be considered in planning a method validation according to recommendations from the Codex Alimentarius Commission[21,25] includes:

- Analyte stability (see Sections 8.5.4, 8.5.6, 8.5.7)
- Ruggedness testing/robustness (see Section 8.5.8)
- Calibration curve (see Section 8.7.1)
- Analytical Range (see Section 8.7.1)
- Linearity (see Section 8.7.1)
- Sensitivity (see Section 8.7.2)
- Selectivity (specificity) (see Section 8.7.3)
- Accuracy (see Section 8.7.4)
- Recovery (see Section 8.7.5)
- Precision: repeatability and reproducibility (see Section 8.7.6)
- Measurement uncertainty (see Section 8.7.8)
- Sample stability (see Section 8.5.7)
- Method comparisons, CRMs (see Section 8.7.4)
- Limit of detection (see Section 8.7.9)
- Limit of quantification (see Section 8.7.9)

The EC requirements do not include limits of detection and quantification or MU, but instead require the determination of other statistical indicators of result reliability, that is, the decision limit (CCα) and the detection capability (CCβ)[35] as discussed in Section 8.7.10.

Most of these terms have clear and specific definitions, many of which are found in guidance documents from IUPAC[6–8] and referenced in other documents, including CAC/GL 72–2009, the guideline on analytical terminology issued by the CAC in 2009.[21] When available, an authoritative definition issued by a recognized independent scientific authority, such as IUPAC or ISO, or a definition contained in regulations of a regulatory organization, such as the USFDA or the EC, should be referenced. In the absence of such internationally accepted definitions, a laboratory's protocols should include a definition of any term used in defining method performance and cite the source of such definitions.

Much has been done to harmonize analytical terminology, particularly by IUPAC's working parties in association with members of the CCMAS. Some of these definitions are included in the current Codex Guideline CAC/GL 72–2009.[21] Other sources of definitions include ISO,[15] the Joint Committee for Guides in Metrology, or JCGM (of which ISO is a member),[43] the National Association of Testing Authorities, Australia (NATA),[44] Eurachem,[16] European Commission Decision 2002/657/EC,[35] and documents issued by ICH[24] and VICH.[19] Use of terms as defined by these organizations prevents confusion during audits and also prevents wasted time and effort during validation. Not using an appropriate definition and, as a result, designing experiments around such a definition, may lead to work that must be repeated so that it conforms to the accepted definition used in regulatory analysis. For example, auditors under an ISO/IEC-17025 : 2005 accreditation audit will reference definitions from the VIM,[15,43] while auditors performing an audit related to laboratory test equivalency for export of food may first reference definitions used by the CAC[21] or the EU.[16] This is seldom a major issue, as the VIM is the source of many definitions used by the CAC. However, it is necessary to be aware that there may be different definitions in use by different organizations, and these may, in turn, have an effect on the manner in which method development and validation are carried out, or at least require some explanation of the different use of the terminology in the laboratory's protocols.

8.5 COMPONENTS OF METHOD DEVELOPMENT

Codex Guideline CAC/GL 72-2009[21] defines the term "fitness for purpose" as the "degree to which data produced by a measurement process enables a user to make technically and administratively correct decisions for a stated purpose." In the EU, the guidance on method validation and quality control procedures published in 2009[45] advises that "within laboratory validation should be performed to provide evidence that a method is fit for the purpose for which it is to be used," thus confirming the importance of within laboratory validation to end users. Similar views are expressed by regulatory authorities in regards to residues of veterinary drugs to ensure that analytical methods are capable of providing robust results at prescribed legislative limits. However, if the purpose of the analysis is to provide quantitative data below legislative limits, such as for dietary intake assessments, then it is important that the analytical method must be adequately validated, normally at concentrations substantially below legislative limits, to permit data generated from the analysis to be used for this purpose.

8.5.1 Identification of "Fitness for Purpose" of an Analytical Method

The regulatory world dealing with food safety is governed by risk analysis and risk management. Validated analytical methods are required in toxicology studies to generate data on trace quantities of chemicals resulting from their use in food production or from their presence as contaminants so that decisions on safe or acceptable limits may be made. Validated methods are also required for pharmacokinetic, metabolic, and environmental degradation studies to identify metabolic and other degradation pathways, identify suitable marker residues and target tissues or matrices for sampling programs, and to understand the mechanisms involved in the presence of residues and contaminants in foods. These data are also used to establish MRLs and withdrawal periods or withholding times between treatment and harvest of crops, collection of milk or eggs, or slaughter of food animals. Validated methods of analysis are required to monitor compliance of food products with these limits for national survey samples and also for dietary intake studies conducted by health authorities. The reliability of the data generated in all of these types of studies that are used to set and enforce food safety standards hinges on the reliability of the analytical methods used and therefore on the validation of these methods of analysis. The consequences of faulty data can be measured in millions of dollars, lost employment, and lawsuits.

When engaged in developing a new method or validating an existing method, there are some fundamental issues to consider. First, it is necessary to know precisely what the regulatory requirements are that must be met to ensure that the method is "fit for purpose." This includes sufficient information on the compound, the metabolic or other degradation pathways followed when it has been used, the matrices that will be included in the testing program, and the concentration range to be targeted to support any MRLs or other regulatory limits that have been established. The compound targeted in the analysis as evidence of use of a particular drug is termed the *marker residue*.[46] This may be the parent compound, a metabolite or degradation product that can be uniquely identified with the parent compound, or a transformation product of the parent compound and/or metabolites. Information on the pharmacokinetics, elimination pathways, and distribution of the chemical in treated animals is required to identify the appropriate marker residue for the matrix (or matrices) that is (are) to be analyzed and, in the case of drugs or pesticides applied or administered to food animals, the tissue in which evidence of use is most likely to be detected. In the case of animal tissues, the tissue containing the highest and/or most persistent residues is typically preferred for sampling and is usually referred to as the *target tissue*.

Selection of the appropriate marker residue may be challenging for some antibiotics, particularly those that typically are prepared from fermentation media and may contain multiple active constituents in variable proportions according to the particular manufacturer, such as gentamicin.[47] For these substances, there is batch-to-batch variability, so the ratios of the components are not necessarily consistent, and therefore standards of individual components are preferable for reliable quantification. The rates at which certain components may be more readily metabolized or eliminated from tissues may differ from those of other components. As a result, the residue profile found can vary according to sample collection time since the last treatment, as well as sample storage time and temperature if particular components are subject to degradation on storage prior to analyses.

The performance characteristics, or figures of merit, which are typically included in the validation of a quantitative analytical method for residues, have traditionally included the accuracy of the method, typically expressed in terms of recovery for residue methods, selectivity, sensitivity, precision, analytical range, and linearity of the calibration curve. Regulatory agencies have imposed additional criteria that have become widely accepted in the analytical community. These include demonstrating method ruggedness, in part to facilitate transfer of methods to other analysts or laboratories, as well as to identify critical steps that may not have been identified during method development. Many regulatory methods at one time were subjected to inter-laboratory studies to determine reproducibility; more recently, other approaches to statistically define method performance and establish the appropriate result that should lead to regulatory action have been developed.

Any other factors that may define the analytical requirements should also be considered. For example, when dealing with veterinary drug residue analysis, while the target tissue for domestically produced animals may be liver or kidney, these organ tissues are less commonly available as imported products. The majority of imported meat products are muscle tissue. Therefore, although the method has been validated for analysis of kidney for domestic samples, it is not fit for purpose for use on most import samples until it has also been validated for muscle tissue and possibly even some processed meat products. A method validated for the analysis of an aquaculture drug or natural toxin in oysters from domestic production may also, for example, require validation for shrimp or tilapia for application to imports. In addition, the requirement may include development of a test that is capable of detecting and quantifying residues of multiple compounds, analysis time considerations to facilitate rapid reporting of results, or ease of transfer to other laboratories. In regulatory analysis, cost, time constraints, and transfer of methods to other laboratories are frequent considerations. Few residue control authorities have the resources to conduct separate analytical methods for each of the hundreds of compounds that may be included in a national residue control program.

An additional consideration is comparability with an existing regulatory method, particularly if that method has "official" status, which requires that analytical results must be traceable to that method when used in legal action. For example, as discussed in Chapter 3, the analytical method submitted as part of a new animal drug application to the USFDA, once accepted, links directly with the tolerances established for residues of the drug in food matrices. When the "reference" or "official" regulatory method for an antimicrobial drug is based on a microbiological growth inhibition test, the determination of test equivalency can be more problematic, as the microbial growth inhibition test will detect all inhibitors (whether parent compounds or metabolites) present, while a chemical assay may target only the parent compound or another marker residue that does not include all the microbiologically active residues present. Recoveries from the extraction processes used in the tests may also differ significantly, adding further complexity to the comparison of the results achieved with the two tests. A bridging study is therefore required so that the performance of the two tests can be directly compared and factors derived so that the results of the new test can be made equivalent to the results from the original reference method. This is, essentially, one of the procedures recommended by IUPAC[11] and the CAC[25] for the validation of a new method of analysis—comparison with a reference method. However, in the context of the development of an alternative test method for regulatory use in a situation such as that in the United States, where there is an official method directly linked to the tolerance (or MRL), the validation by this approach takes on additional significance. Several examples of bridging studies for antimicrobial compounds may be found in the scientific literature.[48,49]

8.5.2 Screening versus Confirmation

The intended use of the method must be clearly defined. A screening method is generally considered to be a method used to detect the presence of a compound, usually at or above a specific concentration. In CAC guidelines for veterinary drug residue analysis, a *screening method* is defined as "qualitative or semi-quantitative in nature and are used as screening methods to identify the presence (or absence) of samples from a herd or lot which may contain residues which exceed an MRL or other regulatory action limit established by a competent authority."[32] A *quantitative method* is defined as a method that provides "quantitative information which may be used to determine if residues in a particular sample exceed an MRL or other regulatory action limit, but do not provide unequivocal confirmation of the identity of the residue" and it is considered that "such methods which provide quantitative results must perform in good statistical control within the analytical range that brackets the MRL or regulatory action limit."[32] Finally, *confirmatory methods* "provide unequivocal confirmation of the identity of the residue and may also confirm the quantity present."[32] In some cases, particularly those involving methods using combined chromatographic–mass spectral techniques, such as LC-MS/MS, the method may be developed and validated for use in any of these applications. Screening may be accomplished by monitoring for a single characteristic ion for each of a large number of possible residues in an initial analysis. Any positive findings may then be quantified by reference to appropriate calibration standards and controls, while confirmation may be achieved by a subsequent analysis targeting multiple characteristic ions or transitions.

8.5.3 Purity of Analytical Standards

Method development is undertaken to meet a requirement for analysis that cannot be met by existing methods and also is the initial step at which a fitness-for-purpose statement is required. The fitness-for-purpose statement summarizes the matrices to be analyzed, the analytes that must be included, the concentration range required, and any other specifications that must be met. The next steps in method development typically include obtaining analytical standards and conducting a literature survey to determine what methods are available that may be adaptable to meet the requirement or may at least suggest approaches for investigation during method development. There are two critical considerations concerning the analytical standards: the purity of the standards and information on their stability. First, standards should be obtained from sources that can provide documentation on the date of manufacture, methods used to confirm the identity and purity, a certificate or statement of purity of the standard, and an expiry date for the certification provided. Then, a file should be established which includes the documentation received with the standard.

It is good practice to conduct additional qualification tests with these materials before beginning method development. Unless the amount of standard provided is insufficient to permit additional testing to confirm the identity and purity, some tests should be conducted to provide this confirmation prior to routine use of the standard. Typical checks to verify the identity and purity of the standard include spectral tests to confirm identity and purity checks such as melting-point determination or chromatography. Add this information to the file.

There is also usually a recommended period of time during which the purity of an analytical standard is considered to be assured if the standard is properly stored, typically from 2 to 5 years, and this will normally be given on the certificate of analysis provided with the standard. If no expiry date is provided with the standard, then it is the responsibility of the laboratory to establish the optimum storage conditions and a maximum time period for which the standard may be used. At the end of this period, the standard should be replaced. However, the usage period may be extended if the laboratory establishes a procedure to recertify the standard with tests that show that no changes in the purity and composition of the standard have occurred. Procedures used to establish or extend expiry dates for standards should be documented.

8.5.4 Analyte Stability in Solution

Once the purity of the analytical standard and a time period during which it can be stored as "pure standard" without degradation have been established, the stability of the standard in solution should be investigated by laboratory testing if no information is available from other sources. The two common causes of degradation are exposure to light and the temperature of storage. The behavior of the analyte in solution in solvents that may be used either for the preparation and storage of standards or as solvents during the performance of the analytical method should be investigated. First, obtain information on the solubility of the compound from information provided by the manufacturer, such as a *material safety data sheet* or from other reliable sources, such as handbooks containing solubility information to identify solvents that may be used for preparation and storage of analytical standards or as solvents in the analytical method. Check the available literature to see if there is any published information on the stability of the compound in solution in various solvents. The analyst must ensure that working standards are stable and must establish a time period at which new ones must be prepared.

The stability of the standard in solution is assessed by experiments such as comparing the effect of storage of a standard solution at room temperature in the laboratory and storage in a refrigerator. The study period typically extends from several days to weeks or even months to determine whether there is any evidence of change in concentration or evidence of degradation of the standard. If there are indications of degradation, such as reduced detector response or the appearance of additional peaks in a chromatogram, additional tests should be conducted to determine whether it is necessary to protect the standard from exposure to light. This may be achieved by a simple approach, such as wrapping the container in which the standard is stored with foil, or by using tinted storage containers that protect the contents from exposure to light. The results of these tests may then be used to establish conditions for the storage, handling, and use of the standard solution. In extreme cases, or when a large analytical throughput is required for these analyses, it may be necessary to adjust the lighting conditions within the laboratory or to consider setting up a dedicated area with special lighting.

It is important to include sufficient replicate analyses in the experiment to ensure that any variability observed is real and not simply within the range of variability that should be expected for the analysis of replicates. No significant loss of analyte ($p = 0.05$) should be observed for any of the conditions of storage and use that are established for subsequent routine use. In addition, a working life for standard solutions should be established by testing the stability over the time period typically used in the laboratory for both stock standard solutions and working standards, and this information should be included in the method.

Decision 2002/657/EC recommends procedures for assessment of analyte stability in solution.[35] Sufficient standard solution should be prepared to provide at least 40 replicate aliquots of the standard solution for testing. From this stock, 10 replicate portions should be taken to assess each of the proposed conditions of temperature for storage, which typically are temperatures such as -20°C, $+4^\circ$C and $+20^\circ$C (typical temperatures in a freezer, a refrigerator and "room temperature"). Two sets (10 aliquots per set) should be prepared for testing at $+20^\circ$C, one to be stored in light, the other in the dark. The tests should be conducted at time intervals such as:

> "one, two, three and four weeks or longer if necessary, e.g. until the first degradation phenomena are observable during identification and/or quantification. The maximum storing time and the optimum storing conditions have to be recorded."

Analytical results at each time point should be compared with the results obtained for analysis of an aliquot of the standard when it was freshly prepared.

8.5.5 Planning the Method Development

Once the stability and storage conditions of the analytical standard have been established, experiments can be undertaken to develop the analytical method. The preferred strategy is to keep the method as simple as possible, as each step adds to the complexity when training other analysts, and also each added step potentially adds to the uncertainty budget, which is discussed in Chapter 9. A typical method for an antibiotic residue begins with the initial extraction to separate the antibiotic residue from the bulk matrix material. For some substances, this may include several steps. First, there may need for a chemical or enzymatic treatment to free any residues that are chemically bound to tissues or to convert the residues present as metabolites to a single compound. The next step is usually a solvent extraction, which may also include a protein precipitation treatment as part of the initial efforts to separate the target analytes from matrix materials. For some compounds and matrices, these steps may provide a sufficiently clean extract to use for the final instrumental analysis. More commonly, additional clean-up steps, such as may be achieved by solvent partition or solid-phase extraction, such as to remove fat from foods of animal origin or sugar from honey, will be required. It is important to check the recovery of analyte at each step in the procedure. When using solid-phase extraction (SPE) cartridges, a series of aliquots should be collected from the column and analyzed as eluting solutions pass through the cartridge. Ideally, it should be possible to collect the analyte in a 2–3-ml aliquot of elution solution. However, it is first necessary to develop a clear profile of the elution pattern from the cartridge and modify the elution solution or conditions until a profile is obtained where the analytes of interest are eluted with most interfering substances removed and in a discrete fraction. Collection of multiple fractions containing analyte indicates that the elution solution requires modification to reduce this band broadening.

The choice of solvents, SPE materials, and chromatographic column are dictated by the chemistry of the target analytes, In general, knowledge of the solubilities, polarity, and potential interactions of the analytes will direct the selection of solvents and chromatographic media. The spectral properties of the analytes dictate the detection strategy. Compounds that exhibit a strong UV absorbance or native fluorescence may be suitable for analysis by liquid chromatography with a UV or fluorescence detector. If the compound does not contain chromophores or exhibit native fluorescence, then a derivatization step may be required. Currently, many laboratories equipped with LC-MS instruments find it simpler and more cost-effective to avoid derivatization or the use of less specific detectors and use selected-ion monitoring to detect the analyte. The ability to monitor multiple characteristic mass spectral fragments provides added confidence in the analytical result. Typical methods for residues of antibiotic compounds are discussed in more detail elsewhere in this book (Chapters 4, 6, and 7). The intent in this section is simply to provide an overview of the method development strategy.

Method development is complete once the method steps have been formalized in a draft method standard operating procedure (SOP) ideally prepared in a standard format such as set out in ISO 78-2 : 1999,[50] and sufficient replicate experiments have been conducted using the procedure to provide initial evidence that the method should be "fit for purpose" with respect to the method requirements that were identified before the commencement of method development. Some of the performance characteristics that should be assessed before a method can be considered validated, such as analyte stability, are completed during the method development phase, while the remainder, including the experiments used to formally establish method performance standards, such as recovery and precision, are conducted primarily during the method validation phase.

8.5.6 Analyte Stability during Sample Processing (Analysis)

It must be established that the analyte remains stable throughout the analysis. There have been some reports in the literature of conversion of the analytes to apparent metabolites or degradation products, such as in the case of tetracyclines, and this can certainly confuse the interpretation of the analytical results.[51] This has led some jurisdictions to amend MRLs to take account of this phenomenon. For example, in the EU, the MRL for the various tetracyclines is expressed as the sum of the parent tetracycline and the 4-epimer of the veterinary drug.[52]

The pure standard in solution should be taken through the complete analysis once the method has been developed. If this experiment demonstrates significant loss of analyte, the experiment should be repeated for each step in the analysis to determine where loss of analyte occurs, and this step should be modified to eliminate the loss. The cause of the loss or degradation of the analyte should be determined to ensure that similar steps are not included in future methods involving the analyte. Potential causes of degradation or loss include poor solubility or instability in a solvent, instability or lack of solubility under certain conditions of pH, interactions with a chromatographic support material, and adsorption onto glass or other contact materials.

8.5.7 Analyte Stability during Sample Storage

When information on the stability of analyte in matrix during typical conditions of storage is unavailable from the literature or from proficiency testing (PT) providers, such as the food analysis proficiency assessment scheme (FAPAS),[53] this information should be developed in the laboratory as part of the method development and validation. However, a study to obtain information on stability of the analyte first requires the availability of a validated method if results are to be considered reliable. When a method is being developed to introduce a test capability for a new analyte into the laboratory, the new method should first be developed and validated, so that the sample stability investigation becomes the final step in method validation.

The stability of the analyte in typical stored sample material should be investigated over a time period that is representative of the accepted sample storage practice in the laboratory and should consider the possibility of requests for reanalysis in case of dispute (typically 3–12 months). Managers and analysts working in laboratories involved in the import/export testing of foods for regulatory purposes should be aware of the contents of the recent CAC guideline on settlement of disputes on analytical results and ensure that they have data to support their standard procedures for storage and disposal of sample material.[54] The guidance in Decision 2002/657/EC suggests that testing be conducted for storage at −20°C and also at lower temperatures, if indicated, at intervals such as 1, 2, and 4 weeks, in addition to the longer storage times suggested above.[35] A minimum of five replicate test portions should be frozen for testing for each timepoint to provide sufficient data to identify significant changes in concentration.

Storage stability studies should be conducted at several representative concentrations and using known incurred material, if it is available, or otherwise using a homogeneous pool of fortified blank material for each matrix that will be routinely analyzed. The stability study should be conducted at several representative concentrations and should include several freeze–thaw cycles to represent typical conditions that may occur when sample material is shipped, removed from storage and thawed for subsampling, and re-frozen. The information obtained should then be used to establish maximum times and conditions under which samples may be stored for re-analysis without significant degradation.

8.5.8 Ruggedness Testing (Robustness)

In the authors' opinion, ruggedness testing should be conducted during the final phase of method development, as the results of the ruggedness testing may result in some changes to the method SOP. Conducting method validation first, followed by ruggedness testing, may mean that the efforts and resources expended in the method validation were wasted if the ruggedness testing requires significant changes to the method, with the result that the method validation experiments must be repeated.

Eurachem[16] cites a definition from AOAC International's Peer-Verified Methods Program in defining ruggedness testing as an "intra-laboratory study to study the behaviour of an analytical process when small changes in the environment and/or operating conditions are made, akin to those likely to arise in different test environments. Ruggedness testing allows information to be obtained on effects of minor changes in a quick and systematic manner." However, CAC/GL 72–2009 can cause confusion among analysts as to the appropriate terminology to be used with the following definition,[21] citing ICH[24] and IUPAC[11] sources:

> *Robustness (ruggedness):* A measure of the capacity of an analytical procedure to remain unaffected by small but deliberate variations in method parameters and provides an indication of its reliability during normal usage.

Analysts should be aware that the terms *ruggedness* and *robustness* are used interchangeably in the literature. There has been discussion in some laboratories as to whether the term used in method validation protocols should be *robustness* or *ruggedness*. ICH and VICH guidelines recommended *robustness*, while the 18th edition of the Codex *Procedural Manual* referred to *ruggedness*.[55] The priority given to the term *robustness* in the new Codex guideline CAC/GL 72–2009 suggests that there is now a strong argument to adopt this term in laboratory protocols, but further discussions will be necessary before a consensus can be reached on this point. In the meantime, it may be appropriate to refer in laboratory SOPs to "ruggedness testing (robustness)."

Typically, *ruggedness testing* is conducted using a standard factorial design approach to determine any steps in the method that are sensitive to minor changes.[56] A typical design for a ruggedness test, that is the Youden approach, is described in an AOAC publication[56] and is also found in Decision 2002/657/EC.[35] The basic approach involves identifying seven factors, identified by capital letters $A–G$. Minor changes are then introduced for each factor, identified by the lower case letters $a–g$. These provide a possible total of 128 combinations. However, when multiple factors are changed in the same analysis using a matrix design such that the effects of each individual change can be identified, the number of combinations is reduced to eight. From this set of experiments, any change that leads to a significant variation can be identified.

Typical factors to consider in the design of a ruggedness experiment include variations in reagent volumes or concentrations, pH, incubation or reaction time, and temperature, reagent quality, and different batch or source of a reagent or chromatographic material. Ruggedness testing of a confirmatory method may be required if the method differs significantly from the quantitative method previously validated (if the method uses extraction or derivatization procedures different from those used in the quantitative method). A rugged procedure should be unaffected by minor variations, and the analytical results should not alter significantly if such minor variations occur during analysis.

There are four analytical results from analyses conducted for each factor. Determine the average for each factor, as described in Section 3.3 of Decision 2002/657/EC[35] or in the AOAC statistical manual.[56] Significant differences between the average result for a factor designated with an uppercase (capital) letter and the corresponding lower case letter indicate a factor that is affected by the change. These should be appropriately identified in the method, either by a note to the analyst or by a stricter requirement, such as adding a "±" to a quantity specified. Examples of a seven-factor ($n = 8$ experiments) design are contained in several more recent papers on the determination of antimicrobial residues in animal-derived foods,[57,58] and a detailed discussion of a typical seven-factor design may be found in Chapter 9.

The seven-factor approach presented in the above referenced documents is essentially based on work from an earlier paper by Plackett and Burman[59] in which a more complex experimental design is based on multiples of 4; for example, a typical design uses $n = 12$, with 11 factors arranged in a 12×11 matrix, instead of the 8×7 matrix described by Youden and Steiner.[56] Examples may also be found for designs based on $n = 20$, 24, 28, or higher.[60] In a Plackett–Burman experiment, variations are introduced by making an increase or a decrease in a quantity specified in a method. Some of the factors may be designated as dummy factors and left unchanged. The more complex study design for the Plackett–Burman-based evaluation is less commonly used than the Youden approach. An example of the application of a Plackett–Burman design to ruggedness testing of a method for avermectin-class endectocides in milk is contained in a 2009 publication, where a design in which $n = 12$ was employed.[61]

8.5.9 Critical Control Points

"Critical control points" are the steps in the method where any deviation from the procedure may result in failure of the analysis, and these should be clearly identified in the method protocol. Critical control points should usually be identified during method development. The ruggedness testing is a final check on the effectiveness of that process, but also identifies steps where failure to follow instructions may result in a less reliable result or, in extreme cases, in complete failure of the analysis. It is important not to view ruggedness testing as the primary means by which critical control points in a method are identified, although these may be revealed during ruggedness testing. Ruggedness testing attempts to mimic what happens when a method is used over an extended period of time, by multiple analysts and in different laboratories. No one with previous experience in a collaborative study would attempt such a study on a method that has not been subjected to ruggedness testing prior to the collaborative study.

8.6 COMPONENTS OF METHOD VALIDATION

8.6.1 Understanding the Requirements

It is necessary to first understand what we should expect to achieve from method validation. There are three basic considerations: (1) the validation should define the method scope, which includes the analyte/matrix combinations to which the method can be applied and the concentration range within which reliable results have been demonstrated; (2) the validation should enable the establishment of performance standards (such as recovery and precision) for analysts in the developing laboratory or in other laboratories to which the method may subsequently be transferred; and (3) the validation should provide confidence in the reliability of results (analyte and sample stability, statistical evaluation of performance).

Method validation should not be considered as an isolated event. Method validation essentially begins once we develop or select a method for use. The first question to be answered is whether the method has been suitably validated by another analyst or laboratory as "fit for purpose." As discussed earlier, the fitness-for-purpose statement used for this determination should include identification of the analytes and matrices to which the method is to be applied, legal limits or other concentrations that define the concentration range required of the method, plus any other considerations such as required accuracy, precision, or measurement uncertainty (MU). When suitable documentation of such validation work is available, then the method may be used by other analysts or laboratories once they have demonstrated that they can routinely attain the performance standards established in the initial validation. If evidence of such validation is not available, then experiments should be conducted to validate the method before implementation.

Once implemented, method performance is monitored with ongoing QA/QC, as discussed in Chapter 10. If the method fails during validation, then further method development and revision is required prior to another attempt at validation. If at any time following validation and implementation, QA/QC reveals the method is not meeting performance standards, an investigation and identification of cause is required. Again, this may require method modification and re-validation. If a method consistently fails validation experiments despite best efforts to remedy the performance issues, then at some point a decision must be made that the method cannot be considered as likely to meet fit-for-purpose requirements and an alternative method should be sought.

8.6.2 Management of the Method Validation Process

Two aspects of method validation (described here and in Section 8.6.3) must be considered when method validation is undertaken. First, the management aspect of the process must be in place to meet accreditation and quality management requirements. This means that there should be a written protocol that documents the processes to be followed, such as who plans and conducts the work, who is responsible for review and approval of both the experimental plans and validation reports, and what documentation is required. There should be a written plan prepared prior to commencement of validation to outline the experiments that will be conducted; the analytes, matrices, and concentration range(s) to be included; and the identification of roles (who does the analytical work, who prepares blind spikes or incurred materials for the analyst, who prepares the report, etc.). This should be reviewed and approved by a manager and/or QA personnel prior to start of method validation. For small laboratories, where the manager may also be the person directly responsible for the method development and validation work, an alternative approach to provide independent review could be an arrangement with other laboratories in the area or region to provide the review function.

In addition, a method development report should be prepared demonstrating that method development has been completed and the method is ready for validation. This report should include a draft of the "standard operating procedure" (method SOP) that will be used in the validation experiments and should be approved by the responsible manager, preferably following a process that includes review by QA personnel and/or another scientist who is not directly involved in the work to ensure that appropriate peer review has been included in the process. Many national authorities require laboratories to produce SOPs in a recognized standard format, such as that set out in ISO 78-2 (1999).[50] It is critical that there be no changes made in this method SOP or divergence from the SOP during validation as this would invalidate the work.

8.6.3 Experimental Design

The second aspect of method validation is the experimental work. This typically involves initial experiments with analytical standards to confirm the reliability and repeatability of calibration of the system using only standards. The next step usually involves a series of analytical runs, conducted over several days or weeks, in which one or more analysts prepare calibration curves and analyze replicates of the typical analyte/matrix combinations and concentrations that are to be routinely analyzed using the method. The final phase of validation typically includes several runs in which fortified or incurred materials, again representing typical analyte/matrix combinations and concentrations, are provided blind to the analyst(s). The results are then summarized in a validation report, which again should receive appropriate peer review within the laboratory prior to management approval. The results and recommendations contained in this report provide the proof that the method is fit for purpose and provide performance standards for analyst qualification and subsequent internal QA/QC.

It is critical that this validation report be properly identified and available for review by external auditors. A properly designed validation exercise and the accompanying report can also form the basis for a publication in a scientific journal when the contents of the report are sufficiently novel. In subsequent sections of the chapter, we will discuss the different aspects of validation experiments typically conducted for methods intended for screening, quantification, or confirmation purposes.

8.7 PERFORMANCE CHARACTERISTICS ASSESSED DURING METHOD DEVELOPMENT AND CONFIRMED DURING METHOD VALIDATION FOR QUANTITATIVE METHODS

8.7.1 Calibration Curve and Analytical Range

Quantitative methods are usually based on a comparison of the response from an analyte in a sample with the response from standards of the analyte in solution at known concentrations. In method development and validation, the calibration curve should first be determined to assess the detector response to standards over a range of concentration. These concentrations should cover the full range of analytical interest, and, although it is usually recommended practice to include a suitable blank with the calibration samples, this does not imply that it is acceptable to extrapolate into the region of the curve below the lowest calibration standard or to force the curve through the origin.

There are recognized definitions and guidance documents related to the assessment of the analytical response for an analyte in a measurement system. These usually are expressed in terms of the relationship between the concentration of analyte present in the sample material and the associated response from a detector. The relationship between the analytical response and the analyte concentration is plotted and usually expressed as a linear regression equation. IUPAC defines the calibration curve in terms of the "calibration function in analysis"; in the IUPAC *Gold Book*[6] the calibration function in analysis is defined as follows:

> The functional (not statistical) relationship for the chemical measurement process, relating the expected value of the observed (gross) signal or response variable $E(y)$ to the analyte amount x. The corresponding graphical display for a single analyte is referred to as the calibration curve. When extended to additional variables or analytes which occur in multicomponent analysis, the "curve" becomes a calibration surface or hypersurface.

Two earlier publications are cited as original source documents for this definition:[62,63]

In addition, the *analytical function* is defined as follows:[6,62]

> A function which relates the measured value C_a to the instrument reading, X, with the value of all interferants, C_i, remaining constant. This function is expressed by the following regression of the calibration results.
>
> $C_a = f(X)$
>
> The analytical function is taken as equal to the inverse of the calibration function.

The definition of calibration function does not specify that the measurement be made in the presence of potential interferants. This serves as an introduction to a discussion of the appropriate approach to calibration in an analytical method for veterinary drug residues, such as antibiotics. Construction of a calibration curve requires a sufficient number of standard solutions to define the response in relation to concentration, where the number of standard solutions used is a function of the concentration range. In most cases, a minimum of five concentrations (plus a blank, or "zero") is considered appropriate for characterization of the calibration curve during method validation. It is also typically recommended that the curve be statistically tested and expressed, usually through linear regression analysis. However, for LC/ESI-MS analysis of residues, the function tends to be quadratic. The analytical range for the analysis is usually defined by the minimum and maximum concentrations used in establishing the calibration curve.

The regulations for analytical methods as contained in Decision 2002/657/EC specify that[35]

- At least five levels (including zero) should be used in the construction of the curve.
- The working range of the curve should be described.
- The mathematical formula of the curve and the goodness-of-fit of the data to the curve should be described.
- Acceptability ranges for the parameters of the curve should be described.

When preparing the calibration curve experiments in a method characterization or validation experiment, there are some key things to remember. It is important to first examine and plot the detector response to analyte in pure or defined solvent (calibration function) and also at various concentrations in the matrix (i.e., in the presence of interferants, or analytical function). Remember that more standard solutions and more calibration points are required when a non-linear response is observed. Extrapolation above or below the analytical range may be used to approximate behavior at "zero" if a blank is not included in the calibration standards, but the curve should not be forced through the origin. Typically, a curve prepared by fortification of blank matrix in a method for a veterinary drug residue in foods passes through the y axis at a point above zero due to the background response. It also is appropriate to apply extrapolation when using the method of additions, but this approach has not been commonly used in veterinary drug residue methods.

When the calibration points are scattered extensively around the regression line rather than the regression line passing through most points, this may indicate problems with the calibration. Similar curves should be observed for both the pure standards and the matrix extracts if there are no significant losses of analyte during extraction and in the absence of matrix effects. When the curves obtained for "pure standards" and "standards in matrix" are parallel or diverging, potential causes include loss of analyte during extraction from matrix (analytical recovery) or matrix effects (suppression or enhancement). The cause of any such divergence should be examined during method development so that an appropriate approach is used for calibration.

The validation experiments are intended to demonstrate the consistency of the observed effect. Additional sources of information on the assessment of the suitability of calibration curves[64–66] are referenced in the Eurachem guidance document.[16]

Non-linear curves are seldom used in well-validated methods, so the more typical requirements in calibration experiments relate to linearity. The linear range is defined by IUPAC[6,62] as the "concentration range over which the intensity of the signal obtained is directly proportional to the concentration of the species producing the signal."

The linear range should be determined in characterizing an analytical method. Common practice in many laboratories has been to require that an acceptable calibration curve yield a linear regression R value ≥ 0.999. As noted by IUPAC,[11] the correlation coefficient determined from linear regression is "misleading and inappropriate as a test for linearity and should not be used." Instead, this guideline recommends evaluation of the residuals from the linear regression analysis and, if suggested by the evaluation, the application of other models, such as weighted regression. Similar advice is found in the Eurachem guide on the fitness for purpose of analytical methods.[16] The IUPAC document also notes that the analytical range validated is not necessarily the whole range that potentially could be calibrated. For regulatory methods, method validation usually focuses on the concentrations of most interest, such as a regulatory limit. Decision 2002/657/EC[35] defines the level of interest as "the concentration of substance or analyte in a sample that is significant to determine its compliance with legislation." CAC/GL 72–2009 defines *linearity* as:[21]

> The ability of a method of analysis, within a certain range, to provide an instrumental response or results proportional to the quality of analyte to be determined in the laboratory

> sample. This proportionality is expressed by an a priori defined mathematical expression. The linearity limits are the experimental limits of concentrations between which a linear calibration model can be applied with a known confidence level (generally taken to be equal to 1%).

It also adds the following definition relevant to this issue:

> *Calibration:* Operation that, under specified conditions, in a first step, establishes a relation between the values with measurement uncertainties provided by measurement standards and corresponding indications with associated measurement uncertainties and in a second step uses this information to establish a relation for obtaining a measurement result from an indication.
>
> Notes:
>
> A calibration may be expressed by a statement, calibration function, calibration diagram, calibration curve, or calibration table. In some cases it may consist of an additive or multiplicative correction of the indication with associated measurement uncertainty.
>
> Calibration should not be confused with adjustment of a measuring system often mistakenly called "self calibration," or with verification of calibration. Often the first step alone in the above definition is perceived as being calibration.[43]

In validating methods for enforcement of a regulatory limit, regulatory laboratories may focus on the performance of the method over a relatively narrow range that brackets the standard or action limit. These methods are typically validated for linearity and other performance characteristics at four nominal concentrations: $0x$, $0.5x$, $1.0x$, and $2.0x$, on three separate days (where x is the normal, usually found concentration, or target concentration for action). The performance over a broader analytical range, using six or more calibration points as recommended by IUPAC and other authorities, is typically assessed during method development. This additional information should be included in the validation report for those situations where analyses may need to be conducted outside the usual target range. To better utilize both regulatory samples and laboratory resources, it is not unusual for regulatory authorities to utilize the full analytical range available with residue and contaminant methods. The limits of detection and quantification of methods may be an order of magnitude or more below the regulatory limit for which they are used to enforce, but using the full analytical range provides additional information on the prevalence of residues below action limits that can be used in dietary exposure studies and for further risk assessment purposes.

8.7.2 Sensitivity

Sensitivity, another characteristic associated with the calibration curve, is a term that is frequently misused. Analysts frequently speak of "highly sensitive" methods, indicating that the method is capable of detecting very low concentrations of the target analyte(s). However, this is not the sensitivity of the method. The *sensitivity* defines the capability of a method or instrument to differentiate between different concentrations of the target analyte and is usually measured in terms of previously defined parameters such as the slope of the calibration curve. The smaller the concentration change that can be measured quantitatively with a method, the greater is the sensitivity. To have appropriate sensitivity for reliable quantitative measurements, the slope of the curve must be neither too shallow nor too steep.

CAC/GL 72-2009[21] defines sensitivity as the "quotient of the change in the indication of a measuring system and the corresponding change in the value of the quantity being measured" and notes that "the sensitivity can depend on the value of the quantity being measured" and "the change considered in the value of the quantity being measured must be large compared with the resolution of the measurement system," referencing the VIM[43] as the source of the definition.

The key steps, therefore, in determining the suitability of a calibration curve for use in a residue method are as follows:

1. Determine the linear range and slope of the curve (sensitivity) for pure standards.
2. Determine the linear range and slope of the curve for curves prepared by fortification and extraction of blank matrix.
3. Determine whether the detector response is affected by the presence of matrix.
4. Determine whether the full linear range available includes (or exceeds) the range of interest for the analyte and whether the sensitivity is adequate to discriminate between concentrations targeted (e.g., whether is it necessary to discriminate between concentration differences of 1 μg/g or 1 ng/g?).
5. Select the calibration range to be used during method development and method validation.
6. Determine whether an external standard calibration (pure standards) or fortified matrix will be used for the calibration curves in the method. When pure standards are used for calibration, it is highly recommended that an internal standard be included in the method to compensate for loss of analyte during extraction or other phenomena associated with the presence of matrix materials in samples.

8.7.3 Selectivity

8.7.3.1 Definitions

Many analytical chemistry texts and older papers in scientific journals use the term "specificity" for "selectivity," but the term "selectivity" is recommended by IUPAC.[67] *Selectivity* is defined in CAC/GL 72-2009[21] as:

> the extent to which a method can determine particular analyte(s) in mixtures or matrices without interferences from other components of similar behaviour. Selectivity is the recommended term in analytical chemistry to express the extent to which a particular method can determine analyte(s) in the presence of interferences from other components. Selectivity can be graded. The use of the term specificity for the same concept is to be discouraged as this often leads to confusion.

Simply put, a method is either "specific" or "not specific." There are no grades of specificity, but selectivity may range from "uniquely selective" to "non-selective." *Selectivity* may be to a specific compound or element, to a group of compounds, to compounds that contain a particular element or functional group or to compounds that share some common property (or properties). There are, therefore, degrees of selectivity, but not of specificity.

Selectivity experiments should be conducted during method development, typically with pure standards once the initial calibration experiments have been conducted and subsequently on blank matrix and with representative blank matrix fortified with potential interfering compounds once the extraction and clean-up procedures to be used in the method have been experimentally determined. During validation, the major focus of the experimentation is on matrix interferences, not on potential interference from related analytes or other compounds that might be used in conjunction with the target analyte. Repeating all selectivity experiments during validation would make little sense and add costly experiments to prove what has already been demonstrated. During validation, include representative matrices from as many diverse sources as can reasonably be obtained to use as the matrix blanks and for preparation of fortified materials. This is particularly important for preparation of the blind samples provided to the analyst during the final validation experiments and provides additional useful confirmation of method selectivity and freedom from matrix interference. A typical recommendation is that at least six different sources of representative matrix material be used during method development and validation.[32]

The selectivity experiments should characterize the manner in which the method is selective and examine other substances that might reasonably be expected to be detected, in addition to the target analyte(s). The selectivity can be characterized, in part, by understanding and documenting how the various steps in the method provide selectivity.[36] For example, steps in the method may introduce selectivity through differences in solubility, separation based on polarity, the inclusion of a derivatization step that targets specific functional groups, or use of a detector that responds to a particular element, wavelength or mass. The selectivity experiments, therefore, should not be carried out simply by checking some readily available chemicals that may, or may not, be structurally related to the target analyte(s). The contributions made by the various steps in the method to isolate and detect the target analyte(s) in the presence of other compounds that might be present in the matrix should be assessed, and the selectivity experiments should then be directed at compounds that might reasonably be expected to pass through all steps of the analysis and could interfere with the detection of the target analyte(s). Potential compounds for investigation include other compounds from the same chemical class, other compounds that may be used in conjunction with the target compounds or that be present in the environment from which samples will be collected, as well as interference that might be naturally present in sample matrices.

8.7.3.2 Suggested Selectivity Experiments

As a first step, test other pure chemicals, such as compounds from the same chemical class and structurally related compounds, chemicals that might have been administered to the commodity or its environment during production and potential (likely) environmental contaminants. Next, test for interference from co-extractives from target matrices by testing representative matrices collected from various sources that reflect the expected profile of sample submissions. Testing of material from a single source is not sufficient. Check also for interference from known metabolites and degradation products.

The following are key considerations for design of the selectivity experiments:

- Analyze reagent blanks and representative blank matrices to ensure that no interfering compounds are present.
- Test the selectivity under different experimental conditions (different analytical principles or detection techniques). For example, if the method typically uses LC with UV detection, check that there is no interference when an MS detector is used in place of UV.
- A method should be able to distinguish the analyte from known interfering materials. For example, a new method developed to replace an older method should test for (and not be subject to) known interference identified in the method being replaced.
- Determine experimentally whether different results are obtained when the analyte is spiked into a matrix extract, compared with results obtained when it is spiked directly into the matrix prior to extraction.
- Verify that there is no interference from compounds that might reasonably be expected to be present in typical field samples, such as compounds with a

common or complementary use (e.g., compounds that are co-formulated with the target analyte or otherwise approved for veterinary use in the same species), structurally related compounds, and known metabolites and degradation products. Use knowledge of production practices combined with analyst experience to identify other compounds that might also be present in samples.

Decision 2002/657/EC advises the following for selectivity experiments:[35]

> potentially interfering substances shall be chosen and relevant blank samples shall be analysed to detect the presence of possible interferences and to estimate the effect of the interferences:
>
> - select a range of chemically related compounds (metabolites, derivatives, etc.) or other substances likely to be encountered with the compound of interest that may be present in the samples;
> - analyse an appropriate number of representative blank samples ($n \geq 20$) and check for any interferences (signals, peaks, ion traces) in the region of interest where the target analyte is expected to elute;
> - additionally, representative blank samples shall be fortified at a relevant concentration with substances that are likely to interfere with the identification and/or quantification of the analyte;
> - after analysis, investigate whether:
>   - the presence may lead to a false identification,
>   - the identification of the target analyte is hindered by the presence of one or more of the interferences, or
>   - the quantification is influenced notably.

Additional guidance is provided by Eurachem, which recommends that the following experiments be conducted for "confirmation of identity" and "selectivity/specificity":[16]

- "Analyse samples and reference materials by candidate and other independent methods," then "use the results from the confirmatory techniques to assess the ability of the method to confirm analyte identity and its ability to measure the analyte in isolation from other interferences."
- In assessing the available data, "decide how much supporting evidence is reasonably required to give sufficient reliability." It is considered that single experiments should be sufficient for this investigation (repeating an experiment is not required unless anomalous results are obtained requiring additional verification).
- "Analyse samples containing various suspected interferences in the presence of the analytes of interest" and "examine effect of interferences—does the presence of the interferent enhance or inhibit detection or quantification of the measurands." In assessing the data from these experiments, "if detection or quantitation is inhibited by the interferences, further method development will be required." Again, only single experiments should be required unless a need to verify a particular observation through additional experiments is apparent.

It is not possible to test for all potential interferences, and there is a possibility that an interfering substance will subsequently be encountered in analysis of field samples from diverse sources. Detection using tandem mass spectrometers, with multiple- or selected-reaction monitoring (MRM or SRM; see Chapters 6 and 7), reduces the probability of encountering unexpected interferences, but does not totally eliminate the possibility. When interference is discovered once a validated method has been put into routine use, it may be necessary to modify the method (clean-up or chromatography) to eliminate the interference. Such modifications, if they involve significant change, may require re-validation of the method.

8.7.3.3 Additional Selectivity Considerations for Mass Spectral Detection

Mass spectral methods are based on the detection of fragmentation products that are structurally significant, and these must be derived from the original structure of the parent compound, not from the structure of a derivatizing agent that may have been chemically bonded to the target compound to aid in analysis. In addition, all fragments monitored should not originate from the same part of the molecule, but, in addition to the molecular ion, characteristic adducts or fragments, as well as isotopes, are considered suitable choices for selected-ion monitoring or selected-reaction monitoring.[35] Non-characteristic fragments, such as water and carbon dioxide, are not considered as acceptable fragments for regulatory methods. The fragmentation pattern should be included in the validation report, showing the chemical structure of each fragment monitored in the method. It should be demonstrated in the validation report that each fragment identified for monitoring is structurally related to the parent compound. If a high-resolution mass spectrometer is available, it should be used to ensure that accurate mass measurements are the basis for assignment of elemental composition of each fragment (Chapter 6). However, if such equipment is not available, it is usually possible when working with standards of known elemental composition and structure to deduce the elemental composition of the fragments from mass spectral results obtained using lower-resolution instruments if these are not available from published literature.

Sources such as the peer-reviewed scientific literature and reference texts may provide information on mass spectral fragmentation of the compound and eliminate the need to conduct experiments to characterize and map the fragmentation pattern. However, the fragmentation results previously reported in the literature should be verified using the equipment that will be used in validation and subsequent routine implementation of the method. While the same fragments and transitions should be observed on different instruments, variations may be observed in ion ratios, particularly when there are differences in source geometry of the instruments used in the reported study and the instrument on which the new method is being developed and validated. When literature sources are used for selection of the ions to be monitored, these should be referenced in the validation report.

Guidance on the fitness for purpose of mass spectrometric methods from the American Society for Mass Spectrometry (ASMS)[68] includes certain basic principles found in most documents on confirmatory analysis using mass spectrometry:

- Use of reference standard analyzed contemporaneously with unknowns.
- Three or more diagnostic ions (except for exact mass measurements)
- Use of relative abundance matching tolerances for selected ion monitoring (SIM)

ASMS also recommends that a minimum signal-to-noise ratio for any diagnostic mass spectral peak should be not less than 3 : 1. According to Decision 2002/657/EC,[35] "confirmatory methods for organic residues or contaminants shall provide information on the chemical structure of the analyte." For mass spectral analyses using SIM, the EC requirement further states that acceptable diagnostic ions include "the molecular ion, characteristic adducts of the molecular ion, characteristic fragment ions and all their isotope ions" and that the "selected diagnostic ions should not exclusively originate from the same part of the molecule." These factors should all be taken into account in choosing appropriate fragment ions to use in mass spectral confirmation.

The mass spectra produced using typical quadrupole mass spectrometers found in most residue control laboratories, or higher-resolution magnetic sector or time-of-flight mass spectrometers available in some laboratories, yield electron impact (EI) spectra that are generally comparable and consistent, whether samples are introduced via direct probe or through a heated interface to a gas chromatograph. Consistent and comparable chemical ionization (CI) mass spectra are also usually obtained when using such instruments. The EI and CI spectra from GC-MS analyses may be compared with reference spectra found in spectral libraries that are available for use with commercially available mass spectrometer data systems or with spectral libraries generated "in house" using instruments in routine use in the laboratory. Such libraries are usually readily transferable between instruments.

The same is not necessarily the case for spectra generated on LC-MS and LC-MS/MS instruments. The source geometry and operating conditions for each instrument have a significant effect on the spectra produced on a given day for each instrument. The same fragment ions should be generated in most situations, but the relative ion abundances and favored transitions may vary from those obtained in a previous experiment, particularly an experiment performed on a different instrument. While spectral libraries and published papers may identify the characteristic fragment ions or transitions that should be produced from each compound, additional work may be required to fully characterize the optimum conditions for analysis of the compound on the instrument used in the analysis. Currently recommended practices suggest that LC-MS and LC-MS/MS spectra generated in confirmatory analyses should be obtained in conjunction with spectra generated from appropriate standards at approximately the same concentration, contained in the same analytical run, for reliable confirmation.[38]

A minimum set of confirmation criteria are contained in Decision 2002/657/EC,[35] including the minimum number of ions or transitions to be monitored for each technique (dependent also on instrument resolution), agreement of ion ratios and minimum signal strength for each ion monitored. In this EU Decision, which is legally binding on official residue control laboratories in EU member states and also establishes requirements for official laboratories responsible for export control of products destined for markets in the EU, it is recommended that a minimum of four characteristic ions be monitored for confirmation of each analyte when the analyses are conducted on typical low-resolution quadrupole mass spectrometers (GC-MS or LC-MS). These fragments may be obtained either from a single analysis when sufficient fragments are generated or by combining results from several different determinations. For example, results of analyses using different forms of ionization may be combined, such as the fragments obtained from an EI analysis of a compound, followed by a second analysis using CI.

One "identification point" (IP) is assigned for each characteristic ion detected using single-stage quadrupole mass spectrometric detectors. High-resolution methods provide more accurate mass measurement (analyses using magnetic sector and time-of-flight instruments), and therefore it is usually considered sufficient to use only two characteristic ions for identification, with two IPs assigned per ion. For low-resolution MS/MS methods (quadrupole instruments), a precursor (parent) ion plus two characteristic transition ions are required, with a single IP assigned for the precursor (or parent) ion and 1.5 points assigned to each transition. A minimum of 4 IPs are required to meet confirmatory criteria required for laboratories operating under Decision 2002/657/EC[35] and are also found as recommendations in the ASMS guidance document.[68] The identification point system for confirmation is also contained in CAC/GL 71–2009.[32] More detailed discussion on IPs may be found in Chapter 6.

The early work on requirements for confirmatory mass spectrometric methods was undertaken to address a regulatory requirement for the analysis of carcinogens established in 1977 by the US Food and Drug Administration. A paper published in 1978 demonstrated the minimum number of peaks and the specifications for relative ion abundances for each peak required to uniquely identify diethylstilbestrol from a GC-MS spectral library containing EI spectra of 30,000 compounds.[69] The same results were obtained when the work was repeated at an ASMS workshop in 1997, but this time were based on comparison with a library containing 270,000 GC-MS EI spectra.[70] The initial library search looked for any match for an ion with m/z 268 and gave 9995 matches out of the 270,000 spectra in the library. Addition of a second ion, m/z 239, reduced the number of matches to 5536, then requiring a relative abundance of 90–100% for m/z 268 and 10–90% for m/z 239 reduced the number of matches to 46. Reducing the range within which the relative abundance of m/z 239 must fall to 50–70% led to only nine spectral matches from the library. Addition of m/z 145 to the ions to be matched and specifying that the ions must display relative abundances of 90–100% for m/z 268, 50–70% for m/z 239, and 45–65% for m/z 145 produced a match with only the spectrum for DES contained in the library of 270,000 spectra.

Decision 2002/657/EC[35] and CAC/GL 71-200932 contain additional performance requirements for mass spectral methods used in the confirmation of veterinary drug residues in foods. It should be noted that the performance specifications given for GC-MS methods using EI spectra are more stringent than for other techniques, including GC-MS with CI, LC-MS, and LC-MS/MS. It is also recommended that only ions with an intensity >10% of the base peak should be used as analytical peaks for confirmation. The EU and CAC requirements address performance of both the chromatographic system and the mass spectrometer:

- For GC-MS methods, the retention time of the target analyte in a sample should agree within ±0.5% with the retention time of a standard, while a match of ±2.5% is required for retention time of the analyte with the standard for LC-based methods.
- For the mass spectrometric detection, the requirement for GC-MS with EI methods is that, using a ratio to the base peak in the analyte spectrum, peaks with a relative abundance of >50% should match the same ratio in the standard within ±10%, the ratio for peaks with a relative abundance of 20–50% should match the same ratio in a standard within ±15%, and peaks that have a relative abundance of 10–20 should be within ±20% of the equivalent ratio in a standard.
- For all other techniques, including GC-MS with CI, peaks with a relative abundance of >50% should match the same ratio in the standard within ±20%, the ratio for peaks with a relative abundance of 20–50% should match the same ratio in a standard within ±25%, and peaks that have a relative abundance of 10–20% should be within ±30% of the equivalent ratio in a standard.

These requirements, along with other recommendations found in the guidance documents available, constitute the fitness-for-purpose requirements for residue methods using mass spectral detection.[32,35,68,71] An excellent discussion of some issues related to the performance of mass spectrometry methods is contained in a 2010 paper, including the misuse of the term "sensitivity" and the challenges of establishing lower limits for the routine application of mass spectrometric methods.[72]

It should be noted, however, that while the original approach requiring a minimum number of peaks and a match on ion ratios used for GC-MS EI spectra was based on an evaluation of a significant pool of data in spectral libraries,[69,70] the recommendation for two transitions to provide sufficient confirmation in LC-MS/MS methods is not based on the same type of evaluation. At least one instance has been reported where the two transition basis for confirmation proved inadequate and both an improved chromatographic separation and monitoring of a third transition were required to provide reliable confirmation.[73] In terms of method development and validation, this suggests that caution and judgment should be exercised in determining the fit-for-purpose criteria for individual confirmatory methods using low-resolution LC-MS/MS instruments. The possibility of interferences should always be considered, and the identification of additional transitions or alternative confirmatory procedures to deal with such issues that may arise in the analysis of real samples should be taken into account.

8.7.4 Accuracy

Our concept of "accuracy" involves two additional terms, "trueness" and "bias." CAC/GL 72–2009,[32] in defining accuracy, includes a note that states that "when applied to a test method, the term accuracy refers to a combination of trueness and precision." "Trueness" is defined as the "closeness of agreement between the average of an infinite number of replicate measured quantity values and a reference quantity value" and "bias" as the "difference between the expectation of the test result or measurement result and the true value. In practice conventional quantity value can be substituted for true value." In residue analysis, the "true value" is rarely known for the content of the analyte in samples analyzed, so bias is usually determined by comparing typical analytical results obtained when certified reference materials are analyzed, if these are available, through differences in analyst or laboratory results from the consensus values obtained in proficiency tests or by comparison with spiked values.

The basic purpose of a quantitative analysis is to provide a measure of the quantity of a particular analyte (or analytes) present in a given mass of a representative sample, whether the measure is to determine the purity of a substance, the percentage of the product that is represented by a particular constituent or the amount of a residue, contaminant, or other trace constituent present in the product. The definitions used should be applicable to all these types of determinations. "Trueness" is defined in IUPAC guidance as "the closeness of agreement between a test result and the accepted reference value of the property being measured,"[11] while "accuracy of measurement" is defined by IUPAC as the "closeness of the agreement between the result of a measurement and a true value of the measurand."[7] The "measurand" is defined by IUPAC as the "particular quantity subject to measurement"[7] and as the "quantity intended to be measured" by the JCGM.[43] Trueness therefore is a measure of the ability of a method to give a result that closely approximates the true (or accepted) value for the concentration of an analyte in a test material. Referring to the analysis of products for which the CAC has developed standards, CAC/GL 72-2009[32] includes in the definition of "accuracy" the added information that "the term 'accuracy,' when applied to a set of test results or measurement results, involves a combination of random components and a common systematic error or bias component." In Decision 2002/657/EC, the definition is also similar, stating that accuracy "means the closeness of agreement between a test result and the accepted reference value" and that it "is determined by determining trueness and precision."[35] In practice, when we validate an analytical method for accuracy, particularly a method for a residue or a contaminant, the practical approach provided in the CAC *Procedural Manual* for determination of the accuracy of methods when validating using the criteria approach in a single-laboratory validation[25] can be seen to be closely related to the procedures recommended by IUPAC for the determination of accuracy.[11] These recommended procedures include analysis of certified reference materials or, in their absence, other reference materials, comparison of results with those obtained using a recognized reference method, or, when these options are not available, recovery experiments using fortified materials.

Method validation should address the two components of accuracy: the trueness (agreement with the expected result) and the bias, the systematic variation observed for the analyst or laboratory when performing the analysis over a period of time. Typically, different analysts in the same laboratory will achieve different results when analyzing the same materials. One may usually achieve a result lower than the accepted value, while another may typically achieve a higher result than will other analysts. This is known as *analyst bias* and reflects slight differences in the manner in which analysts conduct the analysis or perhaps differences in instruments if the analysts involved use different instruments. The cause of such variability should be investigated, whenever possible, during method development and validation to reduce or eliminate the potential for significant analyst bias in on-going work.

When all analysts in a laboratory consistently achieve results that are either above or below expected values, then this is termed *laboratory bias* and may reflect either a consistent error in the manner in which a method is conducted by all analysts, a problem with an analytical standard, an instrument calibration issue, or a fundamental problem with the analytical method. Again, the source of bias should be investigated and, when it is method-related, further method development or modification and re-validation may be required.

The three methods recommended by IUPAC for assessment of the accuracy of an analytical method in a single-laboratory validation are, in order of preference:[11]

1. Analysis of a certified reference material
2. Comparison with results from a recognized reference method
3. Recovery of known quantities of a standard added to representative sample matrix

For the analysis of antibiotic residues in foods, there is a limited number of methods that have been evaluated by collaborative study[13] and a lack of certified reference materials (see other chapters on specific methods of analysis and quality assurance for additional details). Hence, the assessment of accuracy is most frequently based on analytical recovery from fortified materials when developing and validating these methods.

8.7.5 Recovery

Recovery is usually expressed as the percentage of analyte experimentally determined after fortification of sample material at a known concentration and should be assessed over concentrations that cover the analytical range of the method. CAC/GL 72-2009[32] defines recovery as follows:

> *Recovery/recovery factors:* Proportion of the amount of analyte, present in, added to or present in and added to the

analytical portion of the test material, which is presented for measurement.

Notes:

Recovery is assessed by the ratio $R = C_{obs}/C_{ref}$ of the observed concentration or amount C_{obs} obtained by the application of an analytical procedure to a material containing analyte at a reference level C_{ref}.

C_{ref} will be:

(a) a reference material certified value,
(b) measured by an alternative definitive method,
(c) defined by a spike addition or
(d) marginal recovery.

Most regulatory authorities require analytical results for veterinary drug residues to be corrected for the analytical recovery. However, regardless of whether this is a formal requirement, it is important to quantify this parameter and report it with analytical data. Guidance on recovery correction has been provided by IUPAC.[74] IUPAC defines recovery as a "term used in analytical and preparative chemistry to denote the fraction of the total quantity of a substance recoverable following a chemical procedure."[6] Recovery (R) may be mathematically expressed as:

$$R(\%) = \frac{C_1 - C_2}{C} \times 100$$

where $R(\%)$ is percent recovery, C_1 is the measured concentration in fortified material, C_2 is the measured concentration in unfortified material (the background signal), and C is the known increment in concentration (amount added).

To correct an analytical result for recovery, the equation becomes:

$$C_{actual} = \frac{C_{det}}{R_{det}}$$

where C_{actual} is the calculated actual or true concentration, C_{det} is the concentration as determined in the analysis, and R_{det} is the recovery factor determined for the analysis.

In interpreting recoveries for methods for antibiotics in foods, as for other methods involving the analysis of trace constituents in biological materials, it must be understood that analyte added to a blank sample material is a surrogate for incurred material. Thus, this surrogate may not behave in the same manner as the analyte when present in an equivalent biologically incurred material. Investigations done with radiolabeled drugs as part of the evaluation prior to registration of the drug for veterinary use frequently reveal that the amount of an incurred residue that is extracted (the yield or recovered fraction) is less than the total incurred residues present (see residue monographs produced by JECFA[75]). This may be due to losses during extraction, intracellular binding of residues, the presence of conjugates, or other factors that are not fully represented by recovery experiments conducted with analyte-fortified blank tissues. At relatively high concentrations, analytical recoveries are expected to approach 100%. At lower concentrations, particularly with methods involving extensive extraction, isolation, and concentration steps, recoveries may be lower. Regardless of what average recoveries are observed, recovery with low variability is desirable so that a reliable correction for recovery can be made to the final result, when required. Recovery corrections should be made consistent with the guidance provided by IUPAC[74] and the CAC.[28]

8.7.6 Precision

Precision, which quantifies the variation between replicated measurements on test portions from the same sample material, is also an important consideration in determining when a residue in a sample should be considered to exceed a MRL or other regulatory action limit. Precision of a method is usually expressed in terms of the within-laboratory variation (repeatability) and the between-laboratory variability (reproducibility) when the method has been subjected to a multi-laboratory trial. For a single-laboratory method validation, precision should be determined from experiments conducted on different days, using a minimum of six different tissue pools, different reagent batches, preferably different equipment, and so on, and preferably by different analysts.[32] Repeatability of results when determined within a single laboratory but based on results from multiple analysts is termed *intermediate precision*.[16] Precision of a method is usually expressed as the standard deviation. Another useful term is *relative standard deviation*, or *coefficient of variation* (the standard deviation divided by the absolute value of the arithmetic mean result, multiplied by 100 and expressed as a percentage).

Method precision results, as achieved in a laboratory developing a method, will often be superior to the variability achieved by another laboratory that may later use the method. If a method cannot achieve a suitable standard of performance in the laboratory where it was developed, it cannot be generally expected to do any better in other laboratories. Typically, the intermediate precision determined within a laboratory is 1–2% greater than the precision as determined for a single analyst, while the between-laboratory precision, or reproducibility, is again 1–2% higher than the within-laboratory precision.

8.7.7 Experimental Determination of Recovery and Precision

8.7.7.1 Choice of Experimental Design

The analytical recovery and the method precision (relative standard deviation of individual results) should be determined from experiments conducted:

- On different days
- Using different calibration curves
- Using different batches of reagents
- Using matrices from different sources
- By different analysts, when possible

These experiments are designated as "phases II and III" of a method validation or an analyst familiarization for those using the approach recommended in the USDA/FSIS *Chemistry Laboratories Guidebook* (see QA section of the posted methods).[76] The results provide an assessment of the recovery (trueness) and the analyst precision attained with the method under routine conditions of use. In addition, the data generated may be used to calculate statistical estimates of the reliability of the results, including estimates of MU.[11,17]

The *decision limit* (CCα) and the *detection capability* (CCβ) as described in Decision 2002/657/EC[35] may also be calculated from data generated in these experiments, combined with data from the calibration curve experiments, when calculation of these parameters is required. While the decision limit CCα is calculated from the same data used to calculate the detection limit, the calculation of the detection capability CCβ requires that additional data are generated. The calibration curve generated for each analytical run involved in these experiments should be assessed for sensitivity and linearity. All results should also be inspected for interferences to verify selectivity. The data from all calibration curves generated using the same process (whether using external standard calibration with pure standard curves or matrix-fortified curves, including those generated in phase I experiments, may be combined to improve the estimates of LOD and LOQ when these are calculated from calibration curve data. However, do not pool data from experiments using pure standards and data from curves generated using fortified matrix for such calculations; all the calibration curves used should be prepared using the same basis.

Obvious questions are how many replicates should be included in the design and also how many analytical runs should be completed. Eurachem guidance suggests a minimum of 10 replicates for recovery (accuracy) and precision.[16] A collaborative study design is recommended to include a minimum of five materials (matrices) at three concentrations, in blind duplicate,[9] which means that each participating laboratory produces 10 results for each concentration. However, when one considers the recommendation that at least six different sources of matrix should be used in validation for methods for veterinary drug residues in foods[32] and that the validation should include analyses conducted on multiple days, with inclusion of other variables, such as analyst, equipment, and reagents, it is obvious that 10 replicates will prove insufficient to provide the necessary data for a suitable assessment of method performance at the various concentrations in the various different sources of matrix recommended.

The underlying concept of the validation experiments is to provide a prediction of the performance to be expected over an extended period in routine use, and a minimal dataset will not satisfy this expectation. Therefore, a typical validation design will include the six different sources of matrix, usually at each of three concentrations bracketing the MRL, repeated as analyst spikes in three or four analytical runs, followed by one or two additional runs where the materials are provided as unknowns (blind) to the analyst. The design is usually repeated for each required matrix (e.g., each species–tissue combination) for the initial target species and may be also be required when the method is applied routinely to other species. However, when there are obvious commonalities (such as tissues from different ruminants), method extension may require only a reduced dataset, based on experience with the method.

A typical experimental design to assess method performance, particularly recovery and repeatability for an analyte for which a regulatory limit has been established, would therefore include the following experiments:

- *Analyst-Fortified Samples.* Conduct three or four analytical runs (preferably four for a single-analyst method validation), on separate days, each containing six representative blank matrices (preferably from six different sources), plus six representative blank matrices (test portions) fortified at each of 0.5, 1.0, and 2.0 times the MRL or regulatory action level, plus a calibration curve. Additional concentrations may be added when the method will be routinely applied over a larger analytical working range, such as may be required for dietary intake studies or when there is an expectation that some samples may contain residues in excess of twice the MRL. If the run size will not accommodate this many test portions, distribute the test portions over a larger number of analytical runs to produce a minimum of $n = 18$ (preferably $n = 24$) results for the blank matrices and for each of the test concentrations included in the design. The number of data points that are generated allows true "outliers" to be differentiated from the expected scatter of results reflecting the method precision, and these may be removed from the precision calculation using appropriate statistical tests. When a second analyst is involved in producing data to enable calculation of intermediate precision, the number of runs recommended may be reduced from four to three per analyst.
- *Test Portions Provided Blind to the Analyst(s).* A minimum of two analytical runs should be conducted by each analyst. Each run should include a minimum

of three blind test portions (in duplicate) at concentrations of 0.5, 1.0, and 2 times the MRL (or other representative concentrations within the intended range of applications of the method). Incurred samples should be included in the design when suitable incurred material is available. Incurred samples may be diluted with similar blank matrix material to obtain the desired analyte concentrations for these experiments (ensure that such materials are homogeneous). Samples for this phase of validation may be provided to the analyst by the quality manager, the supervisor, or another analyst not involved in the validation.

The experiments conducted with representative fortified matrix materials should include each of the typical sample materials to which the validated method will be routinely applied. Methods applied to animal tissue are usually validated for the designated target tissue for survey samples for each species tested and also for any additional tissues that may be routinely analyzed as part of additional surveillance or compliance testing. This would usually include muscle, liver, kidney, and fat, plus any other organ tissues that are part of the normal diet in the country where the method is being used. Representative species are usually selected to validate methods for fish, such as fin fish with low fat content (e.g., tilapia), a fin fish with high fat content (salmon), and representative shellfish such as shrimp and scallops. The use of representative matrices for validation of multi-residue methods applied to a wide range of commodities has been recommended in CAC guidelines for pesticide residue analysis,[77] and such advice seems equally applicable to methods for other residues, such as antibiotics. Additional information on suggested representative matrices for various commodities can be found in Table 5 of this CAC guidance document, as well as advice on how to determine whether the method is suitably validated when first applied to an additional member of a commodity group for which has the method has been validated using another member of that group as the "representative matrix."

Ultimately, the validation design should demonstrate that the method is fit for purpose, according to the purposes for which the method will be applied within the testing program. When two or more analysts produce data, intermediate precision may be calculated as a more realistic estimate of ongoing method performance within the laboratory. If the opportunity exists to conduct a sample exchange with other laboratories using the same method or to conduct a collaborative trial, then precision under reproducibility conditions may be determined. In the absence of such multi-laboratory trials, precision under reproducibility conditions may be estimated arithmetically using procedures such as those described by Thompson.[78]

When the preceding design is used, approximately 1–2 weeks are required to complete the experiments with analyst-fortified materials for each sample matrix for a typical residue control method (assume one or two analyst days per analytical run) and an additional week is required to complete the runs with sample material that is blind to the analyst. More complex methods that require 3 days or more per run will require additional time to complete the validation work.

A similar experimental design to assess method repeatability is contained in Decision 2002/657/EC:[35]

- Prepare a set of samples of identical matrices, fortified with the analyte to yield concentrations equivalent to 1, 1.5, and 2 times the minimum required performance limit or 0.5, 1, and 1.5 times the permitted limit.
- At each level the analysis should be performed with at least six replicates.
- Analyse the samples.
- Calculate the concentration detected in each sample.
- Find the mean concentration, standard deviation, and the coefficient of variation (%) of the fortified samples.
- Repeat these steps on at least two other occasions.
- Calculate the overall mean concentrations and CVs for the fortified samples.

This design yields 18 replicates at each of three concentrations and involves one less analyst-fortified run than recommended above and targets validation of the method at the regulatory limit.

In addition, Decision 2002/657/EC contains an experimental design for determination of the intermediate precision, which is referred to as *within-laboratory reproducibility*:[35]

- Prepare a set of samples of specified test material (identical or different matrices), fortified with the analyte(s) to yield concentrations equivalent to 1, 1.5, and 2 times the minimum required performance limit or 0.5, 1, and 1.5 times the permitted limit.
- At each level the analysis should be performed with at least six replicates.
- Repeat these steps on at least two other occasions with different operators and different environmental conditions, e.g., different batches of reagents, solvents etc., different room temperatures, different instruments, etc. if possible.
- Analyse the samples.
- Calculate the concentration detected in each sample.
- Find the mean concentration, standard deviation and the coefficient of variation (%) of the fortified samples.

When conducting the validation experiments to meet the requirements of Decision 2002/657/EC,[35] the data generated in these experiments may be used in calculation of the decision limit and the detection capability. The calculation of CCα requires 20 replicates at either the MRL (for MRL substances) or at a concentration equal to the *y* intercept plus 2.33 standard deviations (non-MRL substances). The design given above (four runs by one analyst or three runs each for two analysts, plus the blind runs if they contain samples at the appropriate concentration) provides the data required for calculation of CCα for analytes for which a MRL has been established. For non-MRL substances, additional experiments may be required. The calculation of CCβ requires analysis of 20 blank samples fortified at the decision limit.[35] If this concentration is not one of those used in the design above, additional analyses must be conducted to generate the necessary 20 data points at the specified concentration. However, Decision 2002/657/EC makes it clear that alternative approaches can also be used to calculate these limits.

The inclusion of intermediate-precision experiments is recommended, whenever feasible, in the validation design, as is the inclusion of "blind" samples that are recommended in validation designs following the USDA/FSIS model,[76] but not specified in Decision 2002/657/EC.[35] However, Decision 2002/657/EC does allow for "alternative approaches" for validation and contains an example of such an approach, in which multiple factors are included, such as animal breed and gender and husbandry conditions, as well as analyst experience. The authors of the alternative model, cited in Decision 2002/657/EC, comment in their paper that "there is no commonly accepted validation procedure" for the assessment of the various performance parameters, such as limit of quantification, and that "there is no consensus about the choice of calibration samples and the number of replicates."[79] Unfortunately, there still is no generally accepted formula to follow, although IUPAC has initiated a project to develop specific guidance on experimental design to support the application of their single-laboratory validation guidelines during the preparation of this chapter.[80] The key things to ensure are that the validation model meets the requirements (such as a national regulation or policy), that sufficient data are generated to provide some confidence in the reliability of the work, that the design is documented (and referenced, when appropriate), and that the validation practices in the laboratory are generally consistent and well documented over time.

In summary, in addition to the primary performance characteristics of recovery and precision, other performance characteristics should also be assessed, including:

- Calibration curves used in each analytical run should meet the requirements for analytical range, linearity, and sensitivity required.
- No interfering substances should be detected.
- Performance at the LOQ or LOD (or CCα/CCβ) should be demonstrated in each run, if applicable.

8.7.7.2 Matrix Issues in Calibration

When planning analytical recovery experiments, it is important to compare observed recoveries against known performance. However, absolute recovery using radiolabeled incurred material is rarely an option. This leaves the options of using either matrix-fortified (method matrix-matched or pre-extraction spiking) or matrix-matched (post-extraction spiking) reference materials. In post-extraction spiking, the extract from the analytical sample is spiked with the analyte of interest at a known concentration immediately after extraction. As a result, the extraction efficiency is unknown and therefore there is an additional uncertainty introduced into the analytical procedure.

If, however, the analytical sample is spiked with a known quantity of analytical standard prior to extraction (pre-extraction spiking), a more accurate assessment of the extraction efficiency is obtained, although the possibility of losses due to tissue–analyte binding cannot be disregarded, and a more accurate appreciation of analytical recovery will be obtained. Comparisons of these spiking methods showed that the latter approach gave acceptable results in the absence of radiolabeled incurred material.[81,82]

The use of pre-extraction spiking is particularly important when the presence of matrix co-extractives modifies the response of the analyte as compared with analytical standards. It is increasingly common in methods for veterinary drug residues in foods to base the quantitative determination on a standard curve prepared by addition of standard to known blank representative matrix material at a range of appropriate concentrations that bracket the target value (the analytical function). Use of such a "tissue standard curve" for calibration incorporates a recovery correction into the analytical results obtained.

All LC-MS techniques tend to be subject to matrix effects, especially suppression, although enhancement effects may also be observed. A procedure has been suggested to systematically investigate matrix effects when developing and validating methods using LC-MS or LC-MS/MS for detection.[83] First, run pure standards to determine the analyte response in the absence of matrices. Next, either prepare standards in a matrix extract or infuse standards in the presence of matrix extract into the mass spectrometer and determine whether the response differs from that observed for pure standards. Differences in response may be attributed to matrix suppression (or enhancement) effects. Finally, fortify blank tissue with standards, perform the extraction and clean-up steps of the method, and then determine the detector response. The difference between the response observed for fortification into matrix extract and fortification into matrix prior to extraction and clean-up is attributed to method recovery. The evaluation of matrix effects is discussed in detail in Chapter 6.

Authors have used the term *matrix-matched* to describe both approaches to fortification—fortification of extracts prepared from blank matrix and fortification of blank matrix, although the two approaches can yield quite different results. To provide more clarity, the terms *matrix-matched standard calibration curve* (MSCC) and *method matrix-matched standard calibration curve* (MMSCC) have been proposed to describe calibration curves prepared by fortification of blank extract and by fortification of blank matrix, respectively.[84] This issue is also discussed in Chapter 6 and is an important consideration for methods using LC-MS and LC-MS/MS detection.

Confirmation of quantity may also require the use of the method of standard addition or the inclusion of isotopically labeled standards, although there are currently few such materials available for antimicrobial drugs, as noted in a review in 2009.[85] Some examples of the application of isotopically labeled internal standards to methods for antimicrobial residues include the determination of chloramphenicol in meat, fish, and other biological matrices;[86] nitrofuran residues in milk;[87] and nitroimidazole residues in eggs;[88] and animal plasma.[89]

8.7.8 Measurement Uncertainty (MU)

MU was defined in the earlier version of the VIM as a "parameter associated with the result of a measurement, that characterizes the dispersion of the values that could reasonably be attributed to the measurand."[90] This definition appears in numerous subsequent documents on MU. However, in the current edition of the VIM, MU is defined as a "non-negative parameter characterizing the dispersion of the quantity values being attributed to a measurand, based on the information used."[43]

The provisions of ISO 17025 require that an accredited laboratory assess the MU associated with each test method listed within the scope of the accreditation and must make this information available to customers.[2] The MU associated with an analytical method should be part of the validation experiments for new methods, as discussed in IUPAC guidelines for single-laboratory validation of analytical methods.[11] CAC/GL 71–2009 also recommends that laboratories provide information on MU to their customers on request, while recognising that there are different ways in which MU can be estimated.[32] However, the determination of MU is not mentioned in Decision 2002/657/EC,[35] where calculations of CCα and CCβ may be considered to serve as alternative statistical performance indicators to MU.

Information on MU may be made available by inclusion in the method SOP or in a memorandum to the residue program manager. Measurement uncertainty includes two major elements: the closeness of the analytical result to the true value (*trueness* or *accuracy*) and the variability associated with the measurement (*precision*). There are two general approaches to the estimation of MU, frequently referred to as "top–down" and "bottom–up."[30] The *top–down* approach makes a direct determination of the combined contributions to uncertainty from method performance data, such as variations in recovery and precision, obtained during method validation experiments or using quality control data. In the *bottom–up* approach, steps in the method are reviewed to identify the individual elements that potentially may contribute to uncertainty. Each contribution should be quantified, even if general knowledge suggests that the contribution may be minimal. Some contributions will be found to be insignificant and may be ignored in calculating overall MU. Detailed guidance and examples of experiments to estimate MU may be found in available guidance documents[17] and in Chapter 9.

8.7.9 Limits of Detection and Limits of Quantification

The determination of the limits at which reliable detection, quantification, or confirmation of the presence of an analyte in matrix can be achieved is usually required in validating an analytical method for a residue of a veterinary drug. IUPAC defines the "limit of detection (in analysis)" as follows:[6]

> The limit of detection, expressed as the concentration, c_L, or the quantity, q_L, is derived from the smallest measure, x_L, that can be detected with reasonable certainty for a given analytical procedure. The value of x_L is given by the equation
>
> $$x_L = x_{bi^-} + k s_{bi}$$
>
> where x_{bi} is the mean of the blank measures, s_{bi} is the standard deviation of the blank measures, and k is a numerical factor chosen according to the confidence level desired.

This definition is cross-referenced with the definition for the term *detection limit in analysis*, defined as the "minimum single result which, with a stated probability, can be distinguished from a suitable blank value. The limit defines the point at which the analysis becomes possible and this may be different from the lower limit of the determinable analytical range." CAC/GL 72-2009[32] defines the *limit of detection* (LOD), as "the true net concentration or amount of the analyte in the material to be analyzed which will lead, with probability (1–β), to the conclusion that the concentration or amount of the analyte in the analyzed material is larger than that in the blank material. It is defined as:

$$\Pr(Ł \leq L_C | L = \text{LOD}) = \beta$$

Where Ł is the estimated value, L is the expectation or true value and L_C is the critical value."

In accompanying "Notes," the CAC guideline states that the limit of detection (LOD)

> is estimated by LOD $\approx 2t_{1-\alpha v}\sigma_0$ [where $\alpha = \beta$],
>
> Where $t_{1-\alpha v}$ is Student's-t, based on v degrees of freedom for a one-sided confidence interval of $1-\alpha$ and σ_0 is the standard deviation of the true value (expectation). [It also notes that] the correct estimation of LOD must take into account degrees of freedom, α and β, and the distribution of L as influenced by factors such as analyte concentration, matrix effects and interference.

The detection limit may be described in practical terms as the lowest concentration where the analyte can be identified in a sample. It can be estimated using the standard deviation ($s_{y/x}$) from the linear regression analysis of the calibration curves generated in the experiments described above.[91] In this approach, which provides a conservative estimate, the limit of detection is calculated using the y intercept (assuming a positive value) of the curve plus 3 times $s_{y/x}$. The detection limit may also be estimated by measurements on representative test materials, using the background signal at the point in the chromatogram where the analyte is measured plus 3 times its standard deviation. It is often necessary to fortify test materials at a concentration resulting in a barely detectable response to obtain an approximation of the standard deviation of the blank when using this approach.

The fundamental problem with determining a detection limit for a method applied to trace quantities of an analyte extracted from a complex matrix is that many variables may affect the result on any given day. These include (but are not limited to) variability in matrix coextractives, variations in the source or the detector response, cleanliness of the detector cell, and variations between instruments if more than one instrument is available for use in the laboratory (particularly an issue for LC-MS and LC-MS/MS). The many issues that render determination of a detection limit more complex than it may initially appear are discussed in the IUPAC guidance on method validation in a single laboratory, where it is also suggested that "for analytical systems where the validation range does not include or approach it, the detection limit does not need to be part of a validation."[11] However, if a detection limit specification is critical to a regulatory result, it becomes important to include a sample fortified at the detection limit in each analytical run to demonstrate that in the set of samples analyzed in that run, detection of a positive result at the claimed detection limit was achieved. To demonstrate that a detection limit is realistic, 19 of 20 typical samples fortified at that concentration should be detected as positive.

The limit of quantification (LOQ) is defined in CAC/GL 72-2009 as follows:[32]

> A method performance characteristic generally expressed in terms of the signal or measurement (true) value that will produce estimates having a specified relative standard deviation (RSD), commonly 10% (or 6%). LOQ is estimated by:
>
> $$\mathrm{LOQ} = k_Q\sigma_Q, \quad k_Q = 1/\mathrm{RSD}_Q$$
>
> Where LOQ is the limit of quantification, σ_Q is the standard deviation at that point and k_Q is the multiplier whose reciprocal equals the selected RSD. (The approximate RSD of an estimated σ, based on v-degrees of freedom is $1/\sqrt{2v}$.)

The CAC definition also notes that when

> σ is known and constant, then $\sigma_Q = \sigma_0$, since the standard deviation of the estimated quantity is independent of concentration. Substituting 10% in for k_Q gives:
>
> $$\mathrm{LOQ} = (10 * \sigma_Q) = 10\sigma_0$$
>
> In this case, the LOQ is just 3.04 times the limit of detection, given normality and $\alpha = \beta = 0.05$.

In addition "at the LOQ, a positive identification can be achieved with reasonable and/or previously determined confidence in a defined matrix using a specific analytical method."

The limit of quantification may be established from the validation experiments described above using the y intercept of the curve plus 10 times $s_{y/x}$.[91] For methods used to support MRLs established by the CAC, the LOQ should meet the criteria for precision and accuracy (recovery) specified by the CAC and should be less than or equal to one-half the MRL.[32]

However, when the LOQ of a method is significantly lower than the actual concentrations monitored for compliance with a MRL, it may be more appropriate to carry out the validation experiments based on a *lowest calibrated level* (LCL), typically 0.5 × the MRL.[77,92] For use in a regulatory program, the limits of detection and quantification are important parameters when the method will be applied to estimate exposure to residues, where there may be an interest in monitoring residues at concentrations below the MRL, or when conducting residue analyses for substances that do not have ADIs or MRLs. For monitoring compliance with a MRL, it is important that an LCL be included in the analysis that adequately demonstrates that the MRL concentration may be reliably determined. The LCL of a method used to support a MRL should not be less than the LOQ.

The IUPAC guidance on validation of analytical methods in a single laboratory comments that the "limit of determination or limit of quantification" is sometimes "arbitrarily defined as 10% RSD" and sometimes as a multiple of the limit of detection.[11] However, it recommends that instead of specifying a limit of quantification, it is "preferable to try to express the uncertainty of measurement as a function of concentration and compare that function with a criterion of fitness for purpose agreed between the laboratory and the client or end-user of the data."

8.7.10 Decision Limit (CCα) and Detection Capability (CCβ)

As noted previously, Decision 2002/657/EC does not specify a requirement for the validation of analytical methods to include limits of detection and quantification, but instead includes the decision limit and the detection capability as required performance criteria.[35] The alpha (α) error is defined in 2002/657/EC as "the probability that the tested sample is compliant, even though a non-compliant measurement has been obtained (false non-compliant decision)," and the "decision limit (CCα) means the limit at and above which it can be concluded with an error probability of α that a sample is non-compliant." Accepted values for this probability are usually in the range 1–5%. For substances with zero action limit (AL), the CCα is the lowest concentration at which a method can discriminate with a statistical probability of $1-\alpha$ whether the identified analyte is present. The CCα is equivalent to the limit of detection (LOD) under some definitions (usually for $\alpha = 1\%$). In the case of substances with an established AL, the CCα is the measured concentration, above which it can be decided with a statistical probability of $1-\alpha$ that the identified analyte content is truly above the AL.

Decision 2002/657/EC defines the detection capability (CCβ) as "the smallest content of the substance that may be detected, identified and/or quantified in a sample with an error probability of β."[35] For substances where no permitted limit has been established, the detection capability is the lowest concentration at which a method is able to detect truly contaminated samples with a statistical certainty of $1-\beta$. In the case of substances with an established permitted limit, the detection capability is the concentration at which the method "is able to detect permitted limit concentrations with a statistical certainty of $1-\beta$." In other words, it is the smallest true concentration of the analyte that may be detected, identified, and quantified in a sample with a beta error (false negative). For banned substances the CCβ is the lowest concentration at which a method is able to determine the analyte in contaminated samples with a statistical probability of $1-\beta$. For substances with a MRL, CCβ is the concentration at which the method is able to detect samples that exceed this MRL with a statistical probability of $1-\beta$.

Decision 2002/657/EC provides guidance on the calculation of the decision limit and the detection capability for banned substances and for substances with established MRLs,[35] and examples are provided in various subsequent reports on validation of methods, including examples for mass spectral methods,[93] quinolones in eggs,[94] tetracyclines and sulfonamides in muscle tissue,[95] tetracyclines in tissue,[96] and estimation from matrix-matched calibration data.[97] Additional information is also included in Chapter 10.

8.8 SIGNIFICANT FIGURES

Authors have a tendency to report numbers that do not accurately reflect the actual performance capability of the method. Modern data systems will provide multiple numbers after the decimal place, but it is the responsibility of the analyst to determine the appropriate number of significant figures that should be reported. As a general approach, the method precision should inform the reporting of results. The precision indicates the number at which uncertainty occurs, and this should be the final digit reported.[91] For example, if the standard deviation for a method is determined to be 0.1, it is not appropriate to report a result as 1.463—the result should be reported as 1.5, as adding additional numbers after the decimal gives a false impression of the measurement capabilities of the method. However, it is important that results never be rounded up so that a regulatory or other limit is breached, as this could be challenged in a court of law.

8.9 FINAL THOUGHTS

Method validation is not an end in itself, but rather is undertaken to provide the analyst with a scientifically-based understanding of the routine performance that should be expected and achieved with an analytical method. It is not a guarantee against error, and it does not mean that problems may not be found in application of the method at some future date, particularly if a change has occurred in the materials used and the conditions that applied when the method was initially validated. The method validation establishes a basis for ongoing method performance through a laboratory quality assurance plan and also a basis for assessing when an analyst has successfully completed method training and is ready to undertake routine analyses using a method.

REFERENCES

1. Pohland A, *The Great Collaboration: 25 Years of Change*, AOAC International, Gaithersburg, MD, 2009.
2. International Standards Organization, *General Requirements for the Competence of Calibration and Testing Laboratories*, ISO/IEC 17025: 2005, Geneva, 2005.

3. CAC/GL 27-1997, Guidelines for the Assessment of the Competence of Testing Laboratories Involved in the Import and Export Control of Food, Joint FAO/WHO Food Standards, Rome, 1997 (available at http://www.codexalimentarius.net/download/standards/355/CXG_027e.pdf; accessed 4/19/10).
4. Thompson M, Wood R, International harmonized protocol for proficiency testing of (chemical) analytical laboratories, *Pure Appl. Chem*. 1993;65:2132–2144.
5. Thompson M, Wood R, Harmonized guidelines for internal quality control in analytical chemistry laboratories, *Pure Appl. Chem*. 1993;67:649–666.
6. *IUPAC Compendium of Chemical Terminology—the Gold Book*; International Union of Pure & Applied Chemistry, Research Triangle Park, NC, 2010 (available at http://goldbook.iupac.org/index.html; accessed 3/09/10).
7. *Compendium of Analytical Terminology*, 3rd ed. (The Orange Book), International Union of Pure & Applied Chemistry, Research Triangle Park, NC, 1997 (available at http://old.iupac.org/publications/analytical_compendium/; accessed 3/10/10).
8. Horwitz W, Nomenclature of interlaboratory analytical studies (IUPAC Recommendations 1994), *Pure Appl. Chem*. 1994;66:1903–1911.
9. Horwitz W, Protocol for the design, conduct and interpretation of method performance studies, *Pure Appl. Chem*. 1995;67:331–343.
10. Thompson M, Ellison SLR, Fajeglj A, Willetts P, Wood R, Harmonized guidelines for the use of recovery information in analytical measurement, *Pure Appl. Chem*. 1999;71:337–348.
11. Thompson M, Ellison SLR, Wood R, Harmonized guidelines for single-laboratory validation of methods of analysis, *Pure Appl. Chem*. 2002;74(5):835–855.
12. Helrich K, *The Great Collaboration*, AOAC International, Gaithersburg, MD, 1984.
13. Horwitz W, Latimer G Jr, eds., *Official Methods of Analysis*, 18th ed. (revised), AOAC International, Gaithersburg, MD, 2010.
14. International Standards Organization, *Control Charts for Arithmetic Average with Warning Limits*, ISO 7873: 1993, Geneva, 1993.
15. International Standards Organization, *International Vocabulary of Metrology—Basic and General Concepts and Associated Terms* (VIM), ISO/IEC Guide 99: 2007, Geneva, 2007.
16. *The Fitness for Purpose of Analytical Methods—a Laboratory Guide to Method Validation and Related Topics*, Eurachem, 1998 (available at http://www.eurachem.org/guides/valid.pdf; accessed 3/24/09).
17. Ellison SLR, Rosslein M, Williams A, eds., Eurachem/CITAC Guide CG4, *Quantifying Uncertainty in Analytical Measurement*, 2nd ed. (QUAM 2000:1), Eurachem, 2000 (available at http://www.eurachem.org/guides/QUAM2000-1.pdf; accessed 3/09/10).
18. *CITAC/Eurachem Guide: Guide to Quality in Analytical Chemistry—an Aid to Accreditation*, Eurachem, 2002 (available at http://www.eurachem.org/guides/pdf/CITAC%20EURACHEM%20GUIDE.pdf; accessed 10/22/10).
19. *Validation of Analytical Procedures: Definition and Terminology*, VICH GL1, International Cooperation on Harmonization of Technical Requirements for Registration of Veterinary Medicinal Products (VICH), Brussels, 1998 (available at http://www.vichsec.org/pdf/gl01_st7.pdf; accessed 3/08/10).
20. *Validation of Analytical Procedures*, VICH GL-2, International Cooperation on Harmonization of Technical Requirements for Registration of Veterinary Medicinal Products (VICH), Brussels, 1998 (available at http://www.vichsec.org/pdf/gl02_st7.pdf; accessed 3/08/10).
21. CAC/GL 72-2009, *Guidelines on Analytical Terminology*, Codex Alimentarius Commission, Joint FAO/WHO Food Standards Program, 2009 (available at http://www.codexalimentarius.net/download/standards/11357/cxg_072e.pdf; accessed 1/28/10).
22. *Guidelines for the Nomination and Submission of New, Revised and Alternative Test Methods*, Interagency Coordinating Committee on the Validation of Alternative Methods (ICCVAM), 2003 (available at http://iccvam.niehs.nih.gov/SuppDocs/SubGuidelines/SD_subg034508.pdf; accessed 2/23/10).
23. *AOAC Guidelines for Single Laboratory Validation of Chemical Methods for Dietary Supplements and Botanicals*, AOAC International, Gaithersburg, MD, 2002 (available at http://www.aoac.org/Official_Methods/slv_guidelines.pdf; accessed 2/25/10).
24. Validation of analytical procedures: Text and methodology, *ICH Harmonized Tripartite Guideline Q2(R1), Proc. Intnatl. Harmonisation of Technical Requirements for Registration of Pharmaceuticals for Human Use*, 2005 (available at http://www.ich.org/LOB/media/MEDIA417.pdf; accessed 2/25/10).
25. *Codex Alimentarius Commission Procedural Manual*, 19th ed., Joint FAO/WHO Food Standards Program, 2009 (available at ftp://ftp.fao.org/codex/Publications/ProcManuals/Manual_19e.pdf; accessed 4/19/10).
26. CAC/GL 64–1995, *Protocol for the Design, Conduct and Interpretation of Method Performance Studies*, Joint FAO/WHO Food Standards Program, 1995 (available at http://www.codexalimentarius.net/download/standards/10918/CXG_064e.pdf; accessed 2/23/10).
27. CAC/GL 28–1995, Rev.1-1997, *Food Control Laboratory Management: Recommendations*, Joint FAO/WHO Food Standards Program, 1997 (available at http://www.codexalimentarius.net/download/standards/356/CXG_028e.pdf; accessed 3/08/10).
28. CAC/GL 37–2001, *Harmonized IUPAC Guiidelines for the Use of Recovery Information in Analytical Measurement*, FAO/WHO Food Standards Program, 2001 (available at http://www.codexalimentarius.net/download/standards/376/CXG_037e.pdf; accessed 2/23/10).

29. CAC/GL 65–1997, *Harmonized Guidelines for Internal Quality Control in Analytical Chemistry Laboratories*, Joint FAO/WHO Food Standards Program, 1997 (available at http://www.codexalimentarius.net/download/standards/10920/CXG_065e.pdf; accessed 2/23/10).
30. CAC/GL 59–2006, *Guidelines on Estimation of Uncertainty of Results*, Joint FAO/WHO Food Standards Program, 2006 (available at http://www.codexalimentarius.net/download/standards/10692/cxg_059e.pdf; accessed 2/23/10).
31. CAC/GL 49–2003, *Harmonized IUPAC Guidelines for Single-Laboratory Validation of Methods of Analysis*, Joint FAO/WHO Food Standards Program, 2003 (available at http://www.codexalimentarius.net/download/standards/10256/CXG_049e.pdf; accessed 2/23/10).
32. CAC/GL 71–2009, *Guidelines for the Design and Implementation of National Regulatory Food Safety Quality Assurance Programme Associated with the Use of Veterinary Drugs in Food Producing Animals*, Joint FAO/WHO Food Standards Program, 2009 (available at http://www.codexalimentarius.net/web/more_info.jsp?id_sta = 11252; accessed 2/15/10).
33. *JECFA Introduction*; Food and Agriculture Organization, Rome, 2010 (available at http://www.fao.org/ag/agn/agns/jecfa_index_en.asp; accessed 10/22/10).
34. MacNeil JD, JECFA requirements for validation of analytical methods, in *Residues of Some Veterinary Drugs in Foods*, FAO Food and Nutrition Paper 41/14, Food and Agriculture Organization of the United Nations, Rome, 2002, pp. 95–101.
35. Commission Decision 2002/657/EC, implementing Council Directive 96/23/EC concerning the performance of analytical methods and the interpretation of results, *Off. J. Eur. Commun*., 2002;L221:8.
36. *Guidelines for the Implementation of Decision 2002/657/EC*, SANCO/2004/2726-rev 4, European Commission, 2008 (available at http://ec.europa.eu/food/food/chemicalsafety/residues/cons_2004-2726rev4_en.pdf; accessed 12/03/10).
37. GF1, *Guidance for Industry: General Principles for Evaluating the Safety of Compounds Used in Food-Producing Animals, IV. Guidance for Approval of a Method of Analysis for Residues*, US Food and Drug Administration, Rockville, MD, 2005 (available at http://www.fda.gov/downloads/AnimalVeterinary/GuidanceComplianceEnforcement/GuidanceforIndustry/UCM052180.pdf; accessed 10/22/10).
38. CVM GFI#118, *Guidance for Industry: Mass Spectrometry for Confirmation of the Identity of Animal Drug Residues—Final Guidance*, Center for Veterinary Medicine, US Food and Drug Administration, Rockville, MD, 2003 (available at http://www.fda.gov/downloads/AnimalVeterinary/GuidanceComplianceEnforcement/GuidanceforIndustry/UCM052658.pdf; accessed 3/17/10).
39. *The Codex Alimentarius*, Vol. 3, *Residues of Veterinary Drugs in Foods*, 2nd ed., Codex Alimentarius Commission, Joint FAO/WHO Food Standards Program, Food and Agriculture Organization of the United Nations, Rome, 1994.
40. *Validation of Analytical Methods for Food Control*, FAO Food and Nutrition Paper 68, Food and Agriculture Organization of the United Nations, Rome, 1998 (available at ftp://ftp.fao.org/docrep/fao/007/w8420e/w8420e00.pdf; accessed 10/22/10).
41. *Report of the Joint FAO/IAEA Expert Consultation on Practical Procedures to Validate Method Performance of Analysis of Pesticide and Veterinary Drug Residues, and Trace Organic Contaminants in Food*, 2000 (available at http://www.iaea.org/trc/pest-qa_val2.htm; accessed 2/16/10).
42. Project 2009-006-1-500, *Experimental Requirements for Single-Laboratory Validation*, Analytical Chemistry Division, International Union of Pure and Applied Chemistry, 2010 (available at http://www.iupac.org/web/ins/2009-006-1-500; accessed 1/19/10).
43. *International Vocabulary of Metrology—Basic and General Concepts and Associated Terms* (VIM), 3rd ed., JCGM 200; Joint Committee on Guides on Metrology, 2008 (available at http://www.iso.org/sites/JCGM/VIM/JCGM_200e.html; accessed 10/23/10).
44. *Guidelines for the Validation and Verification of Chemical Test Methods*, Technical Note 17, April 2009, National Association of Testing Authorities, Australia (NATA) (available at http://www.nata.asn.au/phocadownload/publications/Technical_publications/Technotes_Infopapers/technical_note_17_apr09.pdf; accessed 4/20/10).
45. *Method Validation and Quality Control Procedures for Pesticide Residues Analysis in Food and Feed*, SANCO/10684/2009, Directorate General for Health and Consumers (SANCO), European Commission, Brussels, 2009 (available at http://ec.europa.eu/food/plant/protection/resources/qualcontrol_en.pdf; accessed 10/23/10).
46. CAC/MSIC 5-1993 (amended 2003), *Glossary of Terms and Definitions* (*Veterinary Drugs in Foods*), Codex Alimentarius Commission, Joint FAO/WHO Food Standards Program, Food and Agriculture Organization of the United Nations, Rome, 2003 (available at http://www.codexalimentarius.net/web/more_info.jsp?id_sta=348; accessed 10/23/10).
47. Turnipseed SB, Clark SB, Karbiwnyk CM, Andersen WC, Miller KE, Madson MR, Analysis of aminoglycoside residues in bovine milk by liquid chromatography electrospray ion trap mass spectrometry after derivatization with phenyl isocyanate, *J. Chromatogr. B* 2009;877:1487–1493.
48. Ang CYW, Luo W, Kiessling CR, McKim, K, Lochmann R, Walker CC, Thompson HC Jr, A bridging study between liquid chromatography and microbial inhibition assay methods for determining amoxicillin residues in catfish muscle, *J. AOAC Int*. 1998;81(1):33–39.
49. Stehly GR, Gingerich WH, Kiesllng CR, Cutting JH, A bridging study for oxytetracycline in the edible fillet of rainbow trout: Analysis by a liquid chromatographic method and the official microbial inhibition assay, *J. AOAC Int*. 1999;82(4):866–870.

50. International Standards Organization, *Chemistry—Layouts for Standards*, Part 2, *Methods of Chemical Analysis*, ISO 78–2: 1999, Geneva, 1999.
51. Spisso BF, Gonçalves de Araújo Júnior MA, Monteiro MA, Belém Lima AM, Ulberg Pereira M, Alves Luiz R, Wanderley da Nóbrega A, A liquid chromatography–tandem mass spectrometry confirmatory assay for the simultaneous determination of several tetracyclines in milk considering keto–enol tautomerism and epimerization phenomena, *Anal. Chim. Acta* 2009;656:72–84.
52. Committee for Medicinal Veterinary Medical Products, *Oxytetracycline, Tetracycline, Chlortetracycline Summary Report*, EMEA/MRL/023/95, European Medicines Agency, London, 1995 (available at http://www.ema.europa.eu/docs/en_GB/document_library/Maximum_Residue_Limits_-_Report/2009/11/WC500015378.pdf; accessed 10/24/10).
53. *Food Analysis Proficiency Assessment Scheme*, The Food and Environment Research Agency, Sand Hutton, Yorks, UK, 2010 (available at http://www.fapas.com/; accessed 10/24/10).
54. CAC/GL 70-2009, *Guidelines for Settling Disputes over Analytical (Test) Results*, Joint FAO/WHO Food Standards Program, 2009 (available at http://www.codexalimentarius.net/web/more_info.jsp?id_sta = 11256; accessed 3/11/10).
55. *Codex Alimentarius Commission Procedural Manual*, 18th ed., Codex Alimentarius Commission, Joint FAO/WHO Food Standards Program, Rome, 2008 (available at ftp://ftp.fao.org/codex/Publications/ProcManuals/Manual_18e.pdf; accessed 4/28/10).
56. Youden WJ, Steiner EH, *Statistical Manual of the Association of Official Analytical Chemists*, AOAC International, Gaithersburg, MD, 1975.
57. Boison J, Lee S, Gedir R, Analytical determination of virginiamycin drug residues in edible porcine tissues by LC-MS with confirmation by LC-MS/MS, *J. AOAC Int*. 2009;92:329–339.
58. Bohm DA, Stachel CS, Gowik P, Multi-method for the determination of antibiotics of different substance groups in milk and validation in accordance with Commission Decision 2002/657/EC, *J. Chromatogr. A* 2009;1216:8217–8223.
59. Plackett RL, Burman JP, The design of optimal multifactorial experiments. *Biometrika* 1946;33:305–325.
60. *Engineering Statistics Handbook*, 5.3.3.5 Plackett-Burman Designs, National Institute of Standards & Technology, Dept. Commerce, United States of America, 2010 (available at http://www.itl.nist.gov/div898/handbook/pri/section3/pri335.htm; accessed 7/02/10).
61. Durden DA, Wotske J, Quantitation and validation of macrolide endectocides in raw milk by negative ion electrospray MS/MS, *J. AOAC Int*. 2009;92:580–596.
62. Calvert JG, Glossary of atmospheric chemistry terms, *Pure Appl. Chem*. 1990;62:2167–2219.
63. Currie LA, Nomenclature in evaluation of analytical methods including detection and quantification capabilities, *Pure Appl. Chem*. 1995;67:129–1723.
64. Sahai H, Singh RP, The use of R2 as a measure of goodness of fit: An overview, *Virginia J. Sci*. 1989;40(1):5–9.
65. Analytical Methods Committee, Uses (proper and improper) of correlation coefficients, *Analyst* 1988;113:1469–1471.
66. Miller JN, Basic statistical methods for analytical chemistry Part 2. Calibration and regression methods, a review, *Analyst* 1991;116:3–14.
67. Vessman J, Stefan RI, Van Staden JF, Danzer K, Lidner W, Burns DT, Fajgelj A, Müller H, Selectivity in analytical chemistry (IUPAC Recommendations 2001), *Pure Appl. Chem*. 2001;73(8):1381–1386.
68. Bethem R, Boison J, Gale P, Heller D, Lehotay S, Loo J, Musser S, Price P, Stein S, *J. Am. Soc. Mass Spectrom*. 2003;14:528–541.
69. Sphon J, Use of mass spectrometry for confirmation of animal drug residues, *J. Assoc. Off. Anal. Chem*. 1978;61:1247–1252.
70. Baldwin R, Bethem RA, Boyd RK, Buddle WA, Cairns T, Gibbons RD, Henion JD, Kaiser MA, Lewis DL, Matusik JE, Sphon JA, 1996 ASMS Fall Workshop: Limits to confirmation, quantitation, and detection, *J. Am. Soc. Mass Spectrom*. 1997;8:1180–1190.
71. Guidance for Industry #118, *Mass Spectrometry for Confirmation of the Identity of Animal Drug Residues—final guidance*, Center for Veterinary Medicine, US Food and Drug Administration, Rockville, MD, 2003 (available at http://www.fda.gov/downloads/AnimalVeterinary/GuidanceComplianceEnforcement/GuidanceforIndustry/UCM052658.pdf; accessed 3/17/10).
72. Heller DN, Lehotay SJ, Martos PA, Hammack W, Férnandez-Alba AR, Issues in mass spectrometry between bench chemists and regulatory laboratory managers: Summary of the roundtable on mass spectrometry held at the 123rd AOAC International annual meeting, *J. AOAC Int*. 2010;93(5):1625–1632.
73. Schürmann A, Dvorak V, Crüzer C, Butcher P, Kaufmann A, False-positive liquid chromatography/tandem mass spectrometric confirmation of sebuthylazine residues using the identification points system according to EU Directive 2002/657/EC due to a biogenic insecticide in tarragon, *Rapid Commun. Mass Spectrom*. 2009;23:1196–1200.
74. Thompson M, Ellison SLR, Fajeglj A, Willetts P, Wood R, Harmonized guidelines for the use of recovery information in analytical measurement, *Pure Appl. Chem*. 1999;71:337–348.
75. *Residues of Some Veterinary Drugs in Foods and Animals*, online edition, Food and Agriculture Organization of the United Nations, Rome, 2010 (available at http://www.fao.org/ag/agn/jecfa-vetdrugs/search.html; accessed 3/17/10).
76. *Chemistry Laboratory Guidebook*, Food Safety & Inspection Service, US Dept. Agriculture, 2010 (available at, http://www.fsis.usda.gov/Science/Chemistry_Lab_Guidebook/index.asp; accessed 1/04/10).
77. CAC/GL 40-1993, Rev.1–2003, *Guidelines on Good Laboratory Practice in Residue Analysis*, Codex Alimentarius Commission, Joint FAO/WHO Food Standards Program,

Rome, 2003 (available at http://www.codexalimentarius.net/download/standards/378/cxg_040e.pdf; accessed 1/04/10).

78. Thompson M, Recent trends in inter-laboratory precision at ppb and sub-ppb concentrations in relation to fitness for purpose criteria in proficiency testing, *Analyst*. 2000;125:385–386.

79. Jülicher B, Gowik P, Uhlig S, Assessment of detection methods in trace analysis by means of a statistically based in-house validation concept, *Analyst* 1998;123:173–179.

80. Project 2009-006-1-500, *Experimental Requirements for Single-Laboratory Validation*, Ellison SLR (chairman), International Union of Pure and Applied Chemistry, 2009 (available at http://www.iupac.org/web/ins/2009-006-1-500; accessed 2/01/10).

81. Cooper AD, Tarbin JA, Farrington WHH, Shearer G, Aspects of extraction, spiking and distribution in the determination of incurred residues of chloramphenicol in animal tissues, *Food Addit. Contam. A* 1998;15(6):637–644.

82. Cooper AD, Tarbin JA, Farrington WHH, Shearer G, Effects of extraction and spiking procedures on the determination of incurred residues of oxytetracycline in cattle kidney, *Food Addit. Contam. A* 1998;15(6):645–650.

83. Matuszewski BK, Constanzer ML, Chavez-Eng CM, Strategies for the assessment of matrix effect in quantitative bioanalytical methods based on HPLC-MS/MS, *Anal. Chem*. 2003;75:3019–3030.

84. Wang J, Cheung W, Grant D, Determination of pesticides in apple-based infant foods using liquid chromatography electrospray ionization tandem mass spectrometry, *J. Agric. Food Chem*. 2005;53(3):528–537.

85. Bogialli S, Di Corcia A, Recent applications of liquid chromatography–mass spectrometry to residue analysis of antimicrobials in food of animal origin, *Anal. Bioanal. Chem*. 2009;395:947–966.

86. Rønning HT, Einarsen K, Asp TN, Determination of chloramphenicol residues in meat, seafood, egg, honey, milk, plasma and urine with liquid chromatography-tandem mass spectrometry, and the validation of the method based on 2002/657/EC, *J. Chromatogr. A* 2006;1118:226–233.

87. Chu PS, Lopez MI, Determination of nitrofuran residues in milk of dairy cows using liquid chromatography tandem mass spectrometry, *J. Agric. Food Chem*. 2007;55:2129–2135.

88. Cronly M, Behan P, Foley B, Malone E, Regan L, Rapid confirmatory method for the determination of 11 nitroimidazoles in egg using liquid chromatography tandem mass spectrometry, *J. Chromatogr. A* 2009;1216:8101–8109.

89. Cronly M, Behan P, Foley B, Malone E, Regan L, Development and validation of a rapid method for the determination and confirmation of 10 nitroimidazoles in animal plasma using liquid chromatography tandem mass spectrometry, *J. Chromatogr. B* 2009;877:1494–1500.

90. *International Vocabulary of Basic and General Terms in Metrology*, International Standards Organization, Geneva, 1993.

91. Miller JC, Miller JN, *Statistics for Analytical Chemistry*, 3rd ed., Ellis Horwood, Chichester, UK, 1993.

92. Alder L, Holland PT, Lantos J, Lee, M, MacNeil, JD, O'Rangers J, van Zoonen P, Ambrus A, Guidelines for single-laboratory validation of analytical methods for trace-level concentrations of organic chemicals, in Fajgelj A, Ambrus A, eds., *Principles and Practices of Method Validation*, The Royal Society of Chemistry, Cambridge, UK, 2002, pp. 179–248.

93. Antignac J-P, Le Bizec B, Monteau F, Andre F, Validation of analytical methods based on mass spectrometric detection according to the "2002/657/EC" European decision: Guideline and application, *Anal. Chim. Acta* 2002;483:325–334.

94. Bogialli S, D'Ascenzo G, Di Corcia A, Laganà A, Tramontana G, Simple assay for monitoring seven quinolone antibacterials in eggs: Extraction with hot water and liquid chromatography coupled to tandem mass spectrometry laboratory validation in line with the European Union Commission Decision 657/2002/EC, *J. Chromatogr. A* 2009;1216: 794–800.

95. McDonald M, Mannion C, Rafter P, A confirmatory method for the simultaneous extraction, separation, identification and quantification of tetracycline, sulphonamide, trimethoprim and dapsone residues in muscle by ultra-high-performance liquid chromatography–tandem mass spectrometry according to Commission Decision 2002/657/EC, *J. Chromatogr. A* 2009;1216:8110–8116.

96. Nikolaidou KI, Samanidou VF, Papadoyannis IN, Development and validation of an HPLC confirmatory method for the determination of seven tetracycline antibiotics residues in bovine and porcine muscle tissues according to 2002/657/EC, *J. Liq. Chromatogr. Rel. Technol*. 2008;31:3032–3054.

97. Steliopoulos P, Estimating the decision limit and the detection capability using matrix-matched calibration data, *Accred. Qual. Assur*. 2010;15:105–109.

9

MEASUREMENT UNCERTAINTY

JIAN WANG, ANDREW CANNAVAN, LESLIE DICKSON, AND RICK FEDENIUK

9.1 INTRODUCTION

Analytical results for the determination of antibiotic residues in food, in common with results generated in other laboratories or branches of analytical chemistry, must be reliable and comparable. It is a requirement for laboratories accredited under the ISO/IEC 17025 quality system[1] that the measurement uncertainty associated with a result should be made available and reported if it is required by the client, is relevant to the validity of the test results, or may affect compliance with a specification, for example, compliance with a maximum residue limit (MRL) for antibiotics. The Codex Alimentarius Commission also recommends that laboratories provide their customers on request with information on the measurement uncertainty or a statement of confidence associated with quantitative results for veterinary drug residues.[2] The relevant sections in ISO 17025[1] are quoted below:

> 5.4.6.2 Testing laboratories shall have and shall apply procedures for estimating uncertainty of measurement. In certain cases the nature of the test method may preclude rigorous, metrologically and statistically valid, calculation of uncertainty of measurement. In these cases the laboratory shall at least attempt to identify all the components of uncertainty and make a reasonable estimation, and shall ensure that the form of reporting of the result does not give a wrong impression of the uncertainty. Reasonable estimation shall be based on knowledge of the performance of the method and on the measurement scope and shall make use of, for example, previous experience and validation data.
>
> 5.4.6.3 When estimating the uncertainty of measurement, all uncertainty components which are of importance in the given situation shall be taken into account using appropriate methods of analysis.

It is important to note that ISO 17025 does not prescribe a particular approach or method to determine measurement uncertainty; it only recommends that the approach or method be considered valid within the relevant technical discipline and that it gives a reasonable estimate of measurement uncertainty. The degree of rigor in estimating measurement uncertainty should depend on the needs of the client and the level of risk involved with respect to the use of the test results and any decisions that may be based on those results.

In this chapter, general principles and common approaches for the estimation of measurement uncertainty for antibiotic residue analysis are presented and illustrated with practical worked examples.

9.2 GENERAL PRINCIPLES AND APPROACHES

Measurement uncertainty is defined by the International Standards Organization (ISO)[3] as a parameter associated with the result of a measurement that characterizes the dispersion of the values that could reasonably be attributed to the measurand. The uncertainty is expressed as a range within which the true value of the measurand is believed to lie. In a practical sense, measurement uncertainty can be considered as a measure of the quality of measurement results. It gives an answer to the question "How well does the result represent the value of the quantity being measured?" Therefore, the measurement uncertainty associated with a result is an essential part of quantitative results, and along with traceability, it can allow users to assess the reliability of the result and compare results among different sources or with reference values.

Chemical Analysis of Antibiotic Residues in Food, First Edition. Edited by Jian Wang, James D. MacNeil, and Jack F. Kay.

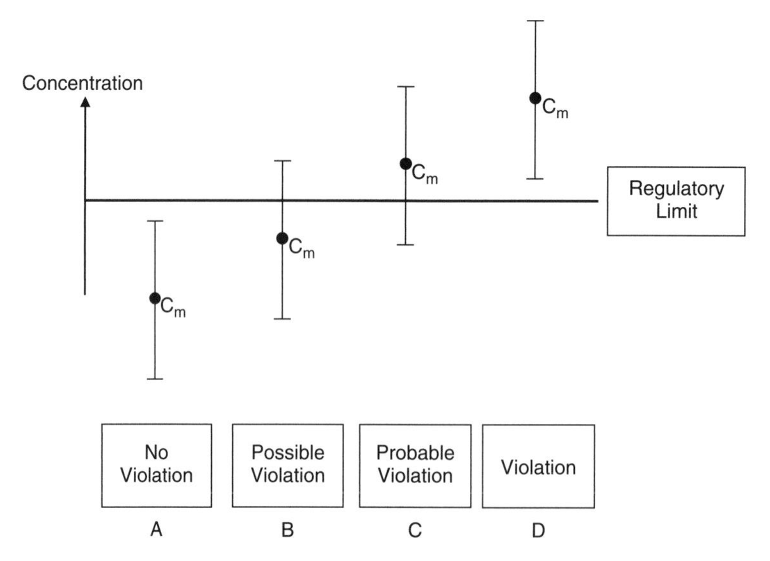

Figure 9.1 Comparison of a measurement result (C_m) with its associated measurement uncertainty interval against a regulatory limit. The error bars represent the uncertainty interval.

Knowledge of the uncertainty allows for reasonable interpretation of results that are close to the regulatory limit. Figure 9.1 illustrates how measurement uncertainty helps interpret a test result. It is clear that scenarios A and D represent no ambiguity in explanation. In scenario A, the measurement result is below the regulatory limit and the uncertainty interval does not overlap the regulatory limit; in scenario D, the measurement result is above the regulatory limit and the uncertainty interval does not overlap the regulatory limit. Scenarios B and C are ambiguous, and in both cases the uncertainty interval overlaps the regulatory limit. It is not possible with sufficient certainty to declare that these measurements exceed the regulatory limit. For scenario B, there is a significant risk of a false-negative result that could lead to the release of food containing truly non-compliant concentrations of a regulated substance into the food supply. For scenario C, there is a significant risk of a false-positive result that could lead to a food producer facing legal charges for not complying with the regulations. Such regulatory decisions cannot be made without knowledge of the measurement uncertainty associated with the testing results.

Measurement uncertainty for an analytical result can arise from many sources, including sampling, preparation of analytical portions, incomplete extraction/clean-up, matrix effects and interference, environmental conditions, uncertainties of balances and volumetric equipment, instrument stability, and random variation. In regulatory testing for some other types of chemical residues and contaminants, such as pesticides and mycotoxins, sampling uncertainty is considered a major contributor to overall measurement uncertainty. However, for antibiotics or veterinary drug residue control in general, sampling uncertainty has generally been considered insignificant and the analytical measurement uncertainty is usually the quoted value. Some recent work,[4] however, has demonstrated that penicillin G residue concentrations are dependent on muscle type, which suggests that sampling procedures and their uncertainty should be considered more closely for veterinary drug residue analysis in the future. From a method performance perspective, the major contributors to measurement uncertainty are overall precision, method recovery, and variation in recovery with matrix and analyte concentration, when gas chromatography (GC) and liquid chromatography (LC) techniques are used as analytical techniques.[5,6]

The overall uncertainty of a measurement may be estimated by combining all of the individual uncertainty components associated with each step of the analytical procedure. Each individual uncertainty component is expressed as a standard deviation or relative standard deviation. In general, the combined standard uncertainty (or combined uncertainty) associated with a result is the estimated standard deviation equal to the positive square root of the total variance obtained by combining all uncertainty components.[3] The expanded uncertainty is obtained by multiplying the combined uncertainty by a coverage factor, k. For most purposes, k is given a value of 2, corresponding to an approximate confidence level of 95%. If a coverage factor of 3 is applied, there is a higher level of confidence (99.7%) that the value will fall within the uncertainty interval. The expanded uncertainty is usually the most appropriate uncertainty estimate to report with a result in analytical chemistry, including antibiotic residues analysis.

It is important to distinguish between measurement uncertainty and error. *Error* is the difference between the true value of the measurand (an unknowable quantity) and a measurement result. It is a single value that could, in principle, be used to correct a measurement result. *Measurement uncertainty*, on the other hand, is a range or interval that encompasses the range of values that could be expected for the given measurement. Errors may be either random or systematic. *Random error* typically arises from unpredictable variations of influencing factors such as thermal effects or electrical noise. Since they are unpredictable, no correction is possible for random errors, but they may be minimized by increasing the number of measurements. *Systematic error*, or *bias*, is defined as a component of error that, in the course of a number of analyses of the same measurand, remains constant or varies in a predictable way. Systematic errors may be constant, for example, those caused by failing to make an allowance for a reagent blank, or non-constant, for example, with changes in results caused by an increase in ambient temperature over the course of measurement. Analytical results can be corrected for recognized systematic errors; however, there will always be some uncertainty associated with the value of the systematic error, so the correction itself is not exact.

There are various ways of estimating measurement uncertainty. Two basic approaches are used, with variations within each. The *bottom–up approach*, as described in the GUM[3] and refined for application to chemical methods by EURACHEM/CITAC[7] and other bodies, derives the uncertainty of a measurement result by combining the uncertainty contributions of all factors influencing the result. This approach was originally developed for physical or metrological measurements, and can be time-consuming and difficult to apply to chemical analytical methods for antibiotic residues in food. Methods for antibiotics in food are frequently complex, involving various extraction and clean-up stages as well as instrumental measurement, and therefore uncertainty sources can be difficult to identify and quantify. An alternative, *top–down approach*, based on the reproducibility estimates from inter-laboratory method performance studies, was adopted by the Analytical Methods Committee of the Royal Society of Chemistry in the mid-1990s.[8] This approach was further developed by the Nordic Committee for Food Analysis[9] to provide uncertainty estimates using only intra-laboratory data. An advantage of this approach is that data produced during the in-house validation of the analytical method can be used to estimate the uncertainty, if the validation protocol is properly designed. A protocol for the design of validation experiments that permits the use of precision, trueness, and ruggedness data to obtain uncertainty estimates was developed by Barwick and Ellison.[10] Other variations and combinations of these approaches applied to the analysis of antibiotic residues in foods have been published, including the use of a nested experimental design to study the measurement uncertainty arising from intermediate precision and recovery in a method for macrolide residues in eggs,[11] and a comparison of top–down uncertainty estimation using reproducibility and repeatability data and a bottom–up approach including repeatability data for an HPLC method for sulfonamides in tissues.[12]

In some cases it may be useful to make a rough estimation of the measurement uncertainty of a method at the target concentration, for example, at the MRL of a veterinary drug, to help to determine whether the method will be fit for purpose before undertaking a full validation and measurement uncertainty estimation exercise. This can be done by applying the Horwitz formula to obtain an estimate applicable to inter-laboratory reproducibility data, or a suitably adjusted version for intra-laboratory data.[13,14] The Horwitz formula, as initially applied to inter-laboratory (between-laboratory) reproducibility data (R) in percentage, and with the concentration C expressed as a mass fraction, is:

$$\mathrm{RSD}_R(\%) = 2 \times C^{-0.1505}$$

or as a standard deviation

$$S_R = 0.02 \times C^{0.8495}$$

To apply to intra-laboratory (within-laboratory) repeatability (r), this expression is divided by 2 to give the estimated standard uncertainty, and to obtain the expanded uncertainty on the basis of intra-laboratory repeatability, the expression is again multiplied by 2:

$$S_r = 0.02 \times C^{0.8495}$$

For example, using this approach a method to test for an antimicrobial residue with a MRL of 100.0 μg/kg would have an estimated expanded uncertainty of:

$$S_r = 0.02 \times (0.0000001)^{0.8495} = 2.26 \times 10^{-8}$$

$$S_r(\%) = 0.02 \times (0.0000001)^{-0.1505} = 22.6\%$$

This is interpreted to mean that 95% of anticipated results for samples with a true value of 100 μg/kg will fall between 77.4 and 122.6 μg/kg.

9.3 WORKED EXAMPLES

9.3.1 EURACHEM/CITAC Approach

The GUM approach consists of identification and quantification of the relevant sources of uncertainty followed by the combination of the individual uncertainty estimates. This is done by means of the law of propagation of uncertainty, namely, a first-order Taylor series, where several parameters are correlated and associated to the combined uncertainty $u_c(y)$[3]

$$u_c(y)=\sqrt{\sum_{i=1}^{n}\left(\frac{\partial f}{\partial x_i}\right)^2 u^2(x_i)+2\sum_{i=1}^{n-1}\sum_{j=i+1}^{n}\frac{\partial f}{\partial x_i}\frac{\partial f}{\partial x_i}u(x_i,x_j)}$$

where y is the measurement result that depends on several parameters x_i, where each x_i is a certain uncertainty source. With x_i the value considered, $u(x_i)$, is the standard uncertainty related to this value, and $\partial f/\partial x_i$ is the partial differential of y with respect to x_i. Under most circumstances for antibiotic residue analysis, independence of the effects is assumed. Therefore, the second term of equation above, which relates to the covariances of dependent variables, can be omitted. The estimation of measurement uncertainty is usually calculated as the summation of the squares of the relative uncertainties:[15,16]

$$u_c(y)/y=\sqrt{\sum_{i=1}^{n}\left(\frac{u(x_i)}{x_i}\right)^2}$$

The GUM approach is most easily achieved using a stepwise strategy, as illustrated below for an HPLC method for the determination of an antimicrobial sulfonamide, for example, sulfamethazine (SMT), in tissues, for evaluation against the MRL of 100 μg/kg. The method has been simplified somewhat for the purpose of demonstration of this approach to uncertainty estimation. The steps of the method are represented schematically in Figure 9.2. For this example, it is supposed that the analysis of six replicates of blank tissue (from various sources) spiked at 100 μg/kg, on three different occasions under within-laboratory reproducibility conditions, gave a mean result of 83.0 μg/kg with a standard deviation s of 5.4 μg/kg (relative standard deviation 0.065 or 6.5%). The within-day precision (repeatability) was calculated using 15 replicate measurements giving a standard deviation of 5.3 μg/kg. The approach used for the calculations is based on that of Leung et al.[17] Results are, in this case, corrected for recovery and reported with recovery information in accordance with Codex Alimentarius Commission recommendations.[18]

The first step is to clearly define what is being measured and to establish the relationship between the measurand and the inputs on which it depends. The following equation for the calculation of the result gives a very good basis for this step:

$$C=\frac{P-a}{b}\times\frac{1}{R} \qquad (9.1)$$

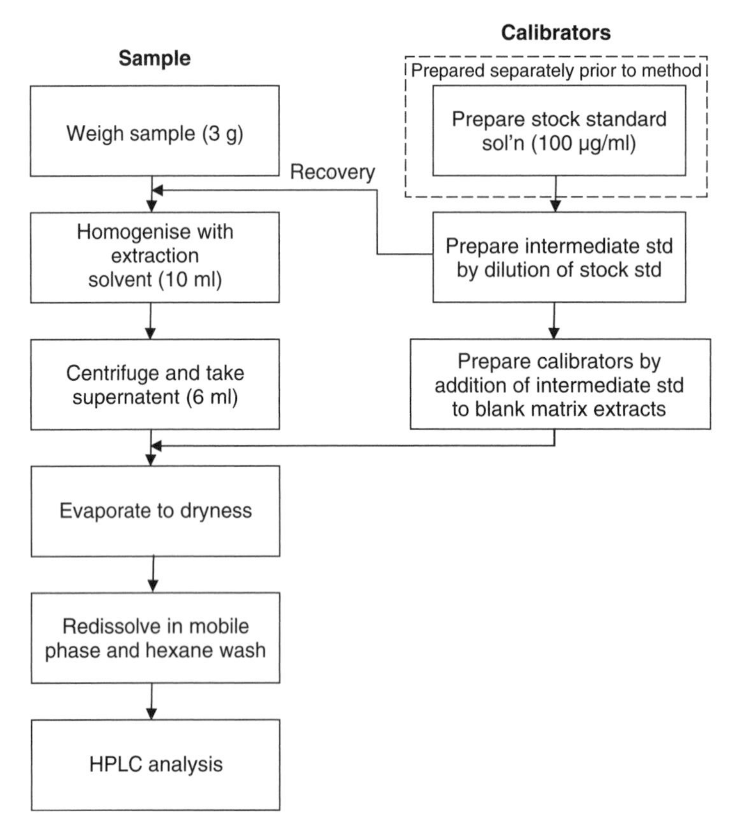

Figure 9.2 Schematic diagram of a method for the determination of sulfamethazine.

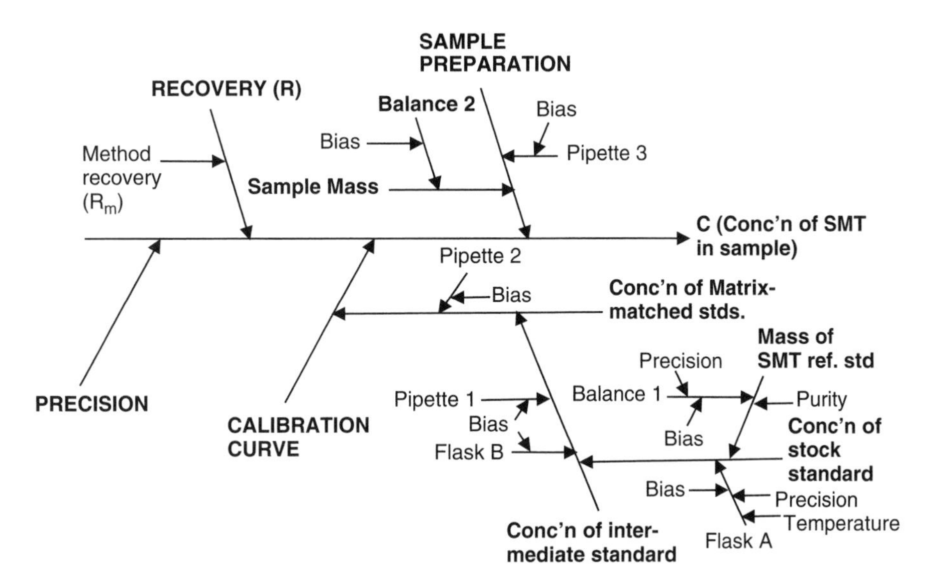

Figure 9.3 "Cause and effect" diagram for a method for the determination of sulfamethazine in tissues.

Here, C is the concentration of SMT, P is the detector response (area of the SMT chromatographic peak), a is the y intercept of the linear regression calibration curve, b is the slope of the calibration curve, and R is the recovery factor (for a recovery-corrected result).

The second step is to identify or to list all possible sources of uncertainty for the method, and to ensure that uncertainty sources of the same nature are grouped together to avoid their inclusion in duplicate in the estimation. A "cause and effect" diagram can serve as a useful tool to elaborate this step. The list, or the cause-and-effect diagram (Fig. 9.3), is developed initially on the basis of factors in the mathematical expression [Eq. (9.1)] that is utilized to calculate the result of the analysis and with reference to the method protocol (Fig. 9.2). The individual components for the uncertainty budget are listed in Table 9.1.

In this case, three main factors are identified: the response (represented by $P-a$), the slope (b), and the recovery factor (R). The sources of uncertainty influencing each of these main branches must then be determined. For example, the branch that represents the slope of the calibration curve has uncertainty contributions from the preparation of the calibrators, and the uncertainty for the peak area branch arises mainly from the sample preparation.

Since no certified reference material was available for this matrix–analyte combination, the method recovery (R) in this example is an estimate of the recovery obtained from spiking a blank sample, as described by Barwick and Ellison.[19] Two uncertainty sources are considered in the estimation of uncertainty associated with the recovery $u(R)$; these are the recovery uncertainty due to the sample preparation method $u(R_m)$, and that due to variation in sample matrices $u(R_s)$, as described by Leung et al.[17] An additional correction factor (R_{rep}) should be considered to account for the fact that a spiked sample may behave differently from a real sample with incurred analyte. However, previous experience from the analysis of SMT in a different tissue matrix for which a certified reference material was available indicated that the spiked samples were representative of the incurred tissue. As a result, it is assumed that the spiked sample in this matrix is also representative of incurred samples and the contribution to the uncertainty of the method is negligible. The uncertainty associated with the recovery is estimated as follows:

$$u(R) = \sqrt{(u(R_m))^2 + (u(R_s))^2} \tag{9.2}$$

The method recovery is calculated as:

$$\overline{R}_m = \frac{\overline{C}_{obs}}{C_{spike}} \tag{9.3}$$

where $\overline{C}_{obs}$ is the mean of replicate analysis of the spiked sample and C_{spike} is the nominal concentration of analyte in the spiked sample. For this example, $\overline{C}_{obs}$ was 83.0 μg/kg and C_{spike} was 100 μg/kg.

The standard uncertainty of the recovery is therefore calculated using the equation:

$$u(\overline{R}_m) = \overline{R}_m \times \sqrt{\left(\frac{s_{obs}^2}{n \times \overline{C}_{obs}^2}\right) + \left(\frac{u(C_{spike})}{C_{spike}}\right)^2} \tag{9.4}$$

where s_{obs} is the standard deviation of replicate analyses of the spiked sample (5.4 μg/kg for this example), n is the number of replicates ($n = 24$), and $u(C_{spike})$ is the standard uncertainty in the concentration of the spiked sample, which is calculated in the same way as $u(\text{cal})$ (Table 9.1) but using uncertainty sources from the procedures used for the recovery study, and was 0.497 μg/kg for this example (calculation not shown).

The recovery uncertainty due to variation in sample matrices $u(R_s)$ is the standard deviation of the mean recovery from different blank spike samples. For the example presented here, the standard deviation of the reproducibility study can be used, since different sources of blank matrix were used for each batch. Other sources of uncertainty that are also associated with the method precision arise from the precision of balances, volumetric glassware, pipettes, and other equipment. In the example given here,

TABLE 9.1 Uncertainty Sources for the EURACHEM/CITAC Approach to Uncertainty Estimation

Contributing Factor	Calculation	Standard Uncertainty $u(y)$	Relative Standard Uncertainty (RSU)
	Uncertainty Arising from Calibrator Preparation, u(cal)		
Calibration of analytical balance, $u(\text{bal}_1\text{bias})$	10 mg weighing, limit of performance of balance (from manufacturer's certificate) = 0.140 mg	0.140 mg	—
Repeatability of weighings $u(\text{bal}_1\text{prec})$	SD from 10 repeated weighings of 10 mg standard mass = 0.024 mg	0.024 mg	—
$u(\text{bal}_1)$	$u(\text{bal}_1) = \sqrt{(u(\text{bal}_1\text{bias})^2 + u(\text{bal}_1 - \text{prec})^2)} = 0.142$ mg $\text{RSU}_{\text{bal}_1} = 0.142 \text{ mg}/10 \text{ mg} = 0.0142$	0.142 mg	0.0142
Purity of SMT standard $u(\text{pur})$	Assumed negligible for the certified standard material used	0	0
Weight of SMT standard $u(\text{weight})$	$\text{RSU}_{\text{weight}} = \sqrt{(\text{RSU}^2_{\text{bal}_1} + \text{RSU}^2_{\text{pur}})} = 0.0142$	—	0.0142
Accuracy of volumetric flask A, $u(\text{flask}_A\text{bias})$	Flask volume 100 ml, uncertainty range ± 0.20 ml; since the range is given without a confidence level, a rectangular distribution is assumed, and SU = half-range/$\sqrt{3}$: $u(\text{flask}_A\text{bias}) = 0.20/\sqrt{3} = 0.1155$ ml	0.1155 ml	—
Precision of volumetric flask A, $u(\text{flask}_A\text{prec})$	SD from weighing 10 repeated fillings of 100 ml = 0.0575 ml	0.0575 ml	—
Uncertainty in volume due to temperature fluctuation $u(\text{flask}_A T)$ (see Ref. 6)	Ambient-temperature specification ±2°C, expansion coefficient (methanol) = $1 \times 10^{-3}\text{C}^{-1}$: Uncertainty range = $100 \times 2 \times (1 \times 10^{-3}) = 0.2$ ml $u(\text{flask}_A T) = 0.2/\sqrt{3} = 0.1155$ ml	0.1155 ml	—
$u(\text{flask}_A)$	$u(\text{flask}_A) = \sqrt{(u(\text{flask}_A\text{bias})^2 + u(\text{flask}_A\text{prec})^2 + u(\text{flask}_A T)^2)}$ = 0.173 ml $\text{RSU}_{\text{flask}A} = 0.173 \text{ ml}/100 \text{ ml} = 0.00173$	0.173 ml	0.00173
Stock SMT standard solution, $u(\text{stock})$	$\text{RSU}_{\text{stock}} = \sqrt{(\text{RSU}^2_{\text{flask}A} + \text{RSU}^2_{\text{weight}})} = 0.0143$	—	0.0143
Pipette 1, $u(\text{pip}_1)$	1000 μl, ±8 μl $u(\text{pip}_1) = 8/\sqrt{3} = 4.62$ μl $\text{RSU}_{\text{pip}_1} = 4.62 \text{ μl}/1000 \text{ μl} = 0.00462$	4.62 μl	0.00462
Accuracy of flask B (100 ml), $u(\text{flask}_B)$	Bias only (precision uncertainty is included in the precision branch) calculated as for flask A	0.1155 ml	0.001155
Intermediate standard solution, $u(\text{inter})$	$\text{RSU}_{\text{inter}} = \sqrt{(\text{RSU}^2_{\text{stock}} + \text{RSU}^2_{\text{pip}_1} + \text{RSU}^2_{\text{flaskB}})} = 0.0151$	—	0.0151
Pipette 2, $u(\text{pip}_2)$	40 μl, ±1 μl $u(\text{pip}_2) = 1/\sqrt{3} = 0.577$ μl $\text{RSU}_{\text{pip}_2} = 0.577 \text{ μl}/40 \text{ μl} = 0.0144$	0.577 μl	0.0144
$u(\text{cal})$	$\text{RSU}_{\text{cal}} = \sqrt{(\text{RSU}^2_{\text{inter}} + \text{RSU}^2_{\text{pip}_2})} = 0.0209$	—	0.0209

TABLE 9.1 (*Continued*)

Contributing Factor	Calculation	Standard Uncertainty $u(y)$	Relative Standard Uncertainty RSU
	Uncertainty Arising from Sample Preparation, u(sample)		
Balance 2 (bias), $u(\text{bal}_2)$	3 g, limit of performance 0.005 g: SU = 0.005 g $\text{RSU}_{\text{bal}_2} = 0.005\text{ g}/3\text{ g} = 0.00166$	0.005 g	0.00166
Pipette 3, $u(\text{pip}_3)$	6 ml, ±0.05 ml $u(\text{pip}_3) = 0.05/\sqrt{3} = 0.0289$ ml $\text{RSU}_{\text{pip}_3} = 0.0289\text{ ml}/6\text{ ml} = 0.00482$	0.0289 ml	0.00482
$u(\text{sample})$	$\text{RSU}_{\text{sample}} = \sqrt{(\text{RSU}_{\text{bal}_2}^2 + \text{RSU}_{\text{pip}_3}^2)} = 0.00509$	—	0.00509
	Uncertainty in Method Precision, u(prec)		
	Mean of 24 controls spiked at 100 μg/kg = 83.0 μg/kg, SD = 5.4 μg/kg $\text{RSU}_{\text{prec}} = \dfrac{5.4\ \mu\text{g/kg}}{83.0\ \mu\text{g/kg}} = 0.065$	5.4 μg/kg	0.065
	Uncertainty in Recovery, u(R)		
Method recovery, $u(R_m)$	Using Eq. (9.4), the standard uncertainty of the recovery is $u(\overline{R}_m) = \overline{R}_m \times \sqrt{\left(\dfrac{s_{\text{obs}}^2}{n \times \overline{C}_{\text{obs}}^2}\right) + \left(\dfrac{u(C_{\text{spike}})}{C_{\text{spike}}}\right)^2}$ where s is the standard deviation of replicate analysis of the spiked sample (5.4 μg/kg for this example), and n is the number of replicates (24 for this example), $u(C_{\text{spike}})$ is the standard uncertainty in the concentration of the spiked sample (0.497 μg/kg for this example) $u(R_m) = 0.83 \times \sqrt{\left(\dfrac{5.4^2}{24 \times 83^2}\right) + \left(\dfrac{0.497}{100}\right)^2} = 0.0118$	0.0118	
Sample recovery, $u(R_s)$	Standard deviation of the mean recovery of the reproducibility study at 100 μg/kg = 0.054	0.054	—
$u(R)$	$u(R) = \sqrt{(u(R_m))^2 + (u(R_s))^2}$ [Eq.(9.2)] = 0.0553	0.0553	0.0666
	The relative standard uncertainty, $u(R)/R_m = 0.0553/0.83 = 0.0666$	—	—

the precision of the analytical balance and the volumetric flask used to prepare the stock standard are included as separate terms because the stock standard is prepared in a separate operation before the analytical method. The best estimate of method precision is the reproducibility, calculated over a period of time (preferably several months) and with different operators and instrumentation.

The third step of the uncertainty estimation procedure is to quantify all the individual components that are identified. It is important that all contributing uncertainty components are quantified as standard deviations, whether they arise from random variability or systematic effects (bias). Estimates derived from experimental data, prior knowledge of the method performance, or professional judgment are treated in the same way and given equal weight, and are also expressed as standard deviations. Where uncertainties are not available as standard deviations, for example, where a range is given without a confidence level for a pipette, they can be converted to standard deviations using the rules given in the ISO and EURACHEM/CITAC guides.[3,7] The uncertainty sources, calculations, standard uncertainties ($u(y)$) and relative standard uncertainties (RSU) for this example are summarized in Table 9.1.

The penultimate step in the procedure is to calculate the combined uncertainty of the method, which is expressed as the relative uncertainty, from the standard uncertainty values of main uncertainty sources. From Table 9.1, the main contributing sources to the overall relative standard uncertainty for the example given are the preparation of calibrators ($RSU_{cal} = 0.0209$), the sample preparation ($RSU_{sample} = 0.00509$), the method precision ($RSU_{prec} = 0.065$), and the recovery ($RSU_{rec} = 0.0666$). The greatest contributions to the uncertainty are the uncertainty due to the recovery (RSU_{rec}) and uncertainty arising from the method precision (RSU_{prec}), which was calculated from the analysis of 24 samples spiked at 100 μg/kg, under reproducibility conditions.

Relative standard uncertainties with values less than 30% of the largest RSU are considered insignificant and can be omitted from calculation of the overall RSU.[7] For this example, therefore, the combined RSU is:

$$\begin{aligned} RSU_{combined} &= \sqrt{(RSU_{cal}^2 + RSU_{prec}^2 + RSU_{rec}^2)} \\ &= \sqrt{0.0209^2 + 0.065^2 + 0.0666^2} = 0.09538 \end{aligned}$$

and the combined standard uncertainty $u_c(y)$ is:

$$u_c(y) = RSU_{combined} \times 83.0 = 0.09538 \times 83.0 = 7.92 \ \mu g/kg$$

The final step is to derive the expanded uncertainty using the coverage factor k. For this example, k is given a value of 2, corresponding to an approximate confidence level of 95%:

$$U(y) = k \times u_c(y) = 15.8 \ \mu g/kg$$

A recovery-corrected result, generated using this analytical method, with a concentration equal to the MRL of 100 μg/kg can therefore be reported as 100 ± 15.8 μg/kg, with 95% confidence that the true SMT content of the sample is between 84.2 and 115.8 μg/kg.

9.3.2 Measurement Uncertainty Based on the Barwick–Ellison Approach Using In-House Validation Data

Although the bottom–up approach can be applied to estimate the uncertainty of analytical methods, as shown in the previous example (in Section 9.3.1), significant difficulties can be encountered in practice. In the analysis of residues and contaminants in foods, the largest contributions to uncertainty frequently arise from the least predictable effects, such as sampling, matrix effects, and interference, which are dependent mainly on individual samples and analytical procedures. The uncertainties associated with such effects can be determined only by experiment. Barwick and Ellison elaborated an approach for uncertainty estimation using three elements of in-house method validation, which included precision, trueness (recovery), and ruggedness.[10,20] The principle of this approach is to account for as many sources of uncertainty as possible in precision and trueness studies and to evaluate any remaining sources through ruggedness studies, or from existing data such as calibration certificates or previous studies. Where there is evidence to suggest that any additional uncertainty sources will be small in comparison with those associated with precision and trueness (<30% of the largest contributor), it is not necessary to evaluate or include them. Planning the validation studies for a new method with uncertainty evaluation in mind enables estimation of the associated uncertainty with very little extra effort. Uncertainty estimates for methods already in routine use can also be calculated using historical validation data.

As for the example in Section 9.3.1, the first step in the procedure is to identify the sources of uncertainty for the method. Using the same example as that used above, of an HPLC method for the determination of SMT, the uncertainty sources are listed in Table 9.1 and Figure 9.2. The individual sources can then be grouped into parameters for consideration in the calculation of the combined uncertainty. Three parameters are considered using the Barwick–Ellison approach:

Precision and Its Uncertainty. For this example, the precision is calculated from the within-laboratory reproducibility results for the method. As shown in Table 9.1, the mean of 24 controls spiked at 100 μg/kg was 83.0 μg/kg, with a standard deviation of 5.4 μg/kg. The relative standard uncertainty (relative standard deviation) was 0.065.

Recovery and Its Uncertainty. The trueness of the method must be considered in the calculation of the uncertainty estimate to cover uncertainties due to method bias. Barwick and Ellison describe several possibilities for estimating uncertainty related to trueness,[19] including the use of data from the analysis of a representative certified reference material, comparison with a reference or standard method, and spiking recovery studies. Since no representative certified reference material was available for this analyte–matrix combination, the trueness was estimated using spiking studies. As stated in the previous section, the spiked samples in this example were assumed to be representative of the incurred tissue, and the uncertainty contribution arising from the difference in behavior of spiked and incurred samples was considered to be negligible. The standard uncertainty of the recovery, calculated as in the previous example (Section 9.3.1 and Table 9.1) was 0.0553, with a relative standard uncertainty of 0.0666.

Ruggedness and Its Uncertainty. Ruggedness tests provide a means of simultaneously investigating the effect of several parameters on the performance of the method by the deliberate introduction of reasonable variations and the observation of their consequences. Factors that could potentially influence the results are identified and are modified by an amount to match the deviations that could reasonably occur within or between laboratory analyses. Examples include factors such as the analyst, the temperature, the source and age of reagents and analytical standards, the pH value at various steps, and the batch number of solvents or SPE cartridges.

The ruggedness test is performed using a fractional factorial design such as that described in the AOAC *Statistical Manual*,[21] which minimizes the number of analyses, time, and effort required to detect influences on the measurement results. Ruggedness tests are advocated by European Commission Decision 2002/657/EC[22] and others as an integral part of the validation of methods of analysis for chemical residues in food. This approach is illustrated in Table 9.2, where the uppercase letters A, B, C, D, E, F, and G represent the nominal values for seven potentially influencing factors and the lowercase letters a, b, c, d, e, f, and g represent those factors with reasonable variations introduced. The results obtained for each of the eight analyses are represented by the uppercase letters S through Z.

The effect of each parameter is calculated by subtracting the mean of the results obtained with the altered parameter from the mean of the results obtained with the nominal value. For example, for parameter A, the difference Dx_A is:

$$Dx_A = \frac{(S + T + U + V)}{4} - \frac{(W + X + Y + Z)}{4} \tag{9.5}$$

For the method in this example, the variables selected were molarity of the hydrochloric acid at the extraction stage (0.1 or 0.11 M), weight of sodium sulfate added at the extraction stage (2 or 2.1 g), homogenization time (45 or 55s), centrifugation speed (650g or 680g), evaporation temperature (55°C or 60°C), evaporation just to dryness before removal from the turbovap compared with allowing the solution to remain in the turbovap for 5 min after evaporation, and the volume of hexane for the final wash step (2 or 2.1 ml). The observed results from the eight experiments are presented in Table 9.3.

The Student's t test is applied to determine whether each parameter has a significant effect on the result, using the formula:

$$t = \frac{\sqrt{n} \times |Dx_i|}{\sqrt{2} \times s} \tag{9.6}$$

where s is an estimate of within-batch (intra-batch) method precision, n is the number of experiments in the ruggedness test for each parameter ($n = 4$ for the design above), and Dx_i is the difference calculated for parameter x_i using Equation (9.5). The precision (s) was calculated using 15 replicate measurements giving a standard deviation of 5.3 μg/kg. The t values for each parameter were calculated using Equation (9.6) and are given in Table 9.4. The t value

TABLE 9.2 A Fractional Factorial Design for Ruggedness Testing

	Combination of Determinations							
Factor Value	1	2	3	4	5	6	7	8
A/*a*	*A*	*A*	*A*	*A*	*a*	*a*	*a*	*a*
B/*b*	*B*	*B*	*b*	*b*	*B*	*B*	*b*	*b*
C/*c*	*C*	*c*	*C*	*c*	*C*	*c*	*C*	*c*
D/*d*	*D*	*D*	*d*	*d*	*d*	*d*	*D*	*D*
E/*e*	*E*	*e*	*E*	*e*	*e*	*E*	*e*	*E*
F/*f*	*F*	*f*	*f*	*F*	*F*	*f*	*f*	*F*
G/*g*	*G*	*g*	*g*	*G*	*g*	*G*	*G*	*g*
Observed Result	*S*	*T*	*U*	*V*	*W*	*X*	*Y*	*Z*

TABLE 9.3 Results from the Ruggedness Test of the SMT HPLC Method (μg/kg)

S	*T*	*U*	*V*	*W*	*X*	*Y*	*Z*
83	74.3	91.7	74.7	78.5	75.1	71.5	81.8

TABLE 9.4 Parameters Varied in the Ruggedness Test for the SMT Method and Values Calculated

Parameter		Nominal Value		Altered Value		Calculated Difference	t Value ($t_{crit} = 2.145$)	Significant Effect	δ_{real}	δ_{test}	$\delta_{real}/\delta_{test}$	$u(y(x_i))$ (μg/kg)	$u(y(x_i))$ (Relative)
Molarity of HCl for extraction	A	0.1 M	a	0.11M	Dx_A	4.2	1.1207	No	0.002	0.01	0.2	0.820	0.00988
Weight of anhydrous Na_2SO_4	B	2 g	b	2.1 g	Dx_B	−2.2	0.5870	No	0.05 g	0.1 g	0.5	2.051	0.02470
Homogenization time	C	45 s	c	55 s	Dx_C	4.7	1.2541	No	5 s	10 s	0.5	2.051	0.02470
Centrifugation	D	650g	d	680g	Dx_D	−2.35	0.6271	No	10g	30g	0.333	1.367	0.01647
Evaporation temperature	E	55°C	e	60°C	Dx_E	8.15	2.1747	Yes	2°C	5°C	—	1.882	0.02268
Evaporation time	F	To dryness	f	+5 min	Dx_F	1.35	0.3602	No	2 min	5 min	0.4	1.640	0.01976
Hexane wash volume	G	2 ml	g	2.1 ml	Dx_G	−5.5	1.4676	No	0.05 ml	0.1 ml	0.5	2.051	0.02470

for each parameter is compared with the two-tailed critical value (t_{crit}) for $N-1$ degrees of freedom at 95% confidence, where $N(=15)$ is the number of determinations used to calculate the within-batch precision s. The value of t_{crit} from t distribution tables is 2.145. If t is less than t_{crit}, variations in that parameter do not have a significant effect on method performance. If t is greater than t_{crit}, variations in the parameter do have a significant effect. In either case there is an uncertainty associated with the parameter. The procedure to calculate the uncertainty associated with each parameter depends on whether varying that parameter has a significant effect.

For cases where the results of the ruggedness test have indicated that variations in a parameter do not have a significant effect on the method, the uncertainty associated with the result y due to the parameter x_i is calculated using the equation:

$$u(y(x_i)) = \frac{\sqrt{2} \times t_{crit} \times s}{\sqrt{n} \times 1.96} \times \frac{\delta_{real}}{\delta_{test}} \tag{9.7}$$

where δ_{real} is the change in parameter that would be expected when the method is operating under control in routine use, δ_{test} is the change in parameter that was introduced in the ruggedness test (Table 9.4); and n is the number of tests carried out for each parameter ($n = 4$ for the typical ruggedness test design shown). This uncertainty estimate is based on the 95% confidence interval, converted to a standard deviation by dividing by 1.96. Substituting the values for t_{crit} and n for this example:

$$u(y(x_i)) = \frac{\sqrt{2} \times 2.145 \times 5.3}{\sqrt{4} \times 1.96} \times \frac{\delta_{real}}{\delta_{test}} = 4.101 \times \frac{\delta_{real}}{\delta_{test}}$$

For the current example, the variations introduced in six of the seven parameters investigated in the ruggedness test did not have a significant effect on the method. The uncertainties for those parameters were calculated using the formula above and are listed in Table 9.4.

For those parameters that are shown to have a significant effect on the method performance, the uncertainty can be estimated using the equations

$$c_i = \frac{\text{observed change in result}}{\text{change in parameter}} \tag{9.8}$$

and

$$u(y(x_i)) = u(x_i) \times c_i \tag{9.9}$$

where c_i is the sensitivity coefficient, $u(x_i)$ is the uncertainty in the parameter, and $u(y(x_i))$ is the uncertainty in the final result.

Of the parameters examined in the current example, only the evaporation temperature was shown to have a significant effect (Table 9.4). The observed change in result (the difference Dx_E) from Table 9.4 was 8.15 μg/kg and the change in parameter was (60°C–55°C) = 5°C. From Equation (9.8), the sensitivity coefficient was calculated as 1.63 μg kg^{-1} °C^{-1}. The uncertainty in the parameter $u(x_E)$ is estimated from the control limit on the evaporation temperature (± 2°C). A rectangular distribution is assumed, and $u(x_E) = 2/\sqrt{3} = 1.1547$°C. Using Equation (9.9), the uncertainty in the final result due to the variation in the evaporation temperature was $u(y(x_E)) = 1.1547 \times 1.63 = 1.882$ μg/kg.

The standard uncertainties for all seven parameters are presented in Table 9.4. The effects of the parameters investigated were considered to be proportional to the analyte concentration. The relative standard uncertainties for each factor were, therefore, calculated by dividing the standard uncertainty by the mean result obtained for samples spiked at 100 μg/kg under reproducibility conditions (83.0 μg/kg).

Finally, the combined uncertainty is calculated from the root sum squares of the relative uncertainties for each of the parameters in the ruggedness test, plus those of the precision and recovery, using an equation of the form

$$\frac{u(y)}{y} = \sqrt{\left(\frac{u(p)}{p}\right)^2 + \left(\frac{u(q)}{q}\right)^2 + \left(\frac{u(r)}{r}\right)^2 + \left(\frac{u(s)}{s}\right)^2 + \cdots} \tag{9.10}$$

For the example given, the combined uncertainty = 0.1085. The combined standard uncertainty $u_c(y)$ is:

$$u_c(y) = 0.1085 \times 83\ \mu\text{g/kg} = 9.00\ \mu\text{g/kg}.$$

The final step is to derive the expanded uncertainty using the coverage factor k. For this example, k is given a value of 2, corresponding to an approximate confidence level of 95%: $U(y) = k \times u_c(y) = 18.0$ μg/kg.

A recovery-corrected result, generated using this analytical method, with a concentration equal to an MRL of 100 μg/kg, can therefore be reported as 100 ± 18.0 μg/kg, with 95% confidence that the true SMT content of the sample is between 82.0 and 118.0 μg/kg.

9.3.3 Measurement Uncertainty Based on Nested Experimental Design Using In-House Validation Data

A *nested design*, which is sometimes referred to as a *hierarchical design*, is used for an experiment in which there is an interest in a set of treatments and the experimental units are subsampled (Fig. 9.4). It has become a practical design or an approach to study and to evaluate method performance criteria that include accuracy expressed as overall recovery, intermediate precision, and

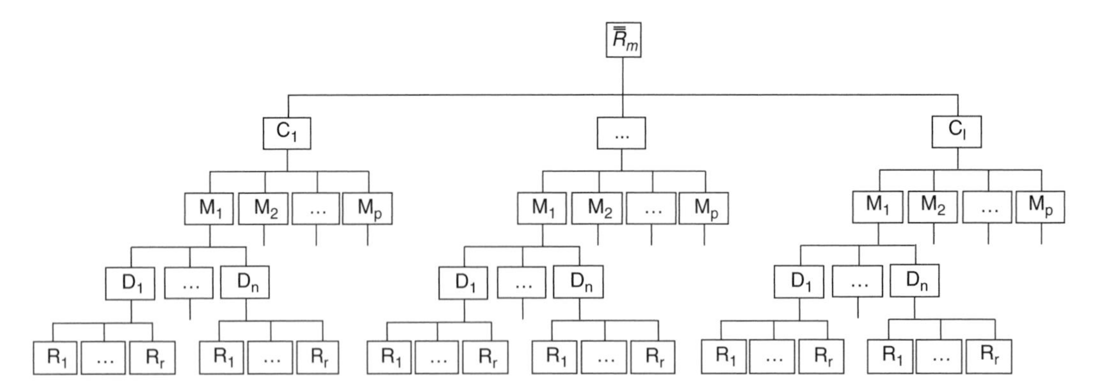

Figure 9.4 Nested experimental design. Main factors associated with uncertainties include: concentration (C) with a number of spiked levels (l), matrix effects (M) with a number of matrices (p), day-to-day variation (D) with a number of days (n), and within-day variation (R) with a number of replicates (r).

TABLE 9.5 ANOVA Table for a Nested Experimental Design and Expression of Uncertainty

Source	Levels	Mean Square (MS)	Variance or Uncertainty Square
Concentrations	$l = 4$	MS_{conc}	$u(C)^2 = \frac{MS_{conc} - MS_m}{pnr}$
Matrices (concentrations)	$p = 3$	MS_m	$u(M)^2 = \frac{MS_m - MS_d}{nr}$
Days (concentrations*matrices)	$n = 2$	MS_d	$u(D)^2 = \frac{MS_d - MS_i}{r}$
Replicates	$r = 3$	MS_i	$u(r)^2 = MS_i$

measurement uncertainty.[11,23–27] To explain the statistics and detailed calculations, an example is presented using the data from LC/ESI-MS/MS analysis of macrolides in milk, for which the analytical method was published elsewhere.[27]

The main factors, which are associated with the method performance criteria or uncertainties of an in-house validated method, consist of concentrations or spike levels, matrix effects, day-to-day variation (between days), and within-day variation (Table 9.5). In this example, four concentrations ($l = 4$; i.e., 5, 40, 50, and 70 μg/kg) are included. For each concentration, the recovery is estimated with three different raw milk matrices ($p = 3$). For each matrix, the analysis is carried out on two different days ($n = 2$) by two analysts using different reagents and analytical columns, and each sample is prepared in triplicate ($r = 3$). Therefore, there are a total of six experiments. Each analyst works on one matrix, which is spiked at four concentrations in triplicate, per day. After completing the experiments, the validation data are organized in nested design format (Fig. 9.4) to perform analysis of variance, from which the overall recovery, intermediate precision and uncertainty are calculated or derived. Table 9.6 lists the experimental data for practice and demonstration. Some of the SAS outputs and the SAS program that is used to perform "PROC GLM" and "PROC VARCOMP" are shown in Scheme 9.1.

9.3.3.1 Recovery (R) and Its Uncertainty [u(R)]

Recovery R of a spiked sample consisted of the sum of three components expressed as follows:

$$R = \overline{\overline{R}}_m + \Delta R_M + \Delta R_C \quad (9.11)$$

The first component ($\overline{\overline{R}}_m$) is the overall recovery; the second one (ΔR_M) considers the variation of recovery caused by different matrices, namely, matrix effects, which is usually the major source of uncertainty for LC-MS quantification; and the last item (ΔR_C) is the variation of recovery due to the amount of an analyte spiked in samples. The uncertainty of the recovery can be calculated from:

$$u(R) = \sqrt{u(\overline{\overline{R}}_m)^2 + u(M)^2 + u(C)^2} \quad (9.12)$$

where $u(\overline{\overline{R}}_m)$ is the uncertainty of the overall recovery, and proportional bias is estimated in terms of the overall recovery; $u(M)$ and $u(C)$ are uncertainties associated with matrix effects (ΔR_M) and concentration variability (ΔR_C). The calculations of $u(M)$ and $u(C)$ are expressed in Table 9.5, and the detailed calculations are described in Scheme 9.1.

The overall recovery $\overline{\overline{R}}_m$ is an estimation of the *method recovery* and is calculated using Equation (9.13) from macrolide recoveries in Scheme 9.1C (mean or average of four spiked concentrations):

$$\overline{\overline{R}}_m = \frac{\sum_{i=1}^{l} \overline{\overline{R}}_i}{l} \quad (9.13)$$

where $\overline{\overline{R}}_i$ is the mean recovery for a given amount $X_{a,i}$ of an analyte added to each of four milk samples and l

TABLE 9.6 LC/ESI-MS/MS Analysis of Macrolides in Milk Experimental Data

Spike Concentration	Matrices	Days	Erythromycin Recovery	Tylosin Recovery
L1 (5 μg/kg)	M1	Analyst 1	0.988	0.894
			0.998	0.924
			0.974	0.932
		Analyst 2	1.110	0.978
			1.020	1.019
			1.000	0.984
	M2	Analyst 1	1.038	1.086
			1.028	1.072
			1.032	1.076
		Analyst 2	1.034	1.091
			1.019	1.009
			1.062	1.053
	M3	Analyst 1	0.976	0.934
			0.924	0.978
			0.950	0.954
		Analyst 2	1.016	0.971
			0.990	0.994
			0.992	0.948
L2 (40 μg/kg)	M1	Analyst 1	0.958	0.899
			0.986	0.937
			0.975	0.912
		Analyst 2	0.958	0.949
			1.031	1.014
			1.015	0.969
	M2	Analyst 1	1.010	1.052
			1.020	1.041
			1.009	1.030
		Analyst 2	1.023	0.991
			1.071	1.036
			1.094	1.090
	M3	Analyst 1	0.969	0.967
			0.917	0.931
			0.961	0.949
		Analyst 2	1.026	0.970
			1.042	0.999
			1.022	1.055
L3 (50 μg/kg)	M1	Analyst 1	0.960	0.907
			0.951	0.908
			0.995	0.931
		Analyst 2	0.979	0.933
			0.956	0.939
			0.981	0.953
	M2	Analyst 1	1.054	1.110
			1.003	1.020
			1.005	1.020
		Analyst 2	0.993	0.992
			1.004	1.088
			0.999	0.999
	M3	Analyst 1	0.967	0.953
			0.962	0.936
			0.952	0.963
		Analyst 2	0.963	0.981
			0.988	0.935
			1.030	0.993

(*continued*)

TABLE 9.6 (*Continued*)

Spike Concentration	Matrices	Days	Erythromycin Recovery	Tylosin Recovery
L4 (70 μg/kg)	M1	Analyst 1	0.950	0.948
			0.946	0.913
			0.937	0.878
		Analyst 2	0.998	0.935
			0.960	0.930
			0.956	0.903
	M2	Analyst 1	1.033	1.001
			1.032	1.029
			1.058	1.043
		Analyst 2	1.027	1.078
			0.950	1.055
			0.981	1.076
	M3	Analyst 1	0.930	0.952
			0.906	0.942
			0.929	0.925
		Analyst 2	1.006	0.994
			1.028	1.011
			0.977	1.006

is the number of concentrations. The overall recoveries of erythromycin and tylosin are 0.994–0.984, respectively (Table 9.7), and they are tested if they are statistically significantly different from one based on t values calculated from Equation (9.14) (Scheme 9.1C):

$$t = \frac{|1 - \overline{\overline{R}}_{\mathrm{m}}|}{u(\overline{\overline{R}}_{\mathrm{m}})} \tag{9.14}$$

where $u(\overline{\overline{R}}_{\mathrm{m}})$ is the uncertainty of the overall recovery expressed in Equation (9.15) (Scheme 9.1):

$$u(\overline{\overline{R}}_{\mathrm{m}}) = \sqrt{\frac{\sum_{i=1}^{l} u(\overline{\overline{R}}_i)^2}{l^2}} \tag{9.15}$$

The uncertainty of mean recoveries $u(\overline{\overline{R}}_i)$ [Eq. (9.16), Scheme 9.1B1–B4], is calculated from relative intermediate precision:

$$u(\overline{\overline{R}}_i)^2 = \frac{u(R_{\mathrm{I}})^2}{pnr} \tag{9.16}$$

The intermediate precision $u(R_{\mathrm{I}})$ is calculated from Equation (9.17) (Scheme 9.1B1–B4), where $u(D)$ represents the variance between days and $u(r)$ is the variance of triplicates. Both $u(D)^2$ and $u(r)^2$, that is, variances (Scheme 9.1B1–B4), are directly output by the SAS program:

$$u(R_{\mathrm{I}})^2 = u(r)^2 + u(D)^2 \tag{9.17}$$

In the significance test of the overall recovery, the two-sided z value ($\alpha = 0.05$; i.e., 1.96) can be used instead of the value of $t_{\mathrm{crit}}(\alpha = 0.05)$ because of the considerable number of degrees of freedom associated with the uncertainty of the overall recovery[24] or the coverage factor k (typically, $k = 2$) is simply applied. The t values of the two macrolides are listed in Table 9.7. The t value (1.28) of erythromycin is less than 1.96, which means that its recovery is not significantly different from one. The t value (3.52) of tylosin (Scheme 9.1C) is higher than 1.96, therefore, its recovery is found to be statistically different from one, the method has a significant bias, and as a result by theory, a correction factor expressed as recovery can be applied to correct the analytical results.[24] When analytical results are not corrected by a correction factor, the uncertainty must be increased using Equation (9.18) (Scheme 9.1C) to account for the fact that the recovery has not been corrected for. The increased uncertainty $u(\overline{\overline{R}}_{\mathrm{m}})''$ is given by:

$$u(\overline{\overline{R}}_{\mathrm{m}})'' = \sqrt{u(\overline{\overline{R}}_{\mathrm{m}})^2 + \left(\frac{1 - \overline{\overline{R}}_{\mathrm{m}}}{k}\right)^2} \tag{9.18}$$

where k is the coverage factor that will be used in the calculation of the expanded uncertainty. In this case, $u(\overline{\overline{R}}_{\mathrm{m}})''$ (Scheme 9.1C) rather than $u(\overline{\overline{R}}_{\mathrm{m}})$ is used to calculate $u(R)$ (Scheme 9.1A).

The uncertainties of the recoveries $[u(R)]$ (Scheme 9.1A) of erythromycin and tylosin are listed in Table 9.7. Whether matrix and concentration effects are in fact statistically significant ($p < 0.05$ or $p \geq 0.05$), $u(M)$ and $u(C)$, which

A. Coumpound = tylosin

Class-level information

Class	Levels	Values
Concentration	$l = 4$	L1 L2 L3 L4
Matrix	$p = 3$	M1 M2 M3
Day	$n = 2$	day1 day2
Replicate	$r = 3$	rep1 rep2 rep3

Type 1 analysis of variance

Source	DF	Sum of squares	Mean square		Expected mean square
Concentration	3	0.003990	0.001330	MS_{conc}	Var(error) + 3 Var(day(concentration*matrix)) + 6 Var(matrix(concentration)) + 18 Var(concentration)
Matrix(concentration)	8	0.160327	0.020041	MS_m	Var(error) + 3 Var(day(concentration*matrix)) + 6 Var(matrix(concentration))
Day(concentration*matrix)	12	0.033014	0.002751	MS_d	Var(error) + 3 Var(day(concentration*matrix))
Error	48	0.038547	0.000803	MS_i	Var(error)
Corrected total	71	0.235878	.		.

Type 1 estimates

Variance component	Estimate	
Var(concentration)	**-0.0010395**	← $u(C)^2 = \dfrac{MS_{conc} - MS_m}{pnr}$
Var(matrix(concentration))	**0.0028816**	← $u(M)^2 = \dfrac{MS_m - MS_d}{nr}$
Var(day(concentration* matrix))	**0.0006494**	← $u(D)^2 = \dfrac{MS_d - MS_i}{r}$
Var(error)	**0.0008031**	← $u(r)^2 = MS_i$

Eq. (9.12)

$$u(R) = \sqrt{u(\bar{\bar{R}}m)^2 + u(M)^2 + u(C)^2} = \sqrt{0.0000846{+}00028816.0{+}0} = \sqrt{0.0029662}$$

Eq. (9.17)

$$u(R_I)^2 = u(r)^2 + u(D)^2 = 0.0006494{+}0.0008031 = 0.0014525$$

Eq. (9.19)

$$u(P) = \frac{u(R_I)}{\bar{\bar{R}}_m} = \frac{\sqrt{0.0014525}}{0.984} = 0.039$$

B1. Compound = tylosin concentration = L1

Class-level Information

Class	Levels	Values
Matrix	3	M1 M2 M3
Day	2	day1 day2

Type 1 analysis of variance

Source	DF	Sum of squares	Mean square		Expected mean square
Matrix	2	0.044572	0.022286	MS_m	Var(error) + 3 Var(day(matrix)) + 6 Var(matrix)
Day(matrix)	3	0.010355	0.003452	MS_d	Var(error) + 3 Var(day(matrix))
Error	12	0.007284	0.000607	MS_i	Var(error)
Corrected total	17	0.062212	.		.

Type 1 estimates

Variance component	Estimate	
Var(matrix)	**0.0031391**	← $u(M)^2 = \dfrac{MS_m - MS_d}{nr}$
Var(day(matrix))	**0.0009482**	← $u(D)^2 = \dfrac{MS_d - MS_i}{r}$
Var(error)	**0.0006070**	← $u(r)^2 = MS_i$

Eq. (9.17)

$$u(R_I)^2 = u(r)^2 + u(D)^2 = 0.0009482{+}0.0006070 = 0.0015552$$

Eq. (9.16)

$$u(\bar{\bar{R}}_1)^2 = \frac{u(R_I)^2}{pnr} = \frac{0.0015552}{3 \times 2 \times 3} = 0.0000864$$

B2. Compound = tylosin concentration = L2

Class-level Information

Class	Levels	Values
Matrix	3	M1 M2 M3
Day	2	day1 day2

Type 1 analysis of variance

Source	DF	Sum of squares	Mean square		Expected mean square
Matrix	2	0.027013	0.013507	MS_m	Var(error) + 3 Var(day(matrix)) + 6 Var(matrix)
Day(matrix)	3	0.010870	0.003623	MS_d	Var(error) + 3 Var(day(matrix))
Error	12	0.012501	0.001042	MS_i	Var(error)
Corrected total	17	0.050384	.		.

Scheme 9.1 SAS output, detailed calculations of method performance criteria, and SAS program.

Type 1 estimates

Variance component	Estimate	
Var(matrix)	**0.0016472**	← $u(M)^2 = \frac{MS_m - MS_d}{nr}$
Var(day(matrix))	**0.0008606**	← $u(D)^2 = \frac{MS_d - MS_i}{r}$
Var(error)	**0.0010417**	← $u(r)^2 = MS_i$

Eq. 9.17

$$u(R_I)^2 = u(r)^2 + u(D)^2 = 0.0008606 + 0.0010417 = 0.0019023$$

Eq. 9.16

$$u(\bar{\bar{R}}_2)^2 = \frac{u(R_I)^2}{pnr} = \frac{0.0019023}{3 \times 2 \times 3} = 0.0001057$$

B3. Compound = tylosin concentration = L3

Class-level Information

Class	Levels	Values
Matrix	3	M1 M2 M3
Day	2	day1 day2

Type 1 analysis of variance

Source	DF	Sum of squares	Mean square		Expected mean square
Matrix	2	0.038227	0.019114	MS_m	Var(error) + 3 Var(day(matrix)) + 6 Var(matrix)
Day(matrix)	3	0.002422	0.000807	MS_d	Var(error) + 3 Var(day(matrix))
Error	12	0.013955	0.001163	MS_i	Var(error)
Corrected total	17	0.054604	.		.

Type 1 estimates

Variance component	Estimate	
Var(matrix)	**0.0030510**	← $u(M)^2 = \frac{MS_m - MS_d}{nr}$
Var(day(matrix))	**−0.0001186**	← $u(D)^2 = \frac{MS_d - MS_i}{r}$
Var(error)	**0.0011629**	← $u(r)^2 = MS_i$

Eq. (9.17)

$$u(R_I)^2 = u(r)^2 + u(D)^2 = 0 + 0.0011629 = 0.0011629$$

Eq. (9.16)

$$u(\bar{\bar{R}}_3)^2 = \frac{u(R_I)^2}{pnr} = \frac{0.0011629}{3 \times 2 \times 3} = 0.0000646$$

B4. Compound = tylosin concentration = L4

Class-level Information

Class	Levels	Values
Matrix	3	M1 M2 M3
Day	2	day1 day2

Type 1 analysis of variance

Source	DF	Sum of squares	Mean square		Expected mean square
Matrix	2	0.050514	0.025257	MS_m	Var(error) + 3 Var(day(matrix)) + 6 Var(matrix)
Day(matrix)	3	0.009367	0.003122	MS_d	Var(error) + 3 Var(day(matrix))
Error	12	0.004807	0.000401	MS_i	Var(error)
Corrected total	17	0.064688	.		.

Type 1 estimates

Variance component	Estimate	
Var(matrix)	**0.0036891**	← $u(M)^2 = \frac{MS_m - MS_d}{nr}$
Var(day(matrix))	**0.0009072**	← $u(D)^2 = \frac{MS_d - MS_i}{r}$
Var(error)	**0.0004006**	← $u(r)^2 = MS_i$

Eq. (9.17)

$$u(R_I)^2 = u(r)^2 + u(D)^2 = 0.0009072 + 0.0004006 = 0.0013078$$

Eq. (9.16)

$$u(\bar{\bar{R}}_4)^2 = \frac{u(R_I)^2}{pnr} = \frac{0.0013078}{3 \times 2 \times 3} = 0.0000727$$

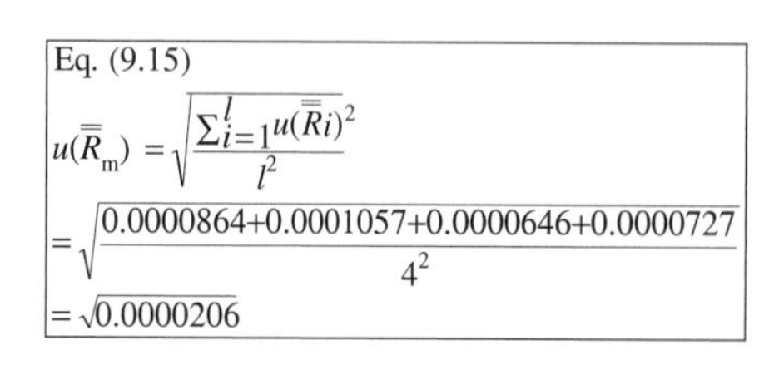

Scheme 9.1 *(Continued)*

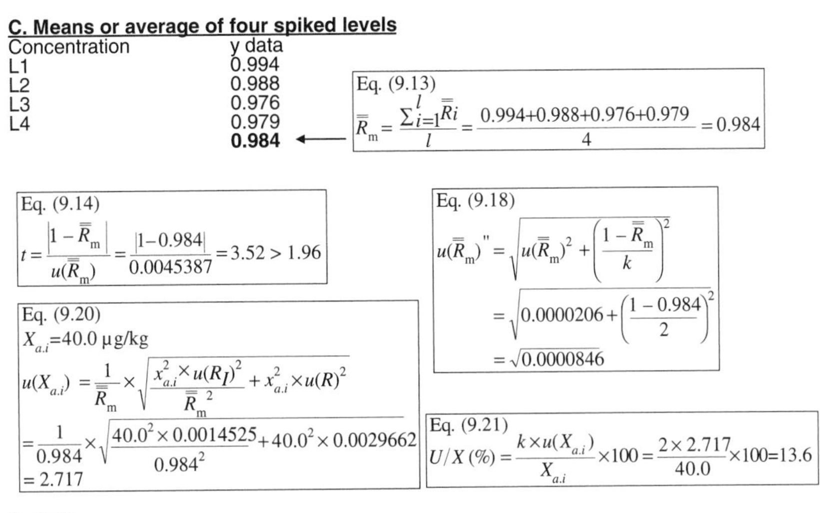

C. Means or average of four spiked levels

Concentration	y data
L1	0.994
L2	0.988
L3	0.976
L4	0.979
	0.984

Eq. (9.13)

$$\overline{\overline{R}}_{\text{m}} = \frac{\sum_{i=1}^{l} \overline{R}i}{l} = \frac{0.994+0.988+0.976+0.979}{4} = 0.984$$

Eq. (9.14)

$$t = \frac{|1 - \overline{\overline{R}}_{\text{m}}|}{u(\overline{\overline{R}}_{\text{m}})} = \frac{|1-0.984|}{0.0045387} = 3.52 > 1.96$$

Eq. (9.18)

$$u(\overline{\overline{R}}_{\text{m}})'' = \sqrt{u(\overline{\overline{R}}_{\text{m}})^2 + \left(\frac{1 - \overline{\overline{R}}_{\text{m}}}{k}\right)^2}$$

$$= \sqrt{0.0000206 + \left(\frac{1 - 0.984}{2}\right)^2}$$

$$= \sqrt{0.0000846}$$

Eq. (9.20)

$X_{a.i} = 40.0\ \mu\text{g/kg}$

$$u(X_{a.i}) = \frac{1}{\overline{\overline{R}}_{\text{m}}} \times \sqrt{\frac{x_{a.i}^2 \times u(R_I)^2}{\overline{\overline{R}}_{\text{m}}^{\,2}} + x_{a.i}^2 \times u(R)^2}$$

$$= \frac{1}{0.984} \times \sqrt{\frac{40.0^2 \times 0.0014525}{0.984^2} + 40.0^2 \times 0.0029662}$$

$$= 2.717$$

Eq. (9.21)

$$U/X\,(\%) = \frac{k \times u(X_{a.i})}{X_{a.i}} \times 100 = \frac{2 \times 2.717}{40.0} \times 100 = 13.6$$

D. SAS program

```
option pagesize=90 linesize=90 formdlim='_';
filename sasm2 dde'Excel|C:\My Documents\[Marolides.xls]Experimental data!R3C4:R74C5';
data sastx;
infile sasm2;
    do concentration = 'L1', 'L2', 'L3', 'L4';
        do matrix = 'M1', 'M2', 'M3';
            do day = 'day1', 'day2';
                do obs =1 to 3;
                    do compounds = 'Erythromycin', 'Tylosin';
                            input ydata @;
                            nobs + 1;
                            output;
                    end;
                end;
            end;
        end;
    end;
proc print;
where nobs lt 21;
title 'Analysis of macrolides in milk by proc glm and varcomp with method=type I';
title2 'Macrolides in milk';
proc sort data=sastx;
    by compounds;
proc glm data=sastx;
    by compounds;
    class concentration matrix day;
    model ydata = concentration matrix(concentration) day(matrix concentration)/ss3;
    test h=concentration e=matrix(concentration);
    test h=matrix(concentration) e=day(matrix concentration);
    lsmeans concentration matrix(concentration) day(matrix concentration)/stderr;
    random concentration matrix(concentration) day(matrix concentration)/test;
proc sort data=sastx;
    by compounds concentration;
proc glm data=sastx;
    by compounds concentration;
    class matrix day;
    model ydata = matrix day(matrix)/ss3;
    test h=matrix e=day(matrix);
    lsmeans matrix day(matrix)/stderr;
    random matrix day(matrix)/test;
proc sort data=sastx;
    by compounds;
proc varcomp method=type1 data=sastx;
    by compounds;
    class concentration matrix day;
    model ydata = concentration matrix(concentration) day(matrix concentration);
proc sort data=sastx;
    by compounds concentration;
proc varcomp method=type1 data=sastx;
    by compounds concentration;
    class matrix day;
    model ydata = matrix day(matrix);
quit;
```

Scheme 9.1 *(Continued)*

are associated with matrix and concentration variability, are included in the uncertainty budget.

9.3.3.2 *Precision and Its Uncertainty [u(P)]*

The uncertainty arising from the precision of the method is expressed as a relative intermediate standard deviation and calculated using Equation (9.19) (Scheme 9.1A); that is, the intermediate precision is divided by the overall recovery:

$$u(P) = \frac{u(R_\mathrm{I})}{\overline{\overline{R}}_\mathrm{m}} \tag{9.19}$$

The results are listed in Table 9.7. The intermediate precisions of erythromycin and tylosin [$u(P)$] are 4.0% and 3.9%, respectively.

9.3.3.3 *Combined Standard Uncertainty and Expanded Uncertainty*

The combined standard uncertainty of the quantitative result $u(X_{\mathrm{a},i})$ of a sample spiked with an amount $X_{\mathrm{a},i}$ is calculated using Equation (9.20) (Scheme 9.1), and this uncertainty from in-house validation data can be applied to future sample testing results:

$$u(X_{\mathrm{a},i}) = \frac{1}{\overline{\overline{R}}_\mathrm{m}} \times \sqrt{\frac{x_{\mathrm{a},i}^2 \times u(R_I)^2}{\overline{\overline{R}}_\mathrm{m}^2} + x_{\mathrm{a},i}^2 \times u(R)^2} \tag{9.20}$$

The first term of Equation (9.20) factors in the uncertainty arising from the experimental variability of the method, that is, the intermediate precision, at fortified concentrations, and the second one takes into account the uncertainty associated with the estimation of recovery including matrix effects and concentration variability as well. The expanded uncertainty U is then calculated using the coverage factor $k = 2$. $U/X(\%)$ is calculated using Eq. (9.21) (Scheme 9.1), and results are listed in Table 9.7:

$$\frac{U}{X}(\%) = \frac{k \times u(X_{\mathrm{a},i})}{X_{\mathrm{a},i}} \times 100 \tag{9.21}$$

The relative uncertainties $U/X(\%)$ obtained at the four fortified concentrations are apparently the same since the uncertainty of constant bias is not included in the budget. For comparison, the between-laboratory relative standard deviations (RSDR, %) according to the Horwitz formula are also calculated, and the within-laboratory relative standard deviations (RSDr, %) should be $\frac{1}{2}$–$\frac{2}{3}$ RSDR (%).[28]

9.3.4 Measurement Uncertainty Based on Inter-laboratory Study Data

Standards ISO 5725-2[29] and ISO 21748[30] provide guidance to carry out properly designed and executed inter-laboratory studies for the estimation of method parameter statistics and subsequently measurement uncertainty.[31] ISO 5725-2 derives measurement uncertainty estimates solely from inter-laboratory data, whereas ISO 21748 expands on ISO 5725-2 by using inter-laboratory data as well as information derived by means outlined in ISO 98-3.[3] ISO 21748 recognizes that even properly designed inter-laboratory studies may not include all relevant uncertainty components, and that they need to be factored in. The model for the determination of measurement uncertainty by this means is expressed as:

$$u^2(y) = u^2(\hat{\delta}) + s_\mathrm{L}^2 + \sum c_i^2 u^2(x_i) + s_\mathrm{r}^2 \tag{9.22}$$

where s_L^2 is an estimate of the between-laboratory variance; s_r^2 is an estimate of the repeatability variance; $u(\hat{\delta})$ is an estimate of the uncertainty of the bias intrinsic to the measurement method; $u(x_i)$ is an estimate of the uncertainty associated with parameter x_i, usually "type B" as defined in ISO 98-3; and c_i^2 is the sensitivity coefficient associated with uncertainty $u(x_i)$. The following equations [Eqs. (9.23)–(9.45)] and their definitions describe the calculations used for this example. Notation in these equations is as follows:

- y_{ijk} — A datum point within the study
- i — The laboratory identifier
- p — The total number of laboratories participating in the study
- j — The level or concentration identifier
- q — The total number of levels or concentrations being tested in the study
- k — The sample replicate identifier done by laboratory i at level j
- n_{ij} — The total number of sample replicates done by laboratory i at level j
- h_{ij} — Mandel's h statistic for assessing and comparing between-laboratory consistency, given in the following equation:

$$h_{ij} = \frac{\bar{y}_{ij} - \bar{\bar{y}}_j}{\sqrt{\frac{1}{(p_j - 1)} \sum_{i=1}^{p_j} (\bar{y}_{ij} - \bar{\bar{y}}_j)^2}} \tag{9.23}$$

The term k_{ij} represents Mandel's k statistic for assessing and comparing within-laboratory consistency, given as follows:

$$k_{ij} = \frac{s_{ij}\sqrt{p_j}}{\sqrt{\sum s_{ij}^2}} \tag{9.24}$$

The term C represents Cochran's test statistic for evaluating within-laboratory consistency, given in

$$C = \frac{s_\mathrm{max}^2}{\sum_{i=1}^{p} s_{ij}^2} \tag{9.25}$$

TABLE 9.7 Overall Recoveries, and Measurement Uncertainty Arising from the Accuracy and Precision of Two Macrolides Spiked in Raw Milk Samples

Compound	$\overline{\overline{R}}_m$	t	$u(P)$	$u(R)$	Spike Concentration (μg/kg)	$\frac{x_{a.i}^2 \times u(R_I)^2}{\overline{\overline{R}}_m^4}$	$\frac{x_{a.i}^2 \times u(R)^2}{\overline{\overline{R}}_m^2}$	U^a $(k = 2)$	U/X (%)	RSDR[b] (%)	$\frac{1}{2}$ RSDR (%)	$\frac{2}{3}$ RSDR (%)
Erythro-mycin	0.994	1.28	4.0×10^{-2}	1.5×10^{-2}	5.0	0.040	0.006	0.4	8.6	35.5	17.8	23.7
					40.0	2.557	0.388	3.4	8.6	26.0	13.0	17.3
					50.0	3.995	0.605	4.3	8.6	25.1	12.6	16.7
					70.0	7.830	1.187	6.0	8.6	23.9	11.9	15.9
Tylosin	0.984	3.52	3.9×10^{-2}	5.4×10^{-2}	5.0	0.039	0.077	0.7	13.6	71.0	17.8	23.7
					40.0	2.476	4.899	5.4	13.6	51.9	13.0	17.3
					50.0	3.869	7.654	6.8	13.6	50.2	12.6	16.7
					70.0	7.583	15.003	9.5	13.6	47.8	11.9	15.9

[a]Using information from the intermediate precision according to the method of Dehouck et al.[25]
[b]RSDR was calculated using the Horwitz equation[28] (RSDR $= 2^{(1-0.5 \log C)}$, where C is the concentration expressed as fractions).

The terms G_p and G_1 represent Grubb's outlier statistics for a single largest or smallest outlying observation respectively, and are given in the following two equations:

$$G_p = \frac{x_p - \bar{x}}{s} \tag{9.26}$$

$$G_1 = \frac{\bar{x} - x_1}{s} \tag{9.27}$$

The term G represents Grubb's outlier statistic for the two largest or two smallest outlying observations. When G represents the two largest outlying observations, it is defined as given in Equation (9.28) [with supporting equations, Eqs. (9.29)–(9.31)]

$$G = \frac{s_{p-1,p}^2}{s_0^2} \tag{9.28}$$

where

$$s_0^2 = \sum_{i=1}^{p} (x_i - \bar{x})^2 \tag{9.29}$$

$$s_{p-1,p}^2 = \sum_{i=1}^{p-2} (x_i - \bar{x}_{p-1,p})^2 \tag{9.30}$$

$$\bar{x}_{p-1,p} = \frac{1}{p-2} \sum_{i=1}^{p-2} x_i \tag{9.31}$$

When G represents the two smallest outlying observations, it is defined as given in Equation (9.32) [with supporting equations, Eqs. (9.29), (9.33), and (9.34)]:

$$G = \frac{s_{1,2}^2}{s_0^2} \tag{9.32}$$

$$s_{1,2}^2 = \sum_{i=3}^{p} (x_i - \bar{x}_{1,2})^2 \tag{9.33}$$

$$\bar{x}_{1,2} = \frac{1}{p-2} \sum_{i=3}^{p} x_i \tag{9.34}$$

The terms T_1, T_2, T_3, T_4, T_5 represent calculations used in ISO 5725-2 to facilitate calculation of estimates of method repeatability and laboratory variability, and are given in the following equations:

$$T_1 = \sum n_i \bar{y}_i \tag{9.35}$$

$$T_2 = \sum n_i (\bar{y}_i)^2 \tag{9.36}$$

$$T_3 = \sum n_i \tag{9.37}$$

$$T_4 = \sum n_i^2 \tag{9.38}$$

$$T_5 = \sum (n_i - 1) s_i^2 \tag{9.39}$$

The term s_r^2 represents the repeatability variance and is calculated as:

$$s_r^2 = \frac{T_5}{T_3 - p} \tag{9.40}$$

The term s_L^2 represents the between-laboratory variance, and is calculated as:

$$s_L^2 = \left[\frac{T_2 T_3 - T_1^2}{T_3(p-1)} - s_r^2\right]\left[\frac{T_3(p-1)}{T_3^2 - T_4}\right] \tag{9.41}$$

The term s_R^2 represents the reproducibility variance, which is given as:

$$s_R^2 = s_L^2 + s_r^2 \tag{9.42}$$

The term $\hat{m}$ represents the average calculated value for a given level, and is calculated as follows:

$$\hat{m} = \frac{T_1}{T_3} \tag{9.43}$$

The term v_i represents the degrees of freedom for uncertainty component i. When v_i represents an uncertainty component of the type B category, as defined in ISO 98-3,[3] it is determined as follows:

$$v_i = \frac{1}{2}\left[\frac{\Delta u(x_i)}{u(x_i)}\right]^{-2} \tag{9.44}$$

where $\Delta u(x_i)/u(x_i)$ is relative uncertainty from the type B component.

The term v_{eff} represents the effective degrees of freedom for an uncertainty expression as determined by the Welch–Satterthwaite formula:[3]

$$v_{\mathrm{eff}} = \frac{u_c^4(y)}{\sum_{i=1}^{N} \frac{u_i^4(y)}{v_i}} \tag{9.45}$$

In the example used to illustrate this model, a data set representing the findings of one analyte from a hypothetical inter-laboratory study is used. It is assumed that the analyte of study has a government regulated maximum residue limit (MRL) of 100 μg/kg in bovine muscle. The study directors evaluated the performance of the method over the range of 0.5–2.0 × MRL (i.e., 50–200 μg/kg) in bovine muscle. The study data were created using the random-number generation function of Microsoft Excel 2002 with the following assumptions:

1. Ten laboratories submitted data for the study.
2. Each laboratory submitted data of four replicates at each concentration (nominal mean) of 50, 100, and 200 μg/kg collected under repeatability conditions.
3. A normal distribution was assumed, with inter-laboratory relative standard deviations (RSD) of 25% assumed at 50 and 100 μg/kg, and 20% at 200 μg/kg. These assumptions are based on the Horwitz equation.[32] The nominal means and RSD assumptions were input into the random-number generator to obtain mean laboratory values.
4. Intra-laboratory RSD (or repeatability standard deviation, as defined in ISO 5725-2) of 0.5 the inter-laboratory standard deviation, and the previously obtained mean laboratory values, were input into the random-number generator to obtain four replicate laboratory values for samples at the three concentrations. The assumption of intra-laboratory standard deviation of ≤0.5 of inter-laboratory standard deviation is based on the publications of Horwitz[32] and the Analytical Methods Committee.[33] Publications by Dehouck et al.[34,35] also support this assumption.
5. The minimum and maximum values obtained for laboratory 5 were decreased and increased respectively by approximately 20% to deliberately increase the range in data for this sample in order to present a situation where less than ideal data were present in a dataset. Raw data as reported by the laboratories and collated as recommended by ISO 5725-2 are given in Table 9.8. Laboratory averages and standard deviations for each laboratory i at level j are given in Tables 9.9 and 9.10, respectively.

Standard ISO 5725-2 recommends that suitable procedures be used to detect and remove outliers in data. The procedures used within the ISO 5725-2 document include Mandel's h and k statistics for overall assessment and comparison of between-laboratory and within-laboratory consistency, respectively; Cochran's test for evaluating within-laboratory consistency; and Grubb's outlier tests for evaluating data. These procedures will also be used for this example. See Equations (9.23)–(9.45) for relevant definitions and equations. Mandel's h and k statistics, given in Tables 9.11 and 9.12, respectively, were calculated using Equations (9.23) and (9.24).

Mandel's h and k statistics are typically used as graphical indicators of between- and within-laboratory consistency. Plots of their values as functions of laboratory or concentration enables visualization of discrepancies for a given laboratory or concentration. Mandel's h statistics (plots not shown) show a pattern of a random spread about the average (0), thereby indicating that there are no issues with between-laboratory consistency. Mandel's k statistics (plots not shown) show that the statistics for laboratory 5 (range 1.97–2.03) are inconsistent with the k statistics of the other laboratories (range 0.14–1.21), suggesting that laboratory 5 has a within-laboratory consistency problem. Table 9.10 also shows that laboratory 5 has the largest standard deviations at all test concentrations for all laboratories. Cochran's test statistics [Eq. 9.25] of laboratory 5 standard deviations at each concentration are given in Table 9.13.

The test statistics at all concentrations exceed the critical values at 5%, but not at 1%. By ISO 5725-2 definition, this classifies them as stragglers, but not as outliers. Application of Grubb's outlier test [Eqs. (9.26) and (9.27)] to the data submitted by laboratory 5 (Table 9.14, where G represents Grubb's outlier statistic) suggests that there are no statistically significant outliers.

Using all of the study's submitted data, Grubb's outlier tests [Eqs. (9.26)–(9.34)] (Tables 9.15–9.18) are performed

TABLE 9.8 Raw Data Collated According to ISO 5725-2 "Form A" Recommendation

	Level (Concentration)		
Laboratory	50 μg/kg	100 μg/kg	200 μg/kg
1	49.7	80.1	216
	37.8	83.5	236
	50.5	80.0	251
	50.8	77.3	252
2	42.0	88.0	224
	28.4	93.8	178
	40.7	90.7	219
	35.4	91.0	222
3	49.7	123	149
	61.4	100	149
	46.6	95.1	167
	60.2	131	163
4	73.4	112	244
	73.3	118	255
	77.9	144	223
	72.6	130	234
5	71.4	166	253
	68.4	128	151
	52.3	107	196
	89.6	88.8	206
6	77.6	136	273
	64.7	136	256
	77.6	113	260
	70.8	128	221
7	18.9	94.8	223
	23.0	63.4	228
	26.4	104	194
	18.7	91.4	209
8	59.6	132	206
	41.4	112	234
	40.4	123	197
	51.7	98.0	193
9	56.2	107	150
	67.4	79.7	155
	48.2	104	162
	67.6	103	134
10	39.0	129	107
	36.2	116	128
	42.1	94.5	144
	41.0	96.0	128

on the cell means to detect any outlying data. All tests were not significant at the 5% critical value.

Outlier tests for cell means as well as for laboratory 5 observations are statistically insignificant. However, the fact that Mandel's k statistics for laboratory 5 were inconsistent with the findings from the other laboratories, and that the Cochran test statistics for all concentrations in laboratory 5 were statistically significant at the 5% but not at the 1% critical value, raises the question as to whether there is a problem with the results reported by laboratory 5. Although

TABLE 9.9 Means Collated According to ISO 5725-2 "Form B" Recommendation

	Level (Concentration)		
Laboratory	50 μg/kg	100 μg/kg	200 μg/kg
1	47.20	80.22	238.8
2	36.63	90.88	210.8
3	54.48	112.3	157.0
4	74.30	126.0	239.0
5	70.42	122.4	201.5
6	72.68	128.2	252.5
7	21.75	88.40	213.5
8	48.28	116.2	207.5
9	59.85	98.42	150.2
10	39.58	108.9	126.8
Grand mean	52.52	107.2	199.8
Standard deviation	17.23	16.87	41.93

TABLE 9.10 Estimated Standard Deviations Collated According to ISO 5725-2 "Form C" Recommendation

	Level (Concentration)		
Laboratory	50 μg/kg	100 μg/kg	200 μg/kg
1	6.284	2.540	16.84
2	6.182	2.371	21.93
3	7.428	17.43	9.381
4	2.426	14.14	13.69
5	15.29	33.16	41.84
6	6.208	10.84	22.22
7	3.679	17.50	15.29
8	9.115	14.66	18.48
9	9.418	12.60	11.90
10	2.590	16.62	15.17

TABLE 9.11 Mandel's *h* Statistics

	Level (Concentration)		
Laboratory	50 μg/kg	100 μg/kg	200 μg/kg
1	−0.31	−1.60	0.93
2	−0.92	−0.97	0.26
3	0.11	0.30	−1.02
4	1.26	1.12	0.94
5	1.04	0.90	0.04
6	1.17	1.25	1.26
7	−1.79	−1.11	0.33
8	−0.25	0.53	0.18
9	0.43	−0.52	−1.18
10	−0.75	0.10	−1.74

ISO 5725-2 Section 7.3.3.2 defines data significant at 5% but not at the 1% critical value as stragglers only, ISO 5725-2 Section 7.3.3.6 states that if several stragglers are reported consistently for a specific laboratory, the whole of

TABLE 9.12 Mandel's *k* Statistics

	Level (Concentration)		
Laboratory	50 μg/kg	100 μg/kg	200 μg/kg
1	0.81	0.15	0.82
2	0.80	0.14	1.07
3	0.96	1.06	0.46
4	0.31	0.86	0.66
5	1.97	2.02	2.03
6	0.80	0.66	1.08
7	0.47	1.07	0.74
8	1.17	0.89	0.90
9	1.21	0.77	0.58
10	0.33	1.01	0.74

TABLE 9.13 Cochran's Test

	Level (Concentration)		
Laboratory	50 μg/kg	100 μg/kg	200 μg/kg
5	0.388	0.409	0.414

Notation: Critical values for Cochran's test, $n = 4$, $p = 10$; 1%—0.447, 5%—0.373.

TABLE 9.14 Grubb's Outlier Test Statistics for Single Smallest or Largest Observation, Laboratory 5 Data

	Level (Concentration)		
G	50 μg/kg	100 μg/kg	200 μg/kg
Smallest	1.19	1.01	1.21
Largest	1.25	1.31	1.23

Notation: Critical values for Grubb's test, single largest or smallest outlier, $n = 4$; 1%—1.496, 5%—1.481.

TABLE 9.15 Grubb's Outlier Test Statistics, Largest Outlying Means

	Level (Concentration)		
Laboratory	50 μg/kg	100 μg/kg	200 μg/kg
4	1.27	—	—
6	—	1.24	—
6	—	—	1.26

Notation: Critical values for Grubb's test, single largest or smallest outlier, $p = 10$; 1%—2.482, 5%—2.290.

the data of the laboratory should be rejected. In keeping with this recommendation, laboratory 5 data are removed prior to calculations of the appropriate variances. Equations (9.35)–(9.43) were used in calculations to obtain the inter-laboratory derived components of uncertainty s_L^2 and s_r^2, results of which are given in Table 9.19.

TABLE 9.16 Grubb's Outlier Test Statistics, Smallest Outlying Means

	Level (Concentration)		
Laboratory	50 μg/kg	100 μg/kg	200 μg/kg
7	1.79	—	—
1	—	1.60	—
10	—	—	1.74

Notation: Critical values for Grubb's test, single largest or smallest outlier, $p = 10$; 1%—2.482, 5%—2.290.

TABLE 9.17 Grubb's Outlier Test Statistics, Two Largest Outlying Means

	Level (Concentration)		
Laboratories	50 μg/kg	100 μg/kg	200 μg/kg
4, 6	0.59	—	—
6, 4	—	0.61	—
6, 4	—	—	0.66
s_0^2	2672	2560	15827
$s_{p-1,p}^2$	1571	1567	10448

Notation: Critical values for Grubb's test, two largest or smallest outliers, $p = 10$; 1%—0.1150, 5%—0.1864.

TABLE 9.18 Grubb's Outlier Test Statistics, Two Smallest Outlying Means

	Level (Concentration)		
Laboratories	50 μg/kg	100 μg/kg	200 μg/kg
7, 2	0.45	—	—
1, 7	—	0.48	—
10, 9	—	—	0.39
s_0^2	2672	2560	15827
$s_{1,2}^2$	1201	1218	6171

Notation: Critical values for Grubb's test, two largest or smallest outliers, $p = 10$; 1%—0.1150, 5%—0.1864.

TABLE 9.19 Calculation of Variances[a]

	Level (Concentration)		
Calculations	50 μg/kg	100 μg/kg	200 μg/kg
T_1	1,819	3,798	7,184
T_2	101,173	409,919	1,497,060
T_3	36	36	36
T_4	144	144	144
T_5	1,108	4,768	7,442
s_r	6.41	13.3	16.6
s_L	16.7	15.6	43.7
s_R	17.9	20.5	46.7
$\hat{m}$	50.5	106	200

[a]Calculations done with the exclusion of all laboratory 5 data.

Determining what other uncertainty components need to be included involves thoughtful evaluation of the method and breaking it down into its constituent parts to see if these parts are in fact included in the inter-laboratory derived

TABLE 9.20 Calculation of Uncertainties and Expanded Uncertainties

Calculations	Level (Concentration)		
	50 μg/kg	100 μg/kg	200 μg/kg
s_L	16.71	15.61	43.69
s_r	6.41	13.29	16.60
$u_{sampling}$	5.05	10.55	19.96
$u_{reference\ standard}$	1.01	2.11	3.99
$u_c(y)$	18.62	23.15	50.98
v_{eff}	12.24	32.54	14.64
$t_{\alpha/2,95\%}, veff$	2.1788	2.0369	2.1448
$U_i(y)95\%$	40.6	47.2	109

uncertainty components. Going through the specifics of this process is beyond the purpose of this example, although some basic assumptions can be made:

1. It can be assumed that with the development and study of new methods, the ability to determine $u^2(\hat{\delta})$, the method bias component of uncertainty, cannot be done given that it can be evaluated only relative to a "true" measure of analyte concentration. This can be achieved by analysis of a certified reference material, which is usually uncommon, or by comparison to a well-characterized/accepted method, which is unlikely to exist for veterinary drug residues of recent interest. Given that method bias is typically corrected using matrix-matched calibration standards, internal standard or recovery spikes, it is considered that the use of these approaches provides correction for the systematic component of method bias.[16] The random error would be considered part of the inter-laboratory derived components of uncertainty.
2. It can be assumed that when determining the $c_i^2 u^2(x_i)$ components of uncertainty as defined in Equation (9.22), two components that most likely needed to be included are the sampling uncertainty and reference standard uncertainty.

Estimating sampling uncertainty goes beyond the scope of this example, although a EURACHEM/CITAC document[36] stated that it can range from a few percent to 84% relative to the measurand. Pharmaceutical-grade reference standards are recommended to have purities exceeding 99.5%.[37] Resultant uncertainties of such products if purity limits were respected would be low; Liu and Hu,[38] using macrolide reference standards, measured relative uncertainties with values ranging from 0% to 2%. Such low uncertainties would not make a significant contribution to overall uncertainties in residue methods.

However, with residue chemistry, the availability of reference standards, particularly if metabolites are the marker residues, may be limited to a few suppliers producing small quantities, thereby precluding the use of rigorous clean-up and/or supplier information on uncertainty. End users of reference substances may be required to obtain this information by themselves over an extended time period by keeping and analyzing records of standard comparisons, a requirement under ISO 17025.[1] For the purpose of this example, it will be assumed that sampling and reference substance uncertainty are the only two additional uncertainty components being accounted for, with relative values of 10% for sampling uncertainty, and 2% for reference substance uncertainty. Sensitivity coefficients are assumed to be 1.

Using the component uncertainties obtained from the inter-laboratory study and the additional uncertainties, and using Equation (9.22), the calculated standard measurement uncertainties for each concentration are given in Table 9.20. To determine the expanded uncertainty, assuming that a 95% coverage factor is desired, Equation (9.44) is required for determining the degrees of freedom contributed by the type B uncertainties sampling and reference standard, and the Welch–Satterthwaite equation [Eq. (9.45)] to determine the effective degrees of freedom of Equation (9.22). Degrees of freedom for s_r^2, s_L^2, sampling and reference standard are 27, 8, 50, and 1250, respectively. The effective degrees of freedom, two-tailed t values at 95% as coverage factor, and expanded uncertainties at the three tested concentrations are given in Table 9.20.

(*Note*: In a simple inter-laboratory study as given in this example, which does not further break down the inter-laboratory derived repeatability variance into its components, output from common spreadsheet software capable of one-way ANOVA such as Microsoft Excel, Corel Quattro, or free-to-download software such as `OpenOffice.org` Calc can also be used.) Table 9.21 and Equation (9.46) shows how output from a one-way ANOVA can be used.

Therefore, the estimate of repeatability variance (s_r^2) is obtained directly from the within-group output. The estimate of laboratory variance (s_L^2) is obtained as follows from the software output:

$$s_L^2 = \frac{MS_L - MS_r}{n} \tag{9.46}$$

9.3.5 Measurement Uncertainty Based on Proficiency Test Data

Given that laboratories accredited to ISO 17025 are encouraged to participate in proficiency testing (PT) programs when available,[1] historical data from PT participation can be used to estimate an individual laboratory's uncertainty.[39] The approaches to using PT data to calculate uncertainties vary; Horwitz models,[40] modifications of the ISO 5725-2 approach,[41] and propagation of uncertainty models with variations depending on reliability of PT participants' uncertainty statements[42] have all been presented in the scientific literature. The overriding principles of any approach are that they should be simple, based on internationally accepted guidelines, and should make use of as much information as possible in the data. ISO 13528's algorithm A[43] adheres to these principles. In brief, algorithm A is an iterative calculation method for calculating the robust average and robust standard deviations of a dataset. As defined in ISO 13528, robust average and robust standard deviation are statistics that are estimated using a robust algorithm. In a robust algorithm, all data are utilized, including those that would be excluded as outliers using standard statistical techniques. Prior to the application of algorithm A, a simple transformation of the dataset as performed by Maroto et al.[39] is done to enable the assessment of method concordance. The transformation is simply a conversion of an individual PT result into a ratio by dividing the individual result by the consensus value for that sample in the respective testing round. As defined by Maroto and co-workers, the average of the ratios, concordance, is then considered an analog of "bias" or "trueness," and it is the average comparability for all PT participants in a given PT round.

It has been proposed that the addition of other uncertainty components not taken into consideration by PT should also be performed.[42] However, unlike a well-designed inter-laboratory test, confidentiality is usually an integral part of PT.[44] It is unlikely that an individual laboratory would have access to the information required to determine the nature of the other uncertainty components. Additionally, given that participant methods are not identical and that robust statistics are used, uncertainty estimates may already be over-estimated. It is therefore assumed that the information derived from PT data are adequate.[41]

The PT data for this example are created using the random-number generation function of Microsoft Excel 2002 with the following assumptions:

1. The data consist of an individual laboratory's results for one analyte from one sample/round for 10 PT rounds (therefore, a total of 10 results for the dataset). In each PT round, the analytical results obtained by the individual laboratory for its sample are divided by the consensus value for that sample from that round; the ratios are used for further calculations.
2. A normal distribution of the ratio is assumed, and that analyte concentration in the PT samples varies from 50 to 200 μg/kg in the samples. At these concentrations, the Horwitz equation predicts inter-laboratory relative standard deviations of 20–25%.[32] An average ratio of 1, with inter-round relative standard deviations (RSDs) of 25%, are input into the random-number generator to obtain individual ratios for the analyte from each of the 10 PT rounds.

With the data, and following algorithm A, ISO 13528, data handling and calculations proceed as follows:

1. Sort the p items of data into increasing order (see column 2, "Ratio," in Table 9.22):

$$x_1, x_2, \ldots, x_i, \ldots, x_p$$

2. Designate the robust average and robust standard deviation of the data as x^* and s^*, respectively.
3. Determine the initial values for x^* and s^* as follows:

$$x^* = \text{median of } x_i \quad \text{(Table 9.22 column 2)} \quad (i = 1, 2, \ldots, p)$$

$$s^* = 1.483 \times \text{ median of } |x_i - x^*| \quad \text{(Table 9.22 column 3)} \quad (i = 1, 2, \ldots, p)$$

4. Update the values of x^* and s^*. First, calculate φ as follows:

$$\varphi = 1.5 \times s^* \tag{9.47}$$

Then, update each original x_i value ($i = 1,2,\ldots,p$), as follows:

$$x_i^* = \begin{cases} x^* - \varphi & \text{if} \quad x_i < x^* - \varphi \\ x^* + \varphi & \text{if} \quad x_i > x^* + \varphi \\ x_i & \text{otherwise} \end{cases} \tag{9.48}$$

TABLE 9.21 One-Way ANOVA Table

Source of Variability	Mean Square Error	Software Output[a]	Variance Estimated
Laboratory	$\text{MS}_\text{L} = \dfrac{n\sum_{i=1}^{p}(\bar{y}_{ij} - \bar{\bar{y}}_j)^2}{p-1}$	Between groups	$s_\text{r}^2 + ns_\text{L}^2$
Repeatability	$\text{MS}_\text{r} = \dfrac{\sum_{i=1}^{p}\sum_{k=1}^{n}(y_{ijk} - \bar{y}_{ij})^2}{p(n-1)}$	Within groups	s_r^2

[a]As given in Microsoft Excel output. Exact nomenclature may vary with software program and version number.

TABLE 9.22 Iterative Process for Determining Robust Average and Robust Standard Deviation According to Algorithm A, ISO 13528

			Update				
Round	Ratio	$\|x_i$-Median$\|$	1	2	3	4	5
7	0.454	0.5465	0.454	0.4625	0.4661	0.4676	0.4682
2	0.681	0.3195	0.681	0.681	0.681	0.681	0.681
10	0.728	0.2725	0.728	0.728	0.728	0.728	0.728
1	0.925	0.0755	0.925	0.925	0.925	0.925	0.925
8	0.941	0.0595	0.941	0.941	0.941	0.941	0.941
3	1.06	0.0595	1.06	1.06	1.06	1.06	1.06
9	1.27	0.2695	1.27	1.27	1.27	1.27	1.27
5	1.30	0.2995	1.30	1.30	1.30	1.30	1.30
4	1.32	0.3195	1.32	1.32	1.32	1.32	1.32
6	1.43	0.4295	1.43	1.43	1.43	1.43	1.43
x^*	1.0005	—	1.0109	1.0117	1.0121	1.0123	1.0123
s^*	—	0.4241	0.3656	0.3638	0.3630	0.3627	0.3625

Source: ISO 13528.[43]

5. Then, calculate the new updated values of x^* and s^* from the updated values of x_i^* as follows:

$$x^* = \frac{\sum_{i=1}^{p} x_i^*}{p} \tag{9.49}$$

$$s^* = 1.134 \times \sqrt{\frac{\sum_{i=1}^{p} (x_i^* - x^*)^2}{p-1}} \tag{9.50}$$

See column 4 ("Update 1") and onward for iterations.

6. Repeat the iterative calculations and updates in Equations (9.47)–(9.50) until values of x^* and s^* converge. Values are assumed to be converged when there is no change in the third significant figure from one iteration to the next in the robust standard deviation and the equivalent figure in the robust average.

Then, as performed by Companyó et al.,[41] the standard relative uncertainty is calculated as follows:

$$u_{\text{rel,lab}} = \frac{s^*}{x^*} \times \sqrt{1 + \frac{1}{p}} \tag{9.51}$$

that is

$$u_{\text{rel,lab}} = \frac{0.3625}{1.0123} \times \sqrt{1 + \frac{1}{10}} = 0.376$$

The expanded relative uncertainty is calculated using a coverage factor that is the two-sided t-tabulated value at 95% and appropriate degrees of freedom as given in the following equation:

$$U_{\text{lab}} = t_{\alpha/2,p-1} \times c \times u_{\text{rel,lab}} \tag{9.52}$$

that is:

$$U_{\text{lab}} = 2.2622 \times c \times 0.376 = 0.8506c$$

where c is the estimated concentration of analyte in the sample. Thus, the expanded relative uncertainty applied to future results is approximately 85% for this example.

9.3.6 Measurement Uncertainty Based on Quality Control Data and Certified Reference Materials

For this top–down approach, it is assumed that the quality control (QC) precision and recovery data have been collected over a sufficiently large number of runs and period of time to allow for natural variation of all factors that can affect the results. These factors include different analysts, analytical instruments, blank tissue lots, lot numbers of reagents, and preparations of standard solutions. Note that other factors that can affect analytical results, such as method bias, variations in the sample matrix, sampling, sample storage and treatment, subsampling, homogeneity, standard purity, and the preparation of standard solutions, are not included in this discussion.

Before using QC data, an appropriate statistical test, such as Grubb's or Dixon's tests, should be applied to test for outliers. Those data points acquired during a period in which the method was not in statistical control should not be included in the calculations. This approach assumes that measurements are being made at concentrations where the relative uncertainty is constant over a defined range, the constant uncertainty that would dominate at concentrations close to the limit of detection or limit of quantification is negligible, and that recovery is independent of concentration.

The overall relative standard uncertainty of the method is calculated by combining the uncertainty contributions from precision and recovery:

$$u_c(y) = \sqrt{\mathrm{RSD}^2 + u_{\mathrm{rel}}(\bar{R}_m)^2} \qquad (9.53)$$

where $u_c(y)$ is the relative combined uncertainty, RSD is relative standard deviation of repeated measurements of a sample, and $u_{\mathrm{rel}}(\bar{R}_m)$ is the relative uncertainty arising from recovery. In Sections 9.3.6.1 and 9.3.6.2, two scenarios to explain step-by-step how $u_c(y)$ is calculated.

In scenario A (see Scheme 9.2), certified reference materials (CRMs) are used to calculate the uncertainties due to both precision and recovery, which are combined to calculate relative uncertainty $u_c(y)$. In scenario B (see Scheme 9.3), QC samples (incurred residues) and fortified blank samples are used to calculate the uncertainties arising from precision and recovery. These examples are adapted from and based in part on literature[7,19,45–47] published elsewhere, especially that of Gluschke et al.,[45] which provides a practical and understandable worked example for estimation of MU from QC data. Some of the data used in these examples were published as part of a worked example submitted by Pantazopoulos to EURACHEM/CITAC,[7] and published in part in Ng et al.[47] and are used with permission.[46]

9.3.6.1 *Scenario A: Use of Certified Reference Material for Estimation of Uncertainty*

Precision and Its Uncertainty The first step is to estimate RSD (Scheme 9.2B), which has been defined as:

$$\mathrm{RSD} = \frac{S_{\mathrm{obs}}}{\bar{C}_{\mathrm{obs}}} \qquad (9.54)$$

where S_{obs} is the standard deviation of series of measurements of a QC sample with mean observed value of $\bar{C}_{\mathrm{obs}}$. In this example, RSD is calculated from analyses of a white wine CRM with a certified concentration of ochratoxin A of 220±14 pg/ml. The data (Scheme 9.2A) are random numbers generated to represent 30 independent analyses of a white wine CRM. These analyses are assumed to have been carried out over sufficient time to allow for natural variation in all factors affecting the results, including inter-run (between-run) variations, with different equipment and different analysts. A Dixon test for outliers was applied to this data set; no outliers were detected at a significance level of $\alpha = 0.01$. From these data, the average observed concentration $\bar{C}_{\mathrm{obs}}$ (187 pg/ml) and the standard deviation of the concentration measurements S_{obs} (21.0 pg/ml) are calculated. Then the RSD, which is equal to 0.113, is determined using Equation 9.54.

Recovery and Its Uncertainty The second step is to estimate uncertainty due to recovery (Scheme 9.2C). In this CRM case, the mean recovery, which is equal to 0.850, is determined by:

$$\bar{R}_m = \frac{\bar{C}_{\mathrm{obs}}}{C_{\mathrm{CRM}}} \qquad (9.55)$$

where $\overline{C}_{\mathrm{obs}}$ is the average of 30 CRM independent analyses and C_{CRM} is the certified concentration of the analyte in the CRM. The relative standard uncertainty of recovery $u_{\mathrm{rel}}(\bar{R}_m)$ of the CRMs, which equals 0.0295, is given by:

$$u_{\mathrm{rel}}(\bar{R}_m) = \sqrt{\frac{S_{\mathrm{obs}}^2}{n\bar{C}_{\mathrm{obs}}^2} + \left(\frac{u(C_{\mathrm{CRM}})}{C_{\mathrm{CRM}}}\right)^2} \qquad (9.56)$$

This last equation pools the relative standard error of the mean recovery $S_{\mathrm{obs}}/(\sqrt{n} \times \overline{C}_{\mathrm{obs}})$ with the relative standard uncertainty, $u(C_{\mathrm{CRM}})/C_{\mathrm{CRM}}$ of the concentration of the analyte in the CRM. The value for the uncertainty $u(C_{\mathrm{CRM}})$ in the concentration of the analyte in the CRM is taken from the certificate of analysis for the CRM. In this example, the white wine CRM has a certified concentration of 220±14 pg/ml. The certificate states that the confidence interval is derived from the reproducibility standard deviation obtained from an inter-laboratory study multiplied by a factor of 3. Therefore, the standard uncertainty of the CRM $u(C_{\mathrm{CRM}})$, which is 4.7, is then the confidence interval (CI) divided by 3:

$$u(C_{\mathrm{CRM}}) = \frac{\mathrm{CI}}{3} \qquad (9.57)$$

The standard uncertainty of recovery $u(\bar{R}_m)$, which is 0.0251, is then determined as follows:

$$u(\bar{R}_m) = u_{\mathrm{rel}}(\bar{R}_m) \times \bar{R}_m \qquad (9.58)$$

Once $\bar{R}_m$ and $u(\bar{R}_m)$ are calculated, a significance test (i.e., Student's t-test) is used to test the assumption that recovery is not significantly different from 1. The t value is calculated as follows:

$$t = \frac{|1 - \bar{R}_m|}{u(\bar{R}_m)} \qquad (9.59)$$

The t value is compared to the coverage factor k (typically, $k = 2$). If $t \le k$ were true, then the assumption that recovery is not significantly different would not be rejected and no increase in $u(\bar{R}_m)$ would be necessary. However, in this example case, t is greater than k. Two cases must then be considered. If the reported analytical results are corrected

A. Data

White wine CRM nominal concentration (pg/ml)	220
White wine CRM confidence Interval (pg/ml)	14.0

CRM analytical results (n=30) (pg/ml) *Note*: Data are generated using a random–number generator.	158 176 156 225 161 190 173 201 188 194 222 160 200 204 231 189 198 194 192 158 207 198 179 160 188 187 163 212 178 168

B. Precision and its uncertainty

CRM summary results

n	$\overline{C}_{obs}$ (pg/ml)	S_{obs} (pg/ml)	RSD
30	**187**	**21**	**0.113**

Eq. (9.54)

$$\text{RSD} = \frac{S_{obs}}{\overline{C}_{obs}} = \frac{21}{187} = 0.113$$

C. Recovery and its uncertainty

CRM data		Concentration. (pg/ml)	Confidence Interval (CI) (pg/ml)
		220	14
Standard uncertainty of the CRM	$u(C_{CRM})$	4.67	
Mean recovery	$\overline{R}_m$	0.850	

Eq. (9.57)

$$u(C_{CRM}) = \frac{CI}{3} = \frac{14}{3} = 4.7$$

Eq. (9.55)

$$\overline{R}_m = \frac{\overline{C}_{obs}}{C_{CRM}} = \frac{187}{220} = 0.850$$

Scheme 9.2 Measurements uncertainties arising from precision and recovery using CRM data.

Relative uncertainty of recovery	$u_{rel}(\overline{R}_m)$	0.0295	Eq. (9.56) $u_{rel}(\overline{R}_m) = \sqrt{\frac{S_{obs}^2}{nC_{obs}^2} + \left[\frac{u(C_{CRM})}{C_{CRM}}\right]^2} = \sqrt{\frac{21^2}{30 \times 187^2} + \left[\frac{4.7}{220}\right]^2} = 0.0295$
Uncertainty of recovery	$u(\overline{R}_m)$	0.0251	Eq. (9.58) $u(\overline{R}_m) = u_{rel}(\overline{R}_m) \times \overline{R}_m = 0.0295 \times 0.850 = 0.0251$
Calculation of *t* value		5.98	Eq. (9.59) $t = \frac{\lvert 1 - \overline{R}_m \rvert}{u(\overline{R}_m)} = \frac{\lvert 1 - 0.850 \rvert}{0.0251} = 5.98$
Extended uncertainty for recovery	$u(\overline{R}_m)''$	0.0791	Eq. (9.60) $u(\overline{R}_m)'' = \sqrt{u(\overline{R}_m)^2 + \left(\frac{1 - \overline{R}_m}{k}\right)^2} = \sqrt{0.0251^2 + \left(\frac{1 - 0.850}{2}\right)^2} = 0.0791$
Extended relative uncertainty for recovery	$u_{rel}(\overline{R}_m)''$	0.0930	Eq. (9.61) $u_{rel}(\overline{R}_m)'' = \frac{u(\overline{R}_m)''}{\overline{R}_m} = \frac{0.0791}{0.850} = 0.0930$

D. Combined uncertainty

Combined relative uncertainty	$u_c(y)$	0.146	Eq. (9.53) $u_c(y) = \sqrt{\text{RSD}^2 + u_{rel}(\overline{R}_m)^2} = \sqrt{0.113^2 + 0.0930^2} = 0.146$
Expanded combined relative uncertainty	$U_{rel}(y)$	0.292	Eq. (9.62) $U_{rel}(y) = k \times u_c(y) = 2 \times 0.146 = 0.292$
Expanded uncertainty for 200 pg/ml	$U(y)$	58	$U(y) = 200 \times 0.292 = 58$
Reported result		(200 ± 58) pg/ml	

Scheme 9.2 (*Continued*)

for recovery, then $u(\bar{R}_m)$ does not need to be extended. However, in this example case, results are not corrected for recovery. As a result, $u(\bar{R}_m)$ has to be extended to include additional uncertainty due to uncorrected recovery, and the additional term is $(1 - \bar{R}_m)/k$ as shown in the following equation:

$$u(\bar{R}_m)'' = \sqrt{u(\bar{R}_m)^2 + \left(\frac{1 - \bar{R}_m}{k}\right)^2} \tag{9.60}$$

Combining terms, the extended standard uncertainty of recovery $u(\bar{R}_m)''$ would then be 0.0791. The extended relative standard uncertainty $u_{rel}(\bar{R}_m)''$ is calculated as 0.0930 according to the following equation:

$$u_{rel}(\bar{R}_m)'' = \frac{u(\bar{R}_m)''}{\bar{R}_m} \tag{9.61}$$

The standard relative uncertainty of recovery increased from 0.0295 to 0.0930, due to the significant uncorrected recovery.

Finally, the combined relative uncertainty $u_c(y)$, which is equal to 0.146, is then calculated using Equation (9.53) (Scheme 9.2C), substituting $u_{rel}(\bar{R}_m)''$ for $u_{rel}(\bar{R}_m)$. Using the coverage factor $k = 2$, the expanded

A. Data

	Analytical Results (pg/ml)
Red wine QC samples (incurred residues)	634.5
	724.1
	768.0
	687.9
	700.2
	659.3
	644.6
	726.8
	580.6
White wine QC sample (incurred residues)	178.2
	215.1
	205.1
	196.2
	222.7

Recoveries from fortified samples	Recoveries
	0.74
	1.11
	1.03
	0.97
	0.89
	0.87
	1.01
	0.93
	0.8
	0.88
	0.79
	1.07
	0.89
	0.72
	0.9
	0.94
	0.98
	0.82

B. Precision and its uncertainty

Wine QC samples	n	$\overline{C}_{obs}$ (pg/ml)	S_{obs} (pg/ml)	RSD
Red	9	680.7	57.0	0.0837
White	5	203.5	17.3	0.0851

RSD pooled for wine control samples — RSD_{pool} 0.0842

Eq. 9.63

$$RSD_{pool} = \sqrt{\frac{RSD_1^2 \times (n_1-1) + RSD_2^2 \times (n_2-1)}{(n_1-1)+(n_2-1)}}$$

$$= \sqrt{\frac{(0.0837^2 \times (9-1) + 0.0851^2 \times (5-1)}{(9-1)+(5-1)}} = 0.0842$$

Scheme 9.3 Calculations of measurement uncertainty using QC data from incurred residues and fortified samples.

C. Recovery and its uncertainty

Analytical runs		n 18	
Mean recovery	$\overline{R}_m$	0.908	
Standard deviation of recovery	S_{rec}	0.109	
Standard uncertainty of recovery	$u(\overline{R}_m)$	0.0257	Eq. 9.64 $u(\overline{R}_m) = \frac{S_{rec}}{\sqrt{n}} = \frac{0.109}{\sqrt{18}}$
Calculation of t value		3.58	$t = \frac{\lvert 1 - \overline{R}_m \rvert}{u(\overline{R}_m)} = \frac{\lvert 1 - 0.908 \rvert}{0.0257} = 3.58$
Expanded uncertainty for recovery	$u(\overline{R}_m)''$	0.0528	Eq. 9.65 $u(\overline{R}_m)'' = \sqrt{\frac{S_{rec}^2}{n} + \left[\frac{(1 - \overline{R}_m)}{k}\right]^2} = \sqrt{\frac{0.109^2}{18} + \left[\frac{(1-0.908)}{2}\right]^2} = 0.0528$
Expanded relative uncertainty for recovery	$u_{rel}(\overline{R}_m)''$	0.0582	$u_{rel}(\overline{R}_m)'' = u(\overline{R}_m)'' / \overline{R}_m = 0.0528/0.908 = 0.0582$

D. Combined uncertainty

Combined relative uncertainty	$u_c(y)$	0.102	Eq. 9.53 $u_c(y) = \sqrt{RSD^2 + u_{rel}(\overline{R}_m)^2} = \sqrt{0.0842^2 + 0.0582^2} = 0.1023$
Expanded combined relative uncertainty:	$u_{rel}(y)$	0.205	Eq. 9.62 $U_{rel}(y) = k \times u_c(y) = 2 \times 0.1023 = 0.205$
Expanded uncertainty for 200 pg/ml	$U(y)$	41	$U(y) = 200 \times 0.205 = 41$
Reported result:		(200 ± 41) pg/ml	

Scheme 9.3 *(Continued)*

relative uncertainty $U_{rel}(y)$ is calculated to be 0.292 as follows:

$$U_{rel}(y) = k \times u_c(y) \tag{9.62}$$

For a determined concentration of 200 pg/ml, the expanded uncertainty would be $U(y) = 200 \times 0.292 = 58$. This result denotes that, for a result of 200 pg/ml, the concentration of ochratoxin A would be expressed as "ochratoxin A: (200±58) pg/ml, where the stated uncertainty is an expanded uncertainty calculated using a coverage factor of 2 that corresponds approximately to the 95% confidence interval."

9.3.6.2 Scenario B. Use of Incurred Residue Samples and Fortified Blank Samples for Estimation of Uncertainty

Precision and Its Uncertainty The first step is to estimate RSD. In this scenario, RSD is calculated from data of two QC samples (Scheme 9.3A), that is, from a red wine QC sample and a white wine QC sample, for residues of ochratoxin A. These analyses were carried out over 9 months and allowed for natural variation in all factors affecting the result, including inter-run variations, different equipment, and different analysts. For each QC sample, the mean or average ($\overline{C}_{obs}$), standard deviation (S_{obs}) and relative standard deviation (RSD) are calculated for red and white wine QC samples, respectively (Scheme 9.3B). Since the RSD values are very similar, it is appropriate to pool the RSD values using Equation (9.63) to obtain $\mathrm{RSD_{pool}}$, which is equal to 0.0842:

$$\mathrm{RSD_{pool}} = \sqrt{\frac{\mathrm{RSD}_1^2 \times (n_1 - 1) + \mathrm{RSD}_2^2 \times (n_2 - 1)}{(n_1 - 1) + (n_2 - 1)}} \tag{9.63}$$

Recovery and Its Uncertainty The second step is to estimate uncertainty due to recovery (Scheme 9.3C). Recovery is calculated from the analysis of blank samples, in which each is fortified with a known mass of analyte. The average recovery ($\bar{R}_\mathrm{m}$) and its standard deviation (S_rec) are determined as 0.908 and 0.109, respectively, from 18 individual recovery QC data. The standard uncertainty of recovery $u(\bar{R}_\mathrm{m})$, which equals 0.0257, is then calculated as follows:

$$u(\bar{R}_\mathrm{m}) = \frac{S_\mathrm{rec}}{\sqrt{n}} \tag{9.64}$$

Once $\bar{R}_\mathrm{m}$ and $u(\bar{R}_\mathrm{m})$ are calculated, a t test is performed, and the t value is equal to 3.58, which is greater than $k(=2)$. Assuming that the results are not corrected for recovery, $u(\bar{R}_\mathrm{m})$ has to be expanded using Equation (9.65) to include additional uncertainty due to uncorrected recovery. The relative uncertainty of recovery $u(\bar{R}_\mathrm{m})''$ is then be calculated as 0.0582:

$$u(\bar{R}_\mathrm{m})'' = \sqrt{\frac{S_\mathrm{rec}^2}{n} + \left[\frac{(1 - \bar{R}_\mathrm{m})}{k}\right]^2} \tag{9.65}$$

Finally, the combined relative uncertainty $u_\mathrm{c}(y)$, which equals 0.102, is then calculated using Equation (9.53). Using the coverage factor $k = 2$, the expanded relative uncertainty $U_\mathrm{rel}(y)$ is calculated as 0.205. For a determined concentration of 200 pg/ml, the expanded uncertainty would be $U(y) = 200 \times 0.205 = 41$. This result denotes that, for a result of 200 pg/ml, the concentration of ochratoxin A would be expressed as "ochratoxin A: (200 ± 41) pg/ml, where the stated uncertainty is an expanded uncertainty calculated using a coverage factor of 2 that corresponds approximately to the 95% confidence interval."

REFERENCES

1. International Standards Organization, ISO/IEC 17025:2005, *General Requirements for the Competence of Testing and Calibration Laboratories*, Geneva, 2005.
2. Codex Alimentarius Commission, CAC/GL 71-2009, *Guidelines for the Design and Implementation of National Regulatory Food Safety Assurance Programme Associated with the Use of Veterinary Drugs in Food Producing Animals*, Rome, 2009.
3. International Standards Organization, ISO Guide 98-3, *Uncertainty of Measurement*—Part 3: *Guide to the Expression of Uncertainty in Measurement* (GUM:1995), Geneva, 2008.
4. Schneider MJ, Mastovska K, Solomon MB, Distribution of penicillin G residues in culled dairy cow muscles: Implications for residue monitoring, *J. Agric. Food Chem*. 2010;58(9):5408–5413.
5. Barwick VJ, Ellison SLR, Estimating measurement uncertainty using a cause and effect and reconciliation approach. Part 2. Measurement uncertainty estimates compared with collaborative trial expectation, *Anal. Commun*. 1998;35(11):377–383.
6. Kolb M, Hippich S, Uncertainty in chemical analysis for the example of determination of caffeine in coffee, *Accred. Qual. Assur*. 2005;10(5):214–218.
7. EURACHEM/CITAC Guide CG4, Quantifying Uncertainty in Analytical Measurement, 2nd ed., 2000 (available at `http://www.eurachem.org/guides/QUAM2000-1.pdf`; accessed 8/11/10).
8. Committee AM, Uncertainty of measurement: Implications of its use in analytical science, *Analyst* 1995;120:2303–2308.
9. NMKL Procedure 5, *Estimation and Expression of Measurement Uncertainty in Chemical Analysis*, 1997 (available at `http://www.nmkl.org/Engelsk/procedures.htm`; accessed 6/02/10).
10. Barwick VJ, Ellison SLR, The evaluation of measurement uncertainty from method validation studies—Part 1: Description of a laboratory protocol, *Accred. Qual. Assur*. 2000;5(2):47–53.
11. Wang J, Leung D, Butterworth F, Determination of five macrolide antibiotic residues in eggs using liquid chromatography/electrospray ionization tandem mass spectrometry, *J. Agric. Food Chem*. 2005;53(6):1857–1865.
12. Dabalus Islam M, Turcu MS, Cannavan A, Comparison of methods for the estimation of measurement uncertainty for an analytical method for sulphonamides, *Food Addit. Contam*. 2008;25(12):1439–1450.
13. Horwitz W, The certainty of uncertainty, *J. AOAC Int*. 2003;86(1):109–111.
14. Horwitz W, Albert R, The Horwitz ratio (HorRat): A useful index of method performance with respect to precision, *J. AOAC Int*. 2006;89(4):1095–1109.
15. Hund E, Massart DL, Smeyers-Verbeke J, Comparison of different approaches to estimate the uncertainty of a liquid chromatographic assay, *Anal. Chim. Acta* 2003;2003:39–52.
16. Hund E, Massart DL, Smeyers-Verbeke J, Operational definitions of uncertainty, *Trends Anal. Chem*. 2001;20(8):394–406.
17. Leung GN, Ho EN, Kwok WH, et al., A bottom-up approach in estimating the measurement uncertainty and other important considerations for quantitative analyses in drug testing for horses, *J. Chromatogr. A* 2007;1163(1–2):237–246.
18. Codex Alimentarius Commission, CAC/GL 37-2001, *Harmonized IUPAC Guidelines for the Use of Recovery Information in Analytical Measurement*, Rome, 2001.

19. Barwick V, Ellison SLR, Measurement uncertainty: Approaches to the evaluation of uncertainties associated with recovery, *Analyst* 1999;124:981–990.
20. VAM Project 3.2.1, *Development and Harmonisation of Measurement Uncertainty* Principles. Part (d): Protocol for Uncertainty Evaluation from Validation Data, LGC/VAM/1998/088, 2000 (available at `http://www.nmschembio.org.uk/GenericHub.aspx?m = 33`; accessed 8/10/05).
21. Youden WJ, Steiner EH, *Statistical Manual of the Association of Official Analytical Chemists*, AOAC International, Gaithersburg; MD, 1975.
22. European Commission, Commission Decision of 12 August 2002 implementing Council Directive 96/23/EC concerning the performance of analytical methods and the interpretation of results. 2002/657/EC, *Off. J. Eur. Commun.* 2002;L221:8–36.
23. International Standards Organization, ISO 5725-3, *Accuracy (Trueness and Precision) of Measurement Methods and Results*—Part 3: *Intermediate Measures of the Precision of a Standard Measurement Method*, Geneva, 1994.
24. Maroto A, Boque R, Riu J, Rius FX, Measurement uncertainty in analytical methods in which trueness is assessed from recovery assays, *Anal. Chim. Acta* 2001;440:171–184.
25. Dehouck P, Van Looy E, Haghedooren E, et al., Analysis of erythromycin and benzoylperoxide in topical gels by liquid chromatography, *J. Chromatogr. B* 2003;794(2):293–302.
26. Vander Heyden Y, De Braekeleer K, Zhu Y, et al., Nested designs in ruggedness testing, *J. Pharm. Biomed. Anal.* 1999;20(6):875–887.
27. Wang J, Leung D, Lenz SP, Determination of five macrolide antibiotic residues in raw milk using liquid chromatography-electrospray ionization tandem mass spectrometry, *J. Agric. Food Chem.* 2006;54(8):2873–2880.
28. Boyer KW, Horwitz W, Albert R, Inter-laboratory variability in trace element analysis, *Anal. Chem.* 1985;57(2): 454–459.
29. International Standards Organization, ISO 5725-2, *Accuracy (Trueness and Precision) of Measurement Methods and Results*—Part 2: *Basic Method for the Determination of Repeatability and Reproducibility of a Standard Measurement Method*, Geneva, 1994.
30. International Standards Organization, ISO 21748, *Guidance for the Use of Repeatability, Reproducibility and Trueness Estimates in Measurement Uncertainty Estimation*, Geneva, 2004.
31. Robertson M, Chan TSS, APLAC interpretation and guidance on the estimation of uncertainty of measurement in testing, *J. AOAC Int.* 2003;86(5):1070–1076.
32. Horwitz W, Kamps LR, Boyer KW, Quality assurance in the analysis of foods and trace constituents, *J. Assoc. Off. Anal. Chem.* 1980;63(6):1344–1354.
33. Analytical Method Committee, Is My Uncertainty Estimate Realistic? AMC Technical Brief 15, Royal Society of Chemsty, Dec. 2003 (available at `http://www.rsc.org/images/brief15_tcm18-25958.pdf`; accessed 8/11/10).
34. Dehouck P, Vander Heyden Y, Smeyers Verbeke J, et al., Determination of uncertainty in analytical measurements from collaborative study results on the analysis of a phenoxymethylpenicillin sample, *Anal. Chim. Acta* 2003;481(2):261–272.
35. Dehouck P, Vander Heyden Y, Smeyers Verbeke J, et al., Inter-laboratory study of a liquid chromatography method for erythromycin: Deter-mination of uncertainty, *J. Chromatogr. A* 2003;1010(1):63–74.
36. EURACHEM/CITAC Guide Measurement Uncertainty Arising from Sampling. A Guide to Methods and Approaches, 2007 (available at `http://www.eurachem.org/guides/pdf/UfS_2007.pdf`; accessed 8/11/10).
37. World Health Organization, General Guidelines for the Establishment, Maintenance and Distribution of Chemical Reference Substances, WHO Technical Report Series 885, Part A, p. 5, 1999 (available at `http://apps.who.int/medicinedocs/en/d/Jwhozip21e/3.html`; accessed 8/11/10).
38. Liu SY, Hu CQ, A comparative uncertainty study of the calibration of macrolide antibiotic reference standards using quantitative nuclear magnetic resonance and mass balance methods, *Anal. Chim. Acta* 2007;602(1):114–121.
39. Maroto A, Boqué R, Riu J, Ruisánchez I, Òdena M, Uncertainty in aflatoxin B1 analysis using information from proficiency tests, *Anal. Bioanal. Chem.* 2005;382(7):1562–1566.
40. Alder L, Korth W, Patey AL, van der Schee HA, Schoeneweiss S, Estimation of measurement uncertainty in pesticide residue analysis, *J. AOAC Int.* 2001;84(5):1569–1578.
41. Companyó R, Rubio R, Sahuquillo A, Boqué R, Maroto A, Riu J, Uncertainty estimation in organic elemental analysis using information from proficiency tests, *Anal. Bioanal. Chem.* 2008;392(7–8):1497–1505.
42. Van der Veen AMH, Uncertainty evaluation in proficiency testing: state-of-the-art, challenges, and perspectives, *Accred. Qual. Assur.* 2001;6(6):160–163.
43. International Standards Organization, ISO 13528, *Statistical Methods for Use in Proficiency Testing by Inter-laboratory Comparisons*, Geneva, 2005.
44. International Standards Organization, ISO 43-1, *Proficiency Testing by Inter-laboratory Comparisons*—Part 1: *Development and Operation of Proficiency Testing Schemes*, Geneva, 1997.
45. Gluschke M, Wellmitz J, Lepom P, A case study in the practical estimation of measurement uncertainty, *Accred. Qual. Assur.* 2005;10:107–111.
46. Pantazopoulos P, uncertainty estimate for Procedure ONT-FCL-0024, *Determination of ochratoxin A in Wine and Grape Juice by High Performance Liquid Chromatography with Fluorescence Detection*, 2001 (available at `http://www.measurementuncertainty.org/mu/examples/pdf/EncertaintyEstimateExampleFoodLaboratoryDivisionOntarioRegion.pdf`; accessed 8/11/10).
47. Ng W, Mankotia M, Pantazopoulos P, Neil RJ, Scott PM, Ochratoxin A in wine and grape juice sold in Canada, *Food Addit. Contam.* 2004;21(10):971–981.

10

QUALITY ASSURANCE AND QUALITY CONTROL

ANDREW CANNAVAN, JACK F. KAY, AND BRUNO LE BIZEC

10.1 INTRODUCTION

The results of the chemical analysis of antibiotic residues in food are frequently used to test for compliance or non-compliance with regulatory limits. Testing is usually performed as part of a verification program designed to provide an appropriate degree of confidence to ensure that the practices and controls applied in animal production are adequate and being applied to the extent necessary to ensure the health of consumers of animal products.[1] It is vital that a high degree of confidence can be placed on the analytical results, since they can be associated with serious consequences. In the case of non-compliant results at the national level, regulatory actions may be instigated, including investigation of the cause of the violative residues and possible penalties to the producer. In the case of import testing, non-compliant samples may result in the rejection of consignments of food products with possible follow-up effects such as requirements for increased testing frequency for products from the exporting country. These actions can have an enormous economic cost. Where the results of the analyses indicate that the products are compliant with the relevant tolerance limits, those products will enter the market and be consumed by the public; false compliant results in this case may have serious implications for the health of the consumer.

Increasingly, residue laboratories have been adopting quality assurance principles that, whilst not guaranteeing that results are correct, increase the likelihood of the data having a sound scientific basis and being fit for purpose. The implementation of a quality system enables a laboratory to demonstrate to its clients that it has the appropriate facilities, equipment, and technical expertise to perform the analyses, and that the work is carried out in a controlled manner using documented and validated procedures and methods. Laboratories can further reinforce confidence in the quality of their work by formalizing their quality systems through certification or accreditation by a recognized authoritative body.

10.1.1 Quality—What Is It?

The term "quality" can be used in different contexts, with various meanings or interpretations. A product labeled as a quality item or of guaranteed quality, for example, is usually regarded as superior or of higher value.

In the context of quality assurance, quality has been defined in ISO 9000:2005[2] as "the degree to which a set of inherent characteristics fulfils requirements." Quality may also be regarded as "fitness for purpose." Since the user or client requirements must be known in order to produce a service that meets those requirements, quality is user-dependent. The client may be a company paying for the services of the laboratory, or, as is often the case for laboratories testing for antibiotic residues, the competent authority or regulatory body responsible for residue monitoring or surveillance in animal-derived foods. Laboratories can improve their performance and ensure the reliability of their test results by implementing a quality system, encompassing quality assurance and quality control procedures.

Quality assurance (QA) refers to the planned and systematic actions and measures that the laboratory uses to ensure the quality of its operations. These measures include, amongst others, implementation of a quality system, a suitable infrastructure and laboratory environment, appropriately trained and skilled staff, calibrated and well-maintained equipment, quality control procedures,

Chemical Analysis of Antibiotic Residues in Food, First Edition. Edited by Jian Wang, James D. MacNeil, and Jack F. Kay.

documented and validated methods, and participation in proficiency testing programs.

Quality control (QC) refers to the operational techniques and activities that are used to fulfill requirements for quality. Internal quality control comprises the routine practical procedures that enable the analytical chemist to make a decision on whether to accept a result or a group of results as fit for purpose, or reject them and repeat the analysis. Tools for quality control include the use of reference standards and certified reference materials, the use of positive (spiked or incurred) and negative control samples and control charts, replicate analyses, and proficiency tests. Quality control in the laboratory is discussed in more detail in Section 10.5 of this chapter.

10.1.2 Why Implement a Quality System?

A *quality system* may be defined as an organizational structure that encompasses the procedures, processes, and resources needed to implement quality management. There are various reasons why laboratories implement quality systems. Internally, implementing a quality system can increase the efficiency of laboratory operations. Analysis and documentation of the operations within the laboratory can help standardize processes, achieve more transparency, retain knowledge, improve the work environment and staff morale, and reduce costs.

In the case of laboratories engaged in testing for antibiotic residues in food, it may be necessary to implement a quality system to meet contractual, statutory, or regulatory requirements, to comply with international regulations or agreements, or to achieve third-party certification or accreditation, which in some instances is mandatory under those international regulations or agreements (see Section 10.3.4).

An important part of the work of many residue laboratories consists of testing products to provide assurance of the equivalence of food safety systems between countries for international trade. The acceptance of test results between countries should be facilitated if laboratories comply with an international standard such as ISO/IEC 17025:2005,[3] especially if they are accredited by bodies that have entered into mutual recognition agreements with equivalent bodies in other countries.

In designing a strategy for implementing a quality system, it is important to realize that the objective of quality assurance is to manage the frequency of quality failures—the greater the effort, the lower the number of quality failures. However, it is necessary to balance the cost of quality assurance against the benefit in reducing quality failures to an acceptable degree. Due to factors such as inevitable occasional gross errors and deviant results arising from measurement uncertainty (see Chapter 9), it is impossible to guarantee that all individual results will be reliable. Quality assurance should focus on the key issues that determine quality results, costs, and timeliness and avoid diversion of energy into less important issues.[4]

10.1.3 Quality System Requirements for the Laboratory

Quality systems are based on the concepts of quality assurance and quality control, as described above. Although quality assurance and quality control activities are often considered separately, they are closely linked, together forming a complete system.

A laboratory may decide to design and follow its own QA system. However, it may be easier and provide better reassurance to the client to follow the protocols of an established system. Several internationally accepted quality assurance standards exist that are relevant to analytical laboratories. The systems differ according to their intended purpose. Examples include:

- ISO/IEC 17025:2005,[3] which provides a framework for quality management and the technical competence of laboratories to carry out tests and calibrations. This is probably the most appropriate standard for regulatory laboratories engaged in testing for antibiotic residues and is discussed in more detail in Section 10.4.3.
- OECD Good Laboratory Practices (GLP),[5] which defines a system for the organizational processes and documentation to ensure the quality, reliability, and validity of data generated in research studies, typically non-clinical health and environmental safety studies to be submitted to a regulatory agency for hazard assessment in support of the approval of regulated products. GLP focuses primarily on the documentation of how results were obtained and refers to specific studies, whereas ISO/IEC 17025:2005 describes general requirements that testing and calibration laboratories must meet to demonstrate that they operate a quality system, are technically competent, and are able to generate technically valid results.
- The ISO 9000 family,[2] which represents an international consensus on good management practices. It aims at ensuring that the organization can deliver products or services that fulfill customer quality requirements and applicable regulatory requirements, enhance customer satisfaction, and achieve continual improvement of its performance in pursuit of the above objectives. In the ISO 9000 context, *quality* refers to all those features of a product or service that are required by the customer. The ISO 9000 family does not certify the quality of the product itself, but states what the organization must do to manage the processes that influence the quality of the product

or service. The ISO 9001:2008[6] standard provides a set of standardized requirements for a quality management system, regardless of what the user organization does, its size, or whether it is in the private or public sector. It is the only standard in the family against which organizations can be certified, although certification is not a compulsory requirement of the standard.

Laboratories implementing a quality system based on an established system or standard may choose to claim informal compliance, or may have an independent assessment or endorsement by a third party through certification or accreditation. In either case, compliance with a certain standard must be verified independently of the organization's management. A quality assurance unit should be set up to take responsibility for internal evaluation and auditing to check for compliance with the standard, and a member of staff appointed as a quality manager with defined responsibility and authority for ensuring that the quality management system is implemented.

Regardless of the standard on which it is based, some of the essential requirements for establishing a laboratory quality system are as follows:

- *Management Commitment and Resources.* The introduction of a quality system is a lengthy and resource-intensive process. It must be carefully planned and needs the full commitment of all parties involved, particularly management. Once in place, a quality system must be continuously maintained, which also requires management and staff resources.
- *Technically Competent Staff.* Staff with the necessary skills and training, including on QA/QC issues, are necessary.
- *Infrastructure and Conditions Suitable to Perform the Analyses.* The laboratory, equipment, and instrumentation must be suitable for the purpose and properly maintained and calibrated, and the analytical methods used must be validated as fit for purpose and under continuous quality control.
- *The Right "Mind-set."* Management and staff must understand the need for the quality system, its objectives, and the procedures involved in developing and maintaining the system. Staff must be intimately involved as stakeholders in the system and take ownership of it.

The quality system demonstrates in an objective and transparent manner that the results produced are reliable, representative, and reproducible, thereby fulfilling the agreed criteria. The science used, including the selection of appropriate analytical procedures, is another very important aspect of quality in a residues laboratory. In many cases, the analytical methods to be used are agreed on with the client in specifying their requirements. For regulatory testing, standard or reference methods may be mandatory, but in most cases the method used must be validated and meet specified performance criteria, as described in Chapter 8 of this volume, and quality controlled, as discussed in Section 10.5.2 of this chapter.

Whilst a quality system does not provide a scientific justification for the type of tests conducted, it does encompass procedures to document the analytical methods used, to demonstrate that they are valid and under control, and to ensure that the laboratory personnel using the test method are adequately trained and skilled to perform the analyses.

10.2 QUALITY MANAGEMENT

Quality management has a much broader meaning than controlling the quality of a service or product by inspection and correction of the end product. An overall system for effective quality management is based on the principle that "prevention is better than cure" and involves a conscious analysis of the management system to correct the causes of any quality failures. Quality assurance is part of a quality management system, focused on providing confidence that quality requirements will be fulfilled. The quality management system is not restricted to the technical laboratory activities, but must include all areas of the laboratory or company that are involved in producing a service that is fit for purpose and satisfies the clients' needs.

Some important aspects of quality management are outlined below. More detail is available in documents such as the ISO/IEC 17025:2005 standard.

10.2.1 Total Quality Management

The concept of *total quality management* (TQM) recognizes the importance of the contributions of all departments and individuals to the quality of the service provided and supports and cultivates a "one team" approach. In order to optimize the quality of outputs, staff must be adequately trained, involved in their tasks in such a way that they can contribute their skills and ideas, and must be provided with the necessary resources to do their job effectively and efficiently. All employees, from top management to technicians and support staff, must know the mission of the laboratory, including the role they play and their specific tasks, and must work in harmony with each other and with the laboratory's clients to achieve the organizational objectives.

Under such a system, it is accepted that quality is a dynamic issue and mistakes and failures are inevitable. The lessons learned from failures that do occur can be used to continuously improve the service provided. Emphasis is placed on continuous improvement and the concept that, at the technical level, good practice in analytical QA is independent of the formal QA system adopted.

10.2.2 Organizational Elements of a Quality System

10.2.2.1 Process Management

As discussed above, quality is user-dependent. As such, the clients can be considered as having an input to the overall laboratory process through defining the requirements for the analyses and providing the samples, and also as the final recipients of the outputs—the results of the analyses. The overall laboratory processes involved can be divided into three groups, albeit with some cross-over between them: management processes, core operating processes, and support processes. The relationship between these processes is illustrated in Figure 10.1.

Typical management functions include setting the organizational strategy, policy and objectives, provision of the appropriate human resources and equipment, management monitoring and review, and administration and audit of the quality system. Typical core operating processes are sample reception, performance of the testing, method development and validation, routine maintenance and calibration of equipment, and the preparation and issue of reports. Support processes include functions such as financial services (accounts, purchasing), information technology support, records management, building/infrastructure maintenance, equipment repair, and training.

10.2.2.2 The Quality Manual

Although not a requirement under all quality standards or systems, it is a requirement under ISO/IEC 17025:2005 that the laboratory's management system policies related to quality, including a quality policy statement, are defined by a quality manual. The quality manual is a formal document describing the overall quality system and policies, specifically, the arrangements for ensuring that the quality policy of the laboratory is followed by the staff (including management) at all times. It is approved by management and distributed to all employees concerned. The quality manual should reflect what the laboratory actually does rather than provide an indication of what it would like to do.

The main purpose of the quality manual is to provide a description of the quality management system while serving as a permanent reference in the implementation and maintenance of that system. When developing a quality manual, it is advisable to provide an efficient and effective method of updating it to allow for changes in the system over time.

The structure of the quality manual can vary between laboratories. Many organizations develop their quality manual in the order of the requirements included in the quality management standard, in which case the manual effectively provides a detailed index to the quality system operated by the laboratory. A quality manual structured in this way can also help simplify the auditing process.

10.2.2.3 Documentation

Another extremely important aspect of an effective quality system is documentation. Accurate and full documentation of all activities is required to ensure the integrity of data generated in the laboratory. Standard operating procedures and working instructions should be prepared and used for the laboratory processes, and all such documents should be controlled. Any notes made, calculations, or changes to procedures should be recorded and, if necessary, explained. It is useful to adopt the phrase "If you did not write it

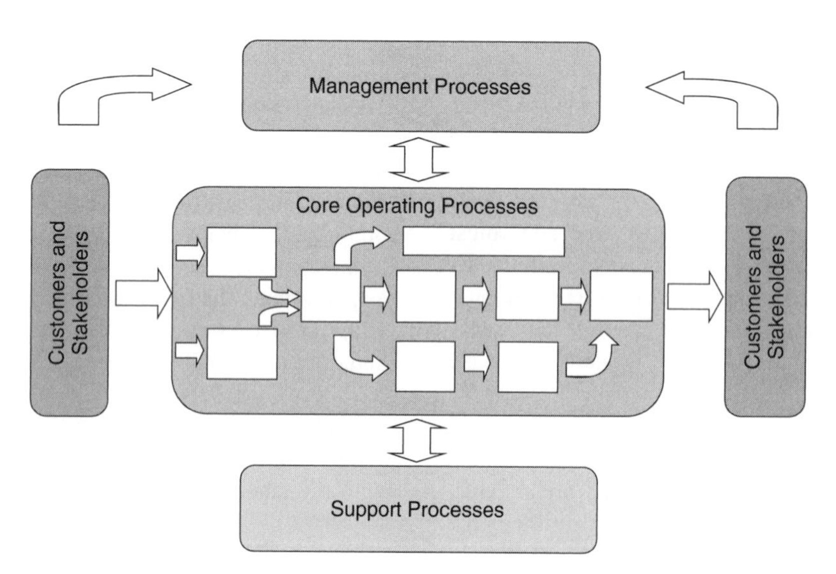

Figure 10.1 Laboratory processes and operations.

down, you did not do it." Data should be recorded directly (not transcribed from a rough copy), promptly, accurately, and legibly, and documents should then be signed and dated. Any corrections must be explained, and the original entry should be crossed out in such a way that it is still legible, rather than deleted, to preserve full transparency. All reports and data must be archived to ensure long-term, secure storage and fast retrieval of data. The archived files should contain all the original scientific data, master documents and reports, and other information to facilitate total traceability of all the events involved in producing an analytical result. The storage of documentation related to analytical samples is discussed in Section 10.5.1.6 of this chapter.

10.2.3 Technical Elements of a Quality System

Setting up a quality system involves working in a systematic way through all the processes involved. A valid approach is to follow the laboratory process step by step, starting with the sample when it arrives at the laboratory and ending when the report is issued. From the moment a sample is received in the laboratory, a defined sequence of actions is necessary to ensure that its identity and integrity are maintained throughout the laboratory process of analysis and reporting. The laboratory process is preceded by sample collection and followed by archiving of data and post-analysis storage of the samples. Sample collection may be outside the direct control of the laboratory. Nevertheless, the condition of the sample has a bearing on the quality of the analytical results, so the laboratory should make every effort to ensure that the samples are received in optimal condition, including specifying sample amounts, types of containers, and transport conditions. The period of storage of the sample after analysis must be agreed with the client, and the laboratory is also responsible for guaranteeing maintenance of sample identity and integrity during this period.

A broad range of technical requirements is important in a laboratory quality system. These include aspects such as the selection of appropriately qualified and experienced personnel; sampling, sample handling and preparation; laboratory accommodation and environmental conditions; equipment and reagents; calibration; reference standards and reference materials; traceability (of standards and of samples); the selection or development, validation, and control of methods; estimation of the uncertainty of measurements; reporting of results; and quality control and proficiency testing.

Detailed guidance on the technical elements of a quality system is available in several published guidelines and standards. Some of those of relevance to residues laboratories are discussed in the following sections of this chapter.

10.3 CONFORMITY ASSESSMENT

Conformity assessment refers to the demonstration that specified requirements relating to a product, process, or system are fulfilled or conform to specific requirements or standards. Conformity assessment helps to ensure that requirements of standards for consistency, compatibility, effectiveness, and safety are met. Conformity to these requirements can be demonstrated through mechanisms such as regular audits and inspections, and is best affirmed through third-party certification or accreditation.

10.3.1 Audits and Inspections

The purpose of an audit is to collect objective evidence to permit an informed judgment on the status of a quality management system. In general, there are two main forms of quality management system audit: external and internal audits.

An audit by an independent external body, usually carried out either as part of the process for initial accreditation or to maintain or expand the scope of accreditation, is frequently termed an *assessment* or *inspection*.

Internal audits are checks carried out within the laboratory to ensure that the quality procedures are in place, are sufficiently well documented to enable adequate and consistent implementation, and are being fully and correctly implemented. The internal audit should address all elements of the management system, including the management, core, and support processes. The purpose is to ensure that the documented system provides adequate evidence to demonstrate the effectiveness of its implementation or to highlight areas in which change is required. The audits should be arranged by the laboratory's quality manager in accordance with a pre-determined schedule and procedure and should be carried out by trained and qualified personnel who are, if possible, independent of the activity to be audited. When an internal audit finds a problem with the effectiveness of the laboratory operations or the validity of the results produced, corrective action must be initiated. If there is a possibility that results already reported may have been affected, the client must be informed in writing.

A second type of internal audit or inspection, usually termed a *review*, is a check by senior laboratory management to ensure that the quality system is effective and is achieving its objectives, and to identify opportunities to improve the system. Over time, client requirements and the needs of the laboratory are likely to change, and the quality system must be sufficiently flexible to allow continuous evolution to fulfill its purpose. Reviews are typically carried out by management on an annual basis, and are coordinated by the laboratory quality manager. Many different sources of information may be used for the review, including information from internal audits, external assessments, proficiency tests, quality control records, client feedback, and complaints.

10.3.2 Certification and Accreditation

In order to satisfy regulatory authority requirements and legal obligations, or customer demands and market forces, independent recognition of laboratory competencies may be necessary or desirable. Certification (or registration) and accreditation are formal procedures assuring conformity to the requirements specified in quality standards and are described below according to the definitions in ISO Guide 2:2004:[7]

- *Certification* is a procedure by which a third party gives written assurance that a product, process, or service conforms to specified requirements. Certification focuses on quality management and technical competence is not specifically addressed. Laboratories and organizations implementing, for example, the ISO 9000 family or OECD GLP may be certified under these standards.
- *Accreditation* is a procedure by which an authoritative body gives formal recognition that a laboratory is competent to carry out specific tasks. In the context of a residues laboratory, accreditation is a formal recognition that the laboratory is competent to carry out specific analytical tests or types of analyses. The core requirements for accreditation are set out in ISO/IEC 17025:2005, which is the most relevant standard for antibiotic residues laboratories. Accreditation under this standard is increasingly required by competent authorities and under national and regional legislation and guidelines, especially with respect to testing of food products for international trade. ISO/IEC 17025:2005 is discussed in more detail in Section 10.4.3 of this chapter.

10.3.3 Advantages of Accreditation

Customers placing work with a laboratory will wish to be assured that the laboratory is competent and capable of producing reliable results. It follows that the laboratory will wish to demonstrate that the faith placed in it by the customer is not misplaced. Accreditation by an independent and authorized body is a clear demonstration by the laboratory of the maintenance of a quality standard, which serves as a guarantee to the client of the quality of the work performed.

Regulators in national and local government may require analyses to be undertaken to ensure that produce complies with regulatory limits for residues of antibiotics or other substances in food. The use of laboratories accredited to ISO/IEC 17025:2005 by a recognized and authorized third party provides an assurance that the results produced by the laboratory are sound and "fit for purpose" and can be used to ensure continued food safety and prosecution for non-compliance if necessary.

Accreditation by an authorized body also has the added benefit that it can reduce the need for national authorities to regulate industry and the professions because it provides an alternative means of ensuring the reliability of activities having the potential to adversely affect public confidence or the national reputation.

Having discussed the benefits of accreditation for regulators, it is important to realize that industry also benefits from the use of accredited laboratories. Companies involved in food production or processing will wish to demonstrate that they have exercised due diligence in respect to monitoring their produce for non-compliant residues. By employing an accredited laboratory to undertake analyses, they can reasonably claim to have used an appropriate laboratory to analyze their samples. It must be borne in mind, however, that the relevance and usefulness of an analytical result depends not only on the sample analysis but also on the appropriateness and implementation of the sampling plan. Questions could still be asked about the sampling program employed if this is not undertaken by an accredited body.

10.3.4 Requirements under Codex Guidelines and EU Legislation

Paragraph 11 of the CAC guidelines (CAC/GL 71–2009)[1] states:

> The reliability of laboratory results is important for the decision making of Competent Authorities. Thus official laboratories should use methods validated as fit for purpose and work under internationally accepted (e.g. ISO 17025) quality management principles.

This is further restated in paragraph 147(c) of the guidelines, which specifically requires laboratories to comply with the general criteria for testing laboratories in ISO/IEC 17025:2005. The CAC guideline does not specify that accredited laboratories must be used, as it is possible for a laboratory to comply with the requirements of ISO/IEC 17025:2005 without formal accreditation. However, if results from a non-accredited laboratory were to be challenged, it might be difficult for the laboratory customer to convince those making the challenge that they had used an appropriate testing laboratory if there is no independent authoritative assessment of the laboratory.

The European Union takes this one stage further. Article 12 of Commission Regulation 882/2004,[8] on official controls performed to ensure the verification of compliance with feed and food law, animal health and animal welfare rules, requires all official food testing laboratories undertaking residue analyses to operate and be assessed and accredited to ISO/IEC 17025:2005. This reduces the potential for a challenge from producers or importers if a non-compliant analytical result is reported. The Regulation also gives "competent authorities" the option to cancel the designation of laboratories as "official laboratories" if they fail to meet the above requirements.

10.4 GUIDELINES AND STANDARDS

The scientific community has long recognized that it is not sufficient to continue producing ever more sophisticated and sensitive analytical methods for the detection and determination of residues of antibiotics and other veterinary medicines in food of animal origin. The methods developed are applied to ensure compliance with national regulations on the use of veterinary medicines and to protect consumers within the producing country and in other countries to which produce may be exported. They are also used in residue monitoring programs in importing countries. In all cases, regulators need to have confidence that the results generated in the monitoring programs are sound and reliable. This has, in turn, led to the development of guidelines and standards, at first in a small number of individual countries but ultimately in major international bodies. This is discussed in more detail below.

10.4.1 Codex Alimentarius

The Codex Alimentarius is a collection of internationally recognized standards, codes of practice, guidelines, and other recommendations relating to food, food production, and food safety. Documents are developed by groups working under the Codex Alimentarius Commission (CAC), a body that was established in 1963 by the Food and Agriculture Organization (FAO) of the United Nations and the World Health Organization (WHO). The main stated aims of the CAC are to protect the health of consumers and to ensure fair practices in international food trade. The CAC is recognized by the World Trade Organization (WTO) as an international reference point for the resolution of disputes concerning food safety and consumer protection.[9]

Whilst the CAC is the final decision-making body in the Codex Alimentarius system, the drafting and recommendation of texts for adoption by the CAC is specialized work undertaken by various committees. A number of these committees have an interest in the field of antibiotic residues in food, particularly the Codex Committee on Residues of Veterinary Drug Residues in Food (CCRVDF), and to a lesser extent the Codex Committee on Pesticide Residues (CCPR), as a small number of plant and crop treatments will have some antibiotic activity.

An international workshop was organized in Miskolc, Hungary in 1999 by the AOAC International, the FAO and the International Atomic Energy Agency. The purpose was to discuss the validation requirements for regulatory analytical methods for the detection and determination of residues, and the workshop covered both pesticides and veterinary medicines as both could ultimately leave detectable residues in food. This consultation produced detailed guidance for both pesticides and veterinary medicines and was published by Fajgelj and Ambrus.[10] These guidelines were subsequently considered by the CCPR and adopted within their guidelines on good laboratory practice in residue analysis, published by the CAC as CAC/GL 40–1993, Rev. 1–2003.[11]

The guidelines above set out to help ensure the reliability of analytical results reported when compliance with maximum residue limits was monitored in international trade. These guidelines were essentially split into three parts and dealt with the analyst, basic resources required in the laboratory (including equipment and supplies), and the analysis. Chapter 8 of this volume discusses the validation requirements for analytical methods in detail, so this will not be covered in this chapter. However, it is worth discussing the importance of the analyst and the basic resources in more detail here.

The analyst is fundamental to the processes and procedures necessary to generate reliable data. Analysts should appreciate the steps involved in the analyses in which they are involved and be appropriately trained and have demonstrated competence before they are expected to conduct analyses. Ideally, they will also have a wider understanding of residue analyses and should be aware of analytical quality assurance systems, especially as implemented in the laboratory in which they work.

CAC/GL 40–1993, Rev. 1–2003,[11] sets out detailed requirements for laboratories working on residue analyses. They must be specifically designed to allow safe operation for staff and free from contamination that could cast doubt on the results generated by staff working in the facility. For instance, sample receipt, storage, and preparation should be in dedicated areas where potential contamination from external sources can be eliminated. It follows from this that analytical standards used in analyses for confirming the identification and quantification of residues should be prepared in secure areas well isolated from routine analytical work.

Equipment used in analyses should be routinely serviced and calibrated. This includes all equipment in use, from refrigerators used to store samples and analytical standards to chromatographic and spectrometric equipment. Solvents and reagents should be stored as appropriate and be of appropriate purity. Where necessary, they should have accompanying certificates of analyses and be used before their expiry date. The laboratory should also operate independently audited analytical quality assurance (AQA) and quality control (QC) systems to ensure that the data reported to customers are dependable and reliable.

10.4.2 Guidelines for the Design and Implementation of a National Regulatory Food Safety Assurance Program Associated with the Use of Veterinary Drugs in Food-Producing Animals

Ellis[12] reviewed in detail the work of the CCRVDF on veterinary drug residue control from when it was founded in 1985. On formation, the immediate priorities of the committee included consideration of the necessary validation of analytical methods to be used for detecting and determining residues of veterinary medicines submitted for evaluation for maximum residue limits. It also considered the need for guidance on the design and implementation of national regulatory food safety assurance programs that should be associated with the controlled use of veterinary drugs. In 1993, this resulted in the publication of Codex guidelines for establishment of a regulatory program for control of veterinary drug residues in foods.[13]

This document was updated and replaced in 2009 by CAC/GL 71–2009.[1] This new document introduced a number of major revisions to previous practice, particularly the introduction of performance criteria for analytical methods and the principle of single-laboratory validation. In the earlier document, analytical methods were accepted by the CCRVDF only if they had been fully validated by a rigorous multi-laboratory collaborative trial. Experience within the CCRVDF had shown that it was becoming increasingly difficult to organise such trials, and many laboratories were adopting performance criteria to validate analytical methods.

The use of performance criteria frees laboratories from the constraints of using prescribed analytical methods that specify all analytical steps, equipment/reagents, and types of instrumentation to be used in the procedure. Rather, the laboratory is permitted to use any analytical method capable of identifying and/or quantifying the residue at the necessary concentration provided that the analytical method can be demonstrated to be "fit for purpose" and meet minimum specified performance criteria. A necessary consequence of adopting this approach is the need to have robust AQA and QC systems in place, and this is discussed in some detail in CAC/GL 71–2009.

The guidelines specify that to satisfy the requirements of the criteria approach, analytical methods must meet the general criteria for the selection of methods of analysis. Methods of analysis must also fulfill the following requirements:

- Methods must be validated according to an internationally recognised protocol;
- Use of the method must be embedded in an ISO/IEC 17025:2005 Standard or Good Laboratory Practice Quality Management System; and
- The analytical method should be complemented with information on accuracy demonstrated for instance by:
  - Regular participation in proficiency testing schemes, where available
  - Calibration using certified reference materials, where applicable
  - Recovery studies performed at the expected concentration of the analytes
  - Verification of the result with another validated method where available.

10.4.3 ISO/IEC 17025:2005

Standard ISO/IEC 17025 was first published in 1999, replacing ISO/IEC Guide 25,[14] and was revised and reissued in May 2005. The standard specifies the general requirements for the competence of laboratories to carry out tests and/or calibrations. It covers testing and calibration performed using standard methods, non-standard methods, and laboratory-developed methods. While ISO/IEC Guide 25 was a widely used document internationally, it did not include all the management requirements outlined in ISO 9001. ISO/IEC 17025:2005 now includes all the management requirements of ISO 9001.

ISO/IEC 17025:2005 contains a total of 15 management requirements and 10 technical requirements, which outline what a laboratory must do to become accredited. The management requirements are related primarily to the operation and effectiveness of the quality management system within the laboratory. The technical requirements address the competence of staff, analytical methods, and test/calibration equipment.

The standard requires laboratories to document their policies, systems, programs, procedures, and instructions to the extent necessary to meet the requirements of customers while ensuring the quality and traceability of measurements, meaning that the laboratory determines the degree of detail found in its documentation. The laboratory must also be able to demonstrate objective evidence that the degree of detail presented in its quality system documentation is generating the desired and required outcome. Documentation must be available in a repeatable form and will normally be in either written or electronic form.

The laboratory should have a systematic quality control program for checking or monitoring the reliability or accuracy of its results for all methods and measurement processes. As the standard covers more than residue testing laboratories, the particular quality control schemes and statistical techniques will vary according to the calibration or testing done within the individual laboratory. Statistical quality control charts or equivalent tabulations will normally be expected to allow monitoring of accuracy and precision of quality control tests (such as reference test materials/standards and replicate tests from the same material source) as is practicable. However, depending on the data, trends may be detectable by a review of data alone or through a regression analysis.

The management requirements imposed by ISO:IEC 17025:2005 require the laboratory to address a range of issues, including:

- Organization
  - An organizational structure and the responsibilities and tasks of management and staff should be clearly defined.
  - The organizational structure should be such that departments having conflicting interests do not adversely influence the laboratory's quality of work.
  - A quality assurance manager should be appointed.
  - All personnel should be free from any commercial and financial pressure that may adversely impact the quality of calibration and test results.
- Management system
  - A management system is implemented, maintained, and continually improved.
  - There should be policies, standard procedures, and work instructions to ensure the quality of test results.
  - There should be a quality manual with policy statements that are issued and communicated by management to all staff.
  - The effectiveness of the management system should be subject to continuous improvement.
- Document control
  - All official documents should be authorized and controlled.
  - Documents should be regularly reviewed and updated if necessary. The review frequency depends on the document.
  - Changes to documents should follow the same review process as for development of initial documents.
- Review of requests, tenders, and contracts
  - The review by the laboratory supervisors should ensure that the laboratory has technical capability and the resources to meet the requirements.
  - Changes in a contract should follow the same process as the initial contract.

Standard ISO/IEC 17025:2005 requires that the laboratory implements routine internal and external audits. Internal audits are managed by the quality manager appointed within the laboratory. The purpose of internal audits is to verify that the laboratory is complying with the requirements of the standard and with company policies, processes, and procedures. These audits are also useful in preparing for external audits. External auditors can come from clients or from accreditation bodies. Their purpose is to verify that the laboratory is operating in compliance with ISO/IEC 17025:2005.

Compliance with the requirements of ISO/IEC 17025:2005 is essential if a laboratory is to become accredited for particular analytical procedures. External audit by a recognized national or international body is a prerequisite for this process. For example, in the United Kingdom, accreditation is awarded by the UK Accreditation Service (UKAS) and in Canada by the Standards Council of Canada. The benefits of accreditation will be discussed in more detail below, but accreditation for a particular function by a recognized body means that the laboratory has been assessed against internationally recognized standards to demonstrate their competence, impartiality, and performance capability.

There should be procedures that clearly outline the roles of those involved in audits before, during, and after internal and external audits. Those with key roles should be identified and their roles defined, and all staff who may be affected by the audit should receive appropriate training.

10.4.4 Method Validation and Quality Control Procedures for Pesticide Residue Analysis in Food and Feed (Document SANCO/10684/2009)

In the European Union, analysts have been considering developing guidance for pesticide residue analyses for a number of years. The latest draft of this guidance is available as Document SANCO/10684/2009.[15] This document (like its immediate precursors) contains valuable and general material that can be readily applied to any residue testing laboratory, whether they are testing for pesticides or antibiotic residues.

10.4.5 EURACHEM/CITAC Guide to Quality in Analytical Chemistry

The EURACHEM/CITAC guide to quality in analytical chemistry[4] provides laboratories with guidance on best practice for the analytical operations that they carry out. The guidance is intended to help management and staff who are implementing quality assurance in laboratories and is useful to laboratories working toward accreditation, certification, or other compliance with particular quality requirements.

The document provides cross references to ISO/IEC 17025, ISO 9000, and OECD GLP requirements.

The guide concentrates on the technical aspects of QA with emphasis on those areas where there is a particular interpretation required for chemical testing, and does not cover general quality assurance topics such as quality systems, reports, and record keeping.

10.4.6 OECD Good Laboratory Practice

The Organization for Economic Co-operation and Development (OECD) principles of good laboratory practice (GLP)[16] were first developed in 1978 through the organization's Environmental Health and Safety Division, in response to concerns regarding the quality, reliability, and validity of product safety data used for determining the safety of chemicals and chemical products. By far the most important aspect of the poor practices that raised concerns was the lack of proper management and organization of studies used to complete regulatory dossiers. Following an OECD Council Decision in 1981,[17] testing facilities were required to follow the GLP principles when conducting studies to be submitted to member countries for the purpose of chemical assessment related to the protection of human health and the environment. An important aspect of GLP was that data produced in a study following the GLP principles would be accepted in other OECD member countries.

The GLP principles are a managerial concept defining the organizational processes and conditions under which laboratory and field research is planned, performed, monitored, recorded, and reported. They were designed to ensure that sampling and analytical procedures and results are complete and of known, documented quality. When a study is GLP compliant, an auditor, regulator, or analyst should be able to review the study some years after its completion and easily determine what work was done, when, where, by whom, and with what equipment and methods; who supervised the study; what results were obtained; and whether there were any problems encountered and, if so, how they were handled. Considerable effort and dedication in planning, support, and implementation is required to achieve this standard of documentation and attention to detail. Successful implementation of GLP provides reliable data for which the precision, accuracy, comparability, and completeness is known.

Laboratories should follow GLP principles when carrying out non-clinical health and environmental safety studies for hazard assessment, including studies on antibiotics and other veterinary pharmaceuticals, and food and feed additives. The GLP principles, while sharing many common principles with ISO/IEC 17025, are less applicable than the ISO standard to laboratories performing routine or regulatory testing.

The OECD Series on Principles of Good Laboratory Practice and Compliance Monitoring has been elaborated and expanded since its inception to reflect technical and scientific developments. The most recent revision of the *Principles on Good Laboratory Practice* (OECD Series on Principles of Good Laboratory Practice and Compliance Monitoring, Vol. 1), published in 1998,[16] is recognized as a sound approach to ensuring high quality data throughout the world.

10.5 QUALITY CONTROL IN THE LABORATORY

This section focuses on the specific parts of the analytical processes that are critical with regard to the quality of the laboratory's performance. Accredited food safety laboratories must ensure the quality of the results of the analysis of samples, in particular by monitoring tests and/or calibration results according to Section 5.9 of ISO/IEC 17025:2005. Standard methods are not systematically available for laboratories involved in the food analysis sector. The laboratory must develop (or adapt), validate, and document methods for the detection of substances that are prohibited, authorized, but subjected to a maximum residue limit, or accepted as contaminants and subjected to a maximum tolerable limit. The methods must be selected and validated so that they are fit for purpose. Internal quality control (IQC) is one of a number of rigorous measures that analytical chemists can take to ensure that the data produced in the laboratory are fit for their intended purpose. IQC covers numerous aspects of the analytical process from the sample reception to the final characterization of the food extract, but is not sufficient to guarantee the quality of the results on its own. Multiple precautions should be taken at every stage from sample reception to final reporting to ensure the global quality of the analytical process.

10.5.1 Sample Reception, Storage, and Traceability throughout the Analytical Process

A schematic of the typical steps in the workflow of an analytical laboratory is presented in Figure 10.2. It is imperative that each sample is fully traceable throughout the various procedures that take place throughout this process.

10.5.1.1 Sample Reception

Samples may be received by any process acceptable within the concepts of ISO/IEC 17025. On receipt, the transport container must first be inspected and its integrity or any irregularities documented. The transfer of the samples from the courier or other person delivering the samples must be documented, including, at a minimum, the date, the time of receipt, and the name and signature of the laboratory

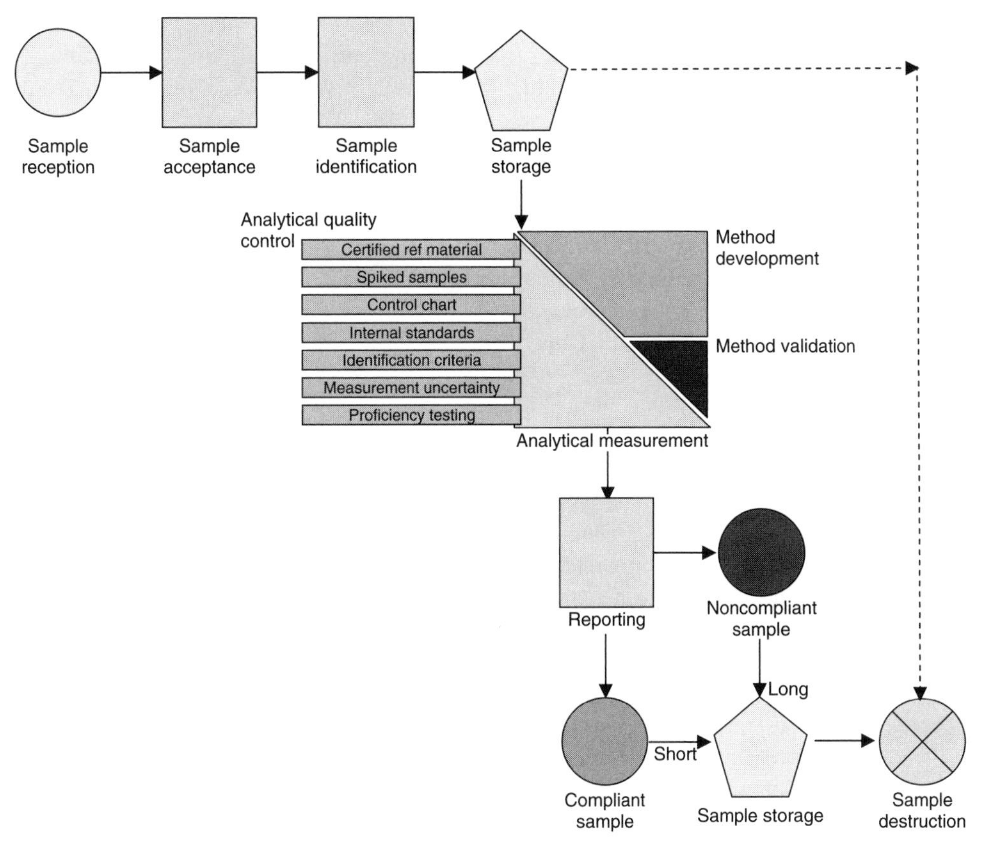

Figure 10.2 Chronological steps in the analytical process.

representative receiving the samples. This information is generally included in the laboratory internal chain of custody (LICC) record, which can be considered as a backbone for maintaining control of, and accountability for, samples from receipt through to their final disposal.

10.5.1.2 Sample Acceptance

The laboratory must observe and document the conditions that exist at the time of receipt and that may adversely impact on the integrity of a sample. Irregularities such as unsealed sample containers on receipt, samples without any identification code, mismatches between the sample number and the sample identification number on the form, inadequate sample volume/weight, unusual condition of the sample (e.g., color, odor, or hemolysis) should be noted. Instructions from competent authorities and/or clients may be required to decide whether the sample should be rejected, and in any case the abnormalities should be documented in the sample logbook for possible future reference.

10.5.1.3 Sample Identification

The laboratory must have a system to uniquely identify the samples and associate each sample with the collection document or other external chain of custody link. Typically, a unique numeric, or alphanumeric code will be assigned to each sample. All information should be stored on a secure computer database system protected by restricted access and passwords. Computer systems should ideally be networked and backed up on a daily basis to an external medium such as a tape drive, ZIP (zone information protocol) disk, or external computer server. Backup copies should be stored in a fireproof safe with a second backup copy located off-site. It is strongly advised to have staff dedicated to maintenance and backup of the database. Manual transfer of information is always liable to error; the less manual transfer of information, the better. Ideally, printed labels with all sample details should be generated by computer at the time of registering a sample in the database. A number of laboratories use barcode printers and readers to record samples and track their progress through the analytical system to enhance sample traceability and minimize the potential for mistaken identity and other problems that may be associated with manual information transfer.

10.5.1.4 Sample Storage (Pre-analysis)

Depending on the nature of the sample and the analytes to be tested for, different conditions of storage must be observed. Whatever the storage temperature chosen, it should be suitable for maintaining the integrity of the sample and the parameters to be measured. Ambient temperatures can be applied for (supposed) stable samples such as hair (the matrix often used for forbidden substances, such as those in group A of Council Directive 96/23/EC in the European Union[18]), refrigeration (+4°C) for samples such as feedstuffs pending their analysis or working aliquots while testing is undertaken, storage in a freezer (−20°C) for most samples (urine, edible tissues, milk, etc.), and in a −80°C freezer for sensitive analyses requiring the absolute stability of the global profile (e.g., metabonomics in blood or urine). Care should be taken to minimize the number of freeze–thaw cycles that a sample undergoes as this may compromise analyte stability.

10.5.1.5 Reporting

All reports should generally be checked independently by (ideally) two suitably qualified staff members. The laboratory should implement a policy regarding the provision of opinions and interpretation of data. The basis on which an opinion has been stated must be documented. An opinion or interpretation may include, but not be limited to, recommendations on how to use results, or information related to the pharmacology, metabolism, and pharmacokinetics of a substance. This is particularly important for classes of compounds for which the client/authority may not be aware of the latest developments or recent knowledge in the field, as is often the case, for instance, for natural hormones such as boldenone and nandrolone in breeding animals, compounds newly discovered to be naturally occurring such as thiouracil,[19] and marker residues that may be unreliable such as semicarbazide as a marker for nitrofurazone abuse.[20]

10.5.1.6 Sample Documentation

Analytical records on negative samples should be retained in secure storage for a period agreed by the client/authorities. In many cases a period of one year may be considered reasonable for annual monitoring plan and accreditation cycles, and this period is often implemented in official laboratories. Analytical records on non-compliant samples must be retained in saferooms for a longer period of time; this may even be for an unlimited period in some laboratories. The raw data supporting these analytical results must be retained in secure storage for at least the same period of time.

10.5.1.7 Sample Storage (Post-Reporting)

The laboratory must retain and store frozen compliant samples for a minimum period of time (typically 1–3 months) after the final analytical report is transmitted to the competent authority. Non-compliant samples must be stored frozen for a longer period of time, typically 1–5 years, following the report to the authorities. If an analytical result is challenged, the storage duration of both the sample and the dossier may be prolonged. If the laboratory wishes to use the samples for other purposes such as research, with the consent, if necessary, of the sample owner, they are generally given new identifiers to make them anonymous.

10.5.2 Analytical Method Requirements

10.5.2.1 Introduction

A major factor affecting the quality of the final result is the suitability of the analytical method applied. Ensuring that the method is fit for purpose can be considered a basic quality control criterion. It is important that laboratories restrict their choice of methods to those that have been characterized as suitable for the matrix and analyte of interest, and at the level of interest. In the EU, and in many other countries and regions, the regulatory limit for authorized veterinary medicinal products is the maximum residue limit (MRL), and for contaminants the maximum permitted limit. For prohibited or unauthorized analytes, there is often a threshold or action limit set; in Europe, for example, the appropriate regulatory limit is the minimum required performance limit (MRPL) or the reference point for action (RPA), as defined in Article 4 of Commission Decision 2002/657/EC,[21] Article 2 of Commission Decision 2005/34/EC,[22] and Articles 18 and 19 of Council Regulation (EEC) 470/2009.[23]

10.5.2.2 Screening Methods

Screening methods, as discussed mainly in Chapter 5, are capable of high sample throughput and are used to identify, in large numbers of samples, those that are potentially non-compliant. The key requirement for a screening method, whether qualitative or quantitative, is its ability to reliably detect the analyte in question at the chosen screening target concentration and to avoid false-compliant results. The screening target concentration should be low enough to ensure that if the analyte in question is present in the sample at the regulatory limit, the sample will be classified as "suspicious." Only those analytical techniques that can be demonstrated to have a false-compliant rate of $<5\%$ (β error) at the level of interest, are suitable for use for screening purposes. In the case of a suspected non-compliant result, the result must be confirmed using a confirmatory method (Chapters 6 and 7). The screening target concentration is the concentration at which a screening test categorizes the sample as "screened positive" (potentially non-compliant). For authorized drugs, the screening target concentration is at or below the maximum regulatory limit (MRL), and should preferably be set at 0.5 MRL wherever possible. For prohibited and unauthorized analytes, the screening target concentration must be at or less than ($\leq$) the threshold or action limit (the MRPL or RPA according to European regulations).

10.5.2.3 Confirmatory Methods

For confirmatory methods, as discussed in Chapters 6 and 7, one of the main quality control elements is demonstration of the selectivity/specificity of the method. Codex Guideline CAC/GL 71-2009[1] states that the selectivity, the ability of the method to unequivocally identify a signal response as being exclusively related to a specific compound, is the primary consideration for confirmatory methods. In the EU, according to Commission Decision 2002/657/EC,[21] confirmatory methods for organic residues or contaminants shall provide information on the chemical structure of the analyte. Consequently, methods based only on chromatographic analysis without the use of spectrometric detection are not suitable on their own for use as confirmatory methods. However, if a single technique lacks sufficient specificity, the desired specificity can be achieved by employing analytical procedures consisting of suitable combinations of clean-up, chromatographic separation(s), and spectrometric detection. The key requirement for a confirmatory method is its ability to reliably identify the analyte at the chosen screening target concentration (at least at the MRPL for banned substances or the MRL for authorized substances) and to avoid false non-compliant results. The term *confirmatory method* refers to methods that provide full or complementary information enabling the target substance to be unequivocally identified and, if necessary, quantified at the level of interest. Only those analytical techniques that can be demonstrated in a documented, traceable manner to be validated and have a false non-compliant rate of $<1\%$ (α error) at the level of interest should be used for confirmatory purposes. Mass spectrometry (MS) coupled to either gas or liquid chromatography (GC or LC) is the analytical technique of choice for confirmation of banned substances, metabolite(s) of a prohibited substance, or marker(s) of the use of a prohibited substance. GC or HPLC coupled with tandem (MS/MS) is acceptable for both initial testing procedures and confirmation procedures.

10.5.2.4 Decision Limit, Detection Capability, Performance Limit, and Sample Compliance

In the EU, and also in other countries that have adopted regulations or guidelines based on the EU approach, the interpretation of analytical results and regulatory decisions is made on the basis of those results that depend on the method performance characteristics known as the *detection capability* and the *decision limit*. The detection capability ($CC\beta$) is defined in point 1.12 of the Annex to Commission Decision 2002/657/EC. $CC\beta$ as the smallest content of the analyte that may be detected, identified, and/or quantified in a sample with an error probability of β. The β error is the probability that the tested sample is truly non-compliant even though a compliant measurement has been obtained. For screening tests the β error (the false-compliant rate) should be $<5\%$.

In the case of analytes for which no regulatory limit has been established, $CC\beta$ is the lowest concentration at which a method is able to detect truly contaminated samples with a statistical certainty of $1-\beta$. In this case, $CC\beta$ must be as low as possible, or lower than recommended concentrations if they exist.

In the case of analytes with an established regulatory limit, $CC\beta$ is the concentration at which the method is able to detect permitted limit concentrations with a statistical certainty of $1-\beta$; in other words, $CC\beta$ is the concentration at which only $\leq 5\%$ false-compliant results remain. In this case, $CC\beta$ must be less than or equal to the regulatory limit.

The decision limit ($CC\alpha$) is defined in point 1.11 of the Annex to Commission Decision 2002/657/EC. The decision limit ($CC\alpha$) means the limit at and above which it can be concluded with an error probability of α that a sample is non-compliant. During a screening process, a substance detected above the $CC\beta$ must be declared suspect (or screening non-compliant). In some laboratories, the $CC\alpha$ is used as a threshold to establish the suspicion, especially for forbidden substances. During a confirmatory process, a substance detected above the $CC\alpha$ must be declared non-compliant, on the condition that the appropriate identification criteria are fulfilled, for example, using the identification point system described in Commission Decision 2002/657/EC[21] and CAC/GL 71–2009.[1]

The use of the decision limit and detection capability characteristics of analytical methods in the cases of authorized and prohibited substances is illustrated graphically in Figures 10.3 and 10.4.

10.5.3 Analytical Standards and Certified Reference Materials

10.5.3.1 Introduction

Standard ISO/IEC 17025:2005 requires a laboratory to have "quality control procedures for monitoring the validity of tests and calibrations undertaken." This means that laboratories must perform internal performance-based quality control checks in accordance with Section 5.9 of ISO/IEC 17025:2005 as it applies to every test, technology, and/or parameter within their scope(s) of accreditation in order to demonstrate compliance with accreditation requirements. Reference or fortified material containing known amounts of analyte, at or near the permitted limit or the decision limit (a non-compliant control sample) as well as compliant control materials and reagent blanks should preferably be carried through the entire procedure simultaneously with each batch of test samples analyzed. Ideally, the control samples should also be very similar to test samples and stable over time. The laboratory should maintain a sufficient amount of control material to last for a significant time period (preferably a number of years) and at suitable analyte concentrations.

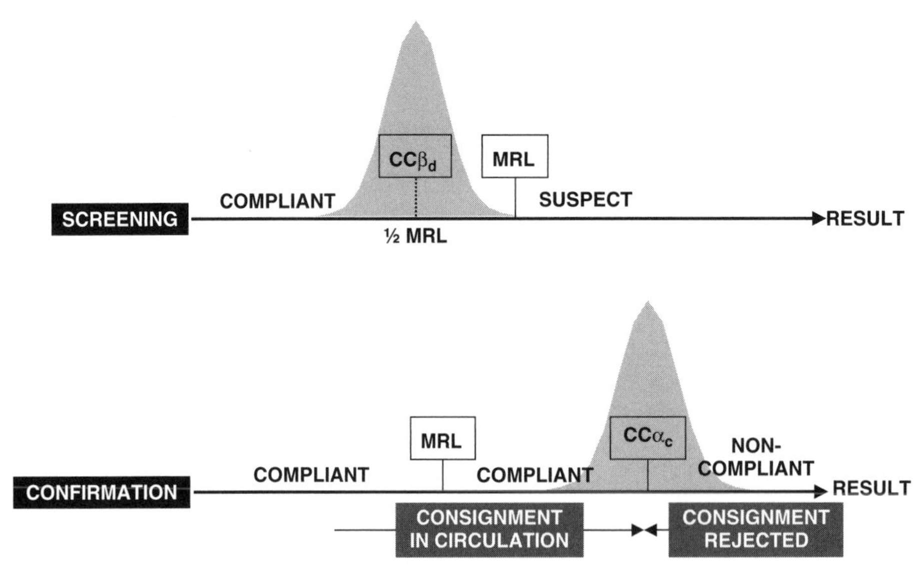

Figure 10.3 Declaration of compliance for authorized substances.

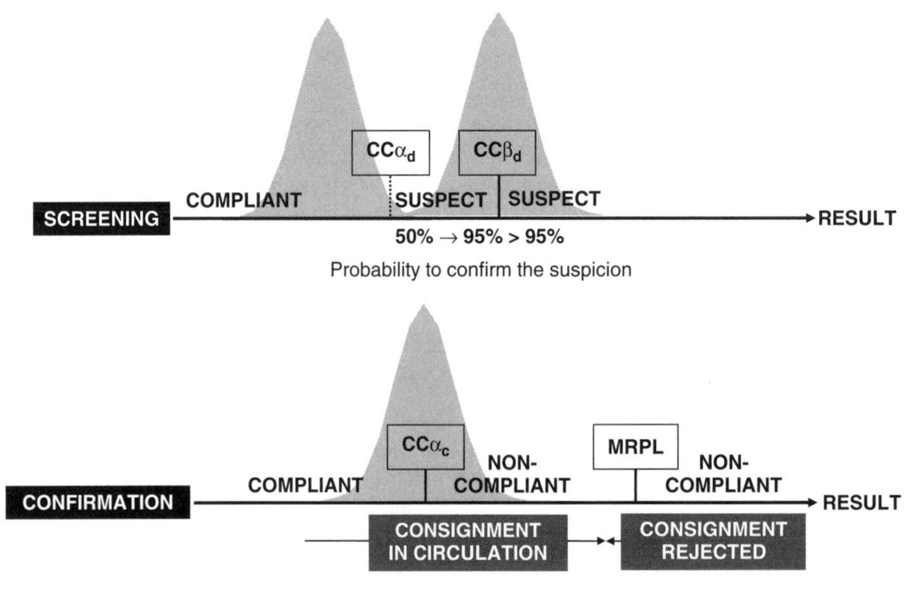

Figure 10.4 Declaration of compliance for banned substances.

10.5.3.2 Certified Reference Materials (CRMs)

Certified reference materials distributed by bodies such as the Institute for Reference Materials and Measurements (IRMM), which is part of the European Commission Joint Research Centre (EC-JRC, Geel, Belgium), give laboratories a means to validate analytical methods, to assess the quality of the measurement results and to demonstrate their traceability to stated references such as SI units. One of the most frequent applications of certified reference materials is in the validation of measurement procedures. To achieve this, measurements on certified reference materials are performed and the results are compared with the certified values. This comparison is often described in a qualitative manner; however, a quantitative approach exists that allows for a statement on the evidence of any bias. This approach takes into account the certified value, the measurement result, and their respective uncertainties. These uncertainties are combined, and the expanded uncertainty is compared to the difference. CRMs are recommended for use as often as practically and economically possible. Unfortunately, a CRM is not always available for the desired sample matrix or concentration range.

10.5.3.3 Blank Samples

Blank samples are taken to ensure that samples have not been contaminated by the data collection process. Any measured value or signal in a blank sample for an analyte that was absent in the blank solution (reagent blank) is believed to be due to contamination. There are many types of blank samples possible, each designed to control a different part of the complete data collection process, including sampling, filtering, preserving, storing, transporting, and analyzing. A blank solution is free of the analyte(s) of interest. Such a solution is used to develop specific types of blank samples. An equipment blank, for example, is a solution that is processed through all equipment used for collecting and processing a sample. A field blank is a solution that is subjected to all aspects of sample collection, field processing, preservation, transportation, and laboratory handling, in the same way as an actual typical sample.

10.5.3.4 Utilization of CRMs and Control Samples

Results from these samples should be recorded continuously, and the data are used to verify that the test works reliably. The choice of analytes to include in routine QC samples should follow the same rules as those selected for the initial or abridged validation exercise, that is, the worst-case analytes that are listed in the method scope or the most relevant analytes in a national control plan. Even if the use of spiked samples as QC is applicable, it is highly preferable to use incurred samples where possible. QC samples should be stored for a period determined by the laboratory according to stability data available for the analyte/matrix. The data obtained with the QC samples should be stored and remain traceable as long as the method is used in the laboratory. The results obtained from the QC samples should be used to supplement the initial and abridged validation data.

10.5.4 Proficiency Testing (PT)

Irrespective of, and in addition to, a laboratory's quality control activities, there is a separate and distinct requirement for all laboratories to participate in relevant and available proficiency testing (PT). PT is one of the important tools used by laboratories and accreditation bodies for monitoring test and calibration results and for verifying the effectiveness of the accreditation process. Results from proficiency testing are an indication of a laboratory's competence and are an integral part of the assessment and accreditation process. When such proficiency testing programs are not available or relevant to the scope of accreditation, performance evaluation should rely on QC checks in accordance with Clause 5.9.1 of ISO/IEC 17025:2005.

Another standard, ISO/IEC 17043:2010,[24] specifies general requirements for the competence of providers of PT schemes and for the development and operation of PT schemes. These requirements are intended to be general for all types of PT schemes, and they can be used as a basis for specific technical requirements for particular fields of application. ISO/IEC 17043:2010 defines PT as the use of inter-laboratory comparisons to determine the performance of individual laboratories for specific tests or measurements and to monitor a laboratory's continuing performance. Proficiency testing is a periodic assessment of the performance of individual laboratories and groups of laboratories that is achieved by the distribution by an independent testing body of typical materials for unsupervised analysis by the participants. Proficiency testing schemes can be regarded as a routine, but relatively infrequent, check on analytical errors.

It is a requirement of accreditation to ISO/IEC 17025:2005 that the laboratory participates in a proficiency testing scheme, if a suitable scheme exists. The analysis of an external quality check sample as part of a laboratory's routine procedures provides objective standards for individual laboratories to perform against and permits them to compare their analytical results with those from other laboratories.

It is important to understand the statistical limitations of this external means of quality assessment when gauging the competence of a laboratory. Typically, the results of a chemical analysis of a sample will have a normal distribution. This means that the majority of results will be centred on a mean value but, inevitably, some results will lie at the extremes of the distribution. The statistics of a normal distribution mean that about 95% of the data points will lie within a z score between -2 and $+2$. Performance in a PT, therefore, is considered satisfactory if a participant's z score lies within this range. It follows that if a participant's z score lies outside this range, that is, $|z| > 2$, there is about a 1 in 20 chance that their result is in fact an acceptable result from the extreme of the distribution. If a participant's z score is greater than 3 ($|z| > 3$), the probability that their result is actually acceptable is only about 1 in 300. A typical PT z-score distribution is illustrated in Figure 10.5.

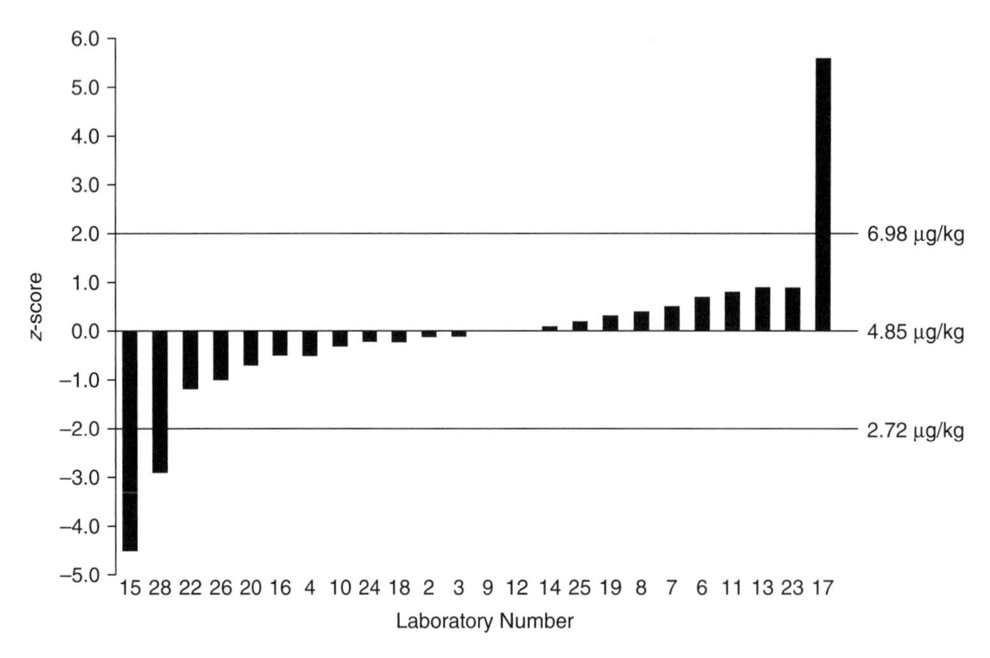

Figure 10.5 Typical presentation of the distribution of z-scores observed in a proficiency test.

The objective of the statistical procedure employed is to obtain a simple and transparent result, which the participant and other interested parties can readily appreciate. The procedure follows that recommended in the IUPAC/ISO/AOAC International Harmonized Protocol for the Proficiency Testing of Analytical Laboratories. The z scores are calculated as:

$$z = \frac{(x - \hat{X})}{\sigma_p}$$

where x is the participant's reported result, $\hat{X}$ is the assigned value, and σ_p is the target standard deviation.

The assigned value corresponds to the best estimate of the true concentration of the analyte and is set as the consensus of the results submitted by participants. The target standard deviation for the proficiency test, σ_p, is derived from the appropriate form of the Horwitz equation[25] and is considered as an appropriate indicator of the best agreement that can be obtained between laboratories. The target relative standard deviation will be set in such a way that:

- A z score between 0 and 2.0, inclusive, is deemed satisfactory performance.
- A z score greater than 2.0 but less than 3.0 is deemed to be questionable performance.
- A z score equal to or greater than 3.0 is deemed to be unsatisfactory performance.

All procedures associated with the handling and testing of the PT samples by the laboratory are, to the greatest extent possible, to be carried out in a manner identical to that expected to be applied to routine samples. No special effort should be made to optimize the instrument (e.g., cleaning ion source, changing multipliers) or method performance prior to analyzing the PT samples. Methods or procedures to be utilized in routine testing should be employed. The laboratory will be aware that the sample is a PT sample, but will not be aware of the content of the sample.

10.5.5 Control of Instruments and Methods in the Laboratory

According to ISO/IEC 17025:2005, the laboratory must have quality control procedures in place for monitoring the validity of tests undertaken. The resulting data should be recorded in such a way that trends are detectable and, where practicable, statistical techniques should be applied to reviewing of the results. The monitoring should include the regular use of internal quality control. Quality control data should be analyzed and, where they are found to be outside pre-defined criteria, planned action should be taken to correct the problem and prevent incorrect results from being reported. Internal quality control in the chemical analytical laboratory involves a continuous, critical evaluation of the laboratory's own analytical methods and working routines. The control encompasses the analytical process starting with the sample entering the laboratory and ending with the analytical report. One of the most important tools for quality control is the use of control charts.

In general, three types of control charts are used in laboratories: the X-chart, the spiked sample chart, and the precision chart (also known as the range or R-chart). The control charts are used to plot variables or data arising from analytical runs over time to help identify trends indicating bias in the analytical results or other anomalies that may require investigation and remediation. The X-chart and the spiked sample chart monitor the process over time, based on the average of a series of observations, called a *sub-group*. The precision chart monitors the variation between replicate observations within the sub-group over time.

The X-chart is based on the use of a standard reference material analyzed preferably with each batch of unknowns. After a reasonable number of analyses of reference material samples (typically $n > 20$), the mean and standard deviation of the data are calculated and a control chart constructed. The center line represents the mean, the two outer lines represent the upper and lower control limits (UCL and LCL), or 99% confidence limits, and the two lines closest to the mean line are the 95% confidence limits, or upper and lower warning limits (UWL and LWL). One analysis outside the 95% confidence limits is not cause for alarm; however, two consecutive analyses falling on one side of the mean line between the 95% and 99% limits would certainly be cause for an investigation. Control charts are very useful in visualizing trends (Fig. 10.6).

Spiked sample control charts are frequently used in cases where check samples of appropriate analyte concentrations are not easily prepared or obtained. The spiked sample control chart is superficially similar to the X-chart, but instead of using a check or reference standard, one of the unknown samples is analyzed and then spiked with a known amount of the analyte of interest. The percentage recovery is calculated and plotted on the control chart. The control chart lines on the spiked sample chart correspond to the mean recovery and the 95% and 99% limits calculated from the standard deviation of the recovery data. The resultant chart can be used and interpreted in the same way as the X-chart.

Control charts can be plotted in the same way using the results from "blind" control samples; that is, blank (or previously analyzed) samples spiked at an appropriate level, unknown to the analyst, by a third party. The use of blind control samples gives an additional degree of quality control, since any bias (intentional or non-intentional) on the part of the analyst is precluded because the analyst is not aware of the expected result. Blind control samples are being used increasingly and are required in some official testing laboratories in the USA. Many laboratories in the UK routinely use blind controls for confirmatory analyses, and in some instances for screening as well.

In precision charts (the range chart or R-chart), the data from duplicates are plotted with the vertical scale (ordinate) in units such as percent, and the horizontal scale (abscissa) in units of batch number or time. Usually the mean of the duplicates is reported and the difference between the duplicates, or range, is examined for acceptability. The mean and standard deviation are calculated from the data. It is common practice in analytical laboratories to run duplicate analyses at frequent intervals as a means of monitoring the precision of analyses and detecting out-of-control situations. This is often done for analyses for which there are no suitable control samples or reference materials available.

The method is in control and the analyst can report the analytical results when the control value is within the warning limits or the control value is between the warning and action limit and the two previous control values were within warning limits.

The method is in control but can be regarded as out of statistical control if all the control values are within the warning limits (maximum one out of the last three between warning and action limits) and if seven consecutive control values are gradually increasing or decreasing, or 10 out of 11 consecutive control values are lying on the same side of the central line. In this case the analyst can report

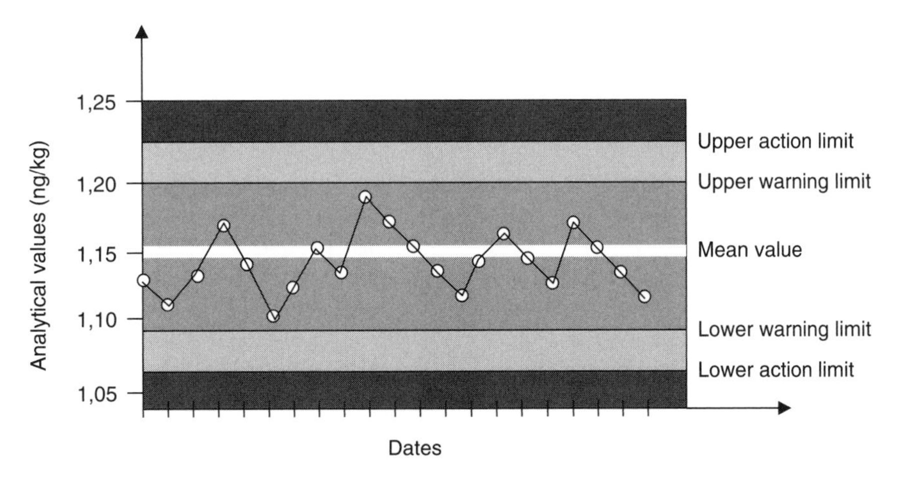

Figure 10.6 Control chart for the determination of dioxins and *d,l*-PCB in edible tissues.

the analytical results, but a problem may be developing. Important trends should be discovered as early as possible in order to avoid serious problems in the future.

The method is out of control and no analytical results can be reported if the control value is outside the action limits or the control value is between the warning and the action limits and at least one of the two previous control values is also between warning and action limit. All results obtained since the last value in control was obtained are suspect, and the samples must be reanalyzed.

10.6 CONCLUSION

Laboratory quality systems must be implemented in the antibiotic residue analytical laboratory in order to ensure that the quality of the results produced meets the requirements of the client. Increasingly in today's global market, quality systems that are formally recognized through accreditation and/or certification are required to facilitate international trade by providing the data that establish equivalence of food safety standards with trading partners. Such systems also provide confidence in domestic food systems when applied in laboratories involved in monitoring and surveillance programs for antibiotic residues in food.

In implementing a quality system, it is essential to define the needs of the laboratory and the customer in order to balance the costs and benefits of the system. Putting in place and maintaining a quality system requires the full commitment of management and staff and the necessary resources in terms of infrastructure, equipment, and appropriately trained and experienced staff. A key issue is development of the right mind-set, in which the laboratory staff accept the system and the procedures involved as necessary and beneficial to both the organization and its clients, and perform the necessary tasks routinely. The system should be implemented on the basis of what is done in the laboratory, rather than what should be done, and should effectively control the laboratory procedures while remaining as simple as possible. It should also retain sufficient flexibility to change in response to varying client demands and to allow continuous improvement.

REFERENCES

1. Codex Alimentarius Commission, CAC/GL 71–2009, *Guidelines for the Design and Implementation of National Regulatory Food Safety Assurance Programme Associated with the Use of Veterinary Drugs in Food Producing animals*, Rome, 2009.
2. International Standards Organization, ISO 9000:2005, *Quality Management Systems—Fundamentals and Vocabulary*, Geneva, 2005.
3. International Standards Organization, ISO/IEC 17025, *General Requirements for the Competence of Testing and Calibration Laboratories*, 2nd ed., Geneva, 2005.
4. EURACHEM/CITAC, *Guide to Quality in Analytical Chemistry—an Aid to Accreditation*, 2002.
5. Organisation for Economic Cooperation and Development, *Good Laboratory Practice* (available at `http://www.oecd.org/department/0,3355,en_2649_34381_1_1_1_1_1,00.html`; accessed 8/12/10).
6. International Standards Organization, ISO 9001:2008, *Quality Management Systems—Requirements*, Geneva, 2008.
7. International Standards Organization. ISO Guide 2:2004, Standardization and related activities - General vocabulary. Geneva; 2004.
8. European Commission. Regulation (EC) 882/2004 of the European Parliament and of The Council of 29 April 2004, on official controls performed to ensure the verification of compliance with feed and food law, animal health and animal welfare rules, *Off. J. Eur. Commun*. 2004;L65:1.
9. World Health Organization and Food and Agriculture Organization of the United Nations, *Understanding the Codex Alimentarius*, Rome, 2005 (available at `http://www.fao.org/docrep/008/y7867e/y7867e00.htm`; accessed 6/20/10).
10. Fajgelj A, Ambrus A, eds., *Principles and Practices of Method Validation*, Royal Society of Chemistry, Cambridge, UK, 2000.
11. Codex Alimentarius Commission, CAC/GL 40–1993, Rev. 1–2003, *Guidelines on Good Laboratory Practice in Residue Analysis*, Rome, 2003.
12. Ellis RL, Development of veterinary drug residue controls by the Codex Alimentarius Commission: A review, *Food Addit. Contam*. 2008;25:1432–1438.
13. Codex Alimentarius Commission, CAC/GL 16–1993, *Codex Guidelines for the Establishment of a Regulatory Programme for Control of Veterinary Drug Residues in Foods*, Rome, 1993.
14. International Standards Organization, ISO/IEC Guide 25, *General Requirements for the Competence of Calibration and Testing Laboratories*, Geneva, 1990.
15. European Commission. SANCO/10684/2009, *Method Validation and Quality Control Procedures for Pesticide Residues Analysis in Food and Feed* (available at `http://ec.europa.eu/food/plant/protection/resources/qualcontrol_en.pdf`; accessed 6/14/10).
16. Organisation for Economic Cooperation and Development, OECD Series on Principles of Good Laboratory Practice and Compliance Monitoring, Vol. 1, *Principles on Good Laboratory Practice* (as revised in 1997), Paris; 1998.
17. Organisation for Economic Cooperation and Development, *Council Decision—Concerning the Mutual Acceptance in the Assessment of Chemicals* [C(81)30(final)], Appendix A, Part 2, Paris, 1981.
18. European Commission. Council Directive 96/23/EC of 29 April 1996 on measures to monitor certain substances and residues thereof in live animals and animal products and

repealing Directives 85/358/EEC and 86/469/EEC and Decisions 89/187/EEC and 91/664/EEC, *Off. J. Eur. Commun.* 1996;L125:10.

19. Pinel G, Mathieu S, Cesbron N, et al., Evidence that urinary excretion of thiouracil in adult bovine submitted to a cruciferous diet can give erroneous indications of the possible illegal use of thyrostats in meat production, *Food Addit. Contam.* 2006;10:974–980.
20. Bendall J, Semicarbazide is non-specific as a marker metabolite to reveal nitrofurazone abuse as it can form under Hofmann conditions, *Food Addit. Contam.* 2009;26:47–56.
21. European Commission. Commission Decision 2002/657/EC of 12 August 2002 implementing Council Directive 96/23/EC concerning the performance of analytical methods and the interpretation of results, *Off. J. Eur. Commun.* 2002;L221:8.
22. European Commission. Commission Decision 2005/34/EC of 11 January 2005 laying down harmonised standards for the testing for certain residues in products of animal origin imported from third countries, *Off. J. Eur. Commun.* 2005;L16:61.
23. European Commission. Regulation (EEC) No 470/2009 of 6 May 2009 laying down Community procedures for the establishment of residue limits of pharmacologically active substances in foodstuffs of animal origin, repealing Council Regulation (EEC) No 2377/90 and amending Directive 2001/82/EC of the European Parliament and of the Council and Regulation (EC) No 726/2004 of the European Parliament and of the Council, *Off. J. Eur. Commun.* 2009;L152:11.
24. International Standards Organization, ISO/IEC 17043:2010, *Conformity Assessment—General Requirements for Proficiency Testing*, Geneva, 2010.
25. Thomson M, Recent trends in interlaboratory precision at ppb and sub-ppb concentrations in relation to fitness for purpose criteria in proficiency testing, *Analyst* 2006;125:385–386.

INDEX

Chemical Analysis of Antibiotic Residues in Food, First Edition. Edited by Jian Wang, James D. MacNeil, and Jack F. Kay.